SPECIAL TRIANGLES

NAME	CHARACTERISTIC	EXAMPLES
Right Triangle	Triangle has a right angle.	90°
Isosceles Triangle	Triangle has two equal sides.	$AB = BC$ B, A, C
Equilateral Triangle	Triangle has three equal sides.	$AB = BC = CA$ B, A, C
Similar Triangles	Corresponding angles are equal; corresponding sides are proportional.	$A = D, B = E, C = F$ $\dfrac{AB}{DE} = \dfrac{AC}{DF} = \dfrac{BC}{EF}$ B, C, A, E, F, D

BEGINNING ALGEBRA

SEVENTH EDITION

MARGARET L. LIAL
AMERICAN RIVER COLLEGE

E. JOHN HORNSBY, JR.
UNIVERSITY OF NEW ORLEANS

CHARLES D. MILLER

HarperCollinsCollegePublishers

Sponsoring Editor: Karin E. Wagner
Developmental Editors: Sandi Goldstein and Ellen Keith
Project Editor: Lisa A. De Mol
Design Administrator: Jess Schaal
Text Design: Lesiak/Crampton Design Inc.: Lucy Lesiak
Cover Design: Lesiak/Crampton Design Inc.: Lucy Lesiak
Photo Researcher: Rosemary Hunter
Production Administrator: Randee Wire
Compositor: Interactive Composition Corporation
Printer and Binder: R. R. Donnelley & Sons
Cover Printer: Phoenix Color Corporation

Beginning Algebra, Seventh Edition

Library of Congress Cataloging-in-Publication Data
Lial, Margaret L.
  Beginning algebra / Margaret L. Lial, E. John Hornsby, Jr.,
  Charles D. Miller. -- 7th ed.
      p.   cm.
  Includes index.
  ISBN 0-673-99139-3—0-673-99543-7 (annotated instructor's edition)
  1. Algebra.   I. Hornsby, E. John. II. Miller, Charles David.
III. Title.
QA152.2.L5   1996

512.9--dc20   95–9797
                                    CIP

95 96 97 98 9 8 7 6 5 4 3 2 1

CONTENTS

CHAPTER 1 THE REAL NUMBER SYSTEM 1

CHAPTER 2 SOLVING EQUATIONS AND INEQUALITIES 81

PREFACE

This seventh edition of *Beginning Algebra* is designed for college students who have not been exposed to algebra or who require further review before taking additional courses in mathematics, science, business, or computer science. The primary objective of the course is for students to gain familiarity with mathematical symbols and operations, and to be able to formulate and solve first- and second-degree equations.

   This edition retains the successful features of previous editions: learning objectives for each section; careful exposition; fully developed examples; cautions and notes; and boxes that set off important definitions, formulas, rules, and procedures. In this new edition, we have made several content changes to follow the guidelines set forth in the *Curriculum and Evaluation Standards for School Mathematics,* published by the National Council of Teachers of Mathematics.

CHANGES IN CONTENT

- For continuing review we have included problems involving fractions and decimals throughout the text.
- More than 80 percent of the exercises are new to this edition and many of these exercises now incorporate real data and graphics.
- Beginning with Chapter 2, cumulative reviews appear after each chapter, covering material learned up to that point.
- In this text we view calculators as a means of allowing students to spend more time on the conceptual nature of mathematics and less time on the mechanics of computation with paper and pencil. We have included an introduction to scientific calculators, and the use of the scientific calculator is discussed throughout the book wherever appropriate.
- Graphics calculator text and exercises are included as appropriate throughout the book. This material is designed so that it can easily be incorporated into the course, treated separately, or omitted, as the instructor chooses.

FEATURES

The following pages illustrate important features. These features are designed to assist students in the learning process and deepen their understanding of the underlying principles and interrelations between topics.

Connections boxes that
provide connections to the
real world or to other mathematical concepts or other disciplines open each chapter and
appear throughout the text—
almost one for every section.
Most of these include thought-
provoking questions for writing or class discussion. The
Connections provide motivation for the topic under discussion, show how mathematics
is used in many different aspects of life, give some historical background, and provide
a larger context for the current
material

CHAPTER 6

EQUATIONS AND INEQUALITIES IN TWO VARIABLES

6.1 Linear Equations in Two Variables

6.2 Graphing Linear Equations in Two Variables

6.3 The Slope of a Line

6.4 Equations of a Line

6.5 Graphing Linear Inequalities in Two Variables

6.6 Functions

CONNECTIONS

It is important in many situations (in business or in science, for example) to be able to make predictions based on known data. An executive may wish to predict next year's costs, profits, or sales, for instance. Scientists are currently trying to predict whether the earth will continue to get increasingly warmer.

The graph below shows actual and predicted Medicare costs since 1967.* It was constructed by plotting points to represent the data for each of the years shown, then connecting the points with lines. It shows the danger of using projections too far into the future. As this graph in dicates, we should be wary about cost estimates in the current national health care debate.

FOR DISCUSSION OR WRITING

In what year did the actual costs begin to diverge sharply from the predicted costs? Estimate the difference between the pre-dicted cost and the actual cost for 1994 (the right end of the graph). What factors may have contributed to the prediction being so far off ?

* Figures per the House Ways and Means Committee: National Center for Policy Analysis, July 1994.

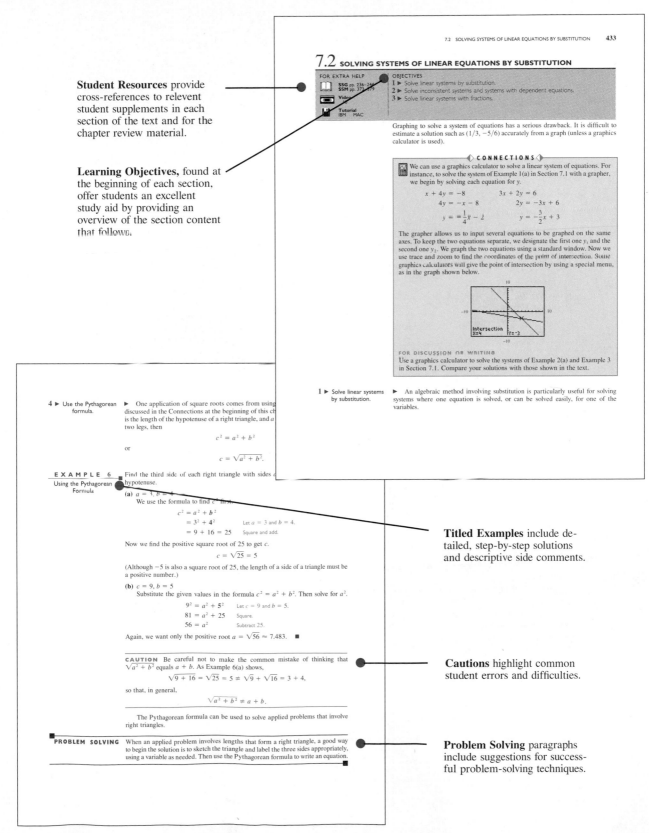

Student Resources provide cross-references to relevent student supplements in each section of the text and for the chapter review material.

Learning Objectives, found at the beginning of each section, offer students an excellent study aid by providing an overview of the section content that follows.

7.2 SOLVING SYSTEMS OF LINEAR EQUATIONS BY SUBSTITUTION

FOR EXTRA HELP

SSG pp. 236–244
SSM pp. 371–379

Video

Tutorial
IBM   MAC

OBJECTIVES

1 ▶ Solve linear systems by substitution.
2 ▶ Solve inconsistent systems and systems with dependent equations.
3 ▶ Solve linear systems with fractions.

Graphing to solve a system of equations has a serious drawback. It is difficult to estimate a solution such as $(1/3, -5/6)$ accurately from a graph (unless a graphics calculator is used).

◆ CONNECTIONS ◆

We can use a graphics calculator to solve a linear system of equations. For instance, to solve the system of Example 1(a) in Section 7.1 with a grapher, we begin by solving each equation for y.

$$x + 4y = -8 \qquad\qquad 3x + 2y = 6$$
$$4y = -x - 8 \qquad\qquad 2y = -3x + 6$$
$$y = -\frac{1}{4}x - 2 \qquad\qquad y = -\frac{3}{2}x + 3$$

The grapher allows us to input several equations to be graphed on the same axes. To keep the two equations separate, we designate the first one y_1 and the second one y_2. We graph the two equations using a standard window. Now we use trace and zoom to find the coordinates of the point of intersection. Some graphics calculators will give the point of intersection by using a special menu, as in the graph shown below.

FOR DISCUSSION OR WRITING
Use a graphics calculator to solve the systems of Example 2(a) and Example 3 in Section 7.1. Compare your solutions with those shown in the text.

1 ▶ Solve linear systems by substitution.

▶ An algebraic method involving substitution is particularly useful for solving systems where one equation is solved, or can be solved easily, for one of the variables.

4 ▶ Use the Pythagorean formula.

▶ One application of square roots comes from using the discussed in the Connections at the beginning of this ch is the length of the hypotenuse of a right triangle, and a two legs, then

$$c^2 = a^2 + b^2$$

or

$$c = \sqrt{a^2 + b^2}.$$

EXAMPLE 6
Using the Pythagorean Formula

Find the third side of each right triangle with sides a hypotenuse.

(a) $a = 3, b = 4$
We use the formula to find c^2 first.

$$c^2 = a^2 + b^2$$
$$= 3^2 + 4^2 \qquad \text{Let } a = 3 \text{ and } b = 4.$$
$$= 9 + 16 = 25 \qquad \text{Square and add.}$$

Now we find the positive square root of 25 to get c.

$$c = \sqrt{25} = 5$$

(Although -5 is also a square root of 25, the length of a side of a triangle must be a positive number.)

(b) $c = 9, b = 5$
Substitute the given values in the formula $c^2 = a^2 + b^2$. Then solve for a^2.

$$9^2 = a^2 + 5^2 \qquad \text{Let } c = 9 \text{ and } b = 5.$$
$$81 = a^2 + 25 \qquad \text{Square.}$$
$$56 = a^2 \qquad \text{Subtract 25.}$$

Again, we want only the positive root $a = \sqrt{56} \approx 7.483$. ∎

CAUTION Be careful not to make the common mistake of thinking that $\sqrt{a^2 + b^2}$ equals $a + b$. As Example 6(a) shows,

$$\sqrt{9 + 16} = \sqrt{25} = 5 \neq \sqrt{9} + \sqrt{16} = 3 + 4,$$

so that, in general,

$$\sqrt{a^2 + b^2} \neq a + b.$$

The Pythagorean formula can be used to solve applied problems that involve right triangles.

PROBLEM SOLVING

When an applied problem involves lengths that form a right triangle, a good way to begin the solution is to sketch the triangle and label the three sides appropriately, using a variable as needed. Then use the Pythagorean formula to write an equation.

Titled Examples include detailed, step-by-step solutions and descriptive side comments.

Cautions highlight common student errors and difficulties.

Problem Solving paragraphs include suggestions for successful problem-solving techniques.

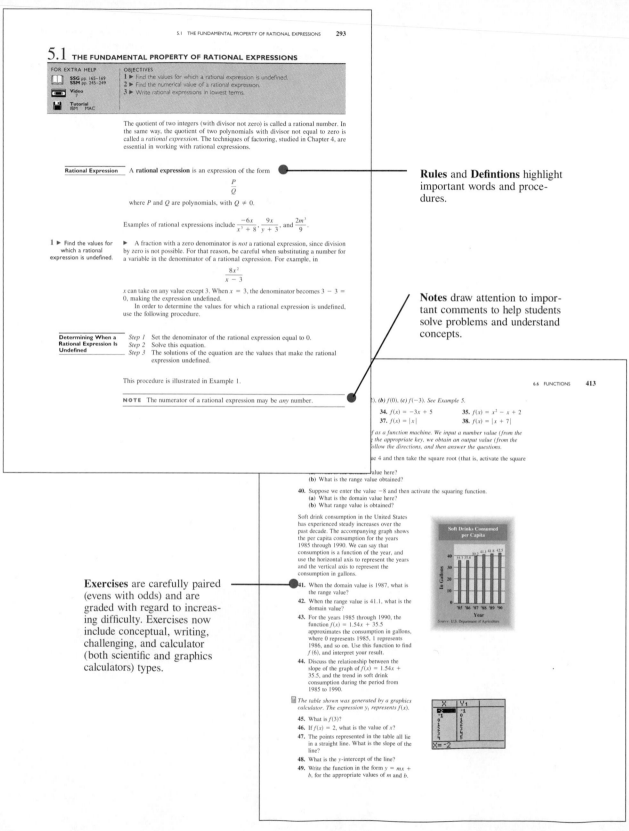

5.1 THE FUNDAMENTAL PROPERTY OF RATIONAL EXPRESSIONS

FOR EXTRA HELP	OBJECTIVES
SSG pp. 165–169 **SSM** pp. 245–249	1 ▶ Find the values for which a rational expression is undefined.
Video 7	2 ▶ Find the numerical value of a rational expression.
Tutorial IBM MAC	3 ▶ Write rational expressions in lowest terms.

The quotient of two integers (with divisor not zero) is called a rational number. In the same way, the quotient of two polynomials with divisor not equal to zero is called a *rational expression*. The techniques of factoring, studied in Chapter 4, are essential in working with rational expressions.

Rational Expression

A **rational expression** is an expression of the form

$$\frac{P}{Q}$$

where P and Q are polynomials, with $Q \neq 0$.

Examples of rational expressions include $\dfrac{-6x}{x^3 + 8}$, $\dfrac{9x}{y + 3}$, and $\dfrac{2m^3}{9}$.

1 ▶ Find the values for which a rational expression is undefined.

▶ A fraction with a zero denominator is *not* a rational expression, since division by zero is not possible. For that reason, be careful when substituting a number for a variable in the denominator of a rational expression. For example, in

$$\frac{8x^2}{x - 3}$$

x can take on any value except 3. When $x = 3$, the denominator becomes $3 - 3 = 0$, making the expression undefined.

In order to determine the values for which a rational expression is undefined, use the following procedure.

Determining When a Rational Expression Is Undefined

Step 1   Set the denominator of the rational expression equal to 0.
Step 2   Solve this equation.
Step 3   The solutions of the equation are the values that make the rational expression undefined.

This procedure is illustrated in Example 1.

NOTE   The numerator of a rational expression may be *any* number.

Rules and **Defintions** highlight important words and procedures.

Notes draw attention to important comments to help students solve problems and understand concepts.

2), **(b)** $f(0)$, **(c)** $f(-3)$. *See Example 5.*

34. $f(x) = -3x + 5$ **35.** $f(x) = x^2 - x + 2$
37. $f(x) = |x|$ **38.** $f(x) = |x + 7|$

f as a function machine. We input a number value (from the the appropriate key, we obtain an output value (from the ollow the directions, and then answer the questions.

ue 4 and then take the square root (that is, activate the square

(b) What is the range value obtained?

40. Suppose we enter the value -8 and then activate the squaring function.
   (a) What is the domain value here?
   (b) What range value is obtained?

Soft drink consumption in the United States has experienced steady increases over the past decade. The accompanying graph shows the per capita consumption for the years 1985 through 1990. We can say that consumption is a function of the year, and use the horizontal axis to represent the years and the vertical axis to represent the consumption in gallons.

41. When the domain value is 1987, what is the range value?

42. When the range value is 41.1, what is the domain value?

43. For the years 1985 through 1990, the function $f(x) = 1.54x + 35.5$ approximates the consumption in gallons, where 0 represents 1985, 1 represents 1986, and so on. Use this function to find $f(6)$, and interpret your result.

44. Discuss the relationship between the slope of the graph of $f(x) = 1.54x + 35.5$, and the trend in soft drink consumption during the period from 1985 to 1990.

The table shown was generated by a graphics calculator. The expression y_1 represents $f(x)$.

45. What is $f(3)$?

46. If $f(x) = 2$, what is the value of x?

47. The points represented in the table all lie in a straight line. What is the slope of the line?

48. What is the y-intercept of the line?

49. Write the function in the form $y = mx + b$, for the appropriate values of m and b.

Exercises are carefully paired (evens with odds) and are graded with regard to increasing difficulty. Exercises now include conceptual, writing, challenging, and calculator (both scientific and graphics calculators) types.

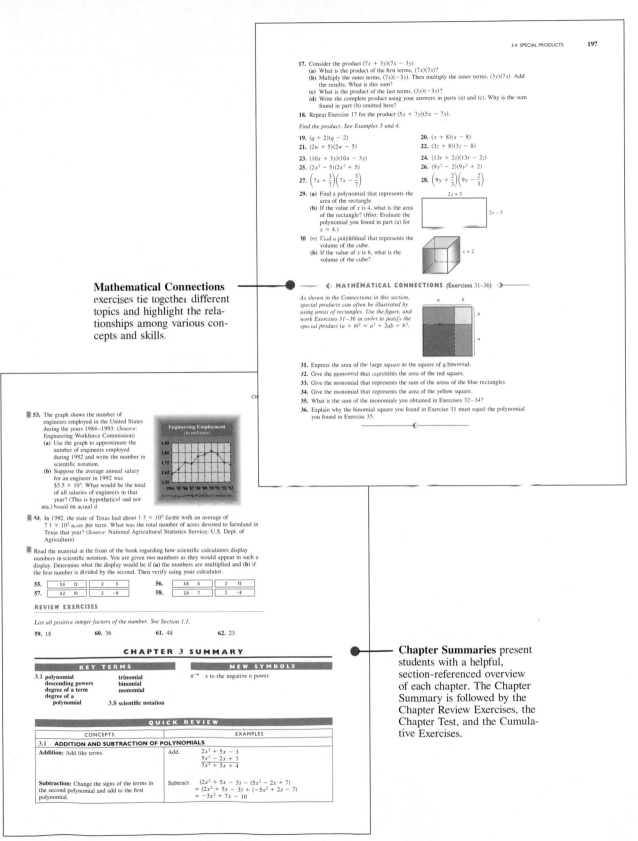

17. Consider the product $(7x + 3y)(7x - 3y)$.
   (a) What is the product of the first terms, $(7x)(7x)$?
   (b) Multiply the outer terms, $(7x)(-3y)$. Then multiply the inner terms, $(3y)(7x)$. Add the results. What is this sum?
   (c) What is the product of the last terms, $(3y)(-3y)$?
   (d) Write the complete product using your answers in parts (a) and (c). Why is the sum found in part (b) omitted here?

18. Repeat Exercise 17 for the product $(5x + 7y)(5x - 7y)$.

Find the product. See Examples 3 and 4.

19. $(q + 2)(q - 2)$ **20.** $(x + 8)(x - 8)$
21. $(2w + 5)(2w - 5)$ **22.** $(3z + 8)(3z - 8)$
23. $(10x + 3y)(10x - 3y)$ **24.** $(13r + 2z)(13r - 2z)$
25. $(2x^2 - 5)(2x^2 + 5)$ **26.** $(9y^2 - 2)(9y^2 + 2)$
27. $\left(7x + \dfrac{3}{7}\right)\left(7x - \dfrac{3}{7}\right)$ **28.** $\left(9y + \dfrac{2}{3}\right)\left(9y - \dfrac{2}{3}\right)$

29. **(a)** Find a polynomial that represents the area of the rectangle.
   (b) If the value of x is 4, what is the area of the rectangle? (*Hint:* Evaluate the polynomial you found in part (a) for $x = 4$.)

30. **(a)** Find a polynomial that represents the volume of the cube.
   (b) If the value of x is 6, what is the volume of the cube?

◆ **MATHEMATICAL CONNECTIONS** (Exercises 31–36) ◇

As shown in the Connections in this section, special products can often be illustrated by using areas of rectangles. Use the figure, and work Exercises 31–36 in order to justify the special product $(a + b)^2 = a^2 + 2ab + b^2$.

31. Express the area of the large square as the square of a binomial.
32. Give the monomial that represents the area of the red square.
33. Give the monomial that represents the sum of the areas of the blue rectangles.
34. Give the monomial that represents the area of the yellow square.
35. What is the sum of the monomials you obtained in Exercises 32–34?
36. Explain why the binomial square you found in Exercise 31 must equal the polynomial you found in Exercise 35.

◆

Mathematical Connections exercises tie together different topics and highlight the relationships among various concepts and skills.

53. The graph shows the number of engineers employed in the United States during the years 1984–1993. (*Source:* Engineering Workforce Commission)
   (a) Use the graph to approximate the number of engineers employed during 1992 and write the number in scientific notation.
   (b) Suppose the average annual salary for an engineer in 1992 was 5.5×10^4. What would be the total of all salaries of engineers in that year? (This is hypothetical and not based on actual data.)

Engineering Employment
(in millions)

1.95
1.85
1.75
1.65
1.55

1984 '85 '86 '87 '88 '89 '90 '91 '92 '93

54. In 1992, the state of Texas had about 1.3×10^5 farms with an average of 7.1×10^2 acres per farm. What was the total number of acres devoted to farmland in Texas that year? (*Source:* National Agricultural Statistics Service, U.S. Dept. of Agriculture)

Read the material at the front of the book regarding how scientific calculators display numbers in scientific notation. You are given two numbers as they would appear in such a display. Determine what the display would be if **(a)** the numbers are multiplied and **(b)** if the first number is divided by the second. Then verify using your calculator.

55. | 3.4 | 12 | | 2 | 5 | **56.** | 4.8 | 6 | | 2 | 15 |
57. | 4.2 | 10 | | 2 | −8 | **58.** | 2.6 | 7 | | 2 | −8 |

REVIEW EXERCISES

List all positive integer factors of the number. See Section 1.1.

59. 18   **60.** 36   **61.** 48   **62.** 23

CHAPTER 3 SUMMARY

KEY TERMS

3.1 polynomial
descending powers
degree of a term
degree of a polynomial

trinomial
binomial
monomial

3.8 scientific notation

NEW SYMBOLS

x^{-n} x to the negative n power

QUICK REVIEW

CONCEPTS	EXAMPLES
3.1 ADDITION AND SUBTRACTION OF POLYNOMIALS	
Addition: Add like terms.	Add. $\begin{aligned} 2x^2 + 5x - 3 \\ 5x^2 - 2x + 7 \\ \hline 7x^2 + 3x + 4 \end{aligned}$
Subtraction: Change the signs of the terms in the second polynomial and add to the first polynomial.	Subtract. $\begin{aligned} &(2x^2 + 5x - 3) - (5x^2 - 2x + 7) \\ &= (2x^2 + 5x - 3) + (-5x^2 + 2x - 7) \\ &= -3x^2 + 7x - 10 \end{aligned}$

Chapter Summaries present students with a helpful, section-referenced overview of each chapter. The Chapter Summary is followed by the Chapter Review Exercises, the Chapter Test, and the Cumulative Exercises.

Connections

Connections boxes that provide connections to the real world or to other mathematical concepts or other disciplines open each chapter and appear throughout the text—almost one for every section. Most of these include thought-provoking questions for writing or class discussion. The Connections provide motivation for the topic under discussion, show how mathematics is used in many different aspects of life, give some historical background, or provide a larger context for the current material.

Calculator Coverage

Connections boxes that focus on scientific and graphics calculators address the growing interest in the use of calculator technology, as well as provide extrinsic motivation. These Connections boxes are included in addition to the calculator passages that are part of the regular text and are optional in nature. Each such Connections box is specially marked with a scientific or graphics calculator symbol. Corresponding exercises are included in the exercise sets.

Graph Reading Activities

A variety of graphs from both popular and professional media sources are used throughout the text to help students visualize mathematics and interpret real data.

EXERCISES

More than 80 percent of the exercises are new to this edition. Care has been taken to pair exercises (evens with odds) and to grade the exercises with regard to increasing difficulty. Exercises now include a number of special types:

Mathematical Connections exercises tie together different topics and highlight the relationships among various concepts and skills. These multiple-skill and multiple-concept exercises sharpen students' problem-solving techniques and improve students' critical thinking abilities. Mathematical Connections exercises are included in most exercise sets and are grouped under a special heading.

Conceptual and writing exercises are designed to require a deeper understanding of concepts. Over 200 of these 600 exercises require the student to respond by writing a few sentences. Answers are not given for the writing exercises because they are open-ended, and instructors may use them in different ways.

Challenging exercises require the student to go beyond the examples in the text.

Calculator exercises are included in some sections and require the student to use a scientific calculator. Other sections include passages on the use of graphics calculators accompanied by appropriately designed exercises. These are grouped and identified with a special symbol so that they can be easily omitted if the instructor prefers.

Applications have been updated and rewritten to include interesting and realistic information. In many cases they use actual data from current events, sports, and other sources.

Cumulative Reviews end each chapter. These begin with Chapter 2 and test the topics covered from the beginning of the text up to that point.

SUPPLEMENTS

Our extensive supplemental package includes an Annotated Instructor's Edition, testing materials, solutions, software, and videotapes.

For the Instructor

Annotated Instructor's Edition   This edition provides instructors with immediate access to the answers to every exercise in the text, with the exception of writing exercises. Each answer is printed in color in the margin or next to the corresponding text exercise.

Symbols are used to identify the conceptual ⊙, writing ✎, and challenging ▲ exercises to assist in making homework assignments. Scientific ▦ and graphics calculator ▦ exercises are marked in both the student and instructor texts. Additional exercises, called Chalkboard Exercises, parallel almost every example, and Teaching Tips corresponding to the text discussions are also included in the Annotated Instructor's Edition.

Instructor's Test Manual   The Instructor's Test Manual includes short-answer and multiple-choice versions of a placement test; six forms of chapter tests for each chapter, including four open response and two multiple-choice forms; short-answer and multiple-choice forms of a final examination; and an extensive set of additional exercises (including more "mixed" exercises) providing 10 to 20 extra exercises for each textbook objective that instructors may use as an additional source of questions for tests, quizzes, or student review of difficult topics. Finally, this manual also includes a list of all conceptual, writing, connection, challenging, and calculator exercises.

Instructor's Solution Manual   This book includes detailed, worked-out solutions to each section exercise in the book, including conceptual *and* writing exercises. This manual also includes a list of all conceptual, writing, connection, challenging, and calculator exercises.

Instructor's Answer Manual   This manual includes answers to all exercises (except writing) and a list of conceptual, writing, connection, challenging, and calculator exercises.

HarperCollins Test Generator/Editor for Mathematics with QuizMaster Available in IBM (both DOS and Windows applications) and Macintosh versions, the Test Generator is fully networkable. The Test Generator enables instructors to select questions by objective, section, or chapter, or to use a ready-made test for each chapter. The Editor allows instructors to edit any preexisting data or to easily create their own questions. The software is algorithm driven, so the instructor may regenerate constants while maintaining problem type, providing a very large number of test or quiz items in multiple-choice and/or open response formats for one or more test forms. The system features printed graphics and accurate mathematics symbols. **QuizMaster** enables instructors to create tests and quizzes using the Test Generator/ Editor and save them to disk so students can take the test or quiz on a stand-alone computer or network. **QuizMaster** then grades the test or quiz and allows the instructor to create reports on individual students or entire classes. CLAST and TASP versions of this package are also available for IBM and Mac machines.

For the Student

Student's Solution Manual   This book contains solutions to every odd-numbered section exercise (including conceptual and writing exercises) as well as solutions to all Connections, chapter review exercises, chapter tests, and cumulative review exercises. (ISBN 0-673-99544-5)

Student's Study Guide   This book provides additional practice and reinforcement for each learning objective in the textbook. Self-tests are included at the end of every chapter.

Interactive Mathematics Tutorial Software with Management System   This innovative package is also available in DOS, Windows, and Macintosh versions and is fully networkable. As with the Test Generator/Editor, this software is algorithm driven, which automatically regenerates constants so that the numbers rarely repeat in a problem type when students revisit any particular section. The tutorial is objective-based, self-paced, and provides unlimited opportunities to review lessons and to practice problem solving. If students give a wrong answer, they can ask to see the problem worked out and get a textbook page reference. Many problems include hints for first incorrect responses. Tools such as an on-line glossary and Quick Reviews provide definitions and examples, and an on-line calculator aids students in computation. The program is menu-driven for ease of use and on-screen help can be obtained at any time with a single keystroke. Students' scores are calculated at the end of each lesson and can be printed for a permanent record. The optional **Management System** lets instructors record student scores on disk and print diagnostic reports for individual students or classes. CLAST and TASP versions of this tutorial are also available for both IBM and Mac machines. This software may also be purchased by students for home use. Student versions include record-keeping and practice tests.

Videotapes   A new videotape series has been developed to accompany *Beginning Algebra,* Seventh Edition. In a separate lesson for each section in the book, the series covers all objectives, topics, and problem-solving techniques discussed in the text.

Overcoming Math Anxiety   This book, written by Randy Davidson and Ellen Levitov, includes step-by-step guides to problem solving, note taking, and applied problems. Students can discover the reasons behind math anxiety and ways to overcome those obstacles. The book also will help them learn relaxation techniques, build better math skills, and improve study habits. (ISBN 0-06-501651-3)

OTHER BOOKS IN THIS SERIES   Other textbooks in this series include: *Intermediate Algebra,* Seventh Edition, *Intermediate Algebra with Early Graphs and Functions,* Seventh Edition, *Algebra for College Students,* Third Edition, and *Beginning and Intermediate Algebra.*

ACKNOWLEDGMENTS   We appreciate the many contributions of users of the sixth edition of the book. We also wish to thank our reviewers for their insightful comments and suggestions:

Barbara R. Allen, *Southeastern Louisiana University*
John Allen, *Saddleback College*
Robert D. Baer, *Miami University—Hamilton Campus*
Gwen M. Barber, *Brunswick College*
Richard B. Basich, *Lakeland Community College*
Elbert Bassham, *Sul Ross State University*
Randall Brian, *Vincennes University*
Stanley Brinton, *Long Beach City College*

Ellen E. Casey, *Massachusetts Bay College and Boston University*
Sandra DeLozier Coleman, *Fayetteville Technical Community College*
James A. Condor, *Manatee Community College*
Linda F. Crabtree, *Longview Community College*
Betsy Dahl, *Glendale Community College*
John DeCoursey, *Vincennes University*

Carol DeVille, *Louisiana Technical University*

Jotilley Dortch, *Paducah Community College*

Lisa E. Downing, *Oakland Community College—Auburn Hills*

Elizabeth Farmer, *Allan Hancock College*

Jeanne Fitzgerald, *Phoenix College*

Donna H. Foster, *Piedmont Technical College*

Herb Gamage, *Alpena Community College*

Elaine M. Hale, *Georgia State University*

Debra Hall, *Indiana University— Purdue University at Fort Wayne*

Elizabeth S. Harold, *University of New Orleans*

Harold Hiken, *University of Wisconsin—Milwaukee*

Everett House, *Nashville State Technical Institute*

Margaret Donovan Hovde, *Grossmont College*

Dale W. Hughes, *Johnson County Community College*

Patricia Ann Jenkins, *South Carolina State University*

Robert Kaiden, *Lorain County Community College*

Margaret Kimbell, *Texas State Technical College*

Troy Kinast, *The Wichita State University*

Linda H. Kodama, *University of Hawaii, Kapiolani Community College*

Jaclyn LeFebvre, *Illinois Central College*

Laila Marei, *San Joaquin Delta College*

Steven E. Martin, *Richard Bland College of the College of William and Mary*

Colleen K. McGraw, *Syracuse University*

Philip J. Metz, *Passaic County Community College*

Elmo Nash, *Butler County Community College*

Magdala Ray, *Palm Beach Community College*

Jane Roads, *Moberly Area Community College*

Sandra V. Rumore, *Liberty University*

Candido K. Sanchez, *Miami-Dade Community College—Downtown Campus*

Samuel Sargis, *Modesto Junior College*

Arnold Schroeder, *Long Beach City College*

Elroy Smith, *Palto Alto College*

Ronald Staszkow, *Ohlone College*

Bill Stoner, *Sacramento City College*

Glynna Strait, *Odessa College*

Charles Sweatt, *Odessa College*

W. Kim Thornsburg, *Sul Ross State University*

Mary K. Vaughn, *Texas State Technical College*

Lenore Vest, *Lower Columbia College*

Gail Wiltse, *St. John's River Community College*

Alice Woolam, *Pensacola Junior College*

As always, Paul Eldersveld, *College of DuPage,* has done an outstanding job of coordinating all the print supplements for us.

We also wish to thank those who did an excellent job checking all the answers:

Gwen M. Barber, *Brunswick College*

Linda Becerra, *University of Houston—Downtown Campus*

Michael Butler, *Mt. San Antonio College*

Betsy Dahl, *Glendale Community College*

Grace DeVelbiss, *Sinclair Community College*

Dale W. Hughes, *Johnson County Community College*

Nancy Johnson, *Broward Community College—North Campus*

Michael Kirby, *Tidewater Community College*

Robert Malena, *Community College of Allegheny County—South Campus*

Kathleen L. Pellissier

Candido K. Sanchez, *Miami Dade Community College—Downtown Campus*

Our appreciation also goes to Tommy Thompson, *Cedar Valley College,* for his suggestions for the feature "To the Student: Success in Algebra."

Special thanks go to the dedicated staff at HarperCollins who have worked so long and hard to make this book a success: Karin Wagner, Sandi Goldstein, Ellen Keith, Lisa De Mol, Anne Kelly, George Duda, Ed Moura, and Linda Youngman.

Margaret L. Lial
E. John Hornsby, Jr.

TO THE STUDENT: SUCCESS IN ALGEBRA

The main reason students have difficulty with mathematics is that they don't know how to study it. Studying mathematics *is* different from studying subjects like English or history. The key to success is regular practice.

This should not be surprising. After all, can you learn to play the piano or to ski well without a lot of regular practice? The same thing is true for learning mathematics. Working problems nearly every day is the key to becoming successful. Here is a list of things you can do to help you succeed in studying algebra.

1. *Attend class regularly.* Pay attention in class to what your teacher says and does, and make careful notes. In particular, note the problems the teacher works on the board and copy the complete solutions. Keep these notes separate from your homework to avoid confusion when you read them over later.

2. Don't hesitate to ask questions in class. It is not a sign of weakness, but of strength. There are always other students with the same question who are too shy to ask.

3. *Read your text carefully.* Many students read only enough to get by, usually only the examples. Reading the complete section will help you to be successful with the homework problems. Most exercises are keyed to specific examples or objectives that will explain the procedures for working them.

4. Before you start on your homework assignment, rework the problems the teacher worked in class. This will reinforce what you have learned. Many students say, "I understand it perfectly when you do it, but I get stuck when I try to work the problem myself."

5. Do your homework assignment only *after* reading the text and reviewing your notes from class. Check your work with the answers in the back of the book. If you get a problem wrong and are unable to see why, mark that problem and ask your instructor about it. Then practice working additional problems of the same type to reinforce what you have learned.

6. Work as neatly as you can. Write your symbols neatly, and make sure the problems are clearly separated from each other. Working neatly will help you to think clearly and also make it easier to review the homework before a test.

7. After you have completed a homework assignment, look over the text again. Try to decide what the main ideas are in the lesson. Often they are clearly highlighted or boxed in the text.

8. Use the chapter test at the end of each chapter as a practice test. Work through the problems under test conditions, without referring to the text or the answers until you are finished. You may want to time yourself to see how long it takes you. When you have finished, check your answers against those in the back of the book and study those problems that you missed. Answers are referenced to the appropriate sections of the text.

9. Keep any quizzes and tests that are returned to you and use them when you study for future tests and the final exam. These quizzes and tests indicate what your instructor considers most important. Be sure to correct any problems on these tests that you missed, so you will have the corrected work to study.

10. Don't worry if you do not understand a new topic right away. As you read more about it and work through the problems, you will gain understanding. Each time you look back at a topic you will understand it a little better. No one understands each topic completely right from the start.

AN INTRODUCTION TO SCIENTIFIC CALCULATORS

There is little doubt that the appearance of handheld calculators over two decades ago and the later development of scientific calculators have changed the methods of learning and studying mathematics forever. Where the study of computations with tables of logarithms and slide rules made up an important part of mathematics courses prior to 1970, today the widespread availability of calculators make their study a topic only of historical significance.

Most consumer models of calculators are inexpensive. At first, however, they were costly. One of the first consumer models available was the Texas Instruments SR-10, which sold for about $150 in 1973. It could perform the four operations of arithmetic and take square roots, but could do very little more.

In the past two decades, the handheld calculator has become an integral part of our everyday existence. Today calculators come in a large array of different types, sizes, and prices. For the course for which this textbook is intended, the most appropriate type is the *scientific calculator,* which costs between ten and twenty dollars. While some scientific calculators have advanced features such as programmability and graphing capability, these two features are not essential for the study of the material in this text.

In this introduction, we explain some of the features of scientific calculators. However, remember that calculators vary among manufacturers and models, and that while the methods explained here apply to many of them, they may not apply to your specific calculator. In particular, they *do not* apply to modern graphics calculators. For this reason, it is important to remember that *this is only a guide, and is not intended to take the place of your owner's manual.* Always refer to the manual in the event you need an explanation of how to perform a particular operation.

FEATURES AND FUNCTIONS OF MOST SCIENTIFIC CALCULATORS

Most scientific calculators use *algebraic logic.* (Models sold by Texas Instruments, Sharp, Casio, and Radio Shack, for example, use algebraic logic.) A notable exception is Hewlett Packard, a company whose calculators use *Reverse Polish Notation* (RPN). In this introduction, we explain the use of calculators with algebraic logic.

Arithmetic Operations

To perform an operation of arithmetic, simply enter the first number, touch the operation key ($+$, $-$, $\times$, or $\div$), enter the second number, and then touch the $=$ key. For example, to add 4 and 3, use the following keystrokes.

(The final answer is displayed in color.)

Change Sign Key

The key marked $\boxed{\pm}$ allows you to change the sign of a display. This is particularly useful when you wish to enter a negative number. For example, to enter -3, use the following keystrokes.

Memory Key

Scientific calculators can hold a number in memory for later use. The label of the memory key varies among models; two of these are $\boxed{M}$ and $\boxed{STO}$. $\boxed{M+}$ and $\boxed{M-}$ allow you to add to or subtract from the value currently in memory. The memory recall key, labeled $\boxed{MR}$, $\boxed{RM}$, or $\boxed{RCL}$, allows you to retrieve the value stored in memory.

Suppose that you wish to store the number 5 in memory. Enter 5, then touch the key for memory. You can then perform other calculations. When you need to retrieve the 5, touch the key for memory recall.

If a calculator has a constant memory feature, the value in memory will be retained even after the power is turned off. Some advanced calculators have more than one memory. It is best to read the owner's manual for your model to see exactly how memory is activated.

Clearing/Clear Entry Keys

These keys allow you to clear the display or clear the last entry entered into the display. They are usually marked $\boxed{C}$ and $\boxed{CE}$. In some models, touching the $\boxed{C}$ key once will clear the last entry, while touching it twice will clear the entire operation in progress.

Second Function Key

This key is used in conjunction with another key to activate a function that is printed *above* an operation key (and not on the key itself). It is usually marked $\boxed{2nd}$. For example, suppose you wish to find the square of a number, and the squaring function (explained in more detail later) is printed above another key. You would need to touch $\boxed{2nd}$ before the desired squaring function can be activated.

Square Root Key

Touching the square root key, $\boxed{\sqrt{x}}$, will give the square root (or an approximation of the square root) of the number in the display. For example, to find the square root of 36, use the following keystrokes.

The square root of 2 is an example of an irrational number (Chapter 8). The calculator will give an approximation of its value, since the decimal for $\sqrt{2}$ never terminates and never repeats. The number of digits shown will vary among models. To find an approximation of $\sqrt{2}$, use the following keystrokes.

   An approximation

Squaring Key

This key, $\boxed{x^2}$, allows you to square the entry in the display. For example, to square 35.7, use the following keystrokes.

The squaring key and the square root key are often found on the same key, with one of them being a second function (that is, activated by the second function key, described above).

Reciprocal Key

The key marked $\boxed{1/x}$ is the reciprocal key. (When two numbers have a product of 1, they are called *reciprocals*.) Suppose that you wish to find the reciprocal of 5. Use the following keystrokes.

$$\boxed{5} \quad \boxed{1/x} \quad \boxed{\qquad 0.2}$$

Inverse Key

Some calculators have an inverse key, marked $\boxed{\text{INV}}$. Inverse operations are operations that "undo" each other. For example, the operations of squaring and taking the square root are inverse operations. The use of the $\boxed{\text{INV}}$ key varies among different models of calculators, so read your owner's manual carefully.

Exponential Key

The key, marked $\boxed{x^y}$ or $\boxed{y^x}$, allows you to raise a number to a power. For example, if you wish to raise 4 to the fifth power (that is, find 4^5), use the following keystrokes.

$$\boxed{4} \quad \boxed{x^y} \quad \boxed{5} \quad \boxed{=} \quad \boxed{\qquad 1024}$$

Root Key

Some calculators have this key specifically marked $\boxed{\sqrt[x]{x}}$ or $\boxed{\sqrt[x]{y}}$; with others, the operation of taking roots is accomplished by using the inverse key in conjunction with the exponential key. Suppose, for example, your calculator is of the latter type and you wish to find the fifth root of 1024. Use the following keystrokes.

$$\boxed{1} \quad \boxed{0} \quad \boxed{2} \quad \boxed{4} \quad \boxed{\text{INV}} \quad \boxed{x^y} \quad \boxed{5} \quad \boxed{=} \quad \boxed{\qquad 4}$$

Notice how this "undoes" the operation explained in the exponential key discussion above.

Pi Key

The number π is an important number in mathematics. It occurs, for example, in the area and circumference formulas for a circle. By touching the $\boxed{\pi}$ key, you can get in the display the first few digits of π. (Because π is irrational, the display shows only an approximation.) One popular model gives the following display when the $\boxed{\pi}$ key is activated: $\boxed{\qquad 3.1415927}$.

log and ln Keys

While logarithms are not covered in this course, they form an important part of subsequent courses. In order to find the common logarithm (base ten logarithm) of a number, enter the number and touch the $\boxed{\text{log}}$ key. To find the natural logarithm, enter the number and touch the $\boxed{\text{ln}}$ key. For example, to find these logarithms of 10, use the following keystrokes.

Common logarithm: $\boxed{1} \quad \boxed{0} \quad \boxed{\text{log}} \quad \boxed{\qquad 1}$

Natural logarithm: $\boxed{1} \quad \boxed{0} \quad \boxed{\text{ln}} \quad \boxed{\qquad 2.3025851}$ An approximation

10^x and e^x Keys

These keys are special exponential keys, and are inverses of the log and ln keys. (On some calculators, they are second functions.) The number e is an irrational number and is the base of the natural logarithm function. Its value is approximately 2.71828. To use these keys, enter the number to which 10 or e is to be raised, and then touch the $\boxed{10^x}$ or $\boxed{e^x}$ key. For example, to raise 10 or e to the 2.5 power, use the following keystrokes.

Base is 10: $\boxed{2} \quad \boxed{.} \quad \boxed{5} \quad \boxed{10^x} \quad \boxed{\qquad 316.22777}$ An approximation

Base is e: $\boxed{2} \quad \boxed{.} \quad \boxed{5} \quad \boxed{e^x} \quad \boxed{\qquad 12.182494}$ An approximation

(Note: If no 10^x key is specifically shown, touching INV followed by log accomplishes raising 10 to the power x. Similarly, if no e^x key is specifically shown, touching INV followed by ln accomplishes raising e to the power x.)

Factorial Key

The factorial key, $x!$, evaluates the factorial of any nonnegative integer within the limits of the calculator. For example, $5! = 1 \cdot 2 \cdot 3 \cdot 4 \cdot 5$. To use the factorial key, just enter the number and touch $x!$. To evaluate 5! on a calculator, use the following keystrokes.

OTHER FEATURES OF SCIENTIFIC CALCULATORS

When decimal approximations are shown on scientific calculators, they are either *truncated* or *rounded.* To see which of these a particular model is programmed to do, evaluate 1/18 as an example. If the display shows .0555555 (last digit 5), it truncates the display. If it shows .0555556 (last digit 6), it rounds off the display.

When very large or very small numbers are obtained as answers, scientific calculators often express these numbers in scientific notation. For example, if you multiply 6,265,804 by 8,980,591, the display might look like this:

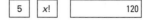

The "13" at the far right means that the number on the left is multiplied by 10^{13}. This means that the decimal point must be moved 13 places to the right if the answer is to be expressed in its usual form. Even then, the value obtained will only be an approximation: 56,270,623,000,000.

Advanced Features

Two features of advanced scientific calculators are programmability and graphics capability. A programmable calculator has the capability of running small programs, much like a mini-computer. A graphics calculator can be used to plot graphs of functions on a small screen. One of the issues in mathematics education today deals with how graphics calculators should be incorporated into the curriculum. Their availability in the 1990s parallels the availability of scientific calculators in the 1980s, and they are responsible for new approaches and attitudes in mathematics education reform.

AN INTRODUCTION TO GRAPHICS CALCULATORS*

CAPABILITIES OF GRAPHICS CALCULATORS

Graphics calculators are a result of the amazingly rapid evolution in computer technology toward packaging more power into smaller "boxes." These machines have powerful graphing capabilities in addition to the full range of features found on programmable scientific calculators. Instead of a one-line display, graphics calculators typically can show up to eight lines of text. This makes it much easier to keep track of the steps of your work, whether you are doing routine computations, entering a long mathematical function, or writing a program. Like programmable scientific calculators, graphics calculators are capable of doing many things we have previously come to expect only from computers. Programs can be written relatively easily, and after they are stored in memory, the programs are always available. Like computers, graphics calculators can be programmed to include graphic displays as part of the program.

It takes some study to learn how to use graphics calculators, but they are much easier to master than most computers—and they can go wherever you do! New models with added features, more memory, greater ease of use, and other improvements are frequently being introduced. The most popular brand names at this time are Casio, Sharp, Texas Instruments, and Hewlett-Packard.

GRAPHICS CALCULATOR FEATURES

Every graphics calculator has keys for the usual operations of arithmetic and all commonly used functions (square root, x^2, log, ln, and so on). All of them can graph functions of the form $y = f(x)$ and have programming capabilities. Except for the cheapest ones, graphics calculators may have a variety of additional features. Before buying one, you should consider which features you are likely to need.

Many of the features of the typical graphics calculator do not play a role in the topics covered in this book, and further mathematics courses will be needed in order to fully appreciate their power. The more advanced models have the capability to perform some symbolic manipulations (such as factoring, adding polynomials, and performing operations with rational algebraic expressions). However, these calculators are more expensive and more difficult to learn to use because of their increased complexity and the fact that they do not use standard algebraic order of operations.

ADVICE ON USING A GRAPHICS CALCULATOR

1. **BASICS**   Graphics calculators have forty-nine or more keys. Most modern desk-top computers have 101 keys on their keyboards. With fewer keys, each key must be used for more actions, so you will find special mode-changing keys such as **"2nd," "shift," "alpha,"** and **"mode".** Become familiar with

* Prepared by Jim Eckerman of *American River College.*

the capabilities of the machine, the layout of the keyboard, how to adjust the screen contrast, and so on. Remember that a graphics calculator is composed of two parts: the machinery *and* the owner's manual!

2. **EDITING**   When keying in expressions, you can pause at any time and use the arrow keys, located at the upper right of the keyboard, to move the cursor to any point in the text. You can then make changes by using the **"DEL"** key to delete and the **"INS"** key to insert material. ["INS" is the "second function" of "DEL."] After an expression has been entered or a calculation made, it can still be edited by using the **edit/replay** feature, available on almost every calculator model. On TI calculators (except TI-81), use "2nd, ENTER" to return to the previously entered expression. On Casio calculators, use the left or right arrow key, and on Sharp models use "2nd, up arrow."

3. **SCIENTIFIC NOTATION**   Learn how to enter and read data in **scientific notation** form. This form is used when the numbers become too large or too small (too many zeros between the decimal point and the first significant digit) for the machine's display.

4. **FUNCTION GRAPHING**

   A. **Setting the Range or Window**   Learn to set **"RANGE"** or **"WINDOW"** values to delineate a window that is appropriate for the function you are graphing before using the **"Graph"** key. This involves keying in the minimum and maximum values of x and y that will be displayed on the screen, along with the distance to be used between tick marks along the axes. If you do not do this, you will often find your graph screen blank! Usually one can quickly find a point on the graph by substitution of either zero or one for x. Use the **"RANGE"** or **"WINDOW"** command to set the x-values to the left and right of the x-coordinate of this point and set the y-values above and below the y-coordinate of this point. Then the **"zoom out"** feature can be used to see more of the graph.

   B. **Using the Function Memory**   Learn how to redraw graphs without reentering the function. Often you will need to change the **"RANGE"** settings several times before you get the "window" that is most appropriate for your function. All graphing calculators allow you to do this without reentering the function. If you plan to graph a particular function often, then you should either store it in the **function memory** (which is labeled "$y = $" on TI and "EQTN" on Sharp models) or store it in **program memory.** Of course, you have to write a program with your function as part of the program in order to make use of the program memory. Once entered into the machine's memory, this function can be used at any time.

   C. **Using the Trace Feature**   With the **"Trace"** feature, the left/right arrows can be used to move the cursor along the last curve plotted, and the values of x and y will be displayed for each point plotted on the screen. If more than one graph was plotted, one can move the cursor vertically between the different graphs by using the up/down arrows.

   D. **Using the Zoom Feature**   The **"zoom"** feature allows a quick redrawing of your graph using smaller ranges of values for x and y (**"zoom in"**) or larger ranges of values (**"zoom out"**). Thus one can easily examine the behavior of the graph of a function within the close vicinity of a particular point or the general behavior as seen from farther away. Using **"zoom box,"** a box can be drawn for a particular region for closer inspection of the graph within that region.

SOLVING EQUATIONS AND SYSTEMS GRAPHICALLY

Some mathematical procedures can be quite difficult or even impossible to do algebraically but can be done easily and to a very high degree of accuracy using a graphics calculator. Listed below are some examples.

1. **SOLVING EQUATIONS OF THE FORM** $f(x) = k$ The quickest way to solve this type of equation is to form the new equation $y = f(x) - k$ and then use the built-in Equation Solver or Root Finder feature. If your calculator does not have this capability or if you prefer to see the solution graphically, you can find the roots of $y = f(x) - k$ by locating the points where the graph crosses the x-axis.

2. **SOLVING SYSTEMS OF TWO EQUATIONS IN TWO UNKNOWNS** Some graphics calculators have built-in programs for solving systems of linear equations, but all of them can be used to solve systems of two equations (linear or not) by finding intersection points. If both equations can be solved explicitly for one variable in terms of the other, then the two equations can be put into the forms $y = f(x)$ and $y = g(x)$. Then we can graph both in the same window and zoom in on the point or points where the two curves intersect.

PROGRAMMING

Many formulas are used often enough to justify automating the process of evaluating them. The distance formula and the quadratic formula are two examples of this. Graphics calculators are able to store programs that will perform such tasks. The realm of *programming* is much different from that of *using* a graphics calculator. Many programs are available from the manufacturers and the literature that they support. Some students like to experiment and program their own calculators. The old adage "The sky's the limit" is certainly applicable to programming, and it is up to the user's imagination as to how far he or she wishes to take the programming capability of the calculator.

SOME SUGGESTIONS FOR REDUCING FRUSTRATION

We all find ways to make even the simplest machines do the wrong things without even trying. One of the more common problems with graphics calculators is getting a blank screen when a graph was expected. This usually results from not setting the **"WINDOW"** values appropriately before graphing the function, although it could also easily result from incorrectly entering the function.

A common problem that is particularly annoying is interpreting cryptic error messages such as **"Syn ERROR"** and **"Ma ERROR."** (Keep that manual handy!) The most common mistakes are made entering formulas and using special functions. For example, on the Casio machines **"Syn ERROR"** means that a mistake was made when entering the function or operation, such as entering "Graph $Y = \log x^y$" instead of "Graph $Y = \log x^2$".

Another common error is having more right parentheses than left parentheses. To confuse us further, these same machines think it is perfectly OK to have more left parentheses than right parentheses. For instance, the expression $5(3 - 4(2 + 7)$ has two left parentheses and one right parenthesis, but it will be evaluated as $5(3 - 4(2 + 7))$. The message **"Ma ERROR"** appears when a number is too large or when a number is not allowed. If you try to find the 1000th power of ten or divide a number by zero you will most certainly see some kind of error message. By pressing one of the **"cursor"** keys on the Casio you will see the cursor blinking at the location of the error in your expression. When the Texas Instruments models detect an error, they display a special menu that lists a code number and a name for the type of error. For certain types of errors the choice **"Go to error"** is offered. The TI-85 will display the number 9.99999999 E999 but shows "ERROR 01 OVERFLOW" for 10 E999 (which means 10 times the 999th power of ten). The latter and other more advanced machines display "(0, 2)" when asked to find the square root of -4. The "(0, 2)" represents the complex number $0 + 2i$.

SOME FINAL COMMENTS

While studying mathematics it is important to learn the mathematical concepts well enough to make intelligent decisions about when to use and when not to use "high tech" aids such as computers and graphics calculators. These machines make it easy to experiment with graphs of mathematical relations. One can learn much about the behavior of different types of functions by playing "what if" games with the formulas. However, in a timed test situation you may find yourself spending too much time working with the graphics calculator when a quick algebraic solution and a rough sketch with pencil and paper are more appropriate.

To get the most return on your investment, learn to use as many features of your machine as possible. Of course, some of the features may not be of use to you, so feel free to ignore them. A first session of two or three hours with your graphics calculator and your user's manual is essential. Be sure to keep your manual handy, referring to it when needed.

A final word of caution: These machines are fun to use, but they can be addictive. So set time limits for yourself, or you may find that your graphics calculator has been more of a detriment than a help!

BEGINNING ALGEBRA

SEVENTH EDITION

THE REAL NUMBER SYSTEM

CONNECTIONS

The Italian astronomer Galileo Galilei (1564–1643) once wrote: *Mathematics is the language with which God has written the Universe.* Can you imagine trying to function in our world without numbers? The numbers studied in arithmetic are positive numbers. In this chapter we extend these numbers to include *negative numbers*, to be introduced in Section 1.4. The Hindus were among the first to see the possibilities in extending the number system to include negative numbers. Leonardo da Pisa (about 1170–1250), an Italian, while working on a financial problem, concluded that the solution must be a negative number, that is, a financial loss. Today, the use of negative numbers is common in business and finance. In 1545, the rules governing operations with negative numbers were published by Girolamo Cardano in his *Ars Magna* (*Great Art*). We will study these rules in this chapter.

In this chapter we examine properties of the real numbers that are used extensively in this book and all more advanced mathematics courses. These properties enable us to solve equations, work with algebraic fractions and radicals, and graph sets of points. They are the basic rules of algebra. As we go through the book, we shall see that algebra allows us to solve a large variety of problems.

1.1 FRACTIONS

FOR EXTRA HELP	OBJECTIVES
📖 **SSG** pp. 1–8 **SSM** pp. 1–6	1 ▶ Learn the definition of *factor*.
📼 **Video** 1	2 ▶ Write fractions in lowest terms.
	3 ▶ Multiply and divide fractions.
💾 **Tutorial** IBM MAC	4 ▶ Add and subtract fractions.
	5 ▶ Solve problems that involve operations with fractions.

As preparation for the study of algebra, this section begins with a brief review of arithmetic. In everyday life the numbers seen most often are the **natural numbers,**

$$1, 2, 3, 4, \ldots,$$

the **whole numbers,**

$$0, 1, 2, 3, 4, \ldots,$$

and the **fractions,** such as

$$\frac{1}{2}, \quad \frac{2}{3}, \quad \text{and} \quad \frac{15}{7}.$$

The parts of a fraction are named as follows.

$$\frac{4}{7} \quad \begin{array}{l} \leftarrow \textbf{numerator} \\ \leftarrow \textbf{denominator} \end{array}$$

◆ **CONNECTIONS** ◆

A common use of fractions is to measure dimensions of tools, amounts of building materials, and so on. We need to understand how to add, subtract, multiply, and divide fractions in order to solve many types of measurement problems.

FOR DISCUSSION OR WRITING

Discuss some situations in your experience where you have needed to perform the operations of addition, subtraction, multiplication, or division on fractions. (*Hint:* To get you started, think of art projects, carpentry projects, adjusting recipes, and working on cars.)

1 ▶ Learn the definition of *factor*.

▶ In the statement $2 \times 9 = 18$, the numbers 2 and 9 are called **factors** of 18. Other factors of 18 include 1, 3, 6, and 18. The result of the multiplication, 18, is called the **product.**

The number 18 is **factored** by writing it as the product of two or more numbers. For example, 18 can be factored in several ways, as $6 \cdot 3$, or $18 \cdot 1$, or $9 \cdot 2$, or $3 \cdot 3 \cdot 2$. In algebra, raised dots are used instead of the $\times$ symbol to indicate multiplication.

A natural number (except 1) is **prime** if it has only itself and 1 as factors. "Factors" are understood here to mean natural number factors. (By agreement, the number 1 is not a prime number.) The first dozen primes are listed here.

$$2, 3, 5, 7, 11, 13, 17, 19, 23, 29, 31, 37$$

A natural number (except 1) that is not prime is a **composite** number.

It is often useful to find all the **prime factors** of a number—those factors that are prime numbers. For example, the only prime factors of 18 are 2 and 3.

EXAMPLE 1
Factoring Numbers

Write the number as the product of prime factors.

(a) 35

Write 35 as the product of the prime factors 5 and 7, or as

$$35 = 5 \cdot 7.$$

(b) 24

One way to begin is to divide by the smallest prime, 2, to get

$$24 = 2 \cdot 12.$$

Now divide 12 by 2 to find factors of 12.

$$24 = 2 \cdot 2 \cdot 6$$

Since 6 can be written as $2 \cdot 3$,

$$24 = 2 \cdot 2 \cdot 2 \cdot 3,$$

where all factors are prime. ■

NOTE It is not necessary to start with the smallest prime factor, as shown in Example 1(b). In fact, no matter which prime factor we start with, we will *always* obtain the same prime factorization.

---

2 ▶ Write fractions in lowest terms.

▶ We use prime numbers to write fractions in *lowest terms*. A fraction is in **lowest terms** when the numerator and denominator have no factors in common (other than 1). By the **basic principle of fractions,** if the numerator and denominator of a fraction are multiplied or divided by the same number, the value of the fraction is unchanged. Thus, the following steps are used to write a fraction in lowest terms.

Writing a Fraction in Lowest Terms

Step 1 Write the numerator and the denominator as the product of prime factors.

Step 2 Divide the numerator and the denominator by the **greatest common factor,** the product of all factors common to both.

EXAMPLE 2 ■

Writing Fractions in
Lowest Terms

Write the fraction in lowest terms.

(a) $\dfrac{10}{15} = \dfrac{2 \cdot 5}{3 \cdot 5} = \dfrac{2 \cdot 1}{3 \cdot 1} = \dfrac{2}{3}$

Since 5 is the greatest common factor of 10 and 15, dividing both numerator and denominator by 5 gives the fraction in lowest terms.

(b) $\dfrac{15}{45} = \dfrac{3 \cdot 5}{3 \cdot 3 \cdot 5} = \dfrac{1 \cdot 3 \cdot 5}{3 \cdot 3 \cdot 5} = \dfrac{1}{3}$

The factored form shows that 3 and 5 are the common factors of both 15 and 45. Dividing both 15 and 45 by $3 \cdot 5 = 15$ gives 15/45 in lowest terms as 1/3. ■

NOTE When you are factoring to write a fraction in lowest terms, you can simplify the process if you can find the greatest common factor in the numerator and denominator by inspection. For instance, in Example 2(b), we can use 15 rather than $3 \cdot 5$.

---

3 ▶ Multiply and divide fractions.

▶ The basic operations on whole numbers are addition, subtraction, multiplication, and division. These same operations apply to fractions. We multiply two fractions by first multiplying their numerators and then multiplying their denominators. This rule is written in symbols as follows.

Multiplying Fractions If $\dfrac{a}{b}$ and $\dfrac{c}{d}$ are fractions, then $\dfrac{a}{b} \cdot \dfrac{c}{d} = \dfrac{a \cdot c}{b \cdot d}$.

EXAMPLE 3 ■

Multiplying Fractions

Find the product of 3/8 and 4/9, and write it in lowest terms.

First, multiply 3/8 and 4/9.

$$\dfrac{3}{8} \cdot \dfrac{4}{9} = \dfrac{3 \cdot 4}{8 \cdot 9} \qquad \text{Multiply numerators; multiply denominators.}$$

It is easiest to write a fraction in lowest terms while the product is in factored form. Factor 8 and 9 and then divide out common factors in the numerator and denominator.

$$\dfrac{3 \cdot 4}{8 \cdot 9} = \dfrac{3 \cdot 4}{2 \cdot 4 \cdot 3 \cdot 3} = \dfrac{1 \cdot 3 \cdot 4}{2 \cdot 4 \cdot 3 \cdot 3} \qquad \text{Factor. Introduce a factor of } 1.$$

$$= \dfrac{1}{2 \cdot 3} \qquad \qquad 3 \text{ and } 4 \text{ are common factors.}$$

$$= \dfrac{1}{6} \qquad \qquad \text{Lowest terms} \quad ■$$

Two fractions are **reciprocals** of each other if their product is 1. For example, 3/4 and 4/3 are reciprocals since

$$\dfrac{3}{4} \cdot \dfrac{4}{3} = \dfrac{12}{12} = 1.$$

Also, 7/11 and 11/7 are reciprocals of each other. We use the reciprocal to divide fractions. To *divide* two fractions, multiply the first fraction and the reciprocal of the second one.

Dividing Fractions   For the fractions $\frac{a}{b}$ and $\frac{c}{d}$, $\frac{a}{b} \div \frac{c}{d} = \frac{a}{b} \cdot \frac{d}{c}$.

(To divide by a fraction, multiply by its reciprocal.)

The reason this method works will be explained in Chapter 5. The answer to a division problem is called a **quotient.** For example, the quotient of 20 and 10 is 2, since $20 \div 10 = 2$.

EXAMPLE 4

Dividing Fractions

Find the following quotients, and write them in lowest terms.

(a) $\frac{3}{4} \div \frac{8}{5} = \frac{3}{4} \cdot \frac{5}{8} = \frac{3 \cdot 5}{4 \cdot 8} = \frac{15}{32}$ Multiply by the reciprocal of $\frac{8}{5}$.

(b) $\frac{3}{4} \div \frac{5}{8} = \frac{3}{4} \cdot \frac{8}{5} = \frac{3 \cdot 8}{4 \cdot 5} = \frac{3 \cdot 4 \cdot 2}{4 \cdot 5} = \frac{6}{5}$ ■

4 ▶ Add and subtract fractions.

▶ The **sum** of two fractions having the same denominator is found by adding the numerators, and keeping the same denominator.

Adding Fractions   If $\frac{a}{b}$ and $\frac{c}{b}$ are fractions, then $\frac{a}{b} + \frac{c}{b} = \frac{a + c}{b}$.

EXAMPLE 5

Adding Fractions with the Same Denominator

Add.

(a) $\frac{3}{7} + \frac{2}{7} = \frac{3 + 2}{7} = \frac{5}{7}$ Add numerators and keep the same denominator.

(b) $\frac{2}{10} + \frac{3}{10} = \frac{2 + 3}{10} = \frac{5}{10} = \frac{1}{2}$ ■

If the fractions to be added do not have the same denominators, the rule above can still be used, but only after the fractions are rewritten with a common denominator. For example, to rewrite 3/4 as a fraction with a denominator of 32,

$$\frac{3}{4} = \frac{?}{32},$$

find the number that can be multiplied by 4 to give 32. Since $4 \cdot 8 = 32$, use the number 8. By the basic principle, we can multiply the numerator and the denominator by 8.

$$\frac{3}{4} = \frac{3 \cdot 8}{4 \cdot 8} = \frac{24}{32}$$

EXAMPLE 6

Changing the Denominator

Write 5/8 as a fraction with a denominator of 72.

Since 8 must be multiplied by 9 to get 72, multiply both numerator and denominator by 9.

$$\frac{5}{8} = \frac{5 \cdot 9}{8 \cdot 9} = \frac{45}{72}$$ ■

E X A M P L E   7 ∎ Add the following fractions.

Adding Fractions with Different Denominators

(a) $\dfrac{4}{15} + \dfrac{5}{9}$

To find the **least common denominator** (**LCD**), we first factor both denominators.

$$15 = 5 \cdot 3 \qquad \text{and} \qquad 9 = 3 \cdot 3$$

The LCD is the smallest multiple of both denominators,

$$5 \cdot 3 \cdot 3.*$$

Write each fraction with the LCD as denominator.

$$\frac{4}{15} = \frac{4 \cdot 3}{15 \cdot 3} = \frac{12}{45} \qquad \text{and} \qquad \frac{5}{9} = \frac{5 \cdot 5}{9 \cdot 5} = \frac{25}{45}$$

Now we add the two equivalent fractions to get the required sum.

$$\frac{4}{15} + \frac{5}{9} = \frac{12}{45} + \frac{25}{45} = \frac{37}{45}$$

(b) $3\dfrac{1}{2} + 2\dfrac{3}{4}$

These numbers are called mixed numbers. A **mixed number** is understood to be the sum of a whole number and a fraction. We can add mixed numbers using either of two methods.

Method 1

Rewrite both numbers as follows.

$$3\frac{1}{2} = 3 + \frac{1}{2} = \frac{3}{1} + \frac{1}{2} = \frac{6}{2} + \frac{1}{2} = \frac{6+1}{2} = \frac{7}{2}$$

$$2\frac{3}{4} = 2 + \frac{3}{4} = \frac{8}{4} + \frac{3}{4} = \frac{8+3}{4} = \frac{11}{4}$$

Now add. The common denominator is 4.

$$3\frac{1}{2} + 2\frac{3}{4} = \frac{7}{2} + \frac{11}{4} = \frac{14}{4} + \frac{11}{4} = \frac{25}{4} \quad \text{or} \quad 6\frac{1}{4}$$

Method 2

Write 3 1/2 as 3 2/4. Then add vertically.

$$
\begin{array}{ccc}
3\dfrac{1}{2} & & 3\dfrac{2}{4} \\[4pt]
 & \longrightarrow & \\[4pt]
+\,2\dfrac{3}{4} & & +\,2\dfrac{3}{4} \\[2pt]
\hline
 & & 5\dfrac{5}{4}
\end{array}
$$

Since 5/4 = 1 1/4,

$$5\frac{5}{4} = 5 + 1\frac{1}{4} = 6\frac{1}{4}, \quad \text{or} \quad \frac{25}{4}. \quad ∎$$

---

*In general, to find the LCD, use every factor that appears in either number. If a factor is repeated (as the $3 \cdot 3$ is), use the largest number of repeats in the LCD.

The **difference** between two numbers is found by subtraction. For example, $9 - 5 = 4$ so the difference between 9 and 5 is 4. Subtraction of fractions is similar to addition. Just subtract the numerators instead of adding them, according to the following definition. Again, keep the same denominator.

Subtracting Fractions

$$\frac{a}{b} - \frac{c}{b} = \frac{a-c}{b}$$

EXAMPLE 8

Subtracting Fractions

Subtract. Write the differences in lowest terms.

(a) $\dfrac{15}{8} - \dfrac{3}{8} = \dfrac{15-3}{8}$ Subtract numerators; keep the same denominator.

$$= \frac{12}{8} = \frac{3}{2} \quad \text{Lowest terms}$$

(b) $\dfrac{7}{18} - \dfrac{4}{15}$

Here, $18 = 2 \cdot 3 \cdot 3$ and $15 = 3 \cdot 5$, so the LCD is $2 \cdot 3 \cdot 3 \cdot 5 = 90$.

$$\frac{7}{18} - \frac{4}{15} = \frac{7 \cdot 5}{2 \cdot 3 \cdot 3 \cdot 5} - \frac{4 \cdot 2 \cdot 3}{2 \cdot 3 \cdot 3 \cdot 5} = \frac{35}{90} - \frac{24}{90} = \frac{11}{90}$$

(c) $\dfrac{15}{32} - \dfrac{11}{45}$

Since $32 = 2 \cdot 2 \cdot 2 \cdot 2 \cdot 2$ and $45 = 3 \cdot 3 \cdot 5$, there are no common factors, and the LCD is $32 \cdot 45 = 1440$.

$$\frac{15}{32} - \frac{11}{45} = \frac{15 \cdot 45}{32 \cdot 45} - \frac{11 \cdot 32}{45 \cdot 32} \quad \text{Get a common denominator.}$$

$$= \frac{675}{1440} - \frac{352}{1440}$$

$$= \frac{323}{1440} \quad \text{Subtract.} \quad \blacksquare$$

5 ▶ Solve problems that involve operations with fractions.

▶ Applied problems often require work with fractions. For example, when a carpenter reads diagrams and plans, he or she often must work with fractions whose denominators are 2, 4, 8, 16, or 32. Therefore, operations with fractions are sometimes necessary to solve such problems. The next example shows a typical diagram.

EXAMPLE 9

Adding Fractions to Solve a Woodworking Problem

The diagram shown on the next page appears in the book *Woodworker's 39 Sure-Fire Projects*. It is the front view of a corner bookcase/desk. Add the fractions shown in the diagram to find the aproximate height of the bookcase/desk.

We must add the following measures (in inches):

$$\frac{3}{4}, \quad 4\frac{1}{2}, \quad 9\frac{1}{2}, \quad \frac{3}{4}, \quad 9\frac{1}{2}, \quad \frac{3}{4}, \quad 4\frac{1}{2}.$$

Begin by changing 4 1/2 to 4 2/4 and 9 1/2 to 9 2/4, since the common denominator is 4. Then, use Method 2 from Example 7(b).

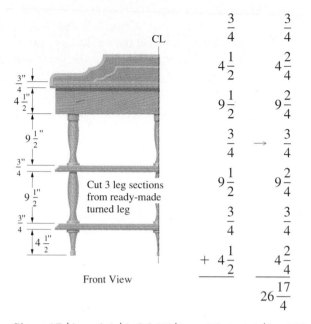

Front View

$$\begin{array}{cc}
& \dfrac{3}{4} \qquad \dfrac{3}{4} \\[6pt]
& 4\dfrac{1}{2} \qquad 4\dfrac{2}{4} \\[6pt]
& 9\dfrac{1}{2} \qquad 9\dfrac{2}{4} \\[6pt]
& \dfrac{3}{4} \;\rightarrow\; \dfrac{3}{4} \\[6pt]
& 9\dfrac{1}{2} \qquad 9\dfrac{2}{4} \\[6pt]
& \dfrac{3}{4} \qquad \dfrac{3}{4} \\[6pt]
+ & 4\dfrac{1}{2} \qquad 4\dfrac{2}{4} \\[6pt]
\hline
& \qquad\quad 26\dfrac{17}{4}
\end{array}$$

Cut 3 leg sections from ready-made turned leg

Since 17/4 = 4 1/4, 26 17/4 = 26 + 4 1/4 = 30 1/4. The approximate height is 30 1/4 inches. It is best to give answers as mixed numbers in applications like this. ∎

1.1 EXERCISES

1. In the fraction $\dfrac{3}{8}$, _____ is the numerator and is _____ the denominator.

2. How may $\dfrac{15}{7}$ be written as a mixed number?

3. If the numerator and the denominator of a fraction are different prime numbers, is the fraction in lowest terms?

4. What is the reciprocal of $\dfrac{9}{8}$?

5. The answer in a multiplication problem is called the _____ , and the answer in a division problem is called the _____ .

6. The answer in an addition problem is called the _____ , and the answer in a subtraction problem is called the _____ .

Identify the number as prime, composite, or neither.

7. 17   **8.** 23   **9.** 54   **10.** 88

11. 3458   **12.** 2895   **13.** 1   **14.** $\dfrac{2}{3}$

▦ *Write the number as the product of prime factors. See Example 1.*

15. 30   **16.** 40   **17.** 500   **18.** 700

19. 124   **20.** 120   **21.** 29   **22.** 31

Write the fraction in lowest terms. See Example 2.

23. $\dfrac{8}{16}$ **24.** $\dfrac{4}{12}$ **25.** $\dfrac{15}{18}$ **26.** $\dfrac{16}{20}$

27. $\dfrac{15}{45}$ **28.** $\dfrac{16}{64}$ **29.** $\dfrac{144}{120}$ **30.** $\dfrac{132}{77}$

31. One of the following is the correct way to write $\dfrac{16}{24}$ in lowest terms. Which one is it?

(a) $\dfrac{16}{24} = \dfrac{8+8}{8+16} = \dfrac{8}{16} = \dfrac{1}{2}$ (b) $\dfrac{16}{24} = \dfrac{4 \cdot 4}{4 \cdot 6} = \dfrac{4}{6}$

(c) $\dfrac{16}{24} = \dfrac{8 \cdot 2}{8 \cdot 3} = \dfrac{2}{3}$ (d) $\dfrac{16}{24} = \dfrac{14+2}{21+3} = \dfrac{2}{3}$

32. For the fractions $\dfrac{p}{q}$ and $\dfrac{r}{s}$, which one of the following can serve as a common denominator?

(a) $q \cdot s$ (b) $q + s$ (c) $p \cdot r$ (d) $p + r$

Find the product or quotient, and write it in lowest terms. See Examples 3 and 4.

33. $\dfrac{4}{5} \cdot \dfrac{6}{7}$ **34.** $\dfrac{5}{9} \cdot \dfrac{10}{7}$ **35.** $\dfrac{1}{10} \cdot \dfrac{12}{5}$ **36.** $\dfrac{6}{11} \cdot \dfrac{2}{3}$

37. $\dfrac{15}{4} \cdot \dfrac{8}{25}$ **38.** $\dfrac{4}{7} \cdot \dfrac{21}{8}$ **39.** $2\dfrac{2}{3} \cdot 5\dfrac{4}{5}$ **40.** $3\dfrac{3}{5} \cdot 7\dfrac{1}{6}$

41. $\dfrac{5}{4} \div \dfrac{3}{8}$ **42.** $\dfrac{7}{6} \div \dfrac{9}{10}$ **43.** $\dfrac{32}{5} \div \dfrac{8}{15}$ **44.** $\dfrac{24}{7} \div \dfrac{6}{21}$

45. $\dfrac{3}{4} \div 12$ **46.** $\dfrac{2}{5} \div 30$ **47.** $2\dfrac{5}{8} \div 1\dfrac{15}{32}$ **48.** $2\dfrac{3}{10} \div 7\dfrac{4}{5}$

49. In your own words, write an explanation of how to divide two fractions.

50. In your own words, write an explanation of how to add two fractions that have different denominators.

Find the sum or difference, and write it in lowest terms. See Examples 5–8.

51. $\dfrac{7}{12} + \dfrac{1}{12}$ **52.** $\dfrac{3}{16} + \dfrac{5}{16}$ **53.** $\dfrac{5}{9} + \dfrac{1}{3}$ **54.** $\dfrac{4}{15} + \dfrac{1}{5}$

55. $3\dfrac{1}{8} + \dfrac{1}{4}$ **56.** $5\dfrac{3}{4} + \dfrac{2}{3}$ **57.** $\dfrac{7}{12} - \dfrac{1}{9}$ **58.** $\dfrac{11}{16} - \dfrac{1}{12}$

59. $6\dfrac{1}{4} - 5\dfrac{1}{3}$ **60.** $8\dfrac{4}{5} - 7\dfrac{4}{9}$ **61.** $\dfrac{5}{3} + \dfrac{1}{6} - \dfrac{1}{2}$ **62.** $\dfrac{7}{15} + \dfrac{1}{6} - \dfrac{1}{10}$

63. A cent is equal to $\dfrac{1}{100}$ of one dollar. Two dimes added to three dimes gives an amount equal to that of one half-dollar. Give an arithmetic problem using fractions that describes this equality, using 100 as a denominator throughout.

64. Three nickels added to twelve nickels gives an amount equal to that of three quarters. Give an arithmetic problem using fractions that describes this equality, using 100 as a denominator throughout.

Work the problem. See Example 9.

65. On Thursday, April 7, 1994, Hewlett Packard stock closed at $83\frac{3}{4}$ dollars (per share). This was $2\frac{1}{4}$ dollars above the price at the start of the day. How much did one share of this stock cost at the start of the day?

66. On Friday, April 8, 1994, the New York Stock Exchange reported that the 52-week high for IBM was 60 (dollars per share) while the low was $40\frac{5}{8}$ (dollars per share). What was the difference between these prices?

67. A hardware store sells a 40-piece socket wrench set. The measure of the largest socket is $\frac{3}{4}$ inch, while the measure of the smallest socket is $\frac{3}{16}$ inch. What is the difference between these measures?

68. Two sockets in a socket wrench set have measures of $\frac{9}{16}$ inch and $\frac{3}{8}$ inch. What is the difference between these two measures?

69. A motel owner has decided to expand his business by buying a piece of property next to the motel. The property has an irregular shape, with five sides as shown in the figure. Find the total distance around the piece of property. (This is called the *perimeter* of the figure.)

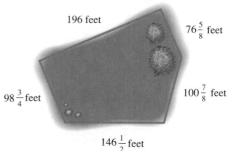

196 feet

$76\frac{5}{8}$ feet

$100\frac{7}{8}$ feet

$98\frac{3}{4}$ feet

$146\frac{1}{2}$ feet

70. A triangle has sides of lengths $5\frac{1}{4}$ feet, $7\frac{1}{2}$ feet, and $10\frac{1}{8}$ feet. Find the perimeter of the triangle. (See Exercise 69.)

71. A piece of board is $15\frac{5}{8}$ inches long. If it must be divided into 3 pieces of equal length, how long must each piece be?

72. If one serving of a macaroni and cheese meal requires $\frac{1}{8}$ cup of chopped onions, how many cups of onions will $7\frac{1}{2}$ servings require?

The following chart appears on a package of Quaker Quick Grits.

	Microwave		Stove Top	
Servings	1	1	4	6
Water	$\frac{3}{4}$ cup	1 cup	3 cups	4 cups
Grits	3 Tbsp	3 Tbsp	$\frac{3}{4}$ cup	1 cup
Salt (optional)	dash	dash	$\frac{1}{4}$ tsp	$\frac{1}{2}$ tsp

Use the chart to answer the questions in Exercises 73–74.

73. How many cups of water would be needed for 6 microwave servings?

74. How many cups of grits would be needed for 5 stove top servings? (*Hint:* 5 is halfway between 4 and 6.)

75. Tex's favorite recipe for barbecue sauce calls for $2\frac{1}{3}$ cups of tomato sauce. The recipe makes enough barbecue sauce to serve 7 people. How much tomato sauce is needed for 1 serving?

76. A cake recipe calls for $1\frac{3}{4}$ cups of sugar. A caterer has $15\frac{1}{2}$ cups of sugar on hand. How many cakes can he make?

A **pie chart** or **circle graph** is often used to give a pictorial representation of data. A circle is used to represent the total of all the categories represented. The circle is divided into sectors, or wedges (like pieces of pie) whose sizes show the relative magnitudes of the categories. Because a complete revolution around a circle measures 360 degrees (360°), the sum of all the measures of the angles at the center must be 360°, while the sum of all the fractional parts represented must be 1 (for 1 whole circle).

77. The pie chart shown depicts the makeup of the population of Mexico. Use the chart to answer the following.
  (a) What fractional part of the population is Caucasian?
  (b) What fractional part of the population is composed of either Mestizo or Indian?

78. The pie chart shown depicts the reactions of a sample of people who were asked to respond to "I feel like people don't take enough time on a day-to-day basis to show that they really care about others." Use the chart to answer the following.
  (a) What fractional part answered "Somewhat disagree"?
  (b) 805 people responded altogether. How many of these answered "Somewhat agree"?

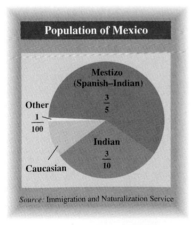

Source: Immigration and Naturalization Service

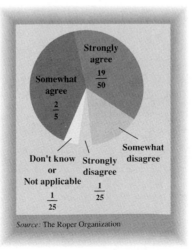

Source: The Roper Organization

79. For each of the following, write a fraction in lowest terms that represents the region described.

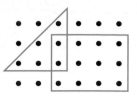

(a) the dots in the rectangle as a part of the dots in the entire figure

(b) the dots in the triangle as a part of the dots in the entire figure

(c) the dots in the overlapping region of the triangle and the rectangle as a part of the dots in the triangle alone

(d) the dots in the overlapping region of the triangle and the rectangle as a part of the dots in the rectangle alone

80. After ten games in the local softball league, the following batting statistics were obtained.

Player	At-bats	Hits	Home Runs
Bishop, Kelly	40	9	2
Carlton, Robert	36	12	3
De Palo, Theresa	11	5	1
Crowe, Vonalaine	16	8	0
Marshall, James	20	10	2

Answer each of the following, using estimation skills as necessary.

(a) Which player got a hit in exactly $\frac{1}{3}$ of his or her at-bats?

(b) Which player got a hit in just less than $\frac{1}{2}$ of his or her at-bats?

(c) Which player got a home run in just less than $\frac{1}{10}$ of his or her at-bats?

(d) Which player got a hit in just less than $\frac{1}{4}$ of his or her at-bats?

(e) Which two players got hits in exactly the same fractional parts of their at-bats? What was the fractional part, reduced to lowest terms?

1.2 EXPONENTS, ORDER OF OPERATIONS, AND INEQUALITY

FOR EXTRA HELP	OBJECTIVES
📖 **SSG** pp. 8–15 **SSM** pp. 6–9	**1 ▶** Use exponents.
	2 ▶ Use the order of operations rules.
📼 **Video** 1	**3 ▶** Use more than one grouping symbol.
	4 ▶ Know the meanings of $\neq$, $<$, $>$, $\leq$, and $\geq$.
💾 **Tutorial** IBM MAC	**5 ▶** Translate word statements to symbols.
	6 ▶ Write statements that change the direction of inequality symbols.

1 ▶ Use exponents.

▶   It is common for a multiplication problem to have the same factor appearing several times. For example, in the product

$$3 \cdot 3 \cdot 3 \cdot 3 = 81$$

the factor 3 appears four times. In algebra, repeated factors are written with an *exponent*. For example, in $3 \cdot 3 \cdot 3 \cdot 3$, the number 3 appears as a factor four times, so the product is written as 3^4, and is read "3 to the **fourth power**."

$$3 \cdot 3 \cdot 3 \cdot 3 = 3^4$$

The number 4 is the **exponent** or **power** and 3 is the **base** in the **exponential expression** 3^4. A natural number exponent, then, tells how many times the base is used as a factor. A number raised to the first power is simply that number. For example, $5^1 = 5$ and $(1/2)^1 = 1/2$.

EXAMPLE 1

Evaluating an Exponential Expression

Find the values of the following.

(a) 5^2

$$\underbrace{5 \cdot 5}_{\text{↑}} = 25$$

—————— 5 is used as a factor 2 times.

Read 5^2 as "5 squared."

(b) 6^3

$$\underbrace{6 \cdot 6 \cdot 6}_{\text{↑}} = 216$$

—————— 6 is used as a factor 3 times.

Read 6^3 as "6 cubed."

(c) $2^5 = 2 \cdot 2 \cdot 2 \cdot 2 \cdot 2 = 32$ 2 is used as a factor 5 times.
Read 2^5 as "2 to the fifth power."

(d) $\left(\dfrac{2}{3}\right)^3 = \dfrac{2}{3} \cdot \dfrac{2}{3} \cdot \dfrac{2}{3} = \dfrac{8}{27}$ $\frac{2}{3}$ is used as a factor 3 times. ■

2 ▶ Use the order of operations rules.

▶   Many problems involve more than one operation. To indicate the order in which the operations should be performed, we often use **grouping symbols.** If no grouping symbols are used, we apply the order of operations rules, which are discussed below.

Suppose we consider the expression $5 + 2 \cdot 3$. If we wish to show that the multiplication should be performed before the addition, parentheses can be used to write

$$5 + (2 \cdot 3) = 5 + 6 = 11.$$

If addition is to be performed first, the parentheses should group $5 + 2$ as follows.

$$(5 + 2) \cdot 3 = 7 \cdot 3 = 21$$

Other grouping symbols used in more complicated expressions are

brackets, $[\quad]$, braces, $\{\quad\}$,

and fraction bars.

The most useful way to work problems with more than one operation is to use the following **order of operations.** This order is used by most calculators and computers.

Order of Operations

If grouping symbols are present, simplify within them, innermost first (and above and below fraction bars separately), in the following order.

Step 1 Apply all exponents.
Step 2 Do any multiplications or divisions in the order in which they occur, working from left to right.
Step 3 Do any additions or subtractions in the order in which they occur, working from left to right.

If no grouping symbols are present, start with Step 1.

A dot has been used to show multiplication; another way to show multiplication is with parentheses. For example, $3(7)$, $(3)7$, and $(3)(7)$ each mean $3 \cdot 7$ or 21. The next example shows the use of parentheses for multiplication.

EXAMPLE 2
Using the Order of Operations

Find the values of the following.

(a) $9(6 + 11)$

Using the order of operations given above, work first inside the parentheses.

$$9(\mathbf{6 + 11}) = 9(\mathbf{17}) \qquad \text{Work inside parentheses.}$$
$$= 153 \qquad \text{Multiply.}$$

(b) $6 \cdot 8 + 5 \cdot 2$

Do any multiplications, working from left to right, and then add.

$$\mathbf{6 \cdot 8 + 5 \cdot 2} = 48 + 10 \qquad \text{Multiply.}$$
$$= 58 \qquad \text{Add.}$$

(c)
$$2(\mathbf{5 + 6}) + 7 \cdot 3 = 2(\mathbf{11}) + 7 \cdot 3 \qquad \text{Work inside parentheses.}$$
$$= 22 + 21 \qquad \text{Multiply.}$$
$$= 43 \qquad \text{Add.}$$

(d) $9 + 2^3 - 5$
Find 2^3 first.

$$9 + 2^3 - 5 = 9 + 2 \cdot 2 \cdot 2 - 5 \qquad \text{Use the exponent.}$$
$$= 9 + 8 - 5 \qquad \text{Multiply.}$$
$$= 17 - 5 \qquad \text{Add.}$$
$$= 12 \qquad \text{Subtract.}$$

(e)
$$16 - 3^2 + 4^2 = 16 - 3 \cdot 3 + \mathbf{4 \cdot 4} \qquad \text{Use the exponents.}$$
$$= 16 - 9 + \mathbf{16} \qquad \text{Multiply.}$$
$$= 7 + 16 \qquad \text{Subtract.}$$
$$= 23 \qquad \text{Add.}$$

Notice that 3^2 and $4 \cdot 4$ must be evaluated before subtracting and adding. ■

NOTE Parentheses and fraction bars are used as grouping symbols to indicate an expression that represents a single number. That is why we must first simplify within parentheses and above and below fraction bars.

3 ▶ Use more than one grouping symbol.

▶ An expression with double parentheses, such as $2(8 + 3(6 + 5))$, can be confusing. To eliminate this, square brackets, [ ], often are used instead of one of the pairs of parentheses, as shown in the next example.

EXAMPLE 3

Using Brackets

Simplify $2[8 + 3(6 + 5)]$.

Work first within the parentheses, and then simplify until a single number is found inside the brackets.

$$2[8 + 3(6 + 5)] = 2[8 + 3(11)]$$
$$= 2[8 + 33]$$
$$= 2[41]$$
$$= 82 \quad \blacksquare$$

Sometimes fraction bars are grouping symbols, as the next example shows.

EXAMPLE 4

Using a Fraction Bar as a Grouping Symbol

Simplify $\dfrac{4(5 + 3) + 3}{2(3) - 1}$.

The expression can be written as the quotient

$$[4(5 + 3) + 3] \div [2(3) - 1],$$

which shows that the fraction bar serves to group the numerator and denominator separately. Simplify both numerator and denominator, then divide, if possible.

$$\frac{4(5 + 3) + 3}{2(3) - 1} = \frac{4(8) + 3}{2(3) - 1} \qquad \text{Work inside parentheses.}$$

$$= \frac{32 + 3}{6 - 1} \qquad \text{Multiply.}$$

$$= \frac{35}{5} \qquad \text{Add and subtract.}$$

$$= 7 \qquad \text{Divide.} \quad \blacksquare$$

4 ▶ Know the meanings of $\neq$, $<$, $>$, $\leq$, and $\geq$.

▶ So far, we have used the symbols for the operations of arithmetic and the symbol for equality ($=$). The equality symbol with a slash through it, $\neq$, means "is not equal to." For example,

$$7 \neq 8$$

indicates that 7 is not equal to 8.

If two numbers are not equal, then one of the numbers must be smaller than the other. The symbol $<$ represents "is less than," so that "7 is less than 8" is written

$$7 < 8.$$

Also, write "6 is less than 9" as $6 < 9$.

The symbol $>$ means "is greater than." Write "8 is greater than 2" as

$$8 > 2.$$

The statement "17 is greater than 11" becomes $17 > 11$.

Keep the meanings of the symbols $<$ and $>$ clear by remembering that the symbol always points to the smaller number. For example, write "8 is less than 15" by pointing the symbol toward the 8:

$$8 < 15.$$

Two other symbols, ≤ and ≥, also represent the idea of inequality. The symbol ≤ means "is less than or equal to," so that

$$5 \leq 9$$

means "5 is less than or equal to 9." This statement is true, since $5 < 9$ is true. If either the $<$ part or the $=$ part is true, then the inequality $\leq$ is true.

The symbol ≥ means "is greater than or equal to." Again,

$$9 \geq 5$$

is true because $9 > 5$ is true. Also, $8 \leq 8$ is true since $8 = 8$ is true. But it is not true that $13 \leq 9$ because neither $13 < 9$ nor $13 = 9$ is true.

EXAMPLE 5

Using Inequality Symbols

Determine whether each statement is true or false.

(a) $6 \neq 6$

The statement is false because 6 *is equal to* 6.

(b) $5 < 19$

Since 5 represents a number that is indeed less than 19, this statement is true.

(c) $15 \leq 20$

The statement $15 \leq 20$ is true, since $15 < 20$.

(d) $25 \geq 30$

Both $25 > 30$ and $25 = 30$ are false. Because of this, $25 \geq 30$ is false.

(e) $12 \geq 12$

Since $12 = 12$, this statement is true.   ■

◇ **C O N N E C T I O N S** ◇

In this section we begin the process of learning to use mathematics to solve real world problems. The most important step in that process, and (unfortunately) the most difficult, is to translate the English statement of the problem into algebraic symbols. The symbols, $=$, $>$, and $<$ are used to write equations and inequalities. In later chapters we learn how to solve equations and inequalities. For now, we are mainly concerned with the process of translation.

FOR DISCUSSION OR WRITING

Write a few English sentences using the words *equals, is less than,* or *is greater than.* An example might be "The number of soft drinks is less than I ordered."

5 ▶ Translate word statements to symbols.

▶   An important part of algebra deals with translating words into algebraic notation.

PROBLEM SOLVING

As we will see throughout this book, the ability to solve problems using mathematics is based on translating the words of the problem into symbols. The next example is the first of many that will be included to illustrate translations from words to symbols.

EXAMPLE 6 ■ Write each word statement in symbols.

Translating From Words to Symbols

(a) Twelve **equals** ten **plus** two.

$$12 = 10 + 2$$

(b) Nine **is less than** ten.

$$9 < 10$$

(c) Fifteen **is not equal to** eighteen.

$$15 \neq 18$$

(d) Seven **is greater than** four.

$$7 > 4$$

(e) Thirteen **is less than or equal to** forty.

$$13 \leq 40$$

(f) Eleven **is greater than or equal to** eleven.

$$11 \geq 11$$ ■

6 ▶ Write statements that change the direction of inequality symbols.

▶ Any statement with $<$ can be converted to one with $>$, and any statement with $>$ can be converted to one with $<$. We do this by reversing the order of the numbers and the direction of the symbol. For example, the statement $6 < 10$ can be written with $>$ as $10 > 6$. Similarly, the statement $4 \leq 10$ can be changed to $10 \geq 4$.

EXAMPLE 7 ■ The following examples show the same statement written in two equally correct ways.

Converting Between Inequality Symbols

(a) $9 < 16$ $16 > 9$

(b) $5 > 2$ $2 < 5$

(c) $3 \leq 8$ $8 \geq 3$

(d) $12 \geq 5$ $5 \leq 12$ ■

Here is a summary of the symbols discussed in this section.

Symbols of Equality and Inequality

$=$ is equal to	$\neq$ is not equal to
$<$ is less than	$>$ is greater than
$\leq$ is less than or equal to	$\geq$ is greater than or equal to

CAUTION The symbols of equality and inequality are used to write mathematical *sentences*. They describe the relationship between two numbers. On the other hand, the symbols for operations ($+$, $-$, $\times$, $\div$) are used to write mathematical *expressions* that represent a single number. For example, compare the sentence $4 < 10$ with the expression $4 + 10$, which represents the number 14.

1.2 EXERCISES

Decide whether the statement is true or false. If it is false, explain why.

1. The exponential expression 3^5 means $3 \cdot 3 \cdot 3 \cdot 3 \cdot 3 \cdot 3$.

2. $7 + 3 \cdot 2$ and $(7 + 3) \cdot 2$ have the same meaning.

3. In an inequality using $>$ or $<$, the inequality symbol should point toward the smaller number for the inequality to be true.

4. $4 + 9 = 13$ is a mathematical sentence, while $4 + 9 - 13$ is a mathematical expression.

Find the value of the exponential expression. See Example 1.

5. 7^2 **6.** 4^2 **7.** 12^2 **8.** 14^2

9. 4^3 **10.** 5^3 **11.** 10^3 **12.** 11^3

13. 3^4 **14.** 6^4 **15.** 4^5 **16.** 3^5

17. $\left(\frac{2}{3}\right)^4$ **18.** $\left(\frac{3}{4}\right)^3$ **19.** $(.04)^3$ **20.** $(.05)^4$

21. Explain in your own words how to evaluate a power of a number, such as 6^3.

22. Explain why any power of 1 must be equal to 1.

Find the value of the expression. See Examples 2–4.

23. $9 \cdot 5 - 13$ **24.** $7 \cdot 6 - 11$ **25.** $\frac{1}{4} \cdot \frac{2}{3} + \frac{2}{5} \cdot \frac{11}{3}$

26. $\frac{9}{4} \cdot \frac{2}{3} + \frac{4}{5} \cdot \frac{5}{3}$ **27.** $9 \cdot 4 - 8 \cdot 3$ **28.** $11 \cdot 4 + 10 \cdot 3$

29. $(4.3)(1.2) + (2.1)(8.5)$ **30.** $(2.5)(1.9) + (4.3)(7.3)$ **31.** $5[3 + 4(2^2)]$

32. $6[2 + 8(3^3)]$ **33.** $3^2[(11 + 3) - 4]$ **34.** $4^2[(13 + 4) - 8]$

35. $\frac{6(3^2 - 1) + 8}{3 \cdot 2 - 2}$ **36.** $\frac{2(8^2 - 4) + 8}{4 \cdot 3 - 10}$

37. $\frac{4(6 + 2) + 8(8 - 3)}{6(4 - 2) - 2^2}$ **38.** $\frac{6(5 + 1) - 9(1 + 1)}{5(8 - 6) - 2^3}$

39. Explain why, in the expression $3 + 4 \cdot 6$, the product $4 \cdot 6$ should be found *before* the addition is performed.

40. When evaluating $(4^2 + 3^3)^4$, what is the *last* exponent that would be applied?

Tell whether the statement is true or false. In Exercises 45–54, first simplify the expression involving an operation. See Example 5.

41. $5 < 6$ **42.** $3 < 7$ **43.** $8 \geq 17$

44. $10 \geq 41$ **45.** $17 \leq 18 - 1$ **46.** $12 \geq 10 + 2$

47. $6 \cdot 8 + 6 \cdot 6 \geq 0$ **48.** $4 \cdot 20 - 16 \cdot 5 \geq 0$

49. $6[5 + 3(4 + 2)] \leq 70$ **50.** $6[2 + 3(2 + 5)] \leq 135$

51. $\frac{9(7 - 1) - 8 \cdot 2}{4(6 - 1)} > 3$ **52.** $\frac{2(5 + 3) + 2 \cdot 2}{2(4 - 1)} > 1$

53. $8 \leq 4^2 - 2^2$ **54.** $10^2 - 8^2 > 6^2$

Write the word statement in symbols. See Example 6.

55. Fifteen is equal to five plus ten.

56. Twelve is equal to twenty minus eight.

57. Nine is greater than five minus four.

58. Ten is greater than six plus one.

59. Sixteen is not equal to nineteen.

60. Three is not equal to four.

61. Two is less than or equal to three.

62. Five is less than or equal to nine.

Write the statement in words and decide whether it is true or false.

63. $7 < 19$ **64.** $9 < 10$ **65.** $3 \neq 6$

66. $9 \neq 13$ **67.** $8 \geq 11$ **68.** $4 \leq 2$

69. Construct a true statement that involves an addition on the left side, the symbol $\geq$, and a multiplication on the right side.

70. Construct a false statement that involves subtraction on the left side, the symbol $\leq$, and a division on the right side. Then tell why the statement is false and how it could be changed to become true.

Write the statement with the inequality symbol reversed while keeping the same meaning. See Example 7.

71. $5 < 30$ **72.** $8 > 4$ **73.** $12 \geq 3$ **74.** $25 \leq 41$

75. What English-language phrase is used to express the fact that one person's age *is less than* another person's age?

76. What English-language phrase is used to express the fact that one person's height *is greater than* another person's height?

77. $12 \geq 12$ is a true statement. Suppose that someone tells you the following: "$12 \geq 12$ is false, because even though 12 is equal to 12, 12 is not greater than 12." How would you respond to this?

78. The symbol $\neq$ means "is not equal to." How do you think we read the symbol $\not>$? How do you think we read $\not<$?

Solve the problem.

A **bar graph,** *as shown in the figure, is a convenient method for depicting data. Here we see that in 1989, $28 million was the figure for direct impact of moviemakers' spending in the state of Louisiana. In 1990, it was $23 million, and so on. Use the bar graph to answer the questions in Exercises 79–84.*

It's Showtime

Moviemakers' spending is a hit in Louisiana

Direct impact (in millions)

'89 $28
'90 $23
'91 $22
'92 $27.9
'93 $37.2

Source: Louisiana Office of Film and Video

79. In what years was the direct impact greater than $23 million?

80. In what years was the direct impact less than $29 million?

81. In what years was the direct impact less than or equal to $29 million?

82. Explain why the answers to Exercises 80 and 81 are the same.

83. The total economic impact is usually determined by multiplying the direct impact by 3. What was the total economic impact from 1989 through 1992?

84. (See Exercise 83.) What was the total economic impact in 1993?

Graph, "It's Showtime: Moviemaker's spending a hit in Louisiana," from *The Times-Picayune,* August 17, 1994. Copyright © 1994 by *The Times-Picayune.* Reprinted by permission.

◆———— ◆ **MATHEMATICAL CONNECTIONS*** (Exercises 85–90) ◆————

Consider the figure that depicts two routes from home to school. Answer Exercises 85–90 in order.

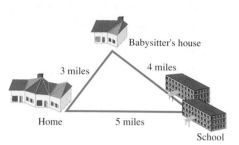

85. What is the distance you must travel from home to school if you must drop off your child at the babysitter's house?

86. What is the distance you must travel from home to school if the babysitter comes to your home to watch your child?

87. Write a sentence using the phrase "is less than" describing the two possible routes described in Exercises 85 and 86.

88. The two possible routes and the distances they cover illustrate an old saying: The _____ distance between two points is a _____
_____ .

89. Where would the babysitter have to live in order for the two routes to be equal?

90. Find the square of the distance from home to the babysitter's and the square of the distance from the babysitter's to school. Add these results. Now find the square of the distance from home to school. How do these two numbers compare? (More about this particular property will be discussed later in the text. It is an example of the Pythagorean relationship for right triangles.)

————————◆————————

1.3 VARIABLES, EXPRESSIONS, AND EQUATIONS

FOR EXTRA HELP	OBJECTIVES
📖 **SSG** pp. 15–19 **SSM** pp. 9–14	**1 ▶** Define *variable*, and find the value of an algebraic expression, given the values of the variables.
📼 **Video** I	**2 ▶** Convert phrases from words to algebraic expressions.
💾 **Tutorial** IBM MAC	**3 ▶** Identify solutions of equations. **4 ▶** Identify solutions of equations from a set of numbers. **5 ▶** Distinguish between an *expression* and an *equation*.

1 ▶ Define *variable*, and find the value of an algebraic expression, given the values of the variables.

▶   A **variable** is a symbol, usually a letter, such as x, y, or z, used to represent any unknown number. An **algebraic expression** is a collection of numbers, variables, symbols for operations, and symbols for grouping (such as parentheses). For example,

$$6(x + 5), \qquad 2m - 9, \qquad \text{and} \qquad 8p^2 + 6p + 2$$

———————

*Some exercise sets will include groups of exercises designated *Mathematical Connections*. The goal of these groups is to relate topics currently being studied to ones studied earlier, to real-life situations, or to universally accepted truths. In general, the exercises in these groups should be worked in numerical order without skipping any. Answers to *all* of these exercises are given in the answer section.

are all algebraic expressions. In the algebraic expression $2m - 9$, the expression $2m$ indicates the product of 2 and m, just as $8p^2$ shows the product of 8 and p^2. Also, $6(x + 5)$ means the product of 6 and $x + 5$. An algebraic expression has different numerical values for different values of the variable.

E X A M P L E   1

Evaluating Expressions

Find the numerical values of the following algebraic expressions when $m = 5$.

(a) $8m$

Replace m with 5, to get

$$8m = 8 \cdot 5 = 40.$$

(b) $3m^2$

For $m = 5$,

$$3m^2 = 3 \cdot 5^2 = 3 \cdot 25 = 75. \quad \blacksquare$$

CAUTION In Example 1(b), it is important to notice that $3m^2$ means $3 \cdot m^2$; it *does not* mean $3m \cdot 3m$. The product $3m \cdot 3m$ is indicated by $(3m)^2$.

E X A M P L E   2

Evaluating Expressions

Find the value of each expression when $x = 5$ and $y = 3$.

(a) $2x + 7y$

Replace x with **5** and y with **3**. Do the multiplication first, and then add.

$$
\begin{aligned}
2x + 7y &= 2 \cdot 5 + 7 \cdot \mathbf{3} &&\text{Let } x = 5 \text{ and } y = 3.\\
&= 10 + 21 &&\text{Multiply.}\\
&= 31 &&\text{Add.}
\end{aligned}
$$

(b) $\dfrac{9x - 8y}{2x - y}$

Replace x with **5** and y with **3**.

$$
\begin{aligned}
\frac{9x - 8y}{2x - y} &= \frac{9 \cdot 5 - 8 \cdot \mathbf{3}}{2 \cdot 5 - \mathbf{3}} &&\text{Let } x = 5 \text{ and } y = 3.\\[2mm]
&= \frac{45 - 24}{10 - 3} &&\text{Multiply.}\\[2mm]
&= \frac{21}{7} &&\text{Subtract.}\\[2mm]
&= 3 &&\text{Divide.}
\end{aligned}
$$

(c)
$$
\begin{aligned}
x^2 - 2y^2 &= 5^2 - 2 \cdot \mathbf{3}^2 &&\text{Let } x = 5 \text{ and } y = 3.\\
&= 25 - 2 \cdot 9 &&\text{Use the exponents.}\\
&= 25 - 18 &&\text{Multiply.}\\
&= 7 &&\text{Subtract.} \quad \blacksquare
\end{aligned}
$$

2 ▶ Convert phrases from words to algebraic expressions.

▶ In Section 1.2 we saw how to translate from words to symbols.

PROBLEM SOLVING Sometimes variables must be used in changing word phrases into algebraic expressions in order to solve problems. The next example illustrates this. Such translations are used extensively in problem solving.

EXAMPLE 3
Changing Word Phrases to Algebraic Expressions

Change the following word phrases to algebraic expressions. Use x as the variable.

(a) The **sum** of a number and 9

"Sum" is the answer to an addition problem. This phrase translates as

$$x + 9 \quad \text{or} \quad 9 + x.$$

(b) 7 **minus** a number

"Minus" indicates subtraction, so the answer is $7 - x$.

CAUTION Here $x - 7$ would *not* be correct; this statement translates as "a number minus 7," not "7 minus a number." The expressions $7 - x$ and $x - 7$ are rarely equal. For example, if $x = 10$, $10 - 7 \neq 7 - 10$. ($7 - 10$ is a *negative number*, discussed in Section 1.4.)

(c) 7 taken from a number

Since 7 is taken *from* a number, write $x - 7$. In this case $7 - x$ would not be correct, because "taken from" means "subtracted from."

(d) The product of 11 and a number

$$11 \cdot x \qquad \text{or} \qquad 11x$$

As mentioned earlier, $11x$ means 11 times x. No symbol is needed to indicate the product of a number and a variable.

(e) 5 divided by a number

This translates as

$$\frac{5}{x}.$$

The expression $\frac{x}{5}$ would *not* be correct here.

(f) The product of 2, and the sum of a number and 8

$$2(x + 8) \quad \blacksquare$$

3 ▶ Identify solutions of equations.

▶ An **equation** is a statement that two algebraic expressions are equal. Therefore, an equation always includes the equality symbol, $=$. Examples of equations are

$$x + 4 = 11, \qquad 2y = 16, \qquad \text{and} \qquad 4p + 1 = 25 - p.$$

Solving an Equation

To **solve** an equation means to find the values of the variable that make the equation true. The values of the variable that make the equation true are called the **solutions** of the equation.

EXAMPLE 4

Deciding Whether a
Number Is a Solution

Decide whether the given number is a solution of the equation.

(a) $5p + 1 = 36$;   7
Replace p with 7.

$$5p + 1 = 36$$
$$5 \cdot 7 + 1 = 36 \qquad ? \qquad \text{Let } p = 7.$$
$$35 + 1 = 36 \qquad ?$$
$$36 = 36 \qquad \qquad \text{True}$$

The number 7 is a solution of the equation.

(b) $9m - 6 = 32$;   4

$$9m - 6 = 32$$
$$9 \cdot 4 - 6 = 32 \qquad ? \qquad \text{Let } m = 4.$$
$$36 - 6 = 32 \qquad ?$$
$$30 = 32 \qquad \qquad \text{False}$$

The number 4 is not a solution of the equation. ■

4 ▶ Identify solutions of equations from a set of numbers.

▶ A **set** is a collection of objects. In mathematics, these objects are most often numbers. The objects that belong to the set, called **elements** of the set, are written between **set braces.** For example, the set containing the numbers 1, 2, 3, 4, and 5 is written as

$$\{1, 2, 3, 4, 5\}.$$

For more information about sets, see Appendix B at the back of this book.

PROBLEM SOLVING

In some cases, the set of numbers from which the solutions of an equation must be chosen is specifically stated. In an application, this set is often determined by the natural restrictions of the problem. For example, if the answer to a problem is a number of people, only whole numbers would make sense, so the set would be the set of whole numbers. In other situations the set of restrictions may be an arbitrary choice.

EXAMPLE 5

Finding a Solution From a
Given Set

Change each word statement to an equation. Use x as the variable. Then find all solutions for the equation from the set

$$\{0, 2, 4, 6, 8, 10\}.$$

(a) The sum of a number and four is six.
      The word "is" suggests "equals." If x represents the unknown number, then translate as follows.

The sum of
a number and four   is   six.
            ↓   ↓   ↓
$$x + 4 \qquad = \qquad 6$$

Try each number from the given set $\{0, 2, 4, 6, 8, 10\}$, in turn, to see that 2 is the only solution of $x + 4 = 6$.

(b) 9 more than five times a number is 49.
Use x to represent the unknown number.

$$\begin{array}{ccccccc} 9 & \text{more than} & \text{five times a number} & \text{is} & 49. \\ \downarrow & \downarrow & \downarrow & \downarrow & \downarrow \\ 9 & + & 5x & = & 49 \end{array}$$

Try each number from $\{0, 2, 4, 6, 8, 10\}$. The solution is 8, since $9 + 5 \cdot 8 = 49$. ■

5 ▶ Distinguish between an *expression* and an *equation*.

▶ Students often have trouble distinguishing between equations and expressions. Remember that an equation is a sentence; an expression is a phrase.

$$\begin{array}{cc} 4x + 5 = 9 & 4x + 5 \\ \uparrow & \uparrow \\ \text{equation} & \text{expression} \\ \text{(to solve)} & \text{(to simplify or evaluate)} \end{array}$$

E X A M P L E 6

Distinguishing Between Equations and Expressions

Decide whether each of the following is an equation or an expression.

(a) $2x - 5y$
There is no equals sign, so this is an expression.

(b) $2x = 5y$
Because of the equals sign, this is an equation. ■

1.3 EXERCISES

Identify as an expression *or an* equation. *See Example 6.*

1. $3x + 2(x - 4)$ 2. $5y - (3y + 6)$ 3. $7t + 2(t + 1) = 4$
4. $9r + 3(r - 4) = 2$ 5. $x + y = 3$ 6. $x + y - 3$

7. Why is $2x^3$ not the same as $2x \cdot 2x \cdot 2x$?

8. Why are "5 less than a number" and "5 is less than a number" translated differently?

9. Explain in your own words why, when evaluating the expression $4x^2$ for $x = 3$, 3 must be squared *before* multiplying by 4.

10. What value of x would cause the expression $2x + 3$ to equal 9?

11. There are many pairs of values of x and y for which $2x + y$ will equal 6. Name two such pairs.

12. Suppose that for the equation $3x - y = 9$, the value of x is given to be 4. What would be the corresponding value of y?

Find the numerical value **(a)** *if* $x = 4$ *and* **(b)** *if* $x = 6$. *See Example 1.*

13. $x + 9$ 14. $x - 1$ 15. $5x$ 16. $7x$ 17. $4x^2$

18. $5x^2$ 19. $\dfrac{x + 1}{3}$ 20. $\dfrac{x - 2}{5}$ 21. $\dfrac{3x - 5}{2x}$ 22. $\dfrac{4x - 1}{3x}$

23. $3x^2 + x$ 24. $2x + x^2$ 25. $6.459x$ 26. $.74x^2$

*Find the numerical value if (**a**) x = 2 and y = 1 and (**b**) x = 1 and y = 5. See Example 2.*

27. $8x + 3y + 5$ **28.** $4x + 2y + 7$ **29.** $3(x + 2y)$ **30.** $2(2x + y)$

31. $x + \dfrac{4}{y}$ **32.** $y + \dfrac{8}{x}$ **33.** $\dfrac{x}{2} + \dfrac{y}{3}$ **34.** $\dfrac{x}{5} + \dfrac{y}{4}$

35. $\dfrac{2x + 4y - 6}{5y + 2}$ **36.** $\dfrac{4x + 3y - 1}{x}$ **37.** $2y^2 + 5x$ **38.** $6x^2 + 4y$

39. $\dfrac{3x + y^2}{2x + 3y}$ **40.** $\dfrac{x^2 + 1}{4x + 5y}$ ▦ **41.** $.841x^2 + .32y^2$ ▦ **42.** $.941x^2 + .2y^2$

Change the word phrase to an algebraic expression. Use x as the variable to represent the number. See Example 3.

43. Twelve times a number

44. Nine times a number

45. Seven added to a number

46. Thirteen added to a number

47. Two subtracted from a number

48. Eight subtracted from a number

49. A number subtracted from seven

50. A number subtracted from fourteen

51. The difference between a number and 6

52. The difference between 6 and a number

53. 12 divided by a number

54. A number divided by 12

55. The product of 6 and four less than a number

56. The product of 9 and five more than a number

57. In the phrase "Four more than the product of a number and 6," does the word *and* signify the operation of addition? Explain.

58. Suppose that the directions on a test read "Solve the following expressions." How would you politely correct the person who wrote these directions?

Decide whether the given number is a solution of the equation. See Example 4.

59. $5m + 2 = 7$;   1

60. $3r + 5 = 8$;   1

61. $2y + 3(y - 2) = 14$;   3

62. $6a + 2(a + 3) = 14$;   2

63. $6p + 4p + 9 = 11$; $\dfrac{1}{5}$

64. $2x + 3x + 8 = 20$; $\dfrac{12}{5}$

65. $3r^2 - 2 = 46$;   4

66. $2x^2 + 1 = 19$;   3

67. $\dfrac{z + 4}{2 - z} = \dfrac{13}{5}$; $\dfrac{1}{3}$

68. $\dfrac{x + 6}{x - 2} = \dfrac{37}{5}$; $\dfrac{13}{4}$

Change the word statement to an equation. Use x as the variable. Find the solutions from the set {0, 2, 4, 6, 8, 10}. See Example 5.

69. The sum of a number and 8 is 18.

70. A number minus three equals 1.

71. Sixteen minus three-fourths of a number is 13.

72. The sum of six-fifths of a number and 2 is 14.

73. Five more than twice a number is 5.

74. The product of a number and 3 is 6.

75. Three times a number is equal to 8 more than twice the number.

76. Twelve divided by a number equals $\dfrac{1}{3}$ times that number.

——————◆ **MATHEMATICAL CONNECTIONS** (Exercises 77–80) ◆——————

Mathematicians who study statistics have developed methods of determining mathematical models. Loosely speaking, a mathematical model is an equation that can be used to determine quantities that may not actually be known. Of course, we cannot always expect a model to give us an answer accurate enough for our purposes, but at least we can obtain a rough estimate of data. For example, based on data obtained from Jupiter Communications, the total revenue of home shopping channels is modeled by the equation $y = .27x + 1.00$, where y is in billions of dollars and $x = 0$ corresponds to the year 1988, $x = 1$ corresponds to 1989, and so on through 1991. Use this model to determine the approximate home shopping channel revenue in the year given.

77. 1988 **78.** 1989 **79.** 1990 **80.** 1991

————————————◇————————————

1.4 REAL NUMBERS AND THE NUMBER LINE

FOR EXTRA HELP	OBJECTIVES
📖 **SSG** pp. 20–24 **SSM** pp. 14–18	**1 ▶** Set up number lines.
📼 Video I	**2 ▶** Identify natural numbers, whole numbers, integers, rational numbers, irrational numbers, and real numbers.
💾 Tutorial IBM MAC	**3 ▶** Tell which of two different real numbers is smaller. **4 ▶** Find additive inverses of real numbers. **5 ▶** Find absolute values of real numbers.

1 ▶ Set up number lines. **▶** In Section 1.1 we introduced two important sets of numbers, the *natural numbers* and the *whole numbers*.

Natural Numbers {1, 2, 3, 4, . . .} is the set of **natural numbers**.

Whole Numbers {0, 1, 2, 3, . . .} is the set of **whole numbers**.

NOTE The three dots show that the list of numbers continues in the same way indefinitely.

These numbers, along with many others, can be represented on **number lines** like the one pictured in Figure 1. We draw a number line by choosing any point on the line and calling it 0. Choose any point to the right of 0 and call it 1. The distance between 0 and 1 gives a unit of measure used to locate other points, as shown in Figure 1. The points labeled in Figure 1 and those continuing in the same way to the right correspond to the set of whole numbers.

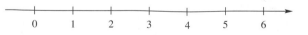

FIGURE 1

All the whole numbers starting with 1 are located to the right of 0 on the number line. But numbers may also be placed to the left of 0. These numbers, written −1, −2, −3, and so on, shown in Figure 2, are called **negative numbers.** (The minus sign is used to show that these numbers are located to the *left* of 0.) The numbers to the *right* of 0 are **positive numbers.** The number 0 itself is neither positive nor negative. Positive numbers and negative numbers are called **signed numbers.**

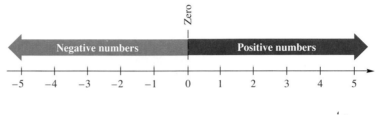

FIGURE 2

◇ **C O N N E C T I O N S** ◇

There are many practical applications of negative numbers. For example, temperatures sometimes fall below zero. The lowest temperature ever recorded in meteorological records was −128.6°F at Vostok, Antarctica, on July 22, 1983. A business that spends more than it takes in has a negative "profit." Altitudes below sea level can be represented by negative numbers. The shore surrounding the Dead Sea is 1312 feet below sea level; this can be represented as −1312 feet.

2 ▶ Identify natural numbers, whole numbers, integers, rational numbers, irrational numbers, and real numbers.

▶ The set of numbers marked on the number line in Figure 2, including positive and negative numbers and zero, is part of the set of *integers.*

Integers   {. . . , −3, −2, −1, 0, 1, 2, 3, . . .} is the set of **integers**.

Not all numbers are integers. For example, 1/2 is not; it is a number halfway between the integers 0 and 1. Also, 3 1/4 is not an integer. Several numbers that are not integers are *graphed* in Figure 3. The **graph** of a number is a point on the number line. The number is called the **coordinate** of the point. Think of the graph of a set of numbers as a picture of the set. All the numbers in Figure 3 can be written as quotients of integers. These numbers are examples of *rational numbers.*

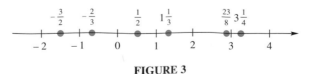

FIGURE 3

Rational Numbers $\{x \mid x$ is a quotient of two integers, with denominator not 0$\}$ is the set of **rational numbers**.

(Read the part in the braces as "the set of all numbers x such that x is a quotient of two integers, with denominator not 0.")

NOTE The set symbolism used in the definition of rational numbers,

$$\{x \mid x \text{ has a certain property}\},$$

is called **set-builder notation.** This notation is convenient to use when it is not possible to list all the elements of the set.

Since any integer can be written as the quotient of itself and 1, all integers also are rational numbers.

All numbers that can be represented by points on the number line are called *real numbers.*

Real Numbers $\{x \mid x$ is a number that can be represented by a point on the number line$\}$ is the set of **real numbers**.

Although a great many numbers are rational, not all are. For example, a floor tile one foot on a side has a diagonal whose length is the square root of 2 (written $\sqrt{2}$). It can be shown that $\sqrt{2}$ cannot be written as a quotient of integers. Because of this, $\sqrt{2}$ is not rational; it is *irrational.*

Irrational Numbers $\{x \mid x$ is a real number that is not rational$\}$ is the set of **irrational numbers**.

Examples of irrational numbers include $\sqrt{3}, \sqrt{7}, -\sqrt{10},$ and π, which is the ratio of the distance around a circle to the distance across it.

Real numbers can be written as decimal numbers. Any rational number will have a decimal that will come to an end (terminate), or repeat in a fixed "block" of digits. For example, $2/5 = .4$ and $27/100 = .27$ are rational numbers with terminating decimals; $1/3 = .3333 \ldots$ and $3/11 = .27272727 \ldots$ are repeating decimals. The decimal representation of an irrational number will neither terminate nor repeat. (A review of decimal numbers can be found in Appendix A.)

An example of a number that is not a real number is the square root of a negative number. These numbers are discussed in the last chapter of this book.

Two ways to represent the relationships among the various types of numbers are shown in Figure 4. Part (a) also gives some examples. Notice that every real number is either a rational number or an irrational number.

EXAMPLE 1

Determining Whether a Number Belongs to a Set

List the numbers in the set

$$\left\{ -5, \quad -\frac{2}{3}, \quad 0, \quad \sqrt{2}, \quad 3\frac{1}{4}, \quad 5, \quad 5.8 \right\}$$

that belong to each of the following sets of numbers.

(a) Natural numbers

The only natural number in the set is 5.

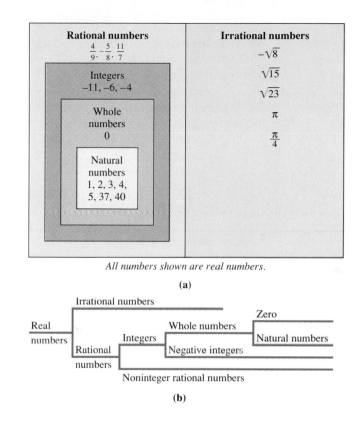

All numbers shown are real numbers.

(a)

(b)

FIGURE 4

(b) Whole numbers
The whole numbers consist of the natural numbers and 0. So the elements of the set that are whole numbers are 0 and 5.

(c) Integers
The integers in the set are -5, 0, and 5.

(d) Rational numbers
The rational numbers are -5, $-2/3$, 0, 3 1/4, 5, 5.8, since each of these numbers *can* be written as the quotient of two integers. For example, $5.8 = 58/10$.

(e) Irrational numbers
The only irrational number in the set is $\sqrt{2}$.

(f) Real numbers
All the numbers in the set are real numbers. ■

3 ▶ Tell which of two different real numbers is smaller.

▶ Given any two whole numbers, you probably can tell which number is smaller. But what happens with negative numbers, as in the set of integers? Positive numbers decrease as the corresponding points on the number line go to the left. For example, $8 < 12$, and 8 is to the left of 12 on the number line. This ordering is extended to all real numbers by definition.

The Ordering of Real Numbers

For any two real numbers a and b, **a is less than b** if a is to the left of b on the number line.

This means that any negative number is smaller than 0, and any negative number is smaller than any positive number. Also, 0 is smaller than any positive number.

EXAMPLE 2

Determining the Order of Real Numbers

Is it true that $-3 < -1$?

To decide whether the statement $-3 < -1$ is true, locate both numbers, -3 and -1, on a number line, as shown in Figure 5. Since -3 is to the left of -1 on the number line, -3 is smaller than -1. The statement $-3 < -1$ is true. ∎

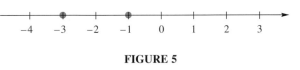

FIGURE 5

NOTE In Section 1.2 we saw how it is possible to rewrite a statement involving $<$ as an equivalent statement involving $>$. The question in Example 2 can also be worded as follows: Is it true that $-1 > -3$? This is, of course, also a true statement.

We can say that for any two real numbers a and b, **a is greater than b** if a is to the right of b on the number line.

4 ▶ Find additive inverses of real numbers.

▶ By a property of the real numbers, for any real number x (except 0), there is exactly one number on the number line the same distance from 0 as x but on the opposite side of 0. For example, Figure 6 shows that the numbers 3 and -3 are each the same distance from 0 but are on opposite sides of 0. The numbers 3 and -3 are called **additive inverses,** or **opposites,** of each other.

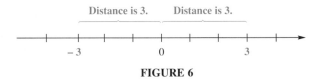

FIGURE 6

Additive Inverse

The **additive inverse** of a number x is the number that is the same distance from 0 on the number line as x, but on the opposite side of 0.

The additive inverse of the number 0 is 0 itself. In fact, 0 is the only real number that is its own additive inverse. Other additive inverses occur in pairs. For example, 4 and -4, and 5 and -5, are additive inverses of each other. Several pairs of additive inverses are shown in Figure 7.

FIGURE 7

The additive inverse of a number can be indicated by writing the symbol $-$ in front of the number. With this symbol, the additive inverse of 7 is written -7. The additive inverse of -4 is written $-(-4)$, and can also be read "the opposite of -4" or "the negative of -4." Figure 7 suggests that 4 is an additive inverse of -4. Since a number can have only one additive inverse, the symbols 4 and $-(-4)$ must represent the same number, which means that

$$-(-4) = 4.$$

This idea can be generalized as follows.

Double Negative Rule   For any real number x,

$$-(-x) = x.$$

E X A M P L E   3 ■ The following chart shows several numbers and their additive inverses.

Finding the Additive
Inverse of a Number

Number	Additive Inverse
-4	$-(-4)$, or 4
0	0
-2	2
19	-19
3	-3 ■

NOTE Example 3 suggests that the additive inverse of a number is found by changing the sign of the number.

An important property of additive inverses will be studied in more detail in a later section of this chapter: $a + (-a) = (-a) + a = 0$ for all real numbers a.

5 ▶ Find absolute values of real numbers.

▶   As mentioned above, additive inverses are numbers that are the same distance from 0 on the number line. This idea can also be expressed by saying that a number and its additive inverse have the same absolute value. The **absolute value** of a real number can be defined as the distance between 0 and the number on the number line. The symbol for the absolute value of the number x is $|x|$, read "the absolute value of x." For example, the distance between 2 and 0 on the number line is 2 units, so that

$$|2| = 2.$$

Because the distance between -2 and 0 on the number line is also 2 units,

$$|-2| = 2.$$

Since distance is a physical measurement, which is never negative, **the absolute value of a number is never negative.** For example, $|12| = 12$ and $|-12| = 12$, since both 12 and -12 lie at a distance of 12 units from 0 on the number line. Also, since 0 is a distance 0 units from 0, $|0| = 0$.

In symbols, the absolute value of x is defined as follows.

Formal Definition of Absolute Value

$$|x| = \begin{cases} x & \text{if } x \geq 0 \\ -x & \text{if } x < 0 \end{cases}$$

By this definition, if x is a positive number or 0, then its absolute value is x itself. For example, since 8 is a positive number, $|8| = 8$. However, if x is a negative number, then its absolute value is the additive inverse of x. This means that if $x = -9$, then $|-9| = -(-9) = 9$, since the additive inverse of -9 is 9.

CAUTION The formal definition of absolute value can be confusing if it is not read carefully. The "$-x$" in the second part of the definition *does not* represent a negative number. Since x is negative in the second part, $-x$ represents the opposite of a negative number, that is, a positive number. Remember that the absolute value of a number is never negative.

EXAMPLE 4

Finding Absolute Value

Simplify by removing absolute value symbols.

(a) $|5| = 5$ **(b)** $|-5| = -(-5) = 5$

(c) $-|5| = -(5) = -5$ **(d)** $-|-14| = -(14) = -14$

(e) $|8 - 2| = |6| = 6$ ■

     Part (e) of Example 4 shows that absolute value bars are also grouping symbols. You must perform any operations that appear inside absolute value symbols before finding the absolute value.

◆ **CONNECTIONS** ◆

Statistical process control is a method of determining when a manufacturing process is out of control, producing defective items. The procedure involves taking samples of a measurement on a production run and calculating the mean (the arithmetic average) and the standard deviation (a measure of the variability) of the measurements. Absolute value can be used to express the control limits, which determine whether the process is out of control. If we let x represent one of the sample measurements, m represent the mean, and s represent the standard deviation, then the process is not out of control as long as $|x - m| \leq ks$, where k is a predetermined constant. For example, if $m = 1.3$, $s = 1.41$, and $k = 1.95$, the process is in control whenever the sample measurements satisfy

$$|x - 1.3| \leq (1.95)(1.41).$$

This means that x (the measurements) must be between -1.4495 and 4.0495.

FOR DISCUSSION OR WRITING
Discuss some processes that might be monitored in this way.

1.4 EXERCISES

Decide whether the statement is true or false.

1. Every whole number is an integer.

2. Every natural number is a whole number.

3. Every rational number is a real number.

4. No number can be both rational and irrational.

5. Every whole number is positive.

6. No natural number is negative.

7. Some real numbers are not rational.

8. Not every rational number is positive.

9. Some whole numbers are not integers.

10. The number 0 is irrational.

For Exercises 11 and 12, see Example 1.

11. List all numbers from the set
$$\left\{ -9, -\sqrt{7}, -1\frac{1}{4}, -\frac{3}{5}, 0, \sqrt{5}, 3, 5.9, 7 \right\}$$
that are
(a) natural numbers;   **(b)** whole numbers;   **(c)** integers;
(d) rational numbers;   **(e)** irrational numbers;   **(f)** real numbers.

12. List all numbers from the set
$$\left\{ 5.3, \quad 5, \quad \sqrt{3}, \quad 1, \quad \frac{1}{9}, 0, 1.2, 1.8, 3, \sqrt{11} \right\}$$
that are
(a) natural numbers;   **(b)** whole numbers;   **(c)** integers;
(d) rational numbers;   **(e)** irrational numbers;   **(f)** real numbers.

Use an integer to express the number in the following applications of numbers.

13. Between the years of 1980 and 1990, the population of Marshalltown, Iowa, decreased by 1760 people.

14. Between 1970 and 1980, the population of the state of New York decreased by 683,226.

15. The city of New Orleans lies 8 feet below sea level.

16. Death Valley lies 282 feet below sea level.

17. The height of Mt. St. Helen's, an active volcano in the state of Washington, is about 8300 feet.

18. Alexander Fedotov, in 1977, flew a jet airplane at an altitude of 123,524 feet.

19. In 1990, New Zealand exported $66,000,000 less than it imported, thus accounting for a negative balance of trade.

20. In 1990, Portugal exported $90,000,000 more than it imported, thus accounting for a positive trade balance.

Graph the group of numbers on a number line. See Figure 3.

21. $0, 3, -5, -6$

22. $2, 6, -2, -1$

23. $-2, -6, -4, 3, 4$

24. $-5, -3, -2, 0, 4$

25. $\frac{1}{4}, 2\frac{1}{2}, -3\frac{4}{5}, -4, -1\frac{5}{8}$

26. $5\frac{1}{4}, 4\frac{5}{9}, -2\frac{1}{3}, 0, -3\frac{2}{5}$

27. A commonly heard statement from students is "Absolute value is always positive." Is this true? If not, explain.

28. If a is a negative number, then is $-|-a|$ positive or negative?

29. Match each expression in Column I with its value in Column II. Some choices in Column II may not be used.

I	II
(a) $\|-7\|$	A. 7
(b) $-(-7)$	B. -7
(c) $-\|-7\|$	C. neither A nor B
(d) $-\|-(-7)\|$	D. both A and B

30. Fill in the blanks with the correct values: The opposite of -2 is _____ , while the absolute value of -2 is _____ . The additive inverse of -2 is _____ , while the additive inverse of the absolute value of -2 is _____ .

Find (a) the opposite (or additive inverse) of the number and (b) the absolute value of the number. See Examples 3 and 4.

31. −2   **32.** −8   **33.** 6   **34.** 11

35. 7 − 4   **36.** 8 − 3   **37.** 7 − 7   **38.** 3 − 3

39. Look at Exercises 35 and 36 and use the results to complete the following: If $a - b > 0$, then the absolute value of $a - b$ in terms of a and b is _____ .

40. Look at Exercises 37 and 38 and use the results to complete the following: If $a - b = 0$, then the absolute value of $a - b$ is _____ .

Select the smaller of the two given numbers. See Examples 2 and 4.

41. −12, −4   **42.** −9, −14   **43.** −8, −1

44. −15, −16   **45.** 3, $|-4|$ **46.** 5, $|-2|$

47. $|-3|, |-4|$ **48.** $|-8|, |-9|$ **49.** $-|-6|, -|-4|$

50. $-|-2|, -|-3|$ **51.** $|5 - 3|, |6 - 2|$ **52.** $|7 - 2|, |8 - 1|$

Decide whether the statement is true or false. See Examples 2 and 4.

53. $6 > -(-2)$ **54.** $-8 > -(-2)$ **55.** $-4 \leq -(-5)$

56. $-6 \leq -(-3)$ **57.** $|-6| < |-9|$ **58.** $|-12| < |-20|$

59. $-|8| > |-9|$ **60.** $-|12| > |-15|$ **61.** $-|-5| \geq -|-9|$

62. $-|-12| \leq -|-15|$ **63.** $|6 - 5| \geq |6 - 2|$ **64.** $|13 - 8| \leq |7 - 4|$

The table shows the annual percent change in the Consumer Price Index for the years 1986 and 1987. Use the information provided to answer Exercises 65–68.

	Percent Change	
Category	1986	1987
Food	3.2	4.1
Shelter	5.5	4.7
Rent, residential	5.8	4.1
Fuel and other utilities	−2.3	−1.1
Apparel and upkeep	.9	4.4
Private transportation	−4.7	−3.0
New cars	4.2	3.6
Gasoline	−21.9	−4.0
Public transportation	5.9	3.5
Medical care	7.5	6.6
Entertainment	3.4	3.3
Commodities	−.9	3.2

Source: Bureau of Labor Statistics, U.S. Dept. of Labor

65. What category of what year represents the greatest drop?

66. Which percent change is represented by a larger number: 1986 fuel and other utilities or 1987 gasoline?

67. True or false? The absolute value of the change in apparel and upkeep in 1986 is less than the absolute value of the change in commodities during the same year.

68. True or false? The absolute value of the change in private transportation in 1986 was less than the absolute value of the change in private transportation in 1987.

For the statement give a pair of values for a and b that make it true, and then give a pair of values that make it false.

69. $|a + b| = |a - b|$

70. $|a - b| = |b - a|$

71. $|a + b| = -|a + b|$

72. $|-(a + b)| = -(a + b)$

Give three numbers that satisfy the given condition.

73. Positive real numbers but not integers

74. Real numbers but not positive numbers

75. Real numbers but not whole numbers

76. Rational numbers but not integers

77. Real numbers but not rational numbers

78. Rational numbers but not negative numbers

1.5 ADDITION OF REAL NUMBERS

FOR EXTRA HELP	OBJECTIVES
📖 **SSG** pp. 25–29 **SSM** pp. 18–21	**1 ▶** Add two numbers with the same sign.
📼 **Video** I	**2 ▶** Add positive and negative numbers.
💾 **Tutorial** IBM MAC	**3 ▶** Use the order of operations with real numbers. **4 ▶** Interpret words and phrases that indicate addition. **5 ▶** Interpret gains and losses as positive and negative numbers.

In this section and the next three sections, we extend the rules for operations with positive numbers to the negative numbers, beginning with addition.

1 ▶ Add two numbers with the same sign.

▶ The number line can be used to explain the addition of real numbers. Later we give the rules for addition.

E X A M P L E   1 ▪
Adding Positive Numbers on the Number Line

Use the number line to find the sum 2 + 3.

Add the positive numbers 2 and 3 on the number line by starting at 0 and drawing an arrow two units to the *right*, as shown in Figure 8. This arrow represents the number 2 in the sum 2 + 3. Then, from the right end of this arrow draw another arrow three units to the right. The number below the end of this second arrow is 5, so 2 + 3 = 5. ▪

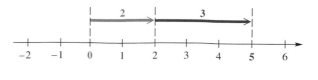

FIGURE 8

EXAMPLE 2

Adding Negative Numbers on the Number Line

Use the number line to find the sum $-2 + (-4)$. (Parentheses are placed around the -4 to avoid the confusing use of $+$ and $-$ next to each other.)

Add the negative numbers -2 and -4 on the number line by starting at 0 and drawing an arrow two units to the *left*, as shown in Figure 9. The arrow is drawn to the left to represent the addition of a *negative* number. From the left end of this first arrow, draw a second arrow four units to the left. The number below the end of this second arrow is -6, so $-2 + (-4) = -6$. ∎

FIGURE 9

In Example 2, the sum of the two negative numbers -2 and -4 is a negative number whose distance from 0 is the sum of the distance of -2 from 0 and the distance of -4 from 0. That is, *the sum of two negative numbers is the negative of the sum of their absolute values.*

$$-2 + (-4) = -(|-2| + |-4|) = -(2 + 4) = -6$$

Adding Numbers with the Same Signs

Add two numbers with the *same* signs by adding the absolute values of the numbers. The sum has the same sign as the numbers being added.

EXAMPLE 3

Adding Two Negative Numbers

Find the sums.

(a) $-2 + (-9) = -(|-2| + |-9|) = -(2 + 9) = -11$

(b) $-8 + (-12) = -20$

(c) $-15 + (-3) = -18$ ∎

2 ▶ Add positive and negative numbers.

▶ We can use the number line again to give meaning to the sum of a positive number and a negative number.

EXAMPLE 4

Adding Numbers with Different Signs

Use the number line to find the sum $-2 + 5$.

Find the sum $-2 + 5$ on the number line by starting at 0 and drawing an arrow two units to the left. From the left end of this arrow, draw a second arrow five units to the right, as shown in Figure 10. The number below the end of the second arrow is 3, so $-2 + 5 = 3$. ∎

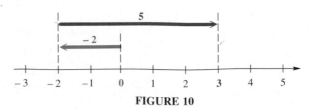

FIGURE 10

Addition of numbers with different signs can also be defined using absolute value.

**Adding Numbers with
Different Signs**

Add two numbers with *different* signs by subtracting the smaller absolute value from the larger absolute value. The answer is given the sign of the number with the larger absolute value.

For example, to add -12 and 5, find their absolute values: $|-12| = 12$ and $|5| = 5$. Then find the difference between these absolute values: $12 - 5 = 7$. Since $|-12| > |5|$, the sum will be negative, so that the final answer is $-12 + 5 = -7$.

While a number line is useful in showing the rules for addition, it is important to be able to do the problems quickly "in your head."

EXAMPLE 5
Adding Mentally

Check each answer, trying to work the addition mentally. If you get stuck, use a number line.

(a) $7 + (-4) = 3$ **(b)** $-8 + 12 = 4$

(c) $-\dfrac{1}{2} + \dfrac{1}{8} = -\dfrac{4}{8} + \dfrac{1}{8} = -\dfrac{3}{8}$ Remember to get a common denominator first.

(d) $\dfrac{5}{6} + \left(-\dfrac{4}{3}\right) = -\dfrac{1}{2}$ **(e)** $-4.6 + 8.1 = 3.5$

(f) $-16 + 16 = 0$ **(g)** $42 + (-42) = 0$ ∎

Parts (f) and (g) in Example 5 suggest that the sum of a number and its additive inverse is 0. This is always true, and this property is discussed further in Section 1.9.

The rules for adding signed numbers are summarized below.

**Adding Signed
Numbers**

Like signs   Add the absolute values of the numbers. The sum has the same sign as the given numbers.
Unlike signs   Find the difference between the larger absolute value and the smaller. The sum has the sign of the number with the larger absolute value.

◆ **CONNECTIONS** ◆

In order to solve many problems, we must be able to add signed numbers. Consider these examples.

1. Like many people, Kareem Dunlap neglects to keep up his checkbook balance. When he finally balanced his account, he found the balance was $-\$23.75$, so he deposited $\$50.00$. What is his new balance?
2. The low temperature in Yellowknife, in the Canadian Northwest Territories, one January was $-26°$F. It rose 16 degrees that day. What was the high temperature?

FOR DISCUSSION OR WRITING
Answer the two questions above. Compare these two problems. How are they alike?

3 ▶ Use the order of operations with real numbers.

▶   Sometimes an addition problem involves adding more than two numbers. As mentioned earlier, do the calculations inside the brackets or parentheses until a single number is obtained. Remember to use the order of operations given in Section 1.2 when adding more than two numbers.

EXAMPLE 6

Adding with Brackets

Find the sums.

(a) $-3 + [4 + (-8)]$

First work inside the brackets. Follow the rules for the order of operations given in Section 1.2.

$$-3 + [4 + (-8)] = -3 + (-4) = -7$$

(b) $8 + [(-2 + 6) + (-3)] = 8 + [4 + (-3)] = 8 + 1 = 9$ ∎

---

4 ▶ Interpret words and phrases that indicate addition.

▶   We now look at the interpretation of words and phrases that involve addition.

PROBLEM SOLVING

As we mentioned earlier, problem solving often requires translating words and phrases into symbols. The word *sum* is one of the words that indicates addition. The chart below lists some of the words and phrases that also signify addition.

Word or Phrase	Example	Numerical Expression and Simplification
Sum of	The *sum of* -3 and 4	$-3 + 4 = 1$
Added to	5 *added to* -8	$-8 + 5 = -3$
More than	12 *more than* -5	$-5 + 12 = 7$
Increased by	-6 *increased by* 13	$-6 + 13 = 7$
Plus	3 *plus* 14	$3 + 14 = 17$

EXAMPLE 7

Interpreting Words and Phrases Involving Addition

Write a numerical expression for each phrase, and simplify the expression.

(a) The *sum of* -8 and 4 and 6

$$-8 + 4 + 6 = [-8 + 4] + 6 = -4 + 6 = 2$$

Notice that brackets were placed around $-8 + 4$ and this addition was done first, using the order of operations given earlier. The same result would be obtained if the brackets were placed around $4 + 6$. (This idea is discussed further in Section 1.9.)

$$-8 + 4 + 6 = -8 + [4 + 6] = -8 + 10 = 2$$

(b) 3 *more than* -5, *increased by* 12

$$-5 + 3 + 12 = [-5 + 3] + 12 = -2 + 12 = 10$$ ∎

---

5 ▶ Interpret gains and losses as positive and negative numbers.

▶   Gains (or increases) and losses (or decreases) sometimes appear in applied problems, such as statements of gains and losses of an investment in stocks.

PROBLEM SOLVING

When problems deal with gains and losses, the gains may be interpreted as positive numbers and the losses as negative numbers. The next example illustrates this idea.

EXAMPLE 8 ■ A football team gained 3 yards on the first play from scrimmage, lost 12 yards on
Interpreting Gains and the second play, and then gained 13 yards on the third play. How many yards did
Losses the team gain or lose altogether?

The gains are represented by positive numbers and the loss by a negative
number.

$$3 + (-12) + 13$$

Add from left to right.

$$3 + (-12) + 13 = [3 + (-12)] + 13 = (-9) + 13 = 4$$

The team gained 4 yards altogether. ■

1.5 EXERCISES

Fill in the blank with the correct response.

1. The sum of two negative numbers will always be a _____
number.   (positive/negative)

2. The sum of a number and its opposite will always be _____ .

3. To simplify the expression $8 + [-2 + (-3 + 5)]$, I should begin by adding
_____ and _____ , according to the rule for order of operations.

4. If I am adding a positive number and a negative number, and the negative number has
the larger absolute value, the sum will be a _____ number.
                                                    (positive/negative)

Find the sum. See Examples 1–6.

5. $6 + (-4)$ **6.** $12 + (-9)$ **7.** $7 + (-10)$ **8.** $4 + (-8)$

9. $-7 + (-3)$ **10.** $-11 + (-4)$ **11.** $-10 + (-3)$ **12.** $-16 + (-7)$

13. $-12.4 + (-3.5)$ **14.** $-21.3 + (-2.5)$

15. $-8 + 7$ **16.** $-12 + 10$

17. $5 + [14 + (-6)]$ **18.** $7 + [3 + (-14)]$

19. $10 + [-3 + (-2)]$ **20.** $13 + [-4 + (-5)]$

21. $-3 + [5 + (-2)]$ **22.** $-7 + [10 + (-3)]$

23. $-8 + [3 + (-1) + (-2)]$ **24.** $-7 + [5 + (-8) + 3]$

25. $-\dfrac{1}{6} + \dfrac{2}{3}$ **26.** $\dfrac{9}{10} + \left(-\dfrac{3}{5}\right)$

27. $\dfrac{5}{8} + \left(-\dfrac{17}{12}\right)$ **28.** $-\dfrac{6}{25} + \dfrac{19}{20}$

29. $2\dfrac{1}{2} + \left(-3\dfrac{1}{4}\right)$ **30.** $-4\dfrac{3}{8} + 6\dfrac{1}{2}$

31. $7.8 + (-9.4)$ **32.** $14.7 + (-10.1)$

33. $-7.1 + [3.3 + (-4.9)]$ **34.** $-9.5 + [-6.8 + (-1.3)]$

35. $[-8 + (-3)] + [-7 + (-7)]$ **36.** $[-5 + (-4)] + [9 + (-2)]$

37. $[-5 + (-7)] + [-4 + (-9)] + [13 + (-12)]$

38. $[-3 + (-11)] + [13 + (-3)] + [19 + (-7)]$

39. Is it possible to add a negative number to another negative number and get a positive number? If so, give an example.

40. Under what conditions will the sum of a positive number and a negative number be a number which is neither negative nor positive?

Perform the operation and then determine whether the statement is true or false. Try to do all work in your head. See Example 5.

41. $-11 + 13 = 13 + (-11)$

42. $16 + (-9) = -9 + 16$

43. $-10 + 6 + 7 = -3$

44. $-12 + 8 + 5 = -1$

45. $18 + (-6) + (-12) = 0$

46. $-5 + 21 + (-16) = 0$

47. $|-8 + 10| = -8 + (-10)$

48. $|-4 + 6| = -4 + (-6)$

49. $\dfrac{11}{5} + \left(-\dfrac{6}{11}\right) = -\dfrac{6}{11} + \dfrac{11}{5}$

50. $-\dfrac{3}{2} + \dfrac{5}{8} = \dfrac{5}{8} + \left(-\dfrac{3}{2}\right)$

51. $-7 + [-5 + (-3)] = [(-7) + (-5)] + 3$

52. $6 + [-2 + (-5)] = [(-4) + (-2)] + 5$

Find all solutions for the equation from the set $\{-3, -2, -1, 0, 1, 2, 3\}$. Guess or use trial and error.

53. $x + 3 = 0$

54. $x + 1 = 0$

55. $x + 2 = 5$

56. $x + 9 = 12$

57. $x + 8 = 7$

58. $x + (-4) = -6$

59. $x + (-2) = -5$

60. $-8 + x = -6$

──────◆ **MATHEMATICAL CONNECTIONS** (Exercises 61–64) ◆──────

Recall the rules for adding signed numbers introduced in this section, and answer Exercises 61–64 in order.

61. Suppose that the sum of two numbers is negative and you know that one of the numbers is positive. What can you conclude about the other number?

62. If you are asked to solve the equation $x + 5 = -7$ from a set of numbers, why could you immediately eliminate any positive numbers as possible solutions? (Remember how you answered Exercise 61.)

63. Suppose that the sum of two numbers is positive, and you know that one of the numbers is negative. What can you conclude about the other number?

64. If you are asked to solve the equation $x + (-8) = 2$ from a set of numbers, why could you immediately eliminate any negative numbers as possible solutions? (Remember how you answered Exercise 63.)

──────────◆──────────

Write a numerical expression for the phrase, and simplify the expression. See Example 7.

65. The sum of -5 and 12 and 6

66. The sum of -3 and 5 and -12

67. 14 added to the sum of -19 and -4

68. -2 added to the sum of -18 and 11

69. The sum of -4 and -10, increased by 12

70. The sum of -7 and -13, increased by 14

71. 4 more than the sum of 8 and -18

72. 10 more than the sum of -4 and -6

Solve the problem by writing a sum of real numbers and adding. No variables are needed. See Example 8.

73. In 1990, the net income of savings instititutions (in millions of dollars) for states in the northeast United States were as follows.

State	Net Income (in millions of dollars)
Maine	0
New Hampshire	−24
Vermont	2
Massachusetts	−212
Rhode Island	−13
Connecticut	−149

What was the total of these net incomes? (*Source:* U.S. Office of Thrift Supervision)

74. The 1991 state general fund balances (in millions of dollars) for states in the Pacific region of the United States were as follows.

State	General Fund Balance (in millions of dollars)
Washington	468
Oregon	380
California	−1259
Alaska	791
Hawaii	347

What was the total of these balances? (*Source:* National Association of State Budget Officers)

75. A college student received a $100 check in the mail from her parents. She then spent $53 for a sociology textbook. How much money does she have left?

76. Shalita's checking account balance is $54.00. She then takes a gamble by writing a check for $89.00. What is her new balance? (Write the balance as a signed number.)

77. The surface, or rim, of a canyon is at altitude 0. On a hike down into the canyon, a party of hikers stops for a rest at 130 meters below the surface. They then descend another 54 meters. What is their new altitude? (Write the altitude as a signed number.)

78. A pilot announces to his passengers that the current altitude of their plane is 34,000 feet. Because of some unexpected turbulence, he is forced to descend 2100 feet. What is the new altitude of the plane? (Write the altitude as a signed number.)

79. The lowest temperature ever recorded in Little Rock, Arkansas, was −5°F. The highest temperature ever recorded there was 117°F more than the lowest. What was this highest temperature?

80. On January 23, 1943, the temperature rose 49°F in two minutes in Spearfish, South Dakota. If the starting temperature was −4°, what was the temperature two minutes later?

81. On a series of three consecutive running plays, Herschel Walker of the Philadelphia Eagles gained 4 yards, lost 3 yards, and lost 2 yards. What positive or negative number represents his total net yardage for the series of plays?

82. On three consecutive passing plays, Troy Aikman of the Dallas Cowboys passed for a gain of 6 yards, was sacked for a loss of 12 yards, and passed for a gain of 43 yards. What positive or negative number represents the total net yardage for the plays?

83. Kim Falgout owes $870.00 on her Master Card account. She returns two items costing $35.90 and $150.00 and receives credits for these on the account. Next, she makes a purchase of $82.50, and then two more purchases of $10.00 each. She finally makes a payment of $500.00. How much does she still owe?

84. A welder working with stainless steel must use precise measurements. Suppose a welder attaches two pieces of steel that are each 3.60 inches in length, and then attaches an additional three pieces that are each 9.10 inches long. She finally cuts off a piece that is 7.60 inches long. Find the length of the welded piece of steel.

The data in Exercises 85 and 86 refer to United States international transactions.

85. The U.S. official reserve assets for the years 1990–1992 are shown in the accompanying bar graph. What is the sum of the assets for these years? (*Source:* Bureau of Economic Analysis)

86. The U.S. Government assets, other than official reserve assets, for the years 1990–1992 are shown in the accompanying bar graph. What is the sum of the assets for these years? (*Source:* Bureau of Economic Analysis)

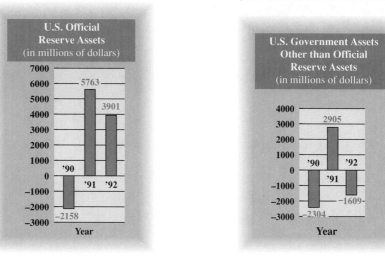

1.6 SUBTRACTION OF REAL NUMBERS

FOR EXTRA HELP

- **SSG** pp. 29–33
- **SSM** pp. 21–25
- **Video** 2
- **Tutorial** IBM   MAC

OBJECTIVES

1 ▶ Find a difference on a number line.
2 ▶ Use the definition of subtraction.
3 ▶ Work subtraction problems that involve grouping.
4 ▶ Interpret words and phrases that involve subtraction.
5 ▶ Solve problems that involve subtraction.

In this section we learn how to subtract with signed numbers.

1 ▶ Find a difference on a number line.

▶ Recall that the answer to a subtraction problem is a *difference*. Differences between signed numbers can be found by using a number line. Since *addition* of a positive number on the number line is shown by drawing an arrow to the *right*, *subtraction* of a positive number is shown by drawing an arrow to the *left*.

EXAMPLE 1

Subtracting with the Number Line

Use the number line to find the difference $7 - 4$.

To find the difference $7 - 4$ on the number line, begin at 0 and draw an arrow seven units to the right. From the right end of this arrow, draw an arrow four units to the left, as shown in Figure 11. The number at the end of the second arrow shows that $7 - 4 = 3$. ■

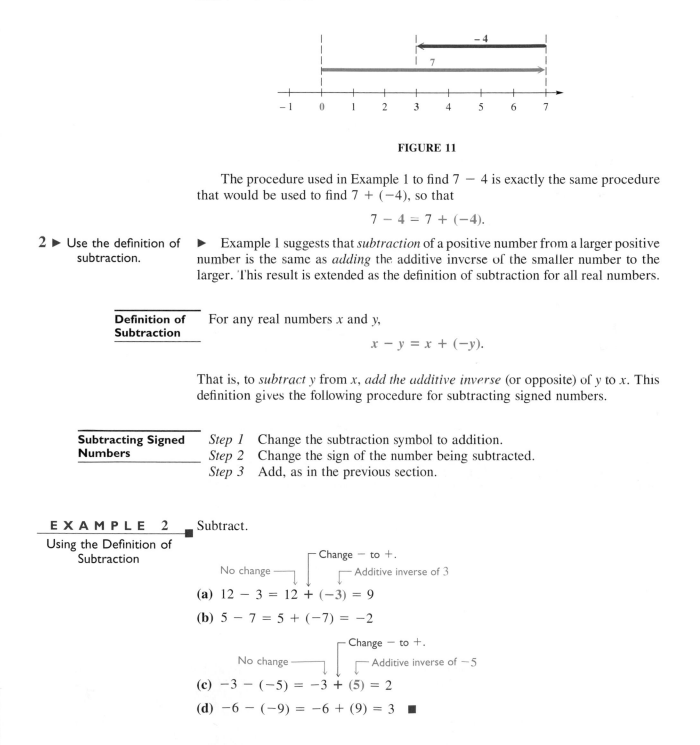

FIGURE 11

The procedure used in Example 1 to find $7 - 4$ is exactly the same procedure that would be used to find $7 + (-4)$, so that

$$7 - 4 = 7 + (-4).$$

2 ▶ Use the definition of subtraction.

▶ Example 1 suggests that *subtraction* of a positive number from a larger positive number is the same as *adding* the additive inverse of the smaller number to the larger. This result is extended as the definition of subtraction for all real numbers.

Definition of Subtraction

For any real numbers x and y,

$$x - y = x + (-y).$$

That is, to *subtract y from x, add the additive inverse* (or opposite) of y to x. This definition gives the following procedure for subtracting signed numbers.

Subtracting Signed Numbers

Step 1 Change the subtraction symbol to addition.
Step 2 Change the sign of the number being subtracted.
Step 3 Add, as in the previous section.

EXAMPLE 2

Using the Definition of Subtraction

Subtract.

┌ Change − to +.
No change ─┐   │   ┌ Additive inverse of 3

(a) $12 - 3 = 12 + (-3) = 9$

(b) $5 - 7 = 5 + (-7) = -2$

┌ Change − to +.
No change ─┐   │   ┌ Additive inverse of −5

(c) $-3 - (-5) = -3 + (5) = 2$

(d) $-6 - (-9) = -6 + (9) = 3$ ■

Subtraction can be used to reverse the result of an addition problem. For example, if 4 is added to a number and then subtracted from the sum, the original number is the result:

$$12 + 4 = 16 \qquad \text{and} \qquad 16 - 4 = 12.$$

The symbol $-$ has now been used for three purposes:

1. to represent subtraction, as in $9 - 5 = 4$;
2. to represent negative numbers, such as -10, -2, and -3;
3. to represent the additive inverse of a number, as in "the additive inverse of 8 is -8."

More than one use may appear in the same problem, such as $-6 - (-9)$, where -9 is subtracted from -6. The meaning of the symbol depends on its position in the algebraic expression.

3 ▶ Work subtraction problems that involve grouping.

▶ As before, with problems that have grouping symbols, first do any operations inside the parentheses and brackets. Work from the inside out.

EXAMPLE 3
Subtracting with Grouping Symbols

Work each problem.

(a)
$$
\begin{aligned}
-6 - [2 - (8 + 3)] &= -6 - [2 - \mathbf{11}] &&\text{Add.} \\
&= -6 - [2 + (\mathbf{-11})] &&\text{Use the definition of subtraction.} \\
&= -6 - (\mathbf{-9}) &&\text{Add.} \\
&= -6 + (9) &&\text{Use the definition of subtraction.} \\
&= 3 &&\text{Add.}
\end{aligned}
$$

(b)
$$
\begin{aligned}
5 - [(-3 - 2) - (4 - 1)] &= 5 - [(-3 + (-2)) - \mathbf{3}] \\
&= 5 - [(\mathbf{-5}) - 3] \\
&= 5 - [(-5) + (-3)] \\
&= 5 - (-8) \\
&= 5 + 8 \\
&= 13
\end{aligned}
$$

(c)
$$
\begin{aligned}
\frac{2}{3} - \left[\frac{1}{12} - \left(-\frac{1}{4}\right)\right] &= \frac{8}{12} - \left[\frac{1}{12} - \left(-\frac{3}{12}\right)\right] &&\text{Get a common denominator.} \\
&= \frac{8}{12} - \left[\frac{1}{12} + \frac{3}{12}\right] &&\text{Use the definition of subtraction.} \\
&= \frac{8}{12} - \frac{4}{12} &&\text{Add.} \\
&= \frac{4}{12} &&\text{Subtract.} \\
&= \frac{1}{3} &&\text{Lowest terms} \quad ∎
\end{aligned}
$$

4 ▶ Interpret words and phrases that involve subtraction.

▶ Let us now look at how we interpret words and phrases that involve subtraction.

PROBLEM SOLVING   In order to solve problems that involve subtraction, we must be able to interpret key words and phrases that indicate subtraction. *Difference* is one of them. Some of these are given in the chart below.

Word or Phrase	Example	Numerical Expression and Simplification
Difference between	The *difference between* -3 and -8	$-3 - (-8) = -3 + 8 = 5$
Subtracted from	12 *subtracted from* 18	$18 - 12 = 6$
Less	6 *less* 5	$6 - 5 = 1$
Less than	6 *less than* 5	$5 - 6 = 5 + (-6) = -1$
Decreased by	9 *decreased by* -4	$9 - (-4) = 9 + 4 = 13$
Minus	8 *minus* 5	$8 - 5 = 3$

CAUTION   When you are subtracting two numbers, it is important that you write them in the correct order, because, in general, $a - b \neq b - a$. For example, $5 - 3 \neq 3 - 5$. For this reason, it is important to *think carefully before interpreting an expression involving subtraction*. (This problem did not arise for addition.)

EXAMPLE 4

Interpreting Words and Phrases Involving Subtraction

Write a numerical expression for each phrase, and simplify the expression.

(a) The **difference between** -8 and 5

It is conventional to write the numbers in the order they are given when "difference between" is used.

$$-8 - 5 = -8 + (-5) = -13$$

(b) 4 **subtracted from** the sum of 8 and -3

Here the operation of addition is also used, as indicated by the word *sum*. First, add 8 and -3. Next, subtract 4 *from* this sum.

$$[8 + (-3)] - 4 = 5 - 4 = 1$$

(c) 4 **less than** -6

Be careful with order here. 4 must be taken *from* -6.

$$-6 - 4 = -6 + (-4) = -10$$

Notice that "4 less than -6" differs from "4 *is less than* -6." The second of these is symbolized as $4 < -6$ (which is a false statement).

(d) 8, **decreased by** 5 **less than** 12

First, write "5 less than 12" as $12 - 5$. Next, subtract $12 - 5$ from 8.

$$8 - (12 - 5) = 8 - 7 = 1 \quad ■$$

5 ▶ Solve problems that involve subtraction.

▶   Recall from Section 1.5 that gains and losses may be interpreted as signed numbers.

PROBLEM SOLVING   Other applications of signed numbers include profit and loss in business, temperatures above and below 0°, and altitudes above and below sea level. The next example illustrates subtraction of signed numbers.

EXAMPLE 5
Solving a Problem
Involving Subtraction

■ The record high temperature of 134°F in the United States was recorded at Death Valley, California, in 1913. The record low was −80°F at Prospect Creek, Alaska, in 1971. What is the difference between the highest and the lowest temperatures?

We must find the value of the highest temperature minus the lowest temperature.

$$134 - (-80) = 134 + 80 \quad \text{Use the definition of subtraction.}$$
$$= 214 \qquad \text{Add.}$$

The difference between the highest and the lowest temperatures is 214°F.   ■

1.6 EXERCISES

Fill in the blank with the correct response.

1. By the definition of subtraction, in order to perform the subtraction problem −6 − (−8), we must add the opposite of _____ to _____ .

2. "The difference between 7 and 12" translates as _____ , while "the difference between 12 and 7" translates as _____ .

3. In order to simplify 6 − [(7 − 8) − (8 − 12)] according to the rules for order of operations, we should begin by subtracting 8 from _____ .

4. −6 − (−3) = −6 + _____

5. −8 − 4 = −8 + _____

6. 0 − 41 = 0 + _____

Find the difference. See Examples 1–3.

7. 4 − 7 **8.** 8 − 13 **9.** 6 − 10 **10.** 9 − 14

11. −7 − 3 **12.** −12 − 5 **13.** −10 − 6 **14.** −13 − 16

15. 7 − (−4) **16.** 9 − (−6) **17.** 6 − (−13) **18.** 13 − (−3)

19. −7 − (−3) **20.** −8 − (−6) **21.** 3 − (4 − 6) **22.** 6 − (7 − 14)

23. −3 − (6 − 9) **24.** −4 − (5 − 12) **25.** $\frac{1}{2} - \left(-\frac{1}{4}\right)$ **26.** $\frac{1}{3} - \left(-\frac{4}{3}\right)$

27. $-\frac{3}{4} - \frac{5}{8}$ **28.** $-\frac{5}{6} - \frac{1}{2}$

29. $\frac{5}{8} - \left(-\frac{1}{2} - \frac{3}{4}\right)$ **30.** $\frac{9}{10} - \left(\frac{1}{8} - \frac{3}{10}\right)$

31. 4.4 − (−9.2) **32.** 6.7 − (−12.6)

33. −7.4 − 4.5 **34.** −5.4 − 9.6

35. −5.2 − (8.4 − 10.8) **36.** −9.6 − (3.5 − 12.6)

37. [(−3.1) − 4.5] − (.8 − 2.1) **38.** [(−7.8) − 9.3] − (.6 − 3.5)

39. Explain in your own words how to subtract signed numbers.

40. We know that in general, $a - b \neq b - a$. Can you give values for a and b so that $a - b$ *is equal to* $b - a$? If so, give two such pairs.

Work the problem. See Example 3.

41. $(-3 - 8) - (7 - 4)$

42. $(-5 - 9) - (7 - 2)$

43. $-10 - [(5 - 4) - (-5 - 8)]$

44. $-12 - [(9 - 2) - (-6 - 3)]$

45. $-4 + [(-6 - 9) - (-7 + 4)]$

46. $-8 + [(-3 - 10) - (-4 + 1)]$

47. $\left(-\dfrac{3}{4} - \dfrac{5}{2}\right) - \left(-\dfrac{1}{8} - 1\right)$

48. $\left(-\dfrac{3}{8} - \dfrac{2}{3}\right) - \left(-\dfrac{9}{8} - 3\right)$

49. $[-34.99 + (6.59 - 12.25)] - 8.33$

50. $[-12.25 - (8.34 + 3.57)] - 17.88$

51. Make up a subtraction problem so that the difference between two negative numbers is a negative number.

52. Make up a subtraction problem so that the difference between two negative numbers is a positive number.

Suppose that a represents a positive number and b represents a negative number. Determine whether the given expression must represent a positive number or a negative number.

53. $a - b$ **54.** $b - a$ **55.** $a + |b|$ **56.** $b - |a|$

──────────◇ **MATHEMATICAL CONNECTIONS** (Exercises 57–62) ◇──────────

The concepts of distance and absolute value are closely related. Exercises 57–62 illustrate how we can express the distance between two points on a number line using absolute value. Work through these exercises in order.

57. Draw a number line and plot the points at -3 and 5. Label the point at -3 as A and the point at 5 as B.

58. Count the number of units between A and B. How many are there?

59. We shall refer to the answer in Exercise 58 as the distance between A and B (or equivalently, between B and A). Subtract -3 from 5. Is this the same as the distance between A and B? If not, why not?

60. Subtract 5 from -3. Is this the same as the distance between A and B? If not, why not?

61. Find the absolute value of each answer in Exercises 59 and 60. Do you get the distance between A and B both times?

62. Based on your results in Exercises 57–61, fill in the blanks with the correct responses: To find the distance between two points on the number line, we can _____ their coordinates in either order, provided we remember to find the _____ _____ of the difference.

──────────◆──────────

Write a numerical expression for the phrase and simplify. See Example 4.

63. The difference between 4 and -8

64. The difference between 7 and -14

65. 8 less than -2

66. 9 less than -13

67. The sum of 9 and -4, decreased by 7

68. The sum of 12 and -7, decreased by 14

69. 12 less than the difference between 8 and -5

70. 19 less than the difference between 9 and -2

Find all solutions for the equation from the set $\{-3, -2, -1, 0, 1, 2, 3\}$. Guess or use trial and error.

71. $x - 1 = -2$ **72.** $x - 2 = -3$ **73.** $3 - x = 6$

74. $2 - x = 4$ **75.** $3 - (-x) = 0$ **76.** $1 - (-x) = 0$

Solve the problem by written a difference between real numbers and subtracting. No variables are needed. See Example 5.

The two charts show the heights of some selected mountains and the depths of some selected trenches. Use the information given to find the answers in Exercises 77–80. (*Source: The World Almanac and Book of Facts,* 1994)

Mountain	Height (in feet)	Trench	Depth in Feet (as a negative number)
Foraker	17,400	Philippine	−32,995
Wilson	14,246	Cayman	−24,721
Pikes Peak	14,110	Java	−23,376

77. What is the difference between the height of Mt. Foraker and the depth of the Philippine Trench?

78. What is the difference between the height of Pikes Peak and the depth of the Java Trench?

79. How much deeper is the Cayman Trench than the Java Trench?

80. How much deeper is the Philippine Trench than the Cayman Trench?

81. On a cold winter day, the temperature reached −5°F in Glenview, Illinois. The following day it was even colder, and the temperature dropped 20°F below −5°F. What was the temperature on the following day?

82. A chemist is running an experiment under precise conditions. At first, she runs it at −174.6°F. She then lowers it by 2.3°F. What is the new temperature for the experiment?

83. The top of Mount Whitney, visible from Death Valley, has an altitude of 14,494 feet above sea level. The bottom of Death Valley is 282 feet below sea level. Using zero as sea level, find the difference between these two elevations.

84. The highest point in Louisiana is Driskill Mountain, at an altitude of 535 feet. The lowest point is at Spanish Fort, 8 feet below sea level. Using zero as sea level, find the difference between these two elevations.

85. Chris owed his brother $10. He later borrowed $70. What positive or negative number represents his present financial status?

86. Francesca has $15 in her purse, and Emilio has a debt of $12. Find the difference between these amounts.

87. During the years 1980–1986, Rickey Henderson was the American League leader in stolen bases each year. The *bar graph* gives a representation of the number of steals each year. Use a signed number to represent the change in the number of steals from one year to the next. For example, from 1980 to 1981 the change was 56 − 100 = −44.
   (a) from 1981 to 1982
   (b) from 1982 to 1983
   (c) from 1983 to 1984
   (d) from 1984 to 1985
   (e) from 1985 to 1986

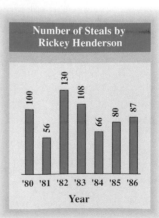

88. During the years 1985–1990, Vince Coleman was the National League leader in stolen bases each year. The *line graph* gives a representation of the number of steals each year. Use a signed number to represent the change in the number of steals from one year to the next. For example, from 1985 to 1986 the change was $107 - 110 = -3$.

(a) from 1986 to 1987

(b) from 1987 to 1988

(c) from 1988 to 1989

(d) from 1989 to 1990

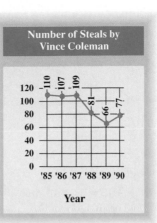

Number of Steals by Vince Coleman

Year

1.7 MULTIPLICATION OF REAL NUMBERS

FOR EXTRA HELP	OBJECTIVES
SSG pp. 33–36 **SSM** pp. 25–28	**1** ▶ Find the product of a positive number and a negative number.
Video 2	**2** ▶ Find the product of two negative numbers.
	3 ▶ Identify factors of integers.
Tutorial IBM MAC	**4** ▶ Use the order of operations in multiplication with signed numbers.
	5 ▶ Evaluate expressions involving variables.
	6 ▶ Interpret words and phrases that indicate multiplication.

Multiplication of real numbers is defined to preserve the properties of multiplication of positive numbers that we are familiar with. In this section we learn how to multiply with positive and negative numbers. We already know the rule for multiplying positive numbers. The product of two positive numbers is positive. We also know that the product of 0 and any positive number is 0, so we extend that property to all real numbers.

Multiplication by Zero   For any real number x, $x \cdot 0 = 0$.

1 ▶ Find the product of a positive number and a negative number.

▶ In order to define the product of a positive and a negative number so that the result is consistent with the multiplication of two positive numbers, look at the following pattern.

$$3 \cdot 5 = 15$$
$$3 \cdot 4 = 12$$
$$3 \cdot 3 = 9$$
$$3 \cdot 2 = 6$$
$$3 \cdot 1 = 3$$
$$3 \cdot 0 = 0$$
$$3 \cdot (-1) = ?$$

Numbers decrease by 3.

What should $3(-1)$ equal? The product $3(-1)$ represents the sum

$$-1 + (-1) + (-1) = -3,$$

so the product should be -3. Also,

$$3(-2) = -2 + (-2) + (-2) = -6$$

and

$$3(-3) = -3 + (-3) + (-3) = -9.$$

These results maintain the pattern in the list, which suggests the following rule.

Multiplying Numbers with Different Signs

For any positive real numbers x and y,

$$x(-y) = -(xy) \qquad \text{and} \qquad (-x)y = -(xy).$$

That is, the product of two numbers with opposite signs is negative.

EXAMPLE 1

Multiplying a Positive Number and a Negative Number

Find the products using the multiplication rule given above.

(a) $8(-5) = -(8 \cdot 5) = -40$ **(b)** $5(-4) = -(5 \cdot 4) = -20$

(c) $(-7)(2) = -(7 \cdot 2) = -14$ **(d)** $(-9)(3) = -(9 \cdot 3) = -27$ ∎

◆ **CONNECTIONS** ◆

Maria Jimenez bet three of her friends $5 each that her team would win the Super Bowl. Unfortunately, the team lost.

FOR DISCUSSION OR WRITING
How much did Maria lose? Write Maria's loss as a multiplication problem using signed numbers. Explain your reasoning. Write another realistic problem that involves multiplication of signed numbers.

---

2 ▶ Find the product of two negative numbers.

▶ The product of two positive numbers is positive, and the product of a positive and a negative number is negative. What about the product of two negative numbers? Look at another pattern.

$$
\begin{array}{c|c}
(-5)(4) = -20 & \text{Numbers} \\
(-5)(3) = -15 & \text{increase} \\
(-5)(2) = -10 & \text{by 5.} \\
(-5)(1) = -5 & \\
(-5)(0) = 0 & \\
(-5)(-1) = ? &
\end{array}
$$

The numbers on the left of the equals sign (in color) decrease by 1 for each step down the list. The products on the right increase by 5 for each step down the list. To maintain this pattern, $(-5)(-1)$ should be 5 more than $(-5)(0)$, or 5 more than 0, so

$$(-5)(-1) = 5.$$

The pattern continues with

$$(-5)(-2) = 10$$
$$(-5)(-3) = 15$$
$$(-5)(-4) = 20$$
$$(-5)(-5) = 25,$$

and so on. This pattern suggests the next rule.

Multiplying Two Negative Numbers

For any positive real numbers x and y,

$$(-x)(-y) = xy.$$

The product of two negative numbers is positive.

E X A M P L E 2
Multiplying Two Negative Numbers

Find the products using the multiplication rule given above.

(a) $(-9)(-2) = 9 \cdot 2 = 18$ **(b)** $(-6)(-12) = 6 \cdot 12 = 72$

(c) $(-8)(-1) = 8 \cdot 1 = 8$ **(d)** $(-15)(-2) = 15 \cdot 2 = 30$ ∎

A summary of the results for multiplying signed numbers is given here.

Multiplying Signed Numbers

The product of two numbers having the *same* sign is *positive,* and the product of two numbers having *different* signs is *negative.*

3 ▶ Identify factors of integers.

▶ In Section 1.1 the definition of a *factor* was given for whole numbers. (For example, since $9 \cdot 5 = 45$, both 9 and 5 are factors of 45.) The definition can now be extended to integers.

If the product of two integers is a third integer, then each of the two integers is a *factor* of the third. For example, $(-3)(-4) = 12$, so -3 and -4 are both factors of 12. The factors of 12 are the numbers -12, -6, -4, -3, -2, -1, 1, 2, 3, 4, 6, and 12.

E X A M P L E 3
Identifying Factors of an Integer

The following chart shows several integers and the factors of those integers.

Integer	Factors
18	$-18, -9, -6, -3, -2, -1, 1, 2, 3, 6, 9, 18$
20	$-20, -10, -5, -4, -2, -1, 1, 2, 4, 5, 10, 20$
15	$-15, -5, -3, -1, 1, 3, 5, 15$
7	$-7, -1, 1, 7$
1	$-1, 1$ ∎

4 ▶ Use the order of operations in multiplication with signed numbers.

▶ In the next example we use the order of operations discussed earlier with multiplication of positive and negative numbers.

EXAMPLE 4

Using the Order of Operations

Perform the indicated operations.

(a) $(-9)(2) - (-3)(2)$

First find all products, working from left to right.

$$(-9)(2) - (-3)(2) = -18 - (-6)$$

Now perform the subtraction.

$$-18 - (-6) = -18 + 6 = -12$$

(b) $(-6)(-2) - (3)(-4) = 12 - (-12) = 12 + 12 = 24$

(c) $-5(-2 - 3) = -5(-5) = 25$ ∎

5 ▶ Evaluate expressions involving variables.

▶ The next examples show numbers substituted for variables where the rules for multiplying signed numbers must be used.

EXAMPLE 5

Evaluating an Expression for Numerical Values

Evaluate the expression $(3x + 4y)(-2m)$ given $x = -1$, $y = -2$, $m = -3$.

First substitute the given values for the variables. Then find the value of the expression.

$$(3x + 4y)(-2m) = [3(-1) + 4(-2)][-2(-3)] \quad \text{Put parentheses around the number for each variable.}$$

$$= [-3 + (-8)][6] \quad \text{Find the products.}$$

$$= (-11)(6) \quad \text{Use order of operations.}$$

$$= -66 \quad ∎$$

EXAMPLE 6

Evaluating an Expression for Numerical Values

Evaluate $2x^2 - 3y^2$ if $x = -3$ and $y = -4$.

Use parentheses as shown.

$$2(-3)^2 - 3(-4)^2 = 2(9) - 3(16) \quad \text{Square} -3 \text{ and} -4.$$

$$= 18 - 48 \quad \text{Multiply.}$$

$$= -30 \quad \text{Subtract.} \quad ∎$$

6 ▶ Interpret words and phrases that indicate multiplication.

▶ Just as there are words and phrases that indicate addition and subtraction, certain ones also indicate multiplication.

PROBLEM SOLVING

The word *product* refers to multiplication. The chart below gives other key words and phrases that indicate multiplication.

Word or Phrase	Example	Numerical Expression and Simplification
Product of	The **product of** −5 and −2	$(-5)(-2) = 10$
Times	13 **times** −4	$13(-4) = -52$
Twice (meaning "2 times")	**Twice** 6	$2(6) = 12$
Of (used with fractions)	$\frac{1}{2}$ **of** 10	$\frac{1}{2}(10) = 5$
Percent of	12% **of** −16	$.12(-16) = -1.92$

E X A M P L E   7

Interpreting Words and
Phrases Involving
Multiplication

Write a numerical expression for each phrase and simplify. Use the order of operations.

(a) The **product of** 12 and the sum of 3 and -6
Here 12 is multiplied by "the sum of 3 and -6."

$$12[3 + (-6)] = 12(-3) = -36$$

(b) **Twice** the difference between 8 and -4

$$2[8 - (-4)] = 2[8 + 4] = 2(12) = 24$$

(c) Two-thirds **of** the sum of -5 and -3

$$\frac{2}{3}[-5 + (-3)] = \frac{2}{3}[-8] = -\frac{16}{3}$$

(d) 15% **of** the difference between 14 and -2
Remember that 15% $= .15$.

$$.15[14 - (-2)] = .15(14 + 2) = .15(16) = 2.4 \quad \blacksquare$$

1.7 EXERCISES

Decide whether the statement is true or false.

1. The product of two negative numbers is a positive number.

2. The product of a positive number and a negative number is a negative number.

3. When the sum of two negative numbers is multiplied by a positive number, the product is a positive number.

4. When the sum of two positive numbers is multiplied by a negative number, the product is a negative number.

5. When a negative number is squared, the result is a positive number.

6. When a negative number is raised to the fifth power, the result is a negative number.

Find the product. See Examples 1 and 2.

7. $(-4)(-5)$

8. $(-4)(-6)$

9. $5(-6)$

10. $3(-7)$

11. $(-7)(4)$

12. $(-8)(5)$

13. $(-4)(-20)$

14. $(-8)(-30)$

15. $(-8)(0)$

16. $(-12)(0)$

17. $\left(-\frac{3}{8}\right)\left(-\frac{20}{9}\right)$

18. $\left(-\frac{5}{4}\right)\left(-\frac{6}{25}\right)$

19. $(-6.8)(.35)$

20. $(-4.6)(.24)$

21. $(-6)\left(-\frac{1}{4}\right)$

22. $(-8)\left(-\frac{1}{2}\right)$

Find all integer factors of the given number. See Example 3.

23. 32   24. 36   25. 40   26. 50   27. 31   28. 17

Perform the indicated operations. See Example 4.

29. $7 - 3 \cdot 6$

30. $8 - 2 \cdot 5$

31. $-10 - (-4)(2)$

32. $-11 - (-3)(6)$

33. $15(8 - 12)$

34. $3(4 - 16)$

35. $-7(3 - 8)$

36. $-5(4 - 7)$

37. $(12 - 14)(1 - 4)$

38. $(8 - 9)(4 - 12)$

39. $(7 - 10)(10 - 4)$

40. $(5 - 12)(19 - 4)$

41. $(-2 - 8)(-6) + 7$

42. $(-9 - 4)(-2) + 10$

43. $3(-5) - (-7)$

44. $4(-8) - (-11)$

45. $(-9 - 3)(-5) - (-4)$

46. $(-7 - 4)(-9) - (-2)$

47. Explain the method you would use to evaluate $3x + 2y$ if $x = -3$ and $y = 4$.

48. If x and y are both replaced by negative numbers, is the value of $4x + 8y$ positive or negative?

Evaluate the expression if $x = 6$, $y = -4$, and $a = 3$. See Examples 5 and 6.

49. $5x - 2y + 3a$

50. $6x - 5y + 4a$

51. $(2x + y)(3a)$

52. $(5x - 2y)(-2a)$

53. $\left(\dfrac{1}{3}x - \dfrac{4}{5}y\right)\left(-\dfrac{1}{5}a\right)$

54. $\left(\dfrac{5}{6}x + \dfrac{3}{2}y\right)\left(-\dfrac{1}{3}a\right)$

55. $(-5 + x)(-3 + y)(3 - a)$

56. $(6 - x)(5 + y)(3 + a)$

57. $-2y^2 + 3a$

58. $5x - 4a^2$

59. $3a^2 - x^2$

60. $4y^2 - 2x^2$

Write a numerical expression for the phrase and simplify. See Example 7.

61. The product of -9 and 2, added to 9

62. The product of 4 and -7, added to -12

63. Twice the product of -1 and 6, subtracted from -4

64. Twice the product of -8 and 2, subtracted from -1

65. Nine subtracted from the product of 7 and -12

66. Three subtracted from the product of -2 and 5

67. The product of 12 and the difference between 9 and -8

68. The product of -3 and the difference between 3 and -7

69. Four-fifths of the sum of -8 and -2

70. Three-tenths of the sum of -2 and -28

Find the solution for the equation from the set $\{-3, -2, -1, 0, 1, 2, 3\}$ by guessing or by using trial and error.

71. $-2x = -6$

72. $-3x = -9$

73. $-4y = 0$

74. $-6t = 0$

75. $-5x = 10$

76. $-2x = 2$

77. $7x = -14$

78. $6x = -12$

79. $\dfrac{1}{5}w = -\dfrac{1}{5}$

80. $\dfrac{2}{3}t = -\dfrac{2}{3}$

81. $6x + 10 = -8$

82. $2x + 6 = 4$

Based on data obtained from the Association of American Medical Colleges, the percent of graduates choosing family practice, general internal medicine, or general pediatrics (known collectively as *generalist*) is approximated by the expression $-2.18x + 36.99$, where $x = 0$ represents the year 1982, $x = 1$ corresponds to 1983, and so on, up to $x = 10$ corresponding to 1992. Use the equation $y = -2.18x + 36.99$ to approximate the percent of graduates choosing generalist for the following years.

83. 1983

84. 1986

85. 1989

86. 1990

────── ◆ **MATHEMATICAL CONNECTIONS** (Exercises 87–90) ◇──────

In this section we used a pattern to justify that the product of a positive number and a negative number is a negative number. Work Exercises 87–90 in order to see another way to justify this property.

87. Multiplication of two positive integers can be interpreted as repeated addition. For example, 3×5 can be thought of as using 3 as a term five times: $3 + 3 + 3 + 3 + 3$. If we add five positive numbers what must be the sign of the sum?

88. Multiplication of a negative integer and a positive integer can also be interpreted as repeated addition, with the negative factor being used as the repeated term. For example, -3×5 can be thought of as using -3 as a term five times: $-3 + (-3) + (-3) + (-3) + (-3)$. If we add five negative numbers what must be the sign of the sum?

89. Interpreting -3×5 as repeated addition as shown in Exercise 88, what is the sum when -3 is used as the added term five times?

90. Because there is only one answer to the multiplication problem -3×5, what must this product be, based on your answer to Exercise 89? In general, what can we say about the product of a negative number and a positive number?

─────── ◆ ───────

1.8 DIVISION OF REAL NUMBERS

FOR EXTRA HELP	OBJECTIVES
SSG pp. 36–39 **SSM** pp. 28–33	**1 ▶** Find the reciprocal, or multiplicative inverse, of a number.
Video 2	**2 ▶** Divide using signed numbers. **3 ▶** Simplify numerical expressions involving quotients.
Tutorial IBM MAC	**4 ▶** Interpret words and phrases that indicate division. **5 ▶** Translate simple sentences into equations.

1 ▶ Find the reciprocal, or multiplicative inverse, of a number.

▶ In Section 1.6 we saw that the difference between two numbers is found by adding the additive inverse of the second number to the first. Similarly, the *quotient* of two numbers is found by *multiplying* by the *multiplicative inverse*. By definition, since

$$8 \cdot \frac{1}{8} = \frac{8}{8} = 1 \qquad \text{and} \qquad \frac{5}{4} \cdot \frac{4}{5} = \frac{20}{20} = 1,$$

the multiplicative inverse of 8 is $1/8$, and of $5/4$ is $4/5$.

Multiplicative Inverse

Pairs of numbers whose product is 1 are **multiplicative inverses,** or **reciprocals,** of each other.

EXAMPLE 1

Finding the Multiplicative Inverse

The following chart shows several numbers and their multiplicative inverses.

Number	Multiplicative Inverse (Reciprocal)
4	$\dfrac{1}{4}$
-5	$\dfrac{1}{-5}$ or $-\dfrac{1}{5}$
$-\dfrac{5}{8}$	$-\dfrac{8}{5}$
0	None
1	1
-1	-1

Why is there no multiplicative inverse for the number 0? Suppose that k is to be the multiplicative inverse of 0. Then $k \cdot 0$ should equal 1. But $k \cdot 0 = 0$ for any number k. Since there is no value of k that is a solution of the equation $k \cdot 0 = 1$, the following statement can be made.

0 has no multiplicative inverse.

2 ▸ Divide using signed numbers.

▸ By definition, the quotient of x and y is the product of x and the multiplicative inverse of y.

Definition of Division

For any real numbers x and y, with $y \neq 0$, $\quad \dfrac{x}{y} = x \cdot \dfrac{1}{y}.$

The definition of division indicates that y, the number to divide by, cannot be 0. The reason is that 0 has no multiplicative inverse, so that $1/0$ is not a number. Because 0 has no multiplicative inverse, *division by 0 is undefined*. If a division problem turns out to involve division by 0, write "undefined."

NOTE While division by zero is undefined, we may divide 0 by any nonzero number. In fact, if $a \neq 0$,

$$\frac{0}{a} = 0.$$

Since division is defined in terms of multiplication, all the rules of multiplication of signed numbers also apply to division.

EXAMPLE 2

Using the Definition of Division

Find the quotients using the definition of division.

(a) $\dfrac{12}{3} = 12 \cdot \dfrac{1}{3} = 4$

(b) $\dfrac{-10}{2} = -10 \cdot \dfrac{1}{2} = -5$

(c) $\dfrac{-14}{-7} = -14\left(\dfrac{1}{-7}\right) = 2$

(d) $-\dfrac{2}{3} \div \left(-\dfrac{4}{5}\right) = -\dfrac{2}{3} \cdot \left(-\dfrac{5}{4}\right) = \dfrac{5}{6}$

(e) $\dfrac{0}{13} = 0\left(\dfrac{1}{13}\right) = 0 \quad \dfrac{0}{a} = 0 \ (a \neq 0)$

(f) $\dfrac{-10}{0} \quad$ undefined ■

When dividing fractions, multiplying by the reciprocal works well. However, using the definition of division directly with integers is awkward. It is easier to divide in the usual way, then determine the sign of the answer. The following rule for division can be used instead of multiplying by the reciprocal.

Dividing Signed Numbers   The quotient of two numbers having the same sign is positive; the quotient of two numbers having *different* signs is *negative.*

Note that these are the same as the rules for multiplication.

E X A M P L E   3

Dividing Signed Numbers

■ Find the quotients.

(a) $\dfrac{8}{-2} = -4$

(b) $\dfrac{-4.5}{-.09} = 50$

(c) $-\dfrac{1}{8} \div \left(-\dfrac{3}{4}\right) = -\dfrac{1}{8} \cdot \left(-\dfrac{4}{3}\right) = \dfrac{1}{6}$ ■

From the definitions of multiplication and division of real numbers,

$$\frac{-40}{8} = -40 \cdot \frac{1}{8} = -5,$$

and

$$\frac{40}{-8} = 40\left(\frac{1}{-8}\right) = -5,$$

so that

$$\frac{-40}{8} = \frac{40}{-8}.$$

Based on this example, the quotient of a positive and a negative number can be expressed in any of the following three forms.

For any positive real numbers x and y, $\dfrac{-x}{y} = \dfrac{x}{-y} = -\dfrac{x}{y}.$

The quotient of two negative numbers can be expressed as a quotient of two positive numbers.

For any positive real numbers x and y, $\dfrac{-x}{-y} = \dfrac{x}{y}.$

3 ► Simplify numerical expressions involving quotients.

► In the next example we simplify numerical expressions involving quotients.

EXAMPLE 4
Simplifying Expressions
Involving Division

■ Simplify $\dfrac{5(-2) - (3)(4)}{2(1 - 6)}$.

Simplify the numerator and denominator separately. Then divide or write in lowest terms.

$$\frac{5(-2) - (3)(4)}{2(1 - 6)} = \frac{-10 - 12}{2(-5)} = \frac{-22}{-10} = \frac{11}{5} \quad ■$$

The rules for operations with signed numbers are summarized here.

**Operations with
Signed Numbers**

Addition

Like signs   Add the absolute values of the numbers. The sum has the same sign as the numbers.

Unlike signs   Subtract the number with the smaller absolute value from the one with the larger. Give the sum the sign of the number having the larger absolute value.

Subtraction

Add the additive inverse, or opposite, of the second number.

Multiplication and Division

Like signs   The product or quotient of two numbers with like signs is positive.

Unlike signs   The product or quotient of two numbers with unlike signs is negative.

Division by 0 is undefined.

0 divided by a nonzero number equals 0.

---

4 ▶ Interpret words and phrases that indicate division.

▶   Certain words and phrases indicate the operation of division.

PROBLEM SOLVING   The word *quotient* refers to the result obtained in a division problem. In algebra, quotients are usually represented with a fraction bar. The symbol ÷ is seldom used. When translating stated problems involving division, interpret using the fraction bar. The following chart gives some key phrases associated with division.

Phrase	Example	Numerical Expression and Simplification
Quotient of	The *quotient of* −24 and 3	$\dfrac{-24}{3} = -8$
Divided by	−16 *divided by* −4	$\dfrac{-16}{-4} = 4$
Ratio of	The *ratio of* 2 to 3	$\dfrac{2}{3}$

It is customary to write the first number named as the numerator and the second as the denominator when interpreting a phrase involving division, as shown in the next example.

EXAMPLE 5

Interpreting Words and Phrases Involving Division

Write a numerical expression for each phrase, and simplify the expression.

(a) The **quotient of** 14 and the sum of -9 and 2

"Quotient" indicates division. The number 14 is the numerator and "the sum of -9 and 2" is the denominator.

$$\frac{14}{-9 + 2} = \frac{14}{-7} = -2$$

(b) The product of 5 and -6, **divided by** the difference between -7 and 8

The numerator of the fraction representing the division is obtained by multiplying 5 and -6. The denominator is found by subtracting -7 and 8.

$$\frac{5(-6)}{-7 - 8} = \frac{-30}{-15} = 2 \quad \blacksquare$$

5 ▶ Translate simple sentences into equations.

▶ In this section and the preceding three sections, important words and phrases involving the four operations of arithmetic have been introduced.

PROBLEM SOLVING

We can use words and phrases involving arithmetic operations to interpret sentences that translate into equations. The ability to do this will help us to solve the types of problems found in later chapters.

EXAMPLE 6

Translating Words Into an Equation

Write the following in symbols, using x as the variable, and guess or use trial and error to find the solution. All solutions come from the list of integers between   12 and 12, inclusive.

(a) Three **times** a number **is** -18.

The word *times* indicates multiplication, and the word *is* translates as the equals sign ($=$).

$$3x = -18$$

Since the integer between -12 and 12, inclusive, that makes this statement true is -6, the solution of the equation is -6.

(b) The **sum** of a number and 9 **is** 12.

$$x + 9 = 12$$

Since $3 + 9 = 12$, the solution of this equation is 3.

(c) The **difference between** a number and 5 **is** 0.

$$x - 5 = 0$$

Since $5 - 5 = 0$, the solution of this equation is 5.

(d) The **quotient of** 24 and a number **is** -2.

$$\frac{24}{x} = -2$$

Here, x must be a negative number, since the numerator is positive and the quotient is negative. Since $24/(-12) = -2$, the solution is -12. $\blacksquare$

CAUTION It is important to recognize the distinction between the types of problems found in Example 5 and Example 6. In Example 5, the phrases translate as *expressions,* while in Example 6, the sentences translate as *equations.* Remember that an equation is a sentence with an = sign, while an expression is a phrase.

$$\frac{5(-6)}{-7-8} \quad \text{is an } \textbf{expression,}$$

$$3x = -18 \quad \text{is an } \textbf{equation.}$$

1.8 EXERCISES

Fill in the blank with the correct response.

1. The quotient of two numbers with the same sign is a _____ number.
   (positive/negative)

2. The quotient of two numbers with unlike signs is a _____ number.
   (positive/negative)

3. If two negative numbers are multiplied and their product is then divided by a negative number, the result is _____ zero.
   (greater than/less than)

4. If a positive number is multiplied by a negative number and their product is then divided by a positive number, the result is _____ zero.
   (greater than/less than)

5. The reciprocal of a positive number is a _____ number.
   (positive/negative)

6. The reciprocal of a negative number is a _____ number.
   (positive/negative)

Find the reciprocal (multiplicative inverse), if one exists, of the number. See Example 1.

7. 11   **8.** 12   **9.** -5 **10.** -3

11. $\dfrac{5}{6}$ **12.** $\dfrac{9}{10}$ **13.** $3-3$ **14.** $5+(-5)$

15. $-\dfrac{8}{7}$ **16.** $-\dfrac{9}{7}$ **17.** .4   **18.** .50

19. Which one of the following expressions is undefined?

   (a) $\dfrac{5-5}{5+5}$ **(b)** $\dfrac{5+5}{5+5}$ **(c)** $\dfrac{5-5}{5-5}$ **(d)** $\dfrac{5-5}{5}$

20. Explain why 0 has no reciprocal.

Find the quotient. See Examples 2 and 3.

21. $\dfrac{-15}{5}$ **22.** $\dfrac{-18}{6}$ **23.** $\dfrac{20}{-10}$ **24.** $\dfrac{28}{-4}$

25. $\dfrac{-160}{-10}$ **26.** $\dfrac{-260}{-20}$ **27.** $\dfrac{0}{-3}$ **28.** $\dfrac{0}{-6}$

29. $\dfrac{-10.252}{-.4}$ **30.** $\dfrac{-29.584}{-.8}$ **31.** $\left(-\dfrac{3}{4}\right) \div \left(-\dfrac{1}{2}\right)$

32. $\left(-\dfrac{3}{16}\right) \div \left(-\dfrac{5}{8}\right)$ **33.** $(-6.8) \div (-2)$ **34.** $(-23.5) \div (-5)$

35. $\dfrac{18}{3 - 9}$ **36.** $\dfrac{21}{4 - 7}$ **37.** $\dfrac{-50}{7 - (-3)}$

38. $\dfrac{-36}{4 - (-5)}$ **39.** $\dfrac{-12 - 36}{-12}$ **40.** $\dfrac{-18 - 2}{-5}$

Simplify the numerator and denominator separately. Then find the quotient. See Example 4.

41. $\dfrac{-5(-6)}{9 - (-1)}$ **42.** $\dfrac{-12(-5)}{7 - (-5)}$ **43.** $\dfrac{-21(3)}{-3 - 6}$

44. $\dfrac{-40(3)}{-2 - 3}$ **45.** $\dfrac{-10(2) + 6(2)}{-3 - (-1)}$ **46.** $\dfrac{8(-1) + 6(-2)}{-6 - (-1)}$

47. $\dfrac{-27(-2) - (-12)(-2)}{-2(3) - 2(2)}$ **48.** $\dfrac{-13(-4) - (-8)(-2)}{(-10)(2) - 4(-2)}$ **49.** $\dfrac{1^2 + 4^2}{3^2 + 5^2}$

50. $\dfrac{3^2 + 4^2}{6^2 + 8^2}$ **51.** $\dfrac{2^2 - 8^2}{6(-4 + 3)}$ **52.** $\dfrac{3^2 - 4^2}{7(-8 + 9)}$

Assume that a is negative, b is positive, and c is positive. Tell whether the value of the given expression is positive or negative.

53. $\dfrac{a \cdot b}{c}$ **54.** $\dfrac{a}{b \cdot c}$ **55.** $a^2 \cdot b \cdot c$

56. $a \cdot b^2 \cdot c$ **57.** $b \cdot (c - a)$ **58.** $\dfrac{a - c}{b}$

Write a numerical expression for the phrase and simplify the expression. See Example 5.

59. The quotient of -36 and -9 **60.** The quotient of -48 and -6

61. The quotient of -12 and the sum of -5 and -1

62. The quotient of -20 and the sum of -8 and -2

63. The sum of 15 and -3, divided by the product of 4 and -3

64. The sum of -18 and -6, divided by the product of 2 and -4

65. The product of -34 and 7, divided by -14

66. The product of -25 and 4, divided by -10

Write the statement in symbols, using x as the variable, and find the solution by guessing or by using trial and error. All solutions come from the list of integers between -12 and 12, inclusive. See Example 6.

67. Six times a number is -42.

68. Four times a number is -36.

69. The quotient of a number and 3 is -3.

70. The quotient of a number and 4 is -1.

71. 6 less than a number is 4.

72. 7 less than a number is 2.

73. When 5 is added to a number the result is -5.

74. When 6 is added to a number the result is -3.

To find the average of a group of numbers, we add the numbers and then divide the sum by the number of terms added. For example, to find the average of 14, 8, 3, 9, and 1, we add them and divide by 5:

$$\frac{14 + 8 + 3 + 9 + 1}{5} = \frac{35}{5} = 7.$$

The average of these numbers is 7. Use the procedure described above to find the average of the numbers.

75. 3, 7, 5, 6, 4, 1, 2

76. 5, 13, 2, 12, 3

77. $-8, -5, -3, 0$

78. $-12, -3, -9$

79. Explain why the average of a group of negative numbers must be negative.

80. Suppose there is a group of numbers with some positive and some negative. Under what conditions will the average be a positive number? Under what conditions will the average be negative?

81. The chart shows the 1994 salaries of several major league baseball players. (*Source*: *USA Today*)
What was the average salary for these three players?

Player	Salary
Chris Bosio	$4,000,000
Rob Ducey	$ 275,000
Mike Benjamin	$ 225,000

82. The bar graph shows the 1993 sales in millions of dollars of four of the largest brands in the United States. What was the average of these sales? (*Source:* Information Resources, Inc.)

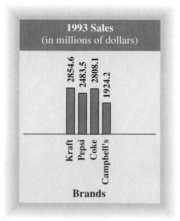

Find the solution of the equation from the set $\{-8, -6, -4, -2, 0, 2, 4, 6, 8\}$ by guessing or by using trial and error.

83. $\dfrac{x}{-4} = 2$

84. $\dfrac{x}{-2} = 3$

85. $\dfrac{n}{-2} = -2$

86. $\dfrac{t}{-2} = -3$

87. $\dfrac{2x}{2} = -4$

88. $\dfrac{2y}{4} = -3$

The operation of division plays an important role in *divisibility tests*. By using a divisibility test, we can determine whether a given natural number is divisible (without remainder) by another natural number. For example, an integer is divisible by 2 if its last digit is divisible by 2, and not otherwise. Exercises 89–96 introduce some of the simpler divisibility tests.

89. Tell why **(a)** 3,473,986 is divisible by 2 and **(b)** 4,336,879 is not divisible by 2.

90. An integer is divisible by 3 if the sum of its digits is divisible by 3, and not otherwise. Show that **(a)** 4,799,232 is divisible by 3 and **(b)** 2,443,871 is not divisible by 3.

91. An integer is divisible by 4 if its last two digits form a number divisible by 4, and not otherwise. Show that **(a)** 6,221,464 is divisible by 4 and **(b)** 2,876,335 is not divisible by 4.

92. An integer is divisible by 5 if its last digit is divisible by 5, and not otherwise. Show that **(a)** 3,774,595 is divisible by 5 and **(b)** 9,332,123 is not divisible by 5.

93. An integer is divisible by 6 if it is divisible by both 2 and 3, and not otherwise. Show that **(a)** 1,524,822 is divisible by 6 and **(b)** 2,873,590 is not divisible by 6.

94. An integer is divisible by 8 if its last three digits form a number divisible by 8, and not otherwise. Show that **(a)** 2,923,296 is divisible by 8 and **(b)** 7,291,623 is not divisible by 8.

95. An integer is divisible by 9 if the sum of its digits is divisible by 9, and not otherwise. Show that **(a)** 4,114,107 is divisible by 9 and **(b)** 2,287,321 is not divisible by 9.

96. An integer is divisible by 12 if it is divisible by both 3 and 4, and not otherwise. Show that **(a)** 4,253,520 is divisible by 12 and **(b)** 4,249,474 is not divisible by 12.

1.9 PROPERTIES OF ADDITION AND MULTIPLICATION

FOR EXTRA HELP	OBJECTIVES
📖 **SSG** pp. 39–46 **SSM** pp. 33–37	1 ▶ Use the commutative properties.
📼 **Video** 2	2 ▶ Use the associative properties.
	3 ▶ Use the identity properties.
💾 **Tutorial** IBM MAC	4 ▶ Use the inverse properties.
	5 ▶ Use the distributive property.

If you were asked to find the sum $3 + 89 + 97$, it is likely that you would mentally add $3 + 97$ to get 100, and then add $100 + 89$ to get 189. While the rule for order of operations says to add from left to right, it is a fact that we may change the order of the terms and group them in any way we choose without affecting the sum. These are examples of shortcuts that we use in everyday mathematics. These shortcuts are justified by the basic properties of addition and multiplication, which are discussed in this section. In the following statements, a, b, and c represent real numbers.

1 ▶ Use the commutative properties.

▶   The word *commute* means to go back and forth. Many people commute to work or to school. If you travel from home to work and follow the same route from work to home, you travel the same distance each time. The **commutative properties** say that if two numbers are added or multiplied in any order, they give the same result.

Commutative Properties	$a + b = b + a$
	$ab = ba$

E X A M P L E   1

Using the Commutative Properties

Use a commutative property to complete each statement.

(a) $-8 + 5 = 5 +$ _____

By the commutative property for addition, the missing number is -8, since $-8 + 5 = 5 + (-8)$.

(b) $(-2)(7) =$ _____ (-2)

By the commutative property for multiplication, the missing number is 7, since $(-2)(7) = (7)(-2)$. ■

2 ▶ Use the associative properties.

▶   When we *associate* one object with another, we tend to think of those objects as being grouped together. The **associative properties** say that when we add or multiply three numbers, we can group the first two together or the last two together and get the same answer.

Associative Properties	$(a + b) + c = a + (b + c)$
	$(ab)c = a(bc)$

E X A M P L E   2

Using the Associative Properties

Use an associative property to complete each statement.

(a) $8 + (-1 + 4) = (8 +$ _____ $) + 4$

The missing number is -1.

(b) $[2 \cdot (-7)] \cdot 6 = 2 \cdot$ _____

The completed expression on the right should be $2 \cdot [(-7) \cdot 6]$. ■

By the associative property of addition, the sum of three numbers will be the same no matter which way the numbers are "associated" in groups. For this reason, parentheses can be left out in many addition problems. For example, both

$$(-1 + 2) + 3 \qquad \text{and} \qquad -1 + (2 + 3)$$

can be written as

$$-1 + 2 + 3.$$

In the same way, parentheses also can be left out of many multiplication problems.

E X A M P L E   3

Distinguishing Between the Associative and Commutative Properties

(a) Is $(2 + 4) + 5 = 2 + (4 + 5)$ an example of the associative property or the commutative property?

The order of the three numbers is the same on both sides of the equals sign. The only change is in the grouping, or association, of the numbers. Therefore, this is an example of the associative property.

(b) Is $6(3 \cdot 10) = 6(10 \cdot 3)$ an example of the associative property or the commutative property?

The same numbers, 3 and 10 are grouped on each side. On the left, the 3 appears first, but on the right, the 10 appears first. Since the only change involves the order of the numbers, this statement is an example of the commutative property.

(c) Is $(8 + 1) + 7 = 8 + (7 + 1)$ an example of the associative property or the commutative property?

In the statement, both the order and the grouping are changed. On the left the order of the three numbers is 8, 1, and 7. On the right it is 8, 7, and 1. On the left the 8 and 1 are grouped, and on the right the 7 and 1 are grouped. Therefore, both the associative and the commutative properties are used   ■

EXAMPLE 4

Using the Commutative
and Associative
Properties

Find the sum: $23 + 41 + 2 + 9 + 25$.

The commutative and associative properties make it possible to choose pairs of numbers whose sums are easy to add.

$$23 + 41 + 2 + 9 + 25 = (41 + 9) + (23 + 2) + 25$$
$$= 50 + 25 + 25$$
$$= 100 \quad ■$$

3 ▶ Use the identity properties.

▶   If a child wears a costume on Halloween, the child's appearance is changed, but his or her *identity* is unchanged. The identity of a real number is left unchanged when identity properties are applied. The **identity properties** say that the sum of 0 and any number equals that number, and the product of 1 and any number equals that number.

Identity Properties

$$a + 0 = a \quad \text{and} \quad 0 + a = a$$
$$a \cdot 1 = a \quad \text{and} \quad 1 \cdot a = a$$

The number 0 leaves the identity, or value, of any real number unchanged by addition. For this reason, 0 is called the **identity element for addition.** Since multiplication by 1 leaves any real number unchanged, 1 is the **identity element for multiplication.**

EXAMPLE 5

Using the Identity
Properties

These statements are examples of the identity properties.

(a) $-3 + 0 = -3$

(b) $1 \cdot \dfrac{1}{2} = \dfrac{1}{2}$ ■

We use the identity property for multiplication to write fractions in lowest terms and to get common denominators.

E X A M P L E   6

Using the Identity Property to Simplify Expressions

Simplify the following expressions.

(a) $\dfrac{49}{35}$

$$\dfrac{49}{35} = \dfrac{7 \cdot 7}{5 \cdot 7} \qquad \text{Factor.}$$

$$= \dfrac{7}{5} \cdot \dfrac{7}{7} \qquad \text{Write as a product.}$$

$$= \dfrac{7}{5} \cdot 1 \qquad \text{Divide.}$$

$$= \dfrac{7}{5} \qquad \text{Identity property}$$

(b) $\dfrac{3}{4} + \dfrac{5}{24}$

$$\dfrac{3}{4} + \dfrac{5}{24} = \dfrac{3}{4} \cdot 1 + \dfrac{5}{24} \qquad \text{Identity property}$$

$$= \dfrac{3}{4} \cdot \dfrac{6}{6} + \dfrac{5}{24} \qquad \text{Get a common denominator.}$$

$$= \dfrac{18}{24} + \dfrac{5}{24} \qquad \text{Multiply.}$$

$$= \dfrac{23}{24} \qquad \text{Add.} \qquad ■$$

4 ▶ Use the inverse properties.

▶ Each day before you go to work or school, you probably put on your shoes before you leave. Before you go to sleep at night, you probably take them off, and this leads to the same situation that existed before you put them on. These operations from everyday life are examples of inverse operations. The **inverse properties** of addition and multiplication lead to the additive and multiplicative identities, respectively. Recall that $-a$ is the **additive inverse** of a and $1/a$ is the **multiplicative inverse** of the nonzero number a. The sum of the numbers a and $-a$ is 0, and the product of the nonzero numbers a and $1/a$ is 1.

Inverse Properties

$$a + (-a) = 0 \qquad \text{and} \qquad -a + a = 0$$

$$a \cdot \dfrac{1}{a} = 1 \qquad \text{and} \qquad \dfrac{1}{a} \cdot a = 1 \qquad (a \neq 0)$$

E X A M P L E   7

Using the Inverse Properties

The following statements are examples of the inverse properties.

(a) $\dfrac{2}{3} \cdot \dfrac{3}{2} = 1$

(b) $(-5)\left(-\dfrac{1}{5}\right) = 1$

(c) $-\dfrac{1}{2} + \dfrac{1}{2} = 0$

(d) $4 + (-4) = 0$ ■

EXAMPLE 8

Using the Additive Inverse Property to Simplify an Expression

Simplify $-2x + 10 + 2x$.

$$\begin{aligned}
-2x + 10 + 2x &= (-2x + 10) + 2x &&\text{Order of operations}\\
&= [10 + (-2x)] + 2x &&\text{Commutative property}\\
&= 10 + [(-2x) + 2x] &&\text{Associative property}\\
&= 10 + 0 &&\text{Inverse property}\\
&= 10 &&\text{Identity property} \quad\blacksquare
\end{aligned}$$

5 ▶ Use the distributive property

▶ The everyday meaning of the word *distribute* is "to give out from one to several." An important property of real number operations involves this idea. Look at the following statements.

$$2(5 + 8) = 2(13) = 26$$
$$2(5) + 2(8) = 10 + 16 = 26$$

Since both expressions equal 26,

$$2(5 + 8) = 2(5) + 2(8).$$

This result is an example of the **distributive property,** the only property involving *both* addition and multiplication. With this property, a product can be changed to a sum or difference.

The distributive property says that multiplying a number a by a sum of numbers $b + c$ gives the same result as multiplying a by b and a by c and then adding the two products.

Distributive Property

$$a(b + c) = ab + ac \qquad \text{and} \qquad (b + c)a = ba + ca$$

As the arrows show, the a outside the parentheses is "distributed" over the b and c inside. Another form of the distributive property is valid for subtraction.

$$a(b - c) = ab - ac \qquad \text{and} \qquad (b - c)a = ba - ca$$

The distributive property also can be extended to more than two numbers.

$$a(b + c + d) = ab + ac + ad$$

EXAMPLE 9

Using the Distributive Property

Use the distributive property to rewrite each expression.

(a) $\begin{aligned}[t] 5(9 + 6) &= 5 \cdot 9 + 5 \cdot 6 &&\text{Distributive property}\\ &= 45 + 30 &&\text{Multiply.}\\ &= 75 &&\text{Add.} \end{aligned}$

(b) $\begin{aligned}[t] 4(x + 5 + y) &= 4x + 4 \cdot 5 + 4y &&\text{Distributive property}\\ &= 4x + 20 + 4y &&\text{Multiply.} \end{aligned}$

(c) $\begin{aligned}[t] -2(x + 3) &= -2x + (-2)(3) &&\text{Distributive property}\\ &= -2x - 6 &&\text{Multiply.} \end{aligned}$

(d) $3(k - 9) = 3k - 3 \cdot 9$ Distributive property

               $= 3k - 27$ Multiply.

(e) $6 \cdot 8 + 6 \cdot 2 = 6(8 + 2)$ Distributive property

                 $= 6(10) = 60$ Add, then multiply.

(f) $4x - 4m = 4(x - m)$ Distributive property

(g) $6x - 12 = 6 \cdot x - 6 \cdot 2 = 6(x - 2)$ Distributive property

(h) $8(3r + 11t + 5z) = 8(3r) + 8(11t) + 8(5z)$ Distributive property

                      $= (8 \cdot 3)r + (8 \cdot 11)t + (8 \cdot 5)z$ Associative property

                      $= 24r + 88t + 40z$ ∎

The symbol $-a$ may be interpreted as $-1 \cdot a$. Similarly, when a negative sign precedes an expression within parentheses, it may also be interpreted as a factor of -1. The distributive property is used to remove parentheses from expressions such as $-(2y + 3)$. We do this by first writing $-(2y + 3)$ as $-1 \cdot (2y + 3)$.

$$-(2y + 3) = -1 \cdot (2y + 3)$$

$$= -1 \cdot (2y) + (-1) \cdot (3)$$ Distributive property

$$= -2y - 3$$ Multiply.

E X A M P L E   10

Using the Distributive Property to Remove Parentheses

Write without parentheses.

(a) $-(7r - 8) = -1(7r) + (-1)(-8)$ Distributive property

              $= -7r + 8$ Multiply.

(b) $-(-9w + 2) = 9w - 2$ ∎

The properties discussed in this section are the basic properties that justify how we do algebra. You should know them by name because we will be referring to them frequently to justify the procedures we introduce in the rest of the book. Here is a summary of these properties.

Properties of Addition and Multiplication

For any real numbers a, b, and c, the following properties hold.

Commutative Properties $a + b = b + a$ $ab = ba$

Associative Properties $(a + b) + c = a + (b + c)$

                            $(ab)c = a(bc)$

Identity Properties   There is a real number 0 such that

              $a + 0 = a$ and $0 + a = a.$

     There is a real number 1 such that

              $a \cdot 1 = a$ and $1 \cdot a = a.$

Inverse Properties

For each real number a, there is a single real number $-a$ such that

$$a + (-a) = 0 \quad \text{and} \quad (-a) + a = 0.$$

For each nonzero real number a, there is a single real number $\frac{1}{a}$ such that

$$a \cdot \frac{1}{a} = 1 \quad \text{and} \quad \frac{1}{a} \cdot a = 1.$$

Distributive Property

$$a(b + c) = ab + ac$$
$$(b + c)a = ba + ca$$

1.9 EXERCISES

Decide whether the statement is true or false.

1. The identity element for addition is 0.

2. The identity element for multiplication is 1.

3. The additive inverse of a is $\dfrac{1}{a}$.

4. The multiplicative inverse of a is $-a$.

5. The sum of a number and its additive inverse is 0.

6. The product of a number and its multiplicative inverse is 1.

7. Every number has a multiplicative inverse.

8. Every number has an additive inverse.

Decide whether the statement is an example of the commutative, associative, identity, inverse, or distributive property. See Examples 1, 2, 3, 5, 7, and 9.

9. $7 + 18 = 18 + 7$

10. $13 + 12 = 12 + 13$

11. $5(13 \cdot 7) = (5 \cdot 13) \cdot 7$

12. $-4(2 \cdot 6) = (-4 \cdot 2) \cdot 6$

13. $\dfrac{2}{3} \cdot (-4) = -4 \cdot \left(\dfrac{2}{3}\right)$

14. $6\left(-\dfrac{5}{6}\right) = \left(-\dfrac{5}{6}\right) \cdot 6$

15. $-6 + (12 + 7) = (-6 + 12) + 7$

16. $(-8 + 13) + 2 = -8 + (13 + 2)$

17. $-6 + 6 = 0$

18. $12 + (-12) = 0$

19. $\left(\dfrac{2}{3}\right)\left(\dfrac{3}{2}\right) = 1$

20. $\left(\dfrac{5}{8}\right)\left(\dfrac{8}{5}\right) = 1$

21. $2.34 \cdot 1 = 2.34$

22. $-8.456 \cdot 1 = -8.456$

23. $(4 + 17) + 3 = 3 + (4 + 17)$

24. $(-8 + 4) + (-12) = -12 + (-8 + 4)$

25. $6(x + y) = 6x + 6y$

26. $14(t + s) = 14t + 14s$

27. $-\dfrac{5}{9} = -\dfrac{5}{9} \cdot \dfrac{3}{3} = -\dfrac{15}{27}$

28. $\dfrac{13}{12} = \dfrac{13}{12} \cdot \dfrac{7}{7} = \dfrac{91}{84}$

29. $5(2x) + 5(3y) = 5(2x + 3y)$

30. $3(5t) - 3(7r) = 3(5t - 7r)$

31. The following conversation actually took place between one of the authors of this book and his son, Jack, when Jack was four years old:

> DADDY: "Jack, what is 3 + 0?"
> JACK: "3."
> DADDY: "Jack, what is 4 + 0?"
> JACK: "4. And Daddy, *string* plus zero equals *string*!"

What property of addition did Jack recognize?

32. The distributive property holds for multiplication with respect to addition. Is there a distributive property for addition with respect to multiplication? If not, give an example to show why.

33. Write a paragraph explaining in your own words the following properties of addition: commutative, associative, identity, inverse.

34. Write a paragraph explaining in your own words the following properties of multiplication: commutative, associative, identity, inverse.

Use the indicated property to write a new expression that is equal to the given expression. Then simplify the new expression if possible. See Examples 1, 2, 5, 7, and 9.

35. $r + 7$; commutative

36. $t + 9$; commutative

37. $s + 0$; identity

38. $w + 0$; identity

39. $-6(x + 7)$; distributive

40. $-5(y + 2)$; distributive

41. $(w + 5) + (-3)$; associative

42. $(b + 8) + (-10)$; associative

Use the properties of this section to simplify the expression. See Examples 7 and 8.

43. $6t + 8 - 6t + 3$

44. $9r + 12 - 9r + 1$

45. $\dfrac{2}{3}x - 11 + 11 - \dfrac{2}{3}x$

46. $\dfrac{1}{5}y + 4 - 4 - \dfrac{1}{5}y$

47. $\left(\dfrac{9}{7}\right)(-.38)\left(\dfrac{7}{9}\right)$

48. $\left(\dfrac{4}{5}\right)(-.73)\left(\dfrac{5}{4}\right)$

49. $t + (-t) + \dfrac{1}{2}(2)$

50. $w + (-w) + \dfrac{1}{4}(4)$

51. Evaluate $25 - (6 - 2)$ and evaluate $(25 - 6) - 2$. Do you think subtraction is associative?

52. Evaluate $180 \div (15 \div 3)$ and evaluate $(180 \div 15) \div 3$. Do you think division is associative?

53. Suppose that a student shows you the following work.

$$-3(4 - 6) = -3(4) - 3(6) = -12 - 18 = -30$$

The student has made a very common error. Write a short paragraph explaining the student's mistake, and work the problem correctly.

54. Explain how the procedure of changing $\dfrac{3}{4}$ to $\dfrac{9}{12}$ requires the use of the multiplicative identity element, 1.

Use the distributive property to rewrite the expression. Simplify if possible. See Example 9.

55. $5x + x$

56. $6q + q$

57. $4(t + 3)$

58. $5(w + 4)$

59. $-8(r + 3)$

60. $-11(x + 4)$

61. $-5(y - 4)$

62. $-9(g - 4)$

63. $-\dfrac{4}{3}(12y + 15z)$

64. $-\dfrac{2}{5}(10b + 20a)$

65. $8 \cdot z + 8 \cdot w$

66. $4 \cdot s + 4 \cdot r$

67. $7(2v) + 7(5r)$

68. $13(5w) + 13(4p)$

69. $8(3r + 4s - 5y)$

70. $2(5u - 3v + 7w)$

71. $q + q + q$

72. $m + m + m + m$

73. $-5x + x$

74. $-9p + p$

Use the distributive property to write the expression without parentheses. See Example 10.

75. $-(4t + 3m)$ **76.** $-(9x + 12y)$ **77.** $-(-5c - 4d)$

78. $-(-13x - 15y)$ **79.** $-(-3q + 5r - 8s)$ **80.** $-(-4z + 5w - 9y)$

81. The operations of "getting out of bed" and "taking a shower" are not commutative. Give an example of another pair of everyday operations that are not commutative.

82. The phrase "dog biting man" has two different meanings, depending on how the words are associated:

$$\text{(dog biting) man} \qquad \text{dog (biting man)}$$

Give another example of a three-word phrase that has different meanings depending on how the words are associated.

◆ **MATHEMATICAL CONNECTIONS** (Exercises 83–86) ◆

In Section 1.7 we used a pattern to see that the product of two negative numbers is a positive number. In the Mathematical Connections in that section we saw how we can justify that the product of a negative number and a positive number is negative, using the idea of repeated addition. In the exercises that follow, we show another justification of the rule for the sign of the product of two negative numbers. Work Exercises 83–86 in order.

83. Evaluate the expression $-3[5 + (-5)]$ by using the rule for order of operations.

84. Write the expression in Exercise 83 using the distributive property. Do not simplify the products.

85. The product -3×5 should be one of the terms you wrote when answering Exercise 84. Based on the results in Section 1.7, what is this product?

86. In Exercise 83, you should have obtained 0 as an answer. Now, consider the following, using the results of Exercises 83 and 85.

$$-3[5 + (-5)] = -3(5) + (-3)(-5)$$
$$0 = -15 + ?$$

The question mark represents the product $(-3)(-5)$. When added to -15, it must give a sum of 0. Therefore, how must we interpret $(-3)(-5)$?

CHAPTER 1 SUMMARY

KEY TERMS

1.1 natural numbers
whole numbers
numerator
denominator
factor
product
factored
prime
composite
greatest common factor
lowest terms
reciprocal
quotient
sum
least common denominator

mixed number
difference

1.2 exponent (power)
exponential expression
base
grouping symbols

1.3 variable
algebraic expression
equation
solution
set
element

NEW SYMBOLS

a^n	n factors of a
$[\ ]$	square brackets (used as grouping symbols)
$=$	is equal to
$\neq$	is not equal to
$<$	is less than (read from left to right)
$>$	is greater than (read from left to right)
$\leq$	is less than or equal to (read from left to right)
$\geq$	is greater than or equal to (read from left to right)
$\{\ \}$	set braces
$\{x \mid x \text{ has a certain property}\}$	set-builder notation
$-x$	the additive inverse, or opposite, of x
$\mid x \mid$	absolute value of x

NEW SYMBOLS

$\frac{1}{x}$ or $1/x$ the multiplicative inverse, or reciprocal, of the nonzero number x

$a(b)$, $(a)(b)$, $a \cdot b$, or ab a times b

$\frac{a}{b}$ or a/b a divided by b

QUICK REVIEW

CONCEPTS	EXAMPLES

1.1 FRACTIONS

Operations with Fractions	Perform the operations.
Addition: To add fractions with the same denominator, add numerators; the denominator is the same.	$$\frac{2}{5} + \frac{7}{5} = \frac{9}{5}$$
Subtraction: To subtract fractions with the same denominator, subtract numerators; the denominator is the same.	$$\frac{5}{4} - \frac{1}{4} = \frac{4}{4} = 1$$
Multiplication: Multiply numerators and multiply denominators.	$$\frac{4}{3} \cdot \frac{5}{6} = \frac{20}{18} = \frac{10}{9}$$
Division: Multiply the first fraction by the reciprocal of the second fraction.	$$\frac{6}{5} \div \frac{1}{4} = \frac{6}{5} \cdot \frac{4}{1} = \frac{24}{5}$$

1.2 EXPONENTS, ORDER OF OPERATIONS, AND INEQUALITY

Order of Operations Simplify within parentheses or above and below fraction bars first, in the following order.	
1. Apply all exponents.	$$36 - 4(2^2 + 3) = 36 - 4(4 + 3)$$
2. Do any multiplications or divisions from left to right.	$$= 36 - 4(7)$$ $$= 36 - 28$$
3. Do any additions or subtractions from left to right. If no grouping symbols are present, start with Step 1.	$$= 8$$

1.3 VARIABLES, EXPRESSIONS, AND EQUATIONS

Evaluate an expression with a variable by substituting a given number for the variable.	Evaluate $2x + y^2$ if $x = 3$ and $y = -4$. $$2x + y^2 = 2(3) + (-4)^2$$ $$= 6 + 16$$ $$= 22$$
Values of a variable that make an equation true are solutions of the equation.	Is 2 a solution of $5x + 3 = 18$? $$5(2) + 3 = 18 \quad ?$$ $$13 = 18 \quad \text{False}$$ 2 is not a solution.

CONCEPTS	EXAMPLES

1.4 REAL NUMBERS AND THE NUMBER LINE

The Ordering of Real Numbers a is less than b if a is to the left of b on the number line.	 $-2 < 3$ $\qquad$ $3 > 0$ $\qquad$ $0 < 3$
The additive inverse of x is $-x$. The absolute value of x, $\lvert x \rvert$, is the distance between x and 0 on the number line.	$-(5) = -5 \qquad -(-7) = 7 \qquad -0 = 0$ $\lvert 13 \rvert = 13 \qquad \lvert 0 \rvert = 0 \qquad \lvert -5 \rvert = 5$

1.5 ADDITION OF REAL NUMBERS

To add two numbers with the same sign, add their absolute values. The sum has that same sign.	$9 + 4 = 13$ $-8 + (-5) = -13$
To add two numbers with different signs, subtract their absolute values. The sum has the sign of the number with larger absolute value.	$7 + (-12) = -5$ $-5 + 13 = 8$

1.6 SUBTRACTION OF REAL NUMBERS

Definition of Subtraction $\qquad x - y = x + (-y)$	$5 - (-2) = 5 + 2 = 7$
Subtracting Signed Numbers **1.** Change the subtraction symbol to addition.	$-3 - 4 = -3 + (-4) = -7$
2. Change the sign of the number being subtracted.	$-2 - (-6) = -2 + 6 = 4$
3. Add, as in the previous section.	$13 - (-8) = 13 + 8 = 21$

1.7, 1.8 MULTIPLICATION AND DIVISION OF REAL NUMBERS

Definition of Division $\qquad \dfrac{x}{y} = x \cdot \dfrac{1}{y}, \qquad y \neq 0$	$\dfrac{10}{2} = 10 \cdot \dfrac{1}{2} = 5$
Multiplying and Dividing Signed Numbers The product (or quotient) of two numbers having the *same sign* is *positive;* the product (or quotient) of two numbers having *different signs* is *negative.*	$6 \cdot 5 = 30 \qquad (-7)(-8) = 56 \qquad \dfrac{20}{4} = 5$ $\dfrac{-24}{-6} = 4 \qquad (-6)(5) = -30 \qquad (6)(-5) = -30$ $\dfrac{-18}{9} = -2 \qquad\qquad\qquad \dfrac{49}{-7} = -7$
Division by 0 is undefined.	$\dfrac{5}{0}$ is undefined.
0 divided by a nonzero number equals 0.	$\dfrac{0}{5} = 0$

CONCEPTS	EXAMPLES
1.9 PROPERTIES OF ADDITION AND MULTIPLICATION	

CONCEPTS	EXAMPLES
Commutative $$a + b = b + a$$ $$ab = ba$$	$$7 + (-1) = -1 + 7$$ $$5(-3) = (-3)5$$
Associative $$(a + b) + c = a + (b + c)$$ $$(ab)c = a(bc)$$	$$(3 + 4) + 8 = 3 + (4 + 8)$$ $$[(-2)(6)](4) = (-2)[(6)(4)]$$
Identity $$a + 0 = a \quad 0 + a = a$$ $$a \cdot 1 = a \quad 1 \cdot a = a$$	$$-7 + 0 = -7 \quad 0 + (-7) = -7$$ $$9 \cdot 1 = 9 \qquad 1 \cdot 9 = 9$$
Inverse $$a + (-a) = 0 \quad -a + a = 0$$ $$a \cdot \frac{1}{a} = 1 \quad \frac{1}{a} \cdot a = 1 \quad (a \neq 0)$$	$$7 + (-7) = 0 \quad -7 + 7 = 0$$ $$-2\left(-\frac{1}{2}\right) = 1 \quad -\frac{1}{2}(-2) = 1$$
Distributive $$a(b + c) = ab + ac$$ $$(b + c)a = ba + ca$$ $$a(b - c) = ab - ac$$	$$5(4 + 2) = 5(4) + 5(2)$$ $$(4 + 2)5 = 4(5) + 2(5)$$ $$9(5 - 4) = 9(5) - 9(4)$$

CHAPTER 1 REVIEW EXERCISES

[1.1] *Perform the operation.* *

1. $\dfrac{8}{5} \div \dfrac{32}{15}$

2. $\dfrac{3}{8} + 3\dfrac{1}{2} - \dfrac{3}{16}$

3. The pie chart illustrates how 800 people responded to a survey that asked "Do you believe that there was a conspiracy to assassinate John F. Kennedy?" What fractional part of the group did not have an opinon?

4. Based on the chart in Exercise 3, how many people responded "yes"?

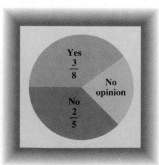

[1.2] *Find the value of the exponential expression.*

5. 5^4

6. $\left(\dfrac{3}{5}\right)^3$

7. $(.02)^5$

8. $(.001)^3$

---

*For help with any of these exercises, refer to the section given in brackets.

Find the value of the expression.

9. $8 \cdot 5 - 13$

10. $7[3 + 6(3^2)]$

11. $\dfrac{9(4^2 - 3)}{4 \cdot 5 - 17}$

12. $\dfrac{6(5 - 4) + 2(4 - 2)}{3^2 - (4 + 3)}$

Tell whether the statement is true or false.

13. $12 \cdot 3 - 6 \cdot 6 \leq 0$ **14.** $3[5(2) - 3] > 20$ **15.** $9 \leq 4^2 - 8$ **16.** $9 \cdot 2 - 6 \cdot 3 \geq 0$

Write the word statement in symbols.

17. Thirteen is less than seventeen.

18. Five plus two is not equal to ten.

[1.3] *Find the numerical value of the expression if $x = 6$ and $y = 3$.*

19. $2x + 6y$

20. $4(3x - y)$

21. $\dfrac{x}{3} + 4y$

22. $\dfrac{x^2 + 3}{3y - x}$

Change the word phrase to an algebraic expression. Use x as the variable to represent the number.

23. Six added to a number

24. A number subtracted from eight

25. Nine subtracted from six times a number

26. Three-fifths of a number added to 12

Decide whether the given number is a solution of the equation.

27. $5x + 3(x + 2) = 22;$ 2

28. $\dfrac{t + 5}{3t} = 1;$ 6

Change the word statement to an equation. Use x as the variable. Then find the solution from the set $\{0, 2, 4, 6, 8, 10\}$.

29. Six less than twice a number is 10.

30. The product of a number and 4 is 8.

[1.4] *Graph the group of numbers on a number line.*

31. $-4, -\dfrac{1}{2}, 0, 2.5, 5$

32. $-2, |-3|, -3, |-1|$

33. $-3\dfrac{1}{4}, 2\dfrac{4}{5}, -1\dfrac{1}{8}, \dfrac{5}{6}$

34. $|-4|, -|-3|, -|-5|, -6$

Select the smaller number in the pair.

35. $-10, 5$

36. $-8, -9$

37. $-\dfrac{2}{3}, -\dfrac{3}{4}$

38. $0, -|23|$

Decide whether the statement is true or false.

39. $12 > -13$

40. $0 > -5$

41. $-9 < -7$

42. $-13 > -13$

*For the following, (**a**) find the opposite of the number and (**b**) find the absolute value of the number.*

43. -9

44. 0

45. 6

46. $-\dfrac{5}{7}$

Simplify by removing absolute value symbols.

47. $|-12|$

48. $-|3|$

49. $-|-19|$

50. $-|9 - 2|$

[1.5] *Find the sum.*

51. $-10 + 4$ **52.** $14 + (-18)$ **53.** $-8 + (-9)$

54. $\dfrac{4}{9} + \left(-\dfrac{5}{4}\right)$ **55.** $-13.5 + (-8.3)$ **56.** $(-10 + 7) + (-11)$

57. $[-6 + (-8) + 8] + [9 + (-13)]$ **58.** $(-4 + 7) + (-11 + 3) + (-15 + 1)$

Write a numerical expression for the phrase, and simplify the expression.

59. 19 added to the sum of -31 and 12   **60.** 13 more than the sum of -4 and -8

Solve the problem.

61. Tri Nguyen has $18 in his checking account. He then writes a check for $26. What negative number represents his balance?

62. The temperature at noon on an August day in Houston was 93°. After a thunderstorm, it dropped 6°. What was the new temperature?

Find the solution of the equation from the set $\{-3, -2, -1, 0, 1, 2, 3\}$ by guessing or using trial and error.

63. $x + (-2) = -4$ **64.** $12 + x = 11$

[1.6] *Find the difference.*

65. $-7 - 4$ **66.** $-12 - (-11)$ **67.** $5 - (-2)$ **68.** $-\dfrac{3}{7} - \dfrac{4}{5}$

69. $2.56 - (-7.75)$ **70.** $(-10 - 4) - (-2)$ **71.** $(-3 + 4) - (-1)$ **72.** $-(-5 + 6) - 2$

Write a numerical expession for the phrase and simplify the expression.

73. The difference between -4 and -6 **74.** Five less than the sum of 4 and -8

Solve the problem.

75. Eric owed his brother $28. He repaid $13 but then borrowed another $14. What positive or negative amount represents his present financial status?

76. If the temperature drops 7° below its previous level of $-3°$, what is the new temperature?

77. Explain in your own words how the subtraction problem $-8 - (-6)$ is performed.

78. Can the difference of two negative numbers be positive? Explain with an example.

[1.7] *Perform the indicated operations.*

79. $(-12)(-3)$ **80.** $15(-7)$ **81.** $\left(-\dfrac{4}{3}\right)\left(-\dfrac{3}{8}\right)$ **82.** $(-4.8)(-2.1)$

83. $5(8 - 12)$ **84.** $(5 - 7)(8 - 3)$ **85.** $2(-6) - (-4)(-3)$**86.** $3(-10) - 5$

Evaluate the expression if $x = -5$, $y = 4$, and $z = -3$.

87. $6x - 4z$ **88.** $5x + y - z$ **89.** $5x^2$ **90.** $z^2(3x - 8y)$

Write a numerical expression for the phrase, and simplify the expression.

91. Nine less than the product of -4 and 5   **92.** Five-sixths of the sum of 12 and -6

[1.8] *Find the quotient.*

93. $\dfrac{-36}{-9}$

94. $\dfrac{220}{-11}$

95. $-\dfrac{1}{2} \div \dfrac{2}{3}$

96. $-33.9 \div (-3)$

97. $\dfrac{-5(3) - 1}{8 - 4(-2)}$

98. $\dfrac{5(-2) - 3(4)}{-2[3 - (-2)] - 1}$

99. $\dfrac{10^2 - 5^2}{8^2 + 3^2 - (-2)}$

100. $\dfrac{(.6)^2 + (.8)^2}{(-1.2)^2 - (-.56)}$

Write a numerical expression for the phrase, and simplify the expression.

101. The quotient of 12 and the sum of 8 and -4

102. The product of -20 and 12, divided by the difference between 15 and -15

Write the sentence in symbols, using x as the variable, and find the solution by guessing or by using trial and error. All solutions come from the list of integers between -12 and 12.

103. 8 times a number is -24.

104. The quotient of a number and 3 is -2.

105. 3 less than a number is -7.

106. The sum of a number and 5 is -6.

[1.9] *Decide whether the statement is an example of the commutative, associative, identity, inverse, or distributive property.*

107. $6 + 0 = 6$

108. $5 \cdot 1 = 5$

109. $-\dfrac{2}{3}\left(-\dfrac{3}{2}\right) = 1$

110. $17 + (-17) = 0$

111. $5 + (-9 + 2) = [5 + (-9)] + 2$

112. $w(xy) = (wx)y$

113. $3x + 3y = 3(x + y)$

114. $(1 + 2) + 3 = 3 + (1 + 2)$

Use the distributive property to rewrite the expression. Simplify if possible.

115. $7y + y$

116. $-12(4 - t)$

117. $3(2s) + 3(5y)$

118. $-(-4r + 5s)$

119. Evaluate $25 - (5 - 2)$ and $(25 - 5) - 2$. Use this example to explain why subtraction is not associative.

120. Evaluate $180 \div (15 \div 5)$ and $(180 \div 15) \div 5$. Use this example to explain why division is not associative.

MIXED REVIEW EXERCISES*

Perform the indicated operations.

121. $[(-2) + 7 - (-5)] + [-4 - (-10)]$

122. $\left(-\dfrac{5}{6}\right)^2$

123. $-|(-7)(-4)| - (-2)$

124. $\dfrac{6(-4) + 2(-12)}{5(-3) + (-3)}$

125. $\dfrac{3}{8} - \dfrac{5}{12}$

126. $\dfrac{12^2 + 2^2 - 8}{10^2 - (-4)(-15)}$

127. $\dfrac{8^2 + 6^2}{7^2 + 1^2}$

128. $-16(-3.5) - 7.2(-3)$

---

*The order of exercises in this final group does not correspond to the order in which topics occur in the chapter. This random ordering should help you prepare for the chapter test in yet another way.

129. $2\dfrac{5}{6} - 4\dfrac{1}{3}$

130. $-8 + [(-4 + 17) - (-3 - 3)]$

131. $-\dfrac{12}{5} \div \dfrac{9}{7}$

132. $(-8 - 3) - 5(2 - 9)$

133. Write a short paragraph explaining the special considerations involving zero when dividing.

134. "Two negatives give a positive" is often heard from students. Is this correct? Use more precise language in explaining what this means.

135. Use x as the variable and write an expression for "the product of 5 and the sum of a number and 7." Then use the distributive property to rewrite the expression.

136. The highest temperature ever recorded in Albany, New York, was 99°F, while the lowest was 112 degrees less than the highest. What was the lowest temperature ever recorded in Albany?

The total revenue of all securities firms in the United States between 1987 and 1990 is illustrated in the accompanying bar graph. Use the graph to answer the questions in Exercises 137 and 138. Use a signed number in your answer.

137. What was the change in revenue from 1988 to 1989?

138. What was the change in revenue from 1989 to 1990?

───────◇ **MATHEMATICAL CONNECTIONS*** (Exercises 139–146) ◇───────

Evaluate the expression $\dfrac{3}{4}x + y^2 - 2z$ for $x = -16$, $y = 2$, and $z = -3$. Then respond to the questions or statements in Exercises 139–146.

139. What is the value of the expression for these particular values of x, y, and z?

140. Is the value of the expression greater than -1 or less than -1?

141. What is the absolute value of the expression?

142. What is the additive inverse (opposite) of the expression?

143. What is the multiplicative inverse of the expression?

144. If parentheses are placed around the first two terms of the expression, will you obtain the same answer? If not, what is the new answer?

145. Is the answer a solution of the equation $x + 5 = 3$?

146. Is the answer an integer?

───────────◇───────────

*Each Chapter Review concludes with a group of exercises designed to show interrelationships among the concepts studied in the current chapter, as well as previous chapters (when applicable).

CHAPTER 1 TEST

1. Write $\dfrac{63}{99}$ in lowest terms.

2. Add: $\dfrac{5}{8} + \dfrac{11}{12} + \dfrac{7}{15}$.

3. Divide: $\dfrac{19}{15} \div \dfrac{6}{5}$.

4. Based on the pie chart shown, what fractional part of total legal gambling revenue in 1992 came from sources other than lotteries, casinos, and pari-mutuel wagering? (*Source:* Gaming and Wagering Business)

Decide whether the statement is true or false.

5. $4[-20 + 7(-2)] \le 135$

6. $(-3)^2 + 2^2 = 5^2$

7. Graph the group of numbers $-1, -3, |-4|, |-1|$ on a number line.

Select the smaller number from the pair.

8. $6, -|-8|$

9. $-.742, -1.277$

10. Write in symbols: The quotient of -6 and the sum of 2 and -8. Simplify the expression.

11. If a and b are both negative, is $\dfrac{a + b}{a \cdot b}$ positive or negative?

Perform the indicated operations.

12. $-2 - (5 - 17) + (-6)$

13. $-5\dfrac{1}{2} + 2\dfrac{2}{3}$

14. $-6 - [-7 + (2 - 3)]$

15. $4^2 + (-8) - (2^3 - 6)$

16. $(-5)(-12) + 4(-4) + (-8)^2$

17. $\dfrac{-7 - (-6 + 2)}{-5 - (-4)}$

18. $\dfrac{30(-1 - 2)}{-9[3 - (-2)] - 12(-2)}$

Find the solution for the equation from the set $\{-6, -4, -2, 0, 2, 4, 6\}$ by guessing or by trial and error.

19. $-3x = -12$

20. $\dfrac{x}{-4} = \dfrac{1}{2}$

Evaluate the expression, given x = −2 and y = 4.

21. $3x - 4y^2$

22. $\dfrac{5x + 7y}{3(x + y)}$

Solve the problem.

23. The highest temperature ever recorded in Wyoming was 114°F, while the lowest was −63°F. What is the difference between the highest and lowest temperatures?

24. The funds available from the National Endowment for the Arts for the years 1986 through 1990 are shown in the chart. Determine the change from one year to the next by subtraction for each year indicated with a blank. For example, from 1986 to 1987, the change was 170.9 − 167.1 = 3.8 million dollars. (*Source:* U.S. National Endowment for the Arts)

Year	Funds Available (in millions of dollars)	Change
1986	167.1	
1987	170.9	3.8
1988	171.1	_____
1989	166.7	_____
1990	170.8	_____

25. The bar graph shows the number of units of elementary/high school texts, in millions, sold during the years 1987 through 1990. Use a signed number to represent the change from one year to the next for **(a)** 1987 to 1988 **(b)** 1988 to 1989 **(c)** 1989 to 1990.

26. Refer to the pie chart in test item 4. If 29.9 billion dollars was the total in gross revenues, how much of it came from lotteries?

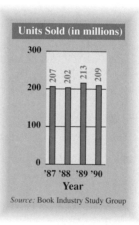

Source: Book Industry Study Group

Match the property in Column I with the example of it in Column II.

I	II
27. Commutative	A. $3x + 0 = 3x$
28. Associative	B. $(5 + 2) + 8 = 8 + (5 + 2)$
29. Inverse	C. $-3(x + y) = -3x + (-3y)$
30. Identity	D. $-5 + (3 + 2) = (-5 + 3) + 2$
31. Distributive	E. $-\dfrac{5}{3}\left(-\dfrac{3}{5}\right) = 1$

32. What property is used to show that $3(x + 1) = 3x + 3$?

33. Consider the expression $-6[5 + (-2)]$.
   (a) Evaluate it by first working within the brackets.
   (b) Evaluate it by using the distributive property.
   (c) Why must the answers in items **(b)** and **(c)** be the same?

CONNECTIONS

The use of algebra to solve equations and applied problems is very old. The 3600-year-old Rhind Papyrus includes the following "word problem." "Aha, its whole, its seventh, it makes 19." This brief sentence describes the equation

$$x + \frac{x}{7} = 19.$$

The word "aha" was used as we would use the words "Let x equal . . . ." The solution of this equation is 16 5/8. The word *algebra* is from the work *Risab al-jabr m'al muquabalah,* written in the ninth century by Muhammed ibn Musa Al-Khowarizmi. The title means "the science of transposition and cancellation." These ideas are used in equation solving. From Latin versions of Khowarizmi's text, "al-jabr" became the broad term covering the art of equation solving.

In this chapter we begin our study of the solution of equations and inequalities, and how they are applied to problem solving. In the second section we learn how to solve the equation mentioned in the Connections box.

2.1 SIMPLIFYING EXPRESSIONS

FOR EXTRA HELP	OBJECTIVES
📖 **SSG** pp. 50–52 **SSM** pp. 51–54	**1 ▶** Simplify expressions.
📼 **Video** 2	**2 ▶** Identify terms and numerical coefficients. **3 ▶** Identify like terms.
💾 **Tutorial** IBM MAC	**4 ▶** Combine like terms. **5 ▶** Simplify expressions from word phrases.

1 ▶ Simplify expressions.   ▶   It is often necessary to simplify the expressions on either side of an equation as the first step in solving it. In this section we show how to simplify expressions using the properties of addition and multiplication introduced in Chapter 1.

EXAMPLE 1
Simplifying Expressions

Simplify the following expressions.

(a) $4x + 8 + 9$
Since $8 + 9 = 17$,

$$4x + 8 + 9 = 4x + 17.$$

(b) $4(3m - 2n)$
Use the distributive property first.

$$4(3m - 2n) = 4(3m) - 4(2n) \qquad \text{Arrows denote distributive property.}$$
$$= (4 \cdot 3)m - (4 \cdot 2)n \qquad \text{Associative property}$$
$$= 12m - 8n$$

(c) $6 + 3(4k + 5) = 6 + 3(4k) + 3(5) \qquad$ Distributive property
$$= 6 + (3 \cdot 4)k + 3(5) \qquad \text{Associative property}$$
$$= 6 + 12k + 15$$
$$= 6 + 15 + 12k \qquad \text{Commutative property}$$
$$= 21 + 12k$$

(d) $5 - (2y - 8) = 5 - 1 \cdot (2y - 8) \qquad$ Replace $-$ with -1.
$$= 5 - 2y + 8 \qquad \text{Distributive property}$$
$$= 5 + 8 - 2y \qquad \text{Commutative property}$$
$$= 13 - 2y \quad ■$$

NOTE In Example 1, parts (c) and (d), a different use of the commutative property would have resulted in answers of $12k + 21$ and $-2y + 13$. These answers also would be acceptable.

The steps using the commutative and associative properties will not be shown in the rest of the examples, but you should be aware that they are usually involved.

2 ▶ Identify terms and numerical coefficients.

▶ A **term** is a number, a variable, or a product or quotient of numbers and variables raised to powers.* Examples of terms include

$$-9x^2, \quad 15y, \quad -3, \quad 8m^2n, \quad \frac{2}{p}, \quad \text{and} \quad k.$$

The **numerical coefficient** of the term $9m$ is 9, the numerical coefficient of $-15x^3y^2$ is -15, the numerical coefficient of x is 1, and the numerical coefficient of 8 is 8.

CAUTION It is important to be able to distinguish between *terms* and *factors*. For example, in the expression $8x^3 + 12x^2$, there are two terms. They are $8x^3$ and $12x^2$. On the other hand, in the expression $(8x^3)(12x^2)$, $8x^3$ and $12x^2$ are *factors*.

EXAMPLE 2

Identifying the Numerical Coefficient of a Term

Give the numerical coefficient of each of the following terms.

Term	Numerical Coefficient
$-7y$	-7
$8p$	8
$34r^3$	34
$-26x^5yz^4$	-26
$-k$	-1

3 ▶ Identify like terms.

▶ Terms with exactly the same variables that have the same exponents are **like terms.** For example, $9m$ and $4m$ have the same variable and are like terms. Also, $6x^3$ and $-5x^3$ are like terms. The terms $-4y^3$ and $4y^2$ have different exponents and are **unlike terms.**

Here are some examples of like terms.

$$5x \text{ and } -12x \qquad 3x^2y \text{ and } 5x^2y$$

Here are some examples of unlike terms.

$$4xy^2 \text{ and } 5xy \qquad -7w^3z^3 \text{ and } 2xz^3$$

4 ▶ Combine like terms.

▶ Recall the distributive property:

$$x(y + z) = xy + xz.$$

This statement can also be written "backward" as

$$xy + xz = x(y + z).$$

This form of the distributive property may be used to find the sum or difference of like terms. For example,

$$3x + 5x = (3 + 5)x = 8x.$$

This process is called **combining like terms.**

NOTE It is important to remember that only like terms may be combined. For example, $5x^2 + 2x \neq 7x^3$.

---

*Another name for certain terms, **monomial,** is introduced in Chapter 3.

EXAMPLE 3
Combining Like Terms

Combine like terms in the following expressions.

(a) $9m + 5m$

Use the distributive property as given above to combine the like terms.

$$9m + 5m = (9 + 5)m = 14m$$

(b) $6r + 3r + 2r = (6 + 3 + 2)r = 11r$ Distributive property

(c) $4x + x = 4x + 1x = (4 + 1)x = 5x$ (Note: $x = 1x$.)

(d) $16y^2 - 9y^2 = (16 - 9)y^2 = 7y^2$

(e) $32y + 10y^2$ cannot be simplified because $32y$ and $10y^2$ are unlike terms. The distributive property cannot be used here to combine coefficients. ■

When an expression involves parentheses, the distributive property is used both "forward" and "backward" to simplify the expression by combining like terms, as shown in the following example.

EXAMPLE 4
Simplifying Expressions
Involving Like Terms

Combine like terms in the following expressions.

(a) $14y + 2(6 + 3y) = 14y + 2(6) + 2(3y)$ Distributive property
$$= 14y + 12 + 6y \qquad \text{Multiply.}$$
$$= 20y + 12 \qquad \text{Combine like terms.}$$

(b) $9k - 6 - 3(2 - 5k) = 9k - 6 - 3(2) - 3(-5k)$ Distributive property
$$= 9k - 6 - 6 + 15k \qquad \text{Multiply.}$$
$$= 24k - 12 \qquad \text{Combine like terms.}$$

(c) $-(2 - r) + 10r = -1(2 - r) + 10r$ Replace $-$ with -1.
$$= -1(2) - 1(-r) + 10r \qquad \text{Distributive property}$$
$$= -2 + r + 10r \qquad \text{Multiply.}$$
$$= -2 + 11r \qquad \text{Combine like terms.}$$

(d) $5(2a - 6) - 3(4a - 9)$
$$= 10a - 30 - 12a + 27 \qquad \text{Distributive property}$$
$$= -2a - 3 \qquad \text{Combine like terms.} \quad ■$$

Example 4(d) shows that the commutative property can be used with subtraction by treating the subtracted terms as the addition of their additive inverses.

NOTE Examples 3 and 4 suggest that like terms may be combined by combining the coefficients of the terms and keeping the same variable factors.

---

5 ▶ Simplify expressions from word phrases.

▶ In Chapter 1 we saw how to translate words, phrases, and statements into expressions and equations. This was done mainly to prepare for solving applied problems. This idea is used extensively in the later sections of this chapter. Now we can simplify translated expressions by combining like terms.

EXAMPLE 5
Converting Words to a
Mathematical Expression

Convert to a mathematical expression, and simplify: The sum of 9, five times a number, four times the number, and six times the number.

The word "sum" indicates that the terms should be added. Use x to represent the number. Then the phrase translates as follows.

$$9 + 5x + 4x + 6x \qquad \text{Write as a mathematical expression.}$$
$$= 9 + 15x \qquad \text{Combine like terms.} \quad \blacksquare$$

CAUTION   In Example 5, we are dealing with an expression to be simplified, and *not* an equation to be solved.

2.1 EXERCISES

Choose the letter of the correct response.

1. Which one of the following is true for all real numbers x?
   (a) $6 + 2x = 8x$ (b) $6 - 2x = 4x$
   (c) $6x - 2x = 4x$ (d) $3 + 8(4t - 6) = 11(4t - 6)$

2. Which one of the following is an example of a pair of like terms?
   (a) $6t, 6w$ (b) $-8x^2y, 9xy^2$ (c) $5ry, 6yr$ (d) $-5x^2, 2x^3$

3. Which one of the following is an example of a term with numerical coefficient 5?
   (a) $5x^3y^7$ (b) x^5 (c) $\dfrac{x}{5}$ (d) 5^2xy^3

4. Which one of the following is a correct translation for "six times a number, subtracted from the product of eleven and the number" (if x represents the number)?
   (a) $6x - 11x$ (b) $11x - 6x$ (c) $(11 + x) - 6x$ (d) $6x - (11 + x)$

Simplify the expression. See Example 1.

5. $4r + 19 - 8$ **6.** $7t + 18 - 4$ **7.** $8(4q - 3t)$

8. $12(9m - 7n)$ **9.** $5 + 2(x - 3y)$ **10.** $8 + 3(s - 6t)$

11. $-2 - (5 - 3p)$ **12.** $-10 - (7 - 14r)$

Give the numerical coefficient of the term. See Example 2.

13. $14x$ **14.** $9x$ **15.** $-12k$ **16.** $-23y$ **17.** $5m^2$ **18.** $-3n^6$

19. xw **20.** pq **21.** $-x$ **22.** $-t$ **23.** 74 **24.** 98

25. Give an example of a pair of like terms in the variable x, such that one of them has a negative numerical coefficient, one has a positive numerical coefficient, and their sum has a positive numerical coefficient.

26. Give an example of a pair of unlike terms such that each term has only x as a variable factor.

Identify the group of terms as like *or* unlike. *See Objective 3.*

27. $8r, -13r$ **28.** $-7a, 12a$ **29.** $5z^4, 9z^3$ **30.** $8x^5, -10x^3$

31. $4, 9, -24$ **32.** $7, 17, -83$ **33.** x, y **34.** t, s

35. There is an old saying "You can't add apples and oranges." Explain how this saying can be applied to the goal of Objective 4 in this section.

36. Explain how the distributive property is used in combining $6t + 5t$ to get $11t$.

Simplify the expression by combining like terms. See Examples 1, 3, and 4.

37. $4k + 3 - 2k + 8 + 7k - 16$

38. $9x + 7 - 13x + 12 + 8x - 15$

39. $-\dfrac{4}{3} + 2t + \dfrac{1}{3}t - 8 - \dfrac{8}{3}t$

40. $-\dfrac{5}{6} + 8x + \dfrac{1}{6}x - 7 - \dfrac{7}{6}$

41. $-5.3r + 4.9 - 2r + .7 + 3.2r$

42. $2.7b + 5.8 - 3b + .5 - 4.4b$

43. $2y^2 - 7y^3 - 4y^2 + 10y^3$

44. $9x^4 - 7x^6 + 12x^4 + 14x^6$

45. $13p + 4(4 - 8p)$

46. $5x + 3(7 - 2x)$

47. $-4(y - 7) - 6$

48. $-5(t - 13) - 4$

49. $-5(5y - 9) + 3(3y + 6)$

50. $-3(2t + 4) + 8(2t - 4)$

51. $-4(-3k + 3) - (6k - 4) - 2k + 1$

52. $-5(8j + 2) - (5j - 3) - 3j + 17$

53. $-7.5(2y + 4) - 2.9(3y - 6)$

54. $8.4(6t - 6) + 2.4(9 - 3t)$

Convert the phrase into a mathematical expression. Use x as the variable. Combine like terms when possible. See Example 5.

55. Five times a number, added to the sum of the number and three

56. Six times a number, added to the sum of the number and six

57. A number multiplied by -7, subtracted from the sum of 13 and six times the number

58. A number multiplied by 5, subtracted from the sum of 14 and eight times the number

59. Six times a number added to -4, subtracted from twice the sum of three times the number and 4 (*Hint*: *Twice* means two times.)

60. Nine times a number added to 6, subtracted from triple the sum of 12 and 8 times the number (*Hint*: *Triple* means three times.)

61. Write the expression $9x - (x + 2)$ using words, as in Exercises 55–60.

62. Write the expression $2(3x + 5) - 2(x + 4)$ using words, as in Exercises 55–60.

──────◆ **MATHEMATICAL CONNECTIONS** (Exercises 63–70) ◇──────

Work Exercises 63–70 in order. They will help prepare you for graphing later in the text.

63. Evaluate the expression $x + 2$ for the values of x shown in the chart.

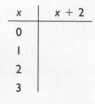

x	$x + 2$
0	
1	
2	
3	

64. Based on your results from Exercise 63, complete the following statement: For every increase of 1 unit for x, the value of $x + 2$ increases by _____ unit(s).

65. Repeat Exercise 63 for these expressions:
   (a) $x + 1$ **(b)** $x + 3$ **(c)** $x + 4$

66. Based on your results from Exercises 63 and 65, make a conjecture (an educated guess) about what happens to the value of an expression of the form $x + b$ for any value of b, as x increases by 1 unit.

67. Repeat Exercise 63 for these expressions:
    (a) $2x + 2$ **(b)** $3x + 2$ **(c)** $4x + 2$

68. Based on your results from Exercise 67, complete the following statement: For every increase of 1 unit for x, the value of $mx + 2$ increases by _____ units.

69. Repeat Exercise 63 and compare your results to those in Exercise 67 for these expressions:
    (a) $2x + 7$ **(b)** $3x + 5$ **(c)** $4x + 1$

70. Based on your results from Exercises 63–69, complete the following statement: For every increase of 1 unit for x, the value of $mx + b$ increases by _____ units.

REVIEW EXERCISES

Most of the exercise sets in the rest of the book end with brief sets of "review exercises." These exercises are designed to help you review ideas introduced earlier, as well as preview ideas needed for the next few sections. If you need help with these review exercises, look in the section or sections indicated.

Find the opposite or additive inverse of the number. See Section 1.4.

71. 5   **72.** 3.4   **73.** -15 **74.** $-\dfrac{3}{4}$

Add a number to the expression so that the final sum is just x. See Section 1.9.

75. $x - 2$ **76.** $x - \dfrac{1}{2}$ **77.** $x + 14$ **78.** $x + 9.6$

Multiply the expression by a number so that the result is just x. What is the number? See Section 1.9.

79. $\dfrac{1}{3}x$ **80.** $\dfrac{2}{5}x$ **81.** $-3x$ **82.** $-7x$

2.2 THE ADDITION AND MULTIPLICATION PROPERTIES OF EQUALITY

FOR EXTRA HELP	OBJECTIVES
📖 **SSG** pp. 52–60 **SSM** pp. 54–59	**1** ▶ Identify linear equations.
📼 **Video** 3	**2** ▶ Use the addition property of equality.
💾 **Tutorial** IBM MAC	**3** ▶ Use the multiplication property of equality. **4** ▶ Simplify equations, and then use the multiplication property of equality.

Quite possibly the most important topic in the study of beginning algebra is the solution of equations. We will investigate many types of equations in this book, and the properties introduced in this section are essential in solving equations.

◇ **CONNECTIONS** ◇

After completing this section you will be able to solve linear equations algebraically. Another method, guessing and testing, was actually used by the early Egyptians to solve equations. This method, called the *rule of false position,* involved making an initial guess of the solution, and then adjusting it in the likely event that the guess was incorrect. For example (using our modern notation) to solve the equation $6x + 2x = 32$, suppose the initial guess was $x = 3$. Replacing x with 3 gives

$$6(3) + 2(3) = 32 \quad ?$$
$$18 + 6 = 32 \quad ?$$
$$24 = 32 \quad \text{False.}$$

The guess, 3, gives 24, a value smaller than that required (32). Since 24 is 24/32 or 3/4 of the required value, the actual solution must be 4, because 3 is 3/4 of 4. A simpler approach is to use trial and error. We see that 3 is too small, so next try 4. In this case, 4 is the correct solution. In other equations, more than two trials will be necessary.

FOR DISCUSSION OR WRITING

Try using the rule of false position and simple trial and error to solve the equation $5x = 33 + 2x$. After completing this section, use algebra to solve the equation. Which method do you prefer?

To solve applied problems, we must be able to solve equations. The simplest type of equation is a *linear equation.* Methods for solving linear equations are introduced in this section. We use the definitions and properties of real numbers that we learned in Chapter 1.

1 ▶ Identify linear equations.

▶ Before we can solve a linear equation we must be able to recognize one.

Linear Equation

A **linear equation** can be written in the form

$$Ax + B = 0$$

for real numbers A and B, with $A \neq 0$.

Methods of solving linear equations are introduced in this section and in Section 2.3. As discussed in Section 1.3, a solution of an equation is a number that when substituted for the variable makes the equation a true statement. Equations that have exactly the same solutions are **equivalent equations.** Linear equations are solved by using a series of steps to produce equivalent equations until an equation of the form

$$x = \text{a number}$$

is obtained.

2 ▶ Use the addition property of equality.

▶ According to the equation $x - 5 = 2$, both $x - 5$ and 2 represent the same number, since this is the meaning of the equals sign. We can solve the equation by changing the left side from $x - 5$ to just x. This is done by adding 5 to $x - 5$. Keep the two sides equal by also adding 5 on the right side.

$$x - 5 = 2 \qquad \text{Given equation}$$
$$x - 5 + 5 = 2 + 5 \qquad \text{Add 5 on both sides.}$$
$$x + 0 = 7 \qquad \text{Additive inverse property}$$
$$x = 7 \qquad \text{Identity property}$$

The solution of the given equation is 7. Check by replacing x with 7 in the given equation.

$$x - 5 = 2 \qquad \text{Given equation}$$
$$7 - 5 = 2 \quad ? \qquad \text{Let } x = 7.$$
$$2 = 2 \qquad \text{True}$$

Since this final result is true, 7 checks as the solution.

To solve the equation above we added the same number to both sides. The **addition property of equality** justifies this step.

Addition Property of Equality

If A, B, and C are mathematical expressions that represent real numbers, then the equations

$$A = B \qquad \text{and} \qquad A + C = B + C$$

have exactly the same solution. In words, the same expression may be added to both sides of an equation without changing the solution.

In the addition property, C represents a real number. That means that numbers or terms with variables, or even sums of terms that represent real numbers, can be added to both sides of an equation.

E X A M P L E   1

Using the Addition Property of Equality

Solve $x - 16 = 7$.

If the left side of this equation were just x, the solution would be found. Get x alone by using the addition property of equality and adding 16 on both sides.

$$x - 16 = 7$$
$$(x - 16) + 16 = 7 + 16 \qquad \text{Add 16 on both sides.}$$
$$x = 23$$

Note that we combined the steps that change $x - 16 + 16$ to $x + 0$ and $x + 0$ to x. We will combine these steps from now on. Check by substituting 23 for x in the original equation.

$$x - 16 = 7 \qquad \text{Given equation}$$
$$23 - 16 = 7 \quad ? \qquad \text{Let } x = 23.$$
$$7 = 7 \qquad \text{True}$$

Since the check results in a true statement, 23 is the solution. ■

In this example, why was 16 added to both sides of the equation $x - 16 = 7$? The equation would be solved if it could be rewritten so that one side contained only the variable and the other side contained only a number. Since $x - 16 + 16 = x + 0 = x$, adding 16 on the left side simplifies that side to just x, the variable, as desired.

The addition property of equality says that the same number may be *added* to both sides of an equation. As was shown in Chapter 1, subtraction is defined in terms of addition. Because of the way subtraction is defined, the addition property also permits *subtracting* the same number on both sides of an equation.

EXAMPLE 2

Subtracting a Variable Expression to Solve an Equation

Solve the equation $3k + 12 + k - 8 - 4 = 15 + 3k + 2$.

Begin by combining like terms on each side of the equation to get

$$4k = 17 + 3k.$$

Next, get all terms that contain variables on the same side of the equation. One way to do this is to subtract $3k$ from both sides.

$$4k = 17 + 3k$$
$$4k - 3k = 17 + 3k - 3k \qquad \text{Subtract } 3k \text{ from both sides.}$$
$$k = 17$$

Check by substituting 17 for k in the original equation.

$$3k + 12 + k - 8 - 4 = 15 + 3k + 2 \qquad \text{Given equation}$$
$$3(\mathbf{17}) + 12 + \mathbf{17} - 8 - 4 = 15 + 3(\mathbf{17}) + 2 \qquad ? \quad \text{Let } k = 17.$$
$$51 + 12 + 17 - 8 - 4 = 15 + 51 + 2 \qquad ? \quad \text{Multiply.}$$
$$68 = 68 \qquad \text{True}$$

The check results in a true statement, so the solution is 17. ■

NOTE Subtracting $3k$ in Example 2 is the same as adding $-3k$, since by the definition of subtraction, $a - b = a + (-b)$.

---

3 ▶ Use the multiplication property of equality.

▶ The addition property of equality by itself is not enough to solve an equation like $3x + 2 = 17$.

$$3x + 2 = 17$$
$$3x + 2 - 2 = 17 - 2 \qquad \text{Subtract 2 from both sides.}$$
$$3x = 15$$

The variable x is not alone on one side of the equation: the equation has $3x$ instead. Another property is needed to change $3x = 15$ to $x = $ a number.

If $3x = 15$, then $3x$ and 15 both represent the same number. Multiplying both $3x$ and 15 by the same number will also result in an equality. The **multiplication property of equality** states that both sides of an equation can be multiplied by the same number.

Multiplication Property of Equality

If A, B, and C are mathematical expressions that represent real numbers, then the equations

$$A = B \qquad \text{and} \qquad AC = BC$$

have exactly the same solution. (Assume that $C \neq 0$.) In words, both sides of an equation may be multiplied by the same nonzero expression without changing the solution.

This property can be used to solve $3x = 15$. The $3x$ on the left must be changed to $1x$, or x, instead of $3x$. To get x, multiply both sides of the equation by $1/3$. Use $1/3$ since it is the reciprocal of the coefficient of x. This works because $3 \cdot 1/3 = 3/3 = 1$.

$$3x = 15$$

$$\frac{1}{3}(3x) = \frac{1}{3} \cdot 15 \qquad \text{Multiply both sides by } \tfrac{1}{3}.$$

$$\left(\frac{1}{3} \cdot 3\right)x = \frac{1}{3} \cdot 15 \qquad \text{Associative property}$$

$$1x = 5 \qquad \text{Multiply.}$$

$$x = 5 \qquad \text{Identity property}$$

The solution of the equation is 5. Check this by substituting 5 for x in the given equation. From now on we shall combine the last two steps shown in this example.

Just as the addition property of equality permits subtracting the same number from both sides of an equation, the multiplication property of equality permits dividing both sides of an equation by the same nonzero number. For example, the equation $3x = 15$, solved above by multiplication, could also be solved by dividing both sides by 3, as follows.

$$3x = 15$$

$$\frac{3x}{3} = \frac{15}{3} \qquad \text{Divide by 3.}$$

$$x = 5 \qquad \text{Simplify.}$$

NOTE In practice, it is usually easier to multiply on each side if the coefficient of the variable is a fraction, and divide on each side if the coefficient is an integer. For example, to solve

$$-\frac{3}{4}x = 12$$

it is easier to multiply by $-4/3$ than to divide by $-3/4$. On the other hand, to solve

$$-5x = -20$$

it is easier to divide by -5 than to multiply by $-1/5$.

EXAMPLE 3

Dividing Each Side of an Equation by a Nonzero Number

Solve $2.1x = 6.09$.

Divide both sides of the equation by 2.1.

$$\frac{2.1x}{2.1} = \frac{6.09}{2.1}$$

$$x = 2.9 \qquad \text{Divide.}$$

You may wish to use a calculator to find the quotient on the right. Check that the solution is 2.9.

$$2.1(2.9) = 6.09 \quad ? \quad \text{Let } x = 2.9.$$

$$6.09 = 6.09 \qquad \text{True.}$$

The solution is 2.9. ■

In the next two examples, multiplication produces the solution more quickly than division.

EXAMPLE 4

Using the Multiplication Property of Equality with a Fraction

■ Solve $\dfrac{a}{4} = 3$.

Replace $\dfrac{a}{4}$ by $\dfrac{1}{4}a$ since division by 4 is the same as multiplication by $1/4$. Get a alone by multiplying both sides by 4, the reciprocal of the coefficient of a.

$$\frac{a}{4} = 3$$

$$\frac{1}{4}a = 3 \qquad \text{Change } \tfrac{a}{4} \text{ to } \tfrac{1}{4}a.$$

$$4 \cdot \frac{1}{4}a = 4 \cdot 3 \qquad \text{Multiply by 4.}$$

$$1a = 12 \qquad \text{Inverse property}$$

$$a = 12 \qquad \text{Identity property}$$

Check the answer.

$$\frac{a}{4} = 3 \qquad \text{Given equation}$$

$$\frac{12}{4} = 3 \quad ? \qquad \text{Let } a = 12.$$

$$3 = 3 \qquad \text{True}$$

The solution 12 is correct. ■

EXAMPLE 5

Using the Multiplication Property of Equality in Two Ways

■ Solve $\dfrac{3}{4}h = 6$.

We will show two ways to solve this equation.
Method 1

To get h alone, multiply both sides of the equation by $4/3$, the reciprocal of $3/4$. Use $4/3$ because $\dfrac{4}{3} \cdot \dfrac{3}{4}h = 1 \cdot h = h$.

$$\frac{3}{4}h = 6$$

$$\frac{4}{3}\left(\frac{3}{4}h\right) = \frac{4}{3} \cdot 6 \qquad \text{Multiply by } \tfrac{4}{3}.$$

$$1 \cdot h = \frac{4}{3} \cdot \frac{6}{1} \qquad \text{Inverse property}$$

$$h = 8 \qquad \text{Identity property}$$

Method 2

Begin by multiplying both sides of the equation by 4 to eliminate the denominator.

$$\frac{3}{4}h = 6$$

$$4\left(\frac{3}{4}h\right) = 4 \cdot 6 \qquad \text{Multiply by 4.}$$

$$3h = 24$$

$$\frac{3h}{3} = \frac{24}{3} \qquad \text{Divide by 3.}$$

$$h = 8$$

Using either method, the solution is 8. Check the answer by substitution in the given equation. ■

4 ▶ Simplify equations, and then use the multiplication property of equality.

▶ The final example of this section requires simplification of one side of the equation before the multiplication property of equality is used.

EXAMPLE 6

Simplifying an Equation Before Solving

Solve $2(3 + 7x) - (1 + 15x) = 2$.

Use the distributive property to first simplify the equation.

$$2(3 + 7x) - (1 + 15x) = 2$$
$$6 + 14x - 1 - 15x = 2 \qquad \text{Distributive property}$$
$$5 - x = 2 \qquad \text{Combine like terms.}$$
$$5 - x - 5 = 2 - 5 \qquad \text{Subtract 5.}$$
$$-x = -3$$

The variable is alone on the left, but its coefficient is -1, since $-x = -1 \cdot x$. When this occurs, simply multiply both sides by -1 (which is the reciprocal of -1).

$$-x = -3$$
$$-1 \cdot x = -3 \qquad -x = -1 \cdot x$$
$$-1(-1 \cdot x) = -1(-3) \qquad \text{Multiply by } -1.$$
$$x = 3$$

The solution is 3. Check by substituting into the original equation. (Incidentally, *dividing* by -1 would also allow us to solve an equation of the form $-x = a$.) ■

NOTE From the final steps in Example 6, we can see that the following is true.

If $-x = a$, then $x = -a$.

2.2 EXERCISES

Decide whether the given item is an expression that can be simplified or an equation to be solved. If it is an expression, simplify it. If it is an equation, solve it.

1. $3x + 7 - 2x + 6$

2. $-8y + 13 + 9y - 4$

3. $3x + 7 - 2x = 6$

4. $-8y + 13 + 9y = -4$

Solve the equation by using the addition property of equality. Check the solution. See Examples 1 and 2.

5. $x - 4 = 8$

6. $x - 8 = 9$

7. $x - 6.5 = -2.3$

8. $y - 5.5 = -1.2$

9. $\frac{9}{7}r - 3 = \frac{2}{7}r$

10. $\frac{8}{5}w - 6 = \frac{3}{5}w$

11. $5.6x + 2 = 4.6x$ **12.** $9.1x - 5 = 8.1x$ **13.** $3p + 6 = 10 + 2p$

14. $8b - 4 = -6 + 7b$ **15.** $1.2y - 4 = .2y - 4$ **16.** $7.7r + 6 = 6.7r + 6$

17. $\frac{1}{2}x + 2 = -\frac{1}{2}x$ **18.** $\frac{1}{5}x - 7 = -\frac{4}{5}x$ **19.** $3x + 7 - 2x = 0$

20. $5x + 4 - 4x = 0$

21. Refer to the definition of *linear equation* given in this section. Why is the restriction $A \neq 0$ necessary?

22. Which of the following are not linear equations?
   (a) $x^2 - 5x + 6 = 0$ **(b)** $x^3 = x$
   (c) $3x - 4 = 0$ **(d)** $7x - 6x = 3 + 9x$

Solve the equation. First simplify each side of the equation as much as possible. You may need to refer to the "Note" following Example 6. Check each solution. See Examples 2 and 6.

23. $5t + 3 + 2t - 6t = 4 + 12$ **24.** $4x + 3x - 6 - 6x = 10 + 3$

25. $10x + 5x + 7 - 8 = 12x + 3 + 2x$ **26.** $7p + 4p + 13 - 7 = 7p + 9 + 3p$

27. $6x + 5 + 7x + 3 = 12x + 4$ **28.** $4x - 3 - 8x + 1 = -5x + 9$

29. $5.2q - 4.6 - 7.1q = -.9q - 4.6$ **30.** $-4.0x + 2.7 - 1.6x = -4.6x + 2.7$

31. $\frac{5}{7}x + \frac{1}{3} = \frac{2}{5} - \frac{2}{7}x + \frac{2}{5}$ **32.** $\frac{6}{7}s - \frac{3}{4} = \frac{4}{5} - \frac{1}{7}s + \frac{1}{6}$

33. $(5y + 6) - (3 + 4y) = 10$ **34.** $(8r - 3) - (7r + 1) = -6$

35. $2(p + 5) - (9 + p) = -3$ **36.** $4(k - 6) - (3k + 2) = -5$

37. $-6(2b + 1) + (13b - 7) = 0$ **38.** $-5(3w - 3) + (1 + 16w) = 0$

39. $10(-2x + 1) = -19(x + 1)$ **40.** $2(2 - 3r) = -5(r - 3)$

41. $-2(8p + 2) - 3(2 - 7p) = 2(4 + 2p)$ **42.** $-5(1 - 2z) + 4(3 - z) = 7(3 + z)$

43. $4(7x - 1) + 3(2 - 5x) = 4(3x + 5) - 6$

44. $9(2m - 3) - 4(5 + 3m) = 5(4 + m) - 3$

45. In the statement of the multiplication property of equality in this section, there is a restriction that $C \neq 0$. What would happen if you should multiply both sides of an equation by 0?

46. Which one of the equations that follow does not require the use of the multiplication property of equality?

   (a) $3x - 5x = 6$ **(b)** $-\frac{1}{4}x = 12$ **(c)** $5x - 4x = 7$ **(d)** $\frac{x}{3} = -2$

Solve the equation and check the solution. See Examples 1–6.

47. $5x = 30$ **48.** $7x = 56$ **49.** $2m = 15$ **50.** $3m = 10$

51. $3a = -15$ **52.** $5k = -70$ **53.** $10t = -36$ **54.** $4s = -34$

55. $-6x = -72$ **56.** $-8x = -64$ **57.** $2r = 0$ **58.** $5x = 0$

59. $\frac{1}{4}y = -12$ **60.** $\frac{1}{5}p = -3$ **61.** $-y = 12$ **62.** $-t = 14$

63. $-x = -\frac{4}{7}$ **64.** $-m = -\frac{9}{5}$ **65.** $.2t = 8$ **66.** $.9x = 18$

67. $4x + 3x = 21$ **68.** $9x + 2x = 121$ **69.** $5m + 6m - 2m = 63$

70. $11r - 5r + 6r = 168$ **71.** $3r - 5r = 10$ **72.** $9p - 13p = 24$

73. $\dfrac{x}{7} = -5$ **74.** $\dfrac{k}{8} = -3$ **75.** $\dfrac{2}{3}t = 6$

76. $\dfrac{4}{3}m = 24$ **77.** $-\dfrac{2}{7}p = -5$ **78.** $-\dfrac{3}{8}y = -2$

79. $-\dfrac{7}{9}c = \dfrac{3}{5}$ **80.** $-\dfrac{5}{6}d = \dfrac{4}{9}$ **81.** $-2.1m = 25.62$

82. $-3.9a = -31.2$

83. Write an equation that requires the use of the multiplication property of equality, where both sides must be multiplied by 2/3, and the solution is a negative number.

84. Write an equation that requires the use of the multiplication property of equality, where both sides must be divided by 100, and the solution is not an integer.

Write an equation using the information given in the problem. Use x as the variable. Then solve the equation.

85. Three times a number is 17 more than twice the number. Find the number.

86. If six times a number is subtracted from seven times the number, the result is −9. Find the number.

87. If five times a number is added to three times the number, the result is the sum of seven times the number and 9. Find the number.

88. When a number is multiplied by 4, the result is 6. Find the number.

89. When a number is divided by −5, the result is 2. Find the number.

90. If twice a number is divided by 5, the result is 4. Find the number.

REVIEW EXERCISES

Simplify the expression. See Sections 1.9 and 2.1.

91. $8(3q + 4)$ **92.** $6(2m - 6)$ **93.** $-7(4p - 3) + 8$

94. $-(3 - 9r) + 12r$ **95.** $6 - 7(2 - 8p)$ **96.** $9(6 + 4y) - 3(3 - 2y)$

2.3 MORE ON SOLVING LINEAR EQUATIONS

FOR EXTRA HELP

📖 **SSG** pp. 60–67
SSM pp. 59–64

📼 **Video**
3

💾 **Tutorial**
IBM   MAC

OBJECTIVES

1 ▶ Learn the four steps for solving a linear equation and how to use them.
2 ▶ Solve equations with fractions or decimals as coefficients.
3 ▶ Recognize equations with no solutions or infinitely many solutions.
4 ▶ Write expressions for two related unknown quantities.

1 ▶ Learn the four steps for solving a linear equation and how to use them.

▶ In this section we learn more about solving linear equations and prepare for solving the types of problems that we will encounter in the next section.

In order to solve linear equations in general, a four-step method can be applied.

Solving a Linear Equation

Step 1 Clear parentheses using the distributive property, if needed; combine terms.

Step 2 Use the addition property to simplify further, if necessary, so that the variable term is on one side of the equation and a number is on the other.

Step 3 Use the multiplication property, if necessary, to get the equation in the form $x =$ a number.

Step 4 Check the solution by substituting into the *original* equation.

EXAMPLE 1

Using the Four Steps to Solve an Equation

Solve $3r + 4 - 2r - 7 = 4r + 3$.

Step 1 $3r + 4 - 2r - 7 = 4r + 3$

$r - 3 = 4r + 3$ Combine like terms.

Step 2 $r - 3 - r = 4r + 3 - r$ Use the addition property of equality. Subtract r.

$-3 = 3r + 3$

$-3 - 3 = -3 + 3r + 3$ Add -3.

$-6 = 3r$

Step 3 $\dfrac{-6}{3} = \dfrac{3r}{3}$ Use the multiplication property of equality. Divide by 3.

$-2 = r$ or $r = -2$

Step 4 Substitute -2 for r in the original equation.

$3r + 4 - 2r - 7 = 4r + 3$

$3(-2) + 4 - 2(-2) - 7 = 4(-2) + 3$?   Let $r = -2$.

$-6 + 4 + 4 - 7 = -8 + 3$?   Multiply.

$-5 = -5$ True

The solution of the equation is -2. ∎

    In Step 2 of Example 1, the terms were added and subtracted in such a way that the variable term ended up on the right. Choosing differently would lead to the variable term being on the left side of the equation. Usually there is no advantage either way.

EXAMPLE 2

Using the Four Steps to Solve an Equation

Solve $4(k - 3) - k = k - 6$.

Step 1 Before combining like terms, use the distributive property to simplify $4(k - 3)$.

$4(k - 3) - k = k - 6$

$4 \cdot k - 4 \cdot 3 - k = k - 6$ Distributive property

$4k - 12 - k = k - 6$

$3k - 12 = k - 6$ Combine like terms.

Step 2 $3k - 12 + 12 = k - 6 + 12$ Add 12.

$3k = k + 6$

$3k - k = k + 6 - k$ Subtract k.

$2k = 6$

Step 3
$$\frac{2k}{2} = \frac{6}{2}$$
Divide by 2.
$$k = 3$$

Step 4   Check your answer by substituting 3 for *k* in the given equation. Remember to do the work inside the parentheses first.

$$4(k - 3) - k = k - 6$$
$$4(3 - 3) - 3 = 3 - 6 \quad ? \qquad \text{Let } k = 3.$$
$$4(0) - 3 = 3 - 6 \quad ?$$
$$0 - 3 = 3 - 6 \quad ?$$
$$-3 = -3 \qquad\qquad \text{True}$$

The solution of the equation is 3.   ■

EXAMPLE 3

Using the Four Steps to Solve an Equation

Solve $8a - (3 + 2a) = 3a + 1$.

Step 1   Simplify.

$$8a - (3 + 2a) = 3a + 1 \qquad\qquad \text{Distributive property}$$
$$8a - 3 - 2a = 3a + 1$$
$$6a - 3 = 3a + 1$$

Step 2
$$6a - 3 + 3 = 3a + 1 + 3 \qquad \text{Add 3.}$$
$$6a = 3a + 4$$
$$6a - 3a = 3a + 4 - 3a \qquad \text{Subtract } 3a.$$
$$3a = 4$$

Step 3
$$\frac{3a}{3} = \frac{4}{3} \qquad\qquad \text{Divide by 3.}$$
$$a = \frac{4}{3}$$

Step 4   Check the solution.

$$8a - (3 + 2a) = 3a + 1$$
$$8\left(\frac{4}{3}\right) - \left[3 + 2\left(\frac{4}{3}\right)\right] = 3\left(\frac{4}{3}\right) + 1 \quad ? \qquad \text{Let } a = \tfrac{4}{3}.$$
$$\frac{32}{3} - \left[3 + \frac{8}{3}\right] = 4 + 1 \quad ?$$
$$\frac{32}{3} - \left[\frac{9}{3} + \frac{8}{3}\right] = 5 \quad ?$$
$$\frac{32}{3} - \frac{17}{3} = 5 \quad ?$$
$$5 = 5 \qquad\qquad \text{True}$$

The check shows that 4/3 is the solution.   ■

CAUTION Be very careful with signs when solving equations like the one in Example 3. When a subtraction sign appears immediately in front of a quantity in parentheses, such as in the expression

$$8 - (3 + 2a),$$

remember that the $-$ sign acts like a factor of -1, and has the effect of changing the sign of *every* term within the parentheses. Thus,

$$8 - (3 + 2a) = 8 - 3 - 2a.$$

$$\uparrow \quad \uparrow$$

Change to $-$ in *both* terms.

EXAMPLE 4

Using the Four Steps to Solve an Equation

■ Solve $4(8 - 3t) = 32 - 8(t + 2)$.

Step 1 Use the distributive property.

$4(8 - 3t) = 32 - 8(t + 2)$	Given equation
$32 - 12t = 32 - 8t - 16$	Distributive property
$32 - 12t = 16 - 8t$	

Step 2

$32 - 12t + \mathbf{12t} = 16 - 8t + \mathbf{12t}$	Add 12t.
$32 = 16 + 4t$	
$32 - \mathbf{16} = 16 + 4t - \mathbf{16}$	Subtract 16.
$16 = 4t$	

Step 3

$\dfrac{16}{4} = \dfrac{4t}{4}$	Divide by 4.
$4 = t \quad \text{or} \quad t = 4$	

Step 4 Check the solution.

$4(8 - 3t) = 32 - 8(t + 2)$		
$4(8 - 3 \cdot \mathbf{4}) = 32 - 8(\mathbf{4} + 2)$	?	Let $t = 4$.
$4(8 - 12) = 32 - 8(6)$	?	
$4(-4) = 32 - 48$	?	
$-16 = -16$		True

The solution, 4, checks. ■

---

2 ▶ Solve equations with fractions or decimals as coefficients.

▶ We can clear an equation of fractions by multiplying both sides by the least common denominator of all denominators in the equation. It is a good idea to do this before starting the four-step method; most students make fewer errors working with integer coefficients.

EXAMPLE 5

Solving an Equation with Fractions as Coefficients

■ Solve $\dfrac{2}{3}x - \dfrac{1}{2}x = -\dfrac{1}{6}x - 2$.

The least common denominator of all the fractions in the equation is 6. Start by multiplying both sides of the equation by 6.

$$\frac{2}{3}x - \frac{1}{2}x = -\frac{1}{6}x - 2$$

$$6\left(\frac{2}{3}x - \frac{1}{2}x\right) = 6\left(-\frac{1}{6}x - 2\right) \qquad \text{Multiply by 6.}$$

$$6\left(\frac{2}{3}x\right) + 6\left(-\frac{1}{2}x\right) = 6\left(-\frac{1}{6}x\right) + 6(-2) \qquad \text{Distributive property}$$

$$4x - 3x = -x - 12$$

Now use the four steps to solve this equivalent equation.

Step 1 $\qquad\qquad\qquad x = -x - 12 \qquad$ Combine like terms.

Step 2 $\qquad\qquad x + x = x - x - 12 \qquad$ Add x.

$$2x = -12$$

Step 3 $\qquad\qquad\qquad \frac{2x}{2} = \frac{-12}{2} \qquad$ Divide by 2.

$$x = -6$$

Step 4 Check the answer.

$$\frac{2}{3}(-6) - \frac{1}{2}(-6) = -\frac{1}{6}(-6) - 2 \qquad ? \qquad \text{Let } x = -6.$$

$$-4 + 3 = 1 - 2 \qquad\qquad ?$$

$$-1 = -1 \qquad\qquad\qquad \text{True}$$

The solution of the equation is -6. ■

CAUTION When clearing equations of fractions be sure to multiply *every* term on both sides of the equation by the least common denominator.

The multiplication property can also be used to clear an equation of decimals.

EXAMPLE 6 ■
Solving an Equation with
Decimal Coefficients

Solve $.20t + .10(20 - t) = .18(20)$.

Since the decimals are all hundredths, start the solution by multiplying both sides of the equation by 100. A number can be multiplied by 100 by moving the decimal point two places to the right.

$$.20t + .10(20 - t) = .18(20) \qquad \text{Multiply by 100.}$$

$$20t + 10(20 - t) = 18(20)$$

Now use the four steps.

Step 1 $\qquad 20t + 10(20) + 10(-t) = 360 \qquad$ Distributive property

$$20t + 200 - 10t = 360$$

$$10t + 200 = 360 \qquad \text{Combine like terms.}$$

Step 2 $\qquad 10t + 200 - 200 = 360 - 200 \qquad$ Subtract 200.

$$10t = 160$$

Step 3

$$\frac{10t}{10} = \frac{160}{10}$$

Divide by 10.

$$t = 16$$

Step 4 Check to see that 16 is the solution of the equation by substituting into the original equation. ∎

3 ▶ Recognize equations with no solutions or infinitely many solutions.

▶ Each equation that we have solved so far has had exactly one solution. As the next examples show, linear equations may also have no solutions or infinitely many solutions. (The four steps are not identified in these examples. See if you can identify them.)

E X A M P L E 7

Solving an Equation that Has Infinitely Many Solutions

Solve $5x - 15 = 5(x - 3)$.

$$5x - 15 = 5(x - 3)$$
$$5x - 15 = 5x - 15 \qquad \text{Distributive property}$$
$$5x - 15 + 15 = 5x - 15 + 15 \qquad \text{Add 15 to each side.}$$
$$5x = 5x \qquad \text{Combine terms.}$$
$$5x - 5x = 5x - 5x \qquad \text{Subtract } 5x \text{ from each side.}$$
$$0 = 0$$

The variable has "disappeared." When this happens look at the resulting statement $(0 = 0)$. Since the statement is a *true* one, *any* real number is a solution. Indicate the solution as "all real numbers." ∎

CAUTION When you are solving an equation like the one in Example 7, do not write "0" as the solution. While 0 is a solution, there are infinitely many other solutions.

E X A M P L E 8

Solving an Equation that Has No Solution

Solve $2x + 3(x + 1) = 5x + 4$.

$$2x + 3(x + 1) = 5x + 4$$
$$2x + 3x + 3 = 5x + 4 \qquad \text{Distributive property}$$
$$5x + 3 = 5x + 4 \qquad \text{Combine terms.}$$
$$5x + 3 - 5x = 5x + 4 - 5x \qquad \text{Subtract } 5x \text{ from each side.}$$
$$3 = 4 \qquad \text{Combine terms.}$$

Again, the variable has disappeared, but this time a *false* statement $(3 = 4)$ results. When this happens, the equation has no solution. Indicate this by writing "no solution." ∎

4 ▶ Write expressions for two related unknown quantities.

▶ We continue our work with translating from words to symbols.

PROBLEM SOLVING

The next example illustrates a type of translation that occurs in many types of problem-solving situations. Very often we are given a problem in which the sum of two quantities is a particular number and we are asked to find the values of the two quantities. Example 9 shows how to express the unknown quantities in terms of a single variable.

EXAMPLE 9 ◾ Two numbers have a sum of 23. If one of the numbers is represented by k, find an
Translating Phrases into
an Algebraic Expression

expression for the other number.

First, suppose that the sum of two numbers is 23, and one of the numbers is **10**.
How would you find the other number? You would subtract 10 from 23 to get 13:
$23 - 10 = 13$. So instead of using **10** as one of the numbers, use k as stated in the
problem. The other number would be obtained in the same way. You must subtract
k from 23. Therefore, an expression for the other number is $23 - k$. ◾

NOTE The approach used in Example 9, first writing an expression using a trial
number, is a useful one that can also be used when translating applied problems to
equations.

2.3 EXERCISES

1. In your own words, give the four steps used to solve a linear equation.

2. Based on the discussion in this section, if an equation has decimals or common
   fractions as coefficients, how can you go about starting the solution to make the work
   easier?

Solve the equation and check the solution. See Examples 1–4.

3. $2h + 4 = 8$ 4. $3x + 5 = 17$

5. $10p + 6 = 12p - 4$ 6. $-5x + 8 = -3x + 10$

7. $x + 3 = -(2x + 2)$ 8. $2x + 1 = -(x + 3)$

9. $4(2x - 1) = -6(x + 3)$ 10. $6(3w + 5) = 2(10w + 10)$

11. $6(4x - 1) = 12(2x + 3)$ 12. $6(2x + 8) = 4(3x - 6)$

13. $3(2x - 4) = 6(x - 2)$ 14. $3(6 - 4x) = 2(-6x + 9)$

15. $7r - 5r + 2 = 5r - r$ 16. $9p - 4p + 6 = 7p - 3p$

17. $11x - 5(x + 3) - 6x = 0$ 18. $6x - 4(x + 2) - 2x = 0$

19. After working correctly through several steps of a linear equation, a student obtains
    the equation $7x = 3x$. Then the student divides both sides by x to get $7 = 3$, and gives
    "no solution" as the answer. Is this correct? If not, explain why.

20. Which one of the following linear equations does not have all real numbers as
    solutions?

    (a) $5x = 4x + x$ **(b)** $2(x + 6) = 2x + 12$ **(c)** $\frac{1}{2}x = .5x$ **(d)** $3x = 2x$

Solve each equation by first clearing of fractions or decimals. See Examples 5 and 6.
do not have to

21. $\frac{3}{5}t - \frac{1}{10}t = t - \frac{5}{2}$ 22. $-\frac{2}{7}r + 2r = \frac{1}{2}r + \frac{17}{2}$

23. $-\frac{1}{4}(x - 12) + \frac{1}{2}(x + 2) = x + 4$ 24. $\frac{1}{9}(y + 18) + \frac{1}{3}(2y + 3) = y + 3$

25. $\frac{2}{3}k - \left(k + \frac{1}{4}\right) = \frac{1}{12}(k + 4)$ 26. $-\frac{5}{6}q - \left(q - \frac{1}{2}\right) = \frac{1}{4}(q + 1)$

27. $.20(60) + .05x = .10(60 + x)$ 28. $.30(30) + .15x = .20(30 + x)$

29. $1.00x + .05(12 - x) = .10(63)$ 30. $.92x + .98(12 - x) = .96(12)$

31. $.06(10,000) + .08x = .072(10,000 + x)$ 32. $.02(5000) + .03x = .025(5000 + x)$

───────── ◆ **MATHEMATICAL CONNECTIONS** (Exercises 33–38) ◆ ─────────

Work Exercises 33–38 in order.

33. Consider the term $100ab$. Evaluate it for $a = 2$ and $b = 4$.

34. Based on your study of Section 1.9, will you get the same answer for Exercise 33 if you evaluate $(100a)b$ for $a = 2$ and $b = 4$? Why or why not?

35. Is the term $(100a)(100b)$ equivalent to $100ab$? Why or why not?

36. If your answer to Exercise 35 is *no*, explain why the distributive property is not involved.

37. Consider the equation

$$.05(x + 2) + .10x = 2.00.$$

To simplify our work, we would begin by multiplying both sides by 100. When applied to the first term on the left, our expression would be

$$100 \cdot .05(x + 2).$$

Is this expression equivalent to $[100 \cdot 05](x + 2)$? (*Hint:* Compare to the expressions in Exercises 33 and 34, letting $a = .05$ and $b = x + 2$.)

38. Students often want to "distribute" the 100 to both .05 and $(x + 2)$ in the expression $100 \cdot .05(x + 2)$. Is this correct? (*Hint:* Compare your answer to those in Exercises 35 and 36.)

───────────────── ◆ ─────────────────

Solve the equation and check the solution. See Examples 1–8.

39. $10(2x - 1) = 8(2x + 1) + 14$

40. $9(3k - 5) = 12(3k - 1) - 51$

41. $-2(2s - 4) - 8 = -3(4s + 4) - 1$

42. $-3(5z + 24) + 2 = 2(3 - 2z) - 4$

43. $-(4y + 2) - (-3y - 5) = 3$

44. $-(6k - 5) - (-5k + 8) = -3$

45. $\frac{1}{2}(x + 2) + \frac{3}{4}(x + 4) = x + 5$

46. $\frac{1}{3}(x + 3) + \frac{1}{6}(x - 6) = x + 3$

47. $.10(x + 80) + .20x = 14$

48. $.30(x + 15) + .40(x + 25) = 25$

49. $4(x + 8) = 2(2x + 6) + 20$

50. $4(x + 3) = 2(2x + 8) - 4$

51. $9(v + 1) - 3v = 2(3v + 1) - 8$

52. $8(t - 3) + 4t = 6(2t + 1) - 10$

53. Explain why the solution of $\frac{1}{10}(x + 80) + \frac{1}{5}x = 14$ is the same as that of the equation in Exercise 47.

54. Explain why an equation involving $\frac{1}{x - 2}$ cannot have 2 as a solution.

Write the answer to the problem as an algebraic expression. See Example 9.

55. Two numbers have a sum of 11. One of the numbers is q. Find the other number.

56. The product of two numbers is 9. One of the numbers is k. What is the other number?

57. Yesterday Walt bought x apples. Today he bought 7 apples. How many apples did he buy altogether?

58. Joann has 15 books. She donated p books to the library. How many books does she have left?

59. Mary is a years old. How old will she be in 12 years? How old was she 5 years ago?

60. Tom has r quarters. Find the value of the quarters in cents.

61. A bank teller has t dollars, all in five-dollar bills. How many five-dollar bills does the teller have?

62. A plane ticket costs b dollars for an adult and d dollars for a child. Find the total cost for 3 adults and 2 children.

REVIEW EXERCISES

Write the phrase as a mathematical expression using x as the variable. See Sections 1.5–1.8.

63. A number added to -6

64. The sum of a number and twice the number

65. A number decreased by 9

66. The difference between -5 and a number

67. The quotient of -6 and a nonzero number

68. A number divided by 17

69. The product of 12 and the difference between a number and 9

70. The quotient of 9 more than a number and 6 less than the number

2.4 AN INTRODUCTION TO APPLICATIONS OF LINEAR EQUATIONS

FOR EXTRA HELP	OBJECTIVES
📖 **SSG** pp. 67–72 **SSM** pp. 65–71 📼 **Video** 3 💾 **Tutorial** IBM MAC	**1 ▶** Learn the six steps to be used to solve an applied problem. **2 ▶** Solve problems involving unknown numbers. **3 ▶** Solve problems involving sums of quantities. **4 ▶** Solve problems involving supplementary and complementary angles. **5 ▶** Solve problems involving consecutive integers.

◆ **CONNECTIONS** ◆

The purpose of algebra is to solve real problems. Since such problems are stated in words, not mathematical symbols, the first step in solving them is to translate the problem into one or more mathematical statements. This is the hardest step for most people. George Polya (1888–1985), a native of Budapest, Hungary, wrote the modern classic *How to Solve It*. In this book he proposed a four-step process for problem solving:

1. Understand the problem.
2. Devise a plan.
3. Carry out the plan.
4. Look back and check.

FOR DISCUSSION OR WRITING

Compare Polya's four-step process with the six steps given in this section. Identify which of the steps in our list match Polya's four steps. In earlier sections of this book, we have mentioned using trial and error or guessing and checking as problem-solving tools. Where do these methods fit into Polya's steps?

1 ▶ Learn the six steps to be used to solve an applied problem.

▶   We now begin to look at how algebra is used to solve applied problems. It must be emphasized that many *meaningful* applications of mathematics require concepts that are beyond the level of this book. Some of the problems you will encounter will seem "contrived," and to some extent they are. But the skills you will develop in solving simple problems will help you in solving more realistic problems in chemistry, physics, biology, business, and other fields.

PROBLEM SOLVING   In earlier sections we learned how to translate words, phrases, and sentences into mathematical expressions and equations. In this section and in many other sections in this book, we use these translations to solve applied problems using algebra. While there is no specific method that enables you to solve all kinds of applied problems, the following general method is suggested. It consists of six steps, and the steps will be specified by number in the examples in this section.

Solving an Applied Problem

Step 1   Read the problem carefully, and choose a variable to represent the numerical value that you are asked to find—the unknown number. *Write down* what the variable represents.

Step 2   *Write down* a mathematical expression using the variable for any other unknown quantities. Draw figures or diagrams if they apply.

Step 3   Translate the problem into an equation.

Step 4   Solve the equation.

Step 5   Answer the question asked in the problem.

Step 6   Check your solution by using the original words of the problem. Be sure that your answer makes sense.

2 ▶ Solve problems involving unknown numbers.

▶   Some of the simplest applied problems involve unknown numbers, and we will look at this type in the next example.

PROBLEM SOLVING   The third step in solving an applied problem is often the hardest. Begin to translate the problem into an equation by writing the given phrases as mathematical expressions. Since equal mathematical expressions are names for the same number, translate any words that mean *equal* or *same* as $=$. The $=$ sign leads to an equation to be solved.

E X A M P L E   1

Finding the Value of an Unknown Number

The product of 4, and a number decreased by 7, is 100. Find the number.

Step 1   After reading the problem carefully, decide on what you are being told to find, and then choose a variable to represent the unknown quantity. In this problem, we are told to find a number, so we write

Let x = the number.

Step 2   There are no other unknown quantities to find.

Step 3   Translate as follows.

The product of 4,	and	a number	decreased by	7,	is	100.
↓		↓	↓	↓	↓	↓
$4 \cdot$	(	x	$-$	7)	$=$	100

Because of the comma in the given sentence, writing the equation as $4x - 7 = 100$ is incorrect. The equation $4x - 7 = 100$ corresponds to the statement "The product of 4 and a number, decreased by 7, is 100."

Step 4 Solve the equation.

$$4(x - 7) = 100$$
$$4x - 28 = 100 \qquad \text{Distributive property}$$
$$4x = 128 \qquad \text{Add 28 to both sides.}$$
$$x = 32 \qquad \text{Divide by 4.}$$

Step 5 The number is 32.

Step 6 Check the solution by using the original words of the problem. When 32 is decreased by 7, we get $32 - 7 = 25$. If 4 is multiplied by 25, we get 100, as the problem required. The answer, 32, is correct. ■

NOTE The commas in the statement of the problem in Example 1 are used in translating correctly.

---

3 ▶ Solve problems involving sums of quantities.

▶ A common type of problem that occurs in elementary algebra is the type that involves finding two quantities when the sum of the quantities is known. In Example 9 of the previous section, we prepared for this type of problem by writing mathematical expressions for two related unknown quantities.

PROBLEM SOLVING

In general, to solve such problems, choose a variable to represent one of the unknowns and then represent the other quantity in terms of the same variable, using information obtained in the problem. Then write an equation based on the words of the problem. The next example illustrates these ideas.

EXAMPLE 2

Finding the Numbers of Men and Women at a Concert

At a concert, there were 25 more women than men. The total number of people at the concert was 139. Find the number of men and the number of women at the concert.

Step 1 Let x = the number of men.

Step 2 Let $x + 25$ = the number of women.

Step 3 Now write an equation.

The total	is	the number of men	plus	the number of women.
↓	↓	↓	↓	↓
139	=	x	+	$(x + 25)$

Step 4 Solve the equation.

$$139 = 2x + 25 \qquad \text{Combine terms.}$$
$$139 - \mathbf{25} = 2x + 25 - \mathbf{25} \qquad \text{Subtract 25.}$$
$$114 = 2x \qquad \text{Simplify.}$$
$$57 = x \qquad \text{Divide by 2.}$$

Step 5   Because x represents the number of men, there were 57 men at the concert. Because $x + 25$ represents the number of women, there were $57 + 25 = 82$ women at the concert.

Step 6   Since there were 57 men and 82 women present, there were $57 + 82 = 139$ people there. Because $82 - 57 = 25$, there were 25 more women than men. This information agrees with what is given in the problem, so the answers check.   ■

NOTE   The problem in Example 2 could also have been solved by letting x represent the number of women. Then $x - 25$ would represent the number of men. The equation would then be

$$139 = x + (x - 25).$$

The solution of this equation is 82, which is the number of women. The number of men would then be $82 - 25 = 57$. You can see that the answers are the same, no matter which approach is used.

EXAMPLE 3

Finding the Number of Orders for Tea

The owner of P. J.'s Coffeehouse found that on one day the number of orders for tea was $1/3$ the number of orders for coffee. If the total number of orders for the two drinks was 76, how many orders were placed for tea?

Step 1   Let x = the number of orders for coffee.

Step 2   Let $\dfrac{1}{3}x$ = the number of orders for tea.

Step 3   Use the fact that the total number of orders was 76 to write an equation.

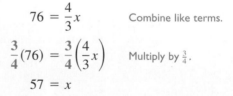

The total	is	orders for coffee	plus	orders for tea.
↓	↓	↓	↓	↓
76	=	x	+	$\dfrac{1}{3}x$

Step 4   Now solve the equation.

$$76 = \frac{4}{3}x \qquad \text{Combine like terms.}$$

$$\frac{3}{4}(76) = \frac{3}{4}\left(\frac{4}{3}x\right) \qquad \text{Multiply by } \tfrac{3}{4}.$$

$$57 = x$$

Step 5   In this problem, *x does not represent the quantity that we are asked to find.* The number of orders placed for tea was $\dfrac{1}{3}x$. So $\dfrac{1}{3}(57) = 19$ is the number of orders for tea.

Step 6   The number of coffee orders (x) was 57 and the number of tea orders was 19. 19 is one-third of 57, and $19 + 57 = 76$. Since this agrees with the information given in the problem, the answer is correct.   ■

PROBLEM SOLVING In Example 3, it was easier to let the variable represent the quantity that was *not* asked for. This required an extra step in Step 5 to find the number of orders for tea. In some cases, this approach is easier than letting the variable represent the quantity that we are asked to find. Experience in solving problems will indicate when this approach is useful, and experience comes only from solving many problems!

Sometimes it is necessary to find three unknown quantities in an applied problem. Frequently the three unknowns are compared in *pairs*. When this happens, it is usually easiest to let the variable represent the unknown found in both pairs.

The next example illustrates how we can find more than two unknown quantities in an application.

E X A M P L E 4

Dividing a Board Into Pieces

The instructions for a woodworking project require three pieces of wood. The longest piece must be twice the length of the middle-sized piece, and the shortest piece must be 10 inches shorter than the middle-sized piece. Maria Gonzoles has a board 70 inches long that she wishes to use. How long can each piece be?

Steps 1 and 2   Since the middle-sized piece appears in both pairs of comparisons, let x represent the length of the middle-sized piece. We have

$$x = \text{the length of the middle-sized piece}$$
$$2x = \text{the length of the longest piece}$$
$$x - 10 = \text{the length of the shortest piece.}$$

A sketch is helpful here. (See Figure 1.)

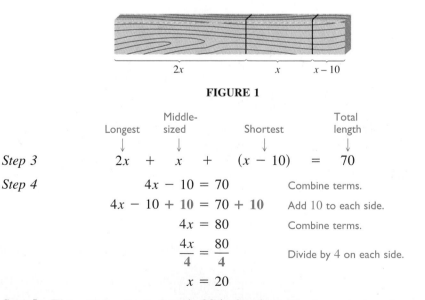

$$2x \qquad x \qquad x - 10$$

FIGURE 1

	Longest	Middle-sized		Shortest		Total length
	↓	↓		↓		↓
Step 3	$2x$	$+$ x	$+$	$(x - 10)$	$=$	70

Step 4

$$4x - 10 = 70 \qquad \text{Combine terms.}$$
$$4x - 10 + 10 = 70 + 10 \qquad \text{Add 10 to each side.}$$
$$4x = 80 \qquad \text{Combine terms.}$$
$$\frac{4x}{4} = \frac{80}{4} \qquad \text{Divide by 4 on each side.}$$
$$x = 20$$

Step 5   The middle-sized piece is 20 inches long, the longest piece is $2(20) = 40$ inches long, and the shortest piece is $20 - 10 = 10$ inches long.

Step 6   Check to see that the sum of the lengths is 70 inches, and that all conditions of the problem are satisfied. ■

E X A M P L E   5

Analyzing a Gasoline/Oil Mixture

A lawn trimmer uses a mixture of gasoline and oil. For each ounce of oil the mixture contains 16 ounces of gasoline. If the tank holds 68 ounces of the mixture, how many ounces of oil and how many ounces of gasoline does it require when it is full?

Step 1   Let x = the number of ounces of oil required when full.
Step 2   Let $16x$ = the number of ounces of gasoline required when full.

Oil

Gasoline   Lawn trimmer

Amount of gasoline		Amount of oil		Total amount in tank
↓		↓		↓

Step 3 $16x$ + x =   68

Step 4 $17x = 68$ Combine terms.

$$\frac{17x}{17} = \frac{68}{17}$$ Divide by 17.

$$x = 4$$

Step 5   The trimmer requires 4 ounces of oil and 16(4) = 64 ounces of gasoline when full.
Step 6   Since 4 + 64 = 68, and 64 is 16 times 4, the answers check.   ∎

4 ▶ Solve problems involving supplementary and complementary angles.

▶   The next example of problem solving in this section deals with concepts from geometry. An angle can be measured by a unit called the **degree** (°). Two angles whose sum is 90° are said to be **complementary,** or complements of each other. Two angles whose sum is 180° are said to be **supplementary,** or supplements of each other. If x represents the degree measure of an angle, then

$90 - x$ represents the degree measure of its complement, and

$180 - x$ represents the degree measure of its supplement.

E X A M P L E   6

Finding the Measure of an Angle

Find the measure of an angle whose supplement is 10 degrees more than twice its complement.

Step 1   Let x = the degree measure of the angle.
Step 2   Let $90 - x$ = the degree measure of its complement;
                $180 - x$ = the degree measure of its supplement.

Step 3   Supplement   is   10   more than   twice   its complement.

$$180 - x = 10 + 2 \cdot (90 - x)$$

Step 4   Solve the equation.

$$180 - x = 10 + 180 - 2x \qquad \text{Distributive property}$$
$$180 - x = 190 - 2x \qquad \text{Combine terms.}$$
$$180 - x + 2x = 190 - 2x + 2x \qquad \text{Add } 2x.$$
$$180 + x = 190 \qquad \text{Combine terms.}$$
$$180 + x - 180 = 190 - 180 \qquad \text{Subtract } 180.$$
$$x = 10$$

Step 5   The measure of the angle is 10 degrees.

Step 6   The complement of 10° is 80° and the supplement of 10° is 170°. 170° is equal to 10° more than twice 80° (170 = 10 + 2(80) is true); therefore, the answer is correct. ∎

5 ▶ Solve problems involving consecutive integers.

▶   We conclude this section with examples of problems involving consecutive integers. Two integers that differ by 1 are called **consecutive integers.** For example, 3 and 4, 6 and 7, and −2 and −1 are pairs of consecutive integers. In general, if x represents an integer, $x + 1$ represents the next larger consecutive integer.

Consecutive even integers, such as 8 and 10, differ by 2. Similarly, consecutive odd integers, such as 9 and 11, also differ by two. In general, if x represents an even integer, $x + 2$ represents the next larger consecutive even integer. The same holds true for odd integers; that is, if x is an odd integer, $x + 2$ is the next larger odd integer.

EXAMPLE 7

Finding Consecutive Integers

Two pages that face each other in this book have 569 as the sum of their page numbers. What are the page numbers?

Because the two pages face each other, they must have page numbers that are consecutive integers.

Step 1   Let $x =$ the smaller page number.

Step 2   Let $x + 1 =$ the larger page number.

Step 3   Because the sum of the page numbers is 569, the equation is

$$x + (x + 1) = 569.$$

Step 4   Solve the equation.

$$x + (x + 1) = 569$$
$$2x + 1 = 569 \qquad \text{Combine like terms.}$$
$$2x = 568 \qquad \text{Subtract } 1.$$
$$x = 284 \qquad \text{Divide by } 2.$$

Step 5   The smaller page number is 284 and the larger page number is $284 + 1 = 285$.

Step 6   The sum of 284 and 285 is 569. Our answer is correct. ∎

In the final example we do not number the steps. See if you are able to identify them.

EXAMPLE 8
Finding Consecutive Odd Integers

If the smaller of two consecutive odd integers is doubled, the result is 7 more than the larger of the two integers. Find the two integers.

Let x be the smaller integer. Since the two numbers are consecutive *odd* integers, then $x + 2$ is the larger. Now write an equation from the statement of the problem:

If the smaller is doubled	the result is	7	more than	the larger.
↓	↓	↓	↓	↓
$2x$	$=$	7	$+$	$x + 2$

Solve the equation.

$$2x = 7 + x + 2$$
$$2x = 9 + x$$
$$2x + (-x) = 9 + x + (-x)$$
$$x = 9$$

The first integer is 9 and the second is $9 + 2 = 11$. To check our answer we see that when 9 is doubled, we get 18, which is 7 more than the larger odd integer, 11. Our answer is correct. ■

2.4 EXERCISES

1. Which one of the following would not be a reasonable answer in an applied problem that requires finding the number of coins in a jar?

    (a) 7   (b) 0   (c) $6\frac{2}{3}$ (d) 80

2. Which one of the following would not be a reasonable answer in an applied problem that requires finding someone's age (in years)?

    (a) $5\frac{1}{2}$ (b) 7   (c) 12.25   (d) -4

3. Explain in your own words the general procedure described for solving applied problems in this section.

4. List some words that will translate as "=" in an applied problem.

Solve the problem. See Example 1.

5. If 1 is added to a number and this sum is doubled, the result is 5 more than the number. Find the number.

6. If 2 is subtracted from a number and this difference is tripled, the result is 4 more than the number. Find the number.

7. If 3 is added to twice a number and this sum is multiplied by 4, the result is the same as if the number is multiplied by 7 and 8 is added to the product. What is the number?

8. The sum of three times a number and 12 more than the number is the same as the difference between -6 and twice the number. What is the number?

Solve the problem. See Examples 2–5.

9. The U.S. Senate has 100 members. After the 1990 election, there were 14 more Democrats than Republicans, with no other parties represented. How many members of each party were there in the Senate?

10. The total number of Democrats and Republicans in the U.S. House of Representatives in 1990 was 434. There were 100 fewer Republicans than Democrats. How many members of each party were there in the House of Representatives?

11. In his coaching career with the Boston Celtics, Red Auerbach had 558 more wins than losses. His total number of games coached was 1516. How many wins did Auerbach have?

12. In the first Super Bowl, played in 1966, Green Bay and Kansas City scored a total of 45 points. Green Bay won by 25 points. What was the score of the first Super Bowl?

13. Nagaraj Nanjappa has a strip of paper 39 inches long. He wants to cut it into two pieces so that one piece will be 9 inches shorter than the other. How long should the two pieces be?

14. On Professor Brandsma's algebra test, the highest grade was 34 points higher than the lowest grade. The sum of the two grades was 160. What were the highest and lowest grades?

15. In one day, Gwen Boyle received 13 packages. Federal Express delivered three times as many as Airborne Express, while Airborne Express delivered two more than United Parcel Service. How many packages did each service deliver to Gwen?

16. In her job at the post office, Janie Quintana works a $6\frac{1}{2}$-hour day. She sorts mail, sells stamps, and does supervisory work. On one day, she sold stamps twice as long as she sorted mail, and sold stamps 1 hour longer than the time she spent doing supervisory work. How many hours did she spend at each task?

17. Venus is 31.2 million miles farther from the sun than Mercury, while Earth is 25.7 million miles farther from the sun than Venus. If the total of the distances for these three planets from the sun is 196.1 million miles, how far away from the sun is Mercury? [All distances given here are *mean (average)* distances.]

18. It is believed that Saturn has 5 more satellites (moons) than the known number of satellites for Jupiter, and 20 more satellites than the known number for Mars. If the total of these numbers is 41, how many satellites does Mars have?

19. During their National League championship year of 1992, Atlanta Braves' hitters Ron Gant, David Justice, and Francisco Cabrera combined for a total of 268 base hits. Gant had 47 times as many hits as Cabrera, while Justice had 17 fewer hits than Gant. How many hits did each player have?

20. During their World Series championship year of 1992, Toronto Blue Jays' pitchers Duane Ward, David Wells, and Jack Morris combined to pitch 462 innings. Together, Morris and Ward pitched 342 innings. Wells pitched $18\frac{2}{3}$ more innings than Ward. How many innings did each player pitch?

21. The sum of the measures of the angles of any triangle is 180 degrees. In triangle ABC, angles A and B have the same measure, while the measure of angle C is 60 degrees larger than each of A and B. What are the measures of the three angles?

22. (See Exercise 21.) In triangle ABC, the measure of angle A is 141 degrees more than the measure of angle B. The measure of angle B is the same as the measure of angle C. Find the measure of each angle.

23. On the 1992 Eagle Premier, the suggested list price for the antilock brake system is $\frac{10}{3}$ the suggested list price of power door locks. Together, these two options cost $1040. What is the suggested list price for each of these options?

24. The 1993 edition of *A Guide Book of United States Coins* lists the value of a Mint State-65 (uncirculated) 1950 Jefferson nickel minted at Denver as $\frac{6}{5}$ the value of a similar condition 1944 nickel minted at Denver. Together the total value of the two coins is $22.00. What is the value of each coin?

25. A pharmacist found that at the end of the day she had $\frac{4}{3}$ as many prescriptions for antibiotics as she did for tranquilizers. She had 42 prescriptions altogether for these two types of drugs. How many did she have for tranquilizers?

26. In a mixture of concrete, there are 3 pounds of cement mix for every 1 pound of gravel. If the mixture contains a total of 140 pounds of these two ingredients, how many pounds of gravel are there?

27. A mixture of nuts contains only peanuts and cashews. For every ounce of cashews there are 5 ounces of peanuts. If the mixture contains a total of 27 ounces, how many ounces of each type of nut does the mixture contain?

28. An insecticide contains 95 centigrams of inert ingredient for every 1 centigram of active ingredient. If a quantity of the insecticide weighs 336 centigrams, how much of each type of ingredient does it contain?

Use the concepts of this section to answer the question.

29. If the sum of two numbers is k, and one of the numbers is m, how can you express the other number?

30. If the product of two numbers is r, and one of the numbers is s $(s \neq 0)$, how can you express the other number?

31. Is there an angle whose supplement is equal to its complement? If so, what is the measure of the angle?

32. Is there an angle that is equal to its supplement? Is there an angle that is equal to its complement? If the answer is yes to either question, give the measure of the angle.

33. If x represents an integer, how can you express the next smaller consecutive integer in terms of x?

34. Express three consecutive even integers, all in terms of x, if x represents the middle of the three.

Solve the problem. See Example 6.

35. Find the measure of an angle whose supplement measures 10 times the measure of its complement.

36. Find the measure of an angle whose supplement measures 4 times the measure of its complement.

37. Find the measure of an angle, if its supplement measures 38° less than three times its complement.

38. Find the measure of an angle, if its supplement measures 39° more than twice its complement.

39. Find the measure of an angle such that the sum of the measures of its complement and its supplement is 160°.

40. Find the measure of an angle such that the difference between the measures of its supplement and three times its complement is 10°.

Solve the problem. See Examples 7 and 8.

41. The sum of two consecutive integers is 137. Find the integers.

42. The sum of two consecutive integers is -357. Find the integers.

43. Find two consecutive even integers such that the smaller added to three times the larger gives a sum of 46.

44. Find two consecutive odd integers such that twice the larger is 17 more than the smaller.

45. Two pages that are back-to-back in this book have 203 as the sum of their page numbers. What are the page numbers?

46. If the sum of three consecutive even integers is 60, what is the smallest even integer?

47. When the smaller of two consecutive integers is added to three times the larger, the result is 43. Find the integers.

48. If five times the smaller of two consecutive integers is added to three times the larger, the result is 59. Find the integers.

49. The smallest of three consecutive integers is added to twice the largest producing a result 15 less than four times the middle integer. Find the integers.

50. If the middle of three consecutive integers is added to 100, the result is 1 less than the sum of the largest and twice the smallest. Find the integers.

51. If 6 is subtracted from the largest of three consecutive odd integers, with this result multiplied by 2, the answer is 23 less than the sum of the first and twice the second of the integers. Find the integers.

52. If the first and third of three consecutive even integers are added, the result is 22 less than three times the second integer. Find the integers.

Apply the ideas of this section to solve the problem based on the graph.

53. In 1991, the funding for Head Start programs increased by .50 billion dollars from the funding in 1990. In 1992, the increase was .25 billion dollars over the funding in 1991. For those three years the total funding was 5.6 billion dollars. How much was funded in each of these years? (*Source:* U.S. Department of Health and Human Services)

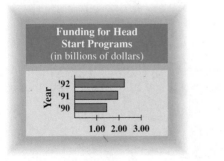

54. According to data provided by the National Safety Council, in 1992 the number of serious injuries per 100,000 participants in football, bicycling, and golf are illustrated in the graph. There were 800 more in bicycling than in golf, and there were 1267 more in football than in bicycling. Altogether there were 3179 serious injuries per 100,000 participants. How many such serious injuries were there in each sport?

REVIEW EXERCISES

Use the given values to evaluate the expression. See Section 1.3.

55. LW; $L = 6, W = 4$

56. rt; $r = 25, t = 4.5$

57. prt; $p = 4000, r = .04, t = 2$

58. $\frac{1}{2}Bh$; $B = 27, h = 6$

2.5 FORMULAS AND APPLICATIONS FROM GEOMETRY

FOR EXTRA HELP	OBJECTIVES
📖 **SSG** pp. 72–79 **SSM** pp. 71–76	**1 ▶** Solve a formula for one variable given the values of the other variables.
📼 **Video** 3	**2 ▶** Use a formula to solve a geometric application.
💾 **Tutorial** IBM MAC	**3 ▶** Solve problems about angle measures.
	4 ▶ Solve a formula for a specified variable.

Many applied problems can be solved with formulas. There are formulas for geometric figures such as squares and circles, for distance, for money earned on bank savings, and for converting English measurements to metric measurements, for example. The formulas used in this book are given in the endsheets of this book.

1 ▶ Solve a formula for one variable given the values of the other variables.

▶ Given the values of all but one of the variables in a formula, the value of the remaining variable can be found by using the methods introduced in this chapter for solving equations.

EXAMPLE 1

Using a Formula to Evaluate a Variable

Find the value of the remaining variable in each of the following.

(a) $A = LW$; $A = 64, L = 10$

As shown in Figure 2, this formula gives the area of a rectangle with length L and width W. Substitute the given values into the formula and then solve for W.

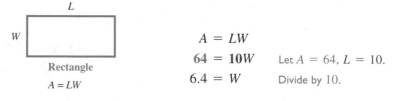

Rectangle
$A = LW$

$$A = LW$$
$$64 = 10W \qquad \text{Let } A = 64, L = 10.$$
$$6.4 = W \qquad \text{Divide by } 10.$$

FIGURE 2

Check that the width of the rectangle is 6.4.

(b) $A = \dfrac{1}{2}(b + B)h$; $A = 210, B = 27, h = 10$

This formula gives the area of a trapezoid with parallel sides of lengths b and B and distance h between the parallel sides. See Figure 3. Again, begin by substituting the given values into the formula.

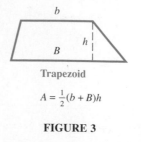

Trapezoid

$A = \frac{1}{2}(b + B)h$

FIGURE 3

$$A = \frac{1}{2}(b + B)h$$

$$\mathbf{210} = \frac{1}{2}(b + \mathbf{27})(\mathbf{10}) \qquad A = 210,\ B = 27,\ h = 10$$

Now solve for b.

$$210 = \frac{1}{2}(10)(b + 27) \qquad \text{Commutative property}$$

$$210 = 5(b + 27)$$

$$210 = 5b + 135 \qquad \text{Distributive property}$$

$$75 = 5b \qquad \text{Subtract 135.}$$

$$15 - b \qquad \text{Divide by 5.}$$

Check that the length of the shorter parallel side, b, is 15. ■

◆ **C O N N E C T I O N S** ◆

In order to solve some applied problems, we can use a ready-made equation, called a formula. For example, the U.S. Postal Service requires that any box sent through the mail have length plus girth (distance around) totaling no more than 108 inches. The maximum volume is obtained if the box has dimensions 18 inches by 18 inches by 72 inches. What is the maximum volume? The volume of a box (a *rectangular solid*) is $V = LWH$, where L is its length, W is its width, and H is its height.

FOR DISCUSSION OR WRITING

Find the maximum volume discussed above. What units are used for volume if length, width, and height are given in inches? What is meant by the volume of a box? (*Hint:* What does it mean to say the volume of a box is 9 cubic inches?)

2 ▶ Use a formula to solve a geometric application.

▶ As the next examples show, formulas can be used to solve applications involving geometric figures. In Step 2, we use a drawing or a diagram to visualize.

Example 2 uses the idea of *perimeter*. The **perimeter** of a figure is the sum of the lengths of its sides.

E X A M P L E   2

Finding the Width of a Rectangular Lot

A rectangular lot is advertised for sale. The advertisement gives the length as 25 meters and the perimeter as 80 meters. To satisfy a potential buyer, the real estate agent must find the width of the lot. (See Figure 4.)

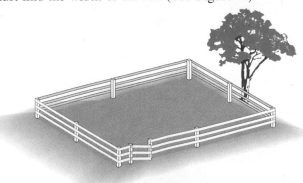

FIGURE 4

Step 1   We want to find the width of the lot, so

let W = the width of the lot in meters.

Step 2   See Figure 4.

Step 3   The formula for the perimeter of a rectangle is

$$P = 2L + 2W.$$

Find the width by substituting 80 for P and 25 for L in the formula.

$$80 = 2(\mathbf{25}) + 2W \qquad P = 80, L = 25$$

Step 4   Solve the equation.

$$80 = 50 + 2W \qquad \text{Multiply.}$$
$$80 - 50 = 50 + 2W - 50 \qquad \text{Subtract 50.}$$
$$30 = 2W$$
$$15 = W \qquad \text{Divide by 2.}$$

Step 5   The width is 15 meters.

Step 6   Check this result. If the width is 15 meters and the length is 25 meters, the distance around the rectangular lot (perimeter) is $2(25) + 2(15) = 50 + 30 = 80$ feet, as required. ∎

The **area** of a geometric figure is a measure of the surface covered by the figure. Example 3 shows an application of area.

E X A M P L E   3

Finding the Height of a Triangular Sail

■ The area of a triangular sail of a sailboat is 126 square meters. The base of the sail is 21 meters. Find the height of the sail.

Step 1   Since we must find the height of the triangular sail,

let h = the height of the sail in meters.

Step 2   See Figure 5.

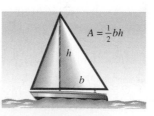

FIGURE 5

Step 3   The formula for the area of a triangle is $A = (1/2)bh$, where A is the area, b is the base, and h is the height. Using the information given in the problem, substitute 126 for A and 21 for b in the formula.

$$A = \frac{1}{2}bh$$

$$\mathbf{126} = \frac{1}{2}(\mathbf{21})h \qquad A = 126, b = 21$$

Step 4   Solve the equation.

$$126 = \frac{21}{2}h$$

$$\frac{2}{21}(126) = \frac{2}{21} \cdot \frac{21}{2}h \qquad\qquad \text{Multiply by } \tfrac{2}{21}.$$

$$12 = h \qquad \text{or} \qquad h = 12$$

Step 5   The height of the sail is 12 meters.

Step 6   Check to see that the values $A = 126$, $b = 21$, and $h = 12$ satisfy the formula for the area of a triangle.   ■

3 ▶ Solve problems about angle measures

▶   Angle measures are vital to the study of *geodesy*, the measurement of the earth's surface. Methods of calculating angle measures are discussed in the next example.

Refer to Figure 6, which shows two intersecting lines forming angles that are numbered ①, ②, ③, and ④. Angles ① and ③ lie "opposite" each other. They are called **vertical angles.** Another pair of vertical angles are ② and ④. In geometry, it is shown that the following property holds.

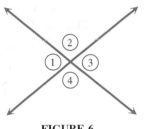

FIGURE 6

Vertical Angles	Vertical angles have equal measures.

Now look at angles ① and ②. When their measures are added, we get the measure of a **straight angle,** which is 180°. There are three other such pairs of angles: ② and ③, ③ and ④, ① and ④.

E X A M P L E   4

Finding Angle Measures

Refer to the appropriate figures in each part.

(a) Find the measure of each marked angle in Figure 7.

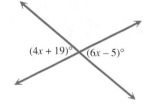

$(4x + 19)°$ $(6x - 5)°$

FIGURE 7

Since the marked angles are vertical angles, they have the same measures. Set $4x + 19$ equal to $6x - 5$ and solve.

$$4x + 19 = 6x - 5$$
$$-4x + 4x + 19 = -4x + 6x - 5 \qquad \text{Add } -4x.$$
$$19 = 2x - 5$$
$$19 + 5 = 2x - 5 + 5 \qquad \text{Add 5.}$$
$$24 = 2x$$
$$12 = x \qquad \text{Divide by 2.}$$

Since $x = 12$, one angle has measure $4(12) + 19 = 67$ degrees. The other has the same measure, since $6(12) - 5 = 67$ as well. Each angle measures $67°$.

(b) Find the measure of each marked angle in Figure 8.

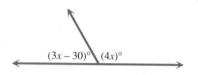

FIGURE 8

The measures of the marked angles must add to $180°$ since together they form a straight angle. The equation to solve is

$$(3x - 30) + 4x = 180.$$
$$7x - 30 = 180 \qquad \text{Combine like terms.}$$
$$7x - 30 + 30 = 180 + 30 \qquad \text{Add 30.}$$
$$7x = 210$$
$$x = 30 \qquad \text{Divide by 7.}$$

To find the measures of the angles, replace x with 30 in the two expressions.

$$3x - 30 = 3(30) - 30 = 90 - 30 = 60$$
$$4x = 4(30) = 120$$

The two angle measures are $60°$ and $120°$. ∎

4 ▶ Solve a formula for a specified variable.

▶ Sometimes it is necessary to solve a large number of problems that use the same formula. For example, a surveying class might need to solve several problems that involve the formula for the area of a rectangle, $A = LW$. Suppose that in each problem the area (A) and the length (L) of a rectangle are given and the width (W) must be found. Rather than solving for W each time the formula is used, it would be simpler to rewrite the *formula* so that it is solved for W. This process is called **solving for a specified variable.** As the following examples show, solving a formula for a specified variable requires the same steps used earlier to solve equations with just one variable.

The formula for converting temperatures given in degrees Celsius to degrees Fahrenheit is

$$F = \frac{9}{5}C + 32.$$

The next example shows how to solve this formula for C.

EXAMPLE 5
Solving for a Specified Variable

Solve $F = \dfrac{9}{5}C + 32$ for C.

First undo the addition of 32 to $(9/5)C$ by subtracting 32 from both sides.

$$F = \frac{9}{5}C + 32$$

$$F - 32 = \frac{9}{5}C + 32 - 32 \qquad \text{Subtract 32.}$$

$$F - 32 = \frac{9}{5}C$$

Now multiply both sides by 5/9. Use parentheses on the left.

$$\frac{5}{9}(F - 32) = \frac{5}{9} \cdot \frac{9}{5}C \qquad \text{Multiply by } \tfrac{5}{9}.$$

$$\frac{5}{9}(F - 32) = C$$

This last result is the formula for converting temperatures from Fahrenheit to Celsius. ■

NOTE When solving a formula for a specified variable, treat that variable as if it were the only variable in the equation, and treat all others as if they were constants. Use the method of solving equations described in Sections 2.2 and 2.3 to solve for the specified variable.

EXAMPLE 6
Solving for a Specified Variable

Solve $A = \dfrac{1}{2}(b + B)h$ for B.

This is the formula for the area of a trapezoid. Begin by multiplying both sides by 2.

$$A = \frac{1}{2}(b + B)h \qquad \text{Given formula}$$

$$2A = 2 \cdot \frac{1}{2}(b + B)h \qquad \text{Multiply by 2.}$$

$$2A = (b + B)h$$

$$2A = bh + Bh \qquad \text{Distributive property}$$

Undo what was done to B by first subtracting bh on both sides. Then divide both sides by h.

$$2A - bh = Bh \qquad\qquad\qquad \text{Subtract } bh.$$

$$\frac{2A - bh}{h} = B \qquad \text{or} \qquad B = \frac{2A - bh}{h} \qquad \text{Divide by } h.$$

The result can be written in a different form as follows.

$$B = \frac{2A - bh}{h} = \frac{2A}{h} - \frac{bh}{h} = \frac{2A}{h} - b$$

Either form is correct. ■

2.5 EXERCISES

1. In your own words, explain what is meant by the *perimeter* of a geometric figure.

2. In your own words, explain what is meant by the *area* of a geometric figure.

Decide whether perimeter or area would be used to solve a problem concerning the measure of the quantity.

3. carpeting for a bedroom
4. sod for a lawn
5. fencing for a yard
6. baseboards for a living room
7. tile for a bathroom
8. fertilizer for a garden
9. determining the cost for replacing a linoleum floor with a wood floor
10. determining the cost for planting rye grass in a lawn for the winter

In the following exercises a formula is given along with the values of all but one of the variables in the formula. Find the value of the variable that is not given. See Example 1.

11. $P = 2L + 2W$ (perimeter of a rectangle); $L = 6, W = 4$

12. $P = 2L + 2W$; $L = 8, W = 5$

13. $P = 4s$ (perimeter of a square); $s = 6$

14. $P = 4s$; $s = 12$

15. $A = \dfrac{1}{2}bh$ (area of a triangle); $b = 10, h = 14$

16. $A = \dfrac{1}{2}bh$; $b = 8, h = 16$

17. $P = a + b + c$ (perimeter of a triangle); $P = 15, a = 3, b = 7$

18. $P = a + b + c$; $P = 12, a = 3, c = 5$

19. $d = rt$ (distance formula); $d = 100, t = 2.5$

20. $d = rt$; $d = 252, r = 45$

21. $I = prt$ (simple interest); $p = 5000, r = .025, t = 7$

22. $I = prt$; $p = 7500, r = .035, t = 6$

23. $A = \dfrac{1}{2}h(b + B)$ (area of a trapezoid); $h = 7, b = 12, B = 14$

24. $A = \dfrac{1}{2}h(b + B)$; $h = 3, b = 19, B = 31$

25. $C = 2\pi r$ (circumference of a circle); $C = 8.164, \pi = 3.14*$

26. $C = 2\pi r$; $C = 16.328, \pi = 3.14$

27. $A = \pi r^2$ (area of a circle); $r = 12, \pi = 3.14$

28. $A = \pi r^2$; $r = 4, \pi = 3.14$

29. If a formula contains exactly five variables, how many values would you need to be given in order to find values for the remaining one?

30. The formula for changing Celsius to Fahrenheit is given in Example 5 as
$F = \dfrac{9}{5}C + 32$. Sometimes it is seen as $F = \dfrac{9C}{5} + 32$. These are both correct. Why is it true that $\dfrac{9}{5}C$ is equal to $\dfrac{9C}{5}$?

---

*Actually, π is approximately equal to 3.14, not *exactly* equal to 3.14.

*The **volume** of a three-dimensional object is a measure of the space occupied by the object. For example, we would need to know the volume of a gasoline tank in order to know how many gallons of gasoline it would take to completely fill the tank. In each of the following exercises, a formula for the volume, V, of a three-dimensional object is given, along with values for the other variables. Solve for V. See Example 1.*

31. $V = LWH$ (volume of a rectangular-sided box); $L = 12, W = 8, H = 4$

32. $V = LWH$; $L = 10, W = 5, H = 3$

33. $V = \frac{1}{3}Bh$ (volume of a pyramid); $B = 36, h = 4$

34. $V = \frac{1}{3}Bh$; $B = 12, h = 13$

35. $V = \frac{4}{3}\pi r^3$ (volume of a sphere); $r = 6, \pi = 3.14$

36. $V = \frac{4}{3}\pi r^3$; $r = 12, \pi = 3.14$

Use a formula to write an equation for the application, and use the problem-solving method of Section 2.4 to solve it. Formulas may be found in the endsheets of this book. See Examples 2 and 3.

37. A radio telescope at the Max Planck Institute in West Germany has a 328-foot diameter. What is the circumference of this telescope? (Use $\pi = 3.14$.)

38. The largest drum ever constructed was played at the Royal Festival Hall in London in 1987. It had a diameter of 13 feet. What was the area of the circular face of the drum? (Use $\pi = 3.14$.)

39. The newspaper *The Constellation*, printed in 1859 in New York City as part of the Fourth of July celebration, had length 51 inches and width 35 inches. What was the perimeter? What was the area?

40. The *Daily Banner*, published in Roseburg, Oregon, in the nineteenth century, had page size 3 inches by 3.5 inches. What was the perimeter? What was the area?

41. A color television set with a liquid crystal display was manufactured by Epson in 1985 and had dimensions 3 inches by $6\frac{3}{4}$ inches by $1\frac{1}{8}$ inches.   What was the volume of this set?

42. A sea lock at Zeebrugge, Belgium, measures 1640 feet by 187 feet by 75.4 feet. What is the volume of the lock? (It is a rectangular solid, so use the formula for the volume of such a figure.)

43. The survey plat shown in the figure shows two lots that form a figure called a trapezoid. The measures of the parallel sides are 115.80 feet and 171.00 feet. The height of the trapezoid is 165.97 feet. Find the combined area of the two lots. Round your answer to the nearest hundredth of a square foot.

44. Lot *A* in the figure is in the shape of a trapezoid. The parallel sides measure 26.84 feet and 82.05 feet. The height of the trapezoid is 165.97 feet. Find the area of Lot *A*. Round your answer to the nearest hundredth of a square foot.

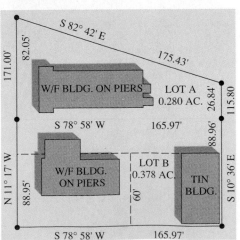

Find the measure of the marked angles. See Example 4.

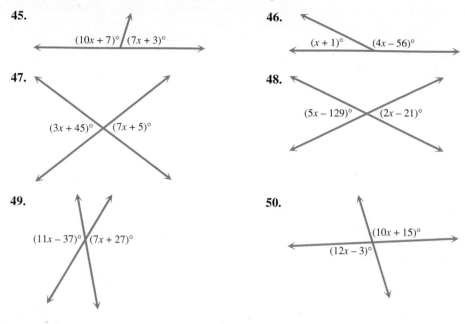

45.

$(10x + 7)°$ $(7x + 3)°$

46.

$(x + 1)°$ $(4x - 56)°$

47.

$(3x + 45)°$ $(7x + 5)°$

48.

$(5x - 129)°$ $(2x - 21)°$

49.

$(11x - 37)°$ $(7x + 27)°$

50.

$(10x + 15)°$ $(12x - 3)°$

Solve the formula for the specified variable. See Examples 5 and 6.

51. $A = LW$ for L

52. $d = rt$ for t

53. $d = rt$ for r

54. $I = prt$ for r

55. $I = prt$ for p

56. $V = LWH$ for H

57. $P = a + b + c$ for a

58. $P = a + b + c$ for b

59. $A = \dfrac{1}{2}bh$ for b

60. $A = \dfrac{1}{2}bh$ for h

61. $A = p + prt$ for r

62. $P = 2L + 2W$ for W

63. $V = \pi r^2 h$ for h

64. $V = \dfrac{1}{3}\pi r^2 h$ for h

65. $y = mx + b$ for m

66. $F = \dfrac{9}{5}C + 32$ for C

◇ **MATHEMATICAL CONNECTIONS** (Exercises 67–70) ◇

Students in beginning algebra are often faced with the following problem: "I know I did the work correctly, but my answer doesn't look like the one given in the book. What's wrong?" In many cases, there are equally acceptable *equivalent* answers for a problem. Work Exercises 67–70 in order to see how two seemingly different answers to a problem can both be correct.

67. Solve the formula $P = 2L + 2W$ for W using the following steps.
   (a) Subtract $2L$ from both sides.
   (b) Divide both sides by 2 to get $\dfrac{P - 2L}{2}$ as the expression for W.

68. Solve the formula $P = 2L + 2W$ for W using the following steps.
   (a) Divide each term on both sides by 2.
   (b) Subtract L from both sides to get $\dfrac{P}{2} - L$ as the expression for W.

69. Compare the results in Exercises 67(b) and 68(b). They are both acceptable answers, but they are written in different forms. To show they are equivalent, follow the steps below and provide the justification for each step.

(a) $\dfrac{P}{2} - L = \dfrac{P}{2} - \dfrac{2}{2} \cdot L$

(b) $\phantom{\dfrac{P}{2} - L} = \dfrac{P}{2} - \dfrac{2}{2} \cdot \dfrac{L}{1}$

(c) $\phantom{\dfrac{P}{2} - L} = \dfrac{P}{2} - \dfrac{2L}{2}$

(d) $\phantom{\dfrac{P}{2} - L} = \dfrac{P - 2L}{2}$

70. Suppose that the expression $\dfrac{5T + 4}{4}$ is obtained in a similar problem. What is another valid answer?

$$\longleftarrow\!\!\!\longrightarrow\;\diamond\;\longleftarrow\!\!\!\longrightarrow$$

REVIEW EXERCISES

Solve the equation. See Section 2.2.

71. $4x = 12$ **72.** $3x = \dfrac{1}{4}$ **73.** $\dfrac{3}{4}y = 21$ **74.** $.06x = 300$

2.6 RATIOS AND PROPORTIONS

FOR EXTRA HELP	OBJECTIVES
📖 **SSG** pp. 79–82 **SSM** pp. 76–82	**1 ▶** Write ratios.
📼 **Video** 3	**2 ▶** Decide whether proportions are true.
💾 **Tutorial** IBM MAC	**3 ▶** Solve proportions. **4 ▶** Solve applied problems using proportions. **5 ▶** Solve problems involving unit pricing.

■

PROBLEM SOLVING   An example of a type of problem that often occurs is given below.

A carpet cleaning service charges $45.00 to clean 2 similarly sized rooms of carpet. How much would it cost to clean 5 rooms of carpet?

Assuming that the cleaning service does not discount its prices for cleaning additional rooms after the first two, the reasoning for solving this problem might be as follows: if it costs $45.00 to clean 2 rooms, then it would cost $45.00/2 = $22.50 per room. So, the total cost for cleaning 5 rooms would be 5 × $22.50 = $112.50.

■

1 ▶ Write ratios.   ▶   The quotient $45.00/2 introduced above is an example of a ratio of price to number of rooms. Ratios provide a way of comparing two numbers or quantities. A **ratio** is a quotient of two quantities. The ratio of the number a to the number b is written as follows.

__Ratio__

$$a \text{ to } b, \qquad \frac{a}{b}, \qquad \text{or} \qquad a:b$$

◆ C O N N E C T I O N S ◆

When you look a long way down a straight road or railroad track, it seems to narrow as it vanishes in the distance. The point where the sides seem to touch is called the **vanishing point.** The same thing occurs in the lens of a camera, as shown in the figure. Suppose I represents the length of the image, O the length of the object, d the distance from the lens to the film, and D the distance from the lens to the object. Then

$$\frac{\text{Image length}}{\text{Object length}} = \frac{\text{Image distance}}{\text{Object distance}}$$

or

$$\frac{I}{O} = \frac{d}{D}.$$

Given the length of the image on the film and its distance from the lens, then the length of the object determines how far away the lens must be from the object to fit on the film.

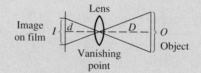

FOR DISCUSSION OR WRITING
How far from the lens should a child 1 meter tall be to fit on 35 mm film (35 mm by 35 mm) if the distance d in the ratio above is 7 mm? Many camera lenses show an infinity symbol, ∞. This symbol is usually used to represent a quantity that grows without bound. Find the definition given for the infinity symbol on a camera.

When ratios are used in comparing units of measure, the units should be the same. This is shown in Example 1.

__E X A M P L E 1__

Writing a Ratio

■ Write a ratio for each word phrase.

(a) The ratio of 5 hours to 3 hours

This ratio can be written as $\frac{5}{3}$.

(b) The ratio of 5 hours to 3 days

First convert 3 days to hours: 3 days $= \mathbf{3 \cdot 24} = 72$ hours. The ratio of 5 hours to 3 days is thus $5/72$. ■

2 ▶ Decide whether proportions are true.

▶ A ratio is used to compare two numbers or amounts. A **proportion** is a statement that two ratios are equal. For example,

$$\frac{3}{4} = \frac{15}{20}$$

is a proportion that says that the ratios 3/4 and 15/20 are equal. In the proportion

$$\frac{a}{b} = \frac{c}{d},$$

a, b, c, and d are the **terms** of the proportion. Beginning with the proportion

$$\frac{a}{b} = \frac{c}{d}$$

and multiplying both sides by the common denominator, bd, gives

$$bd \cdot \frac{a}{b} = bd \cdot \frac{c}{d}$$
$$ad = bc.$$

The products ad and bc can be found by multiplying diagonally.

This is called **cross multiplication,** and ad and bc are called **cross products.**

Cross Products

If $\frac{a}{b} = \frac{c}{d}$, then the cross products ad and bc are equal.

Also, if $ad = bc$, then $\frac{a}{b} = \frac{c}{d}$. $(b, d, \neq 0)$

From the rule given above,

$$\text{if} \quad \frac{a}{b} = \frac{c}{d} \quad \text{then} \quad ad = bc.$$

However, if $\frac{a}{c} = \frac{b}{d}$, then $ad = cb$, or $ad = bc$. This means that the two proportions are equivalent and

the proportion $\frac{a}{b} = \frac{c}{d}$ can also be written as $\frac{a}{c} = \frac{b}{d}$.

Sometimes one form is more convenient to work with than the other.

EXAMPLE 2
Deciding Whether a Proportion Is True

Decide whether the following proportions are true or false.

(a) $\frac{3}{4} = \frac{15}{20}$

Check to see whether the cross products are equal.

$$4 \cdot 15 = 60$$

$$\frac{3}{4} = \frac{15}{20}$$

$$3 \cdot 20 = 60$$

The cross products are equal, so the proportion is true.

(b) $\dfrac{6}{7} = \dfrac{30}{32}$

The cross products are $6 \cdot 32 = 192$ and $7 \cdot 30 = 210$. The cross products are different, so the proportion is false. ■

CAUTION The cross product method cannot be used directly if there is more than one term on either side. For example, you cannot use the method directly to solve the equation

$$\frac{4}{x} + 3 = \frac{1}{9},$$

because there are two terms on the left side.

---

3 ▶ Solve proportions.

▶ Four numbers are used in a proportion. If any three of these numbers are known, the fourth can be found.

E X A M P L E   3

Solving an Equation Using Cross Products

(a) Find x in the proportion

$$\frac{63}{x} = \frac{9}{5}.$$

The cross products must be equal, so

$$63 \cdot 5 = 9x$$

$$315 = 9x.$$

Divide both sides by 9 to get

$$35 = x.$$

(b) Solve the equation

$$\frac{m - 2}{5} = \frac{m + 1}{3}.$$

Find the cross products, and set them equal to each other.

$$3(m - 2) = 5(m + 1) \qquad \text{Be sure to use parentheses.}$$
$$3m - 6 = 5m + 5 \qquad \text{Distributive property}$$
$$3m = 5m + 11 \qquad \text{Add 6.}$$
$$-2m = 11 \qquad \text{Subtract } 5m.$$
$$m = -\frac{11}{2} \qquad \text{Divide by } -2. \quad ■$$

4 ▶ Solve applied problems using proportions.

▶ Proportions occur in many practical applications. For example, if we know the price of a number of similarly priced items, proportions can be used to determine the price of some other number of these items (assuming no discounts are given).

E X A M P L E   4
Applying Proportions ■

A local store is offering 3 packs of toothpicks for $.87. How much would it charge for 10 packs?

Let x = the cost of 10 packs of toothpicks.

Set up a proportion. One ratio in the proportion can involve the number of packs, and the other can involve the costs. Make sure that the corresponding numbers appear in the numerator and the denominator.

$$\frac{\text{Cost of 3}}{\text{Cost of 10}} = \frac{3}{10}$$

$$\frac{.87}{x} = \frac{3}{10}$$

$$3x = .87(10) \qquad \text{Cross products}$$

$$3x = 8.7$$

$$x = 2.90 \qquad \text{Divide by 3.}$$

The 10 packs should cost $2.90. As shown earlier, the proportion could also be written as $\dfrac{3}{.87} = \dfrac{10}{x}$, which would give the same cross products.   ■

NOTE   Many people would solve the problem in Example 4 mentally as follows: Three packs cost $.87, so one pack costs $.87/3 = $.29. Then ten packs will cost 10($.29) = $2.90. If you would do this problem this way, you would be using proportions and probably not even realizing it!

---

5 ▶ Solve problems involving unit pricing.

▶   We now look at the idea of *unit pricing*, an everyday example of the use of a ratio.

We sometimes see shoppers carrying handheld calculators to assist them in their job of budgeting while shopping in supermarkets. While the most common use is to make sure that the shopper does not go over budget, another use is to see which size of an item offered in different sizes produces the best price per unit. In order to do this, simply divide the price of the item by the unit of measure in which the item is measured. The next example illustrates this idea.

E X A M P L E   5
Determining Unit Price to Obtain the Best Buy ■

The local supermarket charges the following prices for a popular brand of pancake syrup:

Size	Price
36-ounce	$3.89
24-ounce	$2.79
12-ounce	$1.89.

Which size is the best buy? That is, which size has the lowest unit price?

To find the best buy, divide the price by the number of units to get the price per ounce. Each result in the following table was found by using a calculator and rounding the answer to three decimal places.

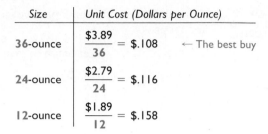

Size	Unit Cost (Dollars per Ounce)	
36-ounce	$\dfrac{\$3.89}{36} = \$.108$	← The best buy
24-ounce	$\dfrac{\$2.79}{24} = \$.116$	
12-ounce	$\dfrac{\$1.89}{12} = \$.158$	

Since the 36-ounce size produces the lowest price per unit, it would be the best buy. (Be careful: Sometimes the largest container *does not* produce the lowest price per unit.)   ■

2.6 EXERCISES

1. Which one of the following ratios is not the same as 3 to 4?
   (a) .75   (b) 6 to 8   (c) 4 to 3   (d) 30 to 40
2. Give three ratios that are equivalent to 3 to 1.

Determine the ratio and write it in lowest terms. See Example 1.

3. 40 miles to 30 miles
4. 60 feet to 70 feet
5. 120 people to 90 people
6. 72 dollars to 220 dollars
7. 20 yards to 8 feet
8. 30 inches to 8 feet
9. 24 minutes to 2 hours
10. 16 minutes to 1 hour
11. 8 days to 40 hours
12. 50 hours to 5 days

13. Explain the distinction between *ratio* and *proportion*.

14. Suppose that someone told you to use cross products in order to multiply fractions. How would you explain to the person what is wrong with his or her thinking?

Decide whether the proportion is true or false. See Example 2.

15. $\dfrac{5}{35} = \dfrac{8}{56}$
16. $\dfrac{4}{12} = \dfrac{7}{21}$
17. $\dfrac{120}{82} = \dfrac{7}{10}$

18. $\dfrac{27}{160} = \dfrac{18}{110}$
19. $\dfrac{\frac{1}{2}}{5} = \dfrac{1}{10}$
20. $\dfrac{\frac{1}{3}}{6} = \dfrac{1}{18}$

Solve the equation. See Example 3.

21. $\dfrac{k}{4} = \dfrac{175}{20}$
22. $\dfrac{49}{56} = \dfrac{z}{8}$
23. $\dfrac{x}{6} = \dfrac{18}{4}$

24. $\dfrac{z}{80} = \dfrac{20}{100}$
25. $\dfrac{3y - 2}{5} = \dfrac{6y - 5}{11}$
26. $\dfrac{2p + 7}{3} = \dfrac{p - 1}{4}$

27. $\dfrac{2r + 8}{4} = \dfrac{3r - 9}{3}$
28. $\dfrac{5k + 1}{6} = \dfrac{3k - 2}{3}$

Solve the problem by setting up and solving a proportion. See Example 4.

29. According to the Home and Garden Bulletin No. 72, four spears of asparagus contain 15 calories. How many spears of asparagus would contain 50 calories?

30. According to the source indicated in Exercise 29, three ounces of bluefish baked with butter or margarine provide 22 grams of protein. How many ounces would provide 242 grams of protein?

31. A chain saw requires a mixture of 2-cycle engine oil and gasoline. According to the directions on a bottle of Oregon 2-cycle Engine Oil, for a 50 to 1 ratio requirement, 2.5 fluid ounces of oil are required for 1 gallon of gasoline. If the tank of the chain saw holds 2.75 gallons, how many fluid ounces of oil are required?

32. The directions on the bottle mentioned in Exercise 31 indicate that if the ratio requirement is 24 to 1, 5.5 ounces of oil are required for 1 gallon of gasoline. If gasoline is to be mixed with 22 ounces of oil, how much gasoline is to be used?

33. In 1992, the average exchange rate between U.S. dollars and United Kingdom pounds was 1 pound to $1.7655. Margaret went to London and exchanged her U.S. currency for U.K. pounds, and received 400 pounds. How much in U.S. money did Margaret exchange?

34. If 3 U.S. dollars can be exchanged for 4.2186 Swiss francs, how many Swiss francs can be obtained for $49.20?

35. If 6 gallons of premium unleaded gasoline cost $3.72, how much would it cost to completely fill a 15-gallon tank?

36. If sales tax on a $16.00 compact disc is $1.32, how much would the sales tax be on a $120.00 compact disc player?

37. The distance between Kansas City, Missouri, and Denver is 600 miles. On a certain wall map, this is represented by a length of 2.4 feet. On the map, how many feet would there be between Memphis and Philadelphia, two cities that are actually 1000 miles apart?

38. The distance between Singapore and Tokyo is 3300 miles. On a certain wall map, this distance is represented by 11 inches. The actual distance between Mexico City and Cairo is 7700 miles. How far apart are they on the same map?

39. A recipe for green salad for 70 people calls for 18 heads of lettuce. How many heads of lettuce would be needed if 175 people were to be served?

40. A recipe for oatmeal macaroons calls for $1\frac{2}{3}$ cups of flour to make four dozen cookies. How many cups of flour would be needed for six dozen cookies?

41. A piece of property assessed at $42,000 requires an annual property tax of $273. How much property tax would be charged for a similar piece of property assessed at $52,000?

42. If 4 pounds of fertilizer will cover 50 square feet of garden, how many pounds would be needed for 225 square feet?

43. Biologists tagged 250 fish in Willow Lake on October 5. On a later date they found 7 tagged fish in a sample of 350. Estimate the total number of fish in Willow Lake to the nearest hundred.

44. On May 13 researchers at Argyle Lake tagged 420 fish. When they returned a few weeks later, their sample of 500 fish contained 9 that were tagged. Give an approximation of the fish population in Argyle Lake to the nearest hundred.

45. The pie chart indicates the number of people who responded in various ways to a survey. Suppose that 1680 people partipated. How many would we expect to answer in each way if 4200 people were surveyed?

46. According to McDonald's Corporation, Canada has 40,113 people for every McDonald's restaurant. The 1992 estimated population of Canada was 27,351,000. According to these figures, about how many McDonald's restaurants would we have expected Canada to have in 1992?

A supermarket was surveyed to find the prices charged for items in various sizes. Find the best buy (based on price per unit) for the particular item. See Example 5.

47. Trash bags
    20-count: $3.09
    30-count: $4.59

48. Black pepper
    1-ounce size: $ .99
    2-ounce size: $1.65
    4-ounce size: $4.39

49. Breakfast cereal
    15-ounce size: $2.99
    25-ounce size: $4.49
    31-ounce size: $5.49

50. Cocoa mix
    8-ounce size: $1.39
    16-ounce size: $2.19
    32-ounce size: $2.99

51. Tomato ketchup
    14-ounce size: $ .89
    32-ounce size: $1.19
    64-ounce size: $2.95

52. Cut green beans
    8-ounce size: $ .45
    16-ounce size: $ .49
    50-ounce size: $1.59

Two triangles are said to be **similar** if they have the same shape (but not necessarily the same size). Similar triangles have sides that are proportional. For example, the figure shows two similar triangles.

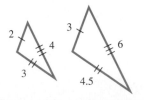

Notice that the ratios of the corresponding sides are all equal to $\frac{3}{2}$:

$$\frac{3}{2} = \frac{3}{2} \qquad \frac{4.5}{3} = \frac{3}{2} \qquad \frac{6}{4} = \frac{3}{2}.$$

If we know that two triangles are similar, we can set up a proportion to solve for the length of an unknown side. Use a proportion to find the length x, given that the pair of triangles are similar.

53.

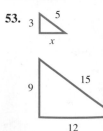

54.

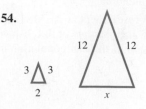

55.

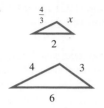

56. 3 △ 3   2 △ 2
      3   x

*For the problems in Exercises 57 and 58, (**a**) draw a sketch consisting of two right triangles, depicting the situation described, and (**b**) solve the problem.*

57. An enlarged version of the chair used by George Washington at the Constitutional Convention casts a shadow 18 feet long at the same time a vertical pole 12 feet high casts a shadow 4 feet long. How tall is the chair?

58 One of the tallest candles ever constructed was exhibited at the 1897 Stockholm Exhibition. If it cast a shadow 5 feet long at the same time a vertical pole 32 feet high cast a shadow 2 feet long, how tall was the candle?

The Consumer Price Index, issued by the U.S. Bureau of Labor Statistics, provides a means of determining the purchasing power of the U.S. dollar from one year to the next. Using the period from 1982 to 1984 as a measure of 100.0, the Consumer Price Index for electricity from 1975 to 1991 is shown here.

Year	Consumer Price Index	Year	Consumer Price Index
1975	50.0	1983	98.9
1976	53.1	1984	105.3
1977	56.6	1985	108.9
1978	60.9	1986	110.4
1979	65.6	1987	110.0
1980	75.8	1988	111.5
1981	87.2	1989	114.7
1982	95.8	1990	117.4
		1991	121.8

To use the Consumer Price Index, we can set up a proportion as follows:

$$\frac{\text{price of electricity in year } a}{\text{Consumer Price Index in year } a} = \frac{\text{price of electricity in year } b}{\text{Consumer Price Index in year } b}.$$

For example, if an electricity bill in 1985 was $150, we could predict the amount of the bill in 1989 by using the following proportion:

$$\begin{array}{l}\text{1985 price} \rightarrow \\ \text{1985 index} \rightarrow\end{array}\ \frac{150}{108.9} = \frac{x}{114.7}.\ \begin{array}{l}\leftarrow \text{1989 price}\\ \leftarrow \text{1989 index}\end{array}$$

By using cross products and a calculator, we can determine that the 1989 price would be approximately $158.

Use the Consumer Price Index figures above to find the amount that would be charged for the use of the same amount of electricity that cost $225 in 1980. Give your answer to the nearest dollar.

59. in 1983   **60.** in 1986   **61.** in 1988   **62.** in 1991

63. The Consumer Price Index figures for shelter for the years 1981 and 1991 are 90.5 and 146.3. If shelter for a particular family cost $3000 in 1981, what would be the comparable cost in 1991? Give your answer to the nearest dollar.

64. Due to a volatile fuel oil market in the early 1980s, the price of fuel decreased during the first three quarters of the decade. The Consumer Price Index figures for 1982 and 1986 were 105.0 and 74.1. If it cost you $21.50 to fill your tank with fuel oil in 1982, how much would it have cost to fill the same tank in 1986? Give your answer to the nearest cent.

———————◊ **MATHEMATICAL CONNECTIONS** (Exercises 65–68) ◊———————

In Section 2.3 we learned that to make the solution process easier, if an equation involves fractions, we can multiply both sides of the equation by the least common denominator of all the fractions in the equation. A proportion consists of two fractions equal to each other, so a proportion is a special case of this kind of equation. Work Exercises 65–68 in order to see how the process of solving by cross products is justified.

65. In the equation $\dfrac{x}{6} = \dfrac{2}{5}$, what is the least common denominator of the two fractions?

66. Solve the equation in Exercise 65 as follows:
    (a) Multiply both sides by the least common denominator. What is the equation that you obtain?
    (b) Solve for x by dividing both sides by the coefficient of x. What is the solution?

67. Solve the equation in Exercise 65 as follows:
    (a) Set the cross products equal. What is the equation you obtain?
    (b) Repeat part (b) of Exercise 66.

68. Compare your results from Exercises 66(a) and 67(a). What do you notice?

————————————◊————————————

REVIEW EXERCISES

Solve using the proper formula. See Section 2.5.

69. If an investment of $8000 earns $1280 in simple interest in 4 years, what rate of interest is being earned?

70. If $200 earned $75 in simple interest at 5%, for how many years was the money earning interest?

71. Seamus earned $5700 in simple interest on a deposit of $19,000 at 3%. For how long was the money deposited?

72. What is the monetary value of 34 quarters? (*Hint:* Monetary value = number of coins × denomination.)

Solve the equation. See Section 2.3.

73. $.15x + .30(3) = .20(3 + x)$

74. $.20(60) + .05x = .10(60 + x)$

75. $.92x + .98(12 - x) = .96(12)$

76. $.10(7) + 1.00x = .30(7 + x)$

2.7 APPLICATIONS OF PERCENT: MIXTURE, INTEREST, AND MONEY

FOR EXTRA HELP

📖 **SSG** pp. 83–88
SSM pp. 82–88

📼 **Video**
4

💾 **Tutorial**
IBM   MAC

OBJECTIVES

1 ▶ Learn how to use percent in problems involving rates.
2 ▶ Learn how to solve problems involving mixtures.
3 ▶ Learn how to solve problems involving simple interest.
4 ▶ Learn how to solve problems involving denominations of money.

◆ **C O N N E C T I O N S** ◆

Experiments done in England, using racing cyclists on stationary bicycles, show that the most efficient saddle height is 109% of a cyclist's inside-leg measurement. You can get the most mileage (or kilometrage) out of your leg work by following these directions.

1. Stand up straight, without shoes. Have someone measure your leg on the inside (from floor to crotch bone.)
2. Find the measure R that is 109% of this length.
3. Adjust your saddle so that the measure R equals the distance between the top of the saddle and the lower pedal spindle when the pedals are positioned as in the diagram.

FOR DISCUSSION OR WRITING

Write a proportion using 109% and the measure R referred to above. Solve the proportion for R. Find your measure R. Compare your results with those for some of your classmates. Is there much difference?

1 ▶ Learn how to use percent in problems involving rates.

▶ Recall that percent means "per hundred." Thus, percents are ratios where the second number is always 100. For example, 50% represents the ratio of 50 to 100 and 27% represents the ratio of 27 to 100.

PROBLEM SOLVING

Percents are often used in problems that involve concentrations or rates. In general, we multiply the rate by the total amount to get the percentage. (The percentage may be an amount of pure substance, or an amount of money, as seen in the examples in this section.) In order to prepare to solve mixture, investment, and money problems, the first example illustrates this basic idea.

E X A M P L E   1
Using Percent to Find a Percentage

(a) If a chemist has 40 liters of a 35% acid solution, then the amount of pure acid in the solution is

$$40 \quad \times \quad .35 \quad = \quad 14 \text{ liters.}$$

Amount of solution   Rate of concentration   Amount of pure acid

(b) If $1300 is invested for one year at 7% simple interest, the amount of interest earned in the year is

$$\$1300 \quad \times \quad .07 \quad = \quad \$91.$$

Principal   interest rate   Interest earned

(c) If a jar contains 37 quarters, the monetary amount of the coins is

$$37 \quad \times \quad \$.25 \quad = \quad \$9.25.$$

$\uparrow$ $\uparrow$ $\uparrow$

Number of coins   Denomination   Monetary value   ■

PROBLEM SOLVING   In the examples that follow, we will use *box diagrams* and charts to organize the information in the problems. Either method enables us to more easily set up the equation for the problem, which is usually the most difficult part of the problem-solving process. The six steps as described in Section 2.4 are used but are not specifically numbered.

---

2 ▶ Learn how to solve problems involving mixtures.

▶   In the next example, we use percent to solve a mixture problem.

E X A M P L E 2

Solving a Mixture Problem

A chemist needs to mix 20 liters of 40% acid solution with some 70% solution to get a mixture that is 50% acid. How many liters of the 70% solution should be used?

Let x = the number of liters of 70% solution that are needed.

Recall from part (a) of Example 1 that the amount of pure acid in this solution will be given by the product of the percent of strength and the number of liters of solution, or

liters of pure acid in x liters of 70% solution = $.70x$.

The amount of pure acid in the 20 liters of 40% solution is

liters of pure acid in the 40% solution = $.40(20) = 8$.

The new solution will contain $20 + x$ liters of 50% solution. The amount of pure acid in this solution is

liters of pure acid in the 50% solution = $.50(20 + x)$.

The given information can be summarized in the box diagram below.

Number of liters of solution	x	20		$20 + x$
		+	=	
Rate of concentration of acid	.70	.40		.50

The number of liters of pure acid in the 70% solution added to the number of liters of pure acid in the 40% solution will equal the number of liters of pure acid in the final mixture, so the equation is

Pure acid in 70%	plus	pure acid in 40%	is	pure acid in 50%.
$\downarrow$	$\downarrow$	$\downarrow$	$\downarrow$	$\downarrow$
$.70x$	$+$	$.40(20)$	$=$	$.50(20 + x)$.

Multiply by 100 to clear decimals.

$$70x + 40(20) = 50(20 + x)$$

Solve for x.

$$70x + 800 = 1000 + 50x \qquad \text{Distributive property}$$
$$20x + 800 = 1000 \qquad \text{Subtract } 50x.$$
$$20x = 200 \qquad \text{Subtract } 800.$$
$$x = 10 \qquad \text{Divide by } 20.$$

Check this solution to see that the chemist needs to use 10 liters of 70% solution. ■

3 ▶ Learn how to solve problems involving simple interest.

▶ The next example uses the formula for simple interest, $I = prt$. Remember that when $t = 1$, the formula becomes $I = pr$, and once again the idea of multiplying the total amount (principal) by the rate (rate of interest) gives the percentage (amount of interest).

NOTE In most real-life applications of interest, compound interest is used; that is, interest is paid on interest. Compound interest involves concepts that are not covered in this book, so our examples must be limited to investments that pay simple interest.

EXAMPLE 3

Solving a Simple Interest Problem

Elizabeth Thornton receives an inheritance. She plans to invest part of it at 9% and $2000 more than this amount at 10%. To earn $1150 per year in interest, how much should she invest at each rate?

Let $\qquad x =$ the amount invested at 9% (in dollars);

$x + 2000 =$ the amount invested at 10% (in dollars).

Use a chart to arrange the information given in the problem.

Amount Invested in Dollars	Rate of Interest	Interest for One Year
x	.09	.09x
$x + 2000$	.10	.10(x + 2000)

We multiply amount by rate to get the interest earned. Since the total interest is to be $1150, the equation is

Interest at 9% plus interest at 10% is total interest.

$$.09x + .10(x + 2000) = 1150.$$

Multiply by 100 to clear decimals.

$$9x + 10(x + 2000) = 115{,}000$$

Now solve for x.

$$9x + 10x + 20{,}000 = 115{,}000 \quad \text{Distributive property}$$
$$19x + 20{,}000 = 115{,}000 \quad \text{Combine terms.}$$
$$19x = 95{,}000 \quad \text{Subtract } 20{,}000.$$
$$x = 5000 \quad \text{Divide by } 19.$$

She should invest \$5000 at 9% and \$5000 + \$2000 = \$7000 at 10%. ■

NOTE Although decimals were cleared in Examples 2 and 3, the equations also can be solved without clearing decimals.

4 ▶ Learn how to solve problems involving denominations of money.

▶ The final example is a problem that can be solved using the same ideas as those in Examples 2 and 3. It deals with different denominations of money.

EXAMPLE 4
Solving a Problem About Money

A bank teller has 25 more five-dollar bills than ten-dollar bills. The total value of the money is \$200. How many of each denomination of bill does he have?
We must find the number of each denomination of bill that the teller has.

Let $x =$ the number of ten-dollar bills;

$x + 25 =$ the number of five-dollar bills.

The information given in the problem can once again be organized in a chart.

Number of Bills	Denomination in Dollars	Dollar Value
x	10	$10x$
$x + 25$	5	$5(x + 25)$

Multiplying the number of bills by the denomination gives the monetary value. The value of the tens added to the value of the fives must be \$200:

Value of fives	plus	value of tens	is	\$200.
↓	↓	↓	↓	↓
$5(x + 25)$	$+$	$10x$	$=$	$200.$

Solve this equation.

$$5x + 125 + 10x = 200 \quad \text{Distributive property}$$
$$15x + 125 = 200 \quad \text{Combine terms.}$$
$$15x = 75 \quad \text{Subtract } 125.$$
$$x = 5 \quad \text{Divide by } 15.$$

Since x represents the number of tens, the teller has 5 tens and $5 + 25 = 30$ fives. Check that the value of this money is $5(\$10) + 30(\$5) = \$200$. ■

CAUTION Most difficulties in solving problems of these types occur because students do not take enough time to read the problem carefully and organize the information given in the problem. By organizing the information in a box diagram or chart, the chances of solving the problem correctly are greatly increased.

2.7 EXERCISES

Solve the problem. See Example 1.

1. How much pure acid is in 250 milliliters of a 14% acid solution?

2. How much pure alcohol is in 150 liters of a 30% alcohol solution?

3. If $10,000 is invested for one year at 3.5% simple interest, how much interest is earned?

4. If $25,000 is invested at 3% simple interest for 2 years, how much interest is earned?

5. What is the monetary amount of 283 nickels?

6. What is the monetary amount of 35 half-dollars?

Use your knowledge of percent to solve the problem involving real-life data.

7. According to a Knight-Ridder Newspapers report, as of May 31, 1992, the nation's "consumer-debt burden" was 16.4%. This means that the average American had consumer debts, such as credit card bills, auto loans, and so on, totaling 16.4% of his or her take-home pay. Suppose that Paul Eldersveld has a take-home pay of $3250 per month. What is 16.4% of his monthly take-home pay?

8. In 1992 General Motors announced that it would raise prices on its 1993 vehicles by an average of 1.6%. If a certain vehicle had a 1992 price of $10,526 and this price was raised 1.6%, what would the 1993 price be?

9. The 1916 dime minted in Denver is quite rare. The 1979 edition of *A Guide Book of United States Coins* listed its value in extremely fine condition as $625.00. The 1991 value had increased to $1750. What was the percent increase in the value of this coin?

10. In 1963, the value of a 1903 Morgan dollar minted in New Orleans in uncirculated condition was $1500. Due to a discovery of a large hoard of these dollars late that year, the value plummeted. Its value as listed in the 1991 edition of *A Guide Book of United States Coins* was $200. What percent of its 1963 value was its 1991 value?

11. At the 1992 Olympic Games in Barcelona, 25% of the medals won by Great Britain were gold medals. The team won a total of 20 medals. How many of the medals were *not* gold?

12. Florence Griffith Joyner has 2 Olympic medals that are not gold medals. These represent 40% of her medals. How many gold medals does she have?

13. The pie chart shows the breakdown, by approximate percents, of age groups buying aerobic shoes in 1990. According to figures of the National Sporting Goods Association, $582,000,000 was spent on aerobic shoes that year. How much was spent by each age group?
    (a) Under 14   (b) 18 to 24
    (c) 45 to 64

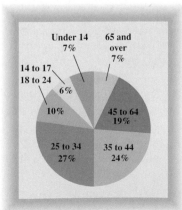

14. (See Exercise 13.) The pie chart shows the breakdown for the year 1990 regarding the sales and purchases of exercise equipment. If $2,295,000,000 was spent that year, determine the amount spent by each age group.
(a) 25 to 34   **(b)** 35 to 44
(c) 65 and over

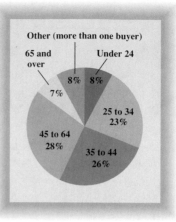

15. Express the amount of alcohol in *r* liters of pure water.

16. Express the amount of alcohol in *k* liters of pure alcohol.

Work the mixture problem. See Example 2.

17. How many gallons of 50% antifreeze must be mixed with 80 gallons of 20% antifreeze to get a mixture that is 40% antifreeze?

18. How many liters of 25% acid solution must be added to 80 liters of 40% solution to get a solution that is 30% acid?

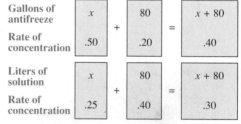

19. A certain metal is 20% tin. How many kilograms of this metal must be mixed with 80 kilograms of a metal that is 70% tin to get a metal that is 50% tin?

20. Ink worth $100 per barrel will be mixed with 30 barrels of ink worth $60 per barrel to get a mixture worth $75 per barrel. How many barrels of $100 ink should be used?

21. How many gallons of milk that is 2% butterfat must be mixed with milk that is 3.5% butterfat to get 10 gallons of milk that is 3% butterfat? (*Hint:* Let *x* represent the number of gallons that are 2% butterfat. Then $10 - x$ represents the number of gallons that are 3.5% butterfat.)

22. A pharmacist has 20 liters of a 10% drug solution. How many liters of 5% solution must be added to get a mixture that is 8%?

23. How many liters of a 60% acid solution must be mixed with a 75% acid solution to get 20 liters of a 72% solution?

24. How many gallons of a 12% indicator solution must be mixed with a 20% indicator solution to get 10 gallons of a 14% solution?

25. Minoxidil is a drug that has recently proven to be effective in treating male pattern baldness. A pharmacist wishes to mix a solution that is 2% minoxidil. She has on hand 50 milliliters of a 1% solution, and she wishes to add some 4% solution to it to obtain the desired 2% solution. How much 4% solution should she add?

26. Water must be added to 20 milliliters of a 4% minoxidil solution to dilute it to a 2% solution. How many milliliters of water should be used?

Use the concepts of this section to answer the question.

27. Suppose that a chemist is mixing two acid solutions, one of 20% concentration and the other of 30% concentration. Which one of the following concentrations could *not* be obtained?
  (a) 22%   **(b)** 24%   **(c)** 28%   **(d)** 32%

28. Suppose that pure alcohol is added to a 24% alcohol mixture. Which one of the following concentrations could *not* be obtained?
  (a) 22%   **(b)** 26%   **(c)** 28%   **(d)** 30%

Work the investment problem. In each case, assume that simple interest is being paid. See Example 3.

29. Li Nguyen invested some money at 3% and $4000 less than that amount at 5%. The two investments produced a total of $200 interest in one year. How much was invested at each rate?

30. LaShondra Williams inherited some money from her uncle. She deposited part of the money in a savings account paying 2%, and $3000 more than that amount in a different account paying 3%. Her annual interest income was $690. How much did she deposit at each rate?

31. Fran Liberto sold a painting that she found at a garage sale to an art dealer. With the $12,000 she got, she invested part at 4% in a certificate of deposit and the rest at 5% in a municipal bond. Her total annual income from the two investments was $515. How much did she invest at each rate? (*Hint:* Let x represent the amount invested at 4%. Then $12,000 - x$ represents the amount invested at 5%.)

32. Two investments produce an annual interest income of $114. The total amount of money invested is $4000, and the two interest rates paid are 2.5% and 3.5%. How much is invested at each rate?

33. With income earned by selling the rights to his life story, an actor invests some of the money at 3% and $30,000 more than twice as much at 4%. The total annual interest earned from the investments is $5600. How much is invested at each rate?

34. An artist invests her earnings in two ways. Some goes into a tax-free bond paying 6%, and $6000 more than three times as much goes into mutual funds paying 5%. Her total annual interest income from the investments is $825. How much does she invest at each rate?

Work the problem involving different monetary rates. See Example 4.

35. A bank teller has some five-dollar bills and some twenty-dollar bills. The teller has 5 more twenties than fives. The total value of the money is $725. Find the number of five-dollar bills that the teller has.

36. A coin collector has $1.70 in dimes and nickels. She has 2 more dimes than nickels. How many nickels does she have?

37. A cashier has a total of 126 bills, made up of fives and tens. The total value of the money is $840. How many of each kind does he have?

38. A convention manager finds that she has $1290, made up of twenties and fifties. She has a total of 42 bills. How many of each kind does she have?

39. For a retirement party, a person buys some 32¢ favors and some 50¢ favors, paying $46 in total. If she buys 10 more of the 50¢ favors, how many of the 32¢ favors were bought?

40. A stamp collector buys some 32¢ stamps and some 29¢ stamps, paying $7.90 for them. He buys 2 more 29¢ stamps than 32¢ stamps. How many 32¢ stamps does he buy?

41. A merchant wishes to mix candy worth $5 per pound with 40 pounds of candy worth $2 per pound to get a mixture that can be sold for $3 per pound. How many pounds of $5 candy should be used?

42. At Vern's Grill, hamburgers cost 90 cents each, and a bag of french fries costs 40 cents. How many hamburgers and how many bags of french fries can a customer buy with $8.80 if he wants twice as many hamburgers as bags of french fries?

Use the concepts of this section to respond.

43. A teacher once commented that the method of solving problems of the type found in this section could be interpreted as "stuff plus stuff equals stuff." Refer to Examples 2, 3, and 4, and determine exactly what the "stuff" is in each problem.

44. Read Example 2. Can a problem of this type have a fraction as an answer? Now read Example 4. Can a problem of this type have a fraction as an answer? Explain.

━━━━━━━━◈ **MATHEMATICAL CONNECTIONS** (Exercises 45–48) ◈━━━━━━━━

Sometimes applied problems that may seem different actually apply the same concepts. Work Exercises 45–48 in order to see how this happens.

45. Consider the following problem: A hoard of coins consists of only nickels and dimes. There are 3400 coins, and the value of the money is $290. How many of each denomination are there?
   (a) Write an equation you would use to solve the problem. Let x represent the number of nickels in the hoard.
   (b) Solve the problem.

46. Consider the following problem: An investor deposits $3400 in two accounts. One account is a passbook savings account that pays 5% interest, and the other is a money market account that pays 10% interest. After one year, the total interest earned is $290. How much did she invest at each rate?
   (a) Write an equation you would use to solve the problem. Let x represent the amount invested in the passbook savings account.
   (b) Solve the problem.

47. Compare the equations you wrote in Exercises 45(a) and 46(a). What do you notice about them?

48. If, in either of the problems in Exercises 45 and 46, you let x represent the other unknown quantity, will you get the same *solution* to the equation? Will you get the same *answers* to the problem?

49. How do you think the pie charts in Exercises 13 and 14 were constructed so that the sections represented the appropriate percentages accurately?

50. Look up the origins of dividing a circle into 360 degrees and discuss the advantages of this choice.

REVIEW EXERCISES

Solve the formula for the indicated variable. See Section 2.5.

51. $d = rt$;   for t

52. $P = a + b + c$;   for c

53. $A = \dfrac{1}{2}bh$;   for b

54. $180 = A + B + C$;   for B

2.8 MORE ABOUT PROBLEM SOLVING

FOR EXTRA HELP

📖 **SSG** pp. 88–92
SSM pp. 88–93

📼 **Video**
4

💾 **Tutorial**
IBM   MAC

OBJECTIVES

1 ▶ Use the formula $d = rt$ to solve problems.
2 ▶ Solve problems involving distance, rate, and time.
3 ▶ Solve problems about geometric figures.

◆ **CONNECTIONS** ◆

The winner of the first Indianapolis 500 race (in 1911) was Ray Harroun driving a Marmon Wasp at an average speed of 74.59 miles per hour. To find his time we need the formula giving the relationship between distance, rate, and time. This formula is used frequently in everyday life. In this section we look at some applications of the distance, rate, and time relationship.

FOR DISCUSSION OR WRITING
Use the formula $d = rt$ to find Harroun's time in the 1911 race. In Section 2.5 we showed two ways to use a formula to solve this type of problem. Explain the two ways and show how this problem can be solved for each method.

1 ▶ **Use the formula $d = rt$ to solve problems.**

▶ If an automobile travels at an average rate of 50 miles per hour for two hours, then it travels $50 \times 2 = 100$ miles. This is an example of the basic relationship between distance, rate, and time:

$$\text{distance} = \text{rate} \times \text{time}.$$

This relationship is given by the formula $d = rt$. By solving, in turn, for r and t in the formula, we obtain two other equivalent forms of the formula. The three forms are given below.

Distance, Rate, Time Relationship

$$d = rt \qquad r = \frac{d}{t} \qquad t = \frac{d}{r}$$

The first example illustrates the uses of these formulas.

EXAMPLE 1
Finding Distance, Rate, or Time

(a) The speed of sound is 1088 feet per second at sea level at 32°F. In 5 seconds under these conditions, sound travels

$$1088 \quad \times \quad 5 \quad = \quad 5440 \text{ feet.}$$
$$\uparrow \qquad\qquad \uparrow \qquad\qquad \uparrow$$
$$\text{Rate} \quad \times \quad \text{Time} \quad = \quad \text{Distance}$$

Here, we found distance given rate and time, using $d = rt$.

(b) Over a short distance, an elephant can travel at a rate of 25 miles per hour. In order to travel 1/4 mile, it would take an elephant

$$\text{Distance} \rightarrow \quad \frac{\frac{1}{4}}{25} = \frac{1}{4} \times \frac{1}{25} = \frac{1}{100} \text{ hour.} \leftarrow \text{Time}$$
$$\text{Rate} \rightarrow$$

Here, we find time given rate and distance, using $t = d/r$. To convert $1/100$ hour to minutes, multiply $1/100$ by 60 to get $60/100$ or $3/5$ minute. To convert $3/5$ minute to seconds, multiply $3/5$ by 60 to get 36 seconds.

(c) In the 1992 Olympic Games in Barcelona, Gwen Torrance of the United States won the women's 200-meter track event in 21.81 seconds. Her rate was

$$\text{Distance} \rightarrow \frac{200}{21.81} = 9.17 \text{ (rounded) meters per second.} \leftarrow \text{Rate}$$
$$\text{Time} \rightarrow$$

This answer was obtained using a calculator. Here, we found rate given distance and time, using $r = d/t$. ■

2 ▶ Solve problems involving distance, rate, and time.

▶ Many applied problems use the formulas just discussed.

PROBLEM SOLVING

The next example shows how to solve a typical application of the formula $d = rt$. A strategy for solving such problems involves two major steps:

Solving Motion Problems

Step 1 Make a sketch showing what is happening in the problem.
Step 2 Make a chart using the information given in the problem, along with the unknown quantities.

The chart will help you organize the information, and the sketch will help you set up the equation.

EXAMPLE 2
Solving a Motion Problem

Two cars leave Baton Rouge, Louisiana, at the same time and travel east on Interstate 12. One travels at a constant speed of 55 miles per hour and the other travels at a constant speed of 63 miles per hour. In how many hours will the distance between them be 24 miles?

Since we are looking for time,

let $t = $ the number of hours until the distance between them is 24 miles.

The sketch in Figure 9 shows what is happening in the problem. Now, construct a chart like the one below. Fill in the information given in the problem, and use t for the time traveled by each car. Multiply rate by time to get the expressions for distances traveled.

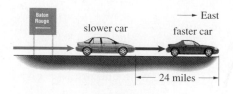

FIGURE 9

	Rate	×	Time	=	Distance	
Faster car	63		t		63t	⎫ Difference is 24 miles.
Slower car	55		t		55t	⎭

The quantities $63t$ and $55t$ represent the different distances. Refer to Figure 9 and notice that the *difference* between the larger distance and the smaller distance is 24 miles. Now write the equation and solve it.

$$63t - 55t = 24$$
$$8t = 24 \qquad \text{Combine terms.}$$
$$t = 3 \qquad \text{Divide by 8.}$$

After 3 hours the faster car will have traveled $63 \times 3 = 189$ miles, and the slower car will have traveled $55 \times 3 = 165$ miles. Since $189 - 165 = 24$, the conditions of the problem are satisfied. It will take 3 hours for the distance between them to be 24 miles. ■

NOTE In motion problems like the one in Example 2, once you have filled in two pieces of information in each row of the chart, you should automatically fill in the third piece of information, using the appropriate form of the formula relating distance, rate, and time. Set up the equation based on your sketch and the information in the chart.

---

3 ▶ Solve problems about geometric figures.

▶ In Section 2.5 we saw some applications of geometric formulas. Example 3 shows another such application.

PROBLEM SOLVING Remember that a sketch is very helpful when solving geometric applications.

EXAMPLE 3
Finding the Length and the Width of a Room

A couple wishes to add a laundry room onto their house. Due to construction limitations the length of the room must be 2 feet more than the width. Find the length and the width of the room if the perimeter is 40 feet.

Start by drawing a rectangle to represent the floor of the room. See Figure 10.

Let x = width of the room in feet;

$x + 2$ = length of the room in feet.

The formula for the perimeter of a rectangle is $P = 2L + 2W$. Substitute x for W, $x + 2$ for L, and 40 for P.

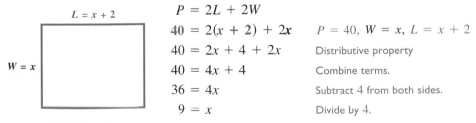

$$P = 2L + 2W$$
$$40 = 2(x + 2) + 2x \qquad P = 40,\ W = x,\ L = x + 2$$
$$40 = 2x + 4 + 2x \qquad \text{Distributive property}$$
$$40 = 4x + 4 \qquad \text{Combine terms.}$$
$$36 = 4x \qquad \text{Subtract 4 from both sides.}$$
$$9 = x \qquad \text{Divide by 4.}$$

FIGURE 10

The width of the room is 9 feet, and the length is $9 + 2 = 11$ feet. Check these answers in the formula for the perimeter. Since $2(11) + 2(9) = 40$, the answers are correct. ■

2.8 EXERCISES

🖩 *In Exercises 1–4, find the time based on the information provided. Use a calculator and round your answer to the nearest thousandth. See Example 1.*

Event and Year	Participant	Distance	Rate
1. Indianapolis 500, 1992	Al Unser, Jr. (Galmer-Chevrolet)	500 miles	134.479 mph
2. Daytona 500, 1992	Davey Allison (Ford)	500 miles	160.256 mph
3. Indianapolis 500, 1980	Johnny Rutherford (Hy-Gain McLaren/ Goodyear)	255 miles (rain-shortened)	148.725 mph
4. Indianapolis 500, 1975	Bobby Unser (Jorgensen Eagle)	435 miles (rain-shortened)	149.213 mph

🖩 *In Exercises 5–8, find the rate based on the information provided. Use a calculator and round your answer to the nearest hundredth.*

Event and Year	Participitant	Distance	Time
5. Summer Olympics 400-meter Hurdles, Women, 1992	Sally Gunnell (Great Britain)	400 meters	53.23 seconds
6. Summer Olympics, 100-meter Dash, Women, 1992	Gail Devers (United States)	100 meters	10.82 seconds
7. Summer Olympics, 100-meter Dash, Men, 1988	Carl Lewis (United States)	100 meters	9.92 seconds
8. Winter Olympics, 500-meter Speed Skating, Women, 1992	Bonnie Blair (United States)	500 meters	40.33 seconds

9. A driver averaged 53 miles per hour and took 10 hours to travel from Memphis to Chicago. What is the distance between Memphis and Chicago?

10. A small plane traveled from Warsaw to Rome, averaging 164 miles per hour. The trip took 2 hours. What is the distance from Warsaw to Rome?

11. Suppose that an automobile averages 45 miles per hour, and travels for 30 minutes. Is the distance traveled $45 \times 30 = 1350$ miles? If not, explain why not, and give the correct distance.

12. Which of the following choices is the best *estimate* for the average speed of a trip of 405 miles that lasted 8.2 hours?
(a) 50 miles per hour   **(b)** 30 miles per hour
(c) 60 miles per hour   **(d)** 40 miles per hour

Solve the problem. See Example 2.

13. St. Louis and Portland are 2060 miles apart. A small plane leaves Portland, traveling toward St. Louis at an average speed of 90 miles per hour. Another plane leaves St. Louis at the same time, traveling toward Portland, averaging 116 miles per hour. How long will it take them to meet?

	r	t	d
Plane leaving Portland	90	t	90t
Plane leaving St. Louis	116	t	116t

14. Atlanta and Cincinnati are 440 miles apart. John leaves Cincinnati, driving toward Atlanta at an average speed of 60 miles per hour. Pat leaves Atlanta at the same time, driving toward Cincinnati in her antique auto, averaging 28 miles per hour. How long will it take them to meet?

	r	t	d
John	60	t	60t
Pat	28	t	28t

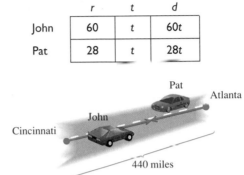

15. Two trains leave a city at the same time. One travels north at 60 miles per hour and the other travels south at 80 miles per hour. In how many hours will they be 280 miles apart?

	r	t	d
Northbound	60	t	
Southbound	80	t	

16. Two planes leave an airport at the same time. One flies east at 300 miles per hour and the other flies west at 450 miles per hour. In how many hours will they be 2250 miles apart?

	r	t	d
Eastbound	300	t	
Westbound	450	t	

17. At a given hour two steamboats leave a city in the same direction on a straight canal. One travels at 18 miles per hour and the other travels at 25 miles per hour. In how many hours will the boats be 35 miles apart?

18. From a point on a straight road, Lupe and Maria ride bicycles in opposite directions. Lupe rides 10 miles per hour and Maria rides 12 miles per hour. In how many hours will they be 55 miles apart?

Work the geometry problem. See Example 3.

19. The largest poster ever constructed was made by the citizens of Obihiri, Hokkaido, Japan. It was in the shape of a square, and its perimeter was 1312 feet. What was the length of a side of the square poster?

20. The largest known map is a relief map of California. It is rectangular in shape, and its length is 25 times its width. The perimeter is 936 feet. Find the length and the width of the map.

21. The largest mosque in use is in Syria. It is rectangular in shape and its length is 121 feet less than twice its width. Its perimeter is 1666 feet. What are the length and the width of the mosque?

22. The perimeter of a rectangle is 16 times its width. The length is 12 centimeters more than the width. Find the width of the rectangle.

23. A lot is in the shape of a triangle. One side is 100 feet longer than the shortest side, while the third side is 200 feet longer than the shortest side. The perimeter of the lot is 1200 feet. Find the lengths of the sides of the lot.

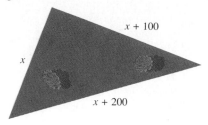

24. A video rental establishment displayed a rectangular cardboard standup advertisement for the movie *The Lion King*. The length was 20 inches more than the width, and the perimeter was 176 inches. What were the dimensions of the rectangle?

25. If the radius of a certain circle is tripled, with 8.2 centimeters then added, the result is the circumference of the circle. Find the radius of the circle. (Use 3.14 as an approximation for π.)

26. The Peachtree Plaza Hotel in Atlanta is in the shape of a cylinder, with a circular foundation. The circumference of the foundation is 6 times the radius, increased by 12.88 feet. Find the radius of the circular foundation. (Use 3.14 as an approximation for π.)

The remaining applications in this exercise set are not organized by type of problem. Use the problem-solving techniques described in this chapter to solve them.

27. According to a survey by Information Resources, Inc., single men spent an average of $10.87 more than single women when buying frozen dinners during the 52 weeks that ended August 22, 1993. Together, a single man and a single woman would spend a total of $92.29, according to the survey. What were the average expenditures for a single man and for a single woman?

28. Team Marketing Report, a sports-business newsletter, computes a Fan Cost Index for a trip to a major league baseball park. The index consists of the cost of four average-priced tickets, two small beers, four small sodas, four hot dogs, parking for one car, two game programs, and two twill baseball caps. For the 1994 season, the New York Yankees had the highest Fan Cost Index, while the Cincinnati Reds had the lowest. The Yankees' index was $43.37 less than twice that of the Reds. What was the Fan Cost Index for each of these teams if the *total* of the two was $194.56?

29. The pie chart shows the approximate percents for different sizes of Health Maintenance Organizations (HMOs) for the year 1991. During that year there were approximately 560 such organizations. Find the approximate number for each size group.

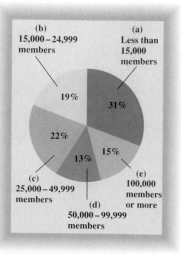

30. At the start of play on June 2, 1994, the standings of the Central Division of the American League were as shown. "Winning percentage" is commonly expressed as a decimal rounded to the nearest thousandth. To find the winning percentage of a team, divide the number of wins by the total number of games played. Find the winning percentage of each of the following teams.
(a) Cleveland   (b) Chicago
(c) Milwaukee

	Won	Lost
Chicago	30	19
Cleveland	27	21
Minnesota	26	24
Kansas City	25	25
Milwaukee	21	30

31. Find the measure of each marked angle.

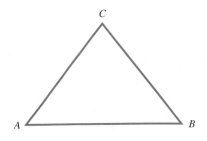

$(3x + 46)°$ $(2x + 29)°$

32. Find the measure of each marked angle.

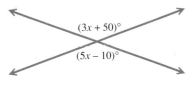

$(3x + 50)°$

$(5x - 10)°$

33. Which size is the best buy for laundry detergent based on unit price?
23-ounce size: $3.07
70-ounce size: $7.09
198-ounce size: $16.47

34. Tulsa and Toledo are 850 miles apart. On a certain map this distance is represented by 14 inches. Houston and Kansas City are 710 miles apart. How far apart are they on the same map? Round your answer to the nearest tenth of an inch.

35. The sum of the measures of the angles of any triangle is 180 degrees. In triangle ABC, angles A and B have the same measure, while the measure of angle C is 24 degrees larger than each of A and B. What are the measures of the three angles?

36. (See Exercise 35.) In triangle ABC, the measure of angle A is 30 degrees more than the measure of angle B. The measure of angle B is the same as the measure of angle C. Find the measure of each angle.

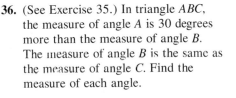

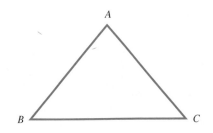

37. In an automobile race, a driver was 240 miles from the finish line after 3 hours. Another driver, who was in a later race, traveled at the same speed as the first driver. After 2.5 hours the second driver was 300 miles from the finish. Find the speed of each driver.

38. Two cars are 400 miles apart. Both start at the same time and travel toward one another. They meet 4 hours later. If the speed of one car is 20 miles per hour faster than the other, what is the speed of each car?

39. A merchant has 80 pounds of candy worth $1.50 per pound. She wishes to upgrade the candy to sell for $3.00 per pound by mixing it with candy worth $4.00 per pound. How many pounds of the $4.00 candy should she use?

40. In the 1992 U.S. presidential election, 1112 votes were cast for Clinton, Bush, and Perot in Sherman County, Oregon. Bush received 62 more votes than Clinton, and Clinton received 36 more votes than Perot. How many votes did each candidate receive?

◇ **MATHEMATICAL CONNECTIONS** (Exercises 41–44) ◇

From our study of angles in Section 2.5 and the fact that the sum of the angles of a triangle must be 180°, we can see how the value of x can be determined in the figure shown in several different ways. Work Exercises 41–44 in order to see how this can be done.

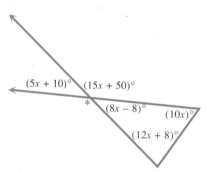

41. Write an equation based on the fact that the sum of the angles of the triangle in the figure is 180°. Solve for x, and then find the measures of the three angles.

42. Based on your results in Exercise 41 and the fact that vertical angles have the same measure, write a different equation that will allow you to solve for x. Does the value of x correspond to the one you found in Exercise 41?

43. Write two equations that are based on the fact that two angles that form a straight angle must have a sum of 180°. Solve each equation. Is the value of x the same as it was in Exercises 41 and 42?

44. Suppose the angle marked * was labeled $10x + 30$. Based on the value of x you have determined, would this be consistent? Explain.

◆

REVIEW EXERCISES

Place < *or* > *in the blank to make a true statement. See Section 1.2.*

45. −7 _____ 3   **46.** −9 _____ 6   **47.** −11 _____ −4   **48.** −10 _____ −12

2.9 THE ADDITION AND MULTIPLICATION PROPERTIES OF INEQUALITY

FOR EXTRA HELP

📖 **SSG** pp. 92–101
SSM pp. 93–99

📼 Video
4

💾 Tutorial
IBM   MAC

OBJECTIVES
1 ▶ Graph intervals on a number line.
2 ▶ Use the addition property of inequality.
3 ▶ Use the multiplication property of inequality.
4 ▶ Solve linear inequalities.
5 ▶ Solve applied problems by using inequalities.
6 ▶ Solve three-part inequalities.

The solution of inequalities is closely related to the methods of equation solving. Just as we will investigate many types of equations, we will do the same with inequalities. In this section we introduce properties that are essential for solving inequalities.

◆ C O N N E C T I O N S ◆

Many mathematical models involve inequalities rather than equations. This is often the case in economics. For example, a company that produces videocassettes has found that revenue from the sales of the cassettes is $5 per cassette less sales costs of $100. Production costs are $125 plus $4 per cassette. Profit (P) is given by revenue (R) less cost (C), so the company must find the production level x that makes

$$P = R - C > 0.$$

FOR DISCUSSION OR WRITING

Write an expression for revenue using the fact that x represents the production level (number of cassettes to be produced). Write an expression for production costs using x. Write an expression for profit and solve the inequality shown above. Describe the solution in terms of the problem.

Inequalities are statements with algebraic expressions related by

$<$ "is less than"

$\le$ "is less than or equal to"

$>$ "is greater than"

$\ge$ "is greater than or equal to."

We solve an inequality by finding all real number solutions for it. For example, the solution of $x \le 2$ includes all real numbers that are less than or equal to 2, and not just the integers less than or equal to 2. For example, -2.5, -1.7, -1, $\frac{7}{4}$, $\frac{1}{2}$, $\sqrt{2}$, and 2 are all real numbers less than or equal to 2, and are therefore solutions of $x \le 2$.

1 ▶ Graph intervals on a number line.

▶ A good way to show the solution of an inequality is by graphing. We graph all the real numbers satisfying $x \le 2$ by placing a dot at 2 on a number line and drawing an arrow extending from the dot to the left (to represent the fact that all numbers less than 2 are also part of the graph). The graph is shown in Figure 11.

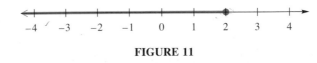

FIGURE 11

E X A M P L E   1

Graphing an Interval on a Number Line

■ Graph $x > -5$.

The statement $x > -5$ says that x can represent any number greater than -5, but x cannot equal -5 itself. Show this on a graph by placing an open circle at -5 and drawing an arrow to the right, as in Figure 12. The open circle at -5 shows that -5 is not part of the graph. ■

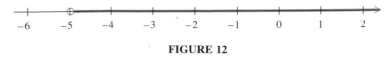

FIGURE 12

NOTE Some texts use parentheses rather than open circles and square brackets rather than closed circles (dots). We have chosen to use open and closed circles throughout this text.

2 ▶ Use the addition property of inequality.

▶ Inequalities such as $x + 4 \leq 9$ can be solved in much the same way as equations. Consider the inequality $2 < 5$. If 4 is added to both sides of this inequality, the result is

$$2 + 4 < 5 + 4$$
$$6 < 9,$$

a true sentence. Now subtract 8 from both sides:

$$2 - 8 < 5 - 8$$
$$-6 < -3.$$

The result is again a true sentence. These examples suggest the following **addition property of inequality,** which states that the same real number can be added to both sides of an inequality without changing the solutions.

Addition Property of Inequality	For any algebraic expressions A, B, and C that represent real numbers, the inequalities

$$A < B \quad \text{and} \quad A + C < B + C$$

have exactly the same solutions. In words, the same expression may be added to both sides of an inequality without changing the solutions.

The addition property of inequality also works with $>$, $\leq$, or $\geq$. Just as with the addition property of equality, the same expression may also be *subtracted* from both sides of an inequality.

The following examples show how the addition property is used to solve inequalities.

EXAMPLE 2

Using the Addition Property of Inequality

■ Solve the inequality $7 + 3k > 2k - 5$.

Use the addition property of inequality twice, once to get the terms containing k on one side of the inequality and a second time to get the integers together on the other side. (These steps can be done in either order.)

$$7 + 3k > 2k - 5$$
$$7 + 3k - 2k > 2k - 5 - 2k \qquad \text{Subtract } 2k.$$
$$7 + k > -5 \qquad \text{Combine terms.}$$
$$7 + k - 7 > -5 - 7 \qquad \text{Subtract 7.}$$
$$k > -12$$

The graph of the solutions $k > -12$ is shown in Figure 13. ■

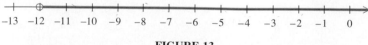

FIGURE 13

3 ▶ Use the multiplication property of inequality.

▶ The addition property of inequality cannot be used to solve inequalities such as $4y \geq 28$. These inequalities require the multiplication property of inequality. To see how this property works, it will be helpful to look at some examples.

First, start with the inequality $3 < 7$ and multiply both sides by the positive number 2.

$$3 < 7$$
$$2(3) < 2(7) \qquad \text{Multiply both sides by } 2.$$
$$6 < 14 \qquad \text{True}$$

Now multiply both sides of $3 < 7$ by the negative number -5.

$$3 < 7$$
$$-5(3) < -5(7) \qquad \text{Multiply both sides by } -5.$$
$$-15 < -35 \qquad \text{False}$$

To get a true statement when multiplying both sides by -5 requires reversing the direction of the inequality symbol.

$$3 < 7$$
$$-5(3) > -5(7) \qquad \text{Multiply by } -5; \text{ reverse the symbol.}$$
$$-15 > -35 \qquad \text{True}$$

Take the inequality $-6 < 2$ as another example. Multiply both sides by the positive number 4.

$$-6 < 2$$
$$4(-6) < 4(2) \qquad \text{Multiply by } 4.$$
$$-24 < 8 \qquad \text{True}$$

Multiplying both sides of $-6 < 2$ by -5 and at the same time reversing the direction of the inequality symbol gives

$$-6 < 2$$
$$(-5)(-6) > (-5)(2) \qquad \text{Multiply by } -5; \text{ change } < \text{ to } >.$$
$$30 > -10. \qquad \text{True}$$

The two parts of the **multiplication property of inequality** are stated below.

Multiplication Property of Inequality

For any algebraic expressions A, B, and C that represent real numbers, with $C \neq 0$,

1. if C is *positive,* then the inequalities

$$A < B \qquad \text{and} \qquad AC < BC$$

have exactly the same solutions;
2. if C is *negative,* then the inequalities

$$A < B \qquad \text{and} \qquad AC > BC$$

have exactly the same solutions.

In other words, both sides of an inequality may be multiplied by the same expression representing a positive number without changing the solutions. If the expression represents a negative number, we must reverse the direction of the inequality symbol.

The multiplication property of inequality works with $>$, $\leq$, or $\geq$, as well. The multiplication property of inequality also permits *division* of both sides of an inequality by the same nonzero expression.

It is important to remember the differences in the multiplication property for positive and negative numbers.

1. When both sides of an inequality are multiplied or divided by a positive number, the direction of the inequality symbol *does not change.* Adding or subtracting terms on both sides also does not change the symbol.
2. When both sides of an inequality are multiplied or divided by a negative number, the direction of the symbol *does change. Reverse the symbol of inequality only when you multiply or divide both sides by a negative number.*

The next examples show how to solve inequalities with the multiplication property.

EXAMPLE 3

Using the Multiplication Property of Inequality

(a) Solve the inequality $3r < -18$.

Simplify this inequality by using the multiplication property of inequality and dividing both sides by 3. Since 3 is a positive number, the direction of the inequality symbol does not change.

$$3r < -18$$
$$\frac{3r}{3} < \frac{-18}{3} \qquad \text{Divide by 3.}$$
$$r < -6$$

The graph of the solutions is shown in Figure 14. ■

FIGURE 14

(b) Solve the inequality $-4t \geq 8$.

Here both sides of the inequality must be divided by -4, a negative number, which *does* change the direction of the inequality symbol.

$$-4t \geq 8$$
$$\frac{-4t}{-4} \leq \frac{8}{-4} \qquad \text{Divide by } -4; \text{ symbol is reversed.}$$
$$t \leq -2$$

The solutions are graphed in Figure 15. ■

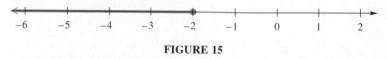

FIGURE 15

CAUTION Even though the number on the right side of the inequality in Example 3(a) is negative (-18), *do not reverse the direction of the inequality symbol.* Reverse the direction only when multiplying or dividing *by* a negative number, as shown in Example 3(b).

4 ▶ Solve linear inequalities.

▶ A **linear inequality** is an inequality that can be written in the form $ax + b < 0$, for real numbers a and b, with $a \neq 0$. ($<$ may be replaced with $>$, $\leq$, or $\geq$ in this definition.) In order to solve a linear inequality go through the following steps.

Solving a Linear Inequality

Step 1 Use the associative, commutative, and distributive properties to combine like terms on each side of the inequality.

Step 2 Use the addition property of inequality to simplify the inequality to one of the form $ax < b$ or $ax > c$, where a, b, and c are real numbers.

Step 3 Use the multiplication property of inequality to simplify further to an inequality of the form $x < d$ or $x > d$, where d is a real number.

Notice how these steps are used in the next example.

E X A M P L E 4

Solving a Linear Inequality

Solve the inequality $3z + 2 - 5 > -z + 7 + 2z$.

Step 1 Combine like terms and simplify.

$$3z + 2 - 5 > -z + 7 + 2z$$
$$3z - 3 > z + 7$$

Step 2 Use the addition property of inequality.

$$3z - 3 + 3 > z + 7 + 3 \qquad \text{Add 3.}$$
$$3z > z + 10$$
$$3z - z > z + 10 - z \qquad \text{Subtract } z.$$
$$2z > 10$$

Step 3 Use the multiplication property of inequality.

$$\frac{2z}{2} > \frac{10}{2} \qquad \text{Divide by 2.}$$
$$z > 5$$

Since 2 is positive, the direction of the inequality symbol was not changed in the third step. A graph of the solutions is shown in Figure 16. ■

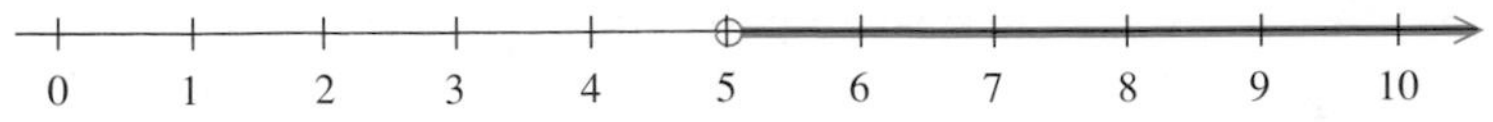

FIGURE 16

E X A M P L E 5

Solving a Linear Inequality

Solve $5(k - 3) - 7k \geq 4(k - 3) + 9$.

Step 1 Simplify and combine like terms.

$$5(k - 3) - 7k \geq 4(k - 3) + 9$$
$$5k - 15 - 7k \geq 4k - 12 + 9 \qquad \text{Distributive property}$$
$$-2k - 15 \geq 4k - 3 \qquad \text{Combine like terms.}$$

Step 2 Use the addition property.

$$-2k - 15 - 4k \geq 4k - 3 - 4k \qquad \text{Subtract } 4k.$$
$$-6k - 15 \geq -3$$
$$-6k - 15 + 15 \geq -3 + 15 \qquad \text{Add 15.}$$
$$-6k \geq 12$$

Step 3 Divide both sides by -6, a negative number. Change the direction of the inequality symbol.

$$\frac{-6k}{-6} \leq \frac{12}{-6} \qquad \begin{array}{l} \text{Divide by } -6; \\ \text{symbol is reversed.} \end{array}$$
$$k \leq -2$$

A graph of the solutions is shown in Figure 17. ■

FIGURE 17

5 ▶ Solve applied problems by using inequalities.

▶ Until now, the applied problems that we have studied have all led to equations.

PROBLEM SOLVING

Inequalities can be used to solve applied problems involving phrases that suggest inequality. The following chart gives some of the more common such phrases along with examples and translations.

Phrase	Example	Inequality
Is more than	A number *is more than* 4	$x > 4$
Is less than	A number *is less than* -12	$x < -12$
Is at least	A number *is at least* 6	$x \geq 6$
Is at most	A number *is at most* 8	$x \leq 8$

CAUTION Do not confuse statements like "5 is more than a number" with the phrase "5 more than a number." The first of these is expressed as "$5 > x$" while the second is expressed with addition, as "$x + 5$."

The next example shows an application of algebra that is important to anyone who has ever asked himself or herself "What score can I make on my next test and have a (particular grade) in this course?" It uses the idea of finding the average of a number of grades. In general, to find the average of n numbers, add the numbers, and divide by n.

EXAMPLE 6
Finding an Average Test Score

Brent has test grades of 86, 88, and 78 on his first three tests in geometry. If he wants an average of at least 80 after his fourth test, what are the possible scores he can make on his fourth test?

Let x = Brent's score on his fourth test. To find his average after 4 tests, add the test scores and divide by 4.

$$\overset{\underset{\displaystyle\downarrow}{\text{Average}}}{\frac{86 + 88 + 78 + x}{4}} \geq \overset{\underset{\displaystyle\downarrow\quad\downarrow}{\text{is at}\ \text{least}\ 80.}}{80}$$

$$\frac{252 + x}{4} \geq 80 \qquad \text{Add the known scores.}$$

$$4\left(\frac{252 + x}{4}\right) \geq 4(80) \qquad \text{Multiply by } 4.$$

$$252 + x \geq 320$$

$$252 - 252 + x \geq 320 - 252 \qquad \text{Subtract } 252.$$

$$x \geq 68 \qquad \text{Combine terms.}$$

He must score 68 or more on the fourth test to have an average of *at least* 80.   ■

CAUTION   Errors often occur when the phrases "at least" and "at most" appear in applied problems. Remember that

	at least	translates as	**greater than or equal to**
and			
	at most	translates as	**less than or equal to**.

---

6 ▶ Solve three-part inequalities.

▶   Inequalities that say that one number is *between* two other numbers are *three-part inequalities*. For example,

$$-3 < 5 < 7$$

says that 5 is between -3 and 7. The inequality translates in words as "-3 is less than 5 *and* 5 is less than 7." It would be *wrong* to write $7 < 5 < -3$, since this would imply that $7 < 5$ and $5 < -3$, which are both false statements.

Three-part inequalities can also be solved by using the addition and multiplication properties of inequality. The idea is to get the inequality in the form

a number $< x <$ **another number**.

The solutions can then easily be graphed.

EXAMPLE 7

Graphing an Interval on a Number Line

Graph $-3 \leq x < 2$.

The statement $-3 \leq x < 2$ is read "-3 is less than or equal to x *and* x is less than 2." Graph this inequality by placing a solid dot at -3 (because -3 is part of the graph) and an open circle at 2 (because 2 is not part of the graph). Then draw a line segment between the two circles, as in Figure 18.   ■

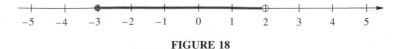

FIGURE 18

EXAMPLE 8 ■

Solving Three-Part
Inequalities

(a) Solve $4 \leq 3x - 5 < 6$ and graph the solutions.

If we were to solve either of the two inequalities alone, the first step would be to add 5 to each part. Do this for each of the three parts.

$$4 \leq 3x - 5 < 6$$
$$4 + 5 \leq 3x - 5 + 5 < 6 + 5 \qquad \text{Add 5.}$$
$$9 \leq 3x < 11$$

Now divide each part by the positive number 3.

$$\frac{9}{3} \leq \frac{3x}{3} < \frac{11}{3} \qquad \text{Divide by 3.}$$

$$3 \leq x < \frac{11}{3}$$

A graph of the solutions is shown in Figure 19.

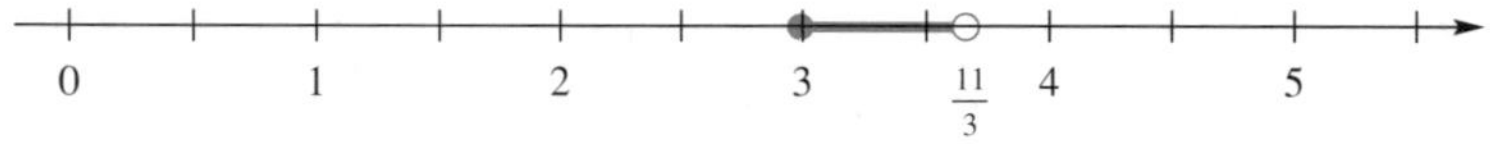

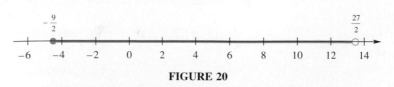

FIGURE 19

(b) Solve $-4 \leq \frac{2}{3}m - 1 < 8$ and graph the solutions.

Recall from Section 2.3 that fractions as coefficients in equations can be eliminated by multiplying both sides by the least common denominator of the fractions. The same is true for inequalities. One way to begin is to multiply all three parts by 3.

$$-4 \leq \frac{2}{3}m - 1 < 8$$
$$3(-4) \leq 3\left(\frac{2}{3}m - 1\right) < 3(8) \qquad \text{Multiply by 3.}$$
$$-12 \leq 2m - 3 < 24 \qquad \text{Distributive property}$$

Now add 3 to each part.

$$-12 + 3 \leq 2m - 3 + 3 < 24 + 3 \qquad \text{Add 3.}$$
$$-9 \leq 2m < 27$$

Finally, divide by 2 to get

$$-\frac{9}{2} \leq m < \frac{27}{2}.$$

A graph of the solutions is shown in Figure 20.

FIGURE 20

This inequality could also have been solved by first adding 1 to each part, and then multiplying each part by 3/2. ■

2.9 EXERCISES

1. Explain how to determine whether to use an open circle or a closed circle at the endpoint when graphing an inequality on a number line.

2. How does the graph of $t \geq -7$ differ from the graph of $t > -7$?

Write an inequality involving the variable x that describes the set of numbers graphed. See Example 1.

3.
$$-4 \ -3 \ -2 \ -1 \ \ 0 \ \ 1 \ \ 2 \ \ 3$$

4.
$$-4 \qquad 0 \qquad 4$$

5.
$$-2 \ -1 \ \ 0 \ \ 1 \ \ 2 \ \ 3 \ \ 4 \ \ 5$$

6.
$$-2 \ -1 \ \ 0 \ \ 1 \ \ 2 \ \ 3 \ \ 4 \ \ 5$$

Graph the inequality on a number line. See Example 1.

7. $k \leq 4$ **8.** $r \leq -11$ **9.** $x < -3$ **10.** $y < 3$

11. $t > 4$ **12.** $m > 5$ **13.** $8 \leq x \leq 10$ **14.** $3 \leq x \leq 5$

15. $0 < y \leq 10$ **16.** $-3 \leq x < 5$

17. Why is it *wrong* to write $3 < x < -2$ to indicate that x is between -2 and 3?

18. If $p < q$ and $r < 0$, which one of the following statements is *false*?
    (a) $pr < qr$ **(b)** $pr > qr$ **(c)** $p + r < q + r$ **(d)** $p - r < q - r$

Solve the inequality and graph the solutions. See Example 2.

19. $z - 8 \geq -7$ **20.** $p - 3 \geq -11$ **21.** $2k + 3 \geq k + 8$

22. $3x + 7 \geq 2x + 11$ **23.** $3n + 5 < 2n - 6$ **24.** $5x - 2 < 4x - 5$

25. Under what conditions must the inequality symbol be reversed when solving an inequality?

26. Explain the steps you would use to solve the inequality $-5x > 20$.

27. Your friend tells you that when solving the inequality $6x < -42$ he reversed the direction of the inequality because of the presence of -42. How would you respond?

28. By what number must you *multiply* both sides of $.2x > 6$ to get just x on the left side?

Solve the inequality and graph the solutions. See Example 3.

29. $3x < 18$

30. $5x < 35$

31. $2y \geq -20$

32. $6m \geq -24$

33. $-8t > 24$

34. $-7x > 49$

35. $-x \geq 0$

36. $-k < 0$

37. $-\frac{3}{4}r < -15$

38. $-\frac{7}{8}t < -14$

39. $-.02x \leq .06$

40. $-.03v \geq -.12$

Solve the inequality and graph the solutions. See Examples 4 and 5.

41. $5r + 1 \geq 3r - 9$

42. $6t + 3 < 3t + 12$

43. $6x + 3 + x < 2 + 4x + 4$

44. $-4w + 12 + 9w \geq w + 9 + w$

45. $-x + 4 + 7x \leq -2 + 3x + 6$

46. $14y - 6 + 7y > 4 + 10y - 10$

47. $5(x + 3) - 6x \leq 3(2x + 1) - 4x$

48. $2(x - 5) + 3x < 4(x - 6) + 1$

49. $\frac{2}{3}(p + 3) > \frac{5}{6}(p - 4)$

50. $\frac{7}{9}(y - 4) \leq \frac{4}{3}(y + 5)$

51. $4x - (6x + 1) \leq 8x + 2(x - 3)$

52. $2y - (4y + 3) > 6y + 3(y + 4)$

53. $5(2k + 3) - 2(k - 8) > 3(2k + 4) + k - 2$

54. $2(3z - 5) + 4(z + 6) \geq 2(3z + 2) + 3z - 15$

Write a three-part inequality involving the variable x that describes the set of numbers graphed. See Example 7.

55.

56.

57.

58.

Solve the inequality and graph the solutions. See Example 8.

59. $-5 \leq 2x - 3 \leq 9$

60. $-7 \leq 3x - 4 \leq 8$

61. $5 < 1 - 6m < 12$

62. $-1 \leq 1 - 5q \leq 16$

63. $10 < 7p + 3 < 24$ **64.** $-8 \leq 3r - 1 \leq -1$ **65.** $-12 \leq \frac{1}{2}z + 1 \leq 4$ **66.** $-6 \leq 3 + \frac{1}{3}a \leq 5$

67. $1 \leq 3 + \frac{2}{3}p \leq 7$ **68.** $2 < 6 + \frac{3}{4}y < 12$ **69.** $-7 \leq \frac{5}{4}r - 1 \leq -1$ **70.** $-12 \leq \frac{3}{7}a + 2 \leq -4$

──────────── ◈ **MATHEMATICAL CONNECTIONS** (Exercises 71–76) ◈ ────────────

The methods for solving linear equations and linear inequalities are quite similar. In Exercises 71–76, we show how the solutions of an inequality are closely connected to the solution of the corresponding equation. Work though these exercises in order.

71. Solve the equation $3x + 2 = 14$ and graph the solution as a single point on the number line.

72. Solve the inequality $3x + 2 > 14$ and graph the solutions as an interval on the number line. How does this result compare to the one in Exercise 71?

73. Solve the inequality $3x + 2 < 14$ and graph the solutions as an interval on the number line. How does this result compare to the one in Exercise 72?

74. If you were to graph all the solutions from Exercises 71–73 on the same number line, what would the graph be? (This is called the *union* of all the solutions.)

75. Based on your results from Exercises 71–74, if you were to graph the union of the solutions of

$$-4x + 3 = -1, \qquad -4x + 3 > -1, \qquad \text{and} \qquad -4x + 3 < -1,$$

what do you think the graph would be?

76. Comment on the following statement: *Equality* is the boundary between *less than* and *greater than*.

──────────── ◈ ────────────

Solve the problem by writing and solving an inequality. See Example 6.

77. When 8 is subtracted from the sum of three times a number and 6, the result is less than 4 more than the number. Find all such numbers.

78. When 2 is added to the difference between six times a number and 5, the result is greater than 13 added to 5 times the number. Find all such numbers.

79. Inkie Landry has grades of 76 and 81 on her first two algebra tests. If she wants an average of at least 80 after her third test, what possible scores can she make on her third test?

80. Mabimi Pampo has grades of 96 and 86 on his first two geometry tests. What possible scores can he make on his third test so that his average is at least 90?

81. The formula for converting Fahrenheit temperature to Celsius is

$$C = \frac{5}{9}(F - 32).$$

If the Celsius temperature on a certain summer day in Toledo is never more than 30 degrees, how would you describe the corresponding Fahrenheit temperatures?

82. The formula for converting Celsius temperature to Fahrenheit is

$$F = \frac{9}{5}C + 32.$$

The Fahrenheit temperature of Key West, Florida, has never exceeded 95 degrees. How would you describe this using Celsius temperature?

83. A product will break even or produce a profit if the revenue R from selling the product is at least equal to the cost C of producing it. Suppose that the cost C (in dollars) to produce x units of bicycle helmets is $C = 50x + 5000$, while the revenue R (in dollars) collected from the sale of x units is $R = 60x$. For what values of x does the product break even or produce a profit?

84. (See Exercise 83.) If the cost to produce x units of basketball cards is $C = 100x + 6000$ (in dollars), and the revenue collected from selling x units is $R = 500x$ (in dollars), for what values of x does the product break even or produce a profit?

85. For what values of x would the rectangle have perimeter of at least 400?

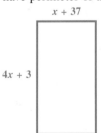

$x + 37$

$4x + 3$

86. For what values of x would the triangle have perimeter of at least 72?

x $x + 5$

$2x + 5$

87. A long-distance phone call costs $2.00 for the first three minutes plus $.30 per minute for each minute or fractional part of a minute after the first three minutes. If x represents the number of minutes of the length of the call after the first three minutes, then $2 + .30x$ represents the cost of the call. If Jorge has $5.60 to spend on a call, what is the maximum total time he can use the phone?

88. If the call described in Exercise 87 costs between $5.60 and $6.50, what are the possible total time lengths for the call?

REVIEW EXERCISES

Evaluate the expression for $x = 3$. See Sections 1.3 and 1.7.

89. $2x^2 - 3x + 9$ **90.** $3x^2 - 3x + 2$ **91.** $4x^3 - 5x^2 + 2x - 6$ **92.** $-4x^3 + 2x^2 - 9x - 3$

Simplify, combining like terms. See Section 2.1.

93. $-3(2x + 4) + 4(2x - 6)$

94. $-8(-3x + 7) - 4(2x + 3)$

95. $5x^3 + 2x^2 + 3x - 10 - 2x^3 + 9x^2 - 3x + 12$

96. $-8x^3 - 4x^2 + 12x - 3 + 9x^3 + 8x^2 + 6x - 14$

CHAPTER 2 SUMMARY

KEY TERMS

2.1 term
  numerical coefficient
  like terms
  combining like
    terms

2.2 linear equation
  equivalent equations

2.4 degree
  complementary
    angles
  supplementary
    angles
  consecutive integers

2.5 perimeter
  area
  vertical angles
  straight angle
  volume

2.6 ratio
  proportion

2.9 linear inequality
  three-part inequality

NEW SYMBOLS

$1°$ one degree

a to b, $a : b$, or $\dfrac{a}{b}$ the ratio of a to b

QUICK REVIEW

CONCEPTS	EXAMPLES
2.1 SIMPLIFYING EXPRESSIONS	
Only like terms may be combined.	$-3y^2 + 6y^2 + 14y^2 = 17y^2$ $4(3 + 2x) - 6(5 - x)$ $\quad = 12 + 8x - 30 + 6x$ Distributive property $\quad = 14x - 18$
2.2 THE ADDITION AND MULTIPLICATION PROPERTIES OF EQUALITY	
The same expression may be added to (or subtracted from) each side of an equation without changing the solution.	Solve $x - 6 = 12$. $\quad x - 6 + 6 = 12 + 6$ Add 6. $\quad\quad\quad\quad x = 18$ Combine terms.
Each side of an equation may be multiplied (or divided) by the same nonzero expression without changing the solution.	Solve $\dfrac{3}{4}x = -9$. $\dfrac{4}{3} \cdot \dfrac{3}{4}x = \dfrac{4}{3}(-9)$ Multiply by $\frac{4}{3}$. $\quad\quad\quad x = -12$
2.3 MORE ON SOLVING LINEAR EQUATIONS	
Solving a Linear Equation **1.** Clear parentheses and combine like terms to simplify each side. **2.** Get the variable term on one side, a number on the other.	Solve the equation $2x + 3(x + 1) = 38$. **1.** $2x + 3x + 3 = 38$ Clear parentheses. $\quad\quad\quad 5x + 3 = 38$ Combine like terms. **2.** $5x + 3 - 3 = 38 - 3$ Subtract 3. $\quad\quad\quad 5x = 35$ Combine terms.

CONCEPTS	EXAMPLES
3. Get the equation into the form $x = $ a number.	**3.** $\dfrac{5x}{5} = \dfrac{35}{5}$ Divide by 5. $x = 7$
4. Check by substituting the result into the original equation.	**4.** $2x + 3(x + 1) = 38$ Check. $2(7) + 3(7 + 1) = 38$? Let $x = 7$. $14 + 24 = 38$? Multiply. $38 = 38$ True

2.4 AN INTRODUCTION TO APPLICATIONS OF LINEAR EQUATIONS

Solving an Applied Problem	One number is 5 more than another. Their sum is 21. Find both numbers.
1. Choose a variable to represent the unknown.	**1.** Let x be the smaller number.
2. Determine expressions for any other unknown quantities, using the variable. Draw figures or diagrams if they apply.	**2.** Let $x + 5$ be the larger number.
3. Translate the problem into an equation.	**3.** $x + (x + 5) = 21$
4. Solve the equation.	**4.** $2x + 5 = 21$ Combine terms. $2x + 5 - 5 = 21 - 5$ Subtract 5. $2x = 16$ Combine terms. $\dfrac{2x}{2} = \dfrac{16}{2}$ Divide by 2. $x = 8$
5. Answer the question asked in the problem.	**5.** The numbers are 8 and 13.
6. Check your solution by using the original words of the problem. Be sure that the answer is appropriate and makes sense.	**6.** 13 is 5 more than 8, and $8 + 13 = 21$. It checks.

2.5 FORMULAS AND APPLICATIONS FROM GEOMETRY

To find the values of one of the variables in a formula given values for the others, substitute the known values into the formula.	Find L if $A = LW$, given that $A = 24$ and $W = 3$. $24 = L \cdot 3$ $A = 24, W = 3$ $\dfrac{24}{3} = \dfrac{L \cdot 3}{3}$ Divide by 3. $8 = L$
To solve a formula for one of the variables, isolate that variable by treating the other variables as numbers and using the steps for solving equations.	Solve $A = \dfrac{1}{2}bh$ for b. $2A = 2\left(\dfrac{1}{2}bh\right)$ Multiply by 2. $2A = bh$ $\dfrac{2A}{h} = b$ Divide by h.

CONCEPTS	EXAMPLES

2.6 RATIOS AND PROPORTIONS

To write a ratio, express quantities in the same units.

Express as a ratio: 4 feet to 8 inches.

$$4 \text{ feet to } 8 \text{ inches} = 48 \text{ inches to } 8 \text{ inches}$$

$$= \frac{48}{8} = \frac{6}{1} \quad \text{or} \quad 6 \text{ to } 1 \quad \text{or} \quad 6:1$$

To solve a proportion, use the method of cross products.

Solve $\frac{x}{12} = \frac{35}{60}$.

$$60x = 12 \cdot 35 \qquad \text{Cross products}$$
$$60x = 420 \qquad \text{Multiply.}$$
$$\frac{60x}{60} = \frac{420}{60} \qquad \text{Divide by 60.}$$
$$x = 7$$

2.7 APPLICATIONS OF PERCENT: MIXTURE, INTEREST, AND MONEY

Problems involving applications of percent can be solved using box diagrams or charts.

A sum of money is invested at simple interest in two ways. Part is invested at 12%, and $20,000 less than that amount is invested at 10%. If the total interest for one year is $9000, find the amount invested at each rate.

Let $\quad x =$ amount invested at 12%;

$\quad\quad x - 20,000 =$ amount invested at 10%.

Amount invested	x		$x - 20,000$		
Rate	.12	$+$	.10	$=$	9000

$$.12x + .10(x - 20,000) = 9000$$
$$12x + 10(x - 20,000) = 900,000 \qquad \text{Multiply by 100.}$$
$$12x + 10x - 200,000 = 900,000 \qquad \text{Distributive property}$$
$$22x - 200,000 = 900,000 \qquad \text{Combine terms.}$$
$$22x = 1,100,000 \qquad \text{Add 200,000.}$$
$$x = 50,000 \qquad \text{Divide by 22.}$$

$50,000 is invested at 12% and $30,000 is invested at 10%.

2.8 MORE ABOUT PROBLEM SOLVING

The three forms of the formula relating distance, rate, and time, are $d = rt$, $r = \frac{d}{t}$, and $t = \frac{d}{r}$.

Two cars leave from the same point, traveling in opposite directions. One travels at 45 miles per hour and the other at 60 miles per hour. How long will it take them to be 210 miles apart?

Let $t =$ time it takes for them to be 210 miles apart.

To solve a problem about distance, set up a sketch showing what is happening in the problem.

210 miles

(continued)

CONCEPTS	EXAMPLES
Make a chart using the information given in the problem, along with the unknown quantities.	The chart gives the information from the problem, with expressions for distance obtained by using $d = rt$. <table><tr><td></td><td>r</td><td>t</td><td>d</td></tr><tr><td>One car</td><td>45</td><td>t</td><td>$45t$</td></tr><tr><td>Other car</td><td>60</td><td>t</td><td>$60t$</td></tr></table> The sum of the distances, $45t$ and $60t$, must be 210 miles. $45t + 60t = 210$ $105t = 210$ Combine like terms. $t = 2$ Divide by 2. It will take them 2 hours to be 210 miles apart.

2.9 THE ADDITION AND MULTIPLICATION PROPERTIES OF INEQUALITY

To solve an inequality: **1.** Clear parentheses and combine like terms. **2.** Add or subtract the same expression on each side to get the variable term on one side and a number on the other side. **3.** Multiply or divide by the same expression on each side to get the form $x > a$ or $x < a$. (When multiplying or dividing by a negative expression, reverse the direction of the inequality symbol.) To solve an inequality such as $$4 < 2x + 6 < 8$$ work with all three expressions at the same time.	Solve $3(1 - x) + 5 - 2x > 9 - 6$. $3 - 3x + 5 - 2x > 9 - 6$ Clear parentheses. $8 - 5x > 3$ Combine terms. $8 - 5x - 8 > 3 - 8$ Subtract 8. $-5x > -5$ Combine terms. $\dfrac{-5x}{-5} < \dfrac{-5}{-5}$ Divide by -5; change $>$ to $<$. $x < 1$ Lowest terms Solve $4 < 2x + 6 < 8$. $4 - 6 < 2x + 6 - 6 < 8 - 6$ Subtract 6. $-2 < 2x < 2$ Combine terms. $\dfrac{-2}{2} < \dfrac{2x}{2} < \dfrac{2}{2}$ Divide by 2. $-1 < x < 1$

CHAPTER 2 REVIEW EXERCISES

[2.1] *Combine terms whenever possible.*

1. $2m + 9m$ **2.** $15p^2 - 7p^2 + 8p^2$ **3.** $5p^2 - 4p + 6p + 11p^2$

4. $-2(3k - 5) + 2(k + 1)$ **5.** $7(2m + 3) - 2(8m - 4)$ **6.** $-(2k + 8) - (3k - 7)$

[2.2–2.3] *Solve the equation.*

7. $m - 5 = 1$ **8.** $y + 8 = -4$ **9.** $3k + 1 = 2k + 8$

10. $5k = 4k + \dfrac{2}{3}$

11. $(4r - 2) - (3r + 1) = 8$

12. $3(2y - 5) = 2 + 5y$

13. $7k = 35$

14. $12r = -48$

15. $2p - 7p + 8p = 15$

16. $\dfrac{m}{12} = -1$

17. $\dfrac{5}{8}k = 8$

18. $12m + 11 = 59$

19. $3(2x + 6) - 5(x + 8) = x - 22$

20. $5x + 9 - (2x - 3) = 2x - 7$

21. $\dfrac{1}{2}r - \dfrac{r}{3} = \dfrac{r}{6}$

22. $.10(x + 80) + .20x = 14$

23. $3x - (-2x + 6) = 4(x - 4) + x$

24. $2(y - 3) - 4(y + 12) = -2(y + 27)$

[2.4] *Solve the problem.*

25. In a recent year, the state of Florida had a total of 120 members In its House of Representatives, consisting of only Democrats and Republicans. There were 30 more Democrats than Republicans. How many representatives from each party were there?

26. Captain Tom Tupper gives deep-sea fishing trips. One day he noticed that the boat contained 2 fewer men than women (not counting himself). If he had 40 customers on the boat, how many men and how many women were there?

27. The land area of Hawaii is 5213 square miles greater than the area of Rhode Island. Together, the areas total 7637 square miles. What is the area of each of the two states?

28. The height of Seven Falls in Colorado is 5/2 the height of Twin Falls in Idaho. The sum of the heights is 420 feet. Find the height of each.

29. The supplement of an angle measures 10 times the measure of its complement. What is the measure of the angle?

[2.5] *A formula is given along with the values for all but one of the variables. Find the value of the variable that is not given.*

30. $A = \dfrac{1}{2}bh;$ $\quad A - 44, b = 8$

31. $A = \dfrac{1}{2}h(b + B);$ $\quad h = 3, B = 4, h = 8$

32. $C = 2\pi r;$ $\quad C = 29.83, \pi = 3.14$

33. $V = \dfrac{4}{3}\pi r^3;$ $\quad r = 6, \pi = 3.14$

Solve the formula for the specified variable.

34. $A = LW$ for L

35. $A = \dfrac{1}{2}h(b + B)$ for h

Find the measure of each marked angle.

36.

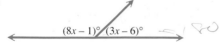

$(8x - 1)°$ $(3x - 6)°$

37.

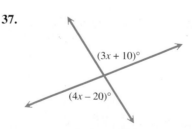

$(3x + 10)°$

$(4x - 20)°$

Solve the application of geometry.

38. A cinema screen in Indonesia has length 92.75 feet and width 70.5 feet. What is the perimeter? What is the area?

39. The Ziegfield Room in Reno, Nevada, has a circular turntable on which its showgirls dance. The circumference of the table is 62.5 feet. What is the diameter? What is the radius? What is the area? (Use $\pi = 3.14$.)

40. What is wrong with the following problem? "The formula for the area of a trapezoid is $A = \frac{1}{2}h(b + B)$. If $h = 12$ and $b = 14$, find the value of A."

Give a ratio for the word phrase writing fractions in lowest terms.

41. 60 centimeters to 40 centimeters

42. 5 days to 2 weeks

43. 90 inches to 10 feet

44. 3 months to 3 years

Solve the equation.

45. $\dfrac{p}{21} = \dfrac{5}{30}$

46. $\dfrac{5 + x}{3} = \dfrac{2 - x}{6}$

47. $\dfrac{y}{5} = \dfrac{6y - 5}{11}$

Solve the problem involving proportion.

48. If 2 pounds of fertilizer will cover 150 square feet of lawn, how many pounds would be needed to cover 500 square feet?

49. If 8 ounces of medicine must be mixed with 20 ounces of water, how many ounces of medicine should be mixed with 90 ounces of water?

50. The tax on a $24.00 item is $2.04. How much tax would be paid on a $36.00 item?

51. The distance between two cities on a road map is 32 centimeters. The two cities are actually 150 kilometers apart. The distance on the map between two other cities is 80 centimeters. How far apart are these cities?

52. In the 1992 Olympic Games held at Barcelona, Hungary earned a total of 30 medals. Two of every five medals were silver. How many silver medals were earned?

53. Which is the best buy for a popular breakfast cereal?
15-ounce size: $2.69
20-ounce size: $3.29
25.5-ounce size: $3.49

[2.7] *Solve the problem.*

54. For a party, Joann bought some 15¢ candy and some 30¢ candy, paying $15 in all. If there are 25 more pieces of 15¢ candy, how many pieces of 30¢ candy did she buy?

55. A person has $250 in fives and tens. He has twice as many tens as fives. How many fives does he have?

56. A nurse must mix 15 liters of a 10% solution of a drug with some 60% solution to get a 20% mixture. How many liters of the 60% solution would be needed?

57. Todd Cardella invested $10,000 from which he earns an annual income of $550 per year. He invested part of it at 5% annual interest and the remainder in bonds paying 6% interest. How much did he invest at each rate?

[2.8] *Solve the problem.*

58. In 1846, the vessel *Yorkshire* traveled from Liverpool to New York, a distance of 3150 miles, in 384 hours. What was the *Yorkshire*'s average speed? Round your answer to the nearest tenth.

59. Sue Fredine drove from Louisville to Dallas, a distance of 819 miles, averaging 63 miles per hour. What was her driving time?

60. Two planes leave St. Louis at the same time. One flies north at 350 miles per hour and the other flies south at 420 miles per hour. In how many hours will they be 1925 miles apart?

61. Jim leaves his house on his bicycle and averages 5 miles per hour. His wife, Annie, leaves 1/2 hour later, following the same path and averaging 8 miles per hour. How long will it take for Annie to catch up to Jim?

62. The perimeter of a rectangle is ten times the width. The length is 9 meters more than the width. Find the width of the rectangle.

63. The longest side of a triangle is 11 meters longer than the shortest side. The medium side is 15 meters long. The perimeter of the triangle is 46 meters. Find the length of the shortest side of the triangle.

64. Write a short paragraph explaining the general method for problem solving as explained in this chapter.

65. Of all the different types of problems presented in this chapter which type is your favorite? Which type is your least favorite? Explain.

[2.9] *Graph the inequality on a number line.*

66. $p \geq -4$ **67.** $x < 7$ **68.** $-5 \leq y < 6$

Solve the inequality and graph the solutions.

69. $y + 6 \geq 3$ **70.** $5t < 4t + 2$

71. $-6x \leq -18$ **72.** $8(k - 5) - (2 + 7k) \geq 4$

73. $4x - 3x > 10 - 4x + 7x$ **74.** $3(2w + 5) + 4(8 + 3w) < 5(3w + 2) + 2w$

75. $-3 \leq 2m + 1 \leq 4$ **76.** $9 < 3m + 5 \leq 20$

Solve the problem by writing an inequality.

77. Carlotta Valdez has grades of 94 and 88 on her first two calculus tests. What possible scores on a third test will give her an average of at least 90?

78. If nine times a number is added to 6 the result is at most 3. Find all such numbers.

MIXED REVIEW EXERCISES*

Solve.

79. $\dfrac{y}{7} = \dfrac{y - 5}{2}$ **80.** $I = prt$ for r

81. $-2x > -4$ **82.** $2k - 5 = 4k + 13$

83. $.05x + .02x = 4.9$ **84.** $2 - 3(y - 5) = 4 + y$

85. $9x - (7x + 2) = 3x + (2 - x)$ **86.** $\dfrac{1}{3}s + \dfrac{1}{2}s + 7 = \dfrac{5}{6}s + 5 + 2$

87. A recipe for biscuit tortoni calls for 2/3 cup of macaroon cookie crumbs. The recipe is for 8 servings. How many cups of macaroon cookie crumbs would be needed for 30 servings?

88. The Golden Gate Bridge in San Francisco is 2605 feet longer than the Brooklyn Bridge. Together, their spans total 5795 feet. How long is each bridge?

89. In the 1988 U.S. presidential election, Sherman County in Oregon registered 120 more votes for George Bush than it did for Michael Dukakis. If the two men together received 990 votes, how many did each receive?

_____

*The order of the exercises in this final group does not correspond to the order in which topics occur in the chapter. This random ordering should help you in your preparation for the chapter test.

90. Which is the best buy for apple juice?
32-ounce size: $1.19
48-ounce size: $1.79
64-ounce size: $1.99

91. If 1 quart of oil must be mixed with 24 quarts of gasoline, how much oil would be needed for 192 quarts of gasoline?

92. Two trains are 390 miles apart. They start at the same time and travel toward one another meeting 3 hours later. If the speed of one train is 30 miles per hour more than the speed of the other train, find the speed of each train.

93. One side of a triangle is 3 centimeters longer than the shortest side. The third side is twice as long as the shortest side. If the perimeter of the triangle cannot exceed 39 centimeters, find all possible lengths for the shortest side.

94. The shorter base of a trapezoid is 42 centimeters long, and the longer base is 48 centimeters long. The area of the trapezoid is 360 square centimeters. Find the height of the trapezoid.

95. The area of a triangle is 25 square meters. The base is 10 meters in length. Find the height.

96. On a test in geometry, the highest grade was 35 points more than the lowest. The sum of the highest and lowest grades was 157. Find the lowest score.

97. The perimeter of a square cannot be greater than 200 meters. Find the possible values for the length of a side.

98. The distance between two cities on a road map is 16 centimeters. The two cities are actually 150 kilometers apart. The distance on the map between two other cities is 40 centimeters. How far apart are these cities?

──────◆ **MATHEMATICAL CONNECTIONS** (Exercises 99–106) ◆──────

Use your knowledge of angle relationships to answer Exercises 99–106. Work through them in order. Refer to the accompanying figure.

$(.5x + 69)°$

$(4x + 7)°$

$\left(\dfrac{3y + 2}{2}\right)°$

$\dfrac{2}{3}(x + 63)°$

99. Write an equation that allows you to solve for x.

100. In order to solve the equation from Exercise 99, you will probably find it helpful to write the decimal coefficient as a fraction. Do this, writing the fraction in lowest terms.

101. What is the least common denominator of all the fractions in the equation?

102. Multiply the terms of the equation by the least common denominator. What property of equality assures you that your new equation has the same solution as the original one?

103. To solve the equation your next step should be to clear parentheses. Do this, and tell what property you used.

104. Solve the equation.

105. What are the measures of the angles of the triangle?

106. Use your result from Exercise 105 to find the value of y in the figure.

──────◆──────

CHAPTER 2 TEST

Simplify by combining like terms.

1. $8x + 4x - 6x + x + 14x$

2. $5(2x - 1) - (x - 12) + 2(3x - 5)$

Solve the equation.

3. $5x + 9 = 7x + 21$

4. $2 - 3(y - 5) = 3 + (y + 1)$

5. $2.3x + 13.7 = 1.3x + 2.9$

6. $7 - (m - 4) = -3m + 2(m + 1)$

7. $-\dfrac{4}{7}x = -12$

8. $.06(x + 20) + .08(x - 10) = 4.6$

9. $-8(2x + 4) = -4(4x + 8)$

Solve the problem.

10. If three is subtracted from four times a number, the result is 10 less than five times the number. What is the number?

11. The three largest islands in the Hawaiian island chain are Hawaii (the Big Island), Maui, and Kauai. Together, their areas total 5300 square miles. The island of Hawaii is 3293 square miles larger than the island of Maui, and Maui is 177 square miles larger than Kauai. What is the area of each island?

12. The formula for the perimeter of a rectangle is $P = 2L + 2W$.
   (a) Solve for W.
   (b) If $P = 116$ and $L = 40$, find the value of W.

13. Suppose that in the equation $\dfrac{a}{b} = \dfrac{c}{d}$, a and d are positive numbers and b is a negative number. Must c be positive or negative?

Find the measure of each marked angle.

14.

. 15.

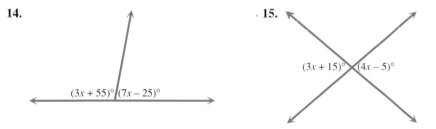

$(3x + 55)°$ $(7x - 25)°$

$(3x + 15)°$ $(4x - 5)°$

16. Find the measure of an angle if its supplement measures 10° more than three times its complement.

Solve the equation.

17. $\dfrac{y + 5}{3} = \dfrac{y - 3}{4}$

Solve the problem.

18. Which is the better buy for processed cheese slices: 8 slices for $2.19 or 12 slices for $3.30?

19. The distance between Milwaukee and Boston is 1050 miles. On a certain map this distance is represented by 21 inches. On the same map Seattle and Cincinnati are 46 inches apart. What is the actual distance between Seattle and Cincinnati?

Solve the problem.

20. Laura Taber invested some money at 3% simple interest and $6000 more than that amount at 4.5% simple interest. After one year her total interest from the two accounts was $870. How much did she invest at each rate?

21. How many liters of a 20% chemical solution must be mixed with 30 liters of a 60% solution to get a 50% mixture?

22. Two cars leave from the same point, traveling in opposite directions. One travels at a constant rate of 50 miles per hour while the other travels at a constant rate of 65 miles per hour. How long will it take for them to be 460 miles apart?

Solve the inequality and graph the solutions.

23. $-4x + 2(x - 3) \geq 4x - (3 + 5x) - 7$ **24.** $-10 < 3k - 4 \leq 14$

Solve the problem.

25. Dee Netzel has grades of 84 and 100 on her first two college algebra tests. What must she score on her third test so that her average is at least 90?

CUMULATIVE REVIEW (Chapters 1–2)

Beginning with this chapter each chapter in the text will conclude with a set of cumulative review exercises designed to cover the major topics from the beginning of the course. This feature allows the student to constantly review topics that have been introduced up to that point.

Write the fraction in lowest terms.

1. $\dfrac{15}{40}$ 　　　　　　　　　　　　　　　　**2.** $\dfrac{108}{144}$

Perform the indicated operation.

3. $\dfrac{5}{6} + \dfrac{1}{4} + \dfrac{7}{15}$ 　**4.** $16\dfrac{7}{8} - 3\dfrac{1}{10}$ 　**5.** $\dfrac{9}{8} \cdot \dfrac{16}{3}$ 　**6.** $\dfrac{3}{4} \div \dfrac{5}{8}$

Translate from words to symbols. Use x as the variable, if necessary.

7. 12 more than twice a number

8. The product of a number and 3

9. The difference between half a number and 18.

10. The quotient of 6 and 12 less than a number

Solve the problem.

11. In making dresses, Earth Works uses $\dfrac{5}{8}$ yard of trim per dress. How many yards of trim would be used to make 56 dresses?

12. A cook wants to increase a recipe that serves 6 to make enough for 20 people. The recipe calls for $1\dfrac{1}{4}$ cups of cheese. How much cheese will be needed to serve 20?

13. John and Gwen are painting a bedroom for their new baby. They painted $\dfrac{1}{4}$ of the room on Saturday and $\dfrac{1}{3}$ of the room on Sunday. How much of the room is still unpainted?

14. A recipe for lemon sherbet calls for $1\frac{1}{3}$ cups of lemon juice. How many cups of lemon juice would be needed for 9 recipes?

Tell whether the statement is true or false.

15. $\dfrac{8(7) - 5(6 + 2)}{3 \cdot 5 + 1} \geq 1$

16. $\dfrac{4(9 + 3) - 8(4)}{2 + 3 - 3} \geq 2$

Perform the indicated operations.

17. $-11 + 20 + (-2)$

18. $13 + (-19) + 7$

19. $9 - (-4)$

20. $-2(-5)(-4)$

21. $\dfrac{4 \cdot 9}{-3}$

22. $\dfrac{8}{7 - 7}$

23. $(-5 + 8) + (-2 - 7)$

24. $(-7 - 1)(-4) + (-4)$

25. $\dfrac{-3 - (-5)}{1 - (-1)}$

26. $\dfrac{6(-4) - (-2)(12)}{3^2 + 7^2}$

27. $\dfrac{(-3)^2 - (-4)(2^4)}{5 \cdot 2 - (-2)^3}$

28. $\dfrac{-2(5^3) - 6}{4^2 + 2(-5) + (-2)}$

Find the value of the expression when $x = -2$, $y = -4$, and $z = 3$.

29. $xz^3 - 5y^2$

30. $\dfrac{3x - y^3}{-4z}$

Name the property illustrated.

31. $7(k + m) = 7k + 7m$

32. $3 + (5 + 2) = 3 + (2 + 5)$

33. $7 + (-7) = 0$

34. $3.5(1) = 3.5$

Simplify the expression by combining terms.

35. $4p - 6 + 3p - 8$

36. $-4(k + 2) + 3(2k - 1)$

Solve the equation and check the solution.

37. $2r - 6 = 8$

38. $2(p - 1) = 3p + 2$

39. $4 - 5(a + 2) = 3(a + 1) - 1$

40. $2 - 6(z + 1) = 4(z - 2) + 10$

41. $-(m - 1) = 3 - 2m$

42. $\dfrac{y - 2}{3} = \dfrac{2y + 1}{5}$

43. $\dfrac{2x + 3}{5} = \dfrac{x - 4}{2}$

44. $\dfrac{2}{3}y + \dfrac{3}{4}y = -17$

Solve the formula for the indicated variable.

45. $P = a + b + c$ for c

46. $P = 4s$ for s

Solve the inequality. Graph the solutions.

47. $-5z \geq 4z - 18$

48. $6(r - 1) + 2(3r - 5) \leq -4$

Solve the problem.

49. Hassan Siddiqui earned $200 at odd jobs during July, $300 during August, and $225 during September. If his average salary for the four months from July through October is to be at least $250, what possible amounts could he earn during October?

50. A piece of string is 40 centimeters long. It is cut into three pieces. The longest piece is 3 times as long as the middle-sized piece, and the shortest piece is 23 centimeters shorter than the longest piece. Find the lengths of the three pieces.

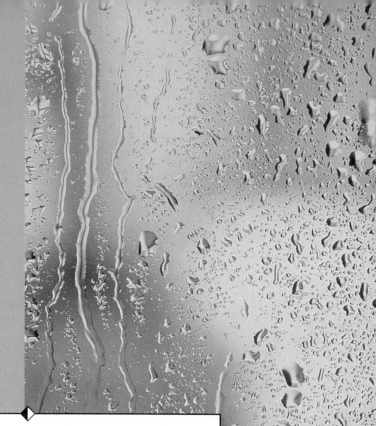

CHAPTER 3

POLYNOMIALS AND EXPONENTS

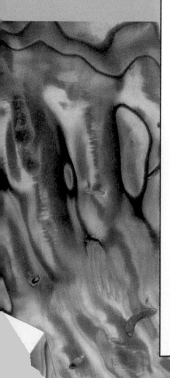

CONNECTIONS

Just as the integers are the basic real numbers, polynomials are the basic algebraic expressions. In this chapter we see how polynomials are added, subtracted, multiplied, and divided, as well as see some of their applications.

Have you ever wondered how your scientific calculator evaluates a number like $\sqrt{.9}$ when you touch the $\sqrt{}$ key? The value of $\sqrt{.9}$ is approximated using the basic operations of addition, subtraction, multiplication, and division of numbers by writing $\sqrt{.9}$ as a *polynomial,* a sum of terms. For example, $\sqrt{.9}$ may be approximated by replacing x in the polynomial

$$1 + \frac{1}{2}x - \frac{1}{8}x^2 + \frac{1}{16}x^3 - \frac{5}{128}x^4$$

with $.9 - 1 = -.1$.

$$\sqrt{.9} \approx 1 + \frac{1}{2}(-.1) - \frac{1}{8}(-.1)^2 + \frac{1}{16}(-.1)^3 - \frac{5}{128}(-.1)^4$$

$$= 1 - .05 - .00125 - .0000625 - .000003906$$

$$= .948683594$$

As more terms are used in the polynomial, the approximation becomes more accurate. The polynomial shown here is most accurate for values of x near 1.

FOR DISCUSSION OR WRITING

Use a calculator to find $\sqrt{.9}$ and compare your result with our approximation. Try using the given polynomial for other values of x "near" 1, such as .5, .7, and so on, and compare the results if you use a value farther from 1, such as 2 or 5. What do you observe?

In Chapter 2 we worked with linear expressions, which are simple polynomials. In this chapter we discuss polynomial expressions that include variables with whole number exponents greater than one. The properties and techniques in this chapter will be important in the subjects to be discussed later in Chapters 4 and 5 and again in Chapters 8 and 9.

3.1 ADDITION AND SUBTRACTION OF POLYNOMIALS

FOR EXTRA HELP	OBJECTIVES
📖 **SSG** pp. 104–110 **SSM** pp. 126–130 📼 **Video** 4 💾 **Tutorial** IBM MAC	**1** ▶ Identify terms and coefficients. **2** ▶ Add like terms. **3** ▶ Know the vocabulary for polynomials. **4** ▶ Evaluate polynomials. **5** ▶ Add polynomials. **6** ▶ Subtract polynomials.

1 ▶ Identify terms and coefficients.

▶ In Chapter 2 we saw that in an expression such as

$$4x^3 + 6x^2 + 5x + 8,$$

the quantities $4x^3$, $6x^2$, $5x$, and 8 are called *terms*. As mentioned earlier, in the term $4x^3$, the number 4 is called the *numerical coefficient,* or simply the *coefficient,* of x^3. In the same way, 6 is the coefficient of x^2 in the term $6x^2$, 5 is the coefficient of x in the term $5x$, and 8 is the coefficient in the term 8. A constant term, like 8 in the polynomial above, can be thought of as $8x^0$, where

$$x^0 \text{ is defined to equal 1.}$$

We explain the reason for this definition later in this chapter.

EXAMPLE 1

Identifying Coefficients

Name the (numerical) coefficient of each term in these expressions.

(a) $4x^3$

The coefficient is 4.

(b) $x - 6x^4$

The coefficient of x is 1 because $x = 1 \cdot x$. The coefficient of x^4 is -6 since $x - 6x^4$ can be written as the sum $x + (-6x^4)$.

(c) $5 - v^3$

The coefficient of the term 5 is 5 because $5 = 5v^0$. By writing $5 - v^3$ as a sum, $5 + (-v^3)$, or $5 + (-1v^3)$, the coefficient of v^3 can be identified as -1. ◾

2 ▶ Add like terms.

▶ Recall from Section 2.1 that *like terms* have exactly the same combination of variables with the same exponents on the variables. Only the coefficients may differ. Examples of like terms are

$$19m^5 \quad \text{and} \quad 14m^5,$$
$$6y^9, \quad -37y^9, \quad \text{and} \quad y^9,$$
$$3pq \quad \text{and} \quad -2pq,$$
$$2xy^2 \quad \text{and} \quad -xy^2.$$

Using the distributive property, we add like terms by adding their coefficients.

EXAMPLE 2
Adding Like Terms

Simplify each expression by adding like terms.

(a) $-4x^3 + 6x^3 = (-4 + 6)x^3 = 2x^3$ Distributive property

(b) $9x^6 - 14x^6 + x^6 = (9 - 14 + 1)x^6 = -4x^6$

(c) $12m^2 + 5m + 4m^2 = (12 + 4)m^2 + 5m = 16m^2 + 5m$

(d) $3x^2y + 4x^2y - x^2y = (3 + 4 - 1)x^2y = 6x^2y$ ■

Example 2(c) shows that it is not possible to combine $16m^2$ and $5m$. These two terms are unlike because the exponents on the variables are different. *Unlike terms* have different variables or different exponents on the same variables.

3 ▶ Know the vocabulary for polynomials.

▶   A **polynomial in x** is a term or the sum of a finite number of terms of the form ax^n, for any real number a and any whole number n. For example,

$$16x^8 - 7x^6 + 5x^4 - 3x^2 + 4$$

is a polynomial in x (the 4 can be written as $4x^0$). This polynomial is written in **descending powers** of the variable, since the exponents on x decrease from left to right. On the other hand,

$$2x^3 - x^2 + \frac{4}{x}$$

is not a polynomial in x, since $4/x$ is not a *product, ax^n,* for a *whole number n.* Of course, we could define *polynomial* using any variable and not just x. In fact, polynomials may have terms with *more* than one variable as in Example 2(d).

The **degree** of a term is the sum of the exponents on the variables. For example, $3x^4$ has degree 4, while $6x^{17}$ has degree 17. The term $5x$ has degree 1, -7 has degree 0 (since -7 can be written as $-7x^0$), and $2x^2y$ has degree $2 + 1 = 3$ (y has an exponent of 1.) The **degree of a polynomial** is the highest degree of any nonzero term of the polynomial. For example, $3x^4 - 5x^2 + 6$ is of degree 4, the polynomial $5x + 7$ is of degree 1, 3 (or $3x^0$) is of degree 0, and $x^2y + xy - 5xy^2$ is of degree 3.

Three types of polynomials are very common and are given special names. A polynomial with exactly three terms is called a **trinomial.** (*Tri-* means "three," as in *tri*angle.) Examples are

$$9m^3 - 4m^2 + 6, \qquad 19y^2 + 8y + 5, \qquad \text{and} \qquad -3m^5n^2 + 2n^3 - m^4.$$

A polynomial with exactly two terms is called a **binomial.** (*Bi-* means "two," as in *bi*cycle.) Examples are

$$-9x^4 + 9x^3, \qquad 8m^2 + 6m, \qquad \text{and} \qquad 3m^5n^2 - 9m^2n^4.$$

A polynomial with only one term is called a **monomial.** (*Mon(o)-* means "one," as in *mono*rail.) Examples are

$$9m, \qquad -6y^5, \qquad a^2b^2, \qquad \text{and} \qquad 6.$$

EXAMPLE 3
Classifying Polynomials

For each polynomial, first simplify if possible by combining like terms. Then give the degree and tell whether it is a monomial, a binomial, a trinomial, or none of these.

(a) $2x^3 + 5$

The polynomial cannot be simplified. The degree is 3. The polynomial is a binomial.

(b) $4xy - 5xy + 2xy$

Add like terms to simplify: $4xy - 5xy + 2xy = xy$ which is a monomial of degree 2. ∎

4 ▶ Evaluate polynomials.

▶ A polynomial usually represents different numbers for different values of the variable, as shown in the next examples.

EXAMPLE 4

Evaluating a Polynomial

Find the value of $3x^4 + 5x^3 - 4x - 4$ when $x = -2$ and when $x = 3$.

First, substitute -2 for x.

$$3x^4 + 5x^3 - 4x - 4 = 3(-2)^4 + 5(-2)^3 - 4(-2) - 4$$
$$= 3 \cdot 16 + 5 \cdot (-8) + 8 - 4$$
$$- 48 - 40 + 8 - 4$$
$$= 12$$

Next, replace x with 3.

$$3x^4 + 5x^3 - 4x - 4 = 3(3)^4 + 5(3)^3 - 4(3) - 4$$
$$= 3 \cdot 81 + 5 \cdot 27 - 12 - 4$$
$$= 362 \quad ∎$$

CAUTION Notice the use of parentheses around the numbers that are substituted for the variable in Example 4. This is particularly important when substituting a negative number for a variable that is raised to a power, so that the sign of the product is correct.

◆ **CONNECTIONS** ◆

Polynomials often give good approximations for real data. According to the U.S. Department of Agriculture, the number of cigarettes sold in the United States from 1981 through 1990 (in billions) decreased according to the polynomial $.043x^5 - .87x^4 + 5.52x^3 - 10.16x^2 - 15.48x + 630$, where x represents the number of years from 1981. Thus, for 1981, $x = 0$; for 1982, $x = 1$; and so on. By replacing x with 9, for example, we find that about 523 billion cigarettes were sold in 1990 (for 1990, $x = 1990 - 1981 = 9$). This was down 107 billion from the 630 billion cigarettes sold in 1981 (replace x with 0 to get 630). Although this "model" of the number of cigarettes sold can be used for years later than 1990, it may not give good approximations for those years, since those data were not used in the construction of the model.

FOR DISCUSSION OR WRITING

Use the given polynomial to approximate the number of cigarettes sold in the U.S. in other years between 1981 and 1990. Evaluate the polynomial for years 1984 and 1985. What do you notice about the number of cigarettes sold in those years?

5 ▶ Add polynomials.   ▶   Polynomials may be added, subtracted, multiplied, and divided. Polynomial addition and subtraction are explained in the rest of this section.

_____
Adding Polynomials   To add two polynomials, we add like terms.

EXAMPLE 5   ▪ Add $6x^3 - 4x^2 + 3$ and $-2x^3 + 7x^2 - 5$.
Adding Polynomials   Write like terms in columns.
Vertically

$$6x^3 - 4x^2 + 3$$
$$\underline{-2x^3 + 7x^2 - 5}$$

Now add, column by column.

$6x^3$	$-4x^2$	3
$-2x^3$	$7x^2$	-5
$4x^3$	$3x^2$	-2

Add the three sums together.

$$4x^3 + 3x^2 + (-2) = 4x^3 + 3x^2 - 2 \quad ▪$$

The polynomials in Example 5 also could be added horizontally, as shown in the next example.

EXAMPLE 6   ▪ Add $6x^3 - 4x^2 + 3$ and $-2x^3 + 7x^2 - 5$.
Adding Polynomials   Write the sum as
Horizontally

$$(6x^3 - 4x^2 + 3) + (-2x^3 + 7x^2 - 5).$$

Use the associative and commutative properties to rewrite this sum with the parentheses removed and with the subtractions changed to additions of inverses.

$$6x^3 + (-4x^2) + 3 + (-2x^3) + 7x^2 + (-5)$$

Place like terms together.

$$6x^3 + (-2x^3) + (-4x^2) + 7x^2 + 3 + (-5)$$

Combine like terms to get

$$4x^3 + 3x^2 + (-2), \quad \text{or simply} \quad 4x^3 + 3x^2 - 2,$$

the same answer found in Example 5.   ▪

6 ▶ Subtract   ▶   Earlier, we defined the difference $x - y$ as $x + (-y)$. (We find the difference **polynomials.** $x - y$ by adding x and the opposite of y.) For example,

$$7 - 2 = 7 + (-2) = 5 \quad \text{and} \quad -8 - (-2) = -8 + 2 = -6.$$

A similar method is used to subtract polynomials.

_____
Subtracting   We subtract two polynomials by changing all the signs on the second
Polynomials   polynomial and adding the result to the first polynomial.

E X A M P L E   7
Subtracting Polynomials

Subtract: $(5x - 2) - (3x - 8)$.

By the definition of subtraction,

$$(5x - 2) - (3x - 8) = (5x - 2) + [-(3x - 8)].$$

As shown in Chapter 1, the distributive property gives

$$-(3x - 8) = -1(3x - 8) = -3x + 8,$$

so

$$(5x - 2) - (3x - 8) = (5x - 2) + (-3x + 8) = 2x + 6. \quad \blacksquare$$

E X A M P L E   8
Subtracting Polynomials

Subtract $6x^3 - 4x^2 + 2$ from $11x^3 + 2x^2 - 8$.

Write the problem.

$$(11x^3 + 2x^2 - 8) - (6x^3 - 4x^2 + 2)$$

Change all the signs in the second polynomial and add the two polynomials.

$$(11x^3 + 2x^2 - 8) + (-6x^3 + 4x^2 - 2) = 5x^3 + 6x^2 - 10$$

We can check a subtraction problem by using the fact that if $a - b = c$, then $a = b + c$. For example, $6 - 2 = 4$, so we check by writing $6 = 2 + 4$, which is correct. We can check the polynomial subtraction above by adding $6x^3 - 4x^2 + 2$ and $5x^3 + 6x^2 - 10$. Since the sum is $11x^3 + 2x^2 - 8$, the subtraction was performed correctly. $\quad \blacksquare$

Subtraction also can be done in columns (vertically). We will use vertical subtraction in Section 3.7 when we study polynomial division.

E X A M P L E   9
Subtracting Polynomials
Vertically

Use the method of subtracting by columns to find

$$(14y^3 - 6y^2 + 2y - 5) - (2y^3 - 7y^2 - 4y + 6).$$

Arrange like terms in columns.

$$14y^3 - 6y^2 + 2y - 5$$
$$\underline{2y^3 - 7y^2 - 4y + 6}$$

Change all signs in the second row, and then add.

$$\begin{array}{r} 14y^3 - 6y^2 + 2y - 5 \\ \underline{-2y^3 + 7y^2 + 4y - 6} \\ 12y^3 + y^2 + 6y - 11 \end{array}$$ Change all signs.
Add. $\blacksquare$

Either the horizontal or the vertical method may be used for adding and subtracting polynomials.

Polynomials in more than one variable are added and subtracted by combining like terms, just as with single variable polynomials.

E X A M P L E   10
Adding and Subtracting
Polynomials with More
Than One Variable

Add or subtract as indicated.

(a) $(4a + 2ab - b) + (3a - ab + b)$

$$(4a + 2ab - b) + (3a - ab + b) = 4a + 2ab - b + 3a - ab + b$$
$$= 7a + ab$$

    (b) $(2x^2y + 3xy + y^2) - (3x^2y - xy - 2y^2)$

$$(2x^2y + 3xy + y^2) - (3x^2y - xy - 2y^2)$$
$$= 2x^2y + 3xy + y^2 - 3x^2y + xy + 2y^2$$
$$= -x^2y + 4xy + 3y^2 \quad \blacksquare$$

3.1 EXERCISES

For the polynomial, determine the number of terms and name the coefficients of the terms. See Example 1.

1. $6x^4$ **2.** $-9y^5$ **3.** t^4 **4.** s^7

5. $-19r^2 - r$ **6.** $2y^3 - y$ **7.** $x + 8x^2$ **8.** $v - 2v^3$

In the polynomial add like terms whenever possible. Write the result in descending powers of the variable. See Examples 2 and 3.

9. $-3m^5 + 5m^5$ **10.** $-4y^3 + 3y^3$ **11.** $2r^5 + (-3r^5)$

12. $(-19y^2) + 9y^2$ **13.** $.2m^5 - .5m^2$ **14.** $-.9y + .9y^2$

15. $-3x^5 + 2x^5 - 4x^5$ **16.** $6x^3 - 8x^3 + 9x^3$ **17.** $-4p^7 + 8p^7 + 5p^9$

18. $-3a^8 + 4a^8 - 3a^2$ **19.** $-4y^2 + 3y^2 - 2y^2 + y^2$ **20.** $3r^5 - 8r^5 + r^5 + 2r^5$

Tell whether the statement is true always, sometimes, or never.

21. A polynomial is a binomial.   **22.** A polynomial is a trinomial.

23. A polynomial has degree 5.   **24.** A binomial has degree 5.

25. A trinomial is a polynomial.   **26.** A binomial is a polynomial.

27. A trinomial is a binomial.   **28.** A binomial is a trinomial.

For the polynomial first simplify, if possible, and write it in descending powers of the variable. Then give the degree of the resulting polynomial and tell whether it is a monomial, a binomial, a trinomial, or none of these. See Example 3.

29. $6x^4 - 9x$ **30.** $7t^3 - 3t$

31. $5m^4 - 3m^2 + 6m^4 - 7m^3$ **32.** $6p^5 + 4p^3 - 8p^5 + 10p^2$

33. $\frac{5}{3}x^4 - \frac{2}{3}x^4$ **34.** $\frac{4}{5}r^6 + \frac{1}{5}r^6$

35. $.8x^4 - .3x^4 - .5x^4 + 7$ **36.** $1.2t^3 - .9t^3 - .3t^3 + 9$

*Find the value of the polynomial when (**a**) $x = 2$ and when (**b**) $x = -1$. See Example 4.*

37. $2x^2 - 4x$ **38.** $8x + 5x^2 + 2$ **39.** $2x^5 - 4x^4 + 5x^3 - x^2$

40. $2x^2 + 5x + 1$ **41.** $-3x^2 + 14x - 2$ **42.** $-2x^2 + 3$

——————◆ **MATHEMATICAL CONNECTIONS** (Exercises 43–46) ◇——————

As explained earlier in this section, the polynomial

$$.043x^5 - .87x^4 + 5.52x^3 - 10.16x^2 - 15.48x + 630$$

gives a good approximation for the number of cigarettes sold in the United States from 1981 through 1990 (in billions), where x represents the number of years from 1981. If we evaluate the polynomial for a specific value of x we will get one and only one value as a result. This idea is basic to the study of functions, *one of the most important concepts in mathematics. Work Exercises 43–46 in order.*

43. If gasoline costs $1.25 per gallon then the monomial $1.25x$ gives the cost of x gallons. Evaluate this monomial for 4, and then use the result to fill in the blanks: If _____ gallons are purchased, the cost is _____ .

44. If it costs $15 to rent a chain saw plus $2 per day, the binomial $2x + 15$ gives the cost to rent the chain saw for x days. Evaluate this polynomial for 6 and then use the result to fill in the blanks: If the saw is rented for _____ days, the cost is _____ .

45. If an object is thrown upward under certain conditions its height in feet is given by the trinomial $-16x^2 + 60x + 80$, where x is in seconds. Evaluate this polynomial for 2.5 and then use the result to fill in the blanks: If _____ seconds have elapsed, the height of the object is _____ feet.

46. Use the polynomial at the beginning of this Connections set to determine the approximate number of cigarettes that were sold in 1986.

——————◆——————

Add or subtract as indicated. See Examples 5 and 9.

47. Add.
$$3m^2 + 5m$$
$$2m^2 - 2m$$

48. Add.
$$4a^3 - 4a^2$$
$$6a^3 + 5a^2$$

49. Subtract.
$$12x^4 - x^2$$
$$8x^4 + 3x^2$$

50. Subtract.
$$13y^5 - y^3$$
$$7y^5 + 5y^3$$

51. Add.
$$\frac{2}{3}x^2 + \frac{1}{5}x + \frac{1}{6}$$
$$\frac{1}{2}x^2 - \frac{1}{3}x + \frac{2}{3}$$

52. Add.
$$\frac{4}{7}y^2 - \frac{1}{5}y + \frac{7}{9}$$
$$\frac{1}{3}y^2 - \frac{1}{3}y + \frac{2}{5}$$

53. Add.
$$9m^3 - 5m^2 + 4m - 8$$
$$-3m^3 + 6m^2 + 8m - 6$$

54. Add.
$$12r^5 + 11r^4 - 7r^3 - 2r^2$$
$$-8r^5 + 10r^4 + 3r^3 + 2r^2$$

55. Subtract.
$$12m^3 - 8m^2 + 6m + 7$$
$$-3m^3 + 5m^2 - 2m - 4$$

56. Subtract.
$$5a^4 - 3a^3 + 2a^2 - a + 6$$
$$-6a^4 + a^3 - a^2 + a - 1$$

57. After reading Examples 5 and 6, explain whether you have a preference regarding horizontal or vertical addition of polynomials.

58. Repeat Exercise 57 but for subtraction of polynomials.

Perform the indicated operations. See Examples 6–8.

59. $(8m^2 - 7m) - (3m^2 + 7m - 6)$

60. $(x^2 + x) - (3x^2 + 2x - 1)$

61. $(16x^3 - x^2 + 3x) + (-12x^3 + 3x^2 + 2x)$ **62.** $(-2b^6 + 3b^4 - b^2) + (b^6 + 2b^4 + 2b^2)$

63. $(7y^4 + 3y^2 + 2y) - (18y^4 - 5y^2 + y)$ **64.** $(8t^5 + 3t^3 + 5t) - (19t^5 - 6t^3 + t)$

65. $(9a^4 - 3a^2 + 2) + (4a^4 - 4a^2 + 2) + (-12a^4 + 6a^2 - 3)$

66. $(4m^2 - 3m + 2) + (5m^2 + 13m - 4) - (16m^2 + 4m - 3)$

67. $[(8m^2 + 4m - 7) - (2m^2 - 5m + 2)] - (m^2 + m + 1)$

68. $[(9b^3 - 4b^2 + 3b + 2) - (-2b^3 - 3b^2 + b)] - (8b^3 + 6b + 4)$

Find the perimeter of the rectangle.

69.

$4x^2 + 3x + 1$

$x + 2$

70.

$5y^2 + 3y + 8$

$y + 4$

Find **(a)** *a polynomial representing the perimeter of the triangle and* **(b)** *the measures of the angles of the triangle.*

71.

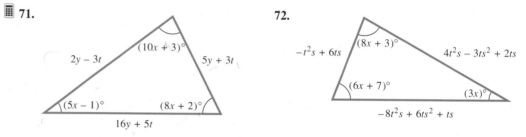

$(10x + 3)°$

$2y - 3t$ $5y + 3t$

$(5x - 1)°$ $(8x + 2)°$

$16y + 5t$

72.

$(8x + 3)°$

$-t^2s + 6ts$ $4t^2s - 3ts^2 + 2ts$

$(6x + 7)°$ $(3x)°$

$-8t^2s + 6ts^2 + ts$

Add or subtract as indicated. See Example 10.

73. $(6b + 3c) + (-2b - 8c)$ **74.** $(-5t + 13s) + (8t - 3s)$

75. $(4x + 2xy - 3) - (-2x + 3xy + 4)$ **76.** $(8ab + 2a - 3b) - (6ab - 2a + 3b)$

77. $(5x^2y - 2xy + 9xy^2) - (8x^2y + 13xy + 12xy^2)$

78. $(16t^3s^2 + 8t^2s^3 + 9ts^4) - (-24t^3s^2 + 3t^2s^3 - 18ts^4)$

The concepts required to work Exercises 79–82 have been covered, but the usual wording of the problem has been changed. Solve the problem.

79. Subtract $9x^2 - 6x + 5$ from $3x^2 - 2$.

80. Find the difference when $9x^4 + 3x^2 + 5$ is subtracted from $8x^4 - 2x^3 + x - 1$.

81. Find the difference between the sum of $5x^2 + 2x - 3$ and $x^2 - 8x + 2$ and the sum of $7x^2 - 3x + 6$ and $-x^2 + 4x - 6$.

82. Subtract the sum of $9t^3 - 3t + 8$ and $t^2 - 8t + 4$ from the sum of $12t + 8$ and $t^2 - 10t + 3$.

83. Explain why the degree of the term 3^4 is not 4. What is its degree?

84. Can the sum of two polynomials in x, both of degree 3, be of degree 2? If so, give an example.

REVIEW EXERCISES

Evaluate each expression. See Section 1.3.

85. $2 \cdot 2 \cdot 2 \cdot 2 \cdot 2 \cdot 2$ **86.** $3 \cdot 3 \cdot 3$

87. $5 \cdot 5 \cdot 5 \cdot 5$ **88.** $4 \cdot 4 \cdot 4 \cdot 4 \cdot 4$

89. $\dfrac{2}{3} \cdot \dfrac{2}{3} \cdot \dfrac{2}{3}$ **90.** $\dfrac{5}{8} \cdot \dfrac{5}{8}$

91. $(2 \cdot 2 \cdot 2)(2 \cdot 2 \cdot 2 \cdot 2)$ **92.** $(3 \cdot 3)(3 \cdot 3 \cdot 3)$

3.2 THE PRODUCT RULE AND POWER RULES FOR EXPONENTS

FOR EXTRA HELP

SSG pp. 111–115
SSM pp. 130–133

Video
4

Tutorial
IBM   MAC

OBJECTIVES

1 ▶ Use exponents.
2 ▶ Use the product rule for exponents.
3 ▶ Use the rule $(a^m)^n = a^{mn}$.
4 ▶ Use the rule $(ab)^m = a^m b^m$.
5 ▶ Use the rule $\left(\dfrac{a}{b}\right)^m = \dfrac{a^m}{b^m}$.

◇ **CONNECTIONS** ◇

Exponential expressions are used to describe many real-life situations. We show one example here. The investment problems discussed in Chapter 2 involved simple interest—interest paid only on the principal. With *compound interest,* interest is paid on the principal and the interest earned earlier. Compound interest is found with an exponential expression. For instance, if $100 is put into an account earning annual (once a year) interest of 3%, then after 1 year the account will have

$$100 + 100(.03) = 100(1 + .03) = 100(1.03)$$

dollars.

Each year thereafter, the interest is found by multiplying the amount in the account by 1.03. At the end of 2 years, the account will have

$$[100(1.03)](1.03) = 100(1.03)^2$$

dollars. After 3 years, the account will have

$$[100(1.03)^2](1.03) = 100(1.03)^3$$

dollars, and so on. A formula for the total amount after n years in an account earning annual compound interest is

$$A = P(1 + r)^n,$$

where P is the original principal deposited, and r is the annual interest rate (as a decimal).

FOR DISCUSSION OR WRITING
Suppose $250 is deposited at 4% annual interest. Use the formula for compound interest to find the amount in the account after 5 years. Find the amount in the account if simple interest was used. (The formula for simple interest is given in the endsheets.) Which method produces more interest? Do you see why?

1 ▶ Use exponents.

▶   Recall from Section 1.2 that in the expression 5^2, the number 5 is the *base* and 2 is the *exponent.* The expression 5^2 is called an *exponential expression.* Usually we do not write the exponent when it is 1; however, sometimes it is convenient to do so. In general, for any quantity a, $a^1 = a$.

EXAMPLE 1

Determining the Base and Exponent in an Exponential Expression

Evaluate each exponential expression. Name the base and the exponent.

(a) $5^4 = 5 \cdot 5 \cdot 5 \cdot 5 = 625$

(b) $-5^4 = -1 \cdot 5^4 = -1 \cdot (5 \cdot 5 \cdot 5 \cdot 5) = -625$

(c) $(-5)^4 = (-5)(-5)(-5)(-5) = 625$

Base	Exponent
5	4
5	4
-5	4

CAUTION It is important to understand the differences between parts (b) and (c) of Example 1. In -5^4 the lack of parentheses shows that the exponent 4 refers only to the base 5, and not -5; in $(-5)^4$ the parentheses show that the exponent 4 refers to the base -5. In summary, $-a^n$ and $(-a)^n$ are not necessarily the same.

Expression	Base	Exponent	Example
$-a^n$	a	n	$-3^2 = -(3 \cdot 3) = -9$
$(-a)^n$	$-a$	n	$(-3)^2 = (-3)(-3) = 9$

---

2 ▶ Use the product rule for exponents.

▶ By the definition of exponents,

$$2^4 \cdot 2^3 = \overbrace{(2 \cdot 2 \cdot 2 \cdot 2)}^{4 \text{ factors}}\overbrace{(2 \cdot 2 \cdot 2)}^{3 \text{ factors}}$$
$$= \underbrace{2 \cdot 2 \cdot 2 \cdot 2 \cdot 2 \cdot 2 \cdot 2}_{4 + 3 = 7 \text{ factors}}$$
$$= 2^7.$$

Also,

$$6^2 \cdot 6^3 = (6 \cdot 6)(6 \cdot 6 \cdot 6)$$
$$= 6 \cdot 6 \cdot 6 \cdot 6 \cdot 6$$
$$= 6^5.$$

Generalizing from these examples, $2^4 \cdot 2^3 = 2^{4+3} = 2^7$ and $6^2 \cdot 6^3 = 6^{2+3} = 6^5$, suggests the **product rule for exponents.**

Product Rule for Exponents

For any positive integers m and n, $\quad a^m \cdot a^n = a^{m+n}$.
(Keep the same base; add the exponents.)
Example: $6^2 \cdot 6^5 = 6^{2+5} = 6^7$

EXAMPLE 2

Using the Product Rule

Use the product rule for exponents to find each product.

(a) $6^3 \cdot 6^5 = 6^{3+5} = 6^8$ by the product rule.

(b) $(-4)^7(-4)^2 = (-4)^{7+2} = (-4)^9$

(c) $x^2 \cdot x = x^2 \cdot x^1 = x^{2+1} = x^3$

(d) $m^4 m^3 m^5 = m^{4+3+5} = m^{12}$

(e) The product rule does not apply to the product $2^3 \cdot 3^2$, since the bases are different.

$$2^3 \cdot 3^2 = 8 \cdot 9 = 72$$

(f) The product rule does not apply to $2^3 + 2^4$, since it is a *sum*, not a *product*.

$$2^3 + 2^4 = 8 + 16 = 24 \quad \blacksquare$$

CAUTION   The bases must be the same before the product rule for exponents can be applied.

EXAMPLE 3

Using the Product Rule

Multiply $2x^3$ and $3x^7$.

Since $2x^3$ means $2 \cdot x^3$ and $3x^7$ means $3 \cdot x^7$, we can use the associative and commutative properties to get

$$2x^3 \cdot 3x^7 = (2 \cdot 3) \cdot (x^3 \cdot x^7) = 6x^{10}. \quad \blacksquare$$

CAUTION   Be sure that you understand the difference between *adding* and *multiplying* exponential expressions. For example,

$$8x^3 + 5x^3 = 13x^3, \qquad \text{but} \qquad (8x^3)(5x^3) = 8 \cdot 5x^{3+3} = 40x^6.$$

---

3 ▶ Use the rule $(a^m)^n = a^{mn}$.

▶   We can simplify an expression such as $(8^3)^2$ with the product rule for exponents as follows.

$$(8^3)^2 = (8^3)(8^3) = 8^{3+3} = 8^6$$

The exponents in $(8^3)^2$ are multiplied to give the exponent in 8^6. As another example,

$$(5^2)^3 = 5^2 \cdot 5^2 \cdot 5^2 = 5^{2+2+2} = 5^6,$$

and $2 \cdot 3 = 6$. These examples suggest **power rule (a) for exponents.**

Power Rule (a) for Exponents

For any positive integers m and n, $(a^m)^n = a^{mn}$.
(Raise a power to a power by multiplying exponents.)
Example: $(3^2)^4 = 3^{2 \cdot 4} = 3^8$

EXAMPLE 4

Using Power Rule (a)

Use power rule (a) for exponents to simplify each expression.

(a) $(2^5)^3 = 2^{5 \cdot 3} = 2^{15}$ 　　　　　**(b)** $(5^7)^2 = 5^{7(2)} = 5^{14}$

(c) $(x^2)^5 = x^{2(5)} = x^{10}$ 　　　　　**(d)** $(n^3)^2 = n^{3(2)} = n^6$

(e) $(-5^6)^3 = (-1 \cdot 5^6)^3 = (-1)^3 \cdot (5^6)^3 = -1 \cdot 5^{18} = -5^{18} \quad \blacksquare$

---

4 ▶ Use the rule $(ab)^m = a^m b^m$.

▶   The properties studied in Chapter 1 can be used to develop two more rules for exponents. Using the definition of an exponential expression and the commutative and associative properties, we can rewrite the expression $(4 \cdot 8)^3$ as follows.

$$
\begin{aligned}
(4 \cdot 8)^3 &= (4 \cdot 8)(4 \cdot 8)(4 \cdot 8) & \text{Definition of exponent} \\
&= (4 \cdot 4 \cdot 4) \cdot (8 \cdot 8 \cdot 8) & \text{Commutative and associative properties} \\
&= 4^3 \cdot 8^3 & \text{Definition of exponent}
\end{aligned}
$$

This example suggests **power rule (b) for exponents.**

Power Rule (b) for Exponents	For any positive integer m, $(ab)^m = a^m b^m$. (Raise a product to a power by raising each factor to the power.) Example: $(2p)^5 = 2^5 p^5$

EXAMPLE 5
Using Power Rule (b)

■ Use power rule (b) for exponents to simplify each expression.

(a) $(3xy)^2 = 3^2 x^2 y^2 = 9x^2 y^2$

(b) $5(pq)^2 = 5(p^2 q^2)$ Power rule (b)

$\qquad = 5p^2 q^2$ Multiply.

(c) $3(2m^2 p^3)^4 = 3[2^4(m^2)^4(p^3)^4]$ Power rule (b)

$\qquad = 3 \cdot 2^4 m^8 p^{12}$ Power rule (a)

$\qquad = 48 m^8 p^{12}$ Multiply. ■

5 ▶ Use the rule $\left(\dfrac{a}{b}\right)^m = \dfrac{a^m}{b^m}$.

▶   Since the quotient $\frac{a}{b}$ can be written as $a(\frac{1}{b})$, we can use power rule (b), together with some of the properties of real numbers, to get **power rule (c) for exponents.**

Power Rule (c) for Exponents	For any positive integer m, $$\left(\frac{a}{b}\right)^m = \frac{a^m}{b^m} \qquad (b \neq 0)$$ (Raise a quotient to a power by raising both numerator and denominator to the power.) Example: $\left(\dfrac{5}{3}\right)^2 = \dfrac{5^2}{3^2}$

EXAMPLE 6
Using Power Rule (c)

■ Use power rule (c) for exponents to simplify each expression.

(a) $\left(\dfrac{2}{3}\right)^5 = \dfrac{2^5}{3^5}$

(b) $\left(\dfrac{m}{n}\right)^3 = \dfrac{m^3}{n^3}, \quad (n \neq 0)$ ■

We list the rules for exponents discussed in this section below. These rules are basic to the study of algebra and should be *memorized*.

Rules for Exponents   For positive integers m and n:

		Examples
Product rule	$a^m \cdot a^n = a^{m+n}$	$6^2 \cdot 6^5 = 6^{2+5} = 6^7$
Power rules (a)	$(a^m)^n = a^{mn}$	$(3^2)^4 = 3^{2 \cdot 4} = 3^8$
(b)	$(ab)^m = a^m b^m$	$(2p)^5 = 2^5 p^5$
(c)	$\left(\dfrac{a}{b}\right)^m = \dfrac{a^m}{b^m} \ (b \neq 0)$	$\left(\dfrac{5}{3}\right)^2 = \dfrac{5^2}{3^2}$

As shown in the next example, more than one rule may be needed to simplify an expression with exponents.

E X A M P L E   7

Using Combinations of Rules

Use the rules for exponents to simplify each expression.

(a) $\left(\dfrac{2}{3}\right)^2 \cdot 2^3 = \dfrac{2^2}{3^2} \cdot \dfrac{2^3}{1}$ Power rule (c)

$= \dfrac{2^2 \cdot 2^3}{3^2 \cdot 1}$ Multiply the fractions.

$= \dfrac{2^5}{3^2}$ Product rule

(b) $(5x)^3(5x)^4 = (5x)^7$ Product rule

$= 5^7 x^7$ Power rule (b)

(c) $(2x^2y^3)^4(3xy^2)^3 = 2^4(x^2)^4(y^3)^4 \cdot 3^3 x^3(y^2)^3$ Power rule (b)

$= 2^4 \cdot 3^3 x^8 y^{12} x^3 y^6$ Power rule (a)

$= 16 \cdot 27 x^{11} y^{18}$ Product rule

$= 432 x^{11} y^{18}$ ■

3.2 EXERCISES

Write the expression using exponents.

1. $3 \cdot 3 \cdot 3 \cdot 3 \cdot 3 \cdot 3 \cdot 3$

2. $8 \cdot 8 \cdot 8 \cdot 8 \cdot 8$

3. $(-6)(-6)(-6)(-6)$

4. $(-2)(-2)(-2)(-2)(-2)$

5. $w \cdot w \cdot w \cdot w \cdot w \cdot w$

6. $t \cdot t \cdot t \cdot t \cdot t \cdot t$

7. $\dfrac{1}{4 \cdot 4 \cdot 4 \cdot 4}$

8. $\dfrac{1}{3 \cdot 3 \cdot 3}$

9. $(-7x)(-7x)(-7x)(-7x)$

10. $(-8p)(-8p)$

11. $\left(\dfrac{1}{2}\right)\left(\dfrac{1}{2}\right)\left(\dfrac{1}{2}\right)\left(\dfrac{1}{2}\right)\left(\dfrac{1}{2}\right)\left(\dfrac{1}{2}\right)$

12. $\left(-\dfrac{1}{4}\right)\left(-\dfrac{1}{4}\right)\left(-\dfrac{1}{4}\right)\left(-\dfrac{1}{4}\right)\left(-\dfrac{1}{4}\right)$

13. Explain how the expressions $(-3)^4$ and -3^4 are different.

14. Explain how the expressions $(5x)^3$ and $5x^3$ are different.

Identify the base and the exponent for the exponential expression. In Exercises 15–18, also evaluate the expression. See Example 1.

15. 3^5

16. 2^7

17. $(-3)^5$

18. $(-2)^7$

19. $(-6x)^4$

20. $(-8x)^4$

21. $-6x^4$

22. $-8x^4$

23. Explain why the product rule does not apply to the expression $5^2 + 5^3$. Then evaluate the expression by finding the individual powers and adding the results.

24. Repeat Exercise 23 for the expression $(-4)^3 + (-4)^4$.

Use the product rule to simplify the expression. Write the answer in exponential form. See Examples 2 and 3.

25. $5^2 \cdot 5^6$

26. $3^6 \cdot 3^7$

27. $4^2 \cdot 4^7 \cdot 4^3$

28. $5^3 \cdot 5^8 \cdot 5^2$

29. $(-7)^3(-7)^6$

30. $(-9)^8(-9)^5$

31. $t^3 \cdot t^8 \cdot t^{13}$

32. $n^5 \cdot n^6 \cdot n^9$

33. $(-8r^4)(7r^3)$

34. $(10a^7)(-4a^3)$

35. $(-6p^5)(-7p^5)$

36. $(-5w^8)(-9w^8)$

37. Explain why the product rule does not apply to the expression $3^2 \cdot 4^3$. Then evaluate the expression by finding the individual powers and multiplying the results.

38. Repeat Exercise 37 for the expression $(-3)^3 \cdot (-2)^5$.

In the following exercises add the given terms. Then, start over and multiply them.

39. $5x^4, 9x^4$ **40.** $8t^5, 3t^5$ **41.** $-7a^2, 2a^2, 10a^2$ **42.** $6x^3, 9x^3, -2x^3$

Use the power rules for exponents to simplify the expression. Write the answer in exponential form. See Examples 4–6.

43. $(4^3)^2$ **44.** $(8^3)^6$ **45.** $(t^4)^5$ **46.** $(y^6)^5$

47. $(7r)^3$ **48.** $(11x)^4$ **49.** $(5xy)^5$ **50.** $(9pq)^6$

51. $(-5^2)^6$ **52.** $(-9^4)^8$ **53.** $(-8^3)^5$ **54.** $(-7^5)^7$

55. $8(qr)^3$ **56.** $4(vw)^5$ **57.** $\left(\dfrac{1}{2}\right)^3$ **58.** $\left(\dfrac{1}{3}\right)^5$

59. $\left(\dfrac{a}{b}\right)^3$ $(b \ne 0)$ **60.** $\left(\dfrac{r}{t}\right)^4$ $(t \ne 0)$ **61.** $\left(\dfrac{9}{5}\right)^8$ **62.** $\left(\dfrac{12}{7}\right)^3$

Use a combination of the rules of exponents introduced in this section to simplify the expression. See Example 7.

63. $\left(\dfrac{5}{2}\right)^3 \cdot \left(\dfrac{5}{2}\right)^2$ **64.** $\left(\dfrac{3}{4}\right)^5 \cdot \left(\dfrac{3}{4}\right)^6$ **65.** $\left(\dfrac{9}{8}\right)^3 \cdot 9^2$

66. $\left(\dfrac{8}{5}\right)^4 \cdot 8^3$ **67.** $(2x)^9(2x)^3$ **68.** $(6y)^5(6y)^8$

69. $(-6p)^4(-6p)$ **70.** $(-13q)^3(-13q)$ **71.** $(6x^2y^3)^5$

72. $(5r^5t^6)^7$ **73.** $(x^2)^3(x^3)^5$ **74.** $(y^4)^5(y^3)^5$

75. $(2w^2x^3y)^2(x^4y)^5$ **76.** $(3x^4y^2z)^3(yz^4)^5$ **77.** $(-r^4s)^2(-r^2s^3)^5$

78. $(-ts^6)^4(-t^3s^5)^3$ **79.** $\left(\dfrac{5a^2b^5}{c^6}\right)^3$ $(c \ne 0)$ **80.** $\left(\dfrac{6x^3y^9}{z^5}\right)^4$ $(z \ne 0)$

81. A student tried to simplify $(10^2)^3$ as 1000^6. Is this correct? If not, how is it simplified using the product rule for exponents?

82. Explain why $(3x^2y^3)^4$ is *not* equivalent to $(3 \cdot 4)x^8y^{12}$.

Find the area of the figure.* (Leave π in the answer for Exercise 86.)

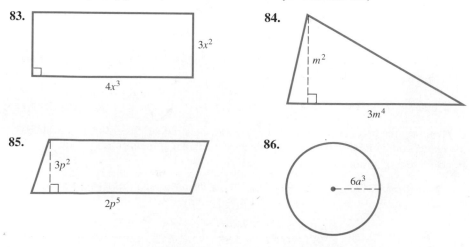

83. $3x^2$, $4x^3$

84. m^2, $3m^4$

85. $3p^2$, $2p^5$

86. $6a^3$

---

*The small square in the figures for Exercises 83–85 indicates a right angle (90°).

Find the volume of the figure.

87.

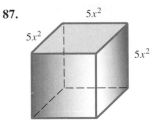

88.

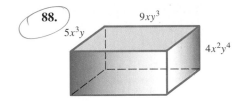

89. Assume a is a positive number greater than 1. Arrange the following terms in order from smallest to largest: $-(-a)^3$, $-a^3$, $(-a)^4$, $-a^4$. Explain how you decided on the order.

90. In your own words, describe a rule to tell whether an exponential expression with a negative base is positive or negative.

Refer to the formula for the total amount of money in an account after n years given in the Connections that introduce this section to find the amount in an account with the given principal (P), annual interest rate (r), and number of years (n). Use a calculator, and round to the nearest cent.

91. $P = \$250$, $r = .04$, $n = 5$ **92.** $P = \$400$, $r = .04$, $n = 3$

93. $P = \$1500$, $r = .035$, $n = 6$ **94.** $P = \$2000$, $r = .025$, $n = 4$

REVIEW EXERCISES

Multiply. See Section 1.9.

95. $5(x + 4)$ **96.** $-3(x^2 + 7)$ **97.** $4(2a + 6b)$ **98.** $\frac{1}{2}(4m - 8n)$

3.3 MULTIPLICATION OF POLYNOMIALS

FOR EXTRA HELP	OBJECTIVES
📖 **SSG** pp. 116–121 **SSM** pp. 133–137	**1** ▶ Multiply a monomial and a polynomial.
📼 **Video** 5	**2** ▶ Multiply two polynomials.
💾 **Tutorial** IBM MAC	**3** ▶ Multiply binomials by the FOIL method.

1 ▶ Multiply a monomial and a polynomial.

▶ As shown in the previous section, the product of two monomials is found by using the rules for exponents and the commutative and associative properties. For example,

$$(-8m^6)(-9m^4) = -8(-9)(m^6)(m^4) = 72m^{6+4} = 72m^{10}.$$

CAUTION It is important not to confuse the *addition* of terms with the *multiplication* of terms. For example,

$$7q^5 + 2q^5 = 9q^5, \quad \text{but} \quad (7q^5)(2q^5) = 7 \cdot 2q^{5+5} = 14q^{10}.$$

To find the product of a monomial and a polynomial with more than one term, we use the distributive property and then the method shown on the previous page.

E X A M P L E   1

Multiplying a Monomial and a Polynomial

Use the distributive property to find each product.

(a) $4x^2(3x + 5)$

$$4x^2(3x + 5) = (4x^2)(3x) + (4x^2)(5) \qquad \text{Distributive property}$$
$$= 12x^3 + 20x^2 \qquad \text{Multiply monomials.}$$

(b) $-8m^3(4m^3 + 3m^2 + 2m - 1)$

$$-8m^3(4m^3 + 3m^2 + 2m - 1)$$
$$= (-8m^3)(4m^3) + (-8m^3)(3m^2) \qquad \text{Distributive property}$$
$$+ (-8m^3)(2m) + (-8m^3)(-1)$$
$$= -32m^6 - 24m^5 - 16m^4 + 8m^3 \qquad \text{Multiply monomials.} \quad \blacksquare$$

2 ▶ Multiply two polynomials.

▶ We can use the distributive property repeatedly to find the product of any two polynomials. For example, to find the product of the polynomials $x + 1$ and $x - 4$, think of $x - 4$ as a single quantity and use the distributive property as follows.

$$(x + 1)(x - 4) = x(x - 4) + 1(x - 4)$$

Now use the distributive property twice to find $x(x - 4)$ and $1(x - 4)$.

$$x(x - 4) + 1(x - 4) = x(x) + x(-4) + 1(x) + 1(-4)$$
$$= x^2 - 4x + x - 4$$
$$= x^2 - 3x - 4$$

(We could have treated $x + 1$ as the single quantity instead to get $(x + 1)\,x + (x + 1)(-4)$ in the first step. Verify that this approach gives the same result, $x^2 - 3x - 4$.)

We give a rule for multiplying any two polynomials below.

Multiplying Polynomials

We multiply two polynomials by multiplying each term of the second polynomial by each term of the first polynomial and adding the products.

E X A M P L E   2

Multiplying Two Polynomials

Find the product of $4m^3 - 2m^2 + 4m$ and $m^2 + 5$.

Multiply each term of the second polynomial by each term of the first. (Either polynomial can be written first in the product.)

$$(m^2 + 5)(4m^3 - 2m^2 + 4m)$$
$$= m^2(4m^3) - m^2(2m^2) + m^2(4m) + 5(4m^3) - 5(2m^2) + 5(4m)$$
$$= 4m^5 - 2m^4 + 4m^3 + 20m^3 - 10m^2 + 20m$$

Now combine like terms.

$$= 4m^5 - 2m^4 + 24m^3 - 10m^2 + 20m \quad \blacksquare$$

When at least one of the factors in a product of polynomials has three or more terms, the multiplication can be simplified by writing one polynomial above the other.

EXAMPLE 3

Multiplying Vertically

Multiply $x^3 + 2x^2 + 4x + 1$ by $3x + 5$.

Start by writing the polynomials as follows.

$$x^3 + 2x^2 + 4x + 1$$
$$3x + 5$$

It is not necessary to line up terms in columns, because any terms may be multiplied (not just like terms). We begin by multiplying each of the terms in the top row by 5.

Step 1

$$
\begin{array}{r}
x^3 + 2x^2 + 4x + 1 \\
3x + 5 \\
\hline
5x^3 + 10x^2 + 20x + 5
\end{array}
\qquad 5(x^3 + 2x^2 + 4x + 1)
$$

Notice how this process is similar to multiplication of whole numbers. Now we multiply each term in the top row by $3x$. Be careful to place the like terms in columns, since the final step will involve addition (as in multiplying two whole numbers).

Step 2

$$
\begin{array}{r}
x^3 + 2x^2 + 4x + 1 \\
3x + 5 \\
\hline
5x^3 + 10x^2 + 20x + 5 \\
3x^4 + 6x^3 + 12x^2 + 3x
\end{array}
\qquad 3x(x^3 + 2x^2 + 4x + 1)
$$

Step 3   Add like terms.

$$
\begin{array}{r}
x^3 + 2x^2 + 4x + 1 \\
3x + 5 \\
\hline
5x^3 + 10x^2 + 20x + 5 \\
3x^4 + 6x^3 + 12x^2 + 3x \\
\hline
3x^4 + 11x^3 + 22x^2 + 23x + 5
\end{array}
$$

The product is $3x^4 + 11x^3 + 22x^2 + 23x + 5$. ■

3 ▶ Multiply binomials by the FOIL method.

▶   In algebra, many of the polynomials to be multiplied are both binomials (with just two terms). For these products a shortcut that eliminates the need to write out all the steps is used. To develop this shortcut, let us first multiply $x + 3$ and $x + 5$ using the distributive property.

$$
\begin{aligned}
(x + 3)(x + 5) &= x(x + 5) + 3(x + 5) \\
&= x(x) + x(5) + 3(x) + 3(5) \\
&= x^2 + 5x + 3x + 15 \\
&= x^2 + 8x + 15
\end{aligned}
$$

The first term in the second line, $(x)(x)$, is the product of the first terms of the two binomials.

$$(x + 3)(x + 5)$$ Multiply the first terms: $(x)(x)$.

The term $(x)(5)$ is the product of the first term of the first binomial and the last term of the second binomial. This is the **outer product.**

$$(x + 3)(x + 5)$$ Multiply the outer terms: $(x)(5)$.

The term $(3)(x)$ is the product of the last term of the first binomial and the first term of the second binomial. The product of these middle terms is the **inner product.**

$$(x + 3)(x + 5) \qquad \text{Multiply the inner terms:} \quad (3)(x).$$

Finally, $(3)(5)$ is the product of the last terms of the two binomials.

$$(x + 3)(x + 5) \qquad \text{Multiply the last terms:} \quad (3)(5).$$

The inner product and the outer product should be added mentally, so that the three terms of the answer can be written without extra steps as

$$(x + 3)(x + 5) = x^2 + 8x + 15.$$

A summary of these steps is given below. This procedure is sometimes called the **FOIL method,** which comes from the abbreviation for *first, outer, inner, last.*

Multiplying Binomials by the FOIL Method	*Step 1* **Multiply the first terms.** Multiply the two first terms of the binomials to get the first term of the answer.
	Step 2 **Find the outer and inner products.** Find the outer product and the inner product and add them (mentally if possible) to get the middle term of the answer.
	Step 3 **Multiply the last terms.** Multiply the two last terms of the binomials to get the last term of the answer.

EXAMPLE 4
Using the FOIL Method

Find the product $(x + 8)(x - 6)$ by the FOIL method.

Step 1 **F** Multiply the *first* terms.

$$x(x) = x^2$$

Step 2 **O** Find the product of the *outer* terms.

$$x(-6) = -6x$$

I Find the product of the *inner* terms.

$$8(x) = 8x$$

Add the outer and inner products mentally.

$$-6x + 8x = 2x$$

Step 3 **L** Multiply the *last* terms.

$$8(-6) = -48$$

The product of $x + 8$ and $x - 6$ is the sum of the terms found in the three steps above, so

$$(x + 8)(x - 6) = x^2 - 6x + 8x - 48 = x^2 + 2x - 48.$$

As a shortcut, this product can be found in the following manner.

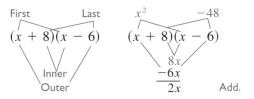

Sometimes it is not possible to add the inner and outer products of the FOIL method, as shown in the next example.

E X A M P L E 5
Using the FOIL Method

Multiply $9x - 2$ and $3y + 1$.

$$
\begin{array}{lll}
First & (9x - 2)(3y + 1) & 27xy \\
Outer & (9x - 2)(3y + 1) & 9x \\
Inner & (9x - 2)(3y + 1) & -6y \\
Last & (9x - 2)(3y + 1) & -2
\end{array}
$$

$$
\begin{array}{cccc}
\text{F} & \text{O} & \text{I} & \text{L} \\
\end{array}
$$
$$(9x - 2)(3y + 1) = 27xy + 9x - 6y - 2 \quad \blacksquare$$

E X A M P L E 6
Using the FOIL Method

Find the following products.

$$
\begin{array}{cccc}
\text{F} & \text{O} & \text{I} & \text{L}
\end{array}
$$
(a) $(2k + 5y)(k + 3y) = (2k)(k) + (2k)(3y) + (5y)(k) + (5y)(3y)$
$$= 2k^2 + 6ky + 5ky + 15y^2$$
$$= 2k^2 + 11ky + 15y^2$$

(b) $(7p + 2q)(3p - q) = 21p^2 - pq - 2q^2 \quad \blacksquare$

N O T E The inner and outer products are often like terms that may be combined, as in Example 6.

3.3 EXERCISES

1. In multiplying two monomials we use (but do not always show) the commutative and associative properties. In each item below, name the property that has been used.

$$
\begin{array}{ll}
(5x^3)(6x^5) = (5x^3 \cdot 6)x^5 & \underline{\hspace{3cm}} \\
= (5 \cdot 6x^3)x^5 & \underline{\hspace{3cm}} \\
= (5 \cdot 6)(x^3 \cdot x^5) & \underline{\hspace{3cm}} \\
= 30x^8
\end{array}
$$

2. In multiplying a monomial by a polynomial, such as in $4x(3x^2 + 7x^3) = 4x(3x^2) + 4x(7x^3)$, the first property that is used is the \underline{\hspace{3cm}} property.

Find the product. See Example 1.

3. $(-5a^9)(-8a^5)$

4. $(-3m^6)(-5m^4)$

5. $-2m(3m + 2)$

6. $-5p(6 + 3p)$

7. $3p(8 - 6p + 12p^3)$

8. $4x(3 + 2x + 5x^3)$

9. $-8z(2z + 3z^2 + 3z^3)$

10. $-7y(3 + 5y^2 - 2y^3)$

11. $7x^2y(2x^3y^2 + 3xy - 4y)$

12. $9xy^3(-3x^2y^4 + 6xy - 2x)$

Find the product. See Examples 2 and 3.

13. $(6x + 1)(2x^2 + 4x + 1)$

14. $(9y - 2)(8y^2 - 6y + 1)$

15. $(4m + 3)(5m^3 - 4m^2 + m - 5)$

16. $(y + 4)(3y^3 - 2y^2 + y + 3)$

17. $(2x - 1)(3x^5 - 2x^3 + x^2 - 2x + 3)$

18. $(2a + 3)(a^4 - a^3 + a^2 - a + 1)$

19. $(5x^2 + 2x + 1)(x^2 - 3x + 5)$

20. $(2m^2 + m - 3)(m^2 - 4m + 5)$

Find the binomial product. Use the FOIL method. See Examples 4–6.

21. $(n - 2)(n + 3)$

22. $(r - 6)(r + 8)$

23. $(x + 6)(x - 6)$

24. $(y + 9)(y - 9)$

25. $(4r + 1)(2r - 3)$

26. $(5x + 2)(2x - 7)$

27. $(3x + 2)(3x - 2)$

28. $(7x + 3)(7x - 3)$

29. $(3q + 1)(3q + 1)$

30. $(4w + 7)(4w + 7)$

31. $(3x + y)(x - 2y)$

32. $(5p + m)(2p - 3m)$

33. $(3t + 4s)(2t + 5s)$

34. $(8v + 5w)(2v + 3w)$

35. $(-3t + 4)(t + 6)$

36. $(-5x + 9)(x - 2)$

37. Find a polynomial that represents the area of this square.

38. Find a polynomial that represents the area of this rectangle.

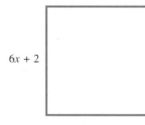

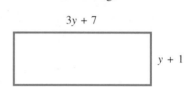

39. Perform the following multiplications:

$$(x + 4)(x - 4); \quad (y + 2)(y - 2); \quad (r + 7)(r - 7).$$

Observe your answers, and explain the pattern that can be found in the answers.

40. Repeat Exercise 39 for the following:

$$(x + 4)(x + 4); \quad (y - 2)(y - 2); \quad (r + 7)(r + 7).$$

Find the product.

41. $\left(3p + \dfrac{5}{4}q\right)\left(2p - \dfrac{5}{3}q\right)$

42. $\left(-x + \dfrac{2}{3}y\right)\left(3x - \dfrac{3}{4}y\right)$

43. $(m^3 - 4)(2m^3 + 3)$

44. $(4a^2 + b^2)(a^2 - 2b^2)$

45. $(2k^3 + h^2)(k^2 - 3h^2)$

46. $(4x^3 - 5y^4)(x^2 + y)$

47. $3p^3(2p^2 + 5p)(p^3 + 2p + 1)$

48. $5k^2(k^2 - k + 4)(k^3 - 3)$

49. $-2x^5(3x^2 + 2x - 5)(4x + 2)$

50. $-4x^3(3x^4 + 2x^2 - x)(-2x + 1)$

Find a polynomial that represents the area of the shaded region. In Exercises 53 and 54 leave π in your answer.

51.

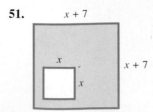

52.

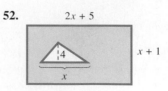

53.

54.

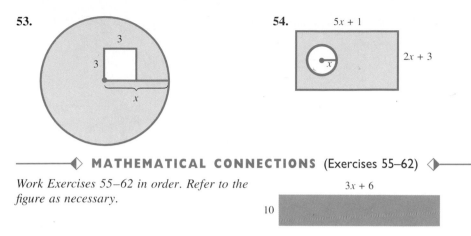

──────◈ **MATHEMATICAL CONNECTIONS** (Exercises 55–62) ◈──────

Work Exercises 55–62 in order. Refer to the figure as necessary.

55. Find a polynomial that represents the area of the rectangle.

56. Suppose you know that the area of the rectangle is 600 square yards. Use this information and the polynomial from Exercise 55 to write an equation that allows you to solve for x.

57. Solve for x.

58. What are the dimensions of the rectangle (assume units are all in yards)?

59. Suppose the rectangle represents a lawn and it costs \$3.50 per square yard to lay sod on the lawn. How much will it cost to sod the entire lawn?

60. Use the result of Exercise 58 to find the perimeter of the lawn.

61. Again, suppose the rectangle represents a lawn and it costs \$9.00 per yard to fence the lawn. How much will it cost to fence the lawn?

62. **(a)** Suppose that it costs k dollars per square yard to sod the lawn. Determine a polynomial in the variables x and k that represents the cost to sod the entire lawn.
(b) Suppose that it costs r dollars per yard to fence the lawn. Determine a polynomial in the variables x and r that represents the cost to fence the lawn.

──────────◈──────────

REVIEW EXERCISES

Apply a power rule for exponents. See Section 3.2.

63. $(3m)^2$ **64.** $(5p)^2$ **65.** $(-2r)^2$ **66.** $(-5a)^2$ **67.** $(4x^2)^2$ **68.** $(8y^3)^2$

3.4 SPECIAL PRODUCTS

FOR EXTRA HELP	OBJECTIVES
📖 **SSG** pp. 122–124 **SSM** pp. 137–141	**1 ▶** Square binomials.
📼 **Video** 5	**2 ▶** Find the product of the sum and difference of two terms.
💾 **Tutorial** IBM MAC	**3 ▶** Find higher powers of binomials.

In the previous section, we saw how the distributive property is used to multiply any two polynomials. In this section, we develop patterns for certain binomial products that occur frequently.

1 ▶ Square binomials.   ▶   The square of a binomial can be found quickly by using the method shown in Example 1.

E X A M P L E   1   ▪Find $(m + 3)^2$.

Squaring a Binomial   Squaring $m + 3$ by the FOIL method gives

$$(m + 3)(m + 3) = m^2 + 3m + 3m + 9 = m^2 + 6m + 9. \quad ▪$$

The result has the square of both the first and the last terms of the binomial:

$$m^2 = m^2 \quad \text{and} \quad 3^2 = 9.$$

The middle term is twice the product of the two terms of the binomial, since both the outer and inner products are $(m)(3)$ and

$$(m)(3) + (m)(3) = 2(m)(3) = 6m.$$

This example suggests the following rule.

Square of a Binomial   The square of a binomial is a trinomial consisting of the square of the first term, plus twice the product of the two terms, plus the square of the last term of the binomial. For x and y,

$$(x + y)^2 = x^2 + 2xy + y^2$$

and

$$(x - y)^2 = x^2 - 2xy + y^2$$

E X A M P L E   2   ▪Use the rule to square each binomial.

Squaring Binomials   **(a)** $(5z - 1)^2 = (5z)^2 - 2(5z)(1) + (1)^2 = 25z^2 - 10z + 1$
Recall that $(5z)^2 = 5^2 z^2 = 25z^2$.

(b) $(3b + 5r)^2 = (3b)^2 + 2(3b)(5r) + (5r)^2 = 9b^2 + 30br + 25r^2$

(c) $(2a - 9x)^2 = 4a^2 - 36ax + 81x^2$

(d) $\left(4m + \dfrac{1}{2} \right)^2 = (4m)^2 + 2(4m)\left(\dfrac{1}{2} \right) + \left(\dfrac{1}{2} \right)^2 = 16m^2 + 4m + \dfrac{1}{4} \quad ▪$

CAUTION   A common error in squaring a binomial is forgetting the middle term of the product. In general, $(a + b)^2 \neq a^2 + b^2$.

---

2 ▶ Find the product of the sum and difference of two terms.   ▶   Binomial products of the form $(x + y)(x - y)$ also occur frequently. In these products, one binomial is the sum of two terms, and the other is the difference of the same two terms. As an example, the product of $a + 2$ and $a - 2$ is

$$(a + 2)(a - 2) = a^2 - 2a + 2a - 4 = a^2 - 4.$$

As we can show with the FOIL method, the product of $x + y$ and $x - y$ is the difference of two squares.

Product of the Sum and Difference of Two Terms	The product of the sum and difference of the two terms x and y is

$$(x + y)(x - y) = x^2 - y^2$$

EXAMPLE 3

Finding the Product of the Sum and Difference of Two Terms

■ Find each product.

(a) $(x + 4)(x - 4)$

Use the pattern for the sum and difference of two terms.

$$(x + 4)(x - 4) = x^2 - 4^2 = x^2 - 16$$

(b) $(3 - w)(3 + w)$

By the commutative property this product is the same as $(3 + w)(3 - w)$.

$$(3 - w)(3 + w) = (3 + w)(3 - w) = 3^2 - w^2 = 9 - w^2$$

(c) $(a - b)(a + b) = a^2 - b^2$. ■

EXAMPLE 4

Finding the Product of the Sum and Difference of Two Terms

■ Find each product.

(a) $(5m + 3)(5m - 3)$

Use the rule for the product of the sum and difference of two terms.

$$(5m + 3)(5m - 3) = (5m)^2 - 3^2 = 25m^2 - 9$$

(b) $(4x + y)(4x - y) = (4x)^2 - y^2 = 16x^2 - y^2$

(c) $\left(z - \dfrac{1}{4}\right)\left(z + \dfrac{1}{4}\right) = z^2 - \dfrac{1}{16}$ ■

◆ **CONNECTIONS** ◆

The algebra of the early Greek period was geometric in nature. For example, a geometric proof for the difference of squares property is possible. Although the property is true for all values of a and b, the proof here is only valid for numbers where $b > a > 0$.

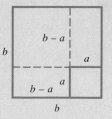

FOR DISCUSSION OR WRITING

Go through the following steps for the proof.

1. Find an expression for the difference of the area of the square with side b and the area of the square with side a.
2. Another way to express this area is as the sum of the three regions that are not colored. Find the area of each of these regions and their sum.
3. Why does this show that $b^2 - a^2 = (b + a)(b - a)$?

The product formulas of this section will be very useful in later work, particularly in Chapters 4 and 5. Therefore, it is important to memorize these formulas and practice using them.

3 ▶ Find higher powers of binomials.

▶ The methods used in the previous section and this section can be combined to find higher powers of binomials.

EXAMPLE 5

Finding Higher Powers of Binomials

Find each product.

(a) $(x + 5)^3$

$$
\begin{array}{ll}
(x + 5)^3 = (x + 5)^2(x + 5) & a^3 = a^2 \cdot a \\
\quad\quad = (x^2 + 10x + 25)(x + 5) & \text{Square the binomial.} \\
\quad\quad = x^3 + 10x^2 + 25x + 5x^2 + 50x + 125 & \text{Multiply polynomials.} \\
\quad\quad = x^3 + 15x^2 + 75x + 125 & \text{Combine terms.}
\end{array}
$$

(b) $(2y - 3)^4$

$$
\begin{array}{ll}
(2y - 3)^4 = (2y - 3)^2(2y - 3)^2 & a^4 = a^2 \cdot a^2 \\
\quad\quad = (4y^2 - 12y + 9)(4y^2 - 12y + 9) & \text{Square each binomial.} \\
\quad\quad = 16y^4 - 48y^3 + 36y^2 - 48y^3 + 144y^2 & \text{Multiply the polynomials.} \\
\quad\quad\quad - 108y + 36y^2 - 108y + 81 & \\
\quad\quad = 16y^4 - 96y^3 + 216y^2 - 216y + 81 & \text{Combine terms.} \quad\blacksquare
\end{array}
$$

3.4 EXERCISES

1. Consider the square $(2x + 3)^2$.
   (a) What is the square of the first term, $(2x)^2$?
   (b) What is twice the product of the two terms, $2(2x)(3)$?
   (c) What is the square of the last term, 3^2?
   (d) Write the final product, which is a trinomial, using your results in parts (a)–(c).

2. Repeat Exercise 1 for the square $(3x - 2)^2$.

Find the square. See Examples 1 and 2.

3. $(p + 2)^2$ 4. $(r + 5)^2$ 5. $(a - c)^2$ 6. $(p - y)^2$

7. $(4x - 3)^2$ 8. $(5y + 2)^2$ 9. $(8t + 7s)^2$ 10. $(7z - 3w)^2$

11. $\left(5x + \dfrac{2}{5}y\right)^2$ 12. $\left(6m - \dfrac{4}{5}n\right)^2$

───────◆ **MATHEMATICAL CONNECTIONS** (Exercises 13–16) ◆───────

To understand how the special product $(a + b)^2 = a^2 + 2ab + b^2$ can be applied to a purely numerical problem, work Exercises 13–16 in order.

13. Evaluate 35^2 using either traditional paper-and-pencil methods or a calculator.

14. The number 35 can be written as $30 + 5$. Therefore, $35^2 = (30 + 5)^2$. Use the special product for squaring a binomial with $a = 30$ and $b = 5$ to write an expression for $(30 + 5)^2$. Do not simplify at this time.

15. Use the rule for order of operations to simplify the expression you found in Exercise 14.

16. How do the answers in Exercises 13 and 14 compare?

────────────────◆────────────────

17. Consider the product $(7x + 3y)(7x - 3y)$.
  (a) What is the product of the first terms, $(7x)(7x)$?
  (b) Multiply the outer terms, $(7x)(-3y)$. Then multiply the inner terms, $(3y)(7x)$. Add the results. What is this sum?
  (c) What is the product of the last terms, $(3y)(-3y)$?
  (d) Write the complete product using your answers in parts (a) and (c). Why is the sum found in part (b) omitted here?

18. Repeat Exercise 17 for the product $(5x + 7y)(5x - 7y)$.

Find the product. See Examples 3 and 4.

19. $(q + 2)(q - 2)$ **20.** $(x + 8)(x - 8)$

21. $(2w + 5)(2w - 5)$ **22.** $(3z + 8)(3z - 8)$

23. $(10x + 3y)(10x - 3y)$ **24.** $(13r + 2z)(13r - 2z)$

25. $(2x^2 - 5)(2x^2 + 5)$ **26.** $(9y^2 - 2)(9y^2 + 2)$

27. $\left(7x + \dfrac{3}{7}\right)\left(7x - \dfrac{3}{7}\right)$ **28.** $\left(9y + \dfrac{2}{3}\right)\left(9y - \dfrac{2}{3}\right)$

29. (a) Find a polynomial that represents the area of the rectangle.
   (b) If the value of x is 4, what is the area of the rectangle? (*Hint:* Evaluate the polynomial you found in part (a) for $x = 4$.)

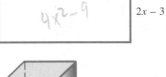

$2x + 3$

$4x^2 - 9$

$2x - 3$

30. (a) Find a polynomial that represents the volume of the cube.
   (b) If the value of x is 6, what is the volume of the cube?

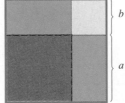

$x + 2$

◆ **MATHEMATICAL CONNECTIONS** (Exercises 31–36) ◆

As shown in the Connections in this section, special products can often be illustrated by using areas of rectangles. Use the figure, and work Exercises 31–36 in order to justify the special product $(a + b)^2 = a^2 + 2ab + b^2$.

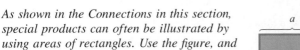

a b

b

a

31. Express the area of the large square as the square of a binomial.

32. Give the monomial that represents the area of the red square.

33. Give the monomial that represents the sum of the areas of the blue rectangles.

34. Give the monomial that represents the area of the yellow square.

35. What is the sum of the monomials you obtained in Exercises 32–34?

36. Explain why the binomial square you found in Exercise 31 must equal the polynomial you found in Exercise 35.

The special product

$$(a + b)(a - b) = a^2 - b^2$$

can be used to perform some multiplication problems. For example,

$$51 \times 49 = (50 + 1)(50 - 1)$$
$$= 50^2 - 1^2 = 2500 - 1$$
$$= 2499$$

$$102 \times 98 = (100 + 2)(100 - 2)$$
$$= 100^2 - 2^2$$
$$= 10{,}000 - 4$$
$$= 9996.$$

Once these patterns are recognized, multiplications of this type can be done mentally.

Use this method to calculate the given product mentally.

37. 101×99 **38.** 103×97 **39.** 201×199

40. 301×299 **41.** $20\frac{1}{2} \times 19\frac{1}{2}$ **42.** $30\frac{1}{3} \times 29\frac{2}{3}$

Find the product. See Example 5.

43. $(m - 5)^3$ **44.** $(p + 3)^3$ **45.** $(2a + 1)^3$ **46.** $(3m - 1)^3$

47. $(y + 4)^4$ **48.** $(z - 2)^4$ **49.** $(3r - 2t)^4$ **50.** $(2z + 5y)^4$

51. Let $a = 2$ and $b = 5$. Evaluate $(a + b)^2$ and $a^2 + b^2$. Does $(a + b)^2 = a^2 + b^2$?

52. Let $p = 7$ and $q = 3$. Evaluate $(p - q)^2$ and $p^2 - q^2$. Does $(p - q)^2 = p^2 - q^2$?

53. Explain how the expressions $x^2 + y^2$ and $(x + y)^2$ differ.

54. Does $a^3 + b^3$ equal $(a + b)^3$? Explain your answer.

Determine a polynomial that represents the area of the figure.

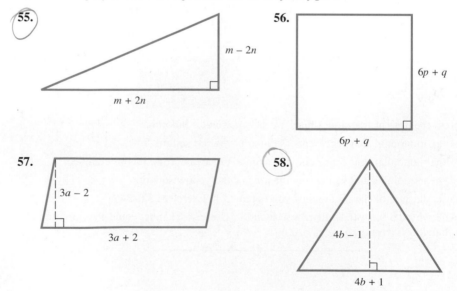

55.

$m - 2n$

$m + 2n$

56.

$6p + q$

$6p + q$

57.

$3a - 2$

$3a + 2$

58.

$4b - 1$

$4b + 1$

59.

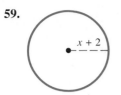

60.

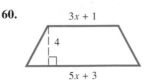

REVIEW EXERCISES

Give the reciprocal of the number. See Section 1.1.

61. 5   **62.** -2 **63.** $-\dfrac{1}{4}$ **64.** .5

Perform the subtraction. See Section 1.6.

65. $8 - (-2)$ **66.** $-2 - 8$
67. Subtract -5 from -2.   **68.** Subtract -2 from -5.

3.5 INTEGER EXPONENTS AND THE QUOTIENT RULE

FOR EXTRA HELP	OBJECTIVES
SSG pp. 124–127 **SSM** pp. 141–143	**1** ▶ Use zero as an exponent.
Video 5	**2** ▶ Use negative numbers as exponents. **3** ▶ Use the quotient rule for exponents.
Tutorial IBM MAC	**4** ▶ Use combinations of rules. **5** ▶ Use variables as exponents.

In an earlier section we studied the product rule for exponents. In all of our work so far, exponents have been positive integers. To develop meanings for exponents other than positive integers (such as 0 and negative integers), we want to define them in such a way that rules for exponents are the same, regardless of the kind of number used for the exponents.

1 ▶ Use zero as an exponent.

▶   Suppose we want to find a meaning for an expression such as

$$6^0,$$

where 0 is used as an exponent. If we were to multiply this by 6^2, for example, we would want the product rule to still be valid. Therefore, we would have

$$6^0 \cdot 6^2 = 6^{0+2} = 6^2.$$

So multiplying 6^2 by 6^0 should give 6^2. Since 6^0 is acting as if it were 1 here, we should define 6^0 to equal 1. This is the definition for 0 used as an exponent with any nonzero base.

Definition of Zero Exponent

For any nonzero real number a, $a^0 = 1$.
Example: $17^0 = 1$.

EXAMPLE 1 ■ Evaluate each exponential expression.

Using Zero Exponents

(a) $60^0 = 1$ **(b)** $(-60)^0 = 1$

(c) $-(60^0) = -(1) = -1$ **(d)** $y^0 = 1$, if $y \neq 0$.

(e) $6y^0 = 6(1) = 6$, if $y \neq 0$ **(f)** $(6y)^0 = 1$, if $y \neq 0$ ■

CAUTION Notice the difference between parts (b) and (c) of Example 1. In Example 1(b) the base is -60 and the exponent is 0. Any nonzero base raised to a zero exponent is 1. But in Example 1(c), the base is 60. Then $60^0 = 1$, and $-60^0 = -1$.

---

2 ▶ Use negative numbers as exponents.

▶   Now let us consider how we can define negative integers as exponents. Suppose that we want to give a meaning to

$$6^{-2}$$

so that the product rule is still valid. If we multiply 6^{-2} by 6^2, we get

$$6^{-2} \cdot 6^2 = 6^{-2+2} = 6^0 = 1.$$

The expression 6^{-2} is acting as if it were the reciprocal of 6^2 since their product is 1. The reciprocal of 6^2 may be written $1/6^2$, leading us to define 6^{-2} as $1/6^2$. This is a particular case of the definition of negative exponents.

Definition of a Negative Exponent

For any nonzero real number a and any integer n, $a^{-n} = \dfrac{1}{a^n}$.

Example: $3^{-2} = \dfrac{1}{3^2}$

By definition, a^{-n} and a^n are reciprocals, since

$$a^n \cdot a^{-n} = a^n \cdot \frac{1}{a^n} = 1.$$

The definition of a^{-n} also can be written as

$$a^{-n} = \frac{1}{a^n} = \left(\frac{1}{a}\right)^n.$$

For example, using the last result above,

$$6^{-3} = \left(\frac{1}{6}\right)^3 \qquad \text{and} \qquad \left(\frac{1}{3}\right)^{-2} = 3^2.$$

EXAMPLE 2 ■ Simplify by using the definition of negative exponents.

Using Negative Exponents

(a) $3^{-2} = \dfrac{1}{3^2} = \dfrac{1}{9}$

(b) $5^{-3} = \dfrac{1}{5^3} = \dfrac{1}{125}$

(c) $\left(\dfrac{1}{2}\right)^{-3} = 2^3 = 8$

(d) $\left(\dfrac{2}{5}\right)^{-4} = \left(\dfrac{5}{2}\right)^{4} = \dfrac{5^4}{2^4} = \dfrac{625}{16}$ $\frac{2}{5}$ and $\frac{5}{2}$ are reciprocals.

(e) $4^{-1} - 2^{-1} = \dfrac{1}{4} - \dfrac{1}{2} = \dfrac{1}{4} - \dfrac{2}{4} = -\dfrac{1}{4}$

(f) $p^{-2} = \dfrac{1}{p^2}, \qquad p \neq 0$

(g) $\dfrac{1}{x^{-4}}, \quad x \neq 0$

$$\dfrac{1}{x^{-4}} = \dfrac{1^{-4}}{x^{-4}} \qquad 1^{-4} = 1$$

$$= \left(\dfrac{1}{x}\right)^{-4} \qquad \text{Power rule (c)}$$

$$= x^4 \qquad \tfrac{1}{x} \text{ and } x \text{ are reciprocals.}$$

(h) $\dfrac{2}{x^{-4}}, \quad x \neq 0$

Write $\dfrac{2}{x^{-4}}$ as $2 \cdot \dfrac{1}{x^{-4}}$. Then use the result from part (g).

$$\dfrac{2}{x^{-4}} = 2 \cdot \dfrac{1}{x^{-4}} = 2 \cdot x^4 \text{ or } 2x^4 \quad \blacksquare$$

CAUTION A negative exponent does not necessarily indicate a negative number; negative exponents lead to reciprocals.

Expression	*Example*	
a^{-n}	$3^{-2} = \dfrac{1}{3^2} = \dfrac{1}{9}$	Not negative
$-a^{-n}$	$-3^{-2} = -\dfrac{1}{3^2} = -\dfrac{1}{9}$	Negative

---

◆ CONNECTIONS ◆

Earlier in this chapter, we gave an example of an exponential expression that modeled the growth of money left at compound interest. Some quantities decay exponentially. That is, the exponential expression that describes the amount present *decreases*. Radioactive substances behave in this way. The amount remaining of 1 gram of the element plutonium 241 after t years is described by the exponential expression $e^{-.053t}$, where e is a constant that is approximately equal to 2.718. By this formula, after 1 year there is about

$$(2.718)^{(-.053)(1)} = .95$$

grams remaining of the original 1 gram. There are many other practical examples of exponential decay.

FOR DISCUSSION OR WRITING

Find the amount of plutonium 241 left after 5 years; then after 10 years. Is the amount getting smaller as expected? Experiment with your calculator to find the number of years when .5 (half) of the plutonium will be left. This value of t is called the *half-life* of the substance.

Examples 2(g) and 2(h) suggest that the definition of negative exponent allows us to move factors in a fraction between the numerator and denominator if we also change the sign of the exponents. For example,

$$\frac{2^{-3}}{3^{-4}} = \frac{\dfrac{1}{2^3}}{\dfrac{1}{3^4}} \qquad \text{Definition of negative exponent}$$

$$= \frac{1}{2^3} \cdot \frac{3^4}{1} \qquad \text{Division by a fraction}$$

$$\frac{2^{-3}}{3^{-4}} = \frac{3^4}{2^3}. \qquad \text{Multiplication of fractions}$$

This fact is generalized below.

Changing from Negative to Positive Exponents

For any nonzero numbers a and b, and any integers m and n,

$$\frac{a^{-m}}{b^{-n}} = \frac{b^n}{a^m}.$$

Example: $\dfrac{3^{-5}}{2^{-4}} = \dfrac{2^4}{3^5}$

EXAMPLE 3

Changing from Negative to Positive Exponents

Write with only positive exponents. Assume all variables represent nonzero real numbers.

(a) $\dfrac{4^{-2}}{5^{-3}} = \dfrac{5^3}{4^2}$

(b) $\dfrac{m^{-5}}{p^{-1}} = \dfrac{p^1}{m^5} = \dfrac{p}{m^5}$

(c) $\dfrac{a^{-2}b}{3d^{-3}} = \dfrac{bd^3}{3a^2}$

(d) $x^3 y^{-4} = \dfrac{x^3 y^{-4}}{1} = \dfrac{x^3}{y^4}$ ∎

CAUTION Be careful. We cannot change negative exponents to positive exponents in this way if they occur in a term. For example,

$$\frac{5^{-2} + 3^{-1}}{7 - 2^{-3}}$$

cannot be written with positive exponents using the rule given here. We would have to use the definition of a negative exponent to rewrite this expression with positive exponents.

---

3 ▶ Use the quotient rule for exponents.

▶ What about the quotient of two exponential expressions with the same base? We know that

$$\frac{6^5}{6^3} = \frac{6 \cdot 6 \cdot 6 \cdot 6 \cdot 6}{6 \cdot 6 \cdot 6} = 6^2.$$

Notice that $5 - 3 = 2$.

Also,

$$\frac{6^2}{6^4} = \frac{6 \cdot 6}{6 \cdot 6 \cdot 6 \cdot 6} = \frac{1}{6^2} = 6^{-2}.$$

Here, $2 - 4 = -2$. These examples suggest the quotient rule for exponents.

Quotient Rule for Exponents

For any nonzero real number a and any integers m and n,

$$\frac{a^m}{a^n} = a^{m-n}.$$

(Keep the base; subtract the exponents.)

Example: $\dfrac{5^8}{5^4} = 5^{8-4} = 5^4$

E X A M P L E   4
Using the Quotient Rule for Exponents

Simplify by using the quotient rule for exponents. Write answers with positive exponents.

(a) $\dfrac{5^8}{5^6} = 5^{8-6} = 5^2$

(b) $\dfrac{4^2}{4^9} = 4^{2-9} = 4^{-7} = \dfrac{1}{4^7}$

(c) $\dfrac{5^{-3}}{5^{-7}} = \dfrac{5^7}{5^3} = 5^{7-3} = 5^4$

(d) $\dfrac{q^5}{q^{-3}} = q^5 q^3 = q^8, \quad q \neq 0$

(e) $\dfrac{3^2 x^5}{3^4 x^3} = \dfrac{3^2}{3^4} \cdot \dfrac{x^5}{x^3} = 3^{2-4} \cdot x^{5-3} = 3^{-2} x^2 = \dfrac{x^2}{3^2} = \dfrac{x^2}{9}, \quad x \neq 0$ ■

Working as shown in Examples 4(c) and 4(d), first changing negative exponents to positive exponents, avoids the potential for errors.

We list all the definitions and rules for exponents given in this section and Section 3.2 in the following summary.

Definitions and Rules for Exponents

If no denominators are zero, for any integers m and n:

		Examples
Product rule	$a^m \cdot a^n = a^{m+n}$	$7^4 \cdot 7^3 = 7^7$
Zero exponent	$a^0 = 1$	$(-3)^0 = 1$
Negative exponent	$a^{-n} = \dfrac{1}{a^n}$	$5^{-3} = \dfrac{1}{5^3}$
Quotient rule	$\dfrac{a^m}{a^n} = a^{m-n}$	$\dfrac{2^2}{2^5} = 2^{-3} = \dfrac{1}{2^3}$
Power rules (a)	$(a^m)^n = a^{mn}$	$(4^2)^3 = 4^6$
(b)	$(ab)^m = a^m b^m$	$(3k)^4 = 3^4 k^4$
(c)	$\left(\dfrac{a}{b}\right)^m = \dfrac{a^m}{b^m}$	$\left(\dfrac{2}{3}\right)^{10} = \dfrac{2^{10}}{3^{10}}$
(d)	$\dfrac{a^{-m}}{b^{-n}} = \dfrac{b^n}{a^m}$	$\dfrac{5^{-3}}{3^{-5}} = \dfrac{3^5}{5^3}$
(e)	$\left(\dfrac{a}{b}\right)^{-m} = \left(\dfrac{b}{a}\right)^m$	$\left(\dfrac{4}{7}\right)^{-2} = \left(\dfrac{7}{4}\right)^2$

4 ▶ Use combinations of rules.

▶ As shown in the next example, sometimes we may need to use more than one rule to simplify an expression.

EXAMPLE 5
Using a Combination of Rules

■ Use a combination of the rules for exponents to simplify each expression.

(a) $\dfrac{(4^2)^3}{4^5}$

Use power rule (a) and then the quotient rule.

$$\frac{(4^2)^3}{4^5} = \frac{4^6}{4^5} = 4^{6-5} = 4^1 = 4$$

(b) $(2x)^3(2x)^2$

Use the product rule first. Then use power rule (b).

$$(2x)^3(2x)^2 = (2x)^5 = 2^5x^5 \quad \text{or} \quad 32x^5$$

(c) $\left(\dfrac{2x^3}{5}\right)^{-4}$

Use power rules (b), (c), and (e).

$$\left(\frac{2x^3}{5}\right)^{-4} = \left(\frac{5}{2x^3}\right)^4 \qquad \text{Change the base to its reciprocal and change the sign of the exponent.}$$

$$= \frac{5^4}{2^4(x^3)^4} \qquad \text{Power rules (b) and (c)}$$

$$= \frac{5^4}{16x^{12}} \qquad \text{Power rule (a)}$$

(d) $\left(\dfrac{3x^{-2}}{4^{-1}y^3}\right)^{-3} = \dfrac{3^{-3}x^6}{4^3y^{-9}} \qquad$ Power rules

$$= \frac{x^6y^9}{3^3 \cdot 4^3} \qquad \text{Power rule (d)}$$

$$= \frac{x^6y^9}{27(64)} \qquad \text{Definition of exponent}$$

$$= \frac{x^6y^9}{1728} \quad ■$$

NOTE Since the steps can be done in several different orders, there are many equally good ways to simplify a problem like Example 5(d).

---

5 ▶ Use variables as exponents.

▶ All the rules and definitions given in the box also apply when variables are used as exponents, as long as the variables represent only integers, an assumption we shall make throughout this chapter.

EXAMPLE 6
Using Variable Exponents

Use the rules for exponents to simplify the following.

(a) $3x^k \cdot 2x^5 = 3 \cdot 2 \cdot x^k \cdot x^5 = 6x^{k+5}$

(b) $\dfrac{y^{7z}}{y^{3z}} = y^{7z-3z} = y^{4z}$

(c) $(2a^4)^r = 2^r \cdot a^{4r} = 2^r a^{4r}$

(d) $z^q \cdot z^{5q} \cdot z^{-4q} = z^{q+5q+(-4q)} = z^{2q}$

(e) $(2m)^a \cdot m^{-a} = 2^a m^a m^{-a} = 2^a m^{a-a} = 2^a m^0 = 2^a \cdot 1 = 2^a$ ∎

3.5 EXERCISES

The given expression is either equal to 0, 1, or −1. Decide which is correct. See Example 1.

1. 9^0

2. 5^0

3. $(-4)^0$

4. $(-10)^0$

5. -9^0

6. -5^0

7. $(-2)^0 - 2^0$

8. $(-8)^0 - 8^0$

9. $\dfrac{0^{10}}{10^0}$

10. $\dfrac{0^5}{5^0}$

11. $x^0 \ (x \neq 0)$

12. $y^0 \ (y \neq 0)$

Evaluate the expression. See Examples 1 and 2.

13. $7^0 + 9^0$

14. $8^0 + 6^0$

15. 4^{-3}

16. 5^{-4}

17. $\left(\dfrac{1}{2}\right)^{-4}$

18. $\left(\dfrac{1}{3}\right)^{-3}$

19. $\left(\dfrac{6}{7}\right)^{-2}$

20. $\left(\dfrac{2}{3}\right)^{-3}$

21. $(-3)^{-4}$

22. $(-4)^{-3}$

23. $5^{-1} + 3^{-1}$

24. $6^{-1} + 2^{-1}$

Decide whether the expression is positive, negative, or zero.

25. $(-2)^{-3}$

26. $(-3)^{-2}$

27. -2^4

28. -3^6

29. $\left(\dfrac{1}{4}\right)^{-2}$

30. $\left(\dfrac{1}{5}\right)^{-2}$

31. $1 - 5^0$

32. $1 - 7^0$

Use the quotient rule to simplify the expression. Write the expression with positive exponents. Assume that all variables represent nonzero real numbers. See Examples 2, 3, and 4.

33. $\dfrac{5^8}{5^5}$

34. $\dfrac{11^6}{11^3}$

35. $\dfrac{9^4}{9^5}$

36. $\dfrac{7^3}{7^4}$

37. $\dfrac{6^{-3}}{6^2}$

38. $\dfrac{4^{-2}}{4^3}$

39. $\dfrac{x^{12}}{x^{-3}}$

40. $\dfrac{y^4}{y^{-6}}$

41. $\dfrac{1}{6^{-3}}$

42. $\dfrac{1}{5^{-2}}$

43. $\dfrac{2}{r^{-4}}$

44. $\dfrac{3}{s^{-8}}$

45. $\dfrac{4^{-3}}{5^{-2}}$

46. $\dfrac{6^{-2}}{5^{-4}}$

47. $p^5 q^{-8}$

48. $x^{-8} y^4$

49. $\dfrac{r^5}{r^{-4}}$

50. $\dfrac{a^6}{a^{-4}}$

51. $\dfrac{6^4 x^8}{6^5 x^3}$

52. $\dfrac{3^8 y^5}{3^{10} y^2}$

—————◆ **MATHEMATICAL CONNECTIONS** (Exercises 53–56) ◆—————

In Objective 1, we showed how 6^0 acts as 1 when it is applied to the product rule, thus motivating the definition for 0 as an exponent. We can also use the quotient rule to motivate this definition. Work Exercises 53–56 in order.

53. Consider the expression $\dfrac{25}{25}$. What is its simplest form?

54. Because $25 = 5^2$, the expression $\dfrac{25}{25}$ can be written as the quotient of powers of 5. Write the expression in this way.

55. Apply the quotient rule for exponents to the expression you wrote in Exercise 54. Give the answer as a power of 5.

56. Your answers in Exercises 53 and 55 must be equal because they both represent $\dfrac{25}{25}$.

Write this equality. What definition does this result support?

—————————————◆—————————————

Problems involving simplification with rules for exponents can often be worked in different ways, yielding the same correct answer. In Example 5(c), we simplified the expression

$$\left(\frac{2x^3}{5}\right)^{-4}$$

in one way. Exercises 57–59 illustrate another way to obtain the same answer. Provide the reason for each step.

57. $\left(\dfrac{2x^3}{5}\right)^{-4} = \dfrac{2^{-4}(x^3)^{-4}}{5^{-4}}$

58. $ = \dfrac{2^{-4}x^{-12}}{5^{-4}}$

59. $ = \dfrac{5^4}{16x^{12}}$

60. If one edge of a cube measures $5x^6$ inches, what is the volume of the cube?

Use a combination of the rules for exponents to simplify the expression. Write answers with only positive exponents. Assume that all variables represent nonzero real numbers. See Example 5.

61. $\dfrac{(7^4)^3}{7^9}$ **62.** $\dfrac{(5^3)^2}{5^2}$ **63.** $x^{-3} \cdot x^5 \cdot x^{-4}$ **64.** $y^{-8} \cdot y^5 \cdot y^{-2}$

65. $\dfrac{(3x)^{-2}}{(4x)^{-3}}$ **66.** $\dfrac{(2y)^{-3}}{(5y)^{-4}}$ **67.** $\left(\dfrac{x^{-1}y}{z^2}\right)^{-2}$ **68.** $\left(\dfrac{p^{-4}q}{r^{-3}}\right)^{-3}$

69. $(6x)^4(6x)^{-3}$ **70.** $(10y)^9(10y)^{-8}$ **71.** $\dfrac{(m^7n)^{-2}}{m^{-4}n^3}$ **72.** $\dfrac{(m^8n^{-4})^2}{m^{-2}n^5}$

Simplify the expression. Assume that all variables represent nonzero integers. Give answers so that the coefficients in the exponents are positive integers. See Example 6.

73. $5^r \cdot 5^{7r} \cdot 5^{-2r}$ **74.** $6^{5p} \cdot 6^p \cdot 6^{-2p}$ **75.** $x^{-a} \cdot x^{-3a} \cdot x^{-7a}$ **76.** $\dfrac{a^{6y}}{a^{2y}}$

77. $\dfrac{q^{-3k}}{q^{-8k}}$ **78.** $\dfrac{z^{-7m}}{z^{-12m}}$ **79.** $(6 \cdot p^{-3})^{-y}$ **80.** $(2 \cdot a^{-p})^4$

REVIEW EXERCISES

Use the distributive property to write each product as a sum of terms. Write answers with positive exponents only. Simplify each term. See Section 1.9.

81. $\dfrac{1}{2p}(4p^2 + 2p + 8)$

82. $\dfrac{1}{5x}(5x^2 - 10x + 45)$

83. $\dfrac{1}{3m}(m^3 + 9m^2 - 6m)$

84. $\dfrac{1}{4y}(y^4 + 6y^2 + 8)$

3.6 THE QUOTIENT OF A POLYNOMIAL AND A MONOMIAL

FOR EXTRA HELP

SSG pp. 127–129
SSM pp. 143–145

Video 5

Tutorial
IBM MAC

OBJECTIVE

1 ▶ Divide a polynomial by a monomial.

1 ▶ Divide a polynomial by a monomial.

▶ We add two fractions with a common denominator as follows.

$$\frac{a}{c} + \frac{b}{c} = \frac{a+b}{c}.$$

Looking at this statement in reverse gives us a rule for dividing a polynomial by a monomial.

Dividing a Polynomial by a Monomial

To divide a polynomial by a monomial, divide each term of the polynomial by the monomial:

$$\frac{a+b}{c} = \frac{a}{c} + \frac{b}{c} \qquad (c \neq 0).$$

The parts of a division problem are named in the diagram.

dividend → $\quad \dfrac{12x^2 + 6x}{6x} = 2x + 1 \quad$ ← quotient
divisor →

EXAMPLE 1
Dividing a Polynomial by a Monomial

Divide $5m^5 - 10m^3$ by $5m^2$.

Use the rule above, with $+$ replaced by $-$. Then use the quotient rule for exponents.

$$\frac{5m^5 - 10m^3}{5m^2} = \frac{5m^5}{5m^2} - \frac{10m^3}{5m^2} = m^3 - 2m$$

Recall from arithmetic that division problems can be checked by multiplication:

$$\frac{63}{7} = 9 \quad \text{because} \quad 7 \cdot 9 = 63.$$

To check the polynomial quotient, multiply $m^3 - 2m$ by $5m^2$. Because

$$5m^2(m^3 - 2m) = 5m^5 - 10m^3,$$

the quotient is correct.

Since division by 0 is undefined, the quotient

$$\frac{5m^5 - 10m^3}{5m^2}$$

is undefined if $m = 0$. In the rest of the chapter we assume that no denominators are 0. ∎

E X A M P L E   2

Dividing a Polynomial by a Monomial

Divide $\dfrac{16a^5 - 12a^4 + 8a^2}{4a^3}$.

Divide each term of $16a^5 - 12a^4 + 8a^2$ by $4a^3$.

$$\frac{16a^5 - 12a^4 + 8a^2}{4a^3} = \frac{16a^5}{4a^3} - \frac{12a^4}{4a^3} + \frac{8a^2}{4a^3}$$

$$= 4a^2 - 3a + \frac{2}{a}$$

The result is not a polynomial because of the expression $2/a$, which has a variable in the denominator. While the sum, difference, and product of two polynomials are always polynomials, the quotient of two polynomials may not be.

Again, check by multiplying.

$$4a^3\left(4a^2 - 3a + \frac{2}{a}\right) = 4a^3(4a^2) - 4a^3(3a) + 4a^3\left(\frac{2}{a}\right)$$

$$= 16a^5 - 12a^4 + 8a^2 \quad ∎$$

E X A M P L E   3

Dividing a Polynomial by a Monomial with a Negative Coefficient

Divide $-7x^3 + 12x^4 - 4x$ by $-4x$.

The polynomial should be written in descending powers before dividing. Write it as $12x^4 - 7x^3 - 4x$; then divide by $-4x$.

$$\frac{12x^4 - 7x^3 - 4x}{-4x} = \frac{12x^4}{-4x} - \frac{7x^3}{-4x} + \frac{-4x}{-4x}$$

$$= -3x^3 + \frac{7x^2}{4} + 1 = -3x^3 + \frac{7}{4}x^2 + 1$$

Check by multiplication. ∎

CAUTION In Example 3, notice that the quotient $\dfrac{-4x}{-4x} = 1$. It is a common error to leave that term out of the answer. Checking by multiplication will show that the answer $-3x^3 + \dfrac{7}{4}x^2$ is not correct.

3.6 EXERCISES

Perform the division.

1. $\dfrac{12x^5}{-2x}$ **2.** $\dfrac{16y^8}{-4y^2}$ **3.** $\dfrac{15r^9y^4}{3x^2y^3}$ **4.** $\dfrac{26w^4z^5}{13w^3z}$

5. Explain why the division problem $\dfrac{16m^3 - 12m^2}{4m}$ can be performed using the methods of this section, while the division problem $\dfrac{4m}{16m^3 - 12m^2}$ cannot.

6. Suppose that a polynomial in the variable x has degree 5 and it is divided by a monomial in the variable x having degree 3. Describe the quotient in mathematical terms giving the degree.

Divide the polynomial by 2m. See Examples 1–3.

7. $60m^4 - 20m^2 + 10m$ **8.** $120m^6 - 60m^3 + 80m^2$ **9.** $10m^5 - 16m^4 + 8m^3$

10. $6m^5 - 4m^3 + 2m^2$ **11.** $8m^5 - 4m^3 + 4m^2$ **12.** $8m^4 - 4m^3 + 6m^2$

13. $m^5 - 4m^2 + 8$ **14.** $m^3 + m^2 + 6$

Divide the polynomial by $3x^2$. See Examples 1–3.

15. $12x^5 - 9x^4 + 6x^3$ **16.** $24x^6 - 12x^5 + 30x^4$ **17.** $3x^2 + 15x^3 - 27x^4$

18. $3x^2 - 18x^4 + 30x^5$ **19.** $36x + 24x^2 + 6x^3$ **20.** $9x - 12x^2 + 9x^3$

21. $4x^4 + 3x^3 + 2x$ **22.** $5x^4 - 6x^3 + 8x$

Perform the division. See Examples 1–3.

23. $\dfrac{27r^4 - 36r^3 - 6r^2 + 26r - 2}{3r}$ **24.** $\dfrac{8k^4 - 12k^3 - 2k^2 + 7k - 3}{2k}$

25. $\dfrac{2m^5 - 6m^4 + 8m^2}{-2m^3}$ **26.** $\dfrac{6r^5 - 8r^4 + 10r^2}{-2r^4}$

27. $(20a^4 - 15a^5 + 25a^3) \div (5a^4)$ **28.** $(16y^5 - 8y^2 + 12y) \div (4y^2)$

29. $(120x^{11} - 60x^{10} + 140x^9 - 100x^8) \div (10x^{12})$

30. $(120x^{12} - 84x^9 + 60x^8 - 36x^7) \div (12x^9)$

31. The quotient in Exercise 21 is $\dfrac{4x^2}{3} + x + \dfrac{2}{3x}$. Notice how the third term is written with x in the denominator. Would $\dfrac{2}{3}x$ be an acceptable form for this term? Explain why or why not.

32. Refer to the quotient given in Example 2 in this section. Write it as an equivalent expression using negative exponents as necessary.

33. Evaluate $\dfrac{5y + 6}{2}$ when $y = 2$. Evaluate $5y + 3$ when $y = 2$. Does $\dfrac{5y + 6}{2}$ equal $5y + 3$?

34. Evaluate $\dfrac{10r + 7}{5}$ when $r = 1$. Evaluate $2r + 7$ when $r = 1$. Does $\dfrac{10r + 7}{5}$ equal $2r + 7$?

Use the appropriate formula to answer the question.

35. The area of the rectangle is given by the polynomial $15x^3 + 12x^2 - 9x + 3$. What is the polynomial that expresses the length?

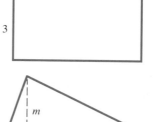

36. The area of the triangle is given by the polynomial $24m^3 + 48m^2 + 12m$. What is the polynomial that expresses the length of the base?

37. The quotient of a certain polynomial and $-7m^2$ is $9m^2 + 3m + 5 - \dfrac{2}{m}$. Find the polynomial.

38. Suppose that a polynomial of degree n is divided by a monomial of degree m to get a *polynomial* quotient. **(a)** How do m and n compare in value? **(b)** What is the expression that gives the degree of the quotient?

———————◆ **MATHEMATICAL CONNECTIONS** (Exercises 39–42) ◆———————

Our system of numeration is called a decimal system. It is based on powers of ten. In a whole number such as 2846, each digit is understood to represent the number of powers of ten for its place value. The 2 represents two thousands (2×10^3), the 8 represents eight hundreds (8×10^2), the 4 represents four tens (4×10^1), and the 6 represents six ones (or units) (6×10^0). In expanded form we write

$$2846 = (2 \times 10^3) + (8 \times 10^2) + (4 \times 10^1) + (6 \times 10^0).$$

Keeping this information in mind, work Exercises 39–42 in order.

39. Divide 2846 by 2, using paper-and-pencil methods: $2\overline{)2846}$.

40. Write your answer in Exercise 39 in expanded form.

41. Use the methods of this section to divide the polynomial $2x^3 + 8x^2 + 4x + 6$ by 2.

42. Compare your answers in Exercises 40 and 41. How are they similar? How are they different? For what value of x does the answer in Exercise 41 equal the answer in Exercise 40?

———————————————◆———————————————

REVIEW EXERCISES

Find the product. See Section 3.3.

43. $-3k(8k^2 - 12k + 2)$ **44.** $(3r + 5)(2r + 1)$ **45.** $(-2k + 1)(8k^2 + 9k + 3)$

Subtract. See Section 3.1.

46. $5t^2 + 2t - 6$
$\underline{5t^2 - 3t - 9}$

47. $-4x^3 + 2x^2 - 3x + 7$
$\underline{-4x^3 - 8x^2 +\ \ x - 4}$

48. $x^4 + 0x^3 - 4x^2 + 3x - 8$
$\underline{x^4 - 2x^3 + 4x^2 + 6x - 7}$

3.7 THE QUOTIENT OF TWO POLYNOMIALS

FOR EXTRA HELP

📖 **SSG** pp. 129–133
SSM pp. 146–150

📼 **Video**
5

💾 **Tutorial**
IBM   MAC

OBJECTIVE

1 ▶ Divide a polynomial by a polynomial.

1 ▶ Divide a polynomial by a polynomial.

▶ We use a method of "long division" to divide a polynomial by a polynomial (other than a monomial). This method is similar to the method of long division used for two whole numbers. For comparison, the division of whole numbers is shown alongside the division of polynomials. Both polynomials must be written with descending powers before beginning the division process.

Step 1

Divide 27 into 6696.

$$27 \overline{)6696}$$

Divide $2x + 3$ into $8x^3 - 4x^2 - 14x + 15$.

$$2x + 3 \overline{)8x^3 - 4x^2 - 14x + 15}$$

Step 2

27 divides into 66 **2** times; $2 \cdot 27 = $ **54**.

$$\begin{array}{r} 2 \\ 27 \overline{)6696} \\ 54 \end{array}$$

$2x$ divides into $8x^3$ **$4x^2$** times; $4x^2(2x + 3) = $ **$8x^3 + 12x^2$**.

$$\begin{array}{r} 4x^2 \\ 2x + 3 \overline{)8x^3 - 4x^2 - 14x + 15} \\ 8x^3 + 12x^2 \end{array}$$

Step 3

Subtract: $66 - 54 = 12$; then bring down the next digit.

$$\begin{array}{r} 2 \\ 27 \overline{)6696} \\ 54 \downarrow \\ \overline{129} \end{array}$$

Subtract: $-4x^2 - 12x^2 = -16x^2$; then bring down the next term.

$$\begin{array}{r} 4x^2 \\ 2x + 3 \overline{)8x^3 - 4x^2 - 14x + 15} \\ 8x^3 + 12x^2 \quad \downarrow \\ \overline{-16x^2 - 14x} \end{array}$$

(To subtract two polynomials, change the sign of the second and then add.)

Step 4

27 divides into 129 **4** times; $4 \cdot 27 = $ **108**.

$$\begin{array}{r} 24 \\ 27 \overline{)6696} \\ 54 \\ \overline{129} \\ 108 \end{array}$$

$2x$ divides into $-16x^2$ **$-8x$** times; $-8x(2x + 3) = $ **$-16x^2 - 24x$**.

$$\begin{array}{r} 4x^2 - 8x \\ 2x + 3 \overline{)8x^3 - 4x^2 - 14x + 15} \\ 8x^3 + 12x^2 \\ \overline{-16x^2 - 14x} \\ -16x^2 - 24x \end{array}$$

Step 5

Subtract: $129 - 108 = 21$; then bring down the next digit.

Subtract: $-14x - (-24x) = 10x$; then bring down the next term.

$$
\begin{array}{r}
24 \\
27\overline{)6696} \\
54 \\
\overline{129} \\
108 \\
\overline{216}
\end{array}
$$

$$
\begin{array}{r}
4x^2 - 8x \phantom{{}+ 15} \\
2x + 3\overline{)8x^3 - 4x^2 - 14x + 15} \\
8x^3 + 12x^2 \phantom{{}- 14x + 15} \\
\overline{-16x^2 - 14x } \\
-16x^2 - 24x \\
\overline{10x + 15}
\end{array}
$$

Step 6

27 divides into 216 **8** times; $8 \cdot 27 = \mathbf{216}$.

$2x$ divides into $10x$ **5** times; $5(2x + 3) = \mathbf{10x + 15}$.

$$
\begin{array}{r}
248 \\
27\overline{)6696} \\
54 \\
\overline{129} \\
108 \\
\overline{216} \\
216 \\
\end{array}
$$

$$
\begin{array}{r}
4x^2 - 8x + 5 \\
2x + 3\overline{)8x^3 - 4x^2 - 14x + 15} \\
8x^3 + 12x^2 \phantom{{}- 14x + 15} \\
\overline{-16x^2 - 14x } \\
-16x^2 - 24x \\
\overline{10x + 15} \\
\mathbf{10x + 15}
\end{array}
$$

6696 divided by 27 is 248. There is no remainder.

$8x^3 - 4x^2 - 14x + 15$ divided by $2x + 3$ is $4x^2 - 8x + 5$. There is no remainder.

Step 7

Check by multiplication.
$27 \cdot 248 = 6696$

Check by multiplication.
$(2x + 3)(4x^2 - 8x + 5)$
$= 8x^3 - 4x^2 - 14x + 15$

Notice that at each step in the polynomial division process, the *first* term was divided into the *first* term.

EXAMPLE 1

Dividing a Polynomial by a Polynomial

Divide $5x + 4x^3 - 8 - 4x^2$ by $2x - 1$.

Both polynomials must be written in descending powers of x. Rewrite the first polynomial as $4x^3 - 4x^2 + 5x - 8$. Then begin the division process.

$$
\begin{array}{r}
2x^2 - x + 2 \\
2x - 1\overline{)4x^3 - 4x^2 + 5x - 8} \\
4x^3 - 2x^2 \phantom{{}+ 5x - 8} \\
\overline{-2x^2 + 5x } \\
-2x^2 + x \\
\overline{4x - 8} \\
4x - 2 \\
\overline{-6}
\end{array}
$$

Step 1 $2x$ divides into $4x^3$ $(\mathbf{2x^2})$ times; $2x^2(2x - 1) = 4x^3 - 2x^2$.

Step 2 Subtract; bring down the next term.

Step 3 $2x$ divides into $-2x^2$ $(\mathbf{-x})$ times; $-x(2x - 1) = -2x^2 + x$.

Step 4 Subtract; bring down the next term.

Step 5 $2x$ divides into $4x$ **2** times; $2(2x - 1) = 4x - 2$.

Step 6 Subtract. The remainder is -6.

Thus, $2x - 1$ divides into $4x^3 - 4x^2 + 5x - 8$ with a quotient of $2x^2 - x + 2$ and a remainder of -6. Write the remainder as a fraction with $2x - 1$ as the denominator. The result is not a polynomial because of the remainder.

$$\frac{4x^3 - 4x^2 + 5x - 8}{2x - 1} = 2x^2 - x + 2 + \frac{-6}{2x - 1}$$

Step 7   Check by multiplication.

$$(2x - 1)\left(2x^2 - x + 2 + \frac{-6}{2x - 1}\right)$$

$$= (2x - 1)(2x^2) + (2x - 1)(-x) + (2x - 1)(2) + (2x - 1)\left(\frac{-6}{2x - 1}\right)$$

$$= 4x^3 - 2x^2 - 2x^2 + x + 4x - 2 - 6$$

$$= 4x^3 - 4x^2 + 5x - 8 \quad \blacksquare$$

E X A M P L E   2

Dividing Into a Polynomial
with Missing Terms

Divide $x^3 + 2x - 3$ by $x - 1$.

Here the polynomial $x^3 + 2x - 3$ is missing the x^2 term. When terms are missing, use 0 as the coefficient for the missing terms.

$$x^3 + 2x - 3 = x^3 + 0x^2 + 2x - 3$$

Now divide.

$$
\begin{array}{r}
x^2 + x + 3 \\
x - 1 \overline{\smash{)}\, x^3 + 0x^2 + 2x - 3} \\
\underline{x^3 - x^2} \\
x^2 + 2x \\
\underline{x^2 - x} \\
3x - 3 \\
\underline{3x - 3} \\
\end{array}
$$

The remainder is 0. The quotient is $x^2 + x + 3$. Check by multiplication.

$$(x^2 + x + 3)(x - 1) = x^3 + 2x - 3 \quad \blacksquare$$

E X A M P L E   3

Dividing by a Polynomial
with Missing Terms

Divide $x^4 + 2x^3 + 2x^2 - x - 1$ by $x^2 + 1$.

Since $x^2 + 1$ has a missing x term, write it as $x^2 + 0x + 1$. Then proceed through the division process.

$$
\begin{array}{r}
x^2 + 2x + 1 \\
x^2 + 0x + 1 \overline{\smash{)}\, x^4 + 2x^3 + 2x^2 - x - 1} \\
\underline{x^4 + 0x^3 + x^2} \\
2x^3 + x^2 - x \\
\underline{2x^3 + 0x^2 + 2x} \\
x^2 - 3x - 1 \\
\underline{x^2 + 0x + 1} \\
-3x - 2 \\
\end{array}
$$

When the result of subtracting ($-3x - 2$, in this case) is a polynomial of smaller degree than the divisor ($x^2 + 0x + 1$), that polynomial is the remainder. Write the result as

$$x^2 + 2x + 1 + \frac{-3x - 2}{x^2 + 1}. \quad \blacksquare$$

◆ CONNECTIONS ◆

In Section 3.1, we found the value of a polynomial in x for a given value of x by substituting that number for x. Surprisingly, we can accomplish the same thing by division. Suppose we want to find the value of $2x^3 - 4x^2 + 3x - 5$ for $x = -3$. Instead of substituting -3 for x in the polynomial, we divide the polynomial by $x - (-3) = x + 3$. The remainder will give the value of the polynomial for $x = -3$.

FOR DISCUSSION OR WRITING
1. Evaluate $2x^3 - 4x^2 + 3x - 5$ for $x = -3$.
2. Divide $2x^3 - 4x^2 + 3x - 5$ by $x + 3$. Give the remainder.
3. Compare the answers to Problems 1 and 2. What do you notice?
4. Choose another polynomial and evaluate it both ways at some value of the variable. Do the answers agree?

3.7 EXERCISES

Perform the division. See Examples 1 and 2.

1. $\dfrac{x^2 - x - 6}{x - 3}$

2. $\dfrac{m^2 - 2m - 24}{m - 6}$

3. $\dfrac{2y^2 + 9y - 35}{y + 7}$

4. $\dfrac{2y^2 + 9y + 7}{y + 1}$

5. $\dfrac{p^2 + 2p + 20}{p + 6}$

6. $\dfrac{x^2 + 11x + 16}{x + 8}$

7. $(r^2 - 8r + 15) \div (r - 3)$

8. $(t^2 + 2t - 35) \div (t - 5)$

9. $\dfrac{12m^2 - 20m + 3}{2m - 3}$

10. $\dfrac{12y^2 + 20y + 7}{2y + 1}$

11. $\dfrac{4a^2 - 22a + 32}{2a + 3}$

12. $\dfrac{9w^2 + 6w + 10}{3w - 2}$

13. $\dfrac{8x^3 - 10x^2 - x + 3}{2x + 1}$

14. $\dfrac{12t^3 - 11t^2 + 9t + 18}{4t + 3}$

15. $\dfrac{14k^2 + 19k - 30}{7k - 8}$

16. $\dfrac{15m^2 + 34m + 28}{5m + 3}$

17. $\dfrac{2x^3 + 3x - x^2 + 2}{2x + 1}$

18. $\dfrac{-11t^2 + 12t^3 + 18 + 9t}{4t + 3}$

19. $\dfrac{8k^4 - 12k^3 - 2k^2 + 7k - 6}{2k - 3}$

20. $\dfrac{27r^4 - 36r^3 - 6r^2 + 26r - 24}{3r - 4}$

21. $\dfrac{5y^4 + 5y^3 + 2y^2 - y - 3}{y + 1}$

22. $\dfrac{2r^3 - 5r^2 - 6r + 15}{r - 3}$

23. $\dfrac{3k^3 - 4k^2 - 6k + 10}{k - 2}$

24. $\dfrac{5z^3 - z^2 + 10z + 2}{z + 2}$

25. $\dfrac{6p^4 - 16p^3 + 15p^2 - 5p + 10}{3p + 1}$

26. $\dfrac{6r^4 - 11r^3 - r^2 + 16r - 8}{2r - 3}$

——————◆ **MATHEMATICAL CONNECTIONS** (Exercises 27–30) ◆——————

Students often would like to know whether the quotient obtained in a polynomial division problem is actually correct or whether an error was made. While the method described here is not a 100% foolproof method, it is quick and will at least give a fairly good idea as to the accuracy of the result. To illustrate, suppose that $4x^4 + 2x^3 - 14x^2 + 19x + 10$ is divided by $2x + 5$. Lakeisha and Stan obtain the following answers:

Lakeisha	*Stan*
$2x^3 - 4x^2 + 3x + 2$	$2x^3 - 4x^2 - 3x + 2.$

As a "quick check" we can evaluate Lakeisha's answer for $x = 1$ and Stan's answer for $x = 1$:

Lakeisha: When $x = 1$, her answer gives $2(1)^3 - 4(1)^2 + 3(1) + 2 = 3$.

Stan: When $x = 1$, his answer gives $2(1)^3 - 4(1)^2 - 3(1) + 2 = -3$.

Now, if the original quotient of the two polynomials is evaluated for $x = 1$, we get

$$\frac{4x^4 + 2x^3 - 14x^2 + 19x + 10}{2x + 5} = \frac{4(1)^4 + 2(1)^3 - 14(1)^2 + 19(1) + 10}{2(1) + 5}$$

$$= \frac{21}{7}$$

$$= 3.$$

Because Stan's answer, -3, is different from the quotient 3 just obtained, Stan can conclude his answer is incorrect. Lakeisha's answer, 3, agrees with the quotient just obtained, and while this does not *guarantee* that she is correct, at least she can feel better and go on to the next problem.

In Exercises 27–29, a division problem is given, along with two possible answers. One is correct and one is incorrect. Use the method just described to determine which one is correct and which one is not.

27. Problem

$$\frac{2x^2 + 3x - 14}{x - 2}$$

Possible answers

(a) $2x + 7$ **(b)** $2x - 7$

28. Problem

$$\frac{x^4 + 4x^3 - 5x^2 - 12x + 6}{x^2 - 3}$$

Possible answers

(a) $x^2 - 4x - 2$ **(b)** $x^2 + 4x - 2$

29. Problem

$$\frac{2y^3 + 17y^2 + 37y + 7}{2y + 7}$$

Possible answers

(a) $y^2 + 5y + 1$ **(b)** $y^2 - 5y + 1$

30. In the explanation preceding Exercise 27 we used 1 for the value of x to check our work. This is due to the fact that a polynomial is easy to evaluate for 1. Explain why this is so. Why would we not be able to use 1 if the divisor is $x - 1$?

————————◆————————

Perform the division. See Examples 2 and 3.

31. $\dfrac{5 - 2r^2 + r^4}{r^2 - 1}$

32. $\dfrac{4t^2 + t^4 + 7}{t^2 + 1}$

33. $\dfrac{y^3 + 1}{y + 1}$

34. $\dfrac{y^3 - 1}{y - 1}$

35. $\dfrac{a^4 - 1}{a^2 - 1}$

36. $\dfrac{a^4 - 1}{a^2 + 1}$

37. $(3a^2 - 11a + 17) \div (2a + 6)$

38. $(4x^2 + 11x - 8) \div (3x + 6)$

39. $\dfrac{x^4 - 4x^3 + 5x^2 - 3x + 2}{x^2 + 3}$

40. $(3t^4 + 5t^3 - 8t^2 - 13t + 2) \div (t^2 - 5)$

41. $\dfrac{2x^5 + 9x^4 + 8x^3 + 10x^2 + 14x + 5}{2x^2 + 3x + 1}$

42. $\dfrac{4t^5 - 11t^4 - 6t^3 + 5t^2 - t + 3}{4t^2 + t - 3}$

43. Suppose that one of your classmates asks you the following question: "How do I know when to stop the division process in a problem like the one in Exercise 39?" How would you respond?

44. Suppose that someone asks you if the following division problem is correct:

$$(6x^3 + 4x^2 - 3x + 9) \div (2x - 3) = 4x^2 + 9x - 3.$$

Explain how, by looking only at the *first term* of the quotient, you can immediately tell the person that the problem has been worked incorrectly.

Find a polynomial that describes the quantity required.

45. Give the length of the rectangle.

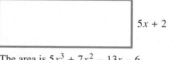

5x + 2

The area is $5x^3 + 7x^2 - 13x - 6$ square units.

46. Find the measure of the base of the parallelogram.

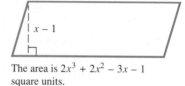

x − 1

The area is $2x^3 + 2x^2 - 3x - 1$ square units.

47. If the distance traveled is $5x^3 - 6x^2 + 3x + 14$ miles and the rate is $x + 1$ miles per hour, what is the time traveled?

48. If it costs $4x^5 + 3x^4 + 2x^3 + 9x^2 - 29x + 2$ dollars to fertilize a garden, and fertilizer costs $x + 2$ dollars per square yard, what is the area of the garden?

REVIEW EXERCISES

Evaluate.

49. $10(6427)$

50. $100(72.69)$

51. $1000(1.23)$

52. $10,000(26.94)$

53. $34 \div 10$

54. $6501 \div 100$

55. $237 \div 1000$

56. $42 \div 10,000$

3.8 AN APPLICATION OF EXPONENTS: SCIENTIFIC NOTATION

FOR EXTRA HELP

📖 **SSG** pp. 133–135
SSM pp. 150–153

📼 **Video**
6

💾 **Tutorial**
IBM   MAC

OBJECTIVES

1 ▶ Express numbers in scientific notation.
2 ▶ Convert numbers in scientific notation to numbers without exponents.
3 ▶ Use scientific notation in calculations.

◈ **CONNECTIONS** ◈

The distance to Earth from the planet Pluto is 4,580,000,000 kilometers. In April 1983, Pioneer 10 transmitted radio signals from Pluto to Earth at the speed of light, 300,000 kilometers per second. As this example indicates, the numbers in science are often extremely large. On the other hand, some numbers used in science are very small: a computer can execute one addition in .00000014 seconds. Because of the difficulty of working with many zeros, these numbers are expressed with powers of 10, as discussed in this section.

FOR DISCUSSION OR WRITING
Write the distance from Earth to Pluto in the following ways.

1. as 458 times an appropriate power of 10
2. as 45.8 times an appropriate power of 10
3. as 4.58 times an appropriate power of 10
4. Your last answer expresses 4,580,000,000 in *scientific notation*, the preferred way to write the number using exponents. Discuss any advantages or disadvantages to writing it this way, rather than as a different product with a power of 10. (*Hint:* Suppose the number was 4,000,000,000.)

In **scientific notation,** a number is written in the form $a \times 10^n$, where n is an integer and $1 \le |a| < 10$. When a number is multiplied by a power of 10, such as $10^1, 10^2 = 100, 10^3 = 1000, 10^{-1} = .1, 10^{-2} = .01, 10^{-3} = .001$, and so on, the net effect is to move the decimal point to the right if it is positive power and to the left if it is a negative power. This is shown in the examples below. (In work with scientific notation, the times symbol, $\times$, is used.)

$23.19 \times 10^1 = 23.19 \times 10 = 231.9$ Decimal point moves 1 place to the right.

$23.19 \times 10^2 = 23.19 \times 100 = 2319.$ Decimal point moves 2 places to the right.

$23.19 \times 10^3 = 23.19 \times 1000 = 23190.$ Decimal point moves 3 places to the right.

$23.19 \times 10^{-1} = 23.19 \times .1 = 2.319$ Decimal point moves 1 place to the left.

$23.19 \times 10^{-2} = 23.19 \times .01 = .2319$ Decimal point moves 2 places to the left.

$23.19 \times 10^{-3} = 23.19 \times .001 = .02319$ Decimal point moves 3 places to the left.

1 ▶ Express numbers in scientific notation.

▶ A number in scientific notation is always written with the decimal point after the first nonzero digit and then multiplied by the appropriate power of 10. For example, 35 is written 3.5×10^1, or 3.5×10; 56,200 is written 5.62×10^4, since

$$56,200 = 5.62 \times \mathbf{10,000} = 5.62 \times \mathbf{10^4}.$$

The steps involved in writing a number in scientific notation are given below.

Writing a Number In Scientific Notation

Step 1   Move the decimal point to the right of the first nonzero digit.

Step 2   Count the number of places you moved the decimal point.

Step 3   The number of places in Step 2 is the absolute value of the exponent on 10.

Step 4   The exponent on 10 is positive if you made the number smaller in Step 1. The exponent is negative if you made the number larger in Step 1. If the decimal point is not moved, the exponent is 0.

EXAMPLE 1

Using Scientific Notation

Write each number in scientific notation.

(a) 93,000,000

The number will be written in scientific notation as 9.3×10^{n}. To find the value of n, first compare 9.3 with the original number, 93,000,000. Is 9.3 larger or smaller? Here, 9.3 is *smaller* than 93,000,000. Therefore, we must multiply by a *positive* power of 10 so the product 9.3×10^{n} will equal the larger number.

Move the decimal point to follow the first nonzero digit (the 9). Count the number of places the decimal point was moved.

$$93,000,000 \qquad \text{7 places}$$

Since the decimal point was moved 7 places, and since n is positive, $93,000,000 = 9.3 \times 10^{7}$.

(b) $63,200,000,000 = 6.3200000000 = 6.32 \times 10^{10}$

$$\text{10 places}$$

(c) .00462

Move the decimal point to the right of the first nonzero digit, and count the number of places the decimal point was moved.

$$.00462 \qquad \text{3 places}$$

Since 4.62 is *larger* than .00462, the exponent must be *negative*.

$$.00462 = 4.62 \times 10^{-3}$$

(d) $.0000762 = 7.62 \times 10^{-5}$ ∎

2 ▶ Convert numbers in scientific notation to numbers without exponents.

▶ To convert a number written in scientific notation to a number without exponents, work in reverse. Multiplying by a positive power of 10 will make the number larger; multiplying by a negative power of 10 will make the number smaller.

EXAMPLE 2

Writing Numbers Without Exponents

Write each number without exponents.

(a) 6.2×10^{3}

Since the exponent is positive, make 6.2 larger by moving the decimal point 3 places to the right.

$$6.2 \times 10^{3} = 6.200 = 6200.$$

(b) $4.283 \times 10^{5} = 4.28300 = 428,300$ Move 5 places to the right.

(c) $7.04 \times 10^{-3} = .00704$ Move 3 places to the left.

As these examples show, the exponent tells the number of places that the decimal point is moved. ∎

3 ▶ Use scientific notation in calculations.

▶ The next example shows how scientific notation can be used with products and quotients.

E X A M P L E   3 ◾

Multiplying and Dividing with Scientific Notation

Write each product or quotient without exponents.

(a) $(6 \times 10^3)(5 \times 10^{-4})$

$$(6 \times 10^3)(5 \times 10^{-4}) = (6 \times 5)(10^3 \times 10^{-4}) \quad \text{Commutative and associative properties}$$

$$= 30 \times 10^{-1} \quad \text{Product rule for exponents}$$

$$= 3.0 \quad \text{Write without exponents.}$$

(b) $\dfrac{4 \times 10^{-5}}{2 \times 10^3} = \dfrac{4}{2} \times \dfrac{10^{-5}}{10^3} = 2 \times 10^{-8} = .00000002$ ◾

NOTE Multiplying or dividing numbers written in scientific notation may produce an answer in the form $a \times 10^0$. Since $10^0 = 1$, $a \times 10^0 = a$. For example,

$$(8 \times 10^{-4})(5 \times 10^4) = 40 \times 10^0 = 40.$$

Some calculators will accept data and display answers in scientific notation. For instance, 6.191736×10^6 is displayed on some calculators as $6.191736 \text{ E} + 6$. See your calculator manual for further information.

3.8 EXERCISES

Determine whether or not the given number is written in scientific notation as defined in Objective 1. If it is not, write it as such.

1. 4.56×10^3 **2.** 7.34×10^5 **3.** $5,600,000$ **4.** $34,000$

5. $.8 \times 10^2$ **6.** $.9 \times 10^3$ **7.** $.004$ **8.** $.0007$

9. Explain in your own words what it means for a number to be written in scientific notation.

10. Explain how to multiply a number by a positive power of ten. Then explain how to multiply a number by a negative power of ten.

Write each number in scientific notation. See Example 1.

11. $5,876,000,000$ **12.** $9,994,000,000$ **13.** $82,350$

14. $78,330$ **15.** $.000007$ **16.** $.0000004$

17. $.00203$ **18.** $.0000578$

Write the number without exponents. See Example 2.

19. 7.5×10^5 **20.** 8.8×10^6 **21.** 5.677×10^{12} **22.** 8.766×10^9

23. 6.21×10^0 **24.** 8.56×10^0 **25.** 7.8×10^{-4} **26.** 8.9×10^{-5}

27. 5.134×10^{-9} **28.** 7.123×10^{-10}

Perform the indicated operations and write the answer without exponents. See Example 3.

29. $(2 \times 10^8) \times (3 \times 10^3)$ **30.** $(4 \times 10^7) \times (3 \times 10^3)$ **31.** $(5 \times 10^4) \times (3 \times 10^2)$

32. $(8 \times 10^5) \times (2 \times 10^3)$ **33.** $(3 \times 10^{-4}) \times (2 \times 10^8)$ **34.** $(4 \times 10^{-3}) \times (2 \times 10^7)$

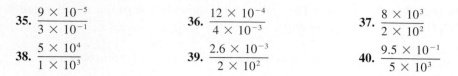

35. $\dfrac{9 \times 10^{-5}}{3 \times 10^{-1}}$

36. $\dfrac{12 \times 10^{-4}}{4 \times 10^{-3}}$

37. $\dfrac{8 \times 10^{3}}{2 \times 10^{2}}$

38. $\dfrac{5 \times 10^{4}}{1 \times 10^{3}}$

39. $\dfrac{2.6 \times 10^{-3}}{2 \times 10^{2}}$

40. $\dfrac{9.5 \times 10^{-1}}{5 \times 10^{3}}$

If the number in the statement is written in scientific notation, write it without exponents. If it is written without exponents, write it in scientific notation. See Examples 1 and 2.

41. The number of possible hands in contract bridge is about 6.35×10^{11}.

42. If there are forty numbers to choose from in a lottery, and a player must choose six different ones, the player has about 3.84×10^{6} ways to make a choice.

43. In a recent year, ESPN'S regular season baseball telecasts averaged just under 1,150,000 viewers.

44. In a recent year, ESPN's estimated losses were at least 36,000,000 dollars.

45. The body of a 150-pound person contains about 2.3×10^{-4} pounds of copper and about 6×10^{-3} pounds of iron.

46. The mean distance from Venus to the sun is about 6.7×10^{6} miles.

The quote is taken from the source cited. Write the number given in scientific notation in the quote without exponents.

47. The muon, a close relative of the electron produced by the bombardment of cosmic rays against the upper atmosphere, has a half-life of 2 millionths of a second (2×10^{-6}s). (Excerpt from *Conceptual Physics,* 6th Edition by Paul G. Hewitt. Copyright © by Paul G. Hewitt. Published by HarperCollins College Publishers.)

48. There are 13 red balls and 39 black balls in a box. Mix them up and draw 13 out one at a time without returning any ball . . . the probability that the 13 drawings each will produce a red ball is . . . 1.6×10^{-12}. (Warren Weaver, *Lady Luck,* New York: Doubleday & Company, Inc., 1963, pp. 298–299.)

The quote is taken from the source cited. Write the given number(s) in scientific notation.

49. An electron and a positron attract each other in two ways: the electromagnetic attraction of their opposite electric charges, and the gravitational attraction of their two masses. The electromagnetic attraction is 4,200,000,000,000,000,000,000,000,000,000,000,000,000,000 times as strong as the gravitational. (Isaac Asimov, *Isaac Asimov's Book of Facts,* New York: Bell Publishing Company, 1981, p. 106.)

50. How is it that the average CEO in Japan receives an income of $300,000, while the average CEO in the United States earns $2.8 million? (Andrew Zimbalist, *Baseball and Billions.* New York: BasicBooks, 1992, p. 78.)

Solve the problem.

51. The distance to Earth from the planet Pluto is 4.58×10^{9} kilometers. In April 1983, Pioneer 10 transmitted radio signals from Pluto to Earth at the speed of light, 3.00×10^{5} kilometers per second. How long (in seconds) did it take for the signals to reach Earth?

52. In Exercise 51, how many hours did it take for the signals to reach Earth?

53. The graph shows the number of engineers employed in the United States during the years 1984–1993. (*Source*: Engineering Workforce Commission)

    (a) Use the graph to approximate the number of engineers employed during 1992 and write the number in scientific notation.

    (b) Suppose the average annual salary for an engineer in 1992 was 5.5×10^4. What would be the total of all salaries of engineers in that year? (This is hypothetical and not based on actual data.)

Engineering Employment
(in millions)

Source: Engineering Workforce Commission

54. In 1992, the state of Texas had about 1.3×10^6 farms with an average of 7.1×10^2 acres per farm. What was the total number of acres devoted to farmland in Texas that year? (*Source*: National Agricultural Statistics Service, U.S. Dept. of Agriculture)

Read the material at the front of the book regarding how scientific calculators display numbers in scientific notation. You are given two numbers as they would appear in such a display. Determine what the display would be if **(a)** the numbers are multiplied and **(b)** if the first number is divided by the second. Then verify using your calculator.

55.	3.4 12	2 5	
57.	4.2 10	2 −8	

56.	4.8 6	2 15	
58.	2.6 7	2 −8	

REVIEW EXERCISES

List all positive integer factors of the number. See Section 1.1.

59. 18   **60.** 36   **61.** 48   **62.** 23

CHAPTER 3 SUMMARY

KEY TERMS

3.1 polynomial
descending powers
degree of a term
degree of a
    polynomial

trinomial
binomial
monomial

3.8 scientific notation

NEW SYMBOLS

x^{-n} x to the negative n power

QUICK REVIEW

CONCEPTS	EXAMPLES
3.1 ADDITION AND SUBTRACTION OF POLYNOMIALS	
Addition: Add like terms.	Add. $$\begin{aligned} 2x^2 + 5x - 3 \\ 5x^2 - 2x + 7 \\ \hline 7x^2 + 3x + 4 \end{aligned}$$
Subtraction: Change the signs of the terms in the second polynomial and add to the first polynomial.	Subtract. $\begin{aligned} &(2x^2 + 5x - 3) - (5x^2 - 2x + 7) \\ &= (2x^2 + 5x - 3) + (-5x^2 + 2x - 7) \\ &= -3x^2 + 7x - 10 \end{aligned}$

CONCEPTS	EXAMPLES

3.2 THE PRODUCT RULE AND POWER RULES FOR EXPONENTS

For any integers m and n with no denominators zero:

Product rule $a^m \cdot a^n = a^{m+n}$

$$2^4 \cdot 2^5 = 2^9$$

Power rules

(a) $(a^m)^n = a^{mn}$

$$(3^4)^2 = 3^8$$

(b) $(ab)^m = a^m b^m$

$$(6a)^5 = 6^5 a^5$$

(c) $\left(\dfrac{a}{b}\right)^m = \dfrac{a^m}{b^m}$ $(b \neq 0)$

$$\left(\dfrac{2}{3}\right)^4 = \dfrac{2^4}{3^4}$$

3.3 MULTIPLICATION OF POLYNOMIALS

Multiply each term of the first polynomial by each term of the second polynomial. Then add like terms.

Multiply

$$
\begin{array}{r}
3x^3 - 4x^2 + 2x - 7 \\
4x + 3 \\
\hline
9x^3 - 12x^2 + 6x - 21 \\
12x^4 - 16x^3 + 8x^2 - 28x \\
\hline
12x^4 - 7x^3 - 4x^2 - 22x - 21
\end{array}
$$

FOIL Method

Find $(2x + 3)(5x - 4)$.

Step 1   Multiply the two first terms to get the first terms of the answer.

$$2x(5x) = 10x^2$$

Step 2   Find the outer product and the inner product and mentally add them, when possible, to get the middle term of the answer.

$$2x(-4) + 3(5x) = 7x$$

Step 3   Multiply the two last terms to get the last term of the answer.

$$3(-4) = -12$$

The product of $(2x + 3)$ and $(5x - 4)$ is $10x^2 + 7x - 12$.

3.4 SPECIAL PRODUCTS

Square of a Binomial

$(a + b)^2 = a^2 + 2ab + b^2$

$(a - b)^2 = a^2 - 2ab + b^2$

$$(3x + 1)^2 = 9x^2 + 6x + 1$$
$$(2m - 5n)^2 = 4m^2 - 20mn + 25n^2$$

Product of the Sum and Difference of Two Terms

$(a + b)(a - b) = a^2 - b^2$

$$(4a + 3)(4a - 3) = 16a^2 - 9$$

3.5 INTEGER EXPONENTS AND THE QUOTIENT RULE

If $a \neq 0$, for integers m and n:

Zero exponent $a^0 = 1$

$$15^0 = 1$$

Negative exponent $a^{-n} = \dfrac{1}{a^n}$

$$5^{-2} = \dfrac{1}{5^2} = \dfrac{1}{25}$$

CONCEPTS	EXAMPLES
Quotient rule $\dfrac{a^m}{a^n} = a^{m-n}$ $$\dfrac{a^{-m}}{b^{-n}} = \dfrac{b^n}{a^m}$$ $$\left(\dfrac{a}{b}\right)^{-m} = \left(\dfrac{b}{a}\right)^{m}$$	$$\dfrac{4^8}{4^3} = 4^5$$ $$\dfrac{4^{-2}}{3^{-5}} = \dfrac{3^5}{4^2}$$ $$\left(\dfrac{6}{5}\right)^{-3} = \left(\dfrac{5}{6}\right)^{3}$$

3.6   THE QUOTIENT OF A POLYNOMIAL AND A MONOMIAL

Divide each term of the polynomial by the monomial: $$\dfrac{a+b}{c} = \dfrac{a}{c} + \dfrac{b}{c}.$$	Divide. $$\dfrac{4x^3 - 2x^2 + 6x - 8}{2x} = 2x^2 - x + 3 - \dfrac{4}{x}$$

3.7   THE QUOTIENT OF TWO POLYNOMIALS

Use "long division."	Divide. $$\begin{array}{r} 2x - 5 + \dfrac{-1}{3x+4} \\ \hline 3x+4\,\overline{)6x^2 - 7x - 21} \\ 6x^2 + 8x \\ \hline -15x - 21 \\ -15x - 20 \\ \hline -1 \end{array}$$

3.8   AN APPLICATION OF EXPONENTS: SCIENTIFIC NOTATION

To write a number in scientific notation (as $a \times 10^n$), move the decimal point to follow the first nonzero digit. The number of places the decimal point is moved is the absolute value of n. If moving the decimal point makes the number smaller, n is positive. Otherwise, n is negative. If the decimal point is not moved, n is 0.	$$247 = 2.47 \times 10^2$$ $$.0051 = 5.1 \times 10^{-3}$$ $$4.8 = 4.8 \times 10^0 = 4.8$$

CHAPTER 3 REVIEW EXERCISES

[3.1] *Combine terms where possible in the polynomial. Write the answer in descending powers of the variable. Give the degree of the answer. Identify the polynomial as a monomial, binomial, trinomial, or none of these.*

1. $9m^2 + 11m^2 + 2m^2$

2. $-4p + p^3 - p^2 + 8p + 2$

3. $12a^5 - 9a^4 + 8a^3 + 2a^2 - a + 3$

4. $-7y^5 - 8y^4 - y^5 + y^4 + 9y$

5. $-7x^5 - 8x - x^5 + x + 9x^3$

6. $-5z^3 + 7 - 6z^2 + 8z$

7. $(12r^4 - 7r^3 + 2r^2) - (5r^4 - 3r^3 + 2r^2 - 1)$

8. Simplify $(5x^3y^2 - 3xy^5 + 12x^2) - (-9x^2 - 8x^3y^2 + 2xy^5)$.

Add or subtract as indicated.

9. Add.
$$-2a^3 + 5a^2$$
$$\underline{3a^3 - a^2}$$

10. Subtract.
$$6y^2 - 8y + 2$$
$$\underline{5y^2 + 2y - 7}$$

11. Subtract.
$$-12k^4 - 8k^2 + 7k$$
$$\underline{k^4 + 7k^2 - 11k}$$

12. $(2m^3 - 8m^2 + 4) + (3m^3 + 2m^2 - 7)$ **13.** $(12r^4 - 7r^3 + 2r^2) - (5r^4 - 3r^3 + 2r^2)$

[3.2]

14. Explain why the product rule for exponents does not apply to the expression $7^2 + 7^4$.

Use the product rule to simplify the expression. Write the answer in exponential form.

15. $4^3 \cdot 4^8$

16. $(-5)^6(-5)^5$

17. $(-8x^4)(9x^3)$

18. $(2x^2)(5x^3)(x^9)$

Use the power rules to simplify the expression. Write the answer in exponential form.

19. $(19x)^5$

20. $(-4y)^7$

21. $5(pt)^4$

22. $\left(\dfrac{7}{5}\right)^6$

Use a combination of rules to simplify the expression.

23. $(3x^2y^3)^3$

24. $(t^4)^8(t^2)^5$

25. $(6x^2z^4)^2(x^3yz^2)^4$

26. $\left(\dfrac{2m^3n}{p^2}\right)^3$

[3.3] *Find the product.*

27. $(-9x^2)(7x^4)$

28. $-m^5(8m^2 - 10m + 6)$

29. $(a + 2)(a^2 - 4a + 1)$

30. $(3r - 2)(2r^2 + 4r - 3)$

31. $(5p^2 + 3p)(p^3 - p^2 + 5)$

32. $(m - 9)(m + 2)$

33. $(3k - 6)(2k + 1)$

34. $(a + 3b)(2a - b)$

35. $(6k + 5q)(2k - 7q)$

36. $(s - 1)^3$

37. Find a polynomial that represents the area of the rectangle shown.

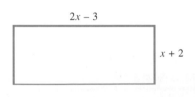

$2x - 3$

$x + 2$

38. If the side of a square has measure represented by $5x^4 + 2x^2$, what polynomial represents its area?

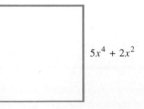

$5x^4 + 2x^2$

[3.4]

39. Explain in your own words how to square a binomial.

40. Explain in your own words how to multiply the sum of two terms by the difference of the same two terms.

Find the product.

41. $(a + 4)^2$

42. $(3p - 2)^2$

43. $(2r + 5t)^2$

44. $(8z - 3y)^2$

45. $(6m - 5)(6m + 5)$

46. $(2z + 7)(2z - 7)$

47. $(5a + 6b)(5a - 6b)$

48. $(9y + 8z)(9y - 8z)$

49. $(r + 2)^3$

50. Choose values for x and y to show that in general,
   (a) $(x + y)^2 \neq x^2 + y^2$ **(b)** $(x + y)^3 \neq x^3 + y^3$.

51. Refer to Exercise 50. Suppose that you happened to choose to let $x = 0$ and $y = 1$. Would your results be sufficient to illustrate the inequalities shown? If not, what would you need to do as your next step in working the exercise?

52. What is the volume of a cube with one side having length $x^2 + 2$ centimeters?

53. What is the volume of a sphere with radius $x + 1$ inches?

[3.5] *Evaluate the expression.*

54. $6^0 + (-6)^0$ **55.** $(-23)^0 - (-23)^0$ **56.** -10^0

Write the expression using only positive exponents.

57. -7^{-2} **58.** $\left(\dfrac{5}{8}\right)^{-2}$ **59.** $2^{-1} + 4^{-1}$ **60.** $(5^{-2})^{-4}$

61. $9^3 \cdot 9^{-5}$ **62.** $(-3)^7(-3)^3$ **63.** $\dfrac{6^{-5}}{6^{-3}}$ **64.** $(9^3)^{-2}$

Simplify. Write the answer with only positive exponents. Assume that all variables are nonzero real numbers.

65. $\dfrac{x^7}{x^{-9}}$ **66.** $\dfrac{y^4 \cdot y^{-2}}{y^{-5}}$ **67.** $(6r^{-2})^{-1}$

68. $(3p)^4(3p^{-7})$ **69.** $\dfrac{ab^{-3}}{a^4b^2}$ **70.** $\dfrac{(6r^{-1})^2(2r^{-4})}{r^{-5}(r^2)^{-3}}$

[3.6] *Perform the division.*

71. $\dfrac{-15y^4}{9y^2}$ **72.** $\dfrac{12x^3y^2}{6xy}$

73. $\dfrac{6y^4 - 12y^2 + 18y}{6y}$ **74.** $(2p^3 - 6p^2 + 5p) \div (2p^2)$

75. $(-10m^4n^2 + 5m^3n^2 + 6m^2n^4) \div (5m^2n)$

76. What polynomial, when multiplied by $6m^2n$, gives the product $12m^3n^2 + 18m^6n^3 - 24m^2n^2$?

77. One of your friends in class simplified $\dfrac{6x^2 - 12x}{6}$ as $x^2 - 12x$. Is this correct? If not, what error did your friend make and how would you explain the correct method of performing the division?

[3.7] *Perform the division.*

78. $\dfrac{2r^2 + 3r - 14}{r - 2}$ **79.** $\dfrac{10a^3 + 9a^2 - 14a + 9}{5a - 3}$

80. $\dfrac{x^4 + 3x^3 - 5x^2 - 3x + 4}{x^2 - 1}$ **81.** $\dfrac{m^4 + 4m^3 - 5m^2 - 12m + 6}{m^2 - 3}$

[3.8] *Write the number in scientific notation.*

82. 48,000,000   **83.** 28,988,000,000

84. .000065   **85.** .0000000824

Write the number without exponents.

86. 2.4×10^4 **87.** 7.83×10^7 **88.** 8.97×10^{-7}

89. 9.95×10^{-12} **90.** 5.6×10^0

Perform the indicated operation and write the answer without exponents.

91. $(2 \times 10^{-3}) \times (4 \times 10^5)$

92. $\dfrac{8 \times 10^4}{2 \times 10^{-2}}$

93. $\dfrac{12 \times 10^{-5} \times 5 \times 10^4}{4 \times 10^3 \times 6 \times 10^{-2}}$

94. $\dfrac{2.5 \times 10^5 \times 4.8 \times 10^{-4}}{7.5 \times 10^8 \times 1.6 \times 10^{-5}}$

Write the number in the statement in scientific notation.

95. Phillip Morris Co., the leading advertiser in the United States, spent $1,558,000 on advertising in a recent year.

96. The distance from Earth to the sun is about 92,900,000 miles.

Write the number in the statement without exponents.

97. A nanosecond is 1×10^{-9} seconds.

98. In the food chain that links the largest sea creature, the whale, to the smallest, the diatom, 4×10^{14} diatoms sustain a medium-sized whale for only a few hours.

99. Based on the accompanying graph, give the approximate federal government expenditure on crime-fighting in 1991 using scientific notation.

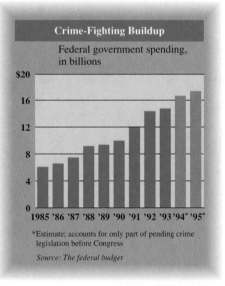

Crime-Fighting Buildup

Federal government spending, in billions

*Estimate; accounts for only part of pending crime legislation before Congress

Source: The federal budget

MIXED REVIEW EXERCISES

Perform the indicated operations. Write with positive exponents only. Assume all variables represent nonzero real numbers.

100. $5^0 + 7^0$

101. $\left(\dfrac{6r^2p}{5}\right)^3$

102. $(12a + 1)(12a - 1)$

103. 2^{-4}

104. $(8^{-3})^4$

105. $\dfrac{12m^2 - 11m - 10}{3m - 5}$

106. $\dfrac{2p^3 - 6p^2 + 5p}{2p^2}$

107. $\dfrac{(2m^{-5})(3m^2)^{-1}}{m^{-2}(m^{-1})^2}$

108. $(3k - 6)(2k^2 + 4k + 1)$

109. $\dfrac{r^9 \cdot r^{-5}}{r^{-2} \cdot r^{-7}}$

110. $(2r + 5s)^2$

111. $(-5y^2 + 3y - 11) + (4y^2 - 7y + 15)$

112. $(2r + 5)(5r - 2)$

113. $(-5m^3)^2$

114. $\dfrac{2y^3 + 17y^2 + 37y + 7}{2y + 7}$

115. $(25x^2y^3 - 8xy^2 + 15x^3y) \div (10x^2y^3)$

116. $(6p^2 - p - 8) - (-4p^2 + 2p - 3)$

117. $\dfrac{5^8}{5^{19}}$

118. $(-7 + 2k)^2$

119. $(5^2)^4$

120. $(3k + 1)(2k - 3)$

121. $\left(\dfrac{x}{y^{-3}}\right)^{-4}$

──────◆ **MATHEMATICAL CONNECTIONS** (Exercises 122–127) ◆──────

In Exercises 122–127 we use the letters P and Q to denote the following polynomials:

$$P: \quad x - 3$$
$$Q: \quad x^3 + x^2 - 11x - 3.$$

Work these exercises in order.

122. **(a)** Evaluate P for $x = 5$.
  (b) Evaluate Q for $x = 5$.
  (c) Find the polynomial represented by $P + Q$.
  (d) Evaluate the polynomial found in part (c) for $x = 5$ and verify that it is equal to the sum of the values you found in parts (a) and (b).

123. **(a)** Evaluate P for $x = 4$.
  (b) Evaluate Q for $x = 4$.
  (c) Find the polynomial represented by $P - Q$.
  (d) Evaluate the polynomial found in part (c) for $x = 4$ and verify that it is equal to the difference between the values you found in parts (a) and (b).

124. **(a)** Evaluate P for $x = -3$.
  (b) Evaluate Q for $x = -3$.
  (c) Find the polynomial represented by $P \cdot Q$.
  (d) Evaluate the polynomial found in part (c) for $x = -3$ and verify that it is equal to the product of the values you found in parts (a) and (b).

125. **(a)** Evaluate P for $x = 6$.
  (b) Evaluate Q for $x = 6$.
  (c) Find the polynomial represented by $\dfrac{Q}{P}$.
  (d) Evaluate the polynomial found in part (c) for $x = 6$ and verify that it is equal to the quotient of the values you found in parts (a) and (b).

126. The concept illustrated in these exercises is this: if we evaluate P for a particular value of x and evaluate Q for the same value of x, and then perform an operation on P and Q, the resulting polynomial will give the appropriate value when evaluated for x. The values used in Exercises 122–125 were chosen arbitrarily. Repeat each exercise for a different value of x.

127. Refer to Exercise 125. Why could we *not* choose 3 as a replacement for x? Is there any other replacement for x that would not be valid?

──────────◆──────────

CHAPTER 3 TEST

For the polynomial, combine terms when possible and write the polynomial in descending powers of the variable. Give the degree of the simplified polynomial. Decide whether the simplified polynomial is a monomial, binomial, trinomial, or none of these.

1. $5x^2 + 8x - 12x^2$

2. $13n^3 - n^2 + n^4 + 3n^4 - 9n^2$

Perform the indicated operations.

3. $(5t^4 - 3t^2 + 7t + 3) - (t^4 - t^3 + 3t^2 + 8t + 3)$

4. $(2y^2 - 8y + 8) + (-3y^2 + 2y + 3) - (y^2 + 3y - 6)$

5. $(-9a^3b^2 + 13ab^5 + 5a^2b^2) - (6ab^5 + 12a^3b^2 + 10a^2b^2)$

Perform the indicated operations.

6. Subtract.
$$9t^3 - 4t^2 + 2t + 2$$
$$9t^3 + 8t^2 - 3t - 6$$

7. $3x^2(-9x^3 + 6x^2 - 2x + 1)$

8. $(t - 8)(t + 3)$

9. $(4x + 3y)(2x - y)$

10. $(5x - 2y)^2$

11. $(10v + 3w)(10v - 3w)$

12. $(2r - 3)(r^2 + 2r - 5)$

13. What polynomial expression represents the area of this square?

$3x + 9$

Evaluate the expression.

14. 5^{-4}

15. $(-3)^0 + 4^0$

16. $4^{-1} + 3^{-1}$

17. Use the rules for exponents to simplify $\dfrac{(3x^2y)^2(xy^3)^2}{(xy)^3}$. Assume x and y are nonzero.

Simplify, and write the answer using only positive exponents. Assume that variables represent nonzero numbers.

18. $\dfrac{8^{-1} \cdot 8^4}{8^{-2}}$

19. $\dfrac{(x^{-3})^{-2}(x^{-1}y)^2}{(xy^{-2})^2}$

20. Do you agree or disagree with the following statement?

$$3^{-4} \text{ represents a negative number.}$$

Justify your answer.

Perform the division.

21. $\dfrac{8y^3 - 6y^2 + 4y + 10}{2y}$

22. $(-9x^2y^3 + 6x^4y^3 + 12xy^3) \div (3xy)$

23. $(3x^3 - x + 4) \div (x - 2)$

24. Write the quotient and write your answer without using scientific notation: $\dfrac{9.5 \times 10^{-1}}{5 \times 10^3}$.

25. The body of a 150-pound person contains about 2.3×10^{-4} pounds of copper. How many pounds of copper would there be in the bodies of 10,000 such persons?

CUMULATIVE REVIEW (Chapters 1–3)

Write the fraction in lowest terms.

1. $\dfrac{28}{16}$

2. $\dfrac{55}{11}$

Work the problem.

3. $\dfrac{2}{3} + \dfrac{1}{8}$

4. $\dfrac{7}{4} - \dfrac{9}{5}$

5. A contractor installs toolsheds. Each requires $1\dfrac{1}{4}$ cubic yards of concrete. How much concrete would be needed for 25 sheds?

6. A retailer has \$34,000 invested in her business. She finds that last year she earned 5.4% on this investment. How much did she earn?

7. List all positive integer factors of 45.

8. If $a < 0$ and $b > 0$, what is the sign of $b - a$?

Find the value of the expression if $x = -2$ and $y = 4$.

9. $\dfrac{4x - 2y}{x + y}$

10. $x^3 - 4xy$

Perform the indicated operations.

11. $\dfrac{(-13 + 15) - (3 + 2)}{6 - 12}$

12. $-7 - 3[2 + (5 - 8)]$

Decide what property justifies the statement.

13. $(9 + 2) + 3 = 9 + (2 + 3)$

14. $-7 + 7 = 0$

15. $6(4 + 2) = 6(4) + 6(2)$

Solve the equation.

16. $2x - 7x + 8x = 30$

17. $2 - 3(t - 5) = 4 + t$

18. $2(5h + 1) = 10h + 4$

19. $d = rt$ for r

20. $\dfrac{x}{5} = \dfrac{x - 2}{7}$

21. $\dfrac{1}{3}p - \dfrac{1}{6}p = -2$

22. $.05x + .15(50 - x) = 5.50$

23. $4 - (3x + 12) = (2x - 9) - (5x - 1)$

Solve the problem.

24. Each month, Janet's allowance is six times as much as Louis' allowance. Together their allowances total \$56. What is the monthly allowance for each?

25. If 8 is subtracted from a number and this difference is tripled, the result is -3 times the number. Find the number.

26. A farmer has 28 more hens than roosters, with 190 chickens in all. Find the number of hens and the number of roosters on this farm.

Solve the inequality.

27. $-8x \le -80$

28. $-2(x + 4) > 3x + 6$

29. $-3 \le 2x + 5 < 9$

Solve the problem.

30. One side of a triangle is twice as long as a second side. The third side of the triangle is 17 feet long. The perimeter of the triangle cannot be more than 50 feet. Find the longest possible values for the other two sides of the triangle.

Evaluate the expression.

31. $4^{-1} + 3^0$

32. $2^{-4} \cdot 2^5$

33. $\dfrac{8^{-5} \cdot 8^7}{8^2}$

34. Write with positive exponents only: $\dfrac{(a^{-3}b^2)^2}{(2a^{-4}b^{-3})^{-1}}$.

35. Write in scientific notation: 34,500.

Perform the indicated operations.

36. $(7x^3 - 12x^2 - 3x + 8) + (6x^2 + 4) - (-4x^3 + 8x^2 - 2x - 2)$

37. $6x^5(3x^2 - 9x + 10)$

38. $(7x + 4)(9x + 3)$

39. $(5x + 8)^2$

40. $\dfrac{y^3 - 3y^2 + 8y - 6}{y - 1}$

CONNECTIONS

Finding prime factors of extremely large numbers has been considered a mere computer exercise—interesting for the improved methods of working with computers, but of no value in its own right. This has changed in recent years with new methods of *computer coding* in which very large numbers are used in an attempt to provide unbreakable codes for computer data. Just as fast as these numbers are used, other people try to find prime factors of them so that the code can be broken. For example, it has been shown that the extremely large number $10^{71} - 1$ has three prime factors: 3, which divides into the number twice, 241,573,142,393,627,673,576,957,439,049, and 45,994,811,347,886,846,310,221,728,895,223,034,301,839. This factorization of $10^{71} - 1$ required 220 minutes of computer time using a Cray computer, one of the fastest of modern supercomputers.*

* From *Mathematical Ideas,* 7th edition, by Charles D. Miller, Vern E. Heeren, and E. John Hornsby, Jr., HarperCollins, 1994, p. 228.

Factoring a polynomial reverses the process of multiplication, discussed in Chapter 3. As shown in Sections 4.5 and 4.7, factoring is used to solve certain equations and inequalities, and therefore, is important in many applications of algebra. We also use factoring to reduce algebraic fractions to lowest terms in Chapter 5.

4.1 THE GREATEST COMMON FACTOR; FACTORING BY GROUPING

FOR EXTRA HELP

📖 **SSG** pp. 138–140
SSM pp. 171–175

📼 **Video**
6

💾 **Tutorial**
IBM   MAC

OBJECTIVES
1 ▶ Find the greatest common factor of a list of terms.
2 ▶ Factor out the greatest common factor.
3 ▶ Factor by grouping.

Recall from Chapter 1 that to **factor** means to write a quantity as a product. That is, factoring is the opposite of multiplication. For example,

Multiplication	*Factoring*
$6 \cdot 2 = 12,$	$12 = 6 \cdot 2.$

Factors   Product   Product   Factors

Other factored forms of 12 are $(-6)(-2), 3 \cdot 4, (-3)(-4), 12 \cdot 1,$ and $(-12)(-1)$. More than two factors may be used, so another factored form of 12 is $2 \cdot 2 \cdot 3$. The positive integer factors of 12 are

$$1, 2, 3, 4, 6, 12.$$

1 ▶ Find the greatest common factor of a list of terms.

▶   An integer that is a factor of two or more integers is called a **common factor** of those integers. For example, 6 is a common factor of 18 and 24 since 6 is a factor of both 18 and 24. Other common factors of 18 and 24 are 1, 2, and 3. The **greatest common factor** of a list of integers is the largest common factor of those integers. Thus, 6 is the greatest common factor of 18 and 24, since it is the largest of the common factors of these numbers.

◆ C O N N E C T I O N S ◆

Many rules have been found for deciding what characteristics determine the numbers that divide into a given number. Here are some divisibility rules for small numbers. The rules are quite useful. It is surprising how many people do not know them.

A Whole Number Divisible by:	Must Have the Following Property:
2	ends in 0, 2, 4, 6, or 8
3	sum of its digits is divisible by 3
4	last two digits form a number divisible by 4
5	ends in 0 or 5
6	divisible by both 2 and 3
8	last three digits form a number divisible by 8
9	sum of its digits is divisible by 9
10	ends in 0

Recall from Chapter 1 that a prime number has only itself and 1 as factors. In Section 1.1 we factored numbers into prime factors. This is the first step in finding the greatest common factor of a list of numbers. We find the greatest common factor of a list of numbers as follows.

Finding the Greatest Common Factor

Step 1 **Factor.** Write each number in prime factored form.

Step 2 **List common factors.** List each prime number that is a factor of every number in the list.

Step 3 **Choose smallest exponents.** Use as exponents on the prime factors the *smallest* exponent from the prime factored forms. (If a prime does not appear in one of the prime factored forms, it cannot appear in the greatest common factor.)

Step 4 **Multiply.** Multiply the primes from Step 3. If there are no primes left after Step 3, the greatest common factor is 1.

EXAMPLE 1

Finding the Greatest Common Factor for Numbers

■ Find the greatest common factor for each list of numbers.

(a) 30, 45

First write each number in prime factored form.

$$30 = 2 \cdot 3 \cdot \mathbf{5}$$
$$45 = 3^2 \cdot \mathbf{5}$$

Now, take each prime the *least* number of times it appears in all the factored forms. There is no 2 in the prime factored form of 45, so there will be no 2 in the greatest common factor. The least number of times 3 appears in all the factored forms is 1, and the least number of times 5 appears is also 1. From this, the greatest common factor is

$$3^1 \cdot \mathbf{5}^1 = 15.$$

(b) 72, 120, 432

Find the prime factored form of each number.

$$72 = 2^3 \cdot 3^2$$
$$120 = 2^3 \cdot 3 \cdot 5$$
$$432 = 2^4 \cdot 3^3$$

The least number of times 2 appears in all the factored forms is 3, and the least number of times 3 appears is 1. There is no 5 in the prime factored form of either 72 or 432, so the greatest common factor is

$$2^3 \cdot 3 = 24$$

(c) 10, 11, 14

Write the prime factored form of each number.

$$10 = 2 \cdot 5$$
$$11 = 11$$
$$14 = 2 \cdot 7$$

There are no primes common to all three numbers, so the greatest common factor is 1. ■

The greatest common factor can also be found for a list of variable terms. For example, the terms x^4, x^5, x^6, and x^7 have x^4 as the greatest common factor because each of these terms can be written with x^4 as a factor.

$$x^4 = 1 \cdot x^4,\ x^5 = x \cdot x^4,\ x^6 = x^2 \cdot x^4,\ x^7 = x^3 \cdot x^4$$

NOTE The exponent on a variable in the greatest common factor is the *smallest* exponent that appears in the factors.

EXAMPLE 2

Finding the Greatest Common Factor for Variable Terms

Find the greatest common factor for each list of terms.

(a) $21m^7,\ -18m^6,\ 45m^8,\ -24m^5$

$$21m^7 = 3 \cdot 7 \cdot \boldsymbol{m^7}$$
$$-18m^6 = -1 \cdot 2 \cdot 3^2 \cdot \boldsymbol{m^6}$$
$$45m^8 = 3^2 \cdot 5 \cdot \boldsymbol{m^8}$$
$$-24m^5 = -1 \cdot 2^3 \cdot 3 \cdot \boldsymbol{m^5}$$

First, 3 is the greatest common factor of the coefficients $21, -18, 45$, and -24. The smallest exponent on m is 5, so the greatest common factor of the terms is $3m^5$.

(b) $x^4y^2,\ x^7y^5,\ x^3y^7,\ y^{15}$

$$x^4y^2 = x^4 \cdot \boldsymbol{y^2}$$
$$x^7y^5 = x^7 \cdot \boldsymbol{y^5}$$
$$x^3y^7 = x^3 \cdot \boldsymbol{y^7}$$
$$y^{15} = \boldsymbol{y^{15}}$$

There is no x in the last term, y^{15}, so x will not appear in the greatest common factor. There is a y in each term, however, and 2 is the smallest exponent on y. The greatest common factor is y^2. ■

2 ▶ Factor out the greatest common factor.

▶ We use the idea of a greatest common factor to write a polynomial (a sum) in factored form as a product. For example, the polynomial

$$3m + 12$$

consists of the two terms $3m$ and 12. The greatest common factor for these two terms is 3. We can write $3m + 12$ so that each term is a product with 3 as one factor.

$$3m + 12 = 3 \cdot m + 3 \cdot 4$$

Now use the distributive property.

$$3m + 12 = 3 \cdot m + 3 \cdot 4 = 3(m + 4)$$

The factored form of $3m + 12$ is $3(m + 4)$. This process is called **factoring out the greatest common factor.**

CAUTION   Notice that the polynomial $3m + 12$ is *not* in factored form when written as

$$3 \cdot m + 3 \cdot 4.$$

The *terms* are factored, but the polynomial is not. The factored form of $3m + 12$ is the *product*

$$3(m + 4).$$

EXAMPLE 3

Factoring Out the Greatest Common Factor

Factor out the greatest common factor.

(a) $20m^5 + 10m^4 + 15m^3$
  The greatest common factor for the terms of this polynomial is $5m^3$.

$$20m^5 + 10m^4 + 15m^3 = (5m^3)(4m^2) + (5m^3)(2m) + (5m^3)3$$
$$= 5m^3(4m^2 + 2m + 3)$$

Check this work by multiplying $5m^3$ and $4m^2 + 2m + 3$. You should get the original polynomial as your answer.

(b) $x^5 + x^3 = (x^3)x^2 + (x^3)1 = x^3(x^2 + 1)$

(c) $20m^7p^2 - 36m^3p^4 = 4m^3p^2(5m^4 - 9p^2)$

(d) $a(a + 3) + 4(a + 3)$
  The binomial $a + 3$ is the greatest common factor here.

$$a(a + 3) + 4(a + 3) = (a + 3)(a + 4)$$ ■

CAUTION   Be careful to avoid the common error of leaving out the 1 in a problem like Example 3(b). Always be sure that the factored form can be multiplied out to give the original polynomial.

---

3 ▶ Factor by grouping.

▶   Common factors are used in **factoring by grouping,** as explained in the next example.

EXAMPLE 4

Factoring by Grouping

Factor by grouping.

(a) $2x + 6 + ax + 3a$
  The first two terms have a common factor of 2, and the last two terms have a common factor of a.

$$2x + 6 + ax + 3a = 2(x + 3) + a(x + 3)$$

The expression is still not in factored form because it is the *sum* of two terms. Now, however, $x + 3$ is a common factor and can be factored out.

$$2x + 6 + ax + 3a = 2(x + 3) + a(x + 3) = (x + 3)(2 + a)$$

The final result is in factored form because it is a *product*. Note that the goal in factoring by grouping is to get a common factor, $x + 3$ here, so that the last step is possible.

Same

(b) $m^2 + 6m + 2m + 12 = m(m + 6) + 2(m + 6)$

$= (m + 6)(m + 2)$

(c) $6xy - 21x - 8y + 28 = 3x(2y - 7) - 4(2y - 7) = (2y - 7)(3x - 4)$

Must be same

Since the quantities in parentheses in the second step must be the same, it was necessary here to factor out -4 rather than 4. ∎

CAUTION Note the careful use of signs in Example 4(c). Sign errors often occur when grouping with negative signs.

Use these steps when factoring four terms by grouping.

Factoring by Grouping

Step 1 Write the four terms so that the first two have a common factor and the last two have a common factor.

Step 2 Use the distributive property to factor each group of two terms.

Step 3 If possible, factor a common binomial factor from the results of Step 2.

Step 4 If Step 2 does not result in a common binomial factor, try grouping the terms of the original polynomial in a different way. (If two groups still have no common binomial factor, the polynomial cannot be factored.)

EXAMPLE 5

Rearranging Terms Before Factoring by Grouping

Factor by grouping.

(a) $10x^2 - 12y^2 + 15xy - 8xy$

Factoring out the common factor of 2 from the first two terms and the common factor of xy terms from the last two terms gives

$$10x^2 - 12y^2 + 15xy - 8xy = 2(5x^2 - 6y^2) + xy(15 - 8).$$

This did not lead to a common factor, so we try rearranging the terms. There is usually more than one way to do this. Let's try

$$10x^2 - 8xy - 12y^2 + 15xy,$$

grouping the first two terms and the last two terms as follows.

$$10x^2 - 8xy - 12y^2 + 15xy = 2x(5x - 4y) + 3y(-4y + 5x)$$
$$= 2x(5x - 4y) + 3y(5x - 4y)$$
$$= (5x - 4y)(2x + 3y)$$

(b) $2xy + 12 - 3y - 8x$

We need to rearrange these terms to get two groups that each have a common factor. Trial and error suggests the following grouping.

$$2xy + 12 - 3y - 8x = (2xy - 3y) + (-8x + 12)$$
$$= y(2x - 3) - 4(2x - 3) \qquad \text{Factor each group.}$$
$$= (2x - 3)(y - 4) \qquad \text{Factor out the common binomial factor.} \quad ∎$$

4.1 EXERCISES

Find the greatest common factor for the list of numbers. See Example 1.

1. 12, 16   **2.** 18, 24   **3.** 40, 20, 4   **4.** 50, 30, 5

5. 18, 24, 36, 48   **6.** 15, 30, 45, 75   **7.** 4, 9, 12   **8.** 9, 16, 24

9. How would you respond to the following? "The numbers 25 and 36 have no greatest common factor."

10. Give an example of three numbers whose greatest common factor is 5.

Find the greatest common factor for the list of terms. See Example 2.

11. $16y$, 24

12. $18w$, 27

13. $30x^3$, $40x^6$, $50x^7$

14. $60z^4$, $70z^8$, $90z^9$

15. $12m^3n^2$, $18m^5n^4$, $36m^8n^3$

16. $25p^5r^7$, $30p^7r^8$, $50p^5r^3$

17. $-x^4y^3$, $-xy^2$

18. $-a^4b^5$, $-a^3b$

19. $42ab^3$, $-36a$, $90b$, $-48ab$

20. $45c^3d$, $75c$, $90d$, $-105cd$

21. Is $-xy$ a common factor of $-x^4y^3$ and $-xy^2$? If so, what is the other factor that when multiplied by $-xy$ gives $-x^4y^3$?

22. Is $-a^5b^2$ a common factor of $-a^4b^5$ and $-a^3b$?

Complete the factoring.

23. $12 = 6($ $)$ **24.** $18 = 9($ $)$ **25.** $3x^2 = 3x($ $)$

26. $8x^3 = 8x($ $)$ **27.** $9m^4 = 3m^2($ $)$ **28.** $12p^5 = 6p^3($ $)$

29. $-8z^9 = -4z^5($ $)$ **30.** $-15k^{11} = -5k^8($ $)$ **31.** $6m^4n^5 = 3m^3n($ $)$

32. $27a^3b^2 = 9a^2b($ $)$ **33.** $-14x^4y^3 = 2xy($ $)$ **34.** $-16m^3n^3 = 4mn^2($ $)$

Factor out the greatest common factor. See Example 3.

35. $12y - 24$ **36.** $18p + 36$ **37.** $10a^7 - 20a$ **38.** $15x^3 - 30x^2$

39. $65y^{10} + 35y^6$ **40.** $100a^5 + 16a^3$ **41.** $11w^3 - 100$ **42.** $13z^5 - 80$

43. $8m^2n^3 + 24m^2n^2$

44. $19p^2y - 38p^2y^3$

45. $13y^8 + 26y^4 - 39y^2$

46. $5x^5 + 25x^4 - 20x^3$

47. $45q^4p^5 + 36qp^6 + 81q^2p^3$

48. $125a^3z^5 + 60a^4z^4 - 85a^5z^2$

49. $a^5 + 2a^3b^2 - 3a^5b^2 + 4a^4b^3$

50. $x^6 + 5x^4y^3 - 6xy^4 + 10xy$

51. $c(x + 2) - d(x + 2)$

52. $r(5 - x) + t(5 - x)$

53. $m(m + 2n) + n(m + 2n)$

54. $3p(1 - 4p) - 2q(1 - 4p)$

Students often have difficulty when factoring by grouping because they are not able to recognize when the polynomial is indeed factored. For example,

$$5y(2x - 3) + 8t(2x - 3)$$

is not in factored form, because it is the *sum* of two terms, $5y(2x - 3)$ and $8t(2x - 3)$. However, because $2x - 3$ is a common factor of these two terms, it can be factored as

$$(2x - 3)(5y + 8t).$$

In this form, the expression is a *product* of two factors, $2x - 3$ and $5y + 8t$.

Determine whether the expression is in factored form or is not in factored form.

55. $8(7t + 4) + x(7t + 4)$ **56.** $3r(5x - 1) + 7(5x - 1)$

57. $(8 + x)(7t + 4)$ **58.** $(3r + 7)(5x - 1)$

59. $18x^2(y + 4) + 7(y + 4)$ **60.** $12k^3(s - 3) + 7(s + 3)$

61. Is it possible to factor the expression in Exercise 59? If so, factor it.

62. Explain why it is not possible to factor the expression in Exercise 60.

Factor by grouping. See Examples 4 and 5.

63. $p^2 + 4p + 3p + 12$ **64.** $m^2 + 2m + 5m + 10$

65. $a^2 - 2a + 5a - 10$ **66.** $y^2 - 6y + 4y - 24$

67. $7z^2 + 14z - az - 2a$ **68.** $5m^2 + 15mp - 2mp - 6p^2$

69. $18r^2 + 12ry - 3xr - 2xy$ **70.** $8s^2 - 4st + 6sy - 3yt$

71. $3a^3 + 3ab^2 + 2a^2b + 2b^3$ **72.** $4x^3 + 3x^2y + 4xy^2 + 3y^3$

73. $1 - a + ab - b$ **74.** $6 - 3x - 2y + xy$

75. $16m^3 - 4m^2p^2 - 4mp + p^3$ **76.** $10t^3 - 2t^2s^2 - 5ts + s^3$

77. $5m + 15 - 2mp - 6p$ **78.** $y^2 - 3y - xy + 3x$

79. $18r^2 + 12ry - 3ry - 2y^2$ **80.** $3a^3 + 3ab^2 + 2a^2b + 2b^3$

81. $a^5 + 2a^5b - 3 - 6b$ **82.** $a^2b - 4a - ab^4 + 4b^3$

◇ **MATHEMATICAL CONNECTIONS** (Exercises 83–88) ◇

In most cases, the choice of which pairs of terms to group when factoring by grouping can be done in several ways to obtain the correct answer. Work Exercises 83–88 in order, and notice how this applies to the polynomial in Example 5(b).

83. Start with the polynomial in Example 5(b), $2xy + 12 - 3y - 8x$, and rearrange the terms as follows: $2xy - 8x + (-3y) + 12$. What properties from Section 1.9 allow us to do this?

84. Group the first pair of terms in the rearranged polynomial. What is the greatest common factor of this pair?

85. Now group the second pair of terms in the rearranged polynomial as well. Is -3 a common factor of this second pair?

86. Factor the greatest common factor from the first pair and -3 from the second pair.

87. Is your result from Exercise 86 in factored form? If not, why?

88. If your answer to Exercise 87 is *no*, factor the polynomial. Is it the same result as the one shown in Example 5(b) in this section?

◆

89. Refer to Exercise 73. The answer given in the back of the book is $(1 - a)(1 - b)$. A student factored this same polynomial and got the result $(a - 1)(b - 1)$.
   (a) Is this student's answer correct?
   (b) If your answer to part (a) is *yes*, explain why these two seemingly different answers are both acceptable.

90. While tutoring someone in introductory algebra, explain how you could make up polynomials like those in Exercises 63–82 that could be factored by grouping.

REVIEW EXERCISES

Find each product. See Section 3.3.

91. $(x + 6)(x - 9)$ **92.** $(x - 3)(x - 6)$

93. $(x + 2)(x + 7)$ **94.** $2x(x + 5)(x - 1)$

4.2 FACTORING TRINOMIALS

FOR EXTRA HELP

📖 **SSG** pp. 141–143
SSM pp. 176–182

📼 **Video**
6

💾 **Tutorial**
IBM   MAC

OBJECTIVES

1 ▶ Factor trinomials with a coefficient of 1 for the squared term.
2 ▶ Factor such polynomials after factoring out the greatest common factor.

Using FOIL we find the product of the polynomials $k - 3$ and $k + 1$ is

$$(k - 3)(k + 1) = k^2 - 2k - 3.$$

Now suppose we are given the polynomial $k^2 - 2k - 3$ and want to rewrite it as the product $(k - 3)(k + 1)$. This product is called the **factored form** of $k^2 - 2k - 3$, and the process of finding the factored form is called **factoring.** The discussion of factoring in this section is limited to trinomials like $x^2 - 2x - 24$ or $y^2 + 2y - 15$, where the coefficient of the squared term is 1.

1 ▶ Factor trinomials with a coefficient of 1 for the squared term.

▶   When factoring polynomials with only integer coefficients, we use only integers for the numerical factors. For example, $x^2 + 5x + 6$ can be factored by finding integers a and b such that

$$x^2 + 5x + 6 = (x + a)(x + b).$$

To find these integers a and b, we first multiply the two factors on the right-hand side of the equation:

$$(x + a)(x + b) = x^2 + ax + bx + ab.$$

By the distributive property,

$$x^2 + ax + bx + ab = x^2 + (a + b)x + ab.$$

Comparing this result with $x^2 + 5x + 6$ shows that we must find integers a and b having a sum of 5 and a product of 6.

$$x^2 + 5x + 6 = x^2 + (a + b)x + ab$$

Sum of a and b is 5.   Product of a and b is 6.

Since many pairs of integers have a sum of 5, it is best to begin by listing those pairs of integers whose product is 6. Both 5 and 6 are positive, so only pairs in which both integers are positive need be considered.

Product	Sum	
$1 \cdot 6 = 6$	$1 + 6 = 7$	
$2 \cdot 3 = 6$	$2 + 3 = 5$	Sum is 5.

Both pairs have a product of 6, but only the pair 2 and 3 has a sum of 5. So 2 and 3 are required integers, and

$$x^2 + 5x + 6 = (x + 2)(x + 3).$$

We can check by multiplying the binomials using FOIL. Make sure that the sum of the outer and inner products produces the correct middle term.

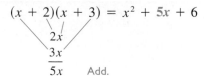

$$(x + 2)(x + 3) = x^2 + 5x + 6$$

$$2x$$
$$\underline{3x}$$
$$5x \qquad \text{Add.}$$

This method of factoring can be used only for trinomials having the coefficient of the squared term equal to 1. Methods for factoring other trinomials are given in the next section.

E X A M P L E   1

Factoring a Trinomial with All Terms Positive

Factor $m^2 + 9m + 14$.

Look for two integers whose product is 14 and whose sum is 9. List the pairs of integers whose product is 14. Then examine the sums. Again, only positive integers are needed because all signs are positive.

$$14, 1 \qquad 14 + 1 = 15$$
$$7, 2 \qquad 7 + 2 = \mathbf{9} \qquad \text{Sum is } 9.$$

From the list, 7 and 2 are the required integers, since $7 \cdot 2 = 14$ and $7 + 2 = 9$. Thus $m^2 + 9m + 14 = (m + 2)(m + 7)$. ∎

NOTE   In Example 1, the answer also could have been written $(m + 7)(m + 2)$. Because of the commutative property of multiplication, the order of the factors does not matter.

The trinomials in the examples so far had all positive terms. If a trinomial has one or more negative terms, both positive and negative factors must be considered.

E X A M P L E   2

Factoring a Trinomial with Two Negative Terms

Factor $p^2 - 2p - 15$.

Find two integers whose product is -15 and whose sum is -2. If these numbers do not come to mind right away, we can find them (if they exist) by listing all the pairs of integers whose product is -15. Because the last term, -15, is negative, we need pairs of integers with different signs.

$$15, -1 \qquad 15 + (-1) = 14$$
$$5, -3 \qquad 5 + (-3) = 2$$
$$-15, 1 \qquad -15 + 1 = -14$$
$$-5, 3 \qquad -5 + 3 = -2 \qquad \text{Sum is } -2.$$

The necessary integers are -5 and 3, and

$$p^2 - 2p - 15 = (p - 5)(p + 3). \quad ∎$$

EXAMPLE 3

Factoring a Trinomial
with One Negative Term

(a) Factor $x^2 - 5x + 12$.

List all pairs of integers whose product is 12. Since the middle term is negative and the last term is positive, we need pairs with both numbers negative. Then examine the sums.

$$-12, -1 \qquad -12 + (-1) = -13$$
$$-6, -2 \qquad -6 + (-2) = -8$$
$$-3, -4 \qquad -3 + (-4) = -7$$

None of the pairs of integers has a sum of -5. Because of this, the trinomial $x^2 - 5x + 12$ *cannot be factored using only integer factors.* A polynomial that cannot be factored using only integer exponents is called a **prime polynomial.**

(b) $k^2 - 8k + 11$

There is no pair of integers whose product is 11 and whose sum is -8, so $k^2 - 8k + 11$ is a prime polynomial. ∎

We can now summarize the procedure for factoring a trinomial of the form $x^2 + bx + c$.

Factoring
$x^2 + bx + c$

Find two integers whose product is c and whose sum is b.

1. Both integers must be positive if b and c are positive.
2. Both integers must be negative if c is positive and b is negative.
3. One integer must be positive and one must be negative if c is negative.

EXAMPLE 4

Factoring a Trinomial
with Two Variables

Factor $z^2 - 2bz - 3b^2$.

To factor $z^2 - 2bz - 3b^2$ look for two expressions whose product is $-3b^2$ and whose sum is $-2b$. The expressions are $-3b$ and b, with

$$z^2 - 2bz - 3b^2 = (z - 3b)(z + b). \quad ∎$$

2 ▶ Factor such
polynomials after
factoring out the greatest
common factor.

▶ The trinomial in the next example does not have a coefficient of 1 for the squared term. (In fact, there is no squared term.) A preliminary step must be taken before using the steps discussed above.

EXAMPLE 5

Factoring a Trinomial
with a Common Factor

Factor $4x^5 - 28x^4 + 40x^3$.

First, factor out the greatest common factor, $4x^3$.

$$4x^5 - 28x^4 + 40x^3 = 4x^3(x^2 - 7x + 10)$$

Now factor $x^2 - 7x + 10$. The integers -5 and -2 have a product of 10 and a sum of -7. The completely factored form is

$$4x^5 - 28x^4 + 40x^3 = 4x^3(x - 5)(x - 2). \quad ∎$$

CAUTION When factoring, always remember to look for a common factor first. Do not forget to include the common factor as part of the answer. Multiplying out the factored form should always give the original polynomial. (This is a good way to check your answer.)

4.2 EXERCISES

In Exercises 1–8, list all pairs of integers with the given product. Then find the pair whose sum is given.

1. Product: 12   Sum: 7   **2.** Product: 18   Sum: 9

3. Product: −24   Sum: −5   **4.** Product: −36   Sum: −16

5. Product: 27   Sum: 28   **6.** Product: 32   Sum: 33

7. Product: −48   Sum: 2   **8.** Product: −48   Sum: 8

9. Which one of the following is the correct factored form of $x^2 - 12x + 32$?
   (a) $(x - 8)(x + 4)$ **(b)** $(x + 8)(x - 4)$ **(c)** $(x - 8)(x - 4)$
   (d) $(x + 8)(x + 4)$

10. What would be the first step you would use in factoring $2x^3 + 8x^2 - 10x$?

Complete the following.

11. $p^2 + 11p + 30 = (p + 5)($ $)$ **12.** $x^2 + 10x + 21 = (x + 7)($ $)$

13. $x^2 + 15x + 44 = (x + 4)($ $)$ **14.** $r^2 + 15r + 56 = (r + 7)($ $)$

15. $x^2 - 9x + 8 = (x - 1)($ $)$ **16.** $t^2 - 14t + 24 = (t - 2)($ $)$

17. $y^2 - 2y - 15 = (y + 3)($ $)$ **18.** $t^2 - t - 42 = (t + 6)($ $)$

19. $x^2 + 9x - 22 = (x - 2)($ $)$ **20.** $x^2 + 6x - 27 = (x - 3)($ $)$

21. $y^2 - 7y - 18 = (y + 2)($ $)$ **22.** $y^2 - 2y - 24 = (y + 4)($ $)$

Factor completely. If the polynomial cannot be factored, write prime. *See Examples 1–3.*

23. $y^2 + 9y + 8$ **24.** $a^2 + 9a + 20$ **25.** $b^2 + 8b + 15$

26. $x^2 + 6x + 8$ **27.** $m^2 + m - 20$ **28.** $p^2 + 4p - 5$

29. $y^2 - 8y + 15$ **30.** $y^2 - 6y + 8$ **31.** $x^2 + 4x + 5$

32. $t^2 + 11t + 12$ **33.** $t^2 - 8t + 16$ **34.** $s^2 - 10s + 25$

35. $r^2 - r - 30$ **36.** $q^2 - q - 42$ **37.** $n^2 - 12n - 35$

38. $x^2 - 4x + 12$

Factor completely. See Examples 4 and 5.

39. $r^2 + 3ra + 2a^2$ **40.** $x^2 + 5xa + 4a^2$

41. $t^2 - tz - 6z^2$ **42.** $a^2 - ab - 12b^2$

43. $x^2 + 4xy + 3y^2$ **44.** $p^2 + 9pq + 8q^2$

45. $v^2 - 11vw + 30w^2$ **46.** $v^2 - 11vx + 24x^2$

47. $4x^2 + 12x - 40$ **48.** $5y^2 - 5y - 30$

49. $2t^3 + 8t^2 + 6t$ **50.** $3t^3 + 27t^2 + 24t$

51. $2x^6 + 8x^5 - 42x^4$ **52.** $4y^5 + 12y^4 - 40y^3$

53. $m^3n - 10m^2n^2 + 24mn^3$ **54.** $y^3z + 3y^2z^2 - 54yz^3$

55. Use the FOIL method from Section 3.3 to show that $(2x + 4)(x - 3) = 2x^2 - 2x - 12$. If you are asked to completely factor $2x^2 - 2x - 12$, why would it be incorrect to give $(2x + 4)(x - 3)$ as your answer?

56. If you are asked to completely factor the polynomial $3x^2 + 9x - 12$, why would it be incorrect to give $(x - 1)(3x + 12)$ as your answer?

Use a combination of the factoring methods discussed in this section to factor the polynomial.

57. $a^5 + 3a^4b - 4a^3b^2$ **58.** $m^3n - 2m^2n^2 - 3mn^3$ **59.** $y^3z + y^2z^2 - 6yz^3$

60. $k^7 - 2k^6m - 15k^5m^2$ **61.** $z^{10} - 4z^9y - 21z^8y^2$ **62.** $x^9 + 5x^8w - 24x^7w^2$

63. $(a + b)x^2 + (a + b)x - 12(a + b)$

64. $(x + y)n^2 + (x + y)n + 16(x + y)$

65. $(2p + q)r^2 - 12(2p + q)r + 27(2p + q)$

66. $(3m - n)k^2 - 13(3m - n)k + 40(3m - n)$

──────◆ **MATHEMATICAL CONNECTIONS** (Exercises 67–74) ◆──────

To check our factoring of a trinomial, remember that when the factors are multiplied, *every* term of the original polynomial must be obtained. In a trinomial such as $x^2 + x - 12$, students often factor so that the first and third terms are correct, but the middle term is incorrect, differing only in the sign. Work Exercises 67–74 in order so that you can see how this problem can be avoided.

67. Add the pair of numbers: -3 and 4.

68. Add the pair of numbers: 3 and -4.

69. In Exercise 68, we added the *opposites* of the numbers in Exercise 67. How does the sum in Exercise 68 compare with the sum in Exercise 67?

70. Consider this factorization, given for $x^2 + x - 12$: $(x + 3)(x - 4)$. Multiply the two binomials. Is this the correct factorization? Why or why not?

71. Consider this factorization, given for $x^2 + x - 12$: $(x - 3)(x + 4)$. Multiply the two binomials. Is this the correct factorization? Why or why not?

72. In Exercises 70 and 71, one of the factorizations is correct and one is not. For the one that is not, how does the sign of the middle term of the product compare with the signs of the middle terms of the correct factorization?

73. Compare the two factorizations shown in Exercises 70 and 71, and use the result of Exercise 72 to complete the following statement: When I factor a trinomial into a product of binomials, and the middle term of the product is different only in sign, I should _____ in order to obtain the correct factorization.

74. Given that the factorization $(x + 5)(x - 3)$ is incorrect only in the sign of the middle term of the product for a particular trinomial, what would be the correct factorization?

──────────◆──────────

REVIEW EXERCISES

Find the product. See Section 3.3.

75. $(2y - 7)(y + 4)$ **76.** $(3a + 2)(2a + 1)$ **77.** $(5z + 2)(3z - 2)$

78. $(4m - 3)(2m + 5)$ **79.** $(4p + 1)(2p - 3)$ **80.** $(6r - 5)(3r + 2)$

4.3 MORE ON FACTORING TRINOMIALS

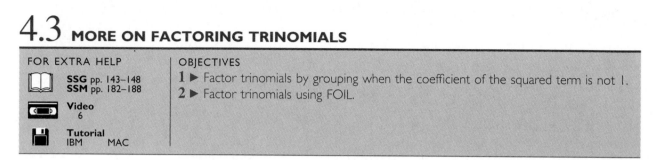

FOR EXTRA HELP	OBJECTIVES
SSG pp. 143–148 **SSM** pp. 182–188	**1 ▶** Factor trinomials by grouping when the coefficient of the squared term is not 1.
Video 6	**2 ▶** Factor trinomials using FOIL.
Tutorial IBM MAC	

Trinomials such as $2x^2 + 7x + 6$, in which the coefficient of the squared term is *not* 1, can be factored with an extension of the method presented in the last section.

1 ▶ Factor trinomials by grouping when the coefficient of the squared term is not 1.

▶ Recall that a trinomial such as $m^2 + 3m + 2$ is factored by finding two numbers whose product is 2 and whose sum is 3. To factor $2x^2 + 7x + 6$, we look for two integers whose product is $2 \cdot 6 = 12$ and whose sum is 7.

$$2x^2 + 7x + 6$$

Sum is 7.

Product is $2 \cdot 6 = 12$.

By considering the pairs of positive integers whose product is 12, the necessary integers are found to be 3 and 4. We use these integers to write the middle term, $7x$, as $7x = 3x + 4x$. With this, the trinomial $2x^2 + 7x + 6$ becomes

$$2x^2 + 7x + 6 = 2x^2 + \underbrace{3x + 4x}_{7x = 3x + 4x} + 6.$$

Factor the new polynomial by grouping as in Section 4.1.

$$2x^2 + 3x + 4x + 6 = x(2x + 3) + 2(2x + 3)$$
$$= (2x + 3)(x + 2)$$

The common factor of $2x + 3$ was factored out to get

$$2x^2 + 7x + 6 = (2x + 3)(x + 2).$$

Check by finding the product of $2x + 3$ and $x + 2$.

We could have written the middle term in the polynomial $2x^2 + 7x + 6$ as $7x = 4x + 3x$ to get

$$2x^2 + 7x + 6 = 2x^2 + 4x + 3x + 6$$
$$= 2x(x + 2) + 3(x + 2)$$
$$= (x + 2)(2x + 3).$$

Either result is correct.

EXAMPLE 1

Factoring Trinomials by Grouping

Factor the trinomial.

(a) $6r^2 + r - 1$

We must find two integers with a product of $6(-1) = -6$ and a sum of 1.

$$\overset{\text{Sum is 1.}}{\underset{\text{Product is } 6(-1) = -6.}{6r^2 + r - 1 = 6r^2 + 1r - 1}}$$

The integers are -2 and 3. We write the middle term, $+r$, as $-2r + 3r$, so that

$$6r^2 + r - 1 = 6r^2 - 2r + 3r - 1.$$

Factor by grouping on the right-hand side.

$$6r^2 + r - 1 = 6r^2 - 2r + 3r - 1$$
$$= 2r(3r - 1) + 1(3r - 1) \qquad \text{The binomials must be the same.}$$
$$= (3r - 1)(2r + 1).$$

(b) $12z^2 - 5z - 2$

Look for two integers whose product is $12(-2) = -24$ and whose sum is -5. The required integers are 3 and -8, and

$$12z^2 - 5z - 2 = 12z^2 + 3z - 8z - 2 \qquad -5z = 3z - 8z$$
$$= 3z(4z + 1) - 2(4z + 1) \qquad \text{Group terms and factor each group.}$$
$$= (4z + 1)(3z - 2). \qquad \text{Factor out } 4z + 1.$$

(c) $10m^2 + mn - 3n^2$

Two integers whose product is $10(-3) = -30$ and whose sum is 1 are -5 and 6. Rewrite the trinomial with four terms.

$$10m^2 + mn - 3n^2 = 10m^2 - 5mn + 6mn - 3n^2 \qquad mn = -5mn + 6mn$$
$$= 5m(2m - n) + 3n(2m - n) \qquad \text{Group terms and factor each group.}$$
$$= (2m - n)(5m + 3n) \qquad \text{Factor out the common factor.} \blacksquare$$

2 ▶ Factor trinomials using FOIL.

▶ The rest of this section shows an alternative method of factoring trinomials in which the coefficient of the squared term is not 1. This method uses trial and error. In the next example, the alternative method is used to factor $2x^2 + 7x + 6$, the same trinomial factored at the beginning of this section.

To factor $2x^2 + 7x + 6$ by trial and error, we must use FOIL backwards. We want to write $2x^2 + 7x + 6$ as the product of two binomials.

$$2x^2 + 7x + 6 = (\quad)(\quad)$$

The product of the two first terms of the binomials is $2x^2$. The possible factors of $2x^2$ are $2x$ and x or $-2x$ and $-x$. Since all terms of the trinomial are positive, only positive factors should be considered. Thus, we have

$$2x^2 + 7x + 6 = (2x\quad)(x\quad).$$

The product of the two last terms, 6, can be factored as $6 \cdot 1$, $1 \cdot 6$, $2 \cdot 3$, or $3 \cdot 2$. Try each pair to find the pair that gives the correct middle term.

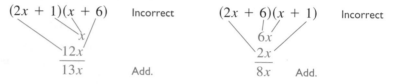

Since $2x + 6 = 2(x + 3)$, the binomial $2x + 6$ has a common factor of 2, while $2x^2 + 7x + 6$ has no common factor other than 1. The product $(2x + 6)(x + 1)$ cannot be correct.

NOTE If the original polynomial has no common factor, then none of its binomial factors will either.

Now try the numbers 2 and 3 as factors of 6. Because of the common factor of 2 in $2x + 2$, $(2x + 2)(x + 3)$ will not work. Try $(2x + 3)(x + 2)$.

$$(2x + 3)(x + 2) = 2x^2 + 7x + 6 \qquad \text{Correct}$$

with the illustration showing:
$$3x$$
$$4x$$
$$\overline{7x} \qquad \text{Add.}$$

Finally, we see that $2x^2 + 7x + 6$ factors as

$$2x^2 + 7x + 6 = (2x + 3)(x + 2).$$

Check by multiplying $2x + 3$ and $x + 2$.

EXAMPLE 2

Factoring a Trinomial with All Terms Positive Using FOIL

Factor $8p^2 + 14p + 5$.

The number 8 has several possible pairs of factors, but 5 has only 1 and 5 or -1 and -5. For this reason, it is easier to begin by considering the factors of 5. Ignore the negative factors since all coefficients in the trinomial are positive. If $8p^2 + 14p + 5$ can be factored, the factors will have the form

$$(\quad + 5)(\quad + 1).$$

The possible pairs of factors of $8p^2$ are $8p$ and p, or $4p$ and $2p$. Try various combinations, checking the middle term in each case.

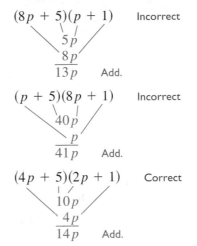

Since $14p$ is the correct middle term, the trinomial $8p^2 + 14p + 5$ factors as $(4p + 5)(2p + 1)$. ■

EXAMPLE 3

Factoring a Trinomial with a Negative Middle Term Using FOIL

Factor $6x^2 - 11x + 3$.

Since 3 has only 1 and 3 or -1 and -3 as factors, it is better here to begin by factoring 3. The last term of the trinomial $6x^2 - 11x + 3$ is positive and the middle term has a negative coefficient, so only negative factors should be considered. Try -3 and -1 as factors of 3:

$$(\quad - 3)(\quad - 1).$$

The factors of $6x^2$ may be either $6x$ and x, or $2x$ and $3x$. Try $2x$ and $3x$.

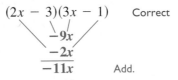

These factors give the correct middle term, so

$$6x^2 - 11x + 3 = (2x - 3)(3x - 1). \quad ■$$

EXAMPLE 4

Factoring a Trinomial with a Negative Last Term Using FOIL

Factor $8x^2 + 6x - 9$.

The integer 8 has several possible pairs of factors, as does -9. Since the last term is negative, one positive factor and one negative factor of -9 are needed. Since the coefficient of the middle term is small, it is wise to avoid large factors such as 8 or 9. Let us try 4 and 2 as factors of 8, and 3 and -3 as factors of -9, and check the middle term.

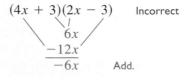

Let us try exchanging 3 and -3.

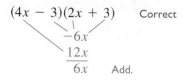

$(4x - 3)(2x + 3)$ Correct

This time we got the correct middle term, so

$$8x^2 + 6x - 9 = (4x - 3)(2x + 3). \quad \blacksquare$$

EXAMPLE 5

Factoring a Trinomial with Two Variables

Factor $12a^2 - ab - 20b^2$.

There are several pairs of factors of $12a^2$, including $12a$ and a, $6a$ and $2a$, and $3a$ and $4a$, just as there are many pairs of factors of $-20b^2$, including $-20b$ and b, $10b$ and $-2b$, $-10b$ and $2b$, $4b$ and $-5b$, and $-4b$ and $5b$. Once again, since the desired middle term is small, we should avoid the larger factors. Let us try as factors $6a$ and $2a$ and $4b$ and $-5b$.

$$(6a + 4b)(2a - 5b)$$

This cannot be correct, as mentioned before, since $6a + 4b$ has a common factor while the given trinomial has none. Let us try $3a$ and $4a$ with $4b$ and $-5b$.

$$(3a + 4b)(4a - 5b) = 12a^2 + ab - 20b^2 \quad \text{Incorrect}$$

Here the middle term has the wrong sign, so we change the signs in the factors.

$$(3a - 4b)(4a + 5b) = 12a^2 - ab - 20b^2 \quad \text{Correct} \quad \blacksquare$$

EXAMPLE 6

Factoring a Trinomial with a Common Factor

Factor $28x^5 - 58x^4 - 30x^3$.

First factor out the greatest common factor, $2x^3$.

$$28x^5 - 58x^4 - 30x^3 = 2x^3(14x^2 - 29x - 15)$$

Now try to factor $14x^2 - 29x - 15$. Try $7x$ and $2x$ as factors of $14x^2$ and -3 and 5 as factors of -15.

$$(7x - 3)(2x + 5) = 14x^2 + 29x - 15 \quad \text{Incorrect}$$

The middle term differs only in sign, so change the signs in the two factors.

$$(7x + 3)(2x - 5) = 14x^2 - 29x - 15 \quad \text{Correct}$$

Finally, the factored form of $28x^5 - 58x^4 - 30x^3$ is

$$28x^5 - 58x^4 - 30x^3 = 2x^3(7x + 3)(2x - 5). \quad \blacksquare$$

CAUTION Do not forget to include the common factor in the final result.

EXAMPLE 7

Factoring a Trinomial with a Negative Common Factor

Factor $-24a^3 - 42a^2 + 45a$

The common factor could be $3a$ or $-3a$. If we factor out $-3a$, the first term of the trinomial factor will be positive, which makes it easier to factor.

$$-24a^3 - 42a^2 + 45a = -3a(8a^2 + 14a - 15) \quad \text{Factor out the greatest common factor.}$$

$$= -3a(4a - 3)(2a + 5) \quad \text{Use trial and error.} \quad \blacksquare$$

4.3 EXERCISES

Decide which is the correct factored form of the given polynomial.

1. $2x^2 - x - 1$ **(a)** $(2x - 1)(x + 1)$ **(b)** $(2x + 1)(x - 1)$
2. $3a^2 - 5a - 2$ **(a)** $(3a + 1)(a - 2)$ **(b)** $(3a - 1)(a + 2)$
3. $4y^2 + 17y - 15$ **(a)** $(y + 5)(4y - 3)$ **(b)** $(2y - 5)(2y + 3)$
4. $12c^2 - 7c - 12$ **(a)** $(6c - 2)(2c + 6)$ **(b)** $(4c + 3)(3c - 4)$
5. $4k^2 + 13mk + 3m^2$ **(a)** $(4k + m)(k + 3m)$ **(b)** $(4k + 3m)(k + m)$
6. $2x^2 + 11x + 12$ **(a)** $(2x + 3)(x + 4)$ **(b)** $(2x + 4)(x + 3)$

Complete the factoring.

7. $6a^2 + 7ab - 20b^2 = (3a - 4b)(\qquad)$
8. $9m^2 - 3mn - 2n^2 = (3m + n)(\qquad)$
9. $2x^2 + 6x - 8 = 2(\qquad) = 2(\qquad)(\qquad)$
10. $3x^2 - 9x - 30 = 3(\qquad) = 3(\qquad)(\qquad)$
11. $4z^3 - 10z^2 - 6z = 2z(\qquad) = 2z(\qquad)(\qquad)$
12. $15r^3 - 39r^2 - 18r = 3r(\qquad) = 3r(\qquad)(\qquad)$

13. For the polynomial $12x^2 + 7x - 12$, 2 is not a common factor. Explain why the binomial $2x - 6$, then, cannot be a factor of the polynomial.

14. Explain how the signs of the last terms of the two binomial factors of a trinomial are determined.

Factor completely. Use either method described in this section. See Examples 1–6.

15. $2x^2 + 7x + 3$
16. $3y^2 + 13y + 4$
17. $3a^2 + 10a + 7$
18. $7r^2 + 8r + 1$
19. $4r^2 + r - 3$
20. $4r^2 + 3r - 10$
21. $15m^2 + m - 2$
22. $6x^2 + x - 1$
23. $8m^2 - 10m - 3$
24. $12s^2 + 11s - 5$
25. $20x^2 + 11x - 3$
26. $20x^2 - 28x - 3$
27. $21m^2 + 13m + 2$
28. $38x^2 + 23x + 2$
29. $20y^2 + 39y - 11$
30. $10x^2 + 11x - 6$
31. $6b^2 + 7b + 2$
32. $6w^2 + 19w + 10$
33. $24x^2 - 42x + 9$
34. $48b^2 - 74b - 10$
35. $40m^2q + mq - 6q$
36. $15a^2b + 22ab + 8b$
37. $2m^3 + 2m^2 - 40m$
38. $3x^3 + 12x^2 - 36x$
39. $15n^4 - 39n^3 + 18n^2$
40. $24a^4 + 10a^3 - 4a^2$
41. $18x^5 + 15x^4 - 75x^3$
42. $32z^5 - 20z^4 - 12z^3$
43. $15x^2y^2 - 7xy^2 - 4y^2$
44. $14a^2b^3 + 15ab^3 - 9b^3$
45. $12p^2 + 7pq - 12q^2$
46. $6m^2 - 5mn - 6n^2$
47. $25a^2 + 25ab + 6b^2$
48. $6x^2 - 5xy - y^2$
49. $6a^2 - 7ab - 5b^2$
50. $25g^2 - 5gh - 2h^2$
51. $6m^6n + 7m^5n^2 + 2m^4n^3$
52. $12k^3q^4 - 4k^2q^5 - kq^6$
53. $5 - 6x + x^2$
54. $7 + 8x + x^2$
55. $16 + 16x + 3x^2$
56. $18 + 65x + 7x^2$
57. $-10x^3 + 5x^2 + 140x$
58. $-18k^3 - 48k^2 + 66k$

If a trinomial has a negative coefficient for the squared term, such as $-2x^2 + 11x - 12$, it may be easier to factor by first factoring out the common factor -1:

$$-2x^2 + 11x - 12 = -1(2x^2 - 11x + 12)$$
$$= -1(2x - 3)(x - 4).$$

Use this method to factor the trinomial. See Example 7.

59. $-x^2 - 4x + 21$
60. $-x^2 + x + 72$
61. $-3x^2 - x + 4$
62. $-5x^2 + 2x + 16$
63. $-2a^2 - 5ab - 2b^2$
64. $-3p^2 + 13pq - 4q^2$

65. The answer given in the back of the book for Exercise 59 is $-1(x + 7)(x - 3)$. Is $(x + 7)(3 - x)$ also a correct answer? Explain why or why not.

66. One answer for Exercise 60 is $-1(x + 8)(x - 9)$. Is $(-x - 8)(-x + 9)$ also a correct answer? Explain.

Factor the polynomial. Remember to factor out the greatest common factor as the first step.

67. $25q^2(m + 1)^3 - 5q(m + 1)^3 - 2(m + 1)^3$

68. $18x^2(y - 3)^2 - 21x(y - 3)^2 - 4(y - 3)^2$

69. $15x^2(r + 3)^3 - 34xy(r + 3)^3 - 16y^2(r + 3)^3$

70. $4t^2(k + 9)^7 + 20ts(k + 9)^7 + 25s^2(k + 9)^7$

Find all integers k so that the trinomial can be factored using the methods of this section.

71. $5x^2 + kx - 1$ **72.** $2c^2 + kc - 3$

73. $2m^2 + km + 5$ **74.** $3y^2 + ky + 3$

◆ **MATHEMATICAL CONNECTIONS** (Exercises 75–82) ◆

One of the most common problems that beginning algebra students face is this: if an answer obtained doesn't look exactly like the one given in the back of the book, is it necessarily incorrect? Very often there are several different equivalent forms of an answer that are all correct. Work Exercises 75–82 in order, so that you can see how and why this is possible for factoring problems.

75. Factor the integer 35 as the product of two prime numbers.

76. Factor the integer 35 as the product of the negatives of two prime numbers.

77. Verify the following factorization: $6x^2 - 11x + 4 = (3x - 4)(2x - 1)$.

78. Verify the following factorization: $6x^2 - 11x + 4 = (4 - 3x)(1 - 2x)$.

79. Compare the two valid factorizations in Exercises 77 and 78. How do the factors in each case compare?

80. Suppose you know that the correct factorization of a particular trinomial is $(7t - 3)(2t - 5)$. Based on your observations in Exercises 77–79, what is another valid factorization?

81. Look at your results in Exercises 75 and 76, and fill in the blank: If an integer factors as the product of a and b, then it also factors as the product of _____ and _____.

82. Look at your results in Exercises 77 and 78 and fill in the blank: If a trinomial factors as the product of the binomials P and Q, then it also factors as the product of the binomials _____ and _____.

◆

REVIEW EXERCISES

Find the product. See Section 3.4.

83. $(7p + 3)(7p - 3)$ **84.** $(3h + 5k)(3h - 5k)$ **85.** $\left(r^2 + \dfrac{1}{2}\right)\left(r^2 - \dfrac{1}{2}\right)$

86. $(x + 6)^2$ **87.** $(3t + 4)^2$ **88.** $\left(c - \dfrac{2}{3}\right)^2$

4.4 SPECIAL FACTORIZATIONS

FOR EXTRA HELP

📖 **SSG** pp. 148–151
SSM pp. 188–193

📼 **Video** 6

💾 **Tutorial** IBM   MAC

OBJECTIVES

1 ▶ Factor the difference of two squares.
2 ▶ Factor a perfect square trinomial.
3 ▶ Factor the difference of two cubes.
4 ▶ Factor the sum of two cubes.

By reversing the rules for the multiplication of binomials that we learned in the last chapter, we get three rules for factoring polynomials in certain forms.

1 ▶ Factor the difference of two squares.

▶   Recall from the last chapter that

$$(x + y)(x - y) = x^2 - y^2.$$

Based on this product, we factor a **difference of two squares** as follows.

Difference of Two Squares

$$x^2 - y^2 = (x + y)(x - y)$$

EXAMPLE 1
Factoring a Difference of Squares

Factor each difference of two squares.

(a) $x^2 - 49 = x^2 - 7^2 = (x + 7)(x - 7)$

(b) $y^2 - m^2 = (y + m)(y - m)$

(c) $z^2 - \dfrac{9}{16} = z^2 - \left(\dfrac{3}{4}\right)^2 = \left(z + \dfrac{3}{4}\right)\left(z - \dfrac{3}{4}\right)$

(d) $p^2 + 16$
   The polynomial is not the *difference* of two squares. Using FOIL,

$$(p + 4)(p - 4) = p^2 - 16$$
$$(p - 4)(p - 4) = p^2 - 8p + 16$$

and
$$(p + 4)(p + 4) = p^2 + 8p + 16$$

so $p^2 + 16$ is a prime polynomial.   ■

CAUTION   As Example 1(d) suggests, after any common factor is removed, the sum of two squares cannot be factored.

EXAMPLE 2
Factoring More Complex Differences of Squares

Factor completely.

(a) $9a^2 - 4b^2$
   This is a difference of two squares because

$$9a^2 - 4b^2 = (3a)^2 - (2b)^2,$$

so $9a^2 - 4b^2 = (3a + 2b)(3a - 2b).$

(b) $81y^2 - 36$

First factor out the common factor of 9.

$$81y^2 - 36 = 9(9y^2 - 4)$$
$$= 9(3y + 2)(3y - 2)$$

(c) $p^4 - 36 = (p^2)^2 - 6^2 = (p^2 + 6)(p^2 - 6)$

Neither $p^2 + 6$ nor $p^2 - 6$ can be factored further.

(d) $m^4 - 16 = (m^2)^2 - 4^2$

$$= (m^2 + 4)(m^2 - 4) \qquad \text{Difference of squares}$$
$$= (m^2 + 4)(m + 2)(m - 2) \qquad \text{Difference of squares} \quad ∎$$

CAUTION A common error is to forget to factor the difference of two squares a second time when several steps are required, as in Example 2(d).

---

2 ▶ Factor a perfect square trinomial.

▶ The expressions 144, $4x^2$, and $81m^6$ are called *perfect squares,* since

$$144 = 12^2, \qquad 4x^2 = (2x)^2, \qquad \text{and} \qquad 81m^6 = (9m^3)^2.$$

A **perfect square trinomial** is a trinomial that is the square of a binomial. For example, $x^2 + 8x + 16$ is a perfect square trinomial since it is the square of the binomial $x + 4$:

$$x^2 + 8x + 16 = (x + 4)^2.$$

For a trinomial to be a perfect square, two of its terms must be perfect squares. For this reason, $16x^2 + 4x + 15$ cannot be a perfect square trinomial since only the term $16x^2$ is a perfect square.

On the other hand, even though two of the terms are perfect squares, the trinomial may not be a perfect square trinomial. For example, $x^2 + 6x + 36$ has two perfect square terms, but it is not a perfect square trinomial. (Try to find a binomial that can be squared to give $x^2 + 6x + 36$.)

We can multiply to see that the square of a binomial gives the following perfect square trinomials.

Perfect Square Trinomial	$x^2 + 2xy + y^2 = (x + y)^2$
	$x^2 - 2xy + y^2 = (x - y)^2$

The middle term of a perfect square trinomial is always twice the product of the two terms in the squared binomial. (This was shown in Section 3.4.) Use this to check any attempt to factor a trinomial that appears to be a perfect square.

While perfect square trinomials can be factored using the procedures of Sections 4.2 and 4.3, it is usually more efficient to recognize the pattern and factor accordingly. The following example illustrates this.

E X A M P L E   3
━━━━━━━━━━━━━━━━━ ▪

Factoring a Perfect
Square Trinomial

Factor the perfect square trinomial.

(a) $x^2 + 10x + 25$.

The term x^2 is a perfect square and so is 25. Try to factor the trinomial as

$$x^2 + 10x + 25 = (x + 5)^2.$$

To check, take twice the product of the two terms in the squared binomial.

$$\text{Twice} \to 2 \cdot x \cdot 5 = 10x$$

First term ⎯⎯⏐   ⏐⎯⎯ Last term
of binomial   of binomial

Since $10x$ is the middle term of the trinomial, the trinomial is a perfect square and can be factored as $(x + 5)^2$.

(b) $x^2 - 22xz + 121z^2$

The first and last terms are perfect squares ($121 = 11^2$). Check to see whether the middle term of $x^2 - 22xz + 121z^2$ is twice the product of the first and last terms of the binomial $x - 11z$.

$$\text{Twice} \cdot 2 \cdot x \cdot 11z = 22xz$$

First term ⎯⏐   ⏐⎯⎯ Last term
of binomial   of binomial

Since twice the product of the first and last terms of the binomial is the middle term, $x^2 - 22xz + 121z^2$ is a perfect square trinomial and

$$x^2 - 22xz + 121z^2 = (x - 11z)^2.$$

(c) $9m^2 - 24m + 16 = (3m)^2 - 2(3m)(4) + 4^2 = (3m - 4)^2$

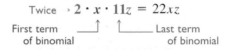

(d) $25y^2 + 20y + 16$

The first and last terms are perfect squares.

$$25y^2 = (5y)^2 \qquad \text{and} \qquad 16 = 4^2$$

Twice the product of the first and last terms of the binomial $5y + 4$ is

$$2 \cdot 5y \cdot 4 = 40y,$$

which is not the middle term of $25y^2 + 20y + 16$. This polynomial is not a perfect square. In fact, the polynomial cannot be factored even with the methods of Section 4.3; it is a prime polynomial. ▪

───

N O T E   The sign of the second term in the squared binomial is always the same as the sign of the middle term in the trinomial. Also, the first and last terms of a perfect square trinomial must be positive, since they are squares. For example, the polynomial $x^2 - 2x - 1$ cannot be a perfect square because the last term is negative.

───

◆ **CONNECTIONS** ◆

In Chapter 3 we saw the connection between multiplication and factoring:

Multiplication →

$$(2x + 3)(5x - 2) = 10x^2 + 11x - 6.$$

← Factoring

We know that multiplication and division are also related: to check a division problem, we multiply the quotient by the divisor to get the dividend. Thus, we can write the example above as a division problem.

$$2x + 3 \overline{)10x^2 + 11x - 6} \quad \overset{5x - 2}{}$$

This means that the factors of a polynomial can also be found by division. Suppose we want to factor $x^3 - 1$. We choose the binomial $x - 1$ as a possible factor and divide.

$$
\begin{array}{r}
x^2 + x + 1 \\
x - 1 \overline{)x^3 + 0x^2 + 0x - 1} \\
\underline{x^3 - x^2} \\
x^2 + 0x \\
\underline{x^2 - x} \\
x - 1 \\
\underline{x - 1} \\
0
\end{array}
$$

Thus, the factored form of $x^3 - 1$ is $(x - 1)(x^2 + x + 1)$.

FOR DISCUSSION OR WRITING

Factor $x^3 + 1$ by dividing it by $x + 1$. What happens when you divide $x^3 + 1$ by $x - 1$? Is $x - 1$ a factor of $x^3 + 1$? What does it mean when the division has a remainder?

3 ▶ Factor the difference of two cubes.

▶ The difference of two squares was factored above; we can also factor the **difference of two cubes.** Use the following pattern.

Difference of Two Cubes

$$x^3 - y^3 = (x - y)(x^2 + xy + y^2)$$

This pattern *should be memorized.* Multiply on the right to see that the pattern gives the correct factors.

$$
\begin{array}{r}
x^2 + xy + y^2 \\
x - y \\
\hline
-x^2y - xy^2 - y^3 \\
x^3 + x^2y + xy^2 \\
\hline
x^3 \qquad\qquad - y^3
\end{array}
$$

CAUTION The polynomial $x^3 - y^3$ is not equivalent to $(x - y)^3$, because $(x - y)^3$ is factored as

$$(x - y)^3 = (x - y)(x - y)(x - y)$$
$$= (x - y)(x^2 - 2xy + y^2)$$

but
$$x^3 - y^3 = (x - y)(x^2 + xy + y^2).$$

EXAMPLE 4

Factoring Differences of Cubes

Factor the following.

(a) $m^3 - 125$

Let $x = m$ and $y = 5$ in the pattern for the difference of two cubes.

$$x^3 - y^3 = (x - y)(x^2 + xy + y^2)$$
$$m^3 - 125 = m^3 - 5^3 = (m - 5)(m^2 + 5m + 5^2) \quad \text{Let } x = m, y = 5.$$
$$= (m - 5)(m^2 + 5m + 25)$$

(b) $8p^3 - 27$

Since $8p^3 = (2p)^3$ and $27 = 3^3$, substitute into the rule using $2p$ for x and 3 for y.

$$8p^3 - 27 = (2p)^3 - 3^3$$
$$= (2p - 3)[(2p)^2 + (2p)3 + 3^2]$$
$$= (2p - 3)(4p^2 + 6p + 9)$$

(c) $4m^3 - 32 = 4(m^3 - 8)$
$$= 4(m^3 - 2^3)$$
$$= 4(m - 2)(m^2 + 2m + 4)$$

(d) $125t^3 - 216s^6 = (5t)^3 - (6s^2)^3$
$$= (5t - 6s^2)[(5t)^2 + (5t)(6s^2) + (6s^2)^2]$$
$$= (5t - 6s^2)(25t^2 + 30ts^2 + 36s^4) \quad ■$$

CAUTION A common error in factoring the difference of two cubes, such as $x^3 - y^3 = (x - y)(x^2 + xy + y^2)$, is to try to factor $x^2 + xy + y^2$. It is easy to confuse this factor with a perfect square trinomial, $x^2 + 2xy + y^2$. Because of the lack of a 2 in $x^2 + xy + y^2$, it is very unusual to be able to further factor an expression of the form $x^2 + xy + y^2$.

4 ▶ Factor the sum of two cubes.

▶ A sum of two squares, such as $m^2 + 25$, cannot be factored using real numbers, but the **sum of two cubes** can be factored by the following pattern, *which should be memorized.*

Sum of Two Cubes

$$x^3 + y^3 = (x + y)(x^2 - xy + y^2)$$

Compare the pattern for the *sum* of two cubes with the pattern for the *difference* of two cubes. The only difference between them is the positive and negative signs.

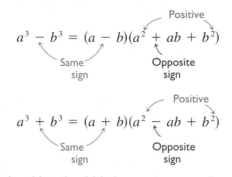

Observing these relationships should help you to remember these patterns.

EXAMPLE 5

Factoring Sums of Cubes

Factor.

(a) $k^3 + 27 = k^3 + 3^3$

$\quad = (k + 3)(k^2 - 3k + 3^2)$

$\quad = (k + 3)(k^2 - 3k + 9)$

(b) $8m^3 + 125 = (2m)^3 + 5^3$

$\quad = (2m + 5)[(2m)^2 - (2m)(5) + 5^2]$

$\quad = (2m + 5)(4m^2 - 10m + 25)$

(c) $1000a^6 + 27b^3 = (10a^2)^3 + (3b)^3$

$\quad = (10a^2 + 3b)[(10a^2)^2 - (10a^2)(3b) + (3b)^2]$

$\quad = (10a^2 + 3b)(100a^4 - 30a^2b + 9b^2)$ ■

The methods of factoring discussed in this section are summarized here. All these rules should be memorized.

Special Factorizations

Difference of two squares

$$x^2 - y^2 = (x + y)(x - y)$$

Perfect square trinomials

$$x^2 + 2xy + y^2 = (x + y)^2$$
$$x^2 - 2xy + y^2 = (x - y)^2$$

Difference of two cubes

$$x^3 - y^3 = (x - y)(x^2 + xy + y^2)$$

Sum of two cubes

$$x^3 + y^3 = (x + y)(x^2 - xy + y^2)$$

CAUTION Remember the *sum* of two *squares* usually cannot be factored.

4.4 EXERCISES

1. In order to help you with the factoring technique for the difference of squares, complete the following list of squares.

   $1^2 =$ _____ $2^2 =$ _____ $3^2 =$ _____ $4^2 =$ _____ $5^2 =$ _____
   $6^2 =$ _____ $7^2 =$ _____ $8^2 =$ _____ $9^2 =$ _____ $10^2 =$ _____
   $11^2 =$ _____ $12^2 =$ _____ $13^2 =$ _____ $14^2 =$ _____ $15^2 =$ _____
   $16^2 =$ _____ $17^2 =$ _____ $18^2 =$ _____ $19^2 =$ _____ $20^2 =$ _____

2. The following powers of x are all perfect squares: $x^2, x^4, x^6, x^8, x^{10}$. Based on this observation, we may make a conjecture (an educated guess) that if the power of a variable is divisible by _____ (with 0 remainder), then we have a perfect square.

3. In order to help you with the factoring techniques for the sum or difference of cubes, complete the following list of cubes.

   $1^3 =$ _____ $2^3 =$ _____ $3^3 =$ _____ $4^3 =$ _____ $5^3 =$ _____
   $6^3 =$ _____ $7^3 =$ _____ $8^3 =$ _____ $9^3 =$ _____ $10^3 =$ _____

4. The following powers of x are all perfect cubes: $x^3, x^6, x^9, x^{12}, x^{15}$. Based on this observation, we may make a conjecture that if the power of a variable is divisible by _____ (with 0 remainder), then we have a perfect cube.

5. Identify the monomial as a perfect square, a perfect cube, both of these, or neither of these.
   (a) $64x^6y^{12}$ **(b)** $125t^6$ **(c)** $49x^{12}$ **(d)** $81r^{10}$

6. What must be true for x^n to be both a perfect square and a perfect cube?

Factor the binomial completely. Use your answers in Exercises 1 and 2 as necessary. See Examples 1 and 2.

7. $y^2 - 25$ **8.** $t^2 - 16$ **9.** $9r^2 - 4$ **10.** $4x^2 - 9$

11. $36m^2 - \dfrac{16}{25}$ **12.** $100b^2 - \dfrac{4}{49}$ **13.** $36x^2 - 16$ **14.** $32a^2 - 8$

15. $196p^2 - 225$ **16.** $361q^2 - 400$ **17.** $16r^2 - 25a^2$ **18.** $49m^2 - 100p^2$

19. $100x^2 + 49$ **20.** $81w^2 + 16$ **21.** $p^4 - 49$ **22.** $r^4 - 25$

23. $x^4 - 1$ **24.** $y^4 - 16$ **25.** $p^4 - 256$ **26.** $16k^4 - 1$

27. When a student was directed to factor $x^4 - 81$ completely, his teacher did not give him full credit for the answer $(x^2 + 9)(x^2 - 9)$. The student argued that since his answer does indeed give $x^4 - 81$ when multiplied out, he should be given full credit. Was the teacher justified in her grading of this item? Why or why not?

28. The binomial $4x^2 + 16$ is a sum of two squares that *can* be factored. How is this binomial factored? When can the sum of two squares be factored?

Factor the trinomial completely. It may be necessary to factor out the greatest common factor first. See Example 3.

29. $w^2 + 2w + 1$ **30.** $p^2 + 4p + 4$ **31.** $x^2 - 8x + 16$

32. $x^2 - 10x + 25$ **33.** $t^2 + t + \dfrac{1}{4}$ **34.** $m^2 + \dfrac{2}{3}m + \dfrac{1}{9}$

35. $x^2 - 1.0x + .25$ **36.** $y^2 - 1.4y + .49$ **37.** $2x^2 + 24x + 72$

38. $3y^2 - 48y + 192$ **39.** $16x^2 - 40x + 25$ **40.** $36y^2 - 60y + 25$

41. $49x^2 - 28xy + 4y^2$ **42.** $4z^2 - 12zw + 9w^2$ **43.** $64x^2 + 48xy + 9y^2$

44. $9t^2 + 24tr + 16r^2$ **45.** $-50h^2 + 40hy - 8y^2$ **46.** $-18x^2 - 48xy - 32y^2$

Factor the binomial completely. Use your answers in Exercises 3 and 4 as necessary. See Examples 4 and 5.

47. $a^3 + 1$ **48.** $m^3 + 8$ **49.** $a^3 - 1$ **50.** $m^3 - 8$

51. $p^3 + q^3$ **52.** $x^3 + z^3$ **53.** $27x^3 - 1$ **54.** $64y^3 - 27$

55. $8p^3 + 729q^3$ **56.** $64x^3 + 125y^3$ **57.** $y^3 - 8x^3$ **58.** $w^3 - 216z^3$

59. $27a^3 - 64b^3$ **60.** $125m^3 - 8p^3$ **61.** $125t^3 + 8s^3$ **62.** $27r^3 + 1000s^3$

———◇ **MATHEMATICAL CONNECTIONS** (Exercises 63–70) ◇———

A binomial may be *both* a difference of squares *and* a difference of cubes. One example of such a binomial is $x^6 - 1$. Using the techniques of this section, one of the factoring methods will give the complete factorization, while the other will not. Work Exercises 63–70 in order to determine which method to use if you have to make such a decision.

63. Factor $x^6 - 1$ as the difference of two squares.

64. The factored form obtained in Exercise 63 consists of a difference of cubes multiplied by a sum of cubes. Factor each binomial further.

65. Now start over and factor $x^6 - 1$ as the difference of two cubes.

66. The factored form obtained in Exercise 65 consists of a binomial which is a difference of squares and a trinomial. Factor the binomial further.

67. Compare your results in Exercises 64 and 66. Which one of these is the completely factored form?

68. Verify that the trinomial in the factored form in Exercise 66 is the product of the two trinomials in the factored form in Exercise 64.

69. Use the results of Exercises 63–68 to complete the following statement: In general, if I must choose between factoring first using the method for difference of squares or the method for difference of cubes, I should choose the _____ method to eventually obtain the complete factorization.

70. Find the *complete* factorization of $x^6 - 729$ using the knowledge you have gained in Exercises 63–69.

————————◆————————

Extend the methods of factoring presented so far in this chapter to factor the polynomial completely.

71. $(m + n)^2 - (m - n)^2$ **72.** $(a - b)^3 - (a + b)^3$

73. $m^2 - p^2 + 2m + 2p$ **74.** $3r - 3k + 3r^2 - 3k^2$

———◇ **MATHEMATICAL CONNECTIONS** (Exercises 75–80) ◇———

It is possible to apply two different methods of factoring a particular polynomial, obtaining the same answer. Work Exercises 75–80 in order, referring to the polynomial

$$(x^2 + 2x + 1) - 4.$$

75. Notice that the first three terms of the polynomial are grouped. Factor that trinomial, which is a perfect square trinomial, to obtain a difference of squares.

76. Factor the difference of squares obtained in Exercise 75.

77. Combine the constant terms within the factors from Exercise 76 to obtain the simplest forms of the factors.

78. Now start over and write the original polynomial $(x^2 + 2x + 1) - 4$ as a trinomial, by combining the constant terms.

79. Factor the trinomial obtained in Exercise 78 using the methods of Sections 4.2 and 4.3.

80. How do your results compare in Exercises 77 and 79? Which method do you prefer?

---◆---

81. For the polynomial $9y^2 + 14y + 25$, the first and last terms are perfect squares. Can the polynomial be factored? If it can, factor it. If it cannot, explain why it is not a perfect square trinomial.

82. Repeat Exercise 81 for $16m^2 + 42m + 49$.

Find the value of the indicated variable.

83. Find a value of b so that $x^2 + bx + 25 = (x + 5)^2$.

84. For what value of c is $4m^2 - 12m + c = (2m - 3)^2$?

85. Find a so that $ay^2 - 12y + 4 = (3y - 2)^2$.

86. Find b so that $100a^2 + ba + 9 = (10a + 3)^2$.

REVIEW EXERCISES

Solve the equation. See Section 2.2.

87. $m - 4 = 0$ **88.** $3t + 2 = 0$ **89.** $4z - 9 = 0$

90. $2t + 10 = 0$ **91.** $9x - 6 = 0$ **92.** $7x = 0$

SUMMARY: EXERCISES ON FACTORING

These mixed exercises are included to give you practice in selecting an appropriate method for factoring a particular polynomial. As you factor a polynomial, ask yourself these questions to decide on a suitable factoring technique.

Factoring a Polynomial	*Step 1*	Is there a common factor?
	Step 2	How many terms are in the polynomial?

Step 2   How many terms are in the polynomial?

Two terms   Check to see whether it is either the difference of two squares or the sum or difference of two cubes.

Three terms   Is it a perfect square trinomial? If the trinomial is not a perfect square, check to see whether the coefficient of the squared term is 1. If so, use the method of Section 4.2. If the coefficient of the squared term of the trinomial is not 1, use the general factoring methods of Section 4.3.

Four terms   Can the polynomial be factored by grouping?

Step 3   Can any factors be factored further?

Factor the polynomial completely.

1. $a^2 - 4a - 12$ **2.** $a^2 + 17a + 72$ **3.** $6y^2 - 6y - 12$

4. $7y^6 + 14y^5 - 168y^4$ **5.** $6a + 12b + 18c$ **6.** $m^2 - 3mn - 4n^2$

7. $p^2 - 17p + 66$

8. $z^2 - 6z + 7z - 42$

9. $10z^2 - 7z - 6$

10. $2m^2 - 10m - 48$

11. $m^2 - n^2 + 5m - 5n$

12. $15y + 5$

13. $8a^5 - 8a^4 - 48a^3$

14. $8k^2 - 10k - 3$

15. $z^2 - 3za - 10a^2$

16. $50z^2 - 100$

17. $x^2 - 4x - 5x + 20$

18. $100n^2r^2 + 30nr^3 - 50n^2r$

19. $6n^2 - 19n + 10$

20. $9y^2 + 12y - 5$

21. $16x + 20$

22. $m^2 + 2m - 15$

23. $6y^2 - 5y - 4$

24. $m^2 - 81$

25. $6z^2 + 31z + 5$

26. $5z^2 + 24z - 5 + 3z + 15$

27. $4k^2 - 12k + 9$

28. $8p^2 + 23p - 3$

29. $54m^2 - 24z^2$

30. $8m^2 - 2m - 3$

31. $3k^2 + 4k - 4$

32. $45a^3b^5 - 60a^4b^2 + 75a^6b^4$

33. $14k^3 + 7k^2 - 70k$

34. $5 + r - 5s - rs$

35. $y^4 - 16$

36. $20y^5 - 30y^4$

37. $8m - 16m^2$

38. $k^2 - 16$

39. $z^3 - 8$

40. $y^2 - y - 56$

41. $k^2 + 9$

42. $27p^{10} - 45p^9 - 252p^8$

43. $32m^9 + 16m^5 + 24m^3$

44. $8m^3 + 125$

45. $16r^2 + 24rm + 9m^2$

46. $z^2 - 12z + 36$

47. $15h^2 + 11hg - 14g^2$

48. $5z^3 - 45z^2 + 70z$

49. $k^2 - 11k + 30$

50. $64p^2 - 100m^2$

51. $3k^3 - 12k^2 - 15k$

52. $y^2 - 4yk - 12k^2$

53. $1000p^3 + 27$

54. $64r^3 - 343$

55. $6 + 3m + 2p + mp$

56. $2m^2 + 7mn - 15n^2$

57. $16z^2 - 8z + 1$

58. $125m^4 - 400m^3n + 195m^2n^2$

59. $108m^2 - 36m + 3$

60. $100a^2 - 81y^2$

61. $64m^2 - 40mn + 25n^2$

62. $4y^2 - 25$

63. $32z^3 + 56z^2 - 16z$

64. $10m^2 + 25m - 60$

65. $20 + 5m + 12n + 3mn$

66. $4 - 2q - 6p + 3pq$

67. $6a^2 + 10a - 4$

68. $36y^6 - 42y^5 - 120y^4$

69. $a^3 - b^3 + 2a - 2b$

70. $16k^2 - 48k + 36$

71. $64m^2 - 80mn + 25n^2$

72. $72y^3z^2 + 12y^2 - 24y^4z^2$

73. $8k^2 - 2kh - 3h^2$

74. $2a^2 - 7a - 30$

75. $(m + 1)^3 + 1$

76. $8a^3 - 27$

77. $10y^2 - 7yz - 6z^2$

78. $m^2 - 4m + 4$

79. $8a^2 + 23ab - 3b^2$

80. $a^4 - 625$

——————◆ **MATHEMATICAL CONNECTIONS** (Exercises 81–86) ◆——————

In this chapter we have presented methods of factoring trinomials, factoring trinomials that are perfect squares, factoring by grouping, and factoring the difference of squares and the sum or difference of cubes. In order to synthesize these ideas, work Exercises 81–86 in order, beginning with the polynomial

$$x^3t^3 + t^3 + 2x^3t^2 + 2t^2 - tx^3 - t - 2x^3 - 2.$$

81. Group the terms in pairs: the first two terms, the next two terms, and so on, until you have four pairs of terms. Be very careful when grouping terms that have negative coefficients.

82. Factor out the greatest common factor from each pair of terms you obtained in Exercise 81 and write the resulting polynomial.

83. You should now have four terms in the polynomial, each with a common factor of $x^3 + 1$. Factor out this common factor.

84. Your factored polynomial should now consist of the sum of cubes $x^3 + 1$ multiplied by a polynomial consisting of four terms. Factor this latter polynomial by grouping.

85. Your factored polynomial should now consist of three factors, two of which can be factored further. Complete the factorization of the polynomial.

86. How do you think that mathematics teachers make up polynomials for students to factor?

——————————————◆——————————————

4.5 SOLVING QUADRATIC EQUATIONS BY FACTORING

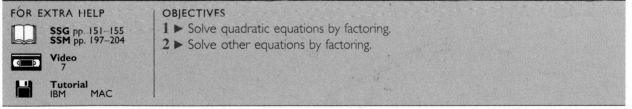

FOR EXTRA HELP

📖 **SSG** pp. 151–155
SSM pp. 197–204

📼 **Video**
7

💾 **Tutorial**
IBM   MAC

OBJECTIVES

1 ▶ Solve quadratic equations by factoring.
2 ▶ Solve other equations by factoring.

In this section we introduce *quadratic equations*, equations that contain a squared term and no terms of higher degree.

Quadratic Equation   An equation that can be put in the form

$$ax^2 + bx + c = 0,$$

where a, b, and c are real numbers, with $a \neq 0$, is a **quadratic equation.**

The form $ax^2 + bx + c = 0$ is the **standard form** of a quadratic equation. For example,

$$x^2 + 5x + 6 = 0, \qquad 2a^2 - 5a = 3, \qquad \text{and} \qquad y^2 = 4$$

are all quadratic equations but only $x^2 + 5x + 6 = 0$ is in standard form.

1 ▶ Solve quadratic equations by factoring.

▶ Some quadratic equations can be solved by factoring. A more general method for solving those equations that cannot be solved by factoring is given in Chapter 9. We use the **zero-factor property** to solve a quadratic equation by factoring.

Zero-Factor Property   If a and b are real numbers and   **if $ab = 0$, then $a = 0$ or $b = 0$.**

In other words, if the product of two numbers is zero, then at least one of the numbers must be zero. This means that one number *must* be 0, but both *may be* 0.

EXAMPLE 1
Using the Zero-Factor Property

Solve the equation $(x + 3)(2x - 1) = 0$.

The product $(x + 3)(2x - 1)$ is equal to zero. By the zero-factor property, the only way that the product of these two factors can be zero is if at least one of the factors is zero. Therefore, either $x + 3 = 0$ or $2x - 1 = 0$. Solve each of these two linear equations as in Chapter 2.

$$x + 3 = 0 \qquad \text{or} \qquad 2x - 1 = 0$$
$$x = -3 \qquad \text{or} \qquad 2x = 1 \qquad \text{Add 1 to both sides.}$$
$$x = \frac{1}{2} \qquad \text{Divide by 2.}$$

Since both of these equations have a solution, the equation $(x + 3)(2x - 1) = 0$ has two solutions, -3 and $1/2$. Check these answers by substituting -3 for x in the original equation. Then start over and substitute $1/2$ for x.

If $x = -3$, then   If $x = 1/2$, then

$$(-3 + 3)[2(-3) - 1] = 0 \quad ? \qquad\qquad \left(\frac{1}{2} + 3\right)\left(2 \cdot \frac{1}{2} - 1\right) = 0 \quad ?$$
$$0(-7) = 0 \quad ? \qquad\qquad\qquad\qquad \frac{7}{2}(1 - 1) = 0 \quad ?$$
$$0 = 0. \quad \text{True} \qquad\qquad\qquad\qquad \frac{7}{2} \cdot 0 = 0 \quad ?$$
$$0 = 0. \quad \text{True}$$

Both -3 and $1/2$ produce true statements, so they are solutions to the original equation. ∎

NOTE   The word "or" as used in Example 1 means "one or the other or both."

◈ C O N N E C T I O N S ◈

Galileo Galilei (1564–1642) developed theories to explain physical phenomena and set up experiments to test his ideas. According to legend, Galileo dropped objects of different weights from the tower of Pisa to disprove the Aristotelian view that heavier objects fall faster than lighter objects. He developed the formula for freely falling objects described by $d = 16t^2$, where d is the distance in feet that an object falls (disregarding air resistance) in t seconds, regardless of weight.*

The equation $d = 16t^2$ is a quadratic equation in the variable t. If we wish to find the number of seconds it would take an object to fall 256 feet, for example, we would substitute 256 for d and solve for t. On the other hand, if we wish to find how far the object would fall in 2 seconds, we would substitute 2 for t and calculate d. This idea of distance traveled depending on the time elapsed is an example of an important mathematical concept, the *function*. Functions are discussed in Chapter 6.

FOR DISCUSSION OR WRITING

1. How long would it take an object to fall 256 feet?
2. How far would an object fall in 2 seconds?

In Example 1 the equation to be solved was presented with the polynomial in factored form. If the polynomial in an equation is not already factored, first make sure that the equation is in standard form. Then factor the polynomial.

EXAMPLE 2 ■
Solving a Quadratic Equation Not in Standard Form

Solve the equation $x^2 - 5x = -6$.

First, rewrite the equation with all terms on one side by adding 6 to both sides.

$$x^2 - 5x = -6$$
$$x^2 - 5x + 6 = 0 \qquad \text{Add 6.}$$

Now factor $x^2 - 5x + 6$. Find two numbers whose product is 6 and whose sum is -5. These two numbers are -2 and -3, so the equation becomes

$$(x - 2)(x - 3) = 0. \qquad \text{Factor.}$$
$$x - 2 = 0 \quad \text{or} \quad x - 3 = 0 \qquad \text{Zero-factor property}$$
$$x = 2 \quad \text{or} \quad x = 3 \qquad \text{Solve each equation.}$$

Check both solutions by substituting first 2 and then 3 for x in the original equation. ■

In summary, we go through the following steps to solve quadratic equations by factoring.

Solving a Quadratic Equation by Factoring

Step 1 **Write in standard form.** Write the equation in standard form: all terms on one side of the equals sign, with 0 on the other side.

Step 2 **Factor.** Factor completely.

Step 3 **Use the zero-factor property.** Set each factor with a variable equal to 0, and solve the resulting equations.

Step 4 **Check.** Check each solution in the original equation.

---

*From *Mathematical Ideas*, 7th edition, by Charles D. Miller, Vern E. Heeren, and E. John Hornsby, Jr., HarperCollins, 1994, page 454.

EXAMPLE 3

Solving a Quadratic Equation with a Common Factor

Solve $4p^2 + 40 = 26p$.

Subtract $26p$ from each side and write in descending powers to get

$$4p^2 - 26p + 40 = 0.$$
$$2(2p^2 - 13p + 20) = 0 \qquad \text{Factor out 2.}$$
$$2(2p - 5)(p - 4) = 0 \qquad \text{Factor the trinomial.}$$

$$2 = 0 \quad \text{or} \quad 2p - 5 = 0 \quad \text{or} \quad p - 4 = 0 \qquad \text{Zero-factor property}$$

The equation $2 = 0$ has no solution. Solve the equation in the middle by first adding 5 on both sides of the equation. Then divide both sides by 2. Solve the equation on the right by adding 4 to both sides.

$$2p - 5 = 0 \qquad \text{or} \qquad p - 4 = 0$$
$$2p = 5 \qquad \text{or} \qquad p = 4$$
$$p = \frac{5}{2}$$

The solutions of $4p^2 + 40 = 26p$ are $5/2$ and 4; check them by substituting in the original equation. ∎

CAUTION A common error is to include 2 as a solution in Example 3.

EXAMPLE 4

Solving Special Quadratic Equations

Solve the equation.

(a) $16m^2 - 25 = 0$

Factor the left-hand side of the equation as the difference of two squares.

$$(4m + 5)(4m - 5) = 0$$

$$4m + 5 = 0 \qquad \text{or} \qquad 4m - 5 = 0 \qquad \text{Zero-factor property}$$
$$4m = -5 \qquad \text{or} \qquad 4m = 5 \qquad \text{Solve each equation.}$$
$$m = -\frac{5}{4} \qquad \text{or} \qquad m = \frac{5}{4}$$

Check the two solutions, $-5/4$ and $5/4$, in the original equation.

(b) $y^2 = 2y$

First write the equation in standard form.

$$y^2 - 2y = 0 \qquad \text{Standard form}$$
$$y(y - 2) = 0 \qquad \text{Factor.}$$
$$y = 0 \quad \text{or} \quad y - 2 = 0 \qquad \text{Zero-factor property}$$
$$y = 2 \qquad \text{Solve.}$$

The solutions are 0 and 2.

(c) $k(2k + 1) = 3$

$$k(2k + 1) = 3$$
$$2k^2 + k = 3 \qquad \text{Distributive property}$$
$$2k^2 + k - 3 = 0 \qquad \text{Subtract 3.}$$
$$(k - 1)(2k + 3) = 0 \qquad \text{Factor.}$$

$$k - 1 = 0 \quad \text{or} \quad 2k + 3 = 0 \qquad \text{Zero-factor property}$$
$$k = 1 \quad \text{or} \quad 2k = -3$$
$$k = -\frac{3}{2}$$

The two solutions are 1 and $-3/2$. ∎

CAUTION In Example 4(b) it is tempting to begin by dividing both sides of the equation by y to get $y = 2$. Note, however, that the other solution, 0, is not found by this method.

In Example 4(c) we could not use the zero-factor property to solve the equation in its given form because of the 3 on the right. Remember that the zero-factor property applies only to a product that equals 0.

---

2 ▶ Solve other equations by factoring.

▶ We can also use the zero-factor property to solve equations that result in more than two factors with variables, as shown in Example 5. (These equations are *not* quadratic equations. Why not?)

EXAMPLE 5

Solving Equations with More Than Two Variable Factors

Solve the equations.

(a) $6z^3 - 6z = 0$

$$6z^3 - 6z = 0$$
$$6z(z^2 - 1) = 0 \qquad \text{Factor out } 6z.$$
$$6z(z + 1)(z - 1) = 0 \qquad \text{Factor } z^2 - 1.$$

By an extension of the zero-factor property this product can equal zero only if at least one of the factors is zero. Write and solve three equations, one for each factor with a variable.

$$6z = 0 \quad \text{or} \quad z + 1 = 0 \quad \text{or} \quad z - 1 = 0$$
$$z = 0 \quad \text{or} \quad z = -1 \quad\quad z = 1$$

Check by substituting, in turn, 0, -1, and 1 in the original equation.

(b) $(3x - 1)(x^2 - 9x + 20) = 0$

$$(3x - 1)(x^2 - 9x + 20) = 0$$
$$(3x - 1)(x - 5)(x - 4) = 0 \qquad \text{Factor } x^2 - 9x + 20.$$
$$3x - 1 = 0 \quad \text{or} \quad x - 5 = 0 \quad \text{or} \quad x - 4 = 0 \qquad \text{Zero-factor property}$$
$$x = \frac{1}{3} \quad \text{or} \quad x = 5 \quad \text{or} \quad x = 4$$

The solutions of the original equation are $1/3$, 4, and 5. Check each solution. ∎

CAUTION In Example 5(b), it would be unproductive to begin by multiplying the two factors together. Keep in mind the zero-factor property requires the product of two or more factors must be equal to zero. Always consider first whether an equation is given in the appropriate form for the zero-factor property.

4.5 EXERCISES

Solve the equation and check the answer. See Example 1.

1. $(x - 4)(x + 5) = 0$ **2.** $(y + 3)(y - 8) = 0$ **3.** $(3k + 8)(k + 7) = 0$

4. $(5r + 9)(r + 6) = 0$ **5.** $t(t + 4) = 0$ **6.** $x(x - 10) = 0$

7. $2x(3x - 4) = 0$ **8.** $6y(4y + 9) = 0$

9. Students often become confused as to how to handle a constant, such as 2 in the equation $2x(3x - 4) = 0$ of Exercise 7. How would you explain to someone how to solve this equation and how to handle the constant 2?

10. As shown in Example 5, the zero-factor property can be extended to more than two factors. For example, to solve $(x - 4)(x + 3)(2x - 7) = 0$, we would set each factor equal to zero and solve three equations. Find the solutions to this equation.

11. Why do you think that 9 is called a *double solution* of the equation $(x - 9)^2 = 0$?

12. Write an equation with the two solutions 5 and $-\dfrac{4}{3}$.

Solve the equation and check the answer. See Examples 2–4.

13. $y^2 + 3y + 2 = 0$ **14.** $p^2 + 8p + 7 = 0$ **15.** $y^2 - 3y + 2 = 0$

16. $r^2 - 4r + 3 = 0$ **17.** $x^2 = 24 - 5x$ **18.** $t^2 = 2t + 15$

19. $x^2 = 3 + 2x$ **20.** $m^2 = 4 + 3m$ **21.** $z^2 = -2 - 3z$

22. $p^2 = 2p + 3$ **23.** $m^2 + 8m + 16 = 0$ **24.** $b^2 - 6b + 9 = 0$

25. $3x^2 + 5x - 2 = 0$ **26.** $6r^2 - r - 2 = 0$ **27.** $6p^2 = 4 - 5p$

28. $6x^2 = 4 + 5x$ **29.** $9s^2 + 12s = -4$ **30.** $36x^2 + 60x = -25$

31. $y^2 - 9 = 0$ **32.** $m^2 - 100 = 0$ **33.** $16k^2 - 49 = 0$

34. $4w^2 - 9 = 0$ **35.** $n^2 = 121$ **36.** $x^2 = 400$

37. What is wrong with this reasoning for solving the equation in Exercise 35? "To solve $n^2 = 121$, I must find a number that, when multiplied by itself, gives 121. Because $11^2 = 121$ the solution of the equation is 11."

38. What is wrong with this reasoning in solving $x^2 = 7x$? "To solve $x^2 = 7x$ first divide both sides by x to get $x = 7$. Therefore, the solution is 7."

Solve the equation and check the answer. See Examples 4 and 5.

39. $x^2 = 7x$ **40.** $t^2 = 9t$ **41.** $6r^2 = 3r$

42. $10y^2 = -5y$ **43.** $g(g - 7) = -10$ **44.** $r(r - 5) = -6$

45. $z(2z + 7) = 4$ **46.** $b(2b + 3) = 9$

47. $2(y^2 - 66) = -13y$ **48.** $3(t^2 + 4) = 20t$

49. $3x(x + 1) = (2x + 3)(x + 1)$ **50.** $2k(k + 3) = (3k + 1)(k + 3)$

51. $(2r + 5)(3r^2 - 16r + 5) = 0$ **52.** $(3m + 4)(6m^2 + m - 2) = 0$

53. $(2x + 7)(x^2 + 2x - 3) = 0$ **54.** $(x + 1)(6x^2 + x - 12) = 0$

55. $9y^3 - 49y = 0$ **56.** $16r^3 - 9r = 0$

57. $r^3 - 2r^2 - 8r = 0$ **58.** $x^3 - x^2 - 6x = 0$

59. $a^3 + a^2 - 20a = 0$ **60.** $y^3 - 6y^2 + 8y = 0$

61. $r^4 = 2r^3 + 15r^2$ **62.** $x^3 = 3x + 2x^2$

63. $6p^2(p + 1) = 4(p + 1) - 5p(p + 1)$ **64.** $6x^2(2x + 3) - 5x(2x + 3) = 4(2x + 3)$

65. $(k + 3)^2 - (2k - 1)^2 = 0$ **66.** $(4y - 3)^3 - 9(4y - 3) = 0$

67. Explain why the solutions of $(x - 3)(x + 2) = 1$ are not found from the two equations

$$x - 3 = 1 \quad \text{and} \quad x + 2 = 1.$$

68. What is wrong with the following solution?

$$4x^2 = 4x$$
$$x = 1 \qquad \text{Divide both sides by } 4x.$$

 **MATHEMATICAL CONNECTIONS** (Exercises 69–74)

Imagine that you are tutoring a student who is studying solving quadratic equations by factoring. You wish to make up a quadratic equation with two different solutions for the student. Let us suppose you want the two solutions to be -8 and $2/5$. Work Exercises 69–74 in order, so that you can see how this equation can be written in several ways. In all cases, use x as your variable

69. Write a binomial such that the coefficient of x is 1 and the number -8 causes the value of the binomial to be 0.

70. Write a binomial such that the coefficient of x is 5 and the number $\frac{2}{5}$ causes the value of the binomial to be 0.

71. Write the *product* of the two binomials you found in Exercises 69 and 70 and set the product equal to 0. What are the solutions of this equation?

72. Multiply the two binomials from Exercise 71 and write an equivalent equation in the form $ax^2 + bx + c = 0$. What are the solutions of this equation?

73. Factor out x from the first two terms in the equation in Exercise 71 and subtract the number represented by c from both sides. What are the solutions of this equation?

74. Use your own ingenuity to transform the equation in Exercise 72 into an equivalent equation different from the others you have written.

REVIEW EXERCISES

Solve the problem. See Sections 2.4 and 2.5.

75. Florida has 9 more counties than California. Together the two states have 125 counties. How many counties does each state have?

76. If a number is doubled and 6 is subtracted from this result, the answer is 3684. The unknown number is the year that Texas was admitted to the Union. What year was Texas admitted?

77. The length of a rectangle is 3 meters more than the width. The perimeter of the rectangle is 34 meters. Find the width of the rectangle.

78. A rectangle has a length 4 meters less than twice the width. The perimeter of the rectangle is 4 meters more than five times the width. Find the width of the rectangle.

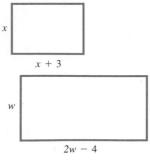

4.6 APPLICATIONS OF QUADRATIC EQUATIONS

FOR EXTRA HELP

SSG pp. 155–159
SSM pp. 204–211

Video
7

Tutorial
IBM   MAC

OBJECTIVES
1 ▶ Solve problems about geometric figures.
2 ▶ Solve problems using the Pythagorean formula.

We are now ready to use factoring to solve quadratic equations that arise from applied problems. Most problems in this section require one of the formulas given in the endsheets. The general approach is the same as in Chapter 2. We still follow the six steps listed in Section 2.4 and continue the work with formulas and geometric problems begun in Section 2.5.

1 ▶ Solve problems about geometric figures.

▶   We begin with a geometry problem.

E X A M P L E   1

Solving an Area Problem

The Goldsteins are planning to add a rectangular porch to their house. The width of the porch will be 4 feet less than its length, and they want it to have an area of 96 square feet. Find the length and width of the porch.

Let x = the length of the porch;

$x - 4$ = the width (the width is 4 less than the length).

See Figure 1. The area of a rectangle is given by the formula

$$\text{area} = LW = \text{length} \times \text{width}.$$

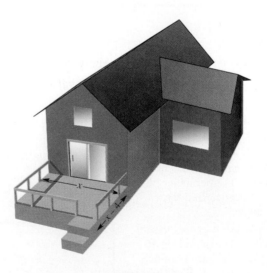

FIGURE 1

Substitute 96 for the area, x for the length, and $x - 4$ for the width into the formula.

$$A = LW$$
$$96 = x(x - 4) \quad \text{Let } A = 96, L = x, W = x - 4.$$
$$96 = x^2 - 4x \qquad \qquad \text{Distributive property}$$
$$0 = x^2 - 4x - 96 \qquad \text{Subtract 96 from both sides.}$$
$$0 = (x - 12)(x + 8) \qquad \text{Factor.}$$
$$x - 12 = 0 \quad \text{or} \quad x + 8 = 0 \qquad \text{Zero-factor property}$$
$$x = 12 \quad \text{or} \quad x = -8 \qquad \text{Solve.}$$

The solutions of the equations are 12 and -8. Since a rectangle cannot have a negative length discard the solution -8. Then 12 feet is the length of the porch and $12 - 4 = 8$ feet is the width. As a check, note that the width is 4 less than the length and the area is $8 \cdot 12 = 96$ square feet as required. ∎

CAUTION   In an applied problem, always be careful to check solutions against physical facts.

The next application involves *perimeter,* the distance around a figure, as well as area.

E X A M P L E   2
Solving an Area and
Perimeter Problem

The length of a rectangular rug is 4 feet more than the width. The area of the rug is numerically 1 more than the perimeter. See Figure 2. Find the length and width of the rug.

$x + 4$

x

FIGURE 2

Let $\quad x =$ the width of the rug;
$\quad x + 4 =$ the length of the rug.

The area is the product of the length and width, so

$$A = LW.$$

Substituting $x + 4$ for the length and x for the width gives

$$A = (x + 4)x.$$

Now substitute into the formula for perimeter.

$$P = 2L + 2W$$
$$P = 2(x + 4) + 2x$$

According to the information given in the problem, the area is numerically 1 more than the perimeter.

The area	is	1	more than	the perimeter.
↓	↓	↓	↓	↓
$(x + 4)x$	$=$	1	$+$	$2(x + 4) + 2x$

Simplify and solve this equation.

$$x^2 + 4x = 1 + 2x + 8 + 2x \qquad \text{Distributive property}$$
$$x^2 + 4x = 9 + 4x \qquad \text{Combine terms.}$$
$$x^2 = 9 \qquad \text{Subtract } 4x \text{ from both sides.}$$
$$x^2 - 9 = 0 \qquad \text{Subtract } 9 \text{ from both sides.}$$
$$(x + 3)(x - 3) = 0 \qquad \text{Factor.}$$
$$x + 3 = 0 \qquad \text{or} \qquad x - 3 = 0 \qquad \text{Zero-factor property}$$
$$x = -3 \qquad \text{or} \qquad x = 3$$

A rectangle cannot have a negative width, so ignore -3. The only valid solution is 3, so the width is 3 feet and the length is $3 + 4 = 7$ feet. Check to see that the area is numerically 1 more than the perimeter. The rug is 3 feet wide and 7 feet long. ∎

2 ▶ Solve problems using the Pythagorean formula. ▶ The next example requires the **Pythagorean formula** from geometry.

Pythagorean Formula   If a right triangle (a triangle with a 90° angle) has longest side of length c and two other sides of lengths a and b, then

$$a^2 + b^2 = c^2.$$

(See the figure.) The longest side, the **hypotenuse,** is opposite the right angle. The two shorter sides are the **legs** of the triangle.

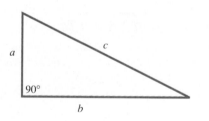

E X A M P L E   3

Using the Pythagorean Formula

Ed and Mark leave their office with Ed traveling north and Mark traveling east. When Mark is 1 mile farther than Ed from the office, the distance between them is 2 miles more than Ed's distance from the office. Find their distances from the office and the distance between them.

Let x represent Ed's distance from the office,

$x + 1$ represent Mark's distance from the office,

$x + 2$ represent the distance between them.

Place these on a right triangle as in Figure 3.

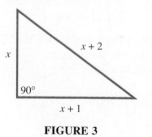

FIGURE 3

Substitute into the Pythagorean formula,

$$a^2 + b^2 = c^2$$
$$x^2 + (x + 1)^2 = (x + 2)^2.$$

Since $(x + 1)^2 = x^2 + 2x + 1$, and since $(x + 2)^2 = x^2 + 4x + 4$, the equation becomes

$$x^2 + x^2 + 2x + 1 = x^2 + 4x + 4.$$

$x^2 - 2x - 3 = 0$	Standard form
$(x - 3)(x + 1) = 0$	Factor.
$x - 3 = 0 \quad$ or $\quad x + 1 = 0$	Zero-factor property
$x = 3 \quad$ or $\quad x = -1$	Solve.

Since -1 cannot be the length of a side of a triangle, 3 is the only possible answer. Therefore, Ed is 3 miles north of the office, Mark is $3 + 1 = 4$ miles east of the office, and they are $3 + 2 = 5$ miles apart. Check to see that $3^2 + 4^2 = 5^2$ is true. ■

CAUTION When solving a problem involving the Pythagorean formula, be sure that the expressions for the sides are properly placed.

$$\mathbf{leg^2 + leg^2 = hypotenuse^2}$$

◆ **CONNECTIONS** ◆

When a carpenter builds a floor for a rectangular room, it is essential that the corners of the floor are at right angles; otherwise, problems will occur when the walls are constructed, when flooring is laid, and so on. To check that the floor is "squared off," the carpenter can use the *converse* of the Pythagorean formula: If $a^2 + b^2 = c^2$, then the angle opposite side c is a right angle.

FOR DISCUSSION OR WRITING
Suppose a carpenter is building an 8-foot by 12-foot room. After the floor is built, the carpenter finds that the length of the diagonal of the floor is 14 feet, 8 inches. Is the floor "squared off" properly? If not, what should the diagonal measure?*

4.6 EXERCISES

In Exercises 1–6, a figure and a corresponding geometric formula are given.
(a) *Write an equation using the formula and the given information.*
(b) *Solve the equation, giving only the solution(s) that make sense in the problem.*
(c) *Use the solution(s) to find the indicated dimensions of the figure.*

1. Area of a rectangle: $A = LW$
The area of this rectangle is 80 square units. Find its length and its width.

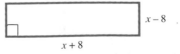

$x - 8$

$x + 8$

* From *Mathematical Ideas,* 7th edition, by Charles D. Miller, Vern E. Heeren, and E. John Hornsby, Jr., HarperCollins, 1994, page 541.

2. Area of a square: $A = s^2$
The area of this square is 196 square units. Find the length of each side.

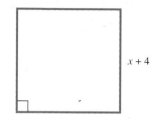

$x + 4$

3. Area of a parallelogram: $A = bh$
The area of this parallelogram is 45 square units. Find its base and its height.

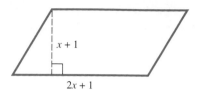

$x + 1$

$2x + 1$

4. Area of a triangle: $A = \frac{1}{2}bh$
The area of this triangle is 60 square units. Find its base and its height.

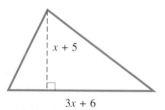

$x + 5$

$3x + 6$

5. Area of a circle: $A = \pi r^2$
The area of this circle is 36π square units. Find its radius.

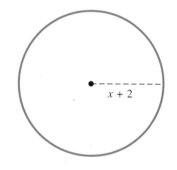

$x + 2$

6. Volume of a rectangular box: $V = LWH$
The volume of this box is 192 cubic units. Find its length and its width.

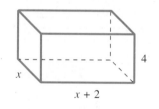

x 4

$x + 2$

Solve the problem. Check your answer to be sure that it is reasonable. Refer to the formulas found in the endsheets. See Examples 1 and 2.

7. The length of a VHS videocassette shell is 3 inches more than its width. The area of the rectangular top side of the shell is 28 square inches. Find the length and the width of the videocassette shell.

8. A plastic box that holds a standard audiocassette has a length 4 centimeters longer than its width. The area of the rectangular top of the box is 77 square centimeters. Find the length and the width of the box.

9. The dimensions of a certain IBM computer monitor screen are such that its length is 3 inches more than its width. If the length is increased by 1 inch while the width remains the same, the area is increased by 8 square inches. What are the dimensions of the screen?

10. The keyboard of the computer mentioned in Exercise 9 is 11 inches longer than it is wide. If both its length and width are increased by 2 inches, the area of the top of the keyboard is increased by 58 square inches. What are the length and the width of the keyboard?

11. A square poster has sides measuring 2 feet less than the sides of a square sign. If the difference between their areas is 32 square feet, find the lengths of the sides of the poster and the sign.

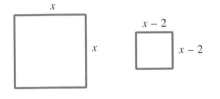

12. The sides of one square have a length 3 meters more than the sides of a second square. If the area of the larger square is subtracted from 4 times the area of the smaller square, the result is 36 square meters. What are the lengths of the sides of each square?

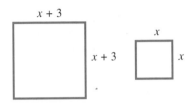

13. The area of a triangle is 30 square inches. The base of the triangle measures 2 inches more than twice the height of the triangle. Find the measures of the base and the height.

14. A certain triangle has its base equal in measure to its height. The area of the triangle is 72 square meters. Find the equal base and height measures.

15. A ten-gallon aquarium holding African cichlids is 3 inches higher than it is wide. Its length is 21 inches, and its volume is 2730 cubic inches. What are the height and width of the aquarium?

16. Nana Nantambu wishes to build a box to hold her tools. It is to be 2 feet high, and the width is to be 3 feet less than its length. If its volume is to be 80 cubic feet, find the length and the width of the box.

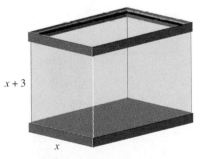

Use the Pythagorean formula to solve the problem. See Example 3.

17. The hypotenuse of a right triangle is 1 centimeter longer than the longer leg. The shorter leg is 7 centimeters shorter than the longer leg. Find the length of the longer leg of the triangle.

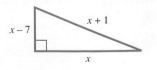

18. The longer leg of a right triangle is 1 meter longer than the shorter leg. The hypotenuse is 1 meter shorter than twice the shorter leg. Find the length of the shorter leg of the triangle.

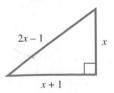

19. A ladder is resting against a wall. The top of the ladder touches the wall at a height of 15 feet. Find the distance from the wall to the bottom of the ladder if the length of the ladder is one foot more than twice its distance from the wall.

20. Two cars leave an intersection. One car travels north; the other travels east. When the car traveling north had gone 24 miles, the distance between the cars was four miles more than three times the distance traveled by the car heading east. Find the distance between the cars at that time.

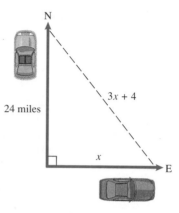

21. A garden has the shape of a right triangle with one leg 2 meters longer than the other. The hypotenuse is two meters less than twice the length of the shorter leg. Find the length of the shorter leg.

22. The hypotenuse of a right triangle is 1 foot more than twice the length of the shorter leg. The longer leg is 1 foot less than twice the length of the shorter leg. Find the length of the shorter leg.

If an object is propelled upward from a height of s feet at an initial velocity of v feet per second, then its height h after t seconds is given by the equation

$$h = -16t^2 + vt + s,$$

where h is in feet. For example, if the object is propelled from a height of 48 feet with an initial velocity of 32 feet per second, its height h is given by the equation $h = -16t^2 + 32t + 48$.

Use this information in Exercises 23–26.

23. After how many seconds is the height 64 feet? (*Hint:* Let $h = 64$ and solve.)

24. After how many seconds is the height 60 feet?

25. After how many seconds does the object hit the ground? (*Hint:* When the object hits the ground, $h = 0$.)

26. The quadratic equation from Exercise 25 has two solutions, yet only one of them is appropriate for answering the question. Why is this so?

If an object is propelled upward from ground level with an initial velocity of 64 feet per second, its height h in feet t seconds later is given by the equation $h = -16t^2 + 64t$. Use this information in Exercises 27–30.

27. After how many seconds is the height 48 feet?

28. It can be shown using concepts developed in other courses that the object reaches its maximum height 2 seconds after it is propelled. What is this maximum height?

29. After how many seconds does the object hit the ground?

30. The quadratic equation from Exercise 29 has two solutions, yet only one of them is appropriate for answering the question. Why is this so?

Exercises 31 and 32 require the formula for the volume of a pyramid,

$$V = \frac{1}{3}Bh,$$

where B is the area of the base.

31. Suppose a pyramid has a rectangular base whose width is 3 centimeters less than the length. If the height is 8 centimeters and the volume is 144 cubic centimeters, find the length of the base.

32. The volume of a pyramid is 32 cubic meters. Suppose the numerical value of the height is 10 less than the numerical value of the area of the base. Find the area of the base.

Work the following problems involving formulas. If an object is dropped, the distance d it falls in t seconds (disregarding air resistance) is given by

$$d = \frac{1}{2}gt^2,$$

where g is approximately 32 feet per second. Find the distance an object would fall in the following times.

33. 4 seconds

34. 8 seconds

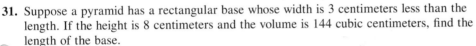 *How long would it take an object to fall from the top of the building described? Use a calculator as necessary, and round your answer to the nearest tenth of a second.*

35. Navarre Building, New York City   512 feet

36. One Canada Square, London   800 feet

37. Central Plaza, Hong Kong   1028 feet

38. Vegas World Tower, Las Vegas   1012 feet

Solve the problem.

39. The product of two consecutive integers is 11 more than their sum. Find the integers.

40. The product of two consecutive integers is 4 less than 4 times their sum. Find the integers.

41. Find three consecutive odd integers such that 3 times the sum of all three is 18 more than the product of the smaller two.

42. Find three consecutive odd integers such that the sum of all three is 42 less than the product of the larger two.

43. Find three consecutive even integers such that the sum of the squares of the smaller two is equal to the square of the largest.

44. Find three consecutive even integers such that the square of the sum of the smaller two is equal to twice the largest.

A square piece of cardboard is to be formed into an open-topped box by cutting 3-inch squares from the corners and folding up the sides. See the figure.

The volume V of the resulting box is given by the formula $V = 3(x - 6)^2$, where x is the original length of each side of the piece of cardboard.

Use this information to work Exercises 45–48.

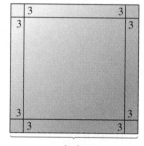

x inches

Original Piece of Cardboard

The Resulting Box

45. What is the volume of the box if the original length of each side of the piece of cardboard is 14 inches?

46. Repeat Exercise 45 if the original length is 20 inches.

47. If the volume is 300 cubic inches, what is the original length of each side of the piece of cardboard?

48. Repeat Exercise 47 if the volume is 1200 cubic inches.

—◆— **MATHEMATICAL CONNECTIONS** (Exercises 49–52) ◆—

One of the many known proofs of the Pythagorean formula is based on the figures shown.

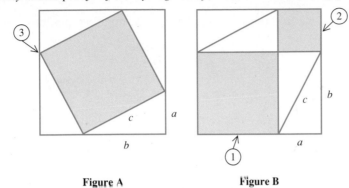

Figure A   **Figure B**

Refer to the appropriate figure and answer Exercises 49–52 in order.

49. What is an expression for the area of the dark square labeled ③ in Figure A?

50. The five regions in Figure A are equal in area to the six regions in Figure B. What is an expression for the area of the square labeled ① in Figure B?

51. What is an expression for the area of the square labeled ② in Figure B?

52. Represent this statement using algebraic expressions: The sum of the areas of the dark regions in Figure B is equal to the area of the dark region in Figure A. What does this equation represent?

—————◆—————

REVIEW EXERCISES

Solve the inequality. See Section 2.9.

53. $-2x + 5 > 0$ **54.** $\frac{1}{2}t - 4 > 6$ **55.** $3 - x \le 5$

Decide whether the inequality is true or false when x is replaced by 3. See Sections 1.4–1.7.

56. $x^2 - x + 4 \le 0$ **57.** $x^2 - 6x + 9 \le 0$ **58.** $x^2 - 6x + 9 < 0$

4.7 SOLVING QUADRATIC INEQUALITIES

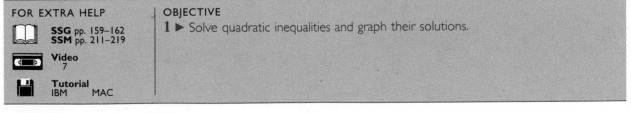

FOR EXTRA HELP

📖 **SSG** pp. 159–162
SSM pp. 211–219

📼 **Video** 7

💾 **Tutorial** IBM   MAC

OBJECTIVE

1 ▶ Solve quadratic inequalities and graph their solutions.

1 ▶ Solve quadratic inequalities and graph their solutions.

▶ A **quadratic inequality** is an inequality that involves a second-degree polynomial. Examples of quadratic inequalities include

$$2x^2 + 3x - 5 < 0, \qquad x^2 \le 4, \qquad \text{and} \qquad x^2 + 5x + 6 > 0.$$

Examples 1 and 2 show how to solve such inequalities.

EXAMPLE 1

Solving a Quadratic Inequality Including Endpoints

Solve $x^2 - 3x - 10 \leq 0$.

To begin, we find the solution of the corresponding quadratic equation,

$$x^2 - 3x - 10 = 0.$$

Factor to get

$$(x - 5)(x + 2) = 0$$

from which

$$x - 5 = 0 \qquad \text{or} \qquad x + 2 = 0$$
$$x = 5 \qquad \text{or} \qquad x = -2.$$

Since 5 and -2 are the only values that satisfy $x^2 - 3x - 10 = 0$, all other values of x will make $x^2 - 3x - 10$ either less than 0 (< 0) or greater than 0 (> 0). The values $x = 5$ and $x = -2$ determine three regions on the number line, as shown in Figure 4. Region A includes all numbers less than -2, region B includes the numbers between -2 and 5, and region C includes all numbers greater than 5.

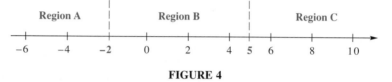

FIGURE 4

All values of x in a given region will cause $x^2 - 3x - 10$ to have the same sign (either positive or negative). Test one value of x from each region to see which regions satisfy $x^2 - 3x - 10 \leq 0$. First, are the points in region A part of the solution? As a trial value, choose any number less than -2, say -6.

$x^2 - 3x - 10 \leq 0$		Original inequality
$(-6)^2 - 3(-6) - 10 \leq 0$	?	Let $x = -6$.
$36 + 18 - 10 \leq 0$	?	Simplify.
$44 \leq 0$		False

Since $44 \leq 0$ is false, the points in Region A do not belong to the solution. What about Region B? Try the value $x = 0$.

$0^2 - 3(0) - 10 \leq 0$	?	Let $x = 0$.
$-10 \leq 0$		True

Since $-10 \leq 0$ is true, the points in Region B do belong to the solution. Try $x = 6$ to check Region C.

$6^2 - 3(6) - 10 \leq 0$	?	Let $x = 6$.
$36 - 18 - 10 \leq 0$	?	Simplify.
$8 \leq 0$		False

Since $8 \leq 0$ is false, the points in Region C do not belong to the solution.

The points in Region B are the only ones that satisfy $x^2 - 3x - 10 \leq 0$. As shown in Figure 5, the solution includes the points in Region B together with the endpoints -2 and 5. The solution is written

$$-2 \leq x \leq 5. \quad \blacksquare$$

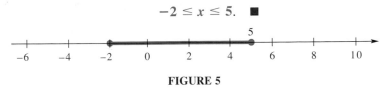

FIGURE 5

To summarize, we use the following steps to solve a quadratic inequality.

Solving a Quadratic Inequality

Step 1 **Write an equation.**   Change the inequality to an equation.

Step 2 **Solve.**   Use the zero-factor property to solve the equation from Step 1.

Step 3 **Determine the regions.**   Use the solutions of the equation in Step 2 to determine regions on the number line.

Step 4 **Test each region.**   Choose a number from each region. Substitute the number into the original inequality. If the number satisfies the inequality, all numbers in that region satisfy the inequality.

Step 5 **Write the solutions.**   Write the solutions as inequalities.

◆ **C O N N E C T I O N S** ◆

The number of items that must be sold for a company's revenue to equal its cost to produce those items is called the *break-even point*. If revenue, R, and cost, C, are given by expressions in x, where x is the number of items produced and sold, then the solution of the equation $R - C = 0$ gives the break-even point. For the company to make a profit, $R - C$ must be greater than 0, and the company would want to solve the inequality $R - C > 0$.

FOR DISCUSSION OR WRITING

Explain why $R - C > 0$ gives values of x that produce a profit. Suppose $R - C = x^2 - x + 12$. What values of x produce a profit? Why is only one part of the solution of the corresponding inequality valid here?

E X A M P L E   2

Solving a Quadratic Inequality Excluding Endpoints

Solve $-x^2 - 5x - 6 < 0$.

It will be easier to factor if we multiply both sides by -1. We must also remember to reverse the direction of the inequality symbol:

$$x^2 + 5x + 6 > 0.$$

Factoring the expression in the corresponding equation $x^2 + 5x + 6 = 0$, we get $(x + 2)(x + 3) = 0$. The solutions of the equation are -2 and -3. These points determine three regions on the number line. See Figure 6. This time, these points will not belong to the solution because only values of x that make $x^2 + 5x + 6$ *greater than* 0 are solutions.

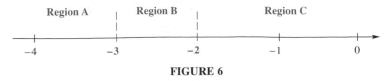

FIGURE 6

Do the points in Region A belong to the solution? Decide by selecting any point in Region A, such as -4. Does -4 satisfy the original inequality?

$$x^2 + 5x + 6 > 0 \qquad \text{Original inequality}$$
$$(-4)^2 + 5(-4) + 6 > 0 \qquad ? \qquad \text{Let } x = -4.$$
$$16 - 20 + 6 > 0 \qquad ? \qquad \text{Simplify.}$$
$$2 > 0 \qquad \text{True}$$

Since $2 > 0$ is true, all the points in Region A belong to the solution of the given inequality.

For Region B, choose a value between -3 and -2, say $-2\ 1/2$, or $-5/2$.

$$\left(-\frac{5}{2}\right)^2 + 5\left(-\frac{5}{2}\right) + 6 > 0 \qquad ? \qquad \text{Let } x = -\frac{5}{2}.$$
$$\frac{25}{4} + \left(-\frac{25}{2}\right) + 6 > 0 \qquad ? \qquad \text{Simplify.}$$
$$-\frac{1}{4} > 0 \qquad \text{False}$$

Since $-1/4 > 0$ is false, no point in Region B belongs to the solution.

For Region C, try the number 0.

$$0^2 + 5(0) + 6 > 0 \qquad ? \qquad \text{Let } x = 0.$$
$$6 > 0 \qquad \text{True}$$

Since $6 > 0$ is true, the points in Region C belong to the solution.

The solution is shown in Figure 7. The graph of $x^2 + 5x + 6 > 0$ includes all values of x less than -3, together with all values of x greater than -2. Write the solution as

$$x < -3 \qquad \text{or} \qquad x > -2. \quad \blacksquare$$

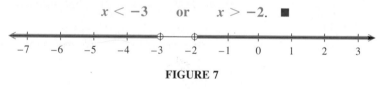

FIGURE 7

CAUTION   There is no shortcut way to write the solution $x < -3$ or $x > -2$.

4.7 EXERCISES

In order to prepare for using test values in the various regions, answer true or false depending on whether the given value of x satisfies or does not satisfy the inequality.

1. $(x - 7)(x + 4) \geq 0$
   (a) $x = 7$ **(b)** $x = -4$ **(c)** $x = 0$ **(d)** $x = 10$

2. $(x + 3)(x - 5) \leq 0$
   (a) $x = -3$ **(b)** $x = 5$ **(c)** $x = 0$ **(d)** $x = -8$

3. $2x^2 - 9x - 5 < 0$

    (a) $x = -\dfrac{1}{2}$ **(b)** $x = 5$ **(c)** $x = 0$ **(d)** $x = -6$

4. $3x^2 - 14x - 24 > 0$

    (a) $x = -\dfrac{4}{3}$ **(b)** $x = 6$ **(c)** $x = 0$ **(d)** $x = -3$

Solve the inequality and graph the solutions. See Examples 1 and 2.

5. $(a + 3)(a - 3) < 0$ **6.** $(b - 2)(b + 2) > 0$ **7.** $(a + 6)(a - 7) \geq 0$ **8.** $(z - 5)(z - 4) \leq 0$

9. $m^2 + 5m + 6 > 0$ **10.** $y^2 - 3y + 2 < 0$ **11.** $z^2 - 4z - 5 \leq 0$ **12.** $3p^2 - 5p - 2 \leq 0$

13. $5m^2 + 3m - 2 < 0$ **14.** $2k^2 + 7k - 4 > 0$ **15.** $6r^2 - 5r < 4$ **16.** $6r^2 + 7r > 3$

17. $q^2 - 7q < -6$ **18.** $2k^2 - 7k \leq 15$ **19.** $6m^2 + m - 1 > 0$ **20.** $30r^2 + 3r - 6 \leq 0$

21. $12p^2 + 11p + 2 < 0$ **22.** $a^2 - 16 < 0$ **23.** $9m^2 - 36 > 0$ **24.** $r^2 - 100 \geq 0$

25. $r^2 > 16$ **26.** $m^2 \geq 25$

The given inequality is not quadratic, but it may be solved in a similar manner. Solve and graph the inequality.

27. $(a + 2)(3a - 1)(a - 4) \geq 0$ **28.** $(2p - 7)(p - 1)(p + 3) \leq 0$

29. $(r - 2)(r^2 - 3r - 4) < 0$ **30.** $(m + 5)(m^2 - m - 6) > 0$

———————◆ **MATHEMATICAL CONNECTIONS** (Exercises 31–36) ◆———————

The method described in this section for solving quadratic inequalities can actually be applied to linear inequalities as well. For example, suppose that we want to solve the linear inequality

$$-3x + 9 < 0.$$

Work Exercises 31–36 in order, and see how this can be done.

31. Solve the corresponding linear equation $-3x + 9 = 0$.

32. Use the solution obtained in Exercise 31 to divide a number line into two regions. Label the region on the left Region A, and the one on the right Region B.

33. The number 0 lies in Region A. Use it as a test value. Does it make $-3x + 9 < 0$ a true statement?

34. The number 6 lies in Region B. Use it as a test value. Does it make $-3x + 9 < 0$ a true statement?

35. Based on your results in Exercises 31, 33, and 34, what is the solution of the linear inequality $-3x + 9 < 0$?

36. Use the method of Section 2.9 to solve this linear inequality. (Your solution should be the same as the one you found in Exercise 35.)

———————————◆———————————

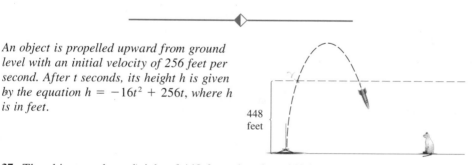

An object is propelled upward from ground level with an initial velocity of 256 feet per second. After t seconds, its height h is given by the equation $h = -16t^2 + 256t$, where h is in feet.

448 feet

37. The object reaches a height of 448 feet when $h = 448$ in the equation. Use the method of Section 4.5 to solve the resulting equation and determine the times at which the object reaches this height.

38. The object is more than 448 feet above the ground when $h > 448$. This is given by the inequality $-16t^2 + 256t > 448$. Use the method of this section to find the time interval over which the object is more than 448 feet above the ground.

39. By solving $-16t^2 + 256t < 448$ and realizing that t must be greater than or equal to 0, we can determine the time intervals over which the object is less than 448 feet above the ground. Use the method of this section to find this time interval. (Note: We must also have $t \le 16$.)

40. Explain the restrictions on t described in Exercise 39.

REVIEW EXERCISES

Write the fraction in lowest terms. See Section 1.1.

41. $\dfrac{50}{72}$ **42.** $\dfrac{26}{156}$ **43.** $\dfrac{26}{13}$ **44.** $\dfrac{-18}{18}$

CHAPTER 4 SUMMARY

KEY TERMS

4.1 factor
common factor
greatest common
factor
factoring by
grouping

4.2 factored form
factoring
prime polynomial

4.4 perfect square
trinomial

4.5 quadratic equation
standard form

4.6 hypotenuse
legs

4.7 quadratic inequality

QUICK REVIEW

CONCEPTS	EXAMPLES
4.1 THE GREATEST COMMON FACTOR; FACTORING BY GROUPING	
Finding the Greatest Common Factor 1. Include the largest numerical factor of every term. 2. Include each variable that is a factor of every term raised to the smallest exponent that appears in a term.	Find the greatest common factor of $$4x^2y, \quad -6x^2y^3, \quad 2xy^2.$$ $$4x^2y = 2^2 \cdot x^2 \cdot y$$ $$-6x^2y^3 = -1 \cdot 2 \cdot 3 \cdot x^2 \cdot y^3$$ $$2xy^2 = 2 \cdot x \cdot y^2$$ The greatest common factor is $2xy$.
Factoring by Grouping 1. Group the terms so that each group has a common factor. 2. Factor out the greatest common factor in each group. 3. Factor a common binomial factor from the result of Step 2. 4. If Step 3 cannot be performed, try a different grouping.	Factor $3x^2 + 5x - 24xy - 40y$. **1.** $(3x^2 + 5x) + (-24xy - 40y)$ **2.** $x(3x + 5) - 8y(3x + 5)$ **3.** $(3x + 5)(x - 8y)$
4.2 FACTORING TRINOMIALS	
To factor $x^2 + bx + c$, find m and n such that $mn = c$ and $m + n = b$. $$\overset{\displaystyle mn = c}{\underset{\displaystyle m + n = b}{x^2 + bx + c}}$$ Then $$x^2 + bx + c = (x + m)(x + n).$$	Factor $x^2 + 6x + 8$. $$\overset{\displaystyle mn = 8}{\underset{\displaystyle m + n = 6}{x^2 + 6x + 8}}$$ $m = 2$ and $n = 4$ $x^2 + 6x + 8 = (x + 2)(x + 4)$

CONCEPTS	EXAMPLES
4.3 MORE ON FACTORING TRINOMIALS	

CONCEPTS	EXAMPLES
To factor $ax^2 + bx + c$: **By Grouping** Find m and n: $$mn = ac$$ $$ax^2 + bx + c$$ $$m + n = b$$	Factor $3x^2 + 14x - 5$. $\overbrace{\qquad -15 \qquad}$ $mn = -15, m + n = 14$
By Trial and Error Use FOIL backwards.	By trial and error or by grouping. $$3x^2 + 14x - 5 = (3x - 1)(x + 5).$$

CONCEPTS	EXAMPLES
4.4 SPECIAL FACTORIZATIONS	

CONCEPTS	EXAMPLES
Difference of Squares $x^2 - y^2 = (x + y)(x - y)$	$4x^2 - 9 = (2x + 3)(2x - 3)$
Perfect Square Trinomial $x^2 + 2xy + y^2 = (x + y)^2$ $x^2 - 2xy + y^2 = (x - y)^2$	$9x^2 + 6x + 1 = (3x + 1)^2$ $4x^2 - 20x + 25 = (2x - 5)^2$
Difference of Cubes $x^3 - y^3 = (x - y)(x^2 + xy + y^2)$	$m^3 - 8 = m^3 - 2^3 = (m - 2)(m^2 + 2m + 4)$
Sum of Cubes $x^3 + y^3 = (x + y)(x^2 - xy + y^2)$	$27 + z^3 = 3^3 + z^3 = (3 + z)(9 - 3z + z^2)$

CONCEPTS	EXAMPLES
4.5 SOLVING QUADRATIC EQUATIONS BY FACTORING	

CONCEPTS	EXAMPLES
Zero-Factor Property If a and b are real numbers and if $ab = 0$, then $a = 0$ or $b = 0$.	If $(x - 2)(x + 3) = 0$, then $x - 2 = 0$ or $x + 3 = 0$.
Solving a Quadratic Equation by Factoring	Solve $2x^2 = 7x + 15$.
1. Write in standard form.	**1.** $2x^2 - 7x - 15 = 0$
2. Factor.	**2.** $(2x + 3)(x - 5) = 0$
3. Use the zero-factor property.	**3.** $2x + 3 = 0$ or $x - 5 = 0$ $2x = -3$ $x = 5$ $x = -\dfrac{3}{2}$
4. Check.	**4.** Both solutions satisfy the original equation.

CONCEPTS	EXAMPLES
4.6 APPLICATIONS OF QUADRATIC EQUATIONS	

CONCEPTS	EXAMPLES
Pythagorean Formula In a right triangle, the square of the hypotenuse equals the sum of the squares of the legs. $$a^2 + b^2 = c^2$$ leg a hypotenuse c leg b	In a right triangle one leg measures 2 feet longer than the other. The hypotenuse measures 4 feet longer than the shorter leg. Find the lengths of the three sides of the triangle. (See next page for solution.)

CONCEPTS	EXAMPLES
	Let $x =$ the length of the shorter leg. Then $$x^2 + (x + 2)^2 = (x + 4)^2$$ Verify that the solutions of this equation are -2 and 6. Discard -2 as a solution. Check that the sides are 6, $6 + 2 = 8$, and $6 + 4 = 10$ feet in length.

4.7   SOLVING QUADRATIC INEQUALITIES

Solving a Quadratic Inequality	Solve $2x^2 + 5x - 3 < 0$.
Step 1 Change the inequality to an equation.	**1.** $2x^2 + 5x - 3 = 0$
Step 2 Use the zero-factor property to solve the equation from Step 1.	**2.** $(2x - 1)(x + 3) = 0$ $2x - 1 = 0$ or $x + 3 = 0$ $2x = 1$ or $x = -3$ $x = \dfrac{1}{2}$
Step 3 Use the solutions of the equation in Step 2 to determine regions on the number line.	**3.** A B C -3 0 $\frac{1}{2}$
Step 4 Choose a number from each region. Substitute the number into the original inequality. If the number satisfies the inequality, all numbers in that region satisfy the inequality.	**4.** Choose -4 from A: $\quad 2(-4)^2 + 5(-4) - 3 < 0 \qquad ?$ $\qquad\qquad\qquad\qquad 9 < 0 \qquad$ False Choose 0 from B: $\quad 2(0)^2 + 5(0) - 3 < 0 \qquad ?$ $\qquad\qquad\qquad\quad -3 < 0 \qquad$ True Choose 1 from C: $\quad 2(1)^2 + 5(1) - 3 < 0 \qquad ?$ $\qquad\qquad\qquad\qquad 4 < 0 \qquad$ False Only the numbers from Region B satisfy the inequality.
Step 5 Write the solutions as inequalities.	**5.** The solution is written $-3 < x < 1/2$.

CHAPTER 4 REVIEW EXERCISES

[4.1] *Factor out the greatest common factor or factor by grouping.*

1. $7t + 14$ 　　　　 **2.** $60z^3 + 30z$ 　　　　 **3.** $35x^3 + 70x^2$

4. $2xy - 8y + 3x - 12$ 　 **5.** $100m^2n^3 - 50m^3n^4 + 150m^2n^2$ 　 **6.** $6y^2 + 9y + 4y + 6$

[4.2] *Factor completely.*

7. $x^2 + 5x + 6$ **8.** $y^2 - 13y + 40$ **9.** $q^2 + 6q - 27$

10. $r^2 - r - 56$ **11.** $r^2 - 4rs - 96s^2$ **12.** $p^2 + 2pq - 120q^2$

13. $8p^3 - 24p^2 - 80p$ **14.** $3x^4 + 30x^3 + 48x^2$ **15.** $m^2 - 3mn - 18n^2$

16. $y^2 - 8yz + 15z^2$ **17.** $p^7 - p^6q - 2p^5q^2$ **18.** $3r^5 - 6r^4s - 45r^3s^2$

19. $x^2 + x + 1$ **20.** $3x^2 + 6x + 6$

[4.3]

21. In order to begin factoring $6r^2 - 5r - 6$, what are the possible first terms of the two binomial factors if we consider only positive integer coefficients?

22. What is the first step you would use to factor $2z^3 + 9z^2 - 5z$?

Factor completely.

23. $2k^2 - 5k + 2$ **24.** $3r^2 + 11r - 4$ **25.** $6r^2 - 5r - 6$

26. $10z^2 - 3z - 1$ **27.** $8v^2 + 17v - 21$ **28.** $24x^5 - 20x^4 + 4x^3$

29. $-6x^2 + 3x + 30$ **30.** $10r^3s + 17r^2s^2 + 6rs^3$

[4.4]

31. Which one of the following is the difference of two squares?
  (a) $32x^2 - 1$ **(b)** $4x^2y^2 - 25z^2$ **(c)** $x^2 + 36$ **(d)** $25y^3 - 1$

32. Which one of the following is a perfect square trinomial?
  (a) $x^2 + x + 1$ **(b)** $y^2 - 4y + 9$ **(c)** $4x^2 + 10x + 25$ **(d)** $x^2 - 20x + 100$

Factor completely.

33. $n^2 - 49$ **34.** $25b^2 - 121$ **35.** $49y^2 - 25w^2$

36. $144p^2 - 36q^2$ **37.** $x^2 + 100$ **38.** $z^2 + 10z + 25$

39. $r^2 - 12r + 36$ **40.** $9t^2 - 42t + 49$ **41.** $m^3 + 1000$

42. $125k^3 + 64x^3$ **43.** $64x^6 - 1$ **44.** $343x^3 - 64$

[4.5] *Solve the equation and check the solutions.*

45. $(4t + 3)(t - 1) = 0$ **46.** $z^2 + 4z + 3 = 0$ **47.** $m^2 - 5m + 4 = 0$

48. $x(x - 8) = -15$ **49.** $3z^2 - 11z - 20 = 0$ **50.** $81t^2 - 64 = 0$

51. $y^2 = 8y$ **52.** $n(n - 5) = 6$ **53.** $t^2 - 14t + 49 = 0$

54. $t^2 = 12(t - 3)$ **55.** $(5z + 2)(z^2 + 3z + 2) = 0$ **56.** $x^2 = 9$

[4.6] *Solve the problem.*

57. The length of a rectangle is 6 meters more than the width. The area is 40 square meters. Find the length and width of the rectangle.

58. The length of a rectangle is three times the width. If the width were increased by 3 meters while the length remained the same, the new rectangle would have an area of 30 square meters. Find the length and width of the original rectangle.

59. The length of a rectangle is 2 centimeters more than the width. The area is numerically 44 more than the perimeter. Find the length and width of the rectangle.

60. The volume of a box is to be 120 cubic meters. The width of the box is to be 4 meters, and the height 1 meter less than the length. Find the length and height of the box.

61. The length of a rectangle is 4 feet more than the width. The area is numerically 1 more than the perimeter. Find the dimensions of the rectangle.

62. The width of a rectangle is 5 inches less than the length. The area is numerically 10 more than the perimeter. Find the dimensions of the rectangle.

63. The product of two consective integers is 29 more than their sum. What are the integers?

64. The sides of a right triangle have lengths (in feet) that are consecutive even integers. What are the lengths of the sides?

If an object is thrown straight up with an initial velocity of 128 feet per second, its height h after t seconds is $h = 128t - 16t^2$.

Find the height of the object after the following periods of time.

65. 1 second

66. 2 seconds

67. 4 seconds

68. When does the object described above return to the ground?

69. A 9-inch by 12-inch picture is to be placed on a cardboard mat so that there is an equal border around the picture. The area of the finished mat and picture is to be 208 square inches. How wide will the border be?

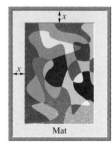

Mat

70. A box is made from a 12-centimeter by 10-centimeter piece of cardboard by cutting equal-sized squares from each corner and folding up the sides. The area of the bottom of the box is to be 48 square centimeters. Find the length of a side of the cutout squares.

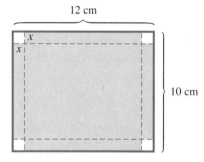

12 cm

10 cm

[4.7] *Solve the inequality.*

71. $(q + 5)(q - 3) > 0$

72. $(2r - 1)(r + 4) \geq 0$

73. $m^2 - 5m + 6 \leq 0$

74. $z^2 - 8z + 15 < 0$

75. $2p^2 + 5p - 12 \geq 0$

76. Suppose you know that the solution of a quadratic inequality involving the $<$ symbol is $-5 < x < 7$. If the symbol is changed to $>$, what is the solution of the new inequality?

MIXED REVIEW EXERCISES

Factor completely.

77. $z^2 - 11zx + 10x^2$

78. $3k^2 + 11k + 10$

79. $15m^2 + 20mp - 12mp - 16p^2$

80. $y^4 - 625$

81. $6m^3 - 21m^2 - 45m$

82. $24ab^3c^2 - 56a^2bc^3 + 72a^2b^2c$

83. $25a^2 + 15ab + 9b^2$

84. $12x^2yz^3 + 12xy^2z - 30x^3y^2z^4$

85. $2a^5 - 8a^4 - 24a^3$

86. $12r^2 + 18rq - 10rq - 15q^2$

87. $1000a^3 + 27$

88. $49t^2 + 56t + 16$

Solve.

89. $t(t - 7) = 0$ **90.** $x(x + 3) = 10$ **91.** $4x^2 - x - 3 \le 0$

92. A lot is shaped like a right triangle. The hypotenuse is 3 meters longer than the longer leg. The longer leg is 6 meters longer than twice the length of the shorter leg. Find the lengths of the sides of the lot.

93. A pyramid has a rectangular base with a length that is 2 meters more than the width. The height of the pyramid is 6 meters, and its volume is 48 cubic meters. Find the length and width of the base.

94. The product of the smaller two of three consecutive integers is equal to 23 plus the largest. Find the integers.

95. The sum of two consecutive even integers is 34 less than their product. Find the integers.

96. The floor plan for a house is a rectangle with length 7 meters more than its width. The area is 170 square meters. Find the width and length of the house.

97. The triangular sail of a schooner has an area of 30 square meters. The height of the sail is 4 meters more than the base. Find the base of the sail.

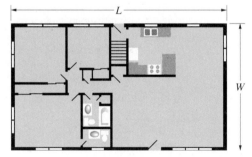

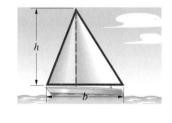

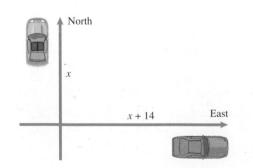

98. Two cars left an intersection at the same time. One traveled north. The other traveled 14 miles farther, but to the east. How far apart were they then, if the distance between them was 4 miles more than the distance traveled east?

99. A ladder is leaning against a building. The distance from the bottom of the ladder to the building is 4 feet less than the length of the ladder. How high up the side of the building is the top of the ladder if that distance is 2 feet less than the length of the ladder?

North

x

$x + 14$ East

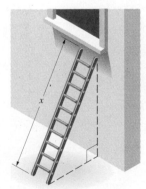

100. A bicyclist heading east and a motorist traveling south left an intersection at the same time. When the motorist had gone 17 miles farther than the bicyclist, the distance between them was 1 mile more than the distance traveled by the motorist. How far apart were they then? (*Hint:* Draw a sketch.)

101. Although $(2x + 8)(3x - 4) = 6x^2 + 16x - 32$ is a true statement, the polynomial is not factored completely. Explain why and give the completely factored form.

102. Explain why $(3x - 15)(x + 2)$ is not the completely factored form of $3x^2 - 9x - 30$ and factor the polynomial completely.

———————◆ **MATHEMATICAL CONNECTIONS** (Exercises 103–108) ◆———————

How would you go about factoring $x^4 + x^2 - 3x - 14$? The methods described in this chapter do not apply to this polynomial. However, methods studied in more advanced courses do allow us to determine the factors of a polynomial such as this. Work Exercises 103–108 in order to see how this is done.

103. A property from more advanced courses says this: *If a polynomial in x is evaluated for x = c, where c is a real number, and the value of the polynomial is 0, then the polynomial has x − c as a factor.* Evaluate $x^4 + x^2 - 3x - 14$ for $x = 2$.

104. Based on your result from Exercise 103, what binomial do we now know is a factor of $x^4 + x^2 - 3x - 14$?

105. To find the other factor, use the long division method described in Section 3.7 to divide $x^4 + x^2 - 3x - 14$ by your answer in Exercise 104. Don't forget to insert $0x^3$ as a placeholder in the dividend.

106. Express $x^4 + x^2 - 3x - 14$ in a factored form.

107. Multiply the two factors from Exercise 106 to assure yourself that their product is indeed $x^4 + x^2 - 3x - 14$.

108. Consider the polynomial $2x^3 - 5x - 39$. Evaluate it for **(a)** $x = 1$ **(b)** $x = 2$ and **(c)** $x = 3$. Based on your results, which one of the following is a factor of the polynomial? **(a)** $x - 1$ **(b)** $x - 2$ **(c)** $x - 3$

CHAPTER 4 TEST

1. Which one of the following is the correct, completely factored form of $2x^2 - 2x - 24$?
  (a) $(2x + 6)(x - 4)$ **(b)** $(x + 3)(2x - 8)$
  (c) $2(x + 4)(x - 3)$ **(d)** $2(x + 3)(x - 4)$

Factor the polynomial completely. If it cannot be factored, write prime.

2. $12x^2 - 30x$ **3.** $2m^3n^2 + 3m^3n - 5m^2n^2$ **4.** $x^2 - 5x - 24$

5. $x^2 - 9x + 14$ **6.** $2x^2 + x - 3$ **7.** $6x^2 - 19x - 7$

8. $3x^2 - 12x - 15$ **9.** $10z^2 - 17z + 3$ **10.** $t^2 + 2t + 3$

11. $x^2 + 36$ **12.** $12 - 6a + 2b - ab$ **13.** $9y^2 - 64$

14. $x^2 + 16x + 64$ **15.** $4x^2 - 28xy + 49y^2$ **16.** $-2x^2 - 4x - 2$

17. $6t^4 + 3t^3 - 108t^2$ **18.** $4r^2 + 10rt + 25t^2$ **19.** $r^3 - 125$

20. $8k^3 + 64$

21. Why is $(p + 3)(p + 3)$ not the correct factorization of $p^2 + 9$?

Solve the equation.

22. $2r^2 - 13r + 6 = 0$

23. $25x^2 - 4 = 0$

24. $x(x - 20) = -100$

25. $t^3 = 9t$

Solve the problem.

26. If an object is propelled from ground level at an initial velocity of 96 feet per second, after t seconds its height h is given by the formula $h = -16t^2 + 96t$. After how many seconds will its height h be 108 feet?

27. A carpenter needs to cut a brace to support a wall stud. See the figure. The brace should be 7 feet less than three times the length of the stud. The brace will be fastened on the floor 1 foot less than twice the length of the stud away from the stud. How long should the brace be?

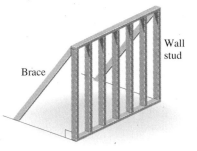

Solve the inequality and graph the solutions.

28. $(3x - 1)(2x + 5) < 0$

29. $x^2 - 2x - 24 \geq 0$

30. Why isn't "$x = \dfrac{2}{3}$" the correct response to "Solve the equation $x^2 = \dfrac{4}{9}$"?

CUMULATIVE REVIEW (Chapters 1–4)

Solve the equation.

1. $3x + 2(x - 4) = 4(x - 2)$

2. $.3x + .9x = .06$

3. $\dfrac{2}{3}y - \dfrac{1}{2}(y - 4) = 3$

4. Solve for P: $A = P + Prt$.

Solve the problem.

5. Four times a number, added to the sum of the number and twelve, gives a result of 3. Find the number.

6. A pharmacist found that at the end of the day she had 2/3 as many prescriptions for antibiotics as she had for tranquilizers. If she had 90 prescriptions in all, how many did she have for each type of drug?

7. In a recent year, the United States exported to the Bahamas 315 million dollars more in goods than it imported. Together, the two amounts totaled 1167 million dollars. How much were the exports and how much were the imports?

8. In a mixture of concrete, there are 3 pounds of cement mix for every 1 pound of gravel. If the mixture contains a total of 140 pounds of these two ingredients, how many pounds of gravel are there?

Evaluate the expression.

9. $2^{-3} \cdot 2^5$

10. $\left(\dfrac{3}{4}\right)^{-2}$

11. $\dfrac{6^5 \cdot 6^{-2}}{6^3}$

12. $\left(\dfrac{4^{-3} \cdot 4^4}{4^5}\right)^{-1}$

Simplify the expression and write the answer using only positive exponents. Assume no denominators are zero.

13. $\dfrac{(p^2)^3 p^{-4}}{(p^{-3})^{-1} p}$

14. $\dfrac{(m^{-2})^3 m}{m^5 m^{-4}}$

Perform the indicated operations.

15. $(2k^2 + 4k) - (5k^2 - 2) - (k^2 + 8k - 6)$

16. $3m^3(2m^5 - 5m^3 + m)$

17. $(y^2 + 3y + 5)(3y - 1)$

18. $(3p + 2)^2$

19. $(2p + 3q)(2p - 3q)$

20. $(9x + 6)(5x - 3)$

21. $\dfrac{8x^4 + 12x^3 - 6x^2 + 20x}{2x}$

22. $(12p^3 + 2p^2 - 12p + 5) \div (2p - 2)$

Factor completely.

23. $2a^2 + 7a - 4$

24. $10m^2 + 19m + 6$

25. $15x^2 - xy - 6y^2$

26. $8t^2 + 10tv + 3v^2$

27. $9x^2 + 6x + 1$

28. $4p^2 - 12p + 9$

29. $-32t^2 - 112tz - 98z^2$

30. $25r^2 - 81t^2$

31. $100x^2 + 25$

32. $6a^2m + am - 2m$

33. $2pq + 6p^3q + 8p^2q$

34. $2ux - 2bx + ay - by$

Solve the equation.

35. $(2p - 3)(p + 2)(p - 6) = 0$

36. $6m^2 + m - 2 = 0$

37. Solve the inequality and graph the solution: $2x^2 + x - 6 \geq 0$.

Solve the problem.

38. The length of a rectangle is 1 centimeter less than twice the width. The area is 15 square centimeters. Find the width of the rectangle.

39. The difference between the squares of two consecutive even integers is 28 less than the square of the smaller integer. Find the two integers.

40. The length of the hypotenuse of a right triangle is twice the length of the shorter leg, plus 3 meters. The longer leg is 7 meters longer than the shorter leg. Find the lengths of the sides.

CHAPTER 5

RATIONAL EXPRESSIONS

CONNECTIONS

Shown here are the winners and their winning jumps in seven Olympic Games Long Jump competitions.

1992 Carl Lewis, United States	28 feet, 5 1/2 inches
1988 Carl Lewis, United States	28 feet, 7 1/4 inches
1984 Carl Lewis, United States	28 feet, 1/4 inch
1980 Lutz Dombrowski, East Germany	28 feet, 1/4 inch
1976 Arnie Robinson, United States	27 feet, 4 1/2 inches
1972 Randy Williams, United States	27 feet, 1/2 inch
1968 Bob Beamon, United States	29 feet, 2 1/2 inches

Notice that each jump was recorded with a fractional part of an inch when it was measured. Whole numbers are not sufficiently accurate to measure with enough precision in many instances in everyday life. Similarly, polynomials are not sufficient in describing certain situations that occur in algebra, and so quotients of polynomials, known as *rational expressions,* are studied. Rational expressions are the algebraic equivalent of fractions in arithmetic.

FOR DISCUSSION OR WRITING
Name several other common uses of fractions in everyday life. What is meant by a fraction, such as 2/3 or 5/4?

5.1 THE FUNDAMENTAL PROPERTY OF RATIONAL EXPRESSIONS

FOR EXTRA HELP

📖 **SSG** pp. 165–169
SSM pp. 245–249

📼 **Video**
7

💾 **Tutorial**
IBM   MAC

OBJECTIVES

1 ▶ Find the values for which a rational expression is undefined.
2 ▶ Find the numerical value of a rational expression.
3 ▶ Write rational expressions in lowest terms.

The quotient of two integers (with divisor not zero) is called a rational number. In the same way, the quotient of two polynomials with divisor not equal to zero is called a *rational expression*. The techniques of factoring, studied in Chapter 4, are essential in working with rational expressions.

Rational Expression

A **rational expression** is an expression of the form

$$\frac{P}{Q}$$

where P and Q are polynomials, with $Q \neq 0$.

Examples of rational expressions include $\dfrac{-6x}{x^3 + 8}$, $\dfrac{9x}{y + 3}$, and $\dfrac{2m^3}{9}$.

1 ▶ **Find the values for which a rational expression is undefined.**

▶ A fraction with a zero denominator is *not* a rational expression, since division by zero is not possible. For that reason, be careful when substituting a number for a variable in the denominator of a rational expression. For example, in

$$\frac{8x^2}{x - 3}$$

x can take on any value except 3. When $x = 3$, the denominator becomes $3 - 3 = 0$, making the expression undefined.

In order to determine the values for which a rational expression is undefined, use the following procedure.

Determining When a Rational Expression Is Undefined

Step 1 Set the denominator of the rational expression equal to 0.
Step 2 Solve this equation.
Step 3 The solutions of the equation are the values that make the rational expression undefined.

This procedure is illustrated in Example 1.

NOTE The numerator of a rational expression may be *any* number.

EXAMPLE 1

Finding Values that Make Rational Expressions Undefined

Find any values for which the following rational expressions are undefined.

(a) $\dfrac{p + 5}{3p + 2}$

Remember that the *numerator* may be any number; we must find any value of p that makes the *denominator* equal to 0. We find these values as described above.

Step 1 Set the denominator equal to 0.

$$3p + 2 = 0$$

Step 2 Solve this equation.

$$3p = -2$$
$$p = -\frac{2}{3}$$

Step 3 Since $p = -2/3$ will make the denominator zero, the given expression is undefined for $-2/3$.

(b) $\dfrac{9m^2}{m^2 - 5m + 6}$

Find the numbers that make the denominator zero by solving the equation

$$m^2 - 5m + 6 = 0.$$
$$(m - 2)(m - 3) = 0 \qquad \text{Factor.}$$
$$m - 2 = 0 \quad \text{or} \quad m - 3 = 0 \qquad \text{Zero-factor property}$$
$$m = 2 \quad \text{or} \quad m = 3 \qquad \text{Solve.}$$

The original expression is undefined for 2 and for 3.

(c) $\dfrac{2r}{r^2 + 1}$

This denominator cannot equal zero for any value of r, since r^2 is always greater than or equal to zero and adding 1 makes the sum greater than zero. Thus, there are no values for which this rational expression is undefined. ■

2 ▶ Find the numerical value of a rational expression.

▶ The following example shows how to find the numerical value of a rational expression for a given value of the variable.

EXAMPLE 2

Evaluating a Rational Expression

Find the numerical value of $\dfrac{3x + 6}{2x - 4}$ for the given values of x.

(a) $x = 1$

Find the value of the rational expression by substituting 1 for x.

$$\frac{3x + 6}{2x - 4} = \frac{3(1) + 6}{2(1) - 4} \qquad \text{Let } x = 1.$$
$$= \frac{9}{-2}$$
$$= -\frac{9}{2}$$

(b) $x = 2$

Substituting 2 for x makes the denominator zero, so the rational expression is undefined when $x = 2$. ■

3 ▶ Write rational expressions in lowest terms.

▶ A fraction such as $2/3$ is said to be in lowest terms. How can "lowest terms" be defined? We use the idea of greatest common factor to give this definition, which applies to all rational expressions.

Lowest Terms

A rational expression P/Q ($Q \neq 0$) is in **lowest terms** if the greatest common factor of its numerator and denominator is 1.

Because a rational expression represents a number for each value of the variable that does not make the denominator zero, the properties of rational numbers also apply to rational expressions. For example, we can use the fundamental property of rational expressions to write a rational expression in lowest terms.

Fundamental Property of Rational Expressions

If P/Q is a rational expression and if K represents any polynomial, where $K \neq 0$, then

$$\frac{PK}{QK} = \frac{P}{Q}.$$

This property is based on the identity property of multiplication, since

$$\frac{PK}{QK} = \frac{P}{Q} \cdot \frac{K}{K} = \frac{P}{Q} \cdot 1 = \frac{P}{Q}.$$

The fundamental property suggests the method used in the next example to write a fraction in lowest terms. We show the procedure with both a rational number and a rational expression. Notice the similarity.

EXAMPLE 3
Writing in Lowest Terms

Write each expression in lowest terms.

(a) $\dfrac{30}{72}$ **(b)** $\dfrac{14k^2}{2k^3}$

Begin by factoring.

$$\frac{30}{72} = \frac{2 \cdot 3 \cdot 5}{2 \cdot 2 \cdot 2 \cdot 3 \cdot 3} \qquad \frac{14k^2}{2k^3} = \frac{2 \cdot 7 \cdot k \cdot k}{2 \cdot k \cdot k \cdot k}$$

Group any factors common to the numerator and denominator.

$$\frac{30}{72} = \frac{5(2 \cdot 3)}{2 \cdot 2 \cdot 3(2 \cdot 3)} \qquad \frac{14k^2}{2k^3} = \frac{7(2 \cdot k \cdot k)}{k(2 \cdot k \cdot k)}$$

Use the fundamental property.

$$\frac{30}{72} = \frac{5}{2 \cdot 2 \cdot 3} = \frac{5}{12} \qquad \frac{14k^2}{2k^3} = \frac{7}{k} \quad ■$$

EXAMPLE 4

Writing in Lowest Terms

Write each rational expression in lowest terms.

(a) $\dfrac{3x - 12}{5x - 20}$

Begin by factoring both numerator and denominator. Then use the fundamental property.

$$\frac{3x - 12}{5x - 20} = \frac{3(x - 4)}{5(x - 4)}$$

$$= \frac{3}{5}$$

The rational expression $\dfrac{3x - 12}{5x - 20}$ is equal to $\dfrac{3}{5}$ for all values of x, where $x \ne 4$ (since the denominator of the original rational expression is 0 when x is 4).

(b) $\dfrac{m^2 + 2m - 8}{2m^2 - m - 6}$

$$\frac{m^2 + 2m - 8}{2m^2 - m - 6} = \frac{(m + 4)(m - 2)}{(2m + 3)(m - 2)} \qquad \text{Factor.}$$

$$= \frac{m + 4}{2m + 3} \qquad \text{Fundamental property}$$

Thus, $\dfrac{m^2 + 2m - 8}{2m^2 - m - 6} = \dfrac{m + 4}{2m + 3}$ for $m \ne -\dfrac{3}{2}$ or 2, since the denominator of the original expression is 0 for these values of m. ■

From now on, we will write statements of equality of rational expressions with the understanding that they apply only to those real numbers that make neither denominator equal to zero.

CAUTION One of the most common errors in algebra occurs when students attempt to write rational expressions in lowest terms *before factoring*. The fundamental property is applied only *after* the numerator and denominator are expressed in factored form. For example, although x appears in both the numerator and denominator in Example 4(a), and 12 and 20 have a common factor of 4, the fundamental property cannot be used before factoring because $3x$, $5x$, 12, and 20 are *terms*, not *factors*. Terms are *added* or *subtracted;* factors are *multiplied* or *divided.* For example,

$$\frac{6 + 2}{3 + 2} = \frac{8}{5}, \quad \textbf{not} \quad \frac{6}{3} + \frac{2}{2} = 2 + 1 = 3.$$

Also, $\qquad \dfrac{2x + 3}{4x + 6} = \dfrac{2x + 3}{2(2x + 3)} = \dfrac{1}{2}, \quad \textbf{but} \quad \dfrac{x^2 + 6}{x + 3} \ne x + 2.$

◇ **CONNECTIONS** ◇

In Chapter 3 we used long division to find the quotient of two polynomials. For example, we can find $(2x^2 + 5x - 12) \div (2x - 3)$ as follows:

$$
\begin{array}{r}
x + 4 \\
2x - 3 \overline{)\smash{2x^2 + 5x - 12}} \\
\underline{2x^2 - 3x} \\
8x - 12 \\
\underline{8x - 12} \\
0
\end{array}
$$

The quotient is $x + 4$. In this section we see how to find this quotient in another way. We do this by writing it as a fraction and reducing it to lowest terms, using the factoring techniques from Chapter 4.

$$\frac{2x^2 + 5x - 12}{2x - 3} = \frac{(2x - 3)(x + 4)}{2x - 3} = x + 4$$

FOR DISCUSSION OR WRITING

What kind of division problem has a quotient that cannot be found by reducing a fraction to lowest terms? Try using that method with the following division problems. Then use long division to compare.

1. $(3x^2 + 11x + 8) \div (x + 2)$
2. $(x^3 - 8) \div (x^2 + 2x + 4)$

EXAMPLE 5

Writing in Lowest Terms (Factors Are Opposites)

■ Write $\dfrac{x - y}{y - x}$ in lowest terms.

At first glance, there does not seem to be any way in which $x - y$ and $y - x$ can be factored to get a common factor. However, one way to approach this is to notice that the numerator can be factored as

$$x - y = -1(-x + y) = -1(y - x).$$

Now the fundamental property can be used to simplify the rational expression.

$$\frac{x - y}{y - x} = \frac{-1(y - x)}{1(y - x)} = \frac{-1}{1} = -1$$

Either the numerator or the denominator could have been factored in the first step. ■

In Example 5, notice that $y - x$ is the opposite of $x - y$. A general rule for this situation follows.

A fraction with numerator and denominator that are opposites equals -1.

EXAMPLE 6

Writing in Lowest Terms
(Factors Are Opposites)

Write each rational expression in lowest terms.

(a) $\dfrac{2 - m}{m - 2}$

Since $2 - m$ and $m - 2$ (or $-2 + m$) are opposites, we use the rule above.

$$\frac{2 - m}{m - 2} = -1$$

(b) $\dfrac{4x^2 - 9}{6 - 4x}$

Factor the numerator and denominator and use the rule above.

$$\frac{4x^2 - 9}{6 - 4x} = \frac{(2x + 3)(2x - 3)}{2(3 - 2x)}$$

$$= \frac{2x + 3}{2}(-1)$$

$$= -\frac{2x + 3}{2}$$

(c) $\dfrac{3 + r}{3 - r}$

The quantity $3 - r$ *is not* the opposite of $3 + r$. This rational expression cannot be written in simpler form. ∎

NOTE The form of the answer given in Example 6(b) is only one of several acceptable forms. The $-$ sign representing the -1 factor is in front of the fraction, on the same line as the fraction bar. The -1 factor may be placed in front of the fraction, in the numerator, or in the denominator. Some other acceptable forms of the answer are

$$\frac{-(2x + 3)}{2}, \qquad \frac{-2x - 3}{2}, \qquad \text{and} \qquad \frac{2x + 3}{-2}.$$

However, can you see why $\dfrac{-2x + 3}{2}$ is *not* an acceptable form?

5.1 EXERCISES

Find any values for which the rational expression is undefined. See Example 1.

1. $\dfrac{2}{5y}$

2. $\dfrac{7}{3z}$

3. $\dfrac{4x^2}{3x - 5}$

4. $\dfrac{2x^3}{3x - 4}$

5. $\dfrac{m + 2}{m^2 + m - 6}$

6. $\dfrac{r - 5}{r^2 - 5r + 4}$

7. $\dfrac{3x}{x^2 + 2}$

8. $\dfrac{4q}{q^2 + 9}$

*Find the numerical value of the rational expression when **(a)** $x = 2$ and **(b)** $x = -3$. See Example 2.*

9. $\dfrac{5x - 2}{4x}$

10. $\dfrac{3x + 1}{5x}$

11. $\dfrac{2x^2 - 4x}{3x}$

12. $\dfrac{4x^2 - 1}{5x}$ **13.** $\dfrac{(-3x)^2}{4x + 12}$ **14.** $\dfrac{(-2x)^3}{3x + 9}$

15. $\dfrac{5x + 2}{2x^2 + 11x + 12}$ **16.** $\dfrac{7 - 3x}{3x^2 - 7x + 2}$

17. If 2 is substituted for x in the rational expression $\dfrac{x - 2}{x^2 - 4}$, the result is $\dfrac{0}{0}$. An often-heard statement is "Any number divided by itself is 1." Does this mean that this expression is equal to 1 for $x = 2$? If not, explain.

18. For $x \neq 2$, the rational expression $\dfrac{2(x - 2)}{x - 2}$ is equal to 2. Can the same be said for $\dfrac{2x - 2}{x - 2}$? Explain.

Write the rational expression in lowest terms. See Examples 3 and 4.

19. $\dfrac{18r^3}{6r}$ **20.** $\dfrac{27p^2}{3p}$ **21.** $\dfrac{4(y - 2)}{10(y - 2)}$

22. $\dfrac{15(m - 1)}{9(m - 1)}$ **23.** $\dfrac{(x + 1)(x - 1)}{(x + 1)^2}$ **24.** $\dfrac{(t + 5)(t - 3)}{(t - 1)(t + 5)}$

25. $\dfrac{7m + 14}{5m + 10}$ **26.** $\dfrac{8z - 24}{4z - 12}$ **27.** $\dfrac{m^2 - n^2}{m + n}$

28. $\dfrac{a^2 - b^2}{a - b}$ **29.** $\dfrac{12m^2 - 3}{8m - 4}$ **30.** $\dfrac{20p^2 - 45}{6p - 9}$

31. $\dfrac{3m^2 - 3m}{5m - 5}$ **32.** $\dfrac{6t^2 - 6t}{2t - 2}$ **33.** $\dfrac{9r^2 - 4s^2}{9r + 6s}$

34. $\dfrac{16x^2 - 9y^2}{12x - 9y}$ **35.** $\dfrac{zw + 4z - 3w - 12}{zw + 4z + 5w + 20}$ **36.** $\dfrac{km + 4k + 4m + 16}{km + 4k + 5m + 20}$

37. $\dfrac{5k^2 - 13k - 6}{5k + 2}$ **38.** $\dfrac{7t^2 - 31t - 20}{7t + 4}$

39. $\dfrac{2x^2 - 3x - 5}{2x^2 - 7x + 5}$ **40.** $\dfrac{3x^2 + 8x + 4}{3x^2 - 4x - 4}$

———————◆ **MATHEMATICAL CONNECTIONS** (Exercises 41–46) ◆———————

If we begin with a common fraction such as $\dfrac{7}{5}$, we can find its additive inverse, or opposite, by changing the sign of the fraction, by changing the sign of its numerator, or by changing the sign of its denominator. Therefore,

$$-\frac{7}{5}, \quad \frac{-7}{5}, \quad \text{and} \quad \frac{7}{-5}$$

all represent the same number. While beginning algebra students seldom have difficulty recognizing this with common fractions, they often have trouble distinguishing between equivalent forms of algebraic fractions.

Consider the rational expression

$$\frac{x - 3}{x + 6},$$

and work Exercises 41–46 in order.

41. Write an expression that represents the additive inverse of the given rational expression by writing a negative sign in front of the entire fraction.

42. Write an expression that represents the additive inverse of the given rational expression by multiplying the numerator by -1 and applying the distributive property.

43. Write an expression that represents the additive inverse of the given rational expression by multiplying the denominator by -1 and applying the distributive property.

44. Compare your final answers in Exercises 41–43. Although they may look different, they all represent the same expression. While the following works with any real number except -6, try it for your favorite positive number: Evaluate each expression obtained in Exercises 41–43 for that number and show that the same result is obtained in each case.

45. Use the concepts of Exercises 41–44 to determine which one of the following rational expressions *is not* equivalent to $-\dfrac{3-x}{x+5}$.

    (a) $\dfrac{-3+x}{x+5}$ **(b)** $\dfrac{3-x}{-x-5}$ **(c)** $\dfrac{x-3}{x+5}$ **(d)** $\dfrac{-3+x}{-x-5}$

46. Use the concepts of Exercises 41–44 to determine which one of the following rational expressions *is* equivalent to $-\dfrac{x^2+1}{x^2+2}$.

    (a) $\dfrac{-x^2-1}{x^2+2}$ **(b)** $\dfrac{-x^2+1}{x^2+2}$ **(c)** $\dfrac{x^2+1}{-x^2+2}$ **(d)** $\dfrac{-x^2-1}{-x^2-2}$

Write the expression in lowest terms. See Examples 5 and 6.

47. $\dfrac{6-t}{t-6}$ **48.** $\dfrac{2-k}{k-2}$ **49.** $\dfrac{m^2-1}{1-m}$ **50.** $\dfrac{a^2-b^2}{b-a}$

51. $\dfrac{q^2-4q}{4q-q^2}$ **52.** $\dfrac{z^2-5z}{5z-z^2}$ **53.** $\dfrac{p+6}{p-6}$ **54.** $\dfrac{5-x}{5+x}$

55. Explain how you can tell, just by looking at a rational expression, if it is equal to -1.

56. Explain why the rational expression $\dfrac{2x+1}{x+1}$ is not equal to 2.

Find the missing dimension.

57. The area of the rectangle is represented by x^4+10x^2+21. What is the width?

$\left(\textit{Hint: Use } W=\dfrac{A}{L}.\right)$

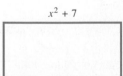

x^2+7

58. The volume of the box is represented by

$$(x^2+8x+15)(x+4).$$

Find the polynomial that represents the area of the bottom of the box.

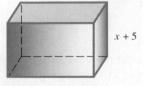

$x+5$

Write the expression in lowest terms.

59. $\dfrac{m^2-n^2-4m-4n}{2m-2n-8}$ **60.** $\dfrac{x^2y+y+x^2z+z}{xy+xz}$ **61.** $\dfrac{b^3-a^3}{a^2-b^2}$

62. $\dfrac{k^3+8}{k^2-4}$ **63.** $\dfrac{z^3+27}{z^3-3z^2+9z}$ **64.** $\dfrac{1-8r^3}{8r^2+4r+2}$

REVIEW EXERCISES

Multiply or divide as indicated. See Section 1.1.

65. $\dfrac{2}{3} \cdot \dfrac{5}{6}$

66. $\dfrac{3}{7} \cdot \dfrac{2}{5}$

67. $\dfrac{6}{15} \cdot \dfrac{25}{3}$

68. $\dfrac{10}{8} \div \dfrac{7}{12}$

69. $\dfrac{10}{3} \div \dfrac{5}{6}$

70. $\dfrac{7}{12} \div \dfrac{15}{4}$

5.2 MULTIPLICATION AND DIVISION OF RATIONAL EXPRESSIONS

FOR EXTRA HELP	OBJECTIVES
📖 **SSG** pp. 169–174 **SSM** pp. 249–253	**1 ▶** Multiply rational expressions.
📼 **Video** 7	**2 ▶** Divide rational expressions.
💾 **Tutorial** IBM MAC	

1 ▶ Multiply rational expressions.

▶ The product of two fractions is found by multiplying the numerators and multiplying the denominators. Rational expressions are multiplied in the same way.

Multiplying Rational Expressions

The product of the rational expressions P/Q and R/S is

$$\frac{P}{Q} \cdot \frac{R}{S} = \frac{PR}{QS}.$$

The next example shows the multiplication of both two rational numbers and two rational expressions. This parallel discussion lets you compare the steps.

EXAMPLE 1

Multiplying Rational Expressions

Multiply. Write answers in lowest terms.

(a) $\dfrac{3}{10} \cdot \dfrac{5}{9}$

(b) $\dfrac{6}{x} \cdot \dfrac{x^2}{12}$

Find the product of the numerators and the product of the denominators.

$$\frac{3}{10} \cdot \frac{5}{9} = \frac{3 \cdot 5}{10 \cdot 9} \qquad \qquad \frac{6}{x} \cdot \frac{x^2}{12} = \frac{6 \cdot x^2}{x \cdot 12}$$

Use the fundamental property to write each product in lowest terms.

$$\frac{3}{10} \cdot \frac{5}{9} = \frac{1 \cdot 3 \cdot 5}{2 \cdot 5 \cdot 3 \cdot 3} = \frac{1}{6} \qquad \qquad \frac{6}{x} \cdot \frac{x^2}{12} = \frac{6 \cdot x \cdot x}{2 \cdot 6 \cdot x} = \frac{x}{2}$$

Notice in the second step above that the products were left in factored form since common factors must be identified to write the product in lowest terms. ∎

NOTE It is also possible to divide out common factors in the numerator and denominator *before* multiplying the rational expressions. Many people use this method with success. For example, to multiply 6/5 and 35/22, the following method can be used.

$$\frac{6}{5} \cdot \frac{35}{22} = \frac{2 \cdot 3}{5} \cdot \frac{5 \cdot 7}{2 \cdot 11} \qquad \text{Identify common factors.}$$

$$= \frac{3 \cdot 7}{11} \qquad \text{Lowest terms}$$

$$= \frac{21}{11} \qquad \text{Multiply in numerator.}$$

EXAMPLE 2
Multiplying Rational Expressions

Multiply and express the product in lowest terms: $\dfrac{x + y}{2x} \cdot \dfrac{x^2}{(x + y)^2}$.

Use the definition of multiplication.

$$\frac{x + y}{2x} \cdot \frac{x^2}{(x + y)^2} = \frac{(x + y)x^2}{2x(x + y)^2} \qquad \text{Multiply numerators; multiply denominators.}$$

$$= \frac{(x + y)x \cdot x}{2x(x + y)(x + y)} \qquad \text{Factor; identify common factors.}$$

$$= \frac{x}{2(x + y)} \cdot \frac{x(x + y)}{x(x + y)} \qquad \text{Definition of multiplication}$$

$$= \frac{x}{2(x + y)} \qquad \text{Lowest terms}$$

Notice how in the third line, the factor $\dfrac{x(x + y)}{x(x + y)}$ appears. Since it is equal to 1, the final product is $\dfrac{x}{2(x + y)}$. ■

EXAMPLE 3
Multiplying Rational Expressions

Multiply and express the product in lowest terms:

$$\frac{x^2 + 3x}{x^2 - 3x - 4} \cdot \frac{x^2 - 5x + 4}{x^2 + 2x - 3}.$$

First factor the numerators and denominators whenever possible. Then use the fundamental property to write the product in lowest terms.

$$\frac{x^2 + 3x}{x^2 - 3x - 4} \cdot \frac{x^2 - 5x + 4}{x^2 + 2x - 3}$$

$$= \frac{x(x + 3)}{(x - 4)(x + 1)} \cdot \frac{(x - 4)(x - 1)}{(x + 3)(x - 1)} \qquad \text{Factor.}$$

$$= \frac{x(x + 3)(x - 4)(x - 1)}{(x - 4)(x + 1)(x + 3)(x - 1)} \qquad \text{Multiply numerators; multiply denominators.}$$

$$= \frac{x}{x + 1} \qquad \text{Lowest terms}$$

The quotients $\dfrac{x+3}{x+3}$, $\dfrac{x-4}{x-4}$, and $\dfrac{x-1}{x-1}$ are all equal to 1, justifying the final product $\dfrac{x}{x+1}$. ■

2 ▶ Divide rational expressions.

▶ To develop a method for dividing rational numbers and rational expressions, consider the following problem. Suppose that you have 7/8 of a gallon of milk and you wish to find how many quarts you have. Since a quart is 1/4 of a gallon, you must ask yourself, "How many 1/4s are there in 7/8?" This would be interpreted as

$$\frac{7}{8} \div \frac{1}{4} \qquad \text{or} \qquad \frac{\dfrac{7}{8}}{\dfrac{1}{4}}$$

since the fraction bar means division.

The fundamental property of rational expressions discussed earlier can be applied to rational number values of P, Q, and K. With $P = 7/8$, $Q = 1/4$, and $K = 4$,

$$\frac{P}{Q} = \frac{P \cdot K}{Q \cdot K} = \frac{\dfrac{7}{8} \cdot 4}{\dfrac{1}{4} \cdot 4} = \frac{\dfrac{7}{8} \cdot 4}{1} = \frac{7}{8} \cdot \frac{4}{1}.$$

So, to divide 7/8 by 1/4, we must multiply 7/8 by the reciprocal of 1/4, namely 4. Since $(7/8)(4) = 7/2$, there are 7/2 or 3 1/2 quarts in 7/8 gallon.

The discussion above illustrates the rule for dividing rational numbers: to divide a/b by c/d, multiply a/b by the reciprocal of c/d. Division of rational expressions is defined in the same way.

Dividing Rational Expressions

If P/Q and R/S are any two rational expressions, with $R/S \neq 0$, then

$$\frac{P}{Q} \div \frac{R}{S} = \frac{P}{Q} \cdot \frac{S}{R} = \frac{PS}{QR}.$$

The next example shows the division of two rational numbers and the division of two rational expressions.

EXAMPLE 4

Dividing Rational Expressions

Divide the following fractions. Write answers in lowest terms.

(a) $\dfrac{5}{8} \div \dfrac{7}{16}$

(b) $\dfrac{y}{y+3} \div \dfrac{4y}{y+5}$

Multiply the first expression and the reciprocal of the second.

$$\frac{5}{8} \div \frac{7}{16} = \frac{5}{8} \cdot \frac{16}{7} \qquad \text{Reciprocal of } \tfrac{7}{16}$$

$$= \frac{5 \cdot 16}{8 \cdot 7}$$

$$= \frac{5 \cdot 8 \cdot 2}{8 \cdot 7}$$

$$= \frac{5 \cdot 2}{7}$$

$$= \frac{10}{7}$$

$$\frac{y}{y+3} \div \frac{4y}{y+5}$$

$$= \frac{y}{y+3} \cdot \frac{y+5}{4y} \qquad \text{Reciprocal of } \tfrac{4y}{y+5}$$

$$= \frac{y(y+5)}{(y+3)(4y)}$$

$$= \frac{y+5}{4(y+3)} \qquad \text{Fundamental property} \blacksquare$$

EXAMPLE 5
Dividing Rational Expressions

■ Divide $\dfrac{(3m)^2}{(2p)^3} \div \dfrac{6m^3}{16p^2}$.

Use the properties of exponents as necessary.

$$\frac{(3m)^2}{(2p)^3} \div \frac{6m^3}{16p^2} = \frac{(3m)^2}{(2p)^3} \cdot \frac{16p^2}{6m^3} \qquad \text{Multiply by reciprocal.}$$

$$= \frac{(3m)(3m)}{(2p)(2p)(2p)} \cdot \frac{16p^2}{6m^3} \qquad \text{Factor.}$$

$$= \frac{9 \cdot 16m^2p^2}{8 \cdot 6p^3m^3} \qquad \begin{array}{l}\text{Multiply numerators.}\\ \text{Multiply denominators.}\end{array}$$

$$= \frac{3}{mp} \qquad \text{Lowest terms} \blacksquare$$

EXAMPLE 6
Dividing Rational Expressions

■ Divide and write the quotient in lowest terms: $\dfrac{x^2-4}{(x+3)(x-2)} \div \dfrac{(x+2)(x+3)}{2x}$.

$$\frac{x^2-4}{(x+3)(x-2)} \div \frac{(x+2)(x+3)}{2x}$$

$$= \frac{x^2-4}{(x+3)(x-2)} \cdot \frac{2x}{(x+2)(x+3)} \qquad \text{Use the definition of division.}$$

$$= \frac{(x+2)(x-2)}{(x+3)(x-2)} \cdot \frac{2x}{(x+2)(x+3)} \qquad \begin{array}{l}\text{Be sure numerators and}\\ \text{denominators are factored.}\end{array}$$

$$= \frac{(x+2)(x-2)(2x)}{(x+3)(x-2)(x+2)(x+3)} \qquad \begin{array}{l}\text{Multiply numerators and}\\ \text{multiply denominators.}\end{array}$$

$$= \frac{2x}{(x+3)^2} \qquad \begin{array}{l}\text{Use the fundamental property}\\ \text{to write in lowest terms.}\end{array} \blacksquare$$

In Example 6, only the numerator had to be factored. Remember that *all* numerators and denominators must be factored before the fundamental property can be applied. The next example requires more factoring than was necessary in Example 6.

EXAMPLE 7
Dividing Rational
Expressions (Factors are
Opposites)

■ Divide and write the quotient in lowest terms: $\dfrac{m^2 - 4}{m^2 - 1} \div \dfrac{2m^2 + 4m}{1 - m}$.

$$\frac{m^2 - 4}{m^2 - 1} \div \frac{2m^2 + 4m}{1 - m}$$

$$= \frac{m^2 - 4}{m^2 - 1} \cdot \frac{1 - m}{2m^2 + 4m}$$ Use the definition of division.

$$= \frac{(m + 2)(m - 2)}{(m + 1)(m - 1)} \cdot \frac{1 - m}{2m(m + 2)}$$ Factor; $1 - m$ and $m - 1$
differ only in sign.

$$= \frac{-1(m - 2)}{2m(m + 1)}$$ From Section 5.1, $\frac{1 - m}{m - 1} = -1$.

$$= \frac{2 - m}{2m(m + 1)}$$ Use the distributive property
in the numerator. ■

The procedure for multiplying or dividing rational expressions is summarized below.

**Multiplying or
Dividing Rational
Expressions**

Step 1 If the operation is division, use the definition of division to rewrite as multiplication.
Step 2 Factor numerators and denominators completely.
Step 3 Multiply numerators and multiply denominators.
Step 4 Write the answer in lowest terms.

◆ **CONNECTIONS** ◆

In Section 1.8 we saw that for positive real numbers x and y,

$$\frac{-x}{y} = \frac{x}{-y} = -\frac{x}{y}.$$

Rational expressions represent real numbers and so may also be written in several different equivalent forms. For example, the final quotient

$$\frac{2 - m}{2m(m + 1)},$$

in Example 7, where x is represented by $2 - m = -m + 2 = -(m - 2)$ and y by $2m(m + 1)$, can be written in the corresponding equivalent forms:

$$\frac{-(m - 2)}{2m(m + 1)}, \quad \frac{m - 2}{-2m(m + 1)}, \quad \text{and} \quad -\frac{m - 2}{2m(m + 1)}.$$

Either $2 - m$ or $-m + 2$ could replace $-(m - 2)$ in each of these forms. So, we can see that there are many equivalent correct ways to represent a rational expression.

FOR DISCUSSION OR WRITING
Write the two forms of the rational expression discussed here using $2 - m$ in place of $-(m - 2)$ and then using $-m + 2$ in place of $-(m - 2)$.

5.2 EXERCISES

Multiply. Write the answer in lowest terms. See Examples 1 and 2.

1. $\dfrac{15a^2}{14} \cdot \dfrac{7}{5a}$

2. $\dfrac{27k^3}{9k} \cdot \dfrac{24}{9k^2}$

3. $\dfrac{12x^4}{18x^3} \cdot \dfrac{-8x^5}{4x^2}$

4. $\dfrac{12m^5}{-2m^2} \cdot \dfrac{6m^6}{28m^3}$

5. $\dfrac{2(c+d)}{3} \cdot \dfrac{18}{6(c+d)^2}$

6. $\dfrac{4(y-2)}{x} \cdot \dfrac{3x}{6(y-2)^2}$

Divide. Write the answer in lowest terms. See Examples 4 and 5.

7. $\dfrac{9z^4}{3z^5} \div \dfrac{3z^2}{5z^3}$

8. $\dfrac{35q^8}{9q^5} \div \dfrac{25q^6}{10q^5}$

9. $\dfrac{4t^4}{2t^5} \div \dfrac{(2t)^3}{-6}$

10. $\dfrac{-12a^6}{3a^2} \div \dfrac{(2a)^3}{27a}$

11. $\dfrac{3}{2y-6} \div \dfrac{6}{y-3}$

12. $\dfrac{4m+16}{10} \div \dfrac{3m+12}{18}$

13. Explain in your own words how to multiply rational expressions.

14. Explain in your own words how to divide rational expressions.

◆ **MATHEMATICAL CONNECTIONS** (Exercises 15–18) ◆

We know that division by 0 is undefined. For example, we know that $5 \div 0$ cannot be a real number because there is no real number that multiplied by 0 gives 5 as a product. In division of rational expressions, we know that no denominator can be zero in either of the expressions in the problem, and furthermore, the numerator of the divisor cannot be zero. Work through Exercises 15–18 in order, referring as necessary to the division problem

$$\frac{x-6}{x+4} \div \frac{x+7}{x+5}.$$

15. In this problem, why must we have the restriction $x \neq -4$?

16. In this problem, why must we have the restriction $x \neq -5$?

17. In this problem, why must we have the restriction $x \neq -7$?

18. In this problem, why is 6 allowed as a replacement for x even though 6 causes the numerator in the first fraction to be 0?

Multiply or divide. Write the answer in lowest terms. See Examples 3, 6, and 7.

19. $\dfrac{5x-15}{3x+9} \cdot \dfrac{4x+12}{6x-18}$

20. $\dfrac{8r+16}{24r-24} \cdot \dfrac{6r-6}{3r+6}$

21. $\dfrac{2-t}{8} \div \dfrac{t-2}{6}$

22. $\dfrac{4}{m-2} \div \dfrac{16}{2-m}$

23. $\dfrac{27-3z}{4} \cdot \dfrac{12}{2z-18}$

24. $\dfrac{5-x}{5+x} \cdot \dfrac{x+5}{x-5}$

25. $\dfrac{6(m-2)^2}{5(m+4)^2} \cdot \dfrac{15(m+4)}{2(2-m)}$

26. $\dfrac{7(q-1)}{3(q+1)^2} \cdot \dfrac{6(q+1)}{3(1-q)^2}$

27. $\dfrac{p^2+4p-5}{p^2+7p+10} \div \dfrac{p-1}{p+4}$

28. $\dfrac{z^2-3z+2}{z^2+4z+3} \div \dfrac{z-1}{z+1}$

29. $\dfrac{2k^2-k-1}{2k^2+5k+3} \div \dfrac{4k^2-1}{2k^2+k-3}$

30. $\dfrac{2m^2-5m-12}{m^2+m-20} \div \dfrac{4m^2-9}{m^2+4m-5}$

31. $\dfrac{2k^2 + 3k - 2}{6k^2 - 7k + 2} \cdot \dfrac{4k^2 - 5k + 1}{k^2 + k - 2}$

32. $\dfrac{2m^2 - 5m - 12}{m^2 - 10m + 24} \div \dfrac{4m^2 - 9}{m^2 - 9m + 18}$

33. $\dfrac{m^2 + 2mp - 3p^2}{m^2 - 3mp + 2p^2} \div \dfrac{m^2 + 4mp + 3p^2}{m^2 + 2mp - 8p^2}$

34. $\dfrac{r^2 + rs - 12s^2}{r^2 - rs - 20s^2} \div \dfrac{r^2 - 2rs - 3s^2}{r^2 + rs - 30s^2}$

35. $\dfrac{m^2 + 3m + 2}{m^2 + 5m + 4} \cdot \dfrac{m^2 + 10m + 24}{m^2 + 5m + 6}$

36. $\dfrac{z^2 - z - 6}{z^2 - 2z - 8} \cdot \dfrac{z^2 + 7z + 12}{z^2 - 9}$

37. $\dfrac{y^2 + y - 2}{y^2 + 3y - 4} \div \dfrac{y + 2}{y + 3}$

38. $\dfrac{2m^2 - 5m - 12}{m^2 - 10m + 24} \div \dfrac{4m^2 - 9}{m^2 - 9m + 18}$

39. $\dfrac{2m^2 + 7m + 3}{m^2 - 9} \cdot \dfrac{m^2 - 3m}{2m^2 + 11m + 5}$

40. $\dfrac{m^2 + 2mp - 3p^2}{m^2 - 3mp + 2p^2} \div \dfrac{m^2 + 4mp + 3p^2}{m^2 + 2mp - 8p^2}$

41. $\dfrac{r^2 + rs - 12s^2}{r^2 - rs - 20s^2} \div \dfrac{r^2 - 2rs - 3s^2}{r^2 + rs - 30s^2}$

42. $\dfrac{(x + 1)^3(x + 4)}{x^2 + 5x + 4} \div \dfrac{x^2 + 2x + 1}{x^2 + 3x + 2}$

43. $\dfrac{(q - 3)^4(q + 2)}{q^2 + 3q + 2} \div \dfrac{q^2 - 6q + 9}{q^2 + 4q + 4}$

In working the exercise, remember how grouping symbols are used (Section 1.2), how to factor sums and differences of cubes (Section 4.4), and how to factor by grouping (Section 4.1).

44. $\left(\dfrac{x^2 + 10x + 25}{x^2 + 10x} \cdot \dfrac{10x}{x^2 + 15x + 50} \right) \div \dfrac{x + 5}{x + 10}$

45. $\left(\dfrac{m^2 - 12m + 32}{8m} \cdot \dfrac{m^2 - 8m}{m^2 - 8m + 16} \right) \div \dfrac{m - 8}{m - 4}$

46. $\dfrac{3a - 3b - a^2 + b^2}{4a^2 - 4ab + b^2} \cdot \dfrac{4a^2 - b^2}{2a^2 - ab - b^2}$

47. $\dfrac{4r^2 - t^2 + 10r - 5t}{2r^2 + rt + 5r} \cdot \dfrac{4r^3 + 4r^2t + rt^2}{2r + t}$

48. $\dfrac{-x^3 - y^3}{x^2 - 2xy + y^2} \div \dfrac{3y^2 - 3xy}{x^2 - y^2}$

49. $\dfrac{b^3 - 8a^3}{4a^3 + 4a^2b + ab^2} \div \dfrac{4a^2 + 2ab + b^2}{-a^3 - ab^3}$

50. If the rational expression $\dfrac{5x^2y^3}{2pq}$ represents the area of a rectangle and $\dfrac{2xy}{p}$ represents the length, what rational expression represents the width?

51. If you are given the problem $\dfrac{4y + 12}{2y - 10} \div \dfrac{?}{y^2 - y - 20} = \dfrac{2(y + 4)}{y - 3}$, what must be the polynomial that is represented by the question mark?

REVIEW EXERCISES

Write the prime factored form of the number. See Section 1.1.

52. 18   **53.** 48   **54.** 108   **55.** 60

Find the greatest common factor of the group of terms. See Section 4.1.

56. $24m$, $18m^2$, 6 **57.** $14x^2$, $28x$, 7 **58.** $84q^3$, $90q^6$ **59.** $54b^3$, $36b^4$

5.3 THE LEAST COMMON DENOMINATOR

FOR EXTRA HELP

📖 **SSG** pp. 174–177
SSM pp. 253–258

📼 Video
8

💾 Tutorial
IBM   MAC

OBJECTIVES
1 ▶ Find least common denominators.
2 ▶ Rewrite rational expressions with given denominators.

1 ▶ Find least common denominators.

▶ In this section, we demonstrate a preliminary step needed to add or subtract rational expressions with different denominators. Just as with rational numbers, adding or subtracting rational expressions (to be discussed in the next section) often requires a **least common denominator,** the least expression that all denominators divide into without a remainder. For example, the least common denominator for $2/9$ and $5/12$ is 36, since 36 is the smallest positive number that both 9 and 12 divide into evenly.

Least common denominators often can be found by inspection. For example, the least common denominator for $1/6$ and $2/(3m)$ is $6m$. In other cases, a least common denominator can be found by a procedure similar to that used in Chapter 4 for finding the greatest common factor.

Finding the Least Common Denominator

Step 1 Factor each denominator into prime factors.
Step 2 List each different denominator factor the *greatest* number of times it appears in any denominator.
Step 3 Multiply the denominator factors from Step 2 to get the least common denominator.

When each denominator is factored into prime factors, every prime factor must divide evenly into the least common denominator. We abbreviate least common denominator as LCD.

In Example 1, the LCD is found both for numerical denominators and algebraic denominators.

EXAMPLE 1

Finding the Least Common Denominator

Find the LCD for each pair of fractions.

(a) $\dfrac{1}{24}, \dfrac{7}{15}$ \qquad\qquad (b) $\dfrac{1}{8x}, \dfrac{3}{10x}$

Write each denominator in factored form with numerical coefficients in prime factored form.

$$24 = 2 \cdot 2 \cdot 2 \cdot 3 \qquad\qquad 8x = 2 \cdot 2 \cdot 2 \cdot x$$
$$= 2^3 \cdot 3 \qquad\qquad\qquad = 2^3 \cdot x$$
$$15 = 3 \cdot 5 \qquad\qquad\qquad 10x = 2 \cdot 5 \cdot x$$

We find the LCD by taking each different factor the greatest number of times it appears as a factor in any of the denominators.

The factor 2 appears three times in one product and not at all in the other, so the greatest number of times 2 appears is three. The greatest number of times both 3 and 5 appear is one.

$$\begin{aligned} \text{LCD} &= 2 \cdot 2 \cdot 2 \cdot 3 \cdot 5 \\ &= 2^3 \cdot 3 \cdot 5 \\ &= 120 \end{aligned}$$

Here 2 appears three times in one product and once in the other, so the greatest number of times the 2 appears is three. The greatest number of times the 5 appears is one, and the greatest number of times x appears in either product is one.

$$\begin{aligned} \text{LCD} &= 2 \cdot 2 \cdot 2 \cdot 5 \cdot x \\ &= 2^3 \cdot 5 \cdot x \\ &= 40x \quad \blacksquare \end{aligned}$$

EXAMPLE 2
Finding the LCD

■ Find the LCD for $\dfrac{5}{6r^2}$ and $\dfrac{3}{4r^3}$.

Factor each denominator.

$$6r^2 = 2 \cdot 3 \cdot r^2$$
$$4r^3 = 2 \cdot 2 \cdot r^3 = 2^2 \cdot r^3$$

The greatest number of times 2 appears is two, the greatest number of times 3 appears is one, and the greatest number of times r appears is three; therefore,

$$\text{LCD} = 2^2 \cdot 3 \cdot r^3 = 12r^3. \quad \blacksquare$$

EXAMPLE 3
Finding the LCD

■ Find the LCD.

(a) $\dfrac{6}{5m}, \dfrac{4}{m^2 - 3m}$

Factor each denominator.

$$5m = 5 \cdot m$$
$$m^2 - 3m = m(m - 3)$$

Take each different factor the greatest number of times it appears as a factor.

$$\text{LCD} = 5 \cdot m \cdot (m - 3) = 5m(m - 3)$$

Since m is not a *factor* of $m - 3$, both factors, m and $m - 3$, must appear in the LCD.

(b) $\dfrac{1}{r^2 - 4r - 5}, \dfrac{1}{r^2 - r - 20}, \dfrac{1}{r^2 - 10r + 25}$

Factor each denominator.

$$r^2 - 4r - 5 = (r - 5)(r + 1)$$
$$r^2 - r - 20 = (r - 5)(r + 4)$$
$$r^2 - 10r + 25 = (r - 5)^2$$

The LCD is $(r - 5)^2(r + 1)(r + 4)$.

(c) $\dfrac{1}{q - 5}, \dfrac{3}{5 - q}$

The expression $5 - q$ can be written as $-1(q - 5)$, since

$$-1(q - 5) = -q + 5 = 5 - q.$$

Because of this, either $q - 5$ or $5 - q$ can be used as the LCD. ■

2 ▶ Rewrite rational expressions with given denominators.

▶ Once we find the LCD, we can use the fundamental property to write equivalent rational expressions with this LCD. The next example shows how to do this with both numerical and algebraic fractions.

E X A M P L E   4

Writing a Fraction with a Given Denominator

Rewrite each expression with the indicated denominator.

(a) $\dfrac{3}{8} = \dfrac{}{40}$

(b) $\dfrac{9k}{25} = \dfrac{}{50k}$

For each example, first factor the denominator on the right. Then compare the denominator on the left with the one on the right to decide what factors are missing.

$$\frac{3}{8} = \frac{}{5 \cdot 8}$$

A factor of 5 is missing.

Multiply by $\dfrac{5}{5}$ to get a denominator of 40.

$$\frac{3}{8} = \frac{3}{8} \cdot \frac{5}{5} = \frac{15}{40}$$
$$\downarrow$$
$$\tfrac{5}{5} = 1$$

$$\frac{9k}{25} = \frac{}{25 \cdot 2k}$$

Factors of 2 and k are missing.

Get a denominator of $50k$ by multiplying by $\dfrac{2k}{2k}$.

$$\frac{9k}{25} = \frac{9k}{25} \cdot \frac{2k}{2k} = \frac{18k^2}{50k}$$
$$\downarrow$$
$$\tfrac{2k}{2k} = 1$$

Notice the use of the multiplicative identity property in each part of this example. ■

E X A M P L E   5

Writing a Fraction with a Given Denominator

Rewrite the following rational expression with the indicated denominator.

$$\frac{12p}{p^2 + 8p} = \frac{}{p^3 + 4p^2 - 32p}$$

Factor $p^2 + 8p$ as $p(p + 8)$. Compare with the denominator on the right which factors as $p(p + 8)(p - 4)$. The factor $p - 4$ is missing, so multiply $\dfrac{12p}{p(p + 8)}$ by $\dfrac{p - 4}{p - 4}$.

$$\frac{12p}{p^2 + 8p} = \frac{12p}{p(p + 8)} \cdot \frac{p - 4}{p - 4} \qquad \text{Multiplicative identity property}$$

$$= \frac{12p(p - 4)}{p(p + 8)(p - 4)} \qquad \text{Multiplication of rational expressions}$$

$$= \frac{12p^2 - 48p}{p^3 + 4p^2 - 32p} \qquad \text{Multiply the factors.} \quad ■$$

NOTE In the next section we learn to add and subtract rational expressions, and this often requires the skill illustrated in Example 5. While it is often beneficial to leave the denominator in factored form, we have multiplied the factors in the denominator in Example 5 to give the answer in the form the original problem was presented.

5.3 EXERCISES

Decide whether the statement is true or false.

1. The LCD for the two fractions $\dfrac{1}{a}$ and $\dfrac{1}{b}$ is ab if the greatest common factor of a and b is 1.

2. If a is a factor of b, then the LCD for $\dfrac{1}{a}$ and $\dfrac{1}{b}$ is b.

3. If x^a and x^b are denominators of two fractions and $a < b$, then x^a is the LCD.

4. If a fraction with denominator $x - 4$ must be written as an equivalent fraction with denominator $(x - 4)^3$, then the original fraction must be multiplied by $x - 4$ in both the numerator and the denominator.

Find the LCD for the fractions in the list. See Examples 1–3.

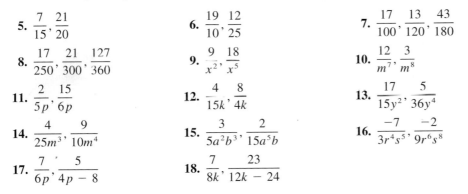

5. $\dfrac{7}{15}, \dfrac{21}{20}$

6. $\dfrac{19}{10}, \dfrac{12}{25}$

7. $\dfrac{17}{100}, \dfrac{13}{120}, \dfrac{43}{180}$

8. $\dfrac{17}{250}, \dfrac{21}{300}, \dfrac{127}{360}$

9. $\dfrac{9}{x^2}, \dfrac{18}{x^5}$

10. $\dfrac{12}{m^7}, \dfrac{3}{m^8}$

11. $\dfrac{2}{5p}, \dfrac{15}{6p}$

12. $\dfrac{4}{15k}, \dfrac{8}{4k}$

13. $\dfrac{17}{15y^2}, \dfrac{5}{36y^4}$

14. $\dfrac{4}{25m^3}, \dfrac{9}{10m^4}$

15. $\dfrac{3}{5a^2b^3}, \dfrac{2}{15a^5b}$

16. $\dfrac{-7}{3r^4s^5}, \dfrac{-2}{9r^6s^8}$

17. $\dfrac{7}{6p}, \dfrac{5}{4p - 8}$

18. $\dfrac{7}{8k}, \dfrac{23}{12k - 24}$

◆ MATHEMATICAL CONNECTIONS (Exercises 19–22) ◆

Suppose we want to find the LCD for the two common fractions

$$\frac{1}{24} \quad \text{and} \quad \frac{1}{20}.$$

In their prime factored forms, the denominators are

$$24 = 2^3 \cdot 3$$
$$\text{and} \quad 20 = 2^2 \cdot 5.$$

Refer to this information as necessary and work through Exercises 19–22 in order.

19. What is the prime factored form of the LCD of the two fractions?

20. Suppose that two algebraic fractions have denominators $(t + 4)^3(t - 3)$ and $(t + 4)^2(t + 8)$. What is the factored form of the LCD of these?

21. What is the similarity between your answers in Exercises 19 and 20?

22. Comment on the following statement: The method for finding the LCD for two algebraic fractions is the same as the method for finding the LCD for two common fractions.

◆

Find the LCD for the fractions in the list. See Examples 1–3.

23. $\dfrac{37}{6r - 12}, \dfrac{5}{9r - 18}$

24. $\dfrac{14}{5p - 30}, \dfrac{5}{6p - 36}$

25. $\dfrac{5}{12p + 60}, \dfrac{7}{p^2 + 5p}, \dfrac{6}{p^2 + 10p + 25}$

26. $\dfrac{13}{r^2 + 7r}, \dfrac{3}{5r + 35}, \dfrac{7}{r^2 + 14r + 49}$

27. $\dfrac{3}{8y + 16}, \dfrac{12}{y^2 + 3y + 2}$

28. $\dfrac{-2}{9m - 18}, \dfrac{-9}{m^2 - 7m + 10}$

29. $\dfrac{12}{m - 3}, \dfrac{4}{3 - m}$

30. $\dfrac{17}{8 - a}, \dfrac{2}{a - 8}$

31. $\dfrac{29}{p - q}, \dfrac{8}{q - p}$

32. $\dfrac{6}{z - x}, \dfrac{8}{x - z}$

33. $\dfrac{6}{a^2 + 6a}, \dfrac{5}{a^2 + 3a - 18}$

34. $\dfrac{8}{y^2 - 5y}, \dfrac{2}{y^2 - 2y - 15}$

35. $\dfrac{5}{k^2 + 2k - 35}, \dfrac{8}{k^2 + 3k - 40}, \dfrac{9}{k^2 - 2k - 15}$

36. $\dfrac{9}{z^2 + 4z - 12}, \dfrac{6}{z^2 + z - 30}, \dfrac{6}{z^2 + 2z - 24}$

37. Suppose that $(2x - 5)^2$ is the LCD for two fractions. Is $(5 - 2x)^2$ also acceptable as an LCD? Why?

38. Suppose that $(4t - 3)(5t - 6)$ is the LCD for two fractions. Is $(3 - 4t)(6 - 5t)$ also acceptable as an LCD? Why?

—————◆ **MATHEMATICAL CONNECTIONS** (Exercises 39–44) ◆—————

Work Exercises 39–44 in order.

39. Suppose that you want to write $\dfrac{3}{4}$ as an equivalent fraction with denominator 28. By what number must you multiply both the numerator and the denominator?

40. If you write $\dfrac{3}{4}$ as an equivalent fraction with denominator 28, by what number are you actually multiplying the fraction?

41. What property of multiplication is being used when we write a common fraction as an equivalent one with a larger denominator? (See Section 1.9.)

42. Suppose that you want to write $\dfrac{2x + 5}{x - 4}$ as an equivalent fraction with denominator $7x - 28$. By what number must you multiply both the numerator and the denominator?

43. If you write $\dfrac{2x + 5}{x - 4}$ as an equivalent fraction with denominator $7x - 28$, by what number are you actually multiplying the fraction?

44. Repeat Exercise 41, changing the word "common" to "algebraic."

————————————◆————————————

Write the rational expression on the left with the indicated denominator. See Examples 4 and 5.

45. $\dfrac{15m^2}{8k} = \dfrac{}{32k^4}$

46. $\dfrac{5t^2}{3y} = \dfrac{}{9y^2}$

47. $\dfrac{19z}{2z - 6} = \dfrac{}{6z - 18}$

48. $\dfrac{2r}{5r - 5} = \dfrac{}{15r - 15}$

49. $\dfrac{-2a}{9a - 18} = \dfrac{}{18a - 36}$

50. $\dfrac{-5y}{6y + 18} = \dfrac{}{24y + 72}$

51. $\dfrac{6}{k^2 - 4k} = \dfrac{}{k(k-4)(k+1)}$

52. $\dfrac{15}{m^2 - 9m} = \dfrac{}{m(m-9)(m+8)}$

53. $\dfrac{36r}{r^2 - r - 6} = \dfrac{}{(r-3)(r+2)(r+1)}$

54. $\dfrac{4m}{m^2 - 8m + 15} = \dfrac{}{(m-5)(m-3)(m+2)}$

55. $\dfrac{a + 2b}{2a^2 + ab - b^2} = \dfrac{a+2b)ab}{2a^3b + a^2b^2 - ab^3}$

56. $\dfrac{m - 4}{6m^2 + 7m - 3} = \dfrac{(m-4)2m}{12m^3 + 14m^2 - 6m}$

57. $\dfrac{4r - t}{r^2 + rt + t^2} = \dfrac{}{t^3 - r^3}$

58. $\dfrac{3x - 1}{x^2 + 2x + 4} = \dfrac{}{x^3 - 8}$

59. $\dfrac{2(z - y)}{y^2 + yz + z^2} = \dfrac{}{y^4 - z^3y}$

60. $\dfrac{2p + 3q}{p^2 + 2pq + q^2} - \dfrac{}{(p+q)(p^3 + q^3)}$

61. Write an explanation of how to find the least common denominator for a group of denominators.

62. Write an explanation of how to write a rational expression as an equivalent rational expression with a given denominator.

REVIEW EXERCISES

Add or subtract as indicated. See Section 1.1.

63. $\dfrac{3}{4} + \dfrac{7}{4}$ **64.** $\dfrac{2}{5} + \dfrac{9}{5}$ **65.** $\dfrac{1}{2} + \dfrac{7}{8}$ **66.** $\dfrac{2}{3} + \dfrac{8}{27}$

67. $\dfrac{7}{5} - \dfrac{3}{4}$ **68.** $\dfrac{11}{6} - \dfrac{2}{5}$ **69.** $\dfrac{4}{3} - \dfrac{1}{4}$ **70.** $\dfrac{7}{8} - \dfrac{10}{3}$

5.4 ADDITION AND SUBTRACTION OF RATIONAL EXPRESSIONS

FOR EXTRA HELP

SSG pp. 177–179
SSM pp. 258–265

Video 8

Tutorial
IBM   MAC

OBJECTIVES

1 ▶ Add rational expressions having the same denominator.
2 ▶ Add rational expressions having different denominators.
3 ▶ Subtract rational expressions.

We are now ready to add and subtract rational expressions. We need the skills we developed in the previous section to find least common denominators and to write fractions with the least common denominator.

1 ▶ Add rational expressions having the same denominator.

▶   We find the sum of two rational expressions with a procedure similar to the one used for adding two fractions.

Adding Rational Expressions

If P/Q and R/Q are rational expressions, then

$$\frac{P}{Q} + \frac{R}{Q} = \frac{P + R}{Q}.$$

Again, the first example shows how the addition of rational expressions compares with that of rational numbers.

EXAMPLE 1

Adding Rational Expressions with the Same Denominator

Add.

(a) $\dfrac{4}{7} + \dfrac{2}{7}$

(b) $\dfrac{3x}{x+1} + \dfrac{2x}{x+1}$

The denominators are the same, so the sum is found by adding the two numerators and keeping the same (common) denominator.

$$\frac{4}{7} + \frac{2}{7} = \frac{4+2}{7}$$

$$= \frac{6}{7}$$

$$\frac{3x}{x+1} + \frac{2x}{x+1} = \frac{3x+2x}{x+1}$$

$$= \frac{5x}{x+1} \quad \blacksquare$$

2 ▶ Add rational expressions having different denominators.

▶ Use the steps given below to add two rational expressions with different denominators. These are the same steps used to add fractions with different denominators.

Adding with Different Denominators

Step 1 Find the least common denominator (LCD).

Step 2 Rewrite each rational expression as an equivalent fraction with the LCD as the denominator.

Step 3 Add the numerators to get the numerator of the sum. The LCD is the denominator of the sum.

Step 4 Write the answer in lowest terms.

EXAMPLE 2

Adding Rational Expressions with Different Denominators

Add.

(a) $\dfrac{1}{12} + \dfrac{7}{15}$

(b) $\dfrac{2}{3y} + \dfrac{1}{4y}$

First find the LCD using the methods of the last section.

$$\text{LCD} = 2^2 \cdot 3 \cdot 5 = 60$$

$$\text{LCD} = 2^2 \cdot 3 \cdot y = 12y$$

Now rewrite each rational expression as a fraction with the LCD (either 60 or 12y) as the denominator.

$$\frac{1}{12} + \frac{7}{15} = \frac{1(5)}{12(5)} + \frac{7(4)}{15(4)}$$

$$= \frac{5}{60} + \frac{28}{60}$$

$$\frac{2}{3y} + \frac{1}{4y} = \frac{2(4)}{3y(4)} + \frac{1(3)}{4y(3)}$$

$$= \frac{8}{12y} + \frac{3}{12y}$$

Since the fractions now have common denominators, add the numerators. (Write in lowest terms if necessary.)

$$\frac{5}{60} + \frac{28}{60} = \frac{5+28}{60}$$

$$= \frac{33}{60} = \frac{11}{20}$$

$$\frac{8}{12y} + \frac{3}{12y} = \frac{8+3}{12y}$$

$$= \frac{11}{12y} \quad \blacksquare$$

In the next example, we show the four steps listed earlier in the section.

EXAMPLE 3

Adding Rational
Expressions

Add and write the sum in lowest terms.

$$\frac{2x}{x^2 - 1} + \frac{-1}{x + 1}$$

Step 1 Since the denominators are different, find the LCD.

$$x^2 - 1 = (x + 1)(x - 1)$$
$$x + 1 \text{ is prime.}$$

The LCD is $(x + 1)(x - 1)$.

Step 2 Rewrite each rational expression as a fraction with denominator $(x + 1)(x - 1)$.

$$\frac{2x}{x^2 - 1} + \frac{-1}{x + 1} = \frac{2x}{(x + 1)(x - 1)} + \frac{1(x - 1)}{(x + 1)(x - 1)} \quad \text{Multiply second fraction by } \frac{x - 1}{x - 1}.$$

$$= \frac{2x}{(x + 1)(x - 1)} + \frac{-x + 1}{(x + 1)(x - 1)} \quad \text{Distributive property}$$

Step 3

$$= \frac{2x - x + 1}{(x + 1)(x - 1)} \quad \text{Add numerators; keep the same denominator.}$$

$$= \frac{x + 1}{(x + 1)(x - 1)} \quad \text{Combine like terms in the numerator.}$$

Step 4

$$= \frac{1(x + 1)}{(x + 1)(x - 1)} \quad \text{Identity property for multiplication}$$

$$= \frac{1}{x - 1} \quad \text{Fundamental property of rational expressions} \quad \blacksquare$$

In the rest of the examples in this section, the steps are not numbered.

EXAMPLE 4

Adding Rational
Expressions

Add and express the answer in lowest terms.

$$\frac{2x}{x^2 + 5x + 6} + \frac{x + 1}{x^2 + 2x - 3}$$

Begin by factoring the denominators completely.

$$\frac{2x}{(x + 2)(x + 3)} + \frac{x + 1}{(x + 3)(x - 1)}$$

The LCD is $(x + 2)(x + 3)(x - 1)$. Use the fundamental property of rational expressions to rewrite each fraction with the LCD.

$$\frac{2x}{(x + 2)(x + 3)} + \frac{x + 1}{(x + 3)(x - 1)}$$

$$= \frac{2x(x - 1)}{(x + 2)(x + 3)(x - 1)} + \frac{(x + 1)(x + 2)}{(x + 3)(x - 1)(x + 2)}$$

$$= \frac{2x(x - 1) + (x + 1)(x + 2)}{(x + 2)(x + 3)(x - 1)} \quad \text{Add numerators; keep the same denominator.}$$

$$= \frac{2x^2 - 2x + x^2 + 3x + 2}{(x + 2)(x + 3)(x - 1)} \quad \text{Distributive property}$$

$$= \frac{3x^2 + x + 2}{(x + 2)(x + 3)(x - 1)} \quad \text{Combine terms.}$$

Since $3x^2 + x + 2$ cannot be factored, the rational expression cannot be reduced. It is usually best to leave the denominator in factored form since it is then easier to identify common factors in the numerator and denominator. ■

In some problems, rational expressions to be added or subtracted have denominators that are opposites of each other. The next example illustrates how to proceed in such a problem.

EXAMPLE 5

Adding Rational Expressions with Denominators that Are Opposites

■ Add and express the answer in lowest terms.

$$\frac{y}{y-2} + \frac{8}{2-y}$$

To get a common denominator of $y - 2$, multiply the second expression by -1 in both the numerator and the denominator.

$$\frac{y}{y-2} + \frac{8}{2-y} = \frac{y}{y-2} + \frac{8(-1)}{(2-y)(-1)} \qquad \text{Fundamental property}$$

$$= \frac{y}{y-2} + \frac{-8}{y-2} \qquad \text{Distributive property}$$

$$= \frac{y-8}{y-2} \qquad \text{Add numerators; keep the same denominator.}$$

If we had chosen to use $2 - y$ as the common denominator, the final answer would be in the form $\frac{8-y}{2-y}$, which is equivalent to $\frac{y-8}{y-2}$. ■

3 ▶ Subtract rational expressions.

▶ To *subtract* rational expressions, use the following rule.

Subtracting Rational Expressions

If P/Q and R/Q are rational expressions, then

$$\frac{P}{Q} - \frac{R}{Q} = \frac{P-R}{Q}.$$

EXAMPLE 6

Subtracting Rational Expressions

■ Subtract: $\dfrac{2m}{m-1} - \dfrac{2}{m-1}$.

By the definition of subtraction,

$$\frac{2m}{m-1} - \frac{2}{m-1} = \frac{2m-2}{m-1} \qquad \text{Subtract numerators; keep the same denominator.}$$

$$= \frac{2(m-1)}{m-1} \qquad \text{Factor the numerator.}$$

$$= 2. \qquad \text{Write in lowest terms.} \quad ■$$

EXAMPLE 7

Subtracting Rational
Expressions with
Different Denominators

■ Subtract: $\dfrac{9}{x-2} - \dfrac{3}{x}$.

The LCD is $x(x-2)$.

$$\frac{9}{x-2} - \frac{3}{x} = \frac{9x}{x(x-2)} - \frac{3(x-2)}{x(x-2)} \qquad \text{Get the least common denominator.}$$

$$= \frac{9x - 3(x-2)}{x(x-2)} \qquad \text{Subtract numerators; keep the same denominator.}$$

$$= \frac{9x - 3x + 6}{x(x-2)} \qquad \text{Distributive property}$$

$$= \frac{6x + 6}{x(x-2)} \qquad \text{Combine like terms in the numerator.} \quad ■$$

NOTE   It would not be wrong to factor the final numerator in Example 7 to get an answer in the form $\dfrac{6(x+1)}{x(x-2)}$; however, the fundamental property would not apply, since there are no common factors that would allow us to write the answer in lower terms. When a rational expression cannot be reduced, we will leave numerators in polynomial form and denominators in factored form.

EXAMPLE 8

Subtracting Rational
Expressions with
Denominators that Are
Opposites

■ Subtract: $\dfrac{3x}{x-5} - \dfrac{2x-25}{5-x}$.

The denominators are opposites, so either may be used as the common denominator. Let us choose $x-5$.

$$\frac{3x}{x-5} - \frac{2x-25}{5-x} = \frac{3x}{x-5} - \frac{2x-25}{5-x} \cdot \frac{-1}{-1} \qquad \text{Fundamental property}$$

$$= \frac{3x}{x-5} - \frac{-2x+25}{x-5} \qquad \text{Multiply.}$$

$$= \frac{3x - (-2x+25)}{x-5} \qquad \text{Subtract numerators.}$$

$$= \frac{3x + 2x - 25}{x-5} \qquad \text{Distributive property}$$

$$= \frac{5x - 25}{x-5} \qquad \text{Combine terms.}$$

$$= \frac{5(x-5)}{x-5} \qquad \text{Factor.}$$

$$= 5 \qquad \text{Lowest terms} \quad ■$$

CAUTION   Sign errors often occur in subtraction problems like the ones in Examples 7 and 8. Remember that the numerator of the fraction being subtracted must be treated as a single quantity. Use parentheses after the subtraction sign to avoid this common error.

EXAMPLE 9 ■

Subtracting Rational Expressions

Subtract: $\dfrac{6x}{x^2 - 2x + 1} - \dfrac{1}{x^2 - 1}$.

Begin by factoring.

$$\frac{6x}{x^2 - 2x + 1} - \frac{1}{x^2 - 1} = \frac{6x}{(x-1)(x-1)} - \frac{1}{(x-1)(x+1)}$$

From the factored denominators, we can identify the common denominator, $(x-1)(x-1)(x+1)$. Use the factor $x - 1$ twice, since it appears twice in the first denominator.

$$\frac{6x}{(x-1)(x-1)} - \frac{1}{(x-1)(x+1)}$$

$$= \frac{6x(x+1)}{(x-1)(x-1)(x+1)} - \frac{1(x-1)}{(x-1)(x-1)(x+1)} \qquad \text{Fundamental property}$$

$$= \frac{6x(x+1) - 1(x-1)}{(x-1)(x-1)(x+1)} \qquad \text{Subtract.}$$

$$= \frac{6x^2 + 6x - x + 1}{(x-1)(x-1)(x+1)} \qquad \text{Distributive property}$$

$$= \frac{6x^2 + 5x + 1}{(x-1)(x-1)(x+1)} \qquad \text{Combine like terms.}$$

The result may be written as $\dfrac{6x^2 + 5x + 1}{(x-1)^2(x+1)}$. ■

NOTE In many problems involving sums and differences of rational expressions, several different equivalent forms of the answer exist. If your answer does not look exactly like the one given in the back of the book, check to see if your answer is an equivalent form.

5.4 EXERCISES

◆———— **MATHEMATICAL CONNECTIONS** (Exercises 1–8) ◇————

As you work through this exercise set, you will probably check your answers against the ones given in the answer section in the back of the book. Work through Exercises 1–8 in order. This will help you later in determining whether your answer is equivalent to the one given if it doesn't match the form shown.

Jill worked a problem involving the sum of two rational expressions correctly. Her answer was given as $\dfrac{5}{x-3}$. Jack worked the same problem and gave his answer as $\dfrac{-5}{3-x}$. We want to decide whether Jack's answer was also correct.

1. Evaluate the fraction $\dfrac{5}{7-3}$ using the rule for order of operations.

2. Multiply the fraction given in Exercise 1 by -1 in both the numerator and the denominator, using the distributive property as necessary. Leave it in unsimplified form.

3. In Exercise 2, what number did you actually multiply the *fraction* by?

4. Simplify the result you found in Exercise 2 and compare it to the one found in Exercise 1. How do they compare?

5. Now look at the answers given by Jill and Jack above. Based on what you learned in Exercises 1–4, determine whether Jack's answer was also correct. Why or why not?

6. Anne Kelly, a perceptive algebra student, made the following comment and was praised by her teacher for her insight: "I can see that if I change the sign of each term in a fraction, the result I get is equivalent to the fraction that I started with." Explain why Anne's observation is correct.

7. Karin Wagner, another perceptive algebra student, followed Anne's comment with one of her own: "I can see that if I put a negative sign in front of a fraction and change the signs of all terms in either the numerator or the denominator, but not both, then the result I get is equivalent to the fraction that I started with." The teacher knew that some real education was happening in the classroom. Explain why Karin's observation is also correct.

8. Use the concepts of Exercises 1–7 to determine whether the rational expression is equivalent or not equivalent to $\dfrac{y-4}{3-y}$.

(a) $\dfrac{y-4}{y-3}$ (b) $\dfrac{-y+4}{-3+y}$ (c) $-\dfrac{y+4}{y-3}$ (d) $-\dfrac{y-4}{3-y}$ (e) $-\dfrac{4-y}{y-3}$ (f) $\dfrac{y+4}{3+y}$

Add or subtract as indicated. Write the answer in lowest terms. See Examples 1 and 6.

9. $\dfrac{4}{m} + \dfrac{7}{m}$

10. $\dfrac{5}{p} + \dfrac{11}{p}$

11. $\dfrac{a+b}{2} - \dfrac{a-b}{2}$

12. $\dfrac{x-y}{2} - \dfrac{x+y}{2}$

13. $\dfrac{x^2}{x+5} + \dfrac{5x}{x+5}$

14. $\dfrac{t^2}{t-3} + \dfrac{-3t}{t-3}$

15. $\dfrac{y^2-3y}{y+3} + \dfrac{-18}{y+3}$

16. $\dfrac{r^2-8r}{r-5} + \dfrac{15}{r-5}$

17. Explain how to add rational expressions with different denominators.

18. Explain how to subtract rational expressions with different denominators.

Add or subtract as indicated. See Examples 2, 3, 4, and 7.

19. $\dfrac{z}{5} + \dfrac{1}{3}$

20. $\dfrac{p}{8} + \dfrac{3}{5}$

21. $\dfrac{5}{7} - \dfrac{r}{2}$

22. $\dfrac{10}{9} - \dfrac{z}{3}$

23. $-\dfrac{3}{4} - \dfrac{1}{2x}$

24. $-\dfrac{5}{8} - \dfrac{3}{2a}$

25. $\dfrac{5+5k}{4} + \dfrac{1+k}{8}$

26. $\dfrac{6-5r}{9} + \dfrac{2-3r}{6}$

27. $\dfrac{b+3}{b} + \dfrac{b+7}{3b}$

28. $\dfrac{3q-1}{q} + \dfrac{q+2}{4q}$

29. $\dfrac{7}{3p^2} - \dfrac{2}{p}$

30. $\dfrac{12}{5m^2} - \dfrac{2}{m}$

31. $\dfrac{-2}{x^2 - 4} + \dfrac{7}{4x + 8}$

32. $\dfrac{-4}{z^2 - 16} + \dfrac{3}{2z + 8}$

33. $\dfrac{8}{m - 2} + \dfrac{3}{5m} + \dfrac{7}{5m(m - 2)}$

34. $\dfrac{-1}{7z} + \dfrac{3}{z + 2} + \dfrac{4}{7z(z + 2)}$

35. What are the two possible LCDs that could be used for the sum

$$\dfrac{10}{m - 2} + \dfrac{5}{2 - m}?$$

36. If one form of the correct answer to a sum or difference of rational expressions is

$\dfrac{4}{k - 3}$, what would an alternate form of the answer be if the denominator is $3 - k$?

Add or subtract as indicated. See Examples 5 and 8.

37. $\dfrac{4}{x - 5} + \dfrac{6}{5 - x}$

38. $\dfrac{10}{m - 2} + \dfrac{5}{2 - m}$

39. $\dfrac{-1}{1 - y} - \dfrac{3}{y - 1}$

40. $\dfrac{-4}{p - 3} - \dfrac{7}{3 - p}$

41. $\dfrac{2}{x - y^2} + \dfrac{7}{y^2 - x}$

42. $\dfrac{-8}{p - q^2} + \dfrac{3}{q^2 - p}$

43. $\dfrac{x}{5x - 3y} - \dfrac{y}{3y - 5x}$

44. $\dfrac{t}{8t - 9s} - \dfrac{s}{9s - 8t}$

45. $\dfrac{3}{4p - 5} + \dfrac{9}{5 - 4p}$

46. $\dfrac{8}{3 - 7y} - \dfrac{2}{7y - 3}$

In these subtraction problems, the rational expression that follows the subtraction sign has a numerator with more than one term. Be very careful with signs and find the difference.

47. $\dfrac{2m}{m - n} - \dfrac{5m + n}{2m - 2n}$

48. $\dfrac{5p}{p - q} - \dfrac{3p + 1}{4p - 4q}$

49. $\dfrac{5}{x^2 - 9} - \dfrac{x + 2}{x^2 + 4x + 3}$

50. $\dfrac{1}{a^2 - 1} - \dfrac{a - 1}{a^2 + 3a - 4}$

51. $\dfrac{2q + 1}{3q^2 + 10q - 8} - \dfrac{3q + 5}{2q^2 + 5q - 12}$

52. $\dfrac{4y - 1}{2y^2 + 5y - 3} - \dfrac{y + 3}{6y^2 + y - 2}$

Perform the indicated operations. See Examples 1–9.

53. $\dfrac{4}{r^2 - r} + \dfrac{6}{r^2 + 2r} - \dfrac{1}{r^2 + r - 2}$

54. $\dfrac{6}{k^2 + 3k} - \dfrac{1}{k^2 - k} + \dfrac{2}{k^2 + 2k - 3}$

55. $\dfrac{x + 3y}{x^2 + 2xy + y^2} + \dfrac{x - y}{x^2 + 4xy + 3y^2}$

56. $\dfrac{m}{m^2 - 1} + \dfrac{m - 1}{m^2 + 2m + 1}$

57. $\dfrac{r + y}{18r^2 + 12ry - 3ry - 2y^2} + \dfrac{3r - y}{36r^2 - y^2}$

58. $\dfrac{2x - z}{2x^2 - 4xz + 5xz - 10z^2} - \dfrac{x + z}{x^2 - 4z^2}$

Perform the indicated operations. Remember the order of operations.

59. $\left(\dfrac{-k}{2k^2 - 5k - 3} + \dfrac{3k - 2}{2k^2 - k - 1}\right)\dfrac{2k + 1}{k - 1}$

60. $\left(\dfrac{3p + 1}{2p^2 + p - 6} - \dfrac{5p}{3p^2 - p}\right)\dfrac{2p - 3}{p + 2}$

61. $\dfrac{k^2 + 4k + 16}{k + 4}\left(\dfrac{-5}{16 - k^2} + \dfrac{2k + 3}{k^3 - 64}\right)$

62. $\dfrac{m - 5}{2m + 5}\left(\dfrac{-3m}{m^2 - 25} - \dfrac{m + 4}{125 - m^3}\right)$

63. Refer to the rectangle in the figure.
     (a) Find an expression that represents its perimeter. Give the simplified form.
     (b) Find an expression that represents its area. Give the simplified form.

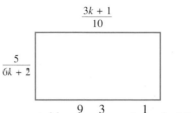

64. If the lengths of the sides of a triangle are represented by $\dfrac{9}{2p}$, $\dfrac{3}{p^2}$, and $\dfrac{1}{p^2}$, find the perimeter.

REVIEW EXERCISES

Perform the indicated operations, using the order of operations as necessary. See Section 1.1.

65. $\dfrac{\dfrac{5}{6} + \dfrac{7}{6}}{\dfrac{2}{3} - \dfrac{1}{3}}$

66. $\dfrac{\dfrac{3}{8} - \dfrac{5}{8}}{\dfrac{1}{4} + \dfrac{7}{4}}$

67. $\dfrac{\dfrac{3}{2} - \dfrac{5}{4}}{\dfrac{7}{4} + \dfrac{1}{3}}$

68. $\dfrac{\dfrac{5}{7} - \dfrac{3}{14}}{\dfrac{5}{3} + \dfrac{1}{2}}$

5.5 COMPLEX FRACTIONS

OBJECTIVES

1 ▶ Simplify a complex fraction by writing it as a division problem (Method 1).

2 ▶ Simplify a complex fraction by multiplying the numerator and the denominator by the LCD of all the fractions within the complex fraction (Method 2).

The quotient of two mixed numbers in arithmetic, such as $2\dfrac{1}{2} \div 3\dfrac{1}{4}$, can be written as a fraction:

$$2\dfrac{1}{2} \div 3\dfrac{1}{4} = \dfrac{2\dfrac{1}{2}}{3\dfrac{1}{4}} = \dfrac{2 + \dfrac{1}{2}}{3 + \dfrac{1}{4}}.$$

The last expression is the quotient of expressions that involve fractions. In algebra, some rational expressions also have fractions in the numerator, or denominator, or both.

Complex Fraction

A rational expression with fractions in the numerator, denominator, or both, is called a **complex fraction.**

Examples of complex fractions include

$$\frac{2 + \dfrac{1}{2}}{3 + \dfrac{1}{4}}, \qquad \frac{\dfrac{3x^2 - 5x}{6x^2}}{2x - \dfrac{1}{x}}, \qquad \text{and} \qquad \frac{3 + x}{5 - \dfrac{2}{x}}.$$

The parts of a complex fraction are named as follows.

$$\frac{\dfrac{2}{p} - \dfrac{1}{q}}{\dfrac{3}{p} + \dfrac{5}{q}}$$

← Numerator of complex fraction
← Main fraction bar
← Denominator of complex fraction

1 ▶ Simplify a complex fraction by writing it as a division problem (Method 1).

▶ Since the main fraction bar represents division in a complex fraction, one method of simplifying a complex fraction involves rewriting it as a division problem.

Method I

To simplify a complex fraction:

Step 1 Write both the numerator and denominator as single fractions.
Step 2 Change the complex fraction to a division problem.
Step 3 Perform the indicated division.

Once again, in this section the first example shows complex fractions from both arithmetic and algebra. Notice the differences and similarities as you read through it.

E X A M P L E   1

Simplifying Complex Fractions by Method I

Simplify each complex fraction.

(a) $\dfrac{\dfrac{2}{3} + \dfrac{5}{9}}{\dfrac{1}{4} + \dfrac{1}{12}}$

(b) $\dfrac{6 + \dfrac{3}{x}}{\dfrac{x}{4} + \dfrac{1}{8}}$

First, write each numerator as a single fraction.

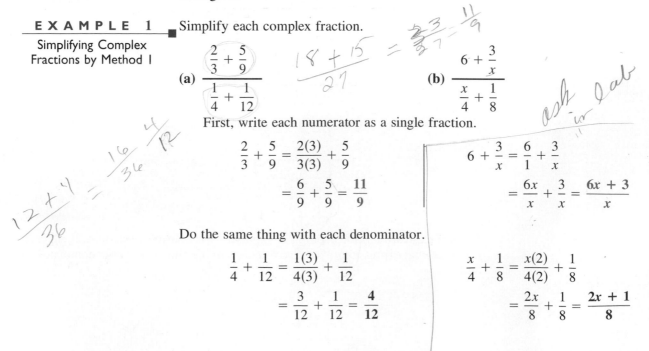

$$\frac{2}{3} + \frac{5}{9} = \frac{2(3)}{3(3)} + \frac{5}{9}$$

$$= \frac{6}{9} + \frac{5}{9} = \frac{11}{9}$$

$$6 + \frac{3}{x} = \frac{6}{1} + \frac{3}{x}$$

$$= \frac{6x}{x} + \frac{3}{x} = \frac{6x + 3}{x}$$

Do the same thing with each denominator.

$$\frac{1}{4} + \frac{1}{12} = \frac{1(3)}{4(3)} + \frac{1}{12}$$

$$= \frac{3}{12} + \frac{1}{12} = \frac{4}{12}$$

$$\frac{x}{4} + \frac{1}{8} = \frac{x(2)}{4(2)} + \frac{1}{8}$$

$$= \frac{2x}{8} + \frac{1}{8} = \frac{2x + 1}{8}$$

The original complex fraction can now be written as follows.

$$\frac{\dfrac{11}{9}}{\dfrac{4}{12}} \qquad\qquad \frac{\dfrac{6x + 3}{x}}{\dfrac{2x + 1}{8}}$$

Now use the rule for division and the fundamental property.

$$\frac{11}{9} \div \frac{4}{12} = \frac{11}{9} \cdot \frac{12}{4} \qquad\qquad \frac{6x + 3}{x} \div \frac{2x + 1}{8} = \frac{6x + 3}{x} \cdot \frac{8}{2x + 1}$$

$$= \frac{11 \cdot 3 \cdot 4}{3 \cdot 3 \cdot 4} \qquad\qquad\qquad = \frac{3(2x + 1)}{x} \cdot \frac{8}{2x + 1}$$

$$= \frac{11}{3} \qquad\qquad\qquad\qquad = \frac{24}{x} \quad\blacksquare$$

EXAMPLE 2

Simplifying a Complex Fraction by Method I

■ Simplify the complex fraction $\dfrac{\dfrac{xp}{q^3}}{\dfrac{p^2}{qx^2}}$.

Here the numerator and denominator are already single fractions, so use the division rule and then the fundamental property: $\dfrac{xp}{q^3} \div \dfrac{p^2}{qx^2} = \dfrac{xp}{q^3} \cdot \dfrac{qx^2}{p^2} = \dfrac{x^3}{q^2 p}$. ■

EXAMPLE 3

Simplifying a Complex Fraction by Method I

■ Simplify $\dfrac{\dfrac{3}{x + 2} - 4}{\dfrac{2}{x + 2} + 1}$.

$$\frac{\dfrac{3}{x + 2} - 4}{\dfrac{2}{x + 2} + 1} = \frac{\dfrac{3}{x + 2} - \dfrac{4(x + 2)}{x + 2}}{\dfrac{2}{x + 2} + \dfrac{1(x + 2)}{x + 2}} \qquad \text{Write both second terms with a denominator of } x + 2.$$

$$= \frac{\dfrac{3 - 4(x + 2)}{x + 2}}{\dfrac{2 + 1(x + 2)}{x + 2}} \qquad \begin{array}{l}\text{Subtract in the numerator.}\\ \text{Add in the denominator.}\end{array}$$

$$= \frac{\dfrac{3 - 4x - 8}{x + 2}}{\dfrac{2 + x + 2}{x + 2}} \qquad \text{Distributive property}$$

$$= \frac{\dfrac{-5 - 4x}{x + 2}}{\dfrac{4 + x}{x + 2}} \qquad \text{Combine terms.}$$

$$= \frac{-5 - 4x}{x + 2} \cdot \frac{x + 2}{4 + x} \qquad \text{Multiply by the reciprocal.}$$

$$= \frac{-5 - 4x}{4 + x} \qquad \text{Lowest terms} \quad\blacksquare$$

2 ► Simplify a complex fraction by multiplying the numerator and the denominator by the LCD of all the fractions within the complex fraction (Method 2).

► As an alternative method, a complex fraction may be simplified by a method that uses the fundamental property of rational expressions. Since any expression can be multiplied by a form of 1 to get an equivalent expression, we may multiply both the numerator and the denominator of a complex fraction by the same nonzero expression to get an equivalent complex fraction. If we choose the expression to be the LCD of all the fractions within the complex fraction, the complex fraction will be simplified. This is Method 2.

Method 2   To simplify a complex fraction:

Step 1   Find the LCD of all fractions within the complex fraction.

Step 2   Multiply both the numerator and the denominator of the complex fraction by this LCD using the distributive property as necessary. Reduce to lowest terms.

In the next example, Method 2 is used to simplify the same complex fractions as in Example 1.

E X A M P L E 4

Simplifying Complex Fractions by Method 2

Simplify each complex fraction.

(a) $\dfrac{\dfrac{2}{3} + \dfrac{5}{9}}{\dfrac{1}{4} + \dfrac{1}{12}}$

(b) $\dfrac{6 + \dfrac{3}{x}}{\dfrac{x}{4} + \dfrac{1}{8}}$

Find the LCD for all denominators in the complex fraction.

The LCD for 3, 9, 4, and 12 is 36. | The LCD for x, 4, and 8 is $8x$.

Multiply numerator and denominator of the complex fraction by the LCD.

$$\dfrac{\dfrac{2}{3} + \dfrac{5}{9}}{\dfrac{1}{4} + \dfrac{1}{12}} = \dfrac{36\left(\dfrac{2}{3} + \dfrac{5}{9}\right)}{36\left(\dfrac{1}{4} + \dfrac{1}{12}\right)}$$

$$\dfrac{6 + \dfrac{3}{x}}{\dfrac{x}{4} + \dfrac{1}{8}} = \dfrac{8x\left(6 + \dfrac{3}{x}\right)}{8x\left(\dfrac{x}{4} + \dfrac{1}{8}\right)}$$

$$= \dfrac{36\left(\dfrac{2}{3}\right) + 36\left(\dfrac{5}{9}\right)}{36\left(\dfrac{1}{4}\right) + 36\left(\dfrac{1}{12}\right)}$$

$$= \dfrac{8x(6) + 8x\left(\dfrac{3}{x}\right)}{8x\left(\dfrac{x}{4}\right) + 8x\left(\dfrac{1}{8}\right)}$$ Distributive property

$$= \dfrac{24 + 20}{9 + 3}$$

$$= \dfrac{48x + 24}{2x^2 + x}$$

$$= \dfrac{44}{12} = \dfrac{4 \cdot 11}{4 \cdot 3}$$

$$= \dfrac{24(2x + 1)}{x(2x + 1)}$$ Factor.

$$= \dfrac{11}{3}$$

$$= \dfrac{24}{x}$$ Lowest terms ■

E X A M P L E 5

Simplifying a Complex Fraction by Method 2

Simplify the complex fraction $\dfrac{\dfrac{3}{5m} - \dfrac{2}{m^2}}{\dfrac{9}{2m} + \dfrac{3}{4m^2}}$.

The LCD for $5m$, m^2, $2m$, and $4m^2$ is $20m^2$. Multiply numerator and denominator by $20m^2$.

$$\frac{\dfrac{3}{5m} - \dfrac{2}{m^2}}{\dfrac{9}{2m} + \dfrac{3}{4m^2}} = \frac{20m^2\left(\dfrac{3}{5m} - \dfrac{2}{m^2}\right)}{20m^2\left(\dfrac{9}{2m} + \dfrac{3}{4m^2}\right)}$$

$$= \frac{20m^2\left(\dfrac{3}{5m}\right) - 20m^2\left(\dfrac{2}{m^2}\right)}{20m^2\left(\dfrac{9}{2m}\right) + 20m^2\left(\dfrac{3}{4m^2}\right)} \quad \text{Distributive property}$$

$$= \frac{12m - 40}{90m + 15} \quad \blacksquare$$

Either of the two methods shown in this section can be applied to a complex fraction to simplify it. You may want to choose one method and stick to it in order to eliminate confusion. However, some students prefer to use Method 1 for problems like Example 2, which is the quotient of two fractions. They prefer Method 2 for problems like Examples 1, 3, 4, and 5, which have sums or differences in the numerators or denominators or both.

◆ **CONNECTIONS** ◇

Some numbers can be expressed as *continued fractions,* which are infinite complex fractions. For example, the irrational number $\sqrt{2}$ can be expressed as follows.

$$\sqrt{2} = 1 + \cfrac{1}{2 + \cfrac{1}{2 + \cfrac{1}{2 + \cfrac{1}{2 + \cdots}}}}$$

Better and better approximations of $\sqrt{2}$ can be found by using more and more terms of the fraction. We give the first three approximations here.

$$1 + \frac{1}{2} = 1.5$$

$$1 + \cfrac{1}{2 + \cfrac{1}{2}} = 1 + \cfrac{1}{\dfrac{5}{2}} = 1 + \frac{2}{5} = \frac{7}{5} = 1.4$$

$$1 + \cfrac{1}{2 + \cfrac{1}{2 + \cfrac{1}{2}}} = 1 + \cfrac{1}{2 + \cfrac{1}{\dfrac{5}{2}}} = 1 + \cfrac{1}{2 + \dfrac{2}{5}} = 1 + \cfrac{1}{\dfrac{12}{5}}$$

$$= 1 + \frac{5}{12} = \frac{17}{12} = 1.41\overline{6}$$

A calculator gives $\sqrt{2} \approx 1.414213562$ to nine decimal places.

FOR DISCUSSION OR WRITING
Give the next approximation of $\sqrt{2}$ and compare it to the calculator value shown above. How many places after the decimal agree?

5.5 EXERCISES

Note: In many problems involving complex fractions, several different equivalent forms of the answer exist. If your answer does not look exactly like the one given in the back of the book, check to see if your answer is an equivalent form.

1. In your own words, describe Method 1 for simplifying complex fractions.

2. In your own words, describe Method 2 for simplifying complex fractions.

3. Which one of the following complex fractions is equivalent to $\dfrac{3 - \dfrac{1}{2}}{2 - \dfrac{1}{4}}$?

Answer this question mentally, without showing any work.

(a) $\dfrac{3 + \dfrac{1}{2}}{2 + \dfrac{1}{4}}$ **(b)** $\dfrac{-3 + \dfrac{1}{2}}{2 - \dfrac{1}{4}}$ **(c)** $\dfrac{-3 - \dfrac{1}{2}}{-2 - \dfrac{1}{4}}$ **(d)** $\dfrac{-3 + \dfrac{1}{2}}{-2 + \dfrac{1}{4}}$

4. Only one of the choices below is equal to $\dfrac{\dfrac{1}{2} + \dfrac{1}{4}}{\dfrac{1}{3} + \dfrac{1}{12}}$. Which one is it?

Answer this question mentally, without showing any work.

(a) $\dfrac{9}{5}$ **(b)** $-\dfrac{9}{5}$ **(c)** $-\dfrac{5}{9}$ **(d)** -12

Simplify the complex fraction. Use either method. See Examples 1–5.

5. $\dfrac{-\dfrac{4}{3}}{\dfrac{2}{9}}$ **6.** $\dfrac{-\dfrac{5}{6}}{\dfrac{5}{4}}$ **7.** $\dfrac{\dfrac{p}{q^2}}{\dfrac{p^2}{q}}$ **8.** $\dfrac{\dfrac{a}{x}}{\dfrac{a^2}{2x}}$

9. $\dfrac{\dfrac{x}{y^2}}{\dfrac{x^2}{y}}$ **10.** $\dfrac{\dfrac{p^4}{r}}{\dfrac{p^2}{r^2}}$ **11.** $\dfrac{\dfrac{4a^4b^3}{3a}}{\dfrac{2ab^4}{b^2}}$ **12.** $\dfrac{\dfrac{2r^4t^2}{3t}}{\dfrac{5r^2t^5}{3r}}$

13. $\dfrac{\dfrac{m + 2}{3}}{\dfrac{m - 4}{m}}$ **14.** $\dfrac{\dfrac{q - 5}{q}}{\dfrac{q + 5}{3}}$ **15.** $\dfrac{\dfrac{2}{x} - 3}{2 - 3x}$ **16.** $\dfrac{6 + \dfrac{2}{r}}{\dfrac{3r + 1}{4}}$

17. $\dfrac{\dfrac{1}{x} + x}{\dfrac{x^2 + 1}{8}}$ **18.** $\dfrac{\dfrac{3}{m} - m}{\dfrac{3 - m^2}{4}}$ **19.** $\dfrac{a - \dfrac{5}{a}}{a + \dfrac{1}{a}}$ **20.** $\dfrac{q + \dfrac{1}{q}}{q + \dfrac{4}{q}}$

21. $\dfrac{\dfrac{5}{8} + \dfrac{2}{3}}{\dfrac{7}{3} - \dfrac{1}{4}}$ **22.** $\dfrac{\dfrac{6}{5} - \dfrac{1}{9}}{\dfrac{2}{5} + \dfrac{5}{3}}$ **23.** $\dfrac{\dfrac{1}{x^2} + \dfrac{1}{y^2}}{\dfrac{1}{x} - \dfrac{1}{y}}$ **24.** $\dfrac{\dfrac{1}{a^2} - \dfrac{1}{b^2}}{\dfrac{1}{a} - \dfrac{1}{b}}$

25. $\dfrac{\dfrac{2}{p^2} - \dfrac{3}{5p}}{\dfrac{4}{p} + \dfrac{1}{4p}}$ 26. $\dfrac{\dfrac{2}{m^2} - \dfrac{3}{m}}{\dfrac{2}{5m^2} + \dfrac{1}{3m}}$ 27. $\dfrac{\dfrac{5}{x^2 y} - \dfrac{2}{xy^2}}{\dfrac{3}{x^2 y^2} + \dfrac{4}{xy}}$ 28. $\dfrac{\dfrac{1}{m^3 p} + \dfrac{2}{mp^2}}{\dfrac{4}{mp} + \dfrac{1}{m^2 p}}$

29. $\dfrac{\dfrac{1}{4} - \dfrac{1}{a^2}}{\dfrac{1}{2} + \dfrac{1}{a}}$ 30. $\dfrac{\dfrac{1}{9} - \dfrac{1}{m^2}}{\dfrac{1}{3} + \dfrac{1}{m}}$ 31. $\dfrac{\dfrac{1}{z + 5}}{\dfrac{4}{z^2 - 25}}$ 32. $\dfrac{\dfrac{1}{a + 1}}{\dfrac{2}{a^2 - 1}}$

33. $\dfrac{\dfrac{1}{m + 1} - 1}{\dfrac{1}{m + 1} + 1}$ 34. $\dfrac{\dfrac{2}{x - 1} + 2}{\dfrac{2}{x - 1} - 2}$

35. In a fraction, what operation does the fraction bar represent?

36. What property of real numbers justifies Method 2 of simplifying complex fractions?

◆ **MATHEMATICAL CONNECTIONS** (Exercises 37–40) ◆

In order to find the average of two numbers, we add them and divide by 2. Suppose that we wish to find the average of 3/8 and 5/6. Work Exercises 37–40, in order, to see how a complex fraction occurs in a problem like this.

37. Write in symbols: the sum of $\dfrac{3}{8}$ and $\dfrac{5}{6}$, divided by 2. Your result should be a complex fraction.

38. Simplify the complex fraction from Exercise 37 using Method 1.

39. Simplify the complex fraction from Exercise 37 using Method 2.

40. Your answers in Exercises 38 and 39 should be the same. Which method did you prefer? Why?

◆

Simplify the continued fraction.

41. $1 + \dfrac{1}{1 + \dfrac{1}{1 + 1}}$ 42. $5 + \dfrac{5}{5 + \dfrac{5}{5 + 5}}$ 43. $7 - \dfrac{3}{5 + \dfrac{2}{4 - 2}}$

44. $3 - \dfrac{2}{4 + \dfrac{2}{4 - 2}}$ 45. $r + \dfrac{r}{4 - \dfrac{2}{6 + 2}}$ 46. $\dfrac{2q}{7} - \dfrac{q}{6 + \dfrac{8}{4 + 4}}$

The given complex fraction is a bit more involved than the ones found earlier in this exercise set. Simplify it using either method.

47. $\dfrac{\dfrac{1}{m - 1} + \dfrac{2}{m + 2}}{\dfrac{2}{m + 2} - \dfrac{1}{m - 3}}$ 48. $\dfrac{\dfrac{5}{r + 3} - \dfrac{1}{r - 1}}{\dfrac{2}{r + 2} + \dfrac{3}{r + 3}}$

REVIEW EXERCISES

Use the distributive property to simplify. See Section 2.1.

49. $9\left(\dfrac{4x}{3} + \dfrac{2}{9}\right)$ 50. $8\left(\dfrac{3r}{4} + \dfrac{9}{8}\right)$ 51. $-12\left(\dfrac{11p^2}{3} - \dfrac{9p}{4}\right)$ 52. $6\left(\dfrac{5z^2}{2} + \dfrac{8z}{3}\right)$

Solve the equation. See Sections 2.3 and 4.5.

53. $3x + 5 = 7x + 3$

54. $9z + 2 = 7z + 6$

55. $6(z - 3) + 5 = 8z - 3$

56. $k^2 + 3k - 4 = 0$

5.6 EQUATIONS INVOLVING RATIONAL EXPRESSIONS

FOR EXTRA HELP	OBJECTIVES
📖 **SSG** pp. 183–188 **SSM** pp. 270–279	**1 ▶** Distinguish between expressions with rational coefficients and equations with terms that are rational expressions.
📼 **Video** 8	**2 ▶** Solve equations with rational expressions.
💾 **Tutorial** IBM MAC	**3 ▶** Solve a formula for a specified variable.

In Section 2.3 we learned how to solve equations with fractions as coefficients. By using the multiplication property of equality we cleared the fractions by multiplying through by the LCD. In this section we examine this procedure in more detail.

1 ▶ Distinguish between expressions with rational coefficients and equations with terms that are rational expressions.

▶   Before solving equations with rational expressions, it is necessary to understand the difference between *sums* and *differences* of terms with rational coefficients, and *equations* with terms that are rational expressions. Sums and differences are *simplified*, while equations are *solved*.

EXAMPLE 1

Distinguishing Between Expressions and Equations

Identify each of the following as an expression or an equation. If it is an expression, simplify it. If it is an equation, solve it.

(a) $\dfrac{3}{4}x - \dfrac{2}{3}x$

This is a difference of two terms, so it is an expression. (There is no equals sign.) Simplify by finding the LCD, writing each cofficient with this LCD, and combining like terms.

$$\frac{3}{4}x - \frac{2}{3}x = \frac{9}{12}x - \frac{8}{12}x \qquad \text{Get a common denominator.}$$

$$= \frac{1}{12}x \qquad \text{Combine like terms.}$$

(b) $\dfrac{3}{4}x - \dfrac{2}{3}x = \dfrac{1}{2}$

Because of the equals sign, this is an equation to be solved. We proceed as in Section 2.3, using the multiplication property of equality to clear fractions. The LCD is 12.

$$\frac{3}{4}x - \frac{2}{3}x = \frac{1}{2}$$

$$12\left(\frac{3}{4}x - \frac{2}{3}x\right) = 12\left(\frac{1}{2}\right) \qquad \text{Multiply by 12.}$$

$$12\left(\frac{3}{4}x\right) - 12\left(\frac{2}{3}x\right) = 12\left(\frac{1}{2}\right) \qquad \text{Distributive property}$$

$$9x - 8x = 6 \qquad \text{Multiply.}$$

$$x = 6 \qquad \text{Combine like terms.}$$

The solution, 6, should be checked in the original equation. ■

The ideas of Example 1 can be summarized as follows.

When adding or subtracting, the LCD must be kept throughout the simplification. When solving an equation, the LCD is used to multiply both sides so that denominators are eliminated.

◆▷ **CONNECTIONS** ◁◆

The plaintiff's rate of success in personal injury cases has been decreasing dramatically in recent years. The expression

$$\frac{893.5}{x + 14.2}$$

gives the success rate, in percent, of plaintiffs in such suits for the years 1989 through 1992. In the expression, x represents the number of years since 1989. In 1989, for example, $x = 0$ and the expression equals

$$\frac{893.5}{0 + 14.2} = \frac{893.5}{14.2} \approx 63 \quad \text{or} \quad 63\%.$$

Rational expressions are often used to model situations like this, where the value of the expression decreases as the variable value increases.

If we call the expression P and write

$$P = \frac{893.5}{x + 14.2},$$

we have an equation. Given values of P, we could then solve for x.

FOR DISCUSSION OR WRITING

1. Find values of the expression for the years 1990, 1991, and 1992. In what year does the equation predict that the rate of success will fall to 40% (let $P = 40$).
2. Use the expression to estimate the success rate for plaintiffs in 1995. What problems do you think might occur when the expression, which was modeled from data from 1989 through 1992, is used for later years? Are the success rates decreasing evenly, year by year? If not, what seems to be happening?

2 ► Solve equations with rational expressions.

► The next few examples illustrate the method of clearing an equation of fractions in order to solve it.

EXAMPLE 2

Solving an Equation Involving Rational Expressions

■ Solve $\dfrac{p}{2} - \dfrac{p-1}{3} = 1$.

Multiply both sides by the LCD, 6.

$$6\left(\frac{p}{2} - \frac{p-1}{3}\right) = 6 \cdot 1$$

$$6\left(\frac{p}{2}\right) - 6\left(\frac{p-1}{3}\right) = 6 \qquad \text{Distributive property}$$

$$3p - 2(p-1) = 6$$

Be careful to put parentheses around $p-1$; otherwise you may get an incorrect solution.

Continue simplifying and solve the equation.

$$3p - 2p + 2 = 6 \qquad \text{Distributive property}$$

$$p + 2 = 6 \qquad \text{Combine like terms.}$$

$$p = 4 \qquad \text{Subtract 2.}$$

Check to see that 4 is correct by replacing p with 4 in the original equation. ■

CAUTION The most common error in equations like the one found in Example 2 occurs when parentheses are not used for the numerator $p-1$ in the second fraction. If parentheses are not used, the sign of the last term on the left side will be incorrect.

The equations in Example 1(b) and Example 2 did not have variables in denominators. When solving equations that have a variable in a denominator, remember that the number 0 cannot be used as a denominator. Therefore, the solution cannot be a number that will make the denominator equal 0. Example 3 illustrates this.

EXAMPLE 3

Solving an Equation Involving Rational Expressions with No Solution

■ Solve $\dfrac{x}{x-2} = \dfrac{2}{x-2} + 2$.

Multiply both sides by the LCD, $x - 2$.

$$(x-2)\left(\frac{x}{x-2}\right) = (x-2)\left(\frac{2}{x-2}\right) + (x-2)(2)$$

$$x = 2 + 2x - 4 \qquad \text{Distributive property}$$

$$x = -2 + 2x \qquad \text{Combine terms.}$$

$$-x = -2 \qquad \text{Subtract } 2x.$$

$$x = 2 \qquad \text{Multiply by } -1.$$

The proposed solution is 2. If we substitute 2 into the original equation, we get

$$\frac{2}{2-2} = \frac{2}{2-2} + 2 \qquad ?$$

$$\frac{2}{0} = \frac{2}{0} + 2. \qquad ?$$

Notice that 2 makes both denominators equal to 0. Because 0 cannot be the denominator of a fraction, *this equation has no solution.* ■

While it is always a good idea to check solutions to guard against arithmetic and algebraic errors, it is *essential* to check proposed solutions when variables appear in denominators in the original equation. Some students like to determine which numbers cannot be solutions *before* solving the equation.

Solving Equations with Rational Expressions

Step 1 Multiply both sides of the equation by the least common denominator. (This clears the equation of fractions.)

Step 2 Solve the resulting equation.

Step 3 Check each proposed solution by substituting it in the original equation. Reject any that cause a denominator to equal 0.

EXAMPLE 4

Solving an Equation Involving Rational Expressions

■ Solve $\dfrac{2}{x^2 - x} = \dfrac{1}{x^2 - 1}$.

Begin by finding a least common denominator. Since $x^2 - x$ can be factored as $x(x - 1)$, and $x^2 - 1$ can be factored as $(x + 1)(x - 1)$, the least common denominator is $x(x + 1)(x - 1)$.

$$\frac{2}{x(x - 1)} = \frac{1}{(x + 1)(x - 1)} \qquad \text{Factor the denominators.}$$

Notice that 0, -1, and 1 cannot be solutions of this equation. Multiply both sides of the equation by $x(x + 1)(x - 1)$.

$$x(x + 1)(x - 1)\frac{2}{x(x - 1)} = x(x + 1)(x - 1)\frac{1}{(x + 1)(x - 1)}.$$

$$2(x + 1) = x$$

$$2x + 2 = x \qquad \text{Distributive property}$$

$$2 = -x \qquad \text{Subtract } 2x.$$

$$x = -2 \qquad \text{Multiply by } -1.$$

The proposed solution is -2, which does not make any denominator equal 0. A check will verify that no arithmetic or algebraic errors have been made. Since -2 satisfies the equation, it is the solution. ■

EXAMPLE 5

Solving an Equation Involving Rational Expressions

■ Solve $\dfrac{1}{x - 1} + \dfrac{1}{2} = \dfrac{2}{x^2 - 1}$.

Factor the denominator on the right.

$$\frac{1}{x - 1} + \frac{1}{2} = \frac{2}{(x + 1)(x - 1)}$$

Notice that 1 and -1 cannot be solutions of this equation. Multiply both sides of the equation by the LCD, $2(x + 1)(x - 1)$.

$$2(x + 1)(x - 1)\left(\frac{1}{x - 1} + \frac{1}{2}\right) = 2(x + 1)(x - 1)\frac{2}{(x + 1)(x - 1)}$$

$$2(x + 1)(x - 1)\frac{1}{x - 1} + 2(x + 1)(x - 1)\frac{1}{2} = 2(x + 1)(x - 1)\frac{2}{(x + 1)(x - 1)}$$

$$2(x + 1) + (x + 1)(x - 1) = 4$$

$$2x + 2 + x^2 - 1 = 4 \qquad \text{Distributive property}$$

$$x^2 + 2x + 1 = 4$$

$$x^2 + 2x - 3 = 0 \qquad \text{Get 0 on the right side.}$$

Factoring gives

$$(x + 3)(x - 1) = 0.$$

$$x + 3 = 0 \qquad \text{or} \qquad x - 1 = 0 \qquad \text{Zero-factor property}$$

$$x = -3 \qquad \text{or} \qquad x = 1$$

-3 and 1 are proposed solutions. However, as noted above, 1 makes an original denominator equal to 0, so 1 is not a solution. However, by substituting -3 for x in the original equation, we find that -3 is a solution. ∎

EXAMPLE 6

Solving an Equation Involving Rational Expressions

∎ Solve $\dfrac{1}{k^2 + 4k + 3} + \dfrac{1}{2k + 2} = \dfrac{3}{4k + 12}$.

Factoring each denominator gives the equation

$$\frac{1}{(k + 1)(k + 3)} + \frac{1}{2(k + 1)} = \frac{3}{4(k + 3)}.$$

The LCD is $4(k + 1)(k + 3)$, indicating that -1 and -3 cannot be solutions of the equation. Multiply both sides by this LCD.

$$4(k + 1)(k + 3)\left(\frac{1}{(k + 1)(k + 3)} + \frac{1}{2(k + 1)}\right) = 4(k + 1)(k + 3)\frac{3}{4(k + 3)}$$

$$4(k + 1)(k + 3)\frac{1}{(k + 1)(k + 3)} + 2 \cdot 2(k + 1)(k + 3)\frac{1}{2(k + 1)}$$

$$= 4(k + 1)(k + 3)\frac{3}{4(k + 3)}$$

$$4 + 2(k + 3) = 3(k + 1)$$

$$4 + 2k + 6 = 3k + 3 \qquad \text{Distributive property}$$

$$2k + 10 = 3k + 3$$

$$10 - 3 = 3k - 2k$$

$$7 = k$$

The proposed solution, 7, does not make an original denominator equal to zero. A check shows that the algebra is correct, so 7 is the solution of the equation. ∎

3 ▶ Solve a formula for a specified variable.

▶ Solving a formula for a specified variable was discussed in Chapter 2. In the next example, this procedure is applied to a formula involving fractions.

EXAMPLE 7
Solving for a Specified Variable

■ Solve the formula $\dfrac{1}{a} = \dfrac{1}{b} + \dfrac{1}{c}$ for c.

The LCD of all the fractions in the equation is abc, so multiply both sides by abc.

$$abc\left(\frac{1}{a}\right) = abc\left(\frac{1}{b} + \frac{1}{c}\right)$$

$$abc\left(\frac{1}{a}\right) = abc\left(\frac{1}{b}\right) + abc\left(\frac{1}{c}\right) \qquad \text{Distributive property}$$

$$bc = ac + ab$$

Since we are solving for c, get all terms with c on one side of the equation. Do this by subtracting ac from both sides.

$$bc - ac = ab \qquad \text{Subtract } ac.$$

Factor out the common factor c on the left.

$$c(b - a) = ab \qquad \text{Factor out } c.$$

Finally, divide both sides by the coefficient of c, which is $b - a$.

$$c = \frac{ab}{b - a} \qquad ■$$

CAUTION Students often have trouble in the step that involves factoring out the variable for which they are solving. In Example 7, we had to factor out c on the left side so that we could divide both sides by $b - a$.

When solving an equation for a specified variable, be sure that the specified variable appears on only one side of the equals sign in the final equation.

5.6 EXERCISES

1. Which one of the following numbers satisfies $\dfrac{y - 1}{2} - \dfrac{y - 3}{4} = 1$?

   (a) 9   **(b)** 0   **(c)** 6   **(d)** 3

2. Which one of the following numbers satisfies $\dfrac{r + 5}{3} - \dfrac{r - 1}{4} = \dfrac{7}{4}$?

   (a) $-\dfrac{61}{4}$ **(b)** $-\dfrac{85}{4}$ **(c)** -2 **(d)** 4

3. What values of x cannot possibly be solutions of the equation $\dfrac{1}{x - 4} = \dfrac{3}{2x}$?

4. What is wrong with the following problem? "Solve $\dfrac{2}{3x} + \dfrac{1}{5x}$."

Identify as an expression or an equation. If it is an expression, simplify it. If it is an equation, solve it. See Example 1.

5. $\dfrac{7}{8}x + \dfrac{1}{5}x$

6. $\dfrac{4}{7}x + \dfrac{3}{5}x$

7. $\dfrac{7}{8}x + \dfrac{1}{5}x = 1$

8. $\dfrac{4}{7}x + \dfrac{3}{5}x = 1$

9. $\dfrac{3}{5}y - \dfrac{7}{10}y$

10. $\dfrac{2}{3}y - \dfrac{7}{4}y$

11. $\dfrac{3}{5}y - \dfrac{7}{10}y = 1$ **12.** $\dfrac{2}{3}y - \dfrac{7}{4}y = -13$

13. Explain how the LCD is used in a different way when adding and subtracting rational expressions compared to solving equations with rational expressions.

14. If we multiply both sides of the equation $\dfrac{6}{x+5} = \dfrac{6}{x+5}$ by $x+5$, we get $6 = 6$. Are all real numbers solutions of this equation? Explain.

Solve the equation and check your answer. See Examples 1(b), 2, and 3.

15. $\dfrac{5}{m} - \dfrac{3}{m} = 8$ **16.** $\dfrac{4}{y} + \dfrac{1}{y} = 2$ **17.** $\dfrac{5}{y} + 4 = \dfrac{2}{y}$ **18.** $\dfrac{11}{q} = 3 - \dfrac{1}{q}$

19. $\dfrac{p}{3} - \dfrac{p}{6} = 4$ **20.** $\dfrac{x}{15} + \dfrac{x}{5} = 4$ **21.** $\dfrac{3x}{5} - 6 = x$ **22.** $\dfrac{5t}{4} + t = 9$

23. $\dfrac{4m}{7} + m = 11$ **24.** $a - \dfrac{3a}{2} = 1$ **25.** $\dfrac{z-1}{4} = \dfrac{z+3}{3}$

26. $\dfrac{r-5}{2} = \dfrac{r+2}{3}$ **27.** $\dfrac{3p+6}{8} = \dfrac{3p-3}{16}$ **28.** $\dfrac{2z+1}{5} = \dfrac{7z+5}{15}$

29. $\dfrac{2x+3}{x} = \dfrac{3}{2}$ **30.** $\dfrac{5-2y}{y} = \dfrac{1}{4}$ **31.** $\dfrac{k}{k-4} - 5 = \dfrac{4}{k-4}$

32. $\dfrac{-5}{a+5} = \dfrac{a}{a+5} + 2$ **33.** $\dfrac{q+2}{3} + \dfrac{q-5}{5} = \dfrac{7}{3}$ **34.** $\dfrac{t}{6} + \dfrac{4}{3} = \dfrac{t-2}{3}$

35. $\dfrac{x}{2} = \dfrac{5}{4} + \dfrac{x-1}{4}$ **36.** $\dfrac{8p}{5} = \dfrac{3p-4}{2} + \dfrac{5}{2}$

Solve the equation. Be very careful with signs. See Example 2.

37. $\dfrac{a+7}{8} - \dfrac{a-2}{3} = \dfrac{4}{3}$ **38.** $\dfrac{x+3}{7} - \dfrac{x+2}{6} = \dfrac{1}{6}$ **39.** $\dfrac{p}{2} - \dfrac{p-1}{4} = \dfrac{5}{4}$

40. $\dfrac{r}{6} - \dfrac{r-2}{3} = -\dfrac{4}{3}$ **41.** $\dfrac{3x}{5} - \dfrac{x-5}{7} = 3$ **42.** $\dfrac{8k}{5} - \dfrac{3k-4}{2} = \dfrac{5}{2}$

Solve the equation and check your answer. See Examples 4–6.

43. $\dfrac{2}{m} = \dfrac{m}{5m+12}$ **44.** $\dfrac{x}{4-x} = \dfrac{2}{x}$

45. $\dfrac{-2}{z+5} + \dfrac{3}{z-5} = \dfrac{20}{z^2-25}$ **46.** $\dfrac{3}{r+3} - \dfrac{2}{r-3} = \dfrac{-12}{r^2-9}$

47. $\dfrac{3}{x-1} + \dfrac{2}{4x-4} = \dfrac{7}{4}$ **48.** $\dfrac{2}{p+3} + \dfrac{3}{8} = \dfrac{5}{4p+12}$

49. $\dfrac{y}{3y+3} = \dfrac{2y-3}{y+1} - \dfrac{2y}{3y+3}$ **50.** $\dfrac{2k+3}{k+1} - \dfrac{3k}{2k+2} = \dfrac{-2k}{2k+2}$

51. $\dfrac{5x}{14x+3} = \dfrac{1}{x}$ **52.** $\dfrac{m}{8m+3} = \dfrac{1}{3m}$

53. $\dfrac{2}{x-1} - \dfrac{2}{3} = \dfrac{-1}{x+1}$ **54.** $\dfrac{5}{p-2} = 7 - \dfrac{10}{p+2}$

55. $\dfrac{x}{2x+2} = \dfrac{-2x}{4x+4} + \dfrac{2x-3}{x+1}$ **56.** $\dfrac{5t+1}{3t+3} = \dfrac{5t-5}{5t+5} + \dfrac{3t-1}{t+1}$

57. If you are solving a formula for the letter k, and your steps lead you to the equation $kr - mr = km$, what would be your next step?

58. If you are solving a formula for the letter k, and your steps lead you to the equation $kr - km = mr$, what would be your next step?

Solve the formula for the specified variable. See Example 7.

59. $m = \dfrac{kF}{a}$ for F

60. $I = \dfrac{kE}{R}$ for E

61. $m = \dfrac{kF}{a}$ for a

62. $I = \dfrac{kE}{R}$ for R

63. $I = \dfrac{E}{R + r}$ for R

64. $I = \dfrac{E}{R + r}$ for r

65. $h = \dfrac{2A}{B + b}$ for A

66. $d = \dfrac{2S}{n(a + L)}$ for S

67. $d = \dfrac{2S}{n(a + L)}$ for a

68. $h = \dfrac{2A}{B + b}$ for B

69. $\dfrac{1}{x} = \dfrac{1}{y} - \dfrac{1}{z}$ for y

70. $\dfrac{3}{k} = \dfrac{1}{p} + \dfrac{1}{q}$ for q

71. $9x + \dfrac{3}{z} = \dfrac{5}{y}$ for z

72. $\dfrac{1}{a} = \dfrac{1}{b} + \dfrac{1}{c}$ for a

Solve the equation and check your answer. See Examples 5 and 6.

73. $\dfrac{8x + 3}{x} = 3x$

74. $\dfrac{2}{y} = \dfrac{y}{5y - 12}$

75. $\dfrac{3y}{y^2 + 5y + 6} = \dfrac{5y}{y^2 + 2y - 3} - \dfrac{2}{y^2 + y - 2}$

76. $\dfrac{m}{m^2 + m - 2} + \dfrac{m}{m^2 - 1} = \dfrac{m}{m^2 + 3m + 2}$

77. $\dfrac{x + 4}{x^2 - 3x + 2} - \dfrac{5}{x^2 - 4x + 3} = \dfrac{x - 4}{x^2 - 5x + 6}$

78. $\dfrac{3}{r^2 + r - 2} - \dfrac{1}{r^2 - 1} = \dfrac{7}{2(r^2 + 3r + 2)}$

◆━━━━ **MATHEMATICAL CONNECTIONS** (Exercises 79–84) ◆━━━━

In Section 5.4 we saw how, after adding or subtracting two rational expressions, the result can often be simplified to lowest terms. In this section we learned that multiplying both sides of an equation by the least common denominator may lead to a solution that must be rejected because it causes a denominator to equal 0.

Work Exercises 79–84 in order to see a connection between these two concepts.

79. Solve the equation $\dfrac{x^2}{x - 3} + \dfrac{2x - 15}{x - 3} = 0$ by multiplying both sides by $x - 3$. What is the solution? What is the number that must be rejected as a solution?

80. Combine the two rational expressions on the left side of the given equation in Exercise 79 and reduce to lowest terms.
 (a) If you set the simplified form equal to 0 and solve, how does your solution compare to the actual solution in Exercise 79?
 (b) If you set the common factor that you divided out equal to 0 and solve, how does your solution compare to the rejected solution in Exercise 79?

81. Consider the equation $\dfrac{1}{x - 1} + \dfrac{1}{2} - \dfrac{2}{x^2 - 1} = 0$, which is equivalent to the equation solved in Example 5. According to Example 5 what is the solution? What is the number that must be rejected as a solution?

82. Combine the three rational expressions on the left side of the given equation in Exercise 81 and reduce to lowest terms. Repeat parts (a) and (b) of Exercise 80, comparing this time to the equation in Exercise 81.

83. Write a short paragraph summarizing what you have learned in working Exercises 79–82.

84. Devise a procedure whereby you could solve an equation involving rational expressions which would not lead to values that must be rejected.

———————◆———————

REVIEW EXERCISES

Write a mathematical expression for each exercise. See Section 2.8.

85. Eryn drives from Philadelphia to Pittsburgh, a distance of 288 miles, in t hours. Find her rate in miles per hour.

86. Chuck drives for 10 hours, traveling from City A to City B, a distance of d kilometers. Find his rate in kilometers per hour.

87. Natalie flies her small plane from St. Louis to Chicago, a distance of 289 miles, at z miles per hour. Find her time in hours.

88. Joshua can do a job in x hours. What portion of the job is done in 1 hour?

SUMMARY: RATIONAL EXPRESSIONS

We have seen how to perform the four operations of arithmetic with rational expressions and how to solve equations with rational expressions. The exercises in this summary provide an opportunity for you to work a mixed variety of problems of these types (as in Example 1 in Section 5.6). In so doing, recall the procedures explained in the earlier sections of this chapter. They are summarized here.

Multiplication of Rational Expressions	Multiply numerators and multiply denominators. Use the fundamental principle to express in lowest terms.
Division of Rational Expressions	First, change the second fraction to its reciprocal; then multiply as described above.
Addition of Rational Expressions	Find the least common denominator (LCD) if necessary. Add numerators, and keep the same denominator. Express in lowest terms.
Subtraction of Rational Expressions	Find the LCD if necessary. Subtract numerators (use parentheses as required), and keep the same denominator. Express in lowest terms.
Solving Equations with Rational Expressions	Multiply both sides of the equation by the LCD of all the rational expressions in the equation. Solve, using methods described in earlier chapters. Be sure to check all proposed solutions and reject any that cause a denominator to equal 0.

A common error that was mentioned in Section 5.6 bears repeating here. Students often confuse *operations* on rational expressions with the *solution of equations* with rational expressions. For example, the four possible operations on the rational expressions $\frac{1}{x}$ and $\frac{1}{x-2}$ can be performed as follows.

Add:

$$\frac{1}{x} + \frac{1}{x-2} = \frac{x-2}{x(x-2)} + \frac{x}{x(x-2)} \qquad \text{Write with a common denominator.}$$

$$= \frac{x-2+x}{x(x-2)} \qquad \text{Add numerators; keep the same denominator.}$$

$$= \frac{2x-2}{x(x-2)} \qquad \text{Combine like terms.}$$

Subtract:

$$\frac{1}{x} - \frac{1}{x-2} = \frac{x-2}{x(x-2)} - \frac{x}{x(x-2)} \qquad \text{Write with a common denominator.}$$

$$= \frac{x-2-x}{x(x-2)} \qquad \text{Subtract numerators; keep the same denominator.}$$

$$= \frac{-2}{x(x-2)} \qquad \text{Combine like terms.}$$

Multiply:

$$\frac{1}{x} \cdot \frac{1}{x-2} = \frac{1}{x(x-2)} \qquad \text{Multiply numerators and multiply denominators.}$$

Divide:

$$\frac{1}{x} \div \frac{1}{x-2} = \frac{1}{x} \cdot \frac{x-2}{1} = \frac{x-2}{x} \qquad \text{Change to multiplication by the reciprocal of the second fraction.}$$

On the other hand, consider the *equation*

$$\frac{1}{x} + \frac{1}{x-2} = \frac{3}{4}.$$

First notice that neither 0 nor 2 could possibly be solutions of this equation, since each will cause a denominator to equal 0. We can use the multiplication property of equality by multiplying both sides by the LCD, $4x(x-2)$, giving an equation with no denominators.

$$4x(x-2)\frac{1}{x} + 4x(x-2)\frac{1}{x-2} = 4x(x-2)\frac{3}{4}$$

$$4x - 8 + 4x = 3x^2 - 6x \qquad \text{Distributive property}$$

$$0 = 3x^2 - 14x + 8 \qquad \text{Get 0 on one side.}$$

$$0 = (3x-2)(x-4) \qquad \text{Factor.}$$

$$3x - 2 = 0 \quad \text{or} \quad x - 4 = 0 \qquad \text{Zero-factor property}$$

$$x = \frac{2}{3} \quad \text{or} \quad x = 4$$

Both 2/3 and 4 are solutions since neither makes a denominator equal to zero.

In conclusion, remember the following points when working exercises involving rational expressions:

1. The fundamental principle is applied only after numerators and denominators have been *factored*.
2. When adding and subtracting rational expressions, the common denominator must be kept throughout the problem and in the final result.
3. Always look to see if the answer is in lowest terms; if it is not, use the fundamental principle.
4. When solving equations with rational expressions, reject any proposed solution that causes an original denominator to equal 0.

EXERCISES

For the given exercise, decide whether an operation should be performed or whether an equation should be solved. Then perform the operation or solve the equation as the case may be.

1. $\dfrac{4}{p} + \dfrac{6}{p}$

2. $\dfrac{x^3 y^2}{x^2 y^4} \cdot \dfrac{y^5}{x^4}$

3. $\dfrac{1}{x^2 + x - 2} \div \dfrac{4x^2}{2x - 2}$

4. $\dfrac{8}{m - 5} = 2$

5. $\dfrac{2y^2 + y - 6}{2y^2 - 9y + 9} \cdot \dfrac{y^2 - 2y - 3}{y^2 - 1}$

6. $\dfrac{2}{k^2 - 4k} + \dfrac{3}{k^2 - 16}$

7. $\dfrac{x - 4}{5} = \dfrac{x + 3}{6}$

8. $\dfrac{3t^2 - t}{6t^2 + 15t} \div \dfrac{6t^2 + t - 1}{2t^2 - 5t - 25}$

9. $\dfrac{4}{p + 2} + \dfrac{1}{3p + 6}$

10. $\dfrac{1}{y} + \dfrac{1}{y - 3} = -\dfrac{5}{4}$

11. $\dfrac{3}{t - 1} + \dfrac{1}{t} = \dfrac{7}{2}$

12. $\dfrac{6}{y} - \dfrac{2}{3y}$

13. $\dfrac{5}{4z} - \dfrac{2}{3z}$

14. $\dfrac{k + 2}{3} = \dfrac{2k - 1}{5}$

15. $\dfrac{1}{m^2 + 5m + 6} + \dfrac{2}{m^2 + 4m + 3}$

16. $\dfrac{2k^2 - 3k}{20k^2 - 5k} \div \dfrac{2k^2 - 5k + 3}{4k^2 + 11k - 3}$

17. $\dfrac{2}{x + 1} + \dfrac{5}{x - 1} = \dfrac{10}{x^2 - 1}$

18. $\dfrac{3}{x + 3} + \dfrac{4}{x + 6} = \dfrac{9}{x^2 + 9x + 18}$

19. $\dfrac{4t^2 - t}{6t^2 + 10t} \div \dfrac{8t^2 + 2t - 1}{3t^2 + 11t + 10}$

20. $\dfrac{x}{x - 2} + \dfrac{3}{x + 2} = \dfrac{8}{x^2 - 4}$

5.7 APPLICATIONS OF RATIONAL EXPRESSIONS

FOR EXTRA HELP

📖 **SSG** pp. 188–195
   SSM pp. 281–289

📼 **Video**
   8

💾 **Tutorial**
   IBM   MAC

OBJECTIVES

1 ▶ Solve problems about numbers using rational expressions.
2 ▶ Solve problems about distance using rational expressions.
3 ▶ Solve problems about work using rational expressions.
4 ▶ Solve problems about variation using rational expressions.

Every time we learn to solve a new type of equation, we are able to apply our knowledge to solving new types of applications. In Section 5.6 we learned how to solve equations involving rational expressions, and now we can solve applications that involve this type of equation. The problem-solving techniques of the earlier chapters still apply. The main difference between the problems of this section and those of earlier sections is that the equations for these problems involve rational expressions.

1 ▶ Solve problems about numbers using rational expressions.

▶ In order to prepare for more meaningful applications, we begin with an example about an unknown number.

EXAMPLE 1

Solving a Problem About an Unknown Number

If the same number is added to both the numerator and the denominator of the fraction 2/5, the result is 2/3. Find the number.

Let x = the number added to the numerator and the denominator. Then

$$\frac{2 + x}{5 + x}$$

represents the result of adding the same number to both the numerator and denominator. Since this result is 2/3,

$$\frac{2 + x}{5 + x} = \frac{2}{3}.$$

Solve this equation by multiplying both sides by the LCD, $3(5 + x)$.

$$3(5 + x)\frac{2 + x}{5 + x} = 3(5 + x)\frac{2}{3}$$
$$3(2 + x) = 2(5 + x)$$
$$6 + 3x = 10 + 2x$$
$$x = 4$$

Check the solution in the words of the original problem: if 4 is added to both the numerator and denominator of 2/5, the result is 6/9 = 2/3, as required. ■

2 ▶ Solve problems about distance using rational expressions.

▶ We now look at more meaningful applications that lead to equations with rational expressions.

PROBLEM SOLVING In Section 2.8 we first saw applications involving distance, rate, and time. Recall the importance of setting up a chart in these problems so that the information is summarized in an organized fashion. The next example shows how to solve a problem when a rational expression appears in the equation.

EXAMPLE 2
Solving a Problem About Distance, Rate, and Time

The Big Muddy River has a current of 3 miles per hour. A motorboat takes the same amount of time to go 12 miles downstream as it takes to go 8 miles upstream. What is the speed of the boat in still water?

This problem requires the distance formula,

$$d = rt \text{ (distance} = \text{rate} \times \text{time)}.$$

Let $x = $ the speed of the boat in still water. Since the current pushes the boat when the boat is going downstream, the speed of the boat downstream will be the sum of the speed of the boat and the speed of the current, or $x + 3$ miles per hour. Also, the boat's speed going upstream is $x - 3$ miles per hour. This, and other information given in the problem, is summarized in the chart.

	d	r	t
Downstream	12	x + 3	x
Upstream	8	x − 3	

Fill in the last column, representing time, by solving the formula $d = rt$ for t.

$$d = rt$$

$$\frac{d}{r} = t \qquad \text{Divide by } r.$$

For the time upstream,

$$\frac{d}{r} = \frac{8}{x - 3},$$

and for the time downstream,

$$\frac{d}{r} = \frac{12}{x + 3}.$$

Now complete the chart.

	d	r	t
Downstream	12	x + 3	$\frac{12}{x + 3}$
Upstream	8	x − 3	$\frac{8}{x - 3}$

Times are equal.

According to the original problem, the time upstream equals the time downstream. The two times from the chart must therefore be equal, giving the equation

$$\frac{12}{x + 3} = \frac{8}{x - 3}.$$

Solve this equation by multiplying both sides by $(x + 3)(x - 3)$.

$$(x + 3)(x - 3)\frac{12}{x + 3} = (x + 3)(x - 3)\frac{8}{x - 3}$$

$$12(x - 3) = 8(x + 3)$$

$$12x - 36 = 8x + 24 \qquad \text{Distributive property}$$
$$4x = 60 \qquad \text{Subtract } 8x; \text{ add } 36.$$
$$x = 15 \qquad \text{Divide by } 4.$$

The speed of the boat in still water is 15 miles per hour. Check this solution by first finding the speed of the boat downstream, which is $15 + 3 = 18$ miles per hour. Traveling 12 miles would take

$$d = rt$$
$$12 = 18t$$
$$t = \frac{2}{3} \text{ hour.}$$

On the other hand, the speed of the boat upstream is $15 - 3 = 12$ miles per hour, and traveling 8 miles would take

$$d = rt$$
$$8 = 12t$$
$$t = \frac{2}{3} \text{ hour.}$$

The time upstream equals the time downstream, as required.   ■

◆ **CONNECTIONS** ◆

Sometimes what seems to be the obvious answer is not correct. Consider the following problem where algebra is used to find the correct solution.

A car travels from A to B at 40 miles per hour and returns at 60 miles per hour. What is its average rate for the entire trip?

The correct answer is not 50 miles per hour! Remembering the distance-rate-time relationship and letting $x =$ the distance between A and B, we can simplify a complex fraction to find the correct answer.

$$\begin{aligned}
\text{Average rate for} &= \frac{\text{Total distance}}{\text{Total time}} \\
\text{entire trip} & \\
&= \frac{x + x}{\dfrac{x}{40} + \dfrac{x}{60}} \\
&= \frac{2x}{\dfrac{3x}{120} + \dfrac{2x}{120}} \\
&= \frac{2x}{\dfrac{5x}{120}} \\
&= 2x \cdot \frac{120}{5x} \\
&= 48
\end{aligned}$$

The average rate for the entire trip is 48 miles per hour.

FOR DISCUSSION OR WRITING

Find the average rate for the entire trip if the speed from A to B is 50 miles per hour. Repeat for 55 miles per hour. Will the result always be less than 1/2 the sum of the two speeds? What might account for this?

3 ▶ Solve problems about work using rational expressions.

▶ Suppose that you can mow your lawn in 4 hours. After 1 hour, you will have mowed 1/4 of the lawn. After 2 hours, you will have mowed 2/4 or 1/2 of the lawn, and so on. This idea is generalized as follows.

Rate of Work If a job can be completed in t units of time, then the rate of work is

$$\frac{1}{t} \text{ job per unit of time.}$$

PROBLEM SOLVING The relationship between problems involving work and problems involving distance is a very close one. Recall that the formula $d = rt$ says that distance traveled is equal to rate of travel multiplied by time traveled. Similarly, the fractional part of a job accomplished is equal to the rate of the work multiplied by the time worked. In the lawn mowing example, after 3 hours, the fractional part of the job done is

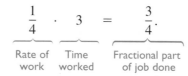

After 4 hours, $(1/4)(4) = 1$ whole job has been done.

These ideas are used in solving problems about the length of time needed to do a job. These problems are often called work problems.

EXAMPLE 3
Solving a Problem About Work

With a riding lawn mower, John, the groundskeeper in a large park, can cut the lawn in 8 hours. With a small mower, his assistant Walt needs 14 hours to cut the same lawn. If both John and Walt work on the lawn, how long will it take to cut it?

Let $x =$ the number of hours it will take for John and Walt to mow the lawn, working together.

Certainly, x will be less than 8, since John alone can mow the lawn in 8 hours. Begin by making a chart as shown. Remember that based on the previous discussion, John's rate alone is 1/8 job per hour, and Walt's rate is 1/14 job per hour.

	Rate	Time Working Together	Fractional Part of the Job Done When Working Together	
John	$\dfrac{1}{8}$	x	$\dfrac{1}{8}x$	← Sum is 1 whole job.
Walt	$\dfrac{1}{14}$	x	$\dfrac{1}{14}x$	←

Since together John and Walt complete 1 whole job, we must add their individual fractional parts and set the sum equal to 1.

$$\underset{\substack{\text{Fractional part}\\\text{done by John}}}{} + \underset{\substack{\text{Fractional part}\\\text{done by Walt}}}{} = 1 \text{ whole job}$$

$$\frac{1}{8}x + \frac{1}{14}x = 1$$

$$56\left(\frac{1}{8}x + \frac{1}{14}x\right) = 56(1) \qquad \text{Multiply by LCD, 56.}$$

$$56\left(\frac{1}{8}x\right) + 56\left(\frac{1}{14}x\right) = 56(1) \qquad \text{Distributive property}$$

$$7x + 4x = 56$$

$$11x = 56 \qquad \text{Combine like terms.}$$

$$x = \frac{56}{11} \qquad \text{Divide by 11.}$$

Working together, John and Walt can mow the lawn in 56/11 hours, or 5 1/11 hours. (Is this answer reasonable?) ■

4 ▶ Solve problems about variation using rational expressions.

▶ Suppose that gasoline costs $1.50 per gallon. Then 1 gallon costs $1.50, 2 gallons cost 2($1.50) = $3.00, 3 gallons cost 3($1.50) = $4.50, and so on. Each time, the total cost is obtained by multiplying the number of gallons by the price per gallon. In general, if k equals the price per gallon and x equals the number of gallons, then the total cost y is equal to kx. Notice that as number of gallons increases, total cost increases.

The preceding discussion is an example of variation. As in the gasoline example, two variables **vary directly** if one is a constant multiple of the other.

Direct Variation

y **varies directly** as x if there exists a constant k such that

$$y = kx.$$

EXAMPLE 4
Using Direct Variation

■ Suppose y varies directly as x, and $y = 20$ when $x = 4$. Find y when $x = 9$.
Since y varies directly as x, there is a constant k such that $y = kx$. Also, $y = 20$ when $x = 4$. Substituting these values into $y = kx$ gives

$$y = kx$$

$$20 = k \cdot 4, \qquad \text{Let } y = 20, x = 4.$$

from which $k = 5$. Since $y = kx$ and $k = 5$, $y = 5x$. When $x = 9$,

$$y = 5x = 5 \cdot 9 = 45.$$

Thus, $y = 45$ when $x = 9$. ■

There are other types of variation that involve fractions. Suppose that a rectangle has area 48 square units and length 24 units. Then its width is 2 units. However, if the area stays the same (48 square units) and the length decreases to 12 units, the width increases to 4 units. As length decreases, width increases. This is an example of **inverse variation:** $W = \dfrac{A}{L}$.

Other types of variation also are defined below, where k represents a constant.

Types of Variation

y varies directly as the square of x $\qquad y = kx^2$

m varies inversely as p $\qquad\qquad\quad m = \dfrac{k}{p}$

r varies inversely as the square of s $\quad r = \dfrac{k}{s^2}$

EXAMPLE 5
Using Inverse Variation

In a certain manufacturing process, the cost of producing an item varies inversely as the number of items produced. If 10,000 items are produced, the cost is $2 per item. Find the cost per item to produce 25,000 items.

Let x = the number of items produced and
c = the cost per item.

Since c varies inversely as x, there is a constant k such that

$$c = \frac{k}{x}.$$

Find k by replacing c with 2 and x with 10,000.

$$2 = \frac{k}{10,000}$$

$$20,000 = k \qquad \text{Multiply by 10,000.}$$

$$c = \frac{20,000}{25,000} = .80 \qquad \text{Let } k = 20,000 \text{ and } x = 25,000.$$

The cost per item to make 25,000 items is $.80. ■

5.7 EXERCISES

Set up the equation you would use to solve the problem. Do not actually solve.

1. One-third of a number is two more than one-sixth of the same number. What is the number? (Let x represent the number.)

2. The numerator of the fraction $\frac{13}{15}$ is increased by an amount so that the value of the resulting fraction is $\frac{7}{5}$. By what amount was the numerator increased? (Let t represent the amount.)

Solve the problem. See Example 1.

3. In a certain fraction, the denominator is 4 less than the numerator. If 3 is added to both the numerator and the denominator, the resulting fraction is equal to $\frac{3}{2}$. Find the original fraction.

4. The denominator of a certain fraction is 3 times the numerator. If 2 is added to the numerator and subtracted from the denominator, the resulting fraction is equal to 1. Find the original fraction.

5. A quantity, its $\frac{2}{3}$, its $\frac{1}{2}$, and its $\frac{1}{7}$, added together, become 33. What is the quantity? (From the *Rhind Mathematical Papyrus*)

6. A quantity, its $\frac{3}{4}$, its $\frac{1}{2}$, and its $\frac{1}{3}$, added together, become 93. What is the quantity?

7. Calgene, a company that sells genetically engineered tomatoes, expects these tomatoes to provide about $\frac{1}{2}$ of its total revenue in 1995. Production of oils will contribute about $\frac{1}{4}$ of its total revenue in 1995. If these two products produce $74.8 million that year, what will the total revenue be?

8. A recent news article stated that Germany invests $\frac{1}{10}$ as much as the United States in Mexico. If the total amount invested in the Mexican economy by these countries is $2 billion, how much does each country invest?

9. Japan invests about $\frac{3}{4}$ as much as Germany in the Mexican economy, according to a recent news article. If Japan and Germany invest a total of $.32 billion, how much does each of these countries invest?

10. Phillip Morris and Ford are among the top ten U.S. companies in sales. In 1992, Phillip Morris's sales were close to $\frac{1}{2}$ of Ford's sales. If total sales for the two companies were $151 billion, what were Ford's sales that year?

Set up the equation you would use to solve the problem. Do not actually solve.

11. Reynaldo flew his airplane 500 miles against the wind in the same time it took him to fly it 600 miles with the wind. If the speed of the wind was 10 miles per hour, what was the average speed of his plane? (Let x = speed of the plane in still air.)

	d	r	t
Against the wind	500	x − 10	
With the wind	600	x + 10	

12. Sam can row 4 miles per hour in still water. It takes as long to row 8 miles upstream as 24 miles downstream. How fast is the current? (Let x = speed of the current.)

	d	r	t
Upstream	8	4 − x	
Downstream	24	4 + x	

Solve the problem. See Example 2.

13. Suppose Amanda walks D miles at R miles per hour in the same time that Kenneth walks d miles at r miles per hour. Give an equation relating D, R, d, and r.

14. If a boat travels m miles per hour in still water, what is its rate when it travels upstream in a river with a current of 5 miles per hour? What is its rate downstream in the river?

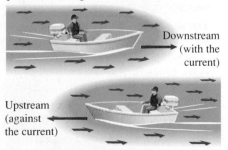

Downstream (with the current)

Upstream (against the current)

▦ **15.** In the 1988 Olympic Nordic Skiing, Gunde Svan of Sweden won the 50 kilometer event in 2.08 hours. What was his average speed?

▦ **16.** Bonnie Blair won the women's 500 meter speed skating in the 1988 Olympics. Her average speed was 12.79 meters per second. What was her time?

▦ **17.** Ibrahim Hussein of Kenya won the 1992 Boston marathon (a 26-mile race) in (about) .8 of an hour less time than the winner of the first Boston marathon in 1897, John J. McDermott of New York. Find each runner's speed, if McDermott's speed was (about) .73 times Hussein's speed. (*Hint:* Don't round until the end of the calculations.)

▦ **18.** Women first ran in the Boston marathon in 1972, when Nina Kuscsik of New York won the race. In 1992, the winner was Olga Markova of Russia, whose time was .8 of an hour less than Kuscsik's in 1972. If Markova ran $\frac{4}{3}$ as fast as Kuscsik, find each runner's speed. (See Exercise 17 for the distance.)

19. A boat can go 20 miles against a current in the same time that it can go 60 miles with the current. The current is 4 miles per hour. Find the speed of the boat in still water.

20. A plane flies 350 miles with the wind in the same time that it can fly 310 miles against the wind. The plane has a still-air speed of 165 miles per hour. Find the speed of the wind.

21. The distance from Seattle, Washington, to Victoria, British Columbia, is about 148 miles by ferry. It takes about 4 hours less to travel by the same ferry to Vancouver, British Columbia, a distance of about 74 miles. What is the average speed of the ferry?

22. Sandi Goldstein flew from Dallas to Indianapolis at 180 miles per hour and then flew back at 150 miles per hour. The trip at the slower speed took 1 hour longer than the trip at the higher speed. Find the distance between the two cities.

Set up the equation you would use to solve the problem. Do not actually solve.

23. Working alone, Jorge can paint a room in 8 hours. Caterina can paint the same room working alone in 6·hours. How long will it take them if they work together? (Let *x* represent the time working together.)

24. Edwin Bedford can tune up his Chevy in 2 hours working alone. His friend, Pat Navarre, can do the job in 3 hours working alone. How long would it take them if they worked together? (Let *t* represent the time working together.)

Solve the problem. See Example 3.

25. If it takes Elayn 10 hours to do a job, what is her rate?

26. If it takes Clay 12 hours to do a job, how much of the job does he do in 8 hours?

27. Geraldo and Luisa Hernandez operate a small cleaners. Luisa, working alone, can clean a day's laundry in 9 hours. Geraldo can clean a day's laundry in 8 hours. How long would it take them if they work together?

28. Lee can clean the house alone in 5 hours, while Tran needs 4 hours. How long will it take them to clean the house if they work together?

29. A pump can pump the water out of a flooded basement in 10 hours. A smaller pump takes 12 hours. How long would it take to pump the water from the basement using both pumps?

30. Doug Todd's copier can do a printing job in 7 hours. Scott's copier can do the same job in 12 hours. How long would it take to do the job using both copiers?

31. An experienced employee can enter tax data into a computer twice as fast as a new employee. Working together, it takes the employees 2 hours. How long would it take the experienced employee working alone?

32. One roofer can put a new roof on a house three times faster than another. Working together they can roof a house in 4 days. How long would it take the faster roofer working alone?

33. One pipe can fill a swimming pool in 6 hours, and another pipe can do it in 9 hours. How long will it take the two pipes working together to fill the pool $\frac{3}{4}$ full?

34. An inlet pipe can fill a swimming pool in 9 hours, and an outlet pipe can empty the pool in 12 hours. Through an error, both pipes are left open. How long will it take to fill the pool?

35. A cold water faucet can fill a sink in 12 minutes, and a hot water faucet can fill it in 15. The drain can empty the sink in 25 minutes. If both faucets are on and the drain is open, how long will it take to fill the sink?

36. Refer to Exercise 34. Assume the error was discovered after both pipes had been running for 3 hours, and the outlet pipe was then closed. How much more time would then be required to fill the pool? (*Hint:* How much of the job had been done when the error was discovered?)

Solve the problem involving variation. See Examples 4 and 5.

37. If x varies directly as y, and $x = 27$ when $y = 6$, find x when $y = 2$.

38. If z varies directly as x, and $z = 30$ when $x = 8$, find z when $x = 4$.

39. If m varies directly as p^2, and $m = 20$ when $p = 2$, find m when p is 5.

40. If a varies directly as b^2, and $a = 48$ when $b = 4$, find a when $b = 7$.

41. If p varies inversely as q^2, and $p = 4$ when $q = \frac{1}{2}$, find p when $q = \frac{3}{2}$.

42. If z varies inversely as x^2, and $z = 9$ when $x = \frac{2}{3}$, find z when $x = \frac{5}{4}$.

43. Assume that the constant of variation, k, is positive.
   (a) If y varies directly as x, then as y increases, x _____?_____ .
   (decreases/increases)
   (b) If y varies inversely as x, then as y increases, x _____?_____ .
   (decreases/increases)

44. (a) The more gasoline you pump, the more you will have to pay. Is this an example of direct or inverse variation?
   (b) The longer the term of your subscription to *Monitoring Times*, the less you will have to pay per year. Is this an example of direct or inverse variation?

Solve the problem. See Examples 4 and 5.

45. The interest on an investment varies directly as the rate of interest. If the interest is $48 when the interest rate is 5%, find the interest when the rate is 4.2%.

46. For a given base, the area of a triangle varies directly as its height. Find the area of a triangle with a height of 6 inches, if the area is 10 square inches when the height is 4 inches.

47. Over a specified distance, speed varies inversely with time. If a car goes a certain distance in one-half hour at 30 miles per hour, what speed is needed to go the same distance in three-fourths of an hour?

48. For a constant area, the length of a rectangle varies inversely as the width. The length of a rectangle is 27 feet when the width is 10 feet. Find the length of a rectangle with the same area if the width is 18 feet.

49. The weight of an object on the moon varies directly as the weight of the object on Earth. According to *The Guiness Book of World Records*, "Shad," a goat owned by a couple in California, is the largest known goat, weighing 352 pounds. Shad would weigh about 59 pounds on the moon. A bull deer weighing 1800 pounds was shot in Canada and is the largest confirmed deer. How much would the deer have weighed on the moon?

50. According to *The Guiness Book of World Records*, the longest recorded voyage in a paddle boat is 2226 miles in 103 days by the foot power of two boaters down the Mississippi River. Assuming a constant rate, how far would they have gone if they had traveled 120 days? (Distance varies directly as time.)

51. The pressure exerted by a certain liquid at a given point varies directly as the depth of the point beneath the surface of the liquid. The pressure at 10 feet is 50 pounds per square inch. What is the pressure at 20 feet?

52. If the volume is constant, the pressure of gas in a container varies directly as the temperature. If the pressure is 5 pounds per square inch at a temperature of 200 Kelvin, what is the pressure at a temperature of 300 Kelvin?

53. If the temperature is constant, the pressure of a gas in a container varies inversely as the volume of the container. If the pressure is 10 pounds per square foot in a container with 3 cubic feet, what is the pressure in a container with 1.5 cubic feet?

54. The force required to compress a spring varies directly as the change in the length of the spring. If a force of 12 pounds is required to compress a certain spring 3 inches, how much force is required to compress the spring 5 inches?

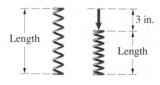

55. For a body falling freely from rest (disregarding air resistance), the distance the body falls varies directly as the square of the time. If an object is dropped from the top of a tower 400 feet high and hits the ground in 5 seconds, how far did it fall in the first 3 seconds?

56. The illumination produced by a light source varies inversely as the square of the distance from the source. If the illumination produced 4 feet from a light source is 75 foot-candles, find the illumination produced 9 feet from the same source.

REVIEW EXERCISES

*Find the value of y when (**a**) x = −2 and (**b**) x = 4. See Sections 1.3 and 2.3.*

57. $y = 5x + 3$ **58.** $y = 4 - 3x$ **59.** $6x - 2 = y$

60. $4x + 7y = 11$ **61.** $2x - 5y = 10$ **62.** $y + 3x = 8$

CHAPTER 5 SUMMARY

KEY TERMS

5.1 rational expression
lowest terms

5.3 least common
denominator
(LCD)

5.5 complex fraction

5.7 direct variation
inverse variation

QUICK REVIEW

CONCEPTS	EXAMPLES

5.1 THE FUNDAMENTAL PROPERTY OF RATIONAL EXPRESSIONS

To find the values for which a rational expression is not defined, set the denominator equal to zero and solve the equation.	Find the values for which the expression $$\frac{x-4}{x^2-16}$$ is not defined. $$x^2-16=0$$ $$(x-4)(x+4)=0$$ $$x-4=0 \quad \text{or} \quad x+4=0$$ $$x=4 \quad \text{or} \quad x=-4$$ The rational expression is not defined for 4 or -4.
To write a rational expression in lowest terms, (1) factor; and (2) use the fundamental property to remove common factors from the numerator and denominator.	Express $\dfrac{x^2-1}{(x-1)^2}$ in lowest terms. $$\frac{x^2-1}{(x-1)^2}=\frac{(x-1)(x+1)}{(x-1)(x-1)}$$ $$=\frac{x+1}{x-1}$$

5.2 MULTIPLICATION AND DIVISION OF RATIONAL EXPRESSIONS

Multiplication	Multiply: $\dfrac{3x+9}{x-5} \cdot \dfrac{x^2-3x-10}{x^2-9}.$
1. Factor.	$$=\frac{3(x+3)}{x-5} \cdot \frac{(x-5)(x+2)}{(x+3)(x-3)}$$
2. Multiply numerators and multiply denominators.	$$=\frac{3(x+3)(x-5)(x+2)}{(x-5)(x+3)(x-3)}$$
3. Write in lowest terms.	$$=\frac{3(x+2)}{x-3}$$
Division	Divide: $\dfrac{2x+1}{x+5} \div \dfrac{6x^2-x-2}{x^2-25}.$
1. Factor.	$$=\frac{2x+1}{x+5} \div \frac{(2x+1)(3x-2)}{(x+5)(x-5)}$$
2. Multiply the first rational expression by the reciprocal of the second.	$$=\frac{2x+1}{x+5} \cdot \frac{(x+5)(x-5)}{(2x+1)(3x-2)}$$
3. Write in lowest terms.	$$=\frac{x-5}{3x-2}$$

5.3 THE LEAST COMMON DENOMINATOR

Finding the LCD	Find the LCD for $\dfrac{3}{k^2-8k+16}$ and $\dfrac{1}{4k^2-16k}.$
1. Factor each denominator into prime factors.	$$k^2-8k+16=(k-4)^2$$ $$4k^2-16k=4k(k-4)$$

CONCEPTS	EXAMPLES
2. List each different factor the greatest number of times it appears. **3.** Multiply the factors from Step 2 to get the LCD.	$$\text{LCD} = (k - 4)^2 \cdot 4 \cdot k$$ $$= 4k(k - 4)^2$$
Writing a Rational Expression with the LCD as Denominator **1.** Factor both denominators.	Find the numerator: $\dfrac{5}{2z^2 - 6z} = \dfrac{}{4z^3 - 12z^2}$. $$\frac{5}{2z(z - 3)} = \frac{1}{4z^2(z - 3)}$$
2. Decide what factors the denominator must be multiplied by to equal the LCD.	$2z(z - 3)$ must be multiplied by $2z$.
3. Multiply the rational expression by that factor over itself (multiply by 1).	$$\frac{5}{2z(z - 3)} \cdot \frac{2z}{2z} = \frac{10z}{4z^2(z - 3)} = \frac{10z}{4z^3 - 12z^2}$$

5.4   ADDITION AND SUBTRACTION OF RATIONAL EXPRESSIONS

Adding Rational Expressions **1.** Find the LCD.	Add: $\dfrac{2}{3m + 6} + \dfrac{m}{m^2 - 4}$. $$3m + 6 = 3(m + 2)$$ $$m^2 - 4 = (m + 2)(m - 2)$$ The LCD is $3(m + 2)(m - 2)$.
2. Rewrite each rational expression with the LCD as denominator. **3.** Add the numerators to get the numerator of the sum. **4.** Write in lowest terms.	$$= \frac{2(m - 2)}{3(m + 2)(m - 2)} + \frac{3m}{3(m + 2)(m - 2)}$$ $$= \frac{2m - 4 + 3m}{3(m + 2)(m - 2)}$$ $$= \frac{5m - 4}{3(m + 2)(m - 2)}$$
Subtracting Rational Expressions Follow the same steps as for addition, but subtract in Step 3.	Subtract: $\dfrac{6}{k + 4} - \dfrac{2}{k}$. The LCD is $k(k + 4)$. $$\frac{6k}{(k + 4)k} - \frac{2(k + 4)}{k(k + 4)} = \frac{6k - 2(k + 4)}{k(k + 4)}$$ $$= \frac{6k - 2k - 8}{k(k + 4)} = \frac{4k - 8}{k(k + 4)}$$

5.5   COMPLEX FRACTIONS

Simplifying Complex Fractions *Method 1* Simplify the numerator and denominator separately. Then divide the simplified numerator by the simplified denominator.	Simplify. **(1)** $\dfrac{\dfrac{1}{a} - a}{1 - a} = \dfrac{\dfrac{1}{a} - \dfrac{a^2}{a}}{1 - a} = \dfrac{\dfrac{1 - a^2}{a}}{1 - a}$ $$= \frac{1 - a^2}{a} \cdot \frac{1}{1 - a}$$ $$= \frac{(1 - a)(1 + a)}{a(1 - a)} = \frac{1 + a}{a}$$

CONCEPTS	EXAMPLES
Method 2 Multiply numerator and denominator of the complex fraction by the LCD of all the denominators in the complex fraction.	(2) $\dfrac{\dfrac{1}{a} - a}{1 - a} = \dfrac{\dfrac{1}{a} - a}{1 - a} \cdot \dfrac{a}{a} = \dfrac{\dfrac{a}{a} - a^2}{(1 - a)a}$ $= \dfrac{1 - a^2}{(1 - a)a} = \dfrac{(1 + a)(1 - a)}{(1 - a)a}$ $= \dfrac{1 + a}{a}$

5.6   EQUATIONS INVOLVING RATIONAL EXPRESSIONS

Solving Equations with Rational Expressions	Solve $\dfrac{2}{x - 1} + \dfrac{3}{4} = \dfrac{5}{x - 1}$.
1. Find the LCD of all denominators in the equation.	The LCD is $4(x - 1)$. Note that 1 cannot be a solution.
2. Multiply each side of the equation by the LCD.	$4(x - 1)\left(\dfrac{2}{x - 1} + \dfrac{3}{4}\right) = 4(x - 1)\left(\dfrac{5}{x - 1}\right)$ $4(x - 1)\left(\dfrac{2}{x - 1}\right) + 4(x - 1)\left(\dfrac{3}{4}\right) = 4(x - 1)\left(\dfrac{5}{x - 1}\right)$
3. Solve the resulting equation which should have no fractions.	$8 + 3(x - 1) = 20$ $8 + 3x - 3 = 20$ $3x = 15$ $x = 5$
4. Check each proposed solution.	The proposed solution, 5, checks.

5.7   APPLICATIONS OF RATIONAL EXPRESSIONS

Solving Problems About Distance	On a trip from Sacramento to Montercy, Marge traveled at an average speed of 60 miles per hour. The return trip, at an average speed of 64 miles per hour, took 1/4 hour less. How far did she travel between the two cities?
1. State what the variable represents.	Let x = the unknown distance.
2. Use a chart to identify distance, rate, and time.	
3. Solve $d = rt$ for the unknown quantity in the chart.	
4. From the wording in the problem, decide the relationship between the quantities. Use those expressions to write an equation.	Since the time for the return trip was 1/4 hour less, the time going equals the time returning plus 1/4. $$\dfrac{x}{60} = \dfrac{x}{64} + \dfrac{1}{4}$$
5. Solve the equation.	The solution of this equation is $x = 240$. She traveled 240 miles.
6. Check the solution.	

	d	r	$t = \dfrac{d}{r}$
Going	x	60	$\dfrac{x}{60}$
Returning	x	64	$\dfrac{x}{64}$

CONCEPTS	EXAMPLES
Solving Problems About Work	It takes the regular mail carrier 6 hours to cover her route. A substitute takes 8 hours to cover the same route. How long would it take them to cover the route together?
1. State what the variable represents.	Let x = the number of hours it would take them working together. The rate of the regular carrier is 1/6 job per hour; the rate of the substitute is 1/8 job per hour. Multiply rate times time to get the fractional part of the job done.
2. Put the information from the problem in a chart. If a job is done in t units of time, the rate is $1/t$.	<table><tr><td></td><td>Rate</td><td>Time</td><td>Part of the job done</td></tr><tr><td>Regular</td><td>$\frac{1}{6}$</td><td>x</td><td>$\frac{1}{6}x$</td></tr><tr><td>Substitute</td><td>$\frac{1}{8}$</td><td>x</td><td>$\frac{1}{8}x$</td></tr></table>
3. Write the equation. The sum of the fractional parts should equal 1 (whole job).	The equation is $$\frac{1}{6}x + \frac{1}{8}x = 1.$$
4. Solve the equation.	The solution of the equation is 24/7.
5. Check the solution.	It would take them 24/7 or 3 3/7 hours to cover the route together.
Solving Variation Problems	If a varies inversely as b, and $a = 4$ when $b = 4$, find a when $b = 6$.
1. Write the variation equation using $y = kx$ or $y = k/x$.	The equation for inverse variation is $a = k/b$.
2. Find k by substituting the given values of x and y into the equation.	Substitute $a = 4$ and $b = 4$. $$4 = \frac{k}{4}.$$
3. Write the equation with the value of k from Step 2 and the given value of x or y. Solve for the remaining variable.	The solution is $k = 16$. Let $k = 16$ and $b = 6$ in the variation equation. $$a = \frac{16}{6} = \frac{8}{3}.$$

CHAPTER 5 REVIEW EXERCISES

[5.1] *Find any values of the variable for which the rational expression is undefined.*

1. $\dfrac{4}{x - 3}$

2. $\dfrac{y + 3}{2y}$

3. $\dfrac{m - 2}{m^2 - 2m - 3}$

4. $\dfrac{2k + 1}{3k^2 + 17k + 10}$

Find the numerical value of the rational expression when **(a)** $x = -2$ *and* **(b)** $x = 4$.

5. $\dfrac{x^2}{x - 5}$ **6.** $\dfrac{4x - 3}{5x + 2}$ **7.** $\dfrac{3x}{x^2 - 4}$ **8.** $\dfrac{x - 1}{x + 2}$

Write the rational expression in lowest terms.

9. $\dfrac{5a^3b^3}{15a^4b^2}$ **10.** $\dfrac{m - 4}{4 - m}$

11. $\dfrac{4x^2 - 9}{6 - 4x}$ **12.** $\dfrac{4p^2 + 8pq - 5q^2}{10p^2 - 3pq - q^2}$

[5.2] *Find the product or quotient and write the answer in lowest terms.*

13. $\dfrac{18p^3}{6} \cdot \dfrac{24}{p^4}$ **14.** $\dfrac{8x^2}{12x^5} \cdot \dfrac{6x^4}{2x}$ **15.** $\dfrac{9m^2}{(3m)^4} \div \dfrac{6m^5}{36m}$

16. $\dfrac{x - 3}{4} \cdot \dfrac{5}{2x - 6}$ **17.** $\dfrac{3q + 3}{5 - 6q} \div \dfrac{4q + 4}{2(5 - 6q)}$ **18.** $\dfrac{2r + 3}{r - 4} \cdot \dfrac{r^2 - 16}{6r + 9}$

19. $\dfrac{6a^2 + 7a - 3}{2a^2 - a - 6} \div \dfrac{a + 5}{a - 2}$ **20.** $\dfrac{y^2 - 6y + 8}{y^2 + 3y - 18} \div \dfrac{y - 4}{y + 6}$

21. $\dfrac{2p^2 + 13p + 20}{p^2 + p - 12} \cdot \dfrac{p^2 + 2p - 15}{2p^2 + 7p + 5}$ **22.** $\dfrac{3z^2 + 5z - 2}{9z^2 - 1} \cdot \dfrac{9z^2 + 6z + 1}{z^2 + 5z + 6}$

[5.3] *Find the least common denominator for the list of fractions.*

23. $\dfrac{1}{8}, \dfrac{5}{12}, \dfrac{7}{32}$ **24.** $\dfrac{4}{9y}, \dfrac{7}{12y^2}, \dfrac{5}{27y^4}$

25. $\dfrac{1}{m^2 + 2m}, \dfrac{4}{m^2 + 7m + 10}$ **26.** $\dfrac{3}{x^2 + 4x + 3}, \dfrac{5}{x^2 + 5x + 4}$

Rewrite the rational expression with the given denominator.

27. $\dfrac{5}{8} = \dfrac{}{56}$ **28.** $\dfrac{10}{k} = \dfrac{}{4k}$

29. $\dfrac{3}{2a^3} = \dfrac{}{10a^4}$ **30.** $\dfrac{9}{x - 3} = \dfrac{}{18 - 6x}$

31. $\dfrac{-3y}{2y - 10} = \dfrac{}{50 - 10y}$ **32.** $\dfrac{4b}{b^2 + 2b - 3} = \dfrac{}{(b + 3)(b - 1)(b + 2)}$

[5.4] *Add or subtract as indicated and write the answer in lowest terms.*

33. $\dfrac{10}{x} + \dfrac{5}{x}$ **34.** $\dfrac{6}{3p} - \dfrac{12}{3p}$ **35.** $\dfrac{9}{k} - \dfrac{5}{k - 5}$

36. $\dfrac{4}{y} + \dfrac{7}{7 + y}$ **37.** $\dfrac{m}{3} - \dfrac{2 + 5m}{6}$ **38.** $\dfrac{12}{x^2} - \dfrac{3}{4x}$

39. $\dfrac{5}{a - 2b} + \dfrac{2}{a + 2b}$ **40.** $\dfrac{4}{k^2 - 9} - \dfrac{k + 3}{3k - 9}$

41. $\dfrac{8}{z^2 + 6z} - \dfrac{3}{z^2 + 4z - 12}$ **42.** $\dfrac{11}{2p - p^2} - \dfrac{2}{p^2 - 5p + 6}$

[5.5]

43. Simplify the complex fraction $\dfrac{\frac{a^4}{b^2}}{\frac{a^3}{b}}$ by

(a) Method 1 as described in Section 5.5.

(b) Method 2 as described in Section 5.5.

(c) Then explain which method you prefer, and why.

Simplify the complex fraction.

44. $\dfrac{\frac{y-3}{y}}{\frac{y+3}{4y}}$

45. $\dfrac{\frac{2}{3}-\frac{1}{6}}{\frac{1}{4}+\frac{2}{5}}$

46. $\dfrac{\frac{1}{p}-\frac{1}{q}}{\frac{1}{q-p}}$

47. $\dfrac{x+\frac{1}{w}}{x-\frac{1}{w}}$

48. $\dfrac{\frac{1}{r+t}-1}{\frac{1}{r+t}+1}$

[5.6]

49. Before even beginning the solution process, how do you know that 2 cannot be a solution to the equation found in Exercise 53 below?

Solve the equation and check your answer.

50. $\dfrac{4-z}{z}+\dfrac{3}{2}=\dfrac{-4}{z}$

51. $\dfrac{x}{2}-\dfrac{x-3}{7}=-1$

52. $\dfrac{3y-1}{y-2}=\dfrac{5}{y-2}+1$

53. $\dfrac{3}{m-2}+\dfrac{1}{m-1}=\dfrac{7}{m^2-3m+2}$

Solve for the specified variable.

54. $m=\dfrac{Ry}{t}$ for t

55. $x=\dfrac{3y-5}{4}$ for y

56. $p^2=\dfrac{4}{3m-q}$ for m

[5.7] *Solve the problem.*

57. When half a number is subtracted from two-thirds of the number, the answer is 2. Find the number.

58. The commission received by a salesperson for selling a small car is $\dfrac{2}{3}$ that received for selling a large car. On a recent day, Linda sold one of each, earning a commission of $300. Find the commission for each type of car.

59. In 1911, at the first Indianapolis 500 (mile) race, Ray Harroun won with an average speed of 74.59 miles per hour. What was his time?

60. A man can plant his garden in 5 hours, working alone. His daughter can do the same job in 8 hours. How long would it take them if they worked together?

61. The head gardener can mow the lawns in the city park twice as fast as his assistant. Working together, they can complete the job in $1\dfrac{1}{3}$ hours. How long would it take the head gardener working alone?

62. The area of a circle varies directly as the square of its radius. A circle with a radius of 5 inches has an area of approximately 78.5 square inches. Find the radius of a circle with an area of 100 square inches.

63. If a parallelogram has a fixed area, the height varies inversely as the base. A parallelogram has a height of 8 centimeters and a base of 12 centimeters. Find the height if the base is changed to 24 centimeters.

64. At a given hour, two steamboats leave a city in the same direction on a straight canal. One travels at 18 miles per hour, and the other travels at 25 miles per hour. In how many hours will the boats be 35 miles apart?

MIXED REVIEW EXERCISES

Perform the indicated operations.

65. $\dfrac{\dfrac{5}{x-y}+2}{3-\dfrac{2}{x+y}}$

66. $\dfrac{4}{m-1}-\dfrac{3}{m+1}$

67. $\dfrac{8p^5}{5} \cdot \dfrac{2p^3}{10}$

68. $\dfrac{r-3}{8} \div \dfrac{3r-9}{4}$

69. $\dfrac{\dfrac{5}{x}-1}{\dfrac{5-x}{3x}}$

70. $\dfrac{4}{z^2-2z+1}-\dfrac{3}{z^2-1}$

Solve.

71. $\dfrac{1}{k}+\dfrac{3}{r}=\dfrac{5}{z}$ for r

72. $\dfrac{5+m}{m}+\dfrac{3}{4}=\dfrac{-2}{m}$

73. About $\dfrac{1}{10}$ as many people in the United States speak French at home as speak Spanish. A total of 19.1 million U.S. residents speak one of these two languages at home. How many speak Spanish?

74. Ann-Marie Buesing flew her plane 400 kilometers with the wind in the same time it took her to go 200 kilometers against the wind. The speed of the wind is 50 kilometers per hour. Find the speed of the plane in still air.

75. If x varies directly as y, and $x = 12$ when $y - 5$, find x when $y = 3$.

76. When Mario and Luigi work together on a job, they can do it in $3\dfrac{3}{7}$ days. Mario can do the job working alone in 8 days. How long would it take Luigi working alone?

──────◆ **MATHEMATICAL CONNECTIONS** (Exercises 77–86) ◇──────

In this chapter we have performed operations with and solved equations involving rational expressions. In these exercises, we summarize the various concepts we have convered. Work Exercises 77–86 in order.

Let P, Q, and R be rational expressions defined as follows.

$$P=\dfrac{6}{x+3} \qquad Q=\dfrac{5}{x+1} \qquad R=\dfrac{4x}{x^2+4x+3}$$

77. Find the value or values for which the expression is undefined.
   (a) P **(b)** Q **(c)** R

78. Find and express in lowest terms: $(P \cdot Q) \div R$.

79. Why is $(P \cdot Q) \div R$ not defined if $x = 0$?

80. Find the LCD for P, Q, and R.

81. Perform the operations and express in lowest terms: $P + Q - R$.

82. Simplify the complex fraction $\dfrac{P+Q}{R}$.

83. Solve the equation $P + Q = R$.

84. How does your answer to Exercise 77 help you in working Exercise 83?

85. Suppose that a car travels 6 miles in $x + 3$ minutes. Explain why P represents the rate of the car (in miles per minute).

86. For what value or values of x is $R = \dfrac{40}{77}$?

CHAPTER 5 TEST

1. Find any values for which $\dfrac{3x - 1}{x^2 - 2x - 8}$ is undefined.

2. Find the numerical value of $\dfrac{6r + 1}{2r^2 - 3r - 20}$ when **(a)** $r = -2$ and **(b)** $r = 4$.

Write the rational expression in lowest terms.

3. $\dfrac{-15x^6y^4}{5x^4y}$

4. $\dfrac{6a^2 + a - 2}{2a^2 - 3a + 1}$

Multiply or divide. Write the answer in lowest terms.

5. $\dfrac{x^6y}{x^3} \cdot \dfrac{y^2}{x^2y^3}$

6. $\dfrac{5(d - 2)}{9} \div \dfrac{3(d - 2)}{5}$

7. $\dfrac{6k^2 - k - 2}{8k^2 + 10k + 3} \cdot \dfrac{4k^2 + 7k + 3}{3k^2 + 5k + 2}$

8. $\dfrac{4a^2 + 9a + 2}{3a^2 + 11a + 10} \div \dfrac{4a^2 + 17a + 4}{3a^2 + 2a - 5}$

Find the least common denominator for the list of fractions.

9. $\dfrac{-3}{10p^2}, \dfrac{21}{25p^3}, \dfrac{-7}{30p^5}$

10. $\dfrac{r + 1}{2r^2 + 7r + 6}, \dfrac{-2r + 1}{2r^2 - 7r - 15}$

Rewrite the rational expression with the given denominator.

11. $\dfrac{15}{4p} = \dfrac{}{64p^3}$

12. $\dfrac{3}{6m - 12} = \dfrac{}{42m - 84}$

Add or subtract as indicated. Write the answer in lowest terms.

13. $\dfrac{4x + 2}{x + 5} + \dfrac{-2x + 8}{x + 5}$

14. $\dfrac{-4}{y + 2} + \dfrac{6}{5y + 10}$

15. $\dfrac{x + 1}{3 - x} + \dfrac{x^2}{x - 3}$

16. $\dfrac{3}{2m^2 - 9m - 5} - \dfrac{m + 1}{2m^2 - m - 1}$

Simplify the complex fraction.

17. $\dfrac{\dfrac{2p}{k^2}}{\dfrac{3p^2}{k^3}}$

18. $\dfrac{\dfrac{1}{x + 3} - 1}{1 + \dfrac{1}{x + 3}}$

19. What values of x could not possibly be solutions of the equation $\dfrac{2}{x + 1} - \dfrac{3}{x - 4} = 6$?

Solve the equation and check your answer.

20. $\dfrac{2x + 8}{9} = \dfrac{10x + 4}{27}$

21. $\dfrac{1}{r + 5} - \dfrac{3}{r - 5} = \dfrac{-10}{r^2 - 25}$

22. Solve the formula $F = \dfrac{k}{d - D}$ for D.

Solve the problem.

23. A boat goes 7 miles per hour in still water. It takes as long to go 20 miles upstream as 50 miles downstream. Find the speed of the current.

24. A man can paint a room in his house, working alone, in 5 hours. His wife can do the job in 4 hours. How long will it take them to paint the room if they work together?

25. The current in a simple electrical circuit varies inversely as the resistance. If the current is 50 amp (an *ampere* is a unit for measuring current) when the resistance is 10 ohm (an *ohm* is a unit for measuring resistance), find the current if the resistance is 5 ohm.

CUMULATIVE REVIEW (Chapters 1–5)

Solve.

1. $\dfrac{5}{8}k = 4$

2. $3(2y - 5) = 2 + 5y$

3. $A = \dfrac{1}{2}bh$ for b

4. $\dfrac{2 + m}{2 - m} = \dfrac{3}{4}$

5. $5y \leq 6y + 8$

6. $5m - 9 > 2m + 3$

Evaluate the expression.

7. $\left(\dfrac{3}{4}\right)^{-2}$

8. $\dfrac{7^{-1}}{7}$

9. $\dfrac{(4^{-2})^3}{4^6 4^{-3}}$

Simplify the expression. Write with only positive exponents.

10. $\dfrac{(2x^3)^{-1} \cdot x}{2^3 x^5}$

11. $\dfrac{(m^{-2})^3 m}{m^5 m^{-4}}$

12. $\dfrac{2p^3 q^4}{8p^5 q^3}$

Perform the indicated operations.

13. $(2k^2 + 3k) - (k^2 + k - 1)$

14. $8x^2 y^2 (9x^4 y^5)$

15. $(2a - b)^2$

16. $(y^2 + 3y + 5)(3y - 1)$

17. $\dfrac{12p^3 + 2p^2 - 12p + 4}{2p - 2}$

Factor completely.

18. $8t^2 + 10tv + 3v^2$

19. $8r^2 - 9rs + 12s^2$

20. $6a^2 m + am - 2m$

Solve the equation.

21. $r^2 = 2r + 15$

22. $8m^2 = 64m$

23. $(r - 5)(2r + 1)(3r - 2) = 0$

Solve the problem.

24. One number is 4 more than another. The product of the numbers is 2 less than the smaller number. Find the smaller number.

25. The length of a rectangle is 2 meters less than twice the width. The area is 60 square meters. Find the width of the rectangle.

26. The length of a rectangle is twice its width. If the width were increased by 2 inches while the length remained the same, the resulting rectangle would have an area of 48 square inches. Find the width of the original rectangle.

27. When four times an integer is subtracted from twice the square of the integer, the result is 16. Find the integer.

28. One of the following is equal to 1 for *all* real numbers. Which one is it?

    (a) $\dfrac{k^2 + 2}{k^2 + 2}$ (b) $\dfrac{4 - m}{4 - m}$ (c) $\dfrac{2x + 9}{2x + 9}$ (d) $\dfrac{x^2 - 1}{x^2 - 1}$

29. Which one of the following rational expressions is *not* equivalent to $\dfrac{4 - 3x}{7}$?

    (a) $-\dfrac{-4 + 3x}{7}$ (b) $-\dfrac{4 - 3x}{-7}$ (c) $\dfrac{-4 + 3x}{-7}$ (d) $\dfrac{-(3x + 4)}{7}$

Perform the operation. Write the expression in lowest terms.

30. $\dfrac{x^6 y^2}{y^5 x^9} \cdot \dfrac{x^2}{y^3}$

31. $\dfrac{5}{q} - \dfrac{1}{q}$

32. $\dfrac{3}{7} + \dfrac{4}{r}$

33. $\dfrac{4}{5q - 20} - \dfrac{1}{3q - 12}$

34. $\dfrac{2}{k^2 + k} - \dfrac{3}{k^2 - k}$

35. $\dfrac{7z^2 + 49z + 70}{16z^2 + 72z - 40} \div \dfrac{3z + 6}{4z^2 - 1}$

36. Simplify the complex fraction $\dfrac{\dfrac{4}{a} + \dfrac{5}{2a}}{\dfrac{7}{6a} - \dfrac{1}{5a}}$.

Solve the equation.

37. $\dfrac{r + 2}{5} = \dfrac{r - 3}{3}$

38. $\dfrac{1}{x} = \dfrac{1}{x + 1} + \dfrac{1}{2}$

Solve the problem.

39. On a business trip, Arlene traveled to her destination at an average speed of 60 miles per hour. Coming home, her average speed was 50 miles per hour, and the trip took $\dfrac{1}{2}$ hour longer. How far did she travel each way?

40. Juanita can weed the yard in 3 hours. Benito can weed the yard in 2 hours. How long would it take them if they worked together?

EQUATIONS AND INEQUALITIES IN TWO VARIABLES

CONNECTIONS

It is important in many situations (in business or in science, for example) to be able to make predictions based on known data. An executive may wish to predict next year's costs, profits, or sales, for instance. Scientists are currently trying to predict whether the earth will continue to get increasingly warmer.

The graph below shows actual and predicted Medicare costs since 1967.* It was constructed by plotting points to represent the data for each of the years shown, then connecting the points with lines. It shows the danger of using projections too far into the future. As this graph indicates, we should be wary about cost estimates in the current national health care debate.

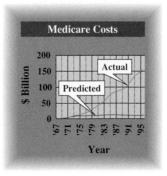

FOR DISCUSSION OR WRITING

In what year did the actual costs begin to diverge sharply from the predicted costs? Estimate the difference between the predicted cost and the actual cost for 1994 (the right end of the graph). What factors may have contributed to the prediction being so far off?

---

* Figures per the House Ways and Means Committee: National Center for Policy Analysis, July 1994.

In this chapter we discuss how the relationship between two variables can be presented pictorially with a graph or algebraically with an equation. These ideas extend the work in earlier chapters where we graphed the solutions of equations or inequalities with one variable on number lines.

6.1 LINEAR EQUATIONS IN TWO VARIABLES

FOR EXTRA HELP	OBJECTIVES
📖 **SSG** pp. 199–206 **SSM** pp. 312–318	**1 ▶** Interpret graphs.
📼 **Video** 9	**2 ▶** Write a solution as an ordered pair. **3 ▶** Decide whether a given ordered pair is a solution of a given equation.
💾 **Tutorial** IBM MAC	**4 ▶** Complete ordered pairs for a given equation. **5 ▶** Complete a table of values. **6 ▶** Plot ordered pairs.

Since graphs are so prevalent in our society, we must be able to read and interpret them intelligently. We begin by looking at some graphs typically seen in a newspaper or a magazine.

1 ▶ Interpret graphs.

▶ There are many ways to represent the relationship between two variables. Bar graphs, pie charts, and line graphs are the most common. A line graph is shown in the Connections at the beginning of this chapter. The first example discusses a bar graph and a pie chart.

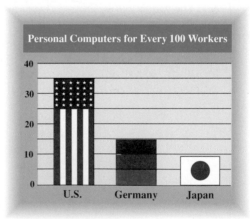

Personal Computers for Every 100 Workers

EXAMPLE 1

Interpreting Graphs

Use the graph to make the interpretations.

(a) The bar graph compares the extent of personal computer ownership in the United States, Germany, and Japan.*

---

*New York Times, February 27, 1994 (Sec. 3, p. 6)

Using the scale given on the left of the graph, we can determine that about 35 of every 100 U.S. workers have a personal computer. Estimate the corresponding numbers for Germany and Japan.

Move horizontally from the top of the bar for Germany over to the scale on the left to see that about 15 of 100 German workers, or 15%, have personal computers. The top of the bar for Japan indicates about 9% of the Japanese have personal computers.

(b) Baby boomers are expected to inherit $10.4 trillion from their parents over the next 45 years, an average of $50,000 each. The following pie chart shows how they plan to spend their inheritance.

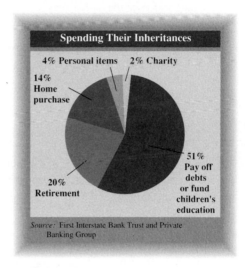

How much of the $50,000 is expected to go toward the purchase of a home? How much to retirement?

The chart shows that 14% will go toward a home. This amounts to .14(50,000) = $7000. They plan to use 20% or .20(50,000) = $10,000 for retirement. ■

Now we consider how to draw a graph that represents a given equation. In most of this chapter we discuss *linear equations*.

Linear Equation

A **linear equation** in two variables is an equation that can be put in the form

$$Ax + By = C,$$

where A, B, and C are real numbers and A and B are not both 0.

2 ▶ Write a solution as an ordered pair.

▶ A solution of a linear equation in two variables requires two numbers, one for each variable. For example, the equation $y = 4x + 5$ is satisfied if x is replaced with 2 and y is replaced with 13, since

13 = 4(2) + 5.   Let $x = 2$; $y = 13$.

The pair of numbers $x = 2$ and $y = 13$ gives a solution of the equation $y = 4x + 5$. The phrase "$x = 2$ and $y = 13$" is abbreviated

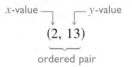

with the x-value, 2, and the y-value, 13, given as a pair of numbers written inside parentheses. *The x-value is always given first.* A pair of numbers such as $(2, 13)$ is called an **ordered pair.** As the name indicates, the order in which the numbers are written is important. The ordered pairs $(\mathbf{2}, \mathbf{13})$ and $(\mathbf{13}, \mathbf{2})$ are not the same. The second pair indicates that $x = 13$ and $y = 2$. (Of course, letters other than x and y may be used in the equation with the numbers.)

3 ▶ Decide whether a given ordered pair is a solution of a given equation.

▶ The next example shows how to decide whether an ordered pair is a solution of an equation. An ordered pair that is a solution of an equation is said to *satisfy* the equation.

E X A M P L E 2

Deciding Whether an Ordered Pair Satisfies an Equation

Decide whether the given ordered pair is a solution of the given equation.

(a) $(3, 2); 2x + 3y = 12$

To see whether $(3, 2)$ is a solution of the equation $2x + 3y = 12$, we substitute 3 for x and 2 for y in the given equation.

$$2x + 3y = 12$$
$$2(3) + 3(2) = 12 \quad ? \quad \text{Let } x = 3; \text{ let } y = 2.$$
$$6 + 6 = 12 \quad ?$$
$$12 = 12 \quad \text{True}$$

This result is true, so $(3, 2)$ satisfies $2x + 3y = 12$.

(b) $(-2, -7); m + 5n = 33$

$$(-2) + 5(-7) = 33 \quad ? \quad \text{Let } m = -2; \text{ let } n = -7.$$
$$-2 + (-35) = 33 \quad ?$$
$$-37 = 33 \quad \text{False}$$

This result is false, so $(-2, -7)$ is *not* a solution of $m + 5n = 33$. ■

4 ▶ Complete ordered pairs for a given equation.

▶ Choosing a number for one variable in a linear equation makes it possible to find the value of the other variable, as shown in the next example.

E X A M P L E 3

Completing an Ordered Pair

Complete the ordered pair $(7, \quad)$ for the equation $y = 4x + 5$.

In this ordered pair, $x = 7$. (Remember that x always comes first.) We find the corresponding value of y by replacing x with 7 in the equation $y = 4x + 5$.

$$y = 4(7) + 5 = 28 + 5 = 33$$

The ordered pair is $(7, 33)$. ■

5 ▶ Complete a table of
values.

▶   Ordered pairs of an equation often are displayed in a **table of values** as in the next example. The table may be written either vertically or horizontally. We will write these tables both horizontally and vertically in this book.

E X A M P L E   4

Completing a Table of
Values

Complete the given table of values for each equation.

(a) $x - 2y = 8$

x	2	10		
y			0	-2

To complete the first two ordered pairs, let $x = 2$ and $x = 10$, respectively.

	If	$x = 2,$		If	$x = 10,$
	then	$x - 2y = 8$		then	$x - 2y = 8$
	becomes	$2 - 2y = 8$		becomes	$10 - 2y = 8$
		$-2y = 6$			$-2y = -2$
		$y = -3.$			$y = 1.$

Now complete the last two ordered pairs by letting $y = 0$ and $y = -2$, respectively.

	If	$y = 0,$		If	$y = -2,$
	then	$x - 2y = 8$		then	$x - 2y = 8$
	becomes	$x - 2(0) = 8$		becomes	$x - 2(-2) = 8$
		$x - 0 = 8$			$x + 4 = 8$
		$x = 8.$			$x = 4.$

The completed table of values is as follows.

x	2	10	8	4
y	-3	1	0	-2

(b) $x = 5$

x			
y	-2	6	3

The given equation is $x = 5$. No matter which value of y might be chosen, the value of x is always the same, 5.

x	5	5	5
y	-2	6	3	■

NOTE   When an equation such as $x = 5$ is discussed along with equations in two variables, think of $x = 5$ as an equation in two variables by rewriting $x = 5$ as $x + 0y = 5$. This form of the equation shows that for any value of y, the value of x is 5. Similarly, $y = -2$ is the same as $y = 0x - 2$.

Every linear equation has an infinite number of solutions, because each choice of a number for one variable leads to a number for the other variable, and there are an infinite number of choices for the first variable. To graph these solutions, represented as the ordered pairs (x, y), we need *two* number lines, one for each variable. These two number lines are drawn as shown in Figure 1. The horizontal number line is called the **x-axis.** The vertical line is called the **y-axis.** Together, the x-axis and y-axis form a **rectangular coordinate system.**

The coordinate system is divided into four regions, called **quadrants.** These quadrants are numbered counterclockwise, as shown in Figure 1. Points on the axes themselves are not in any quadrant. The point at which the x-axis and y-axis meet is called the **origin.** The origin, labeled 0 in Figure 1, is the point corresponding to $(0, 0)$.

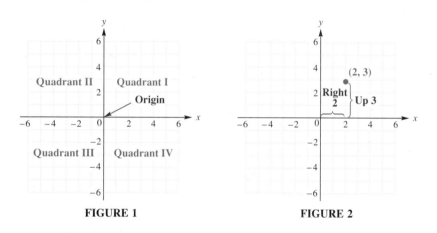

FIGURE 1   FIGURE 2

6 ▶ Plot ordered pairs.

▶ By referring to the two axes, every point on the plane can be associated with an ordered pair. The numbers in the ordered pair are called the **coordinates** of the point. For example, locate the point associated with the ordered pair $(2, 3)$, by starting at the origin. Since the x-coordinate is 2, go 2 units to the right along the x-axis. Then since the y-coordinate is 3, turn and go up 3 units on a line parallel to the y-axis. This is called **plotting** the point $(2, 3)$. (See Figure 2.) From now on we refer to the point with x-coordinate 2 and y-coordinate 3 as the point $(2, 3)$.

EXAMPLE 5
Plotting Ordered Pairs ■

Plot the given points on a coordinate system.

(a) $(1, 5)$ **(b)** $(-2, 3)$ **(c)** $(-1, -4)$

(d) $(7, -2)$ **(e)** $\left(\dfrac{3}{2}, 2\right)$ **(f)** $(5, 0)$

Locate the point $(-1, -4)$, for example, by first going 1 unit to the left along the x-axis. Then turn and go 4 units down, parallel to the y-axis. Plot the point $(3/2, 2)$, by going 3/2 (or 1 1/2) units to the right along the x-axis. Then turn and go 2 units up, parallel to the y-axis. Figure 3 shows the graphs of the points in this example. ■

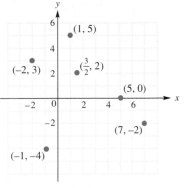

FIGURE 3

EXAMPLE 6

Completing Ordered
Pairs for an Application

A company has found that the cost to produce x calculators is $y = 25x + 250$, where y represents the cost in cents. Complete the following table of values and graph the ordered pairs.

x	1	2	3
y			

To complete the ordered pair $(1, \quad)$ for example, we let $x = 1$.

$$y = 25x + 250$$
$$y = 25(1) + 250 \qquad \text{Let } x = 1.$$
$$y = 25 + 250$$
$$y = 275$$

This gives the ordered pair $(1, 275)$, which says that the cost to produce 1 calculator is 275 cents or \$2.75. Verify that the other ordered pairs are $(2, 300)$ and $(3, 325)$. ∎

These ordered pairs, along with several others that satisfy $y = 25x + 250$, are graphed in Figure 4. Notice how the axes are labeled. In this application x represents the number of calculators and y represents the corresponding cost in cents. Different scales are used on the two axes, since the y-values in the ordered pairs are much larger than the x-values. Here, each square represents 50 units in the vertical direction and $1/2$ unit in the horizontal direction.

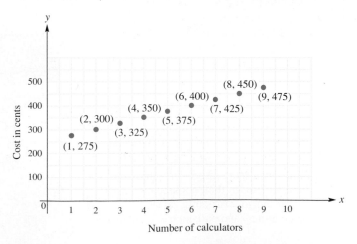

FIGURE 4

6.1 EXERCISES

Use the graph to respond to the statement or question. See Example 1.

1. Between what two consecutive years shown on the graph did the number of jobs in California increase? How much was the increase?

2. How many jobs did California lose between 1990 and 1993?

3. According to Standard & Poor's Corporation, half the losses between 1990 and 1993 were in defense and 40% in construction. How many were lost in defense? How many were lost in construction?

4. Suppose that the same number of jobs were lost between 1993 and 1994 as were lost between 1992 and 1993. How many jobs would there be in California in 1994 based on this assumption?

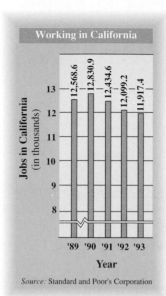

5. Which one of the following would be the best estimate for the total number of units of printer shipments in North America in 1989?
   (a) 5.8 million   (b) 1.5 million
   (c) .2 million   (d) 7.5 million

6. In what year did the number of Inkjet printer shipments first exceed the number of laser printer shipments?

7. What type of printer has shown the most rapid increase in shipments since 1992?

8. Describe how the trend in shipments of Dot Matrix printers has differed from the shipments of Inkjet and Laser printers since 1992.

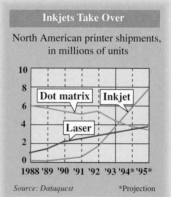

9. In 1992, the gambling industry had $29.9 billion in gross revenues. What percent of this was from lotteries?

10. Suppose that a 4% tax had been levied on gambling revenues in 1992. How much tax would have been collected from casinos?

11. Suppose that casinos paid out 83% of their gross revenues. How much would they have paid out in 1992?

12. How much more in gross revenues did lotteries take in than bookmaking, card rooms, bingo, charitable games, reservations, and pari-mutuel combined?

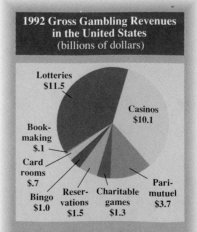

Fill in the blank with the correct response.

13. The symbol (x, y) _____ represent an ordered pair, while the symbols $[x, y]$
        (does/does not)
and $\{x, y\}$ _____ represent ordered pairs.
        (do/do not)

14. The ordered pair $(3, 2)$ is a solution of the equation $2x - 5y = $ _____ .

15. The point whose graph has coordinates $(-4, 2)$ is in quadrant _____ .

16. The point whose graph has coordinates $(0, 5)$ lies along the _____ -axis.

17. The ordered pair $(4,$ _____ $)$ is a solution of the equation $y = 3$.

18. The ordered pair $($ _____ $, -2)$ is a solution of the equation $x = 6$.

Decide whether the given ordered pair is a solution of the given equation. See Example 2.

19. $x + y = 9$; $(0, 9)$ **20.** $x + y = 8$; $(0, 8)$ **21.** $2p - q = 6$; $(4, 2)$

22. $2v + w = 5$; $(3, -1)$ **23.** $4x - 3y = 6$; $(2, 1)$ **24.** $5x - 3y = 15$; $(5, 2)$

25. $y = 3x$; $(2, 6)$ **26.** $x = -4y$; $(-8, 2)$ **27.** $x = -6$; $(-6, 5)$

28. $y = 2$; $(4, 2)$ **29.** $x + 4 = 0$; $(-6, 2)$ **30.** $x - 6 = 0$; $(4, 2)$

31. Explain why there are an infinite number of solutions for a linear equation in two variables.

32. Give the ordered pairs that correspond to the points labeled in the figure.

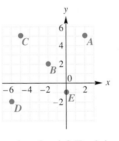

33. Do $(4, -1)$ and $(-1, 4)$ represent the same ordered pair? Explain why.

34. Do the ordered pairs $(3, 4)$ and $(4, 3)$ correspond to the same point on the plane? Explain why.

Complete the ordered pair for the equation $y = 2x + 7$. See Example 3.

35. $(2, \quad)$ **36.** $(5, \quad)$ **37.** $(0, \quad)$

38. $(\quad , 0)$ **39.** $(\quad , -3)$ **40.** $(-6, \quad)$

Complete the ordered pair for the equation $y = -4x - 4$. See Example 3.

41. $(0, \quad)$ **42.** $(\quad , 0)$ **43.** $(\quad , 16)$

44. $(\quad , 24)$ **45.** $(10, \quad)$ **46.** $(5, \quad)$

47. Explain why it would be easier to find the corresponding y-value for $x = \dfrac{1}{3}$ than for $x = \dfrac{1}{7}$ in the equation $y = 6x + 2$.

48. For the equation $y = mx + b$, what is the y-value corresponding to $x = 0$ for *any* value of m?

Complete the table of values. See Example 4.

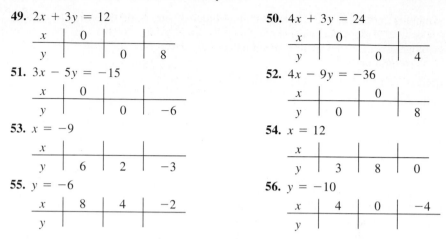

49. $2x + 3y = 12$

x	0		
y		0	8

50. $4x + 3y = 24$

x	0		
y		0	4

51. $3x - 5y = -15$

x	0		
y		0	-6

52. $4x - 9y = -36$

x		0	
y	0		8

53. $x = -9$

x			
y	6	2	-3

54. $x = 12$

x			
y	3	8	0

55. $y = -6$

x	8	4	-2
y			

56. $y = -10$

x	4	0	-4
y			

Plot the ordered pair in a rectangular coordinate system. See Example 5.

57. $(6, 2)$ **58.** $(5, 3)$ **59.** $(-4, 2)$ **60.** $(-3, 5)$ **61.** $\left(-\dfrac{4}{5}, -1\right)$

62. $\left(-\dfrac{3}{2}, -4\right)$ **63.** $(0, 4)$ **64.** $(0, -3)$ **65.** $(4, 0)$ **66.** $(-3, 0)$

Fill in the blank with the word positive *or the word* negative.
The point with coordinates (x, y) *is in*

67. quadrant III if x is _____ and y is _____ .
68. quadrant II if x is _____ and y is _____ .
69. quadrant IV if x is _____ and y is _____ .
70. quadrant I if x is _____ and y is _____ .

Complete the table of values and then plot the ordered pairs. See Examples 4 and 5.

71. $x - 2y = 6$

x	y
0	
	0
2	
	-1

72. $2x - y = 4$

x	y
0	
	0
1	
	-6

73. $3x - 4y = 12$

x	y
0	
	0
-4	
	-4

74. $2x - 5y = 10$

x	y
0	
	0
-5	
	-3

75. $y + 4 = 0$

x	y
0	
5	
-2	
-3	

76. $x - 5 = 0$

x	y
	1
	0
	6
	-4

Solve the problem. See Example 6.

77. Suppose that it costs $5000 to start up a business of selling snow cones. Furthermore, it costs $.50 per cone in labor, ice, syrup, and overhead. Then the cost to make x snow cones is given by y dollars, where $y = .50x + 5000$. Express as an ordered pair each of the following:

   (a) When 100 snow cones are made, the cost is $5050.

   (b) When the cost is $6000, the number of snow cones made is 2000.

78. It costs a flat fee of $20 to rent a pressure washer plus $5 per day. Therefore, the cost to rent the pressure washer for x days is given by $y = 5x + 20$, where y is in dollars. Express as an ordered pair each of the following:

   (a) When the washer is rented for 5 days, the cost is $45.

   (b) I paid $50 when I returned the washer, so I must have rented it for 6 days.

In statistics, ordered pairs are used to decide whether two quantities (such as the height and weight of an individual) are related in such a way that one can be predicted when given the other. Ordered pairs that give these quantities for a number of individuals are plotted on a graph (called a scatter diagram). If the points lie approximately on a line, the variables have a linear relationship.

79. Make a scatter diagram by plotting the following ordered pairs of heights and weights for six women on axes similar to the ones shown here: $(62, 105)$, $(65, 130)$, $(67, 142)$, $(63, 115)$, $(66, 120)$, $(60, 98)$. As shown in the figure provided, the horizontal axis is used to represent the heights, given as the first coordinates, and the vertical axis represents the weights, given as the second coordinates. (We could have assigned the first coordinates to weights and the second coordinates to heights.) Is there a linear relationship between height and weight? (Do the points lie approximately on a straight line?)

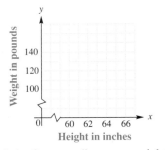

80. The number of U.S. jobs supported by exports to Mexico has steadily increased over the past few years. If we let $x = 0$ represent the year 1986 and let y represent the number of jobs supported (in thousands), then the following ordered pairs are formed: $(0, 275)$, $(1, 300)$, $(2, 400)$, $(3, 500)$, $(4, 425)$, $(5, 610)$, $(6, 725)$. Plot these points.

81. The total for hourly wages and benefits for private industry in Mexico between 1987 and 1992 is shown in the accompanying bar graph. If we let 1987 be represented by 0, 1988 be represented by 1, and so on, write the ordered pairs that describe the monetary amount each year. Write them in the form (x, y), where x is the year and y is the amount.

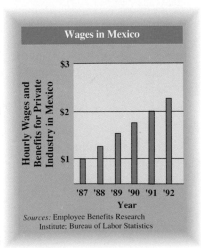

82. The number of all-talk radio stations has risen almost linearly during the past few years. Using 1989 as 0, 1990 as 1, and so on, write the ordered pairs that approximate the number of such stations each year. Write them in the form (x, y), where x is the year and y is the number of stations.

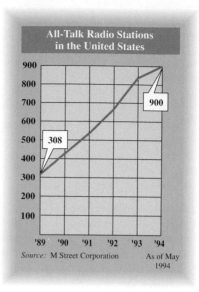

All-Talk Radio Stations in the United States

Source: M Street Corporation   As of May 1994

REVIEW EXERCISES

Solve the equation. See Section 2.2.

83. $3x + 6 = 0$

84. $4 + 2y = 10$

85. $9 - y = -4$

86. $-5 + y = 3$

Solve for y. See Section 2.5.

87. $2x + 3y = 12$

88. $2x - 3y = 12$

6.2 GRAPHING LINEAR EQUATIONS IN TWO VARIABLES

FOR EXTRA HELP

📖 **SSG** pp. 206–212
SSM pp. 318–323

📼 **Video** 9

💾 **Tutorial** IBM   MAC

OBJECTIVES

1 ▶ Graph linear equations.
2 ▶ Find intercepts.
3 ▶ Graph linear equations of the form $Ax + By = 0$.
4 ▶ Graph linear equations of the form $y = k$ or $x = k$.

In this section we see how to use ordered pairs that satisfy an equation to graph the equation.

1 ▶ Graph linear equations.

▶   We know that infinitely many ordered pairs satisfy a linear equation. Some ordered pairs that are solutions of $x + 2y = 7$ are graphed in Figure 5.

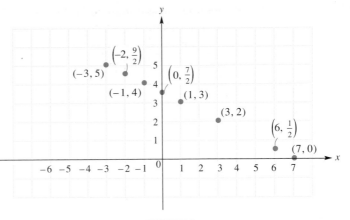

FIGURE 5

Notice that the points plotted in this figure all appear on a straight line, as shown in Figure 6. In fact, all ordered pairs satisfying the equation $x + 2y = 7$ correspond to points that lie on this same straight line. This line gives a "picture" of all the solutions of the equation $x + 2y = 7$. Only a portion of the line is shown here, but it extends indefinitely in both directions, as suggested by the arrowhead on each end of the line. The line is called the **graph** of the equation and the process of plotting the ordered pairs and drawing the line through the corresponding points is called **graphing.**

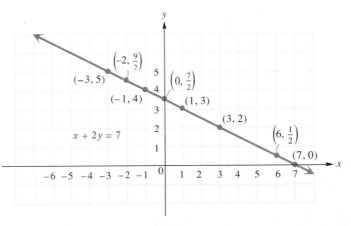

FIGURE 6

The preceding discussion can be generalized.

Graph of a Linear Equation

The graph of any linear equation in two variables is a straight line.

Notice that the word *line* appears in the name "*linear* equation."

Since two distinct points determine a line, we can graph a straight line by finding any two different points on the line. However, it is a good idea to plot a third point as a check.

EXAMPLE 1

Graphing a Linear Equation

■ Graph the linear equation $2y = -3x + 6$.

Although this equation is not in the form $Ax + By = C$, it *could* be put in that form, and so is a linear equation.

For most linear equations, two different points on the graph can be found by first letting $x = 0$, and then letting $y = 0$. Doing this gives the ordered pairs $(0, 3)$ and $(2, 0)$. We get a third ordered pair (as a check) by letting x or y equal some other number. For example, if $x = -2$, we find that $y = 6$, giving the ordered pair $(-2, 6)$. These three ordered pairs are shown in the table of values with Figure 7. Plot the corresponding points, then draw a line through them. This line, shown in Figure 7, is the graph of $2y = -3x + 6$. ■

x	y
0	3
2	0
-2	6

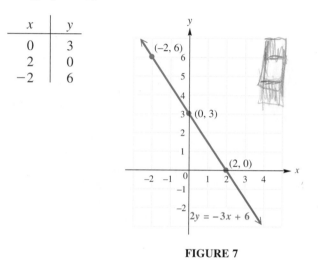

FIGURE 7

2 ► Find intercepts.

► In Figure 7 the graph crosses the y-axis at $(0, 3)$ and the x-axis at $(2, 0)$. For this reason $(0, 3)$ is called the **y-intercept** and $(2, 0)$ is called the **x-intercept** of the graph. The intercepts are particularly useful for graphing linear equations, as in Example 1.

Finding Intercepts

We find the x-intercept by letting $y = 0$ in the given equation and solving for x.

We find the y-intercept by letting $x = 0$ in the given equation and solving for y.

EXAMPLE 2

Finding Intercepts

■ Find the intercepts for the graph of $2x + y = 4$. Draw the graph.

Find the y-intercept by letting $x = 0$; find the x-intercept by letting $y = 0$.

$$
\begin{aligned}
2x + y &= 4 & \qquad 2x + y &= 4 \\
2(0) + y &= 4 & 2x + \mathbf{0} &= 4 \\
0 + y &= 4 & 2x &= 4 \\
y &= 4 & x &= 2
\end{aligned}
$$

The y-intercept is $(0, 4)$. The x-intercept is $(2, 0)$. The graph with the two intercepts shown in color is given in Figure 8. We get a third point as a check. For example, choosing $x = 1$ gives $y = 2$. These three ordered pairs are shown in the table with Figure 8. Plot $(0, 4)$, $(2, 0)$, and $(1, 2)$ and draw a line through them. This line, shown in Figure 8, is the graph of $2x + y = 4$. ■

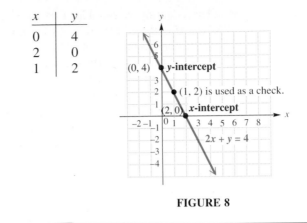

x	y
0	4
2	0
1	2

FIGURE 8

CAUTION When choosing x- or y-values to find ordered pairs to plot, be careful to choose so that the resulting points are not too close together. For example, using $(-1, -1)$, $(0, 0)$, and $(1, 1)$ may result in an inaccurate line. It is better to choose points where the x-values differ by at least 2.

---

3 ▶ Graph linear equations of the form $Ax + By = 0$.

▶ In earlier examples, the x- and y-intercepts were used to help draw the graphs. This is not always possible, as the following examples show. Example 3 shows what to do when the x- and y-intercepts are the same point.

EXAMPLE 3 ■

Graphing an Equation of the Form $Ax + By = 0$

Graph the linear equation $x - 3y = 0$.

If we let $x = 0$, then $y = 0$, giving the ordered pair $(0, 0)$. Letting $y = 0$ also gives $(0, 0)$. This is the same ordered pair, so choose two other values for x or y. Choosing 2 for y gives $x - 3 \cdot 2 = 0$, giving the ordered pair $(6, 2)$. For a check point, we choose -6 for x getting -2 for y. This ordered pair, $(-6, -2)$, along with $(0, 0)$ and $(6, 2)$, was used to get the graph shown in Figure 9. ■

x	y
0	0
6	2
-6	-2

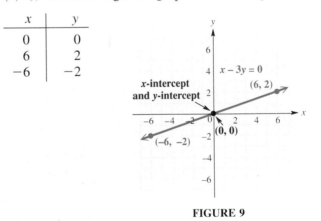

FIGURE 9

Example 3 can be generalized as follows.

Line Through the Origin

If A and B are real numbers, the graph of a linear equation of the form

$$Ax + By = 0$$

goes through the origin $(0, 0)$.

4 ▶ Graph linear equations of the form $y = k$ or $x = k$.

▶ The equation $y = -4$ is a linear equation in which the coefficient of x is 0. (Write $y = -4$ as $0x + y = -4$ to see this.) Also, $x = 3$ is a linear equation in which the coefficient of y is 0. These equations lead to horizontal or vertical straight lines, as the next examples show.

E X A M P L E 4

Graphing an Equation of the Form $y = k$

■ Graph the linear equation $y = -4$.

As the equation states, for any value of x, y is always equal to -4. To get ordered pairs that are solutions of this equation, we choose any numbers for x, always using -4 for y. Three ordered pairs that satisfy the equation are shown in the table of values with Figure 10. Drawing a line through these points gives the horizontal line shown in Figure 10. ■

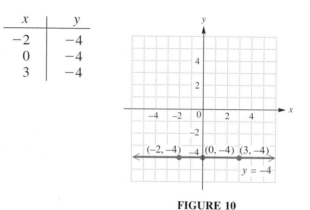

x	y
-2	-4
0	-4
3	-4

FIGURE 10

Horizontal Line

The graph of the linear equation $y = k$, where k is a real number, is the horizontal line going through the point $(0, k)$.

E X A M P L E 5

Graphing an Equation of the Form $x = k$

■ Graph the linear equation $x - 3 = 0$.

First add 3 to each side of the equation $x - 3 = 0$ to get $x = 3$. All the ordered pairs that are solutions of this equation have an x-value of 3. Any number can be used for y. We show three ordered pairs that satisfy the equation in the table of values with Figure 11. Drawing a line through these points gives the vertical line shown in Figure 11. ■

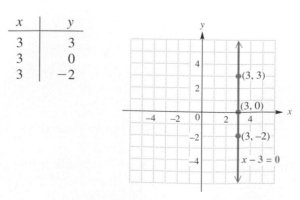

x	y
3	3
3	0
3	-2

FIGURE 11

Vertical Line   The graph of the linear equation $x = k$, where k is a real number, is the vertical line going through the point $(k, 0)$.

In particular, notice that the horizontal line $y = 0$ is the x-axis and the vertical line $x = 0$ is the y-axis.

◈ **C O N N E C T I O N S** ◈

Beginning in this chapter we include information on the basic features of graphics calculators.* The most obvious feature is their ability to graph equations. We must solve the equation for y in order to input it into the calculator. Also, we must select an appropriate "window" for the graph. The window is determined by the minimum and maximum values of x and y. Graphics calculators have a standard window, often from $x = -10$ to $x = 10$ and from $y = -10$ to $y = 10$. These are sometimes called the x-range and the y-range. We indicate this as $[-10, 10]$, $[-10, 10]$, with the range of x-values shown first.

For example, to graph the equation $2x + y = 4$, discussed in Example 2, we first solve for y.

$$2x + y = 4$$
$$y = -2x + 4 \qquad \text{Subtract } 2x.$$

If we input this equation as $y_1 = -2x + 4$ and choose the standard window, the calculator shows the following graph.

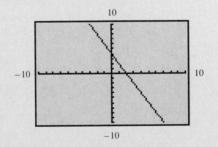

We can also use this graph to solve the equation $-2x + 4 = 0$, by locating the point on the graph where $y = 0$. Of course, this is the x-intercept with an x-value of 2. Thus the solution of the equation is $x = 2$. Most graphics calculators will display this value with a sequence of keystrokes.

FOR DISCUSSION OR WRITING
Use a graphics calculator to solve the following equations from Section 2.3. First rewrite each equation with 0 on one side. You do not need to simplify or combine terms first.

1. $3x + 4 - 2x - 7 = 4x + 3$ (Example 1 with x as the variable)
2. $5x - 15 = 5(x - 3)$ (Example 7)
3. $2x + 3(x + 1) = 5x + 4$ (Example 8)

Compare your results with the examples and describe how each graph gives the solution.

---

*A discussion titled "An Introduction to Graphics Calculators" is included at the front of this book.

The different forms of straight-line equations and the methods of graphing them are given in the following summary.

Graphing Straight Lines

Equation	To Graph	Example
$y = k$	Draw a horizontal line, through $(0, k)$.	
$x = k$	Draw a vertical line, through $(k, 0)$.	
$Ax + By = 0$	Graph goes through $(0, 0)$. Get additional points that lie on the graph by choosing any value of x, or y, except 0.	
$Ax + By = C$ but not of the types above	Find any two points the line goes through. A good choice is to find the intercepts: let $x = 0$, and find the corresponding value of y; then let $y = 0$, and find x. As a check, get a third point by choosing a value of x or y that has not yet been used.	

6.2 EXERCISES

Fill in the blank with the correct response.

1. All ordered pairs that satisfy the equation $2x + 3y = 6$ lie on a straight _line_.

2. To find the x-intercept for the graph of a linear equation, we let __y__ $= 0$.

3. To find the y-intercept for the graph of a linear equation, we let __x__ $= 0$.

4. The graph of an equation of the form $Ax + By = 0$ must go through the __O__ .

5. The graph of an equation of the form $y = k$ is a ____*horizontal*____ line.
   (horizontal/vertical)

6. The graph of an equation of the form $x = k$ is a ____*vertical*____ line.
   (horizontal/vertical)

Complete the given ordered pairs using the given equation. Then graph the equation by plotting the points and drawing a line through them. See Examples 1 and 2.

7. $y = -x + 5$

   $(0,\quad), (\quad , 0), (2,\quad)$

8. $y = x - 2$

   $(0,\quad), (\quad , 0), (5,\quad)$

9. $y = \dfrac{2}{3}x + 1$

   $(0,\quad), (3,\quad), (-3,\quad)$

10. $y = -\dfrac{3}{4}x + 2$

    $(0,\quad), (4,\quad), (-4,\quad)$

11. $3x = -y - 6$

    $(0,\quad), (\quad , 0), \left(-\dfrac{1}{3},\quad \right)$

12. $x = 2y + 3$

    $(\quad , 0), (0,\quad), \left(\quad , \dfrac{1}{2}\right)$

Find the x-intercept and the y-intercept for the graph of the equation. See Example 2.

13. $2x - 3y = 24$

14. $-3x + 8y = 48$

15. $x + 6y = 0$

16. $3x - y = 0$

17. A student attempted to graph $4x + 5y = 0$ by finding intercepts. She first let $x = 0$ and found y; then she let $y = 0$ and found x. In both cases, the resulting point was $(0, 0)$. She knew that she needed at least two points to graph the line, but was unsure what to do next since finding intercepts gave her only one point. How would you explain to her what to do next?

18. What is the equation of the x-axis? What is the equation of the y-axis?

Graph the linear equation. See Examples 1–5.

19. $x = y + 2$

20. $x = -y + 6$

21. $x - y = 4$

22. $x - y = 5$

23. $2x + y = 6$

24. $-3x + y = -6$

25. $3y = 4x + 12$

26. $2y = 5x + 10$

27. $3x + 7y = 14$

28. $6x - 5y = 18$

29. $y - 2x = 0$

30. $y + 3x = 0$

31. $y = -6x$

32. $y = 4x$

33. $y + 1 = 0$

34. $y - 3 = 0$

35. $x = -2$

36. $x = 4$

─────────── ◆ **MATHEMATICAL CONNECTIONS** (Exercises 37–42) ◆ ───────────

In each exercise below, a calculator-generated graph of an equation with one side equal to 0 is shown. Accompanying the graph is the equation itself, where y is expressed in terms of x on the left side. Solve the equation using the methods of Chapter 2, and show that the solution you get is the same as the x-intercept (labeled "Root") on the calculator screen.

37. $8 - 2(3x - 4) - 2x = 0$

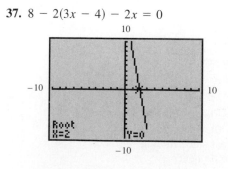

38. $5(2x - 1) - 4(2x + 1) - 7 = 0$

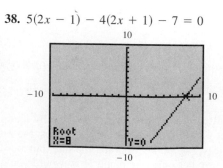

39. $.6x - .1x - x + 2.5 = 0$

40. $-\frac{2}{7}x + 2x - \frac{1}{2}x - \frac{17}{2} = 0$

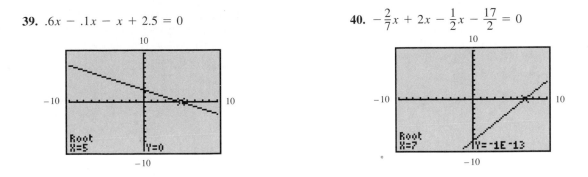

41. Use the results of Exercises 37–40 to explain how the x intercept of a graph corresponds to the solution of the equation.

42. A horizontal line has no x-intercept. If you try to solve $5x - (3x + 2x) + 4 = 0$, you get no solution. What would the graph of $y = 5x - (3x + 2x) + 4$ look like on a graphics calculator?

Solve the problem.

43. Before leveling off in the early 1990s, the number of full-time undergraduate students in engineering programs declined during the years 1984 to 1988. If $x = 0$ represents 1984, $x = 1$ represents 1985, and so on, the number of such students, where y is in thousands, can be approximated by the equation

$$y = -11.1x + 391.2.$$

(This is known as a *linear model* for the data.) The accompanying figure shows the data as a line graph. Use the *equation* to approximate the number of students in each year from 1984 to 1988.

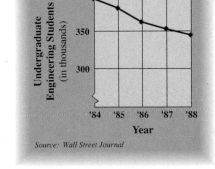

Source: Wall Street Journal

44. The revenue y in billions of dollars generated by the Merck Corporation between 1988 and 1993 can be approximated by the linear equation

$$y = .9x + 5.8,$$

where $x = 0$ corresponds to 1988, $x = 1$ corresponds to 1989, and so on. This information is depicted in the line graph in the accompanying figure. Use the *equation* to approximate the revenue generated in each year from 1988 to 1993.

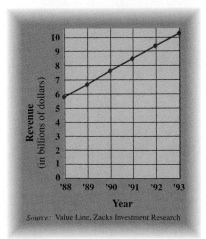

Source: Value Line, Zacks Investment Research

45. The height h of a woman (in centimeters) can be estimated from the length of her radius bone r (from the wrist to the elbow) with the following formula: $h = 73.5 + 3.9r$. Estimate the heights of women with radius bones of the following lengths.
(a) 23 centimeters   (b) 25 centimeters   (c) 20 centimeters
(d) Graph $h = 73.5 + 3.9r$.

46. As a rough estimate, the weight of a man taller than about 60 inches is approximated by $y = 5.5x - 220$, where x is the height of the person in inches, and y is the weight in pounds. Estimate the weights of men whose heights are as follows.
(a) 62 inches   (b) 64 inches   (c) 68 inches   (d) 72 inches
(e) Graph $y = 5.5x - 220$.

47. The graph shows the value of a certain automobile over its first five years. Use the graph to estimate the depreciation (loss in value) during the following years.
(a) First   (b) Second   (c) Fifth
(d) What is the total depreciation over the 5-year period?

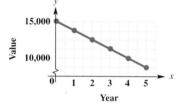

48. The demand for an item is closely related to its price. As price goes up, demand goes down. On the other hand, when price goes down, demand goes up. Suppose the demand for a certain fashionable watch is 1000 when its price is $30 and 8000 when it costs $15.
(a) Let x be the price and y be the demand for the watch. Graph the two given pairs of prices and demands.
(b) Assume the relationship is linear. Draw a line through the two points from part (a). From your graph estimate the demand if the price drops to $10.
(c) Use the graph to estimate the price if the demand is 4000.

REVIEW EXERCISES

Find the quotient. See Section 1.8.

49. $\dfrac{4 - 2}{8 - 5}$

50. $\dfrac{-3 - 5}{2 - 7}$

51. $\dfrac{-2 - (-4)}{3 - (-1)}$

52. $\dfrac{5 - (-7)}{-4 - (-1)}$

6.3 THE SLOPE OF A LINE

FOR EXTRA HELP

📖 **SSG** pp. 213–215
SSM pp. 323–327

📼 **Video**
9

💾 **Tutorial**
IBM   MAC

OBJECTIVES

1 ▶ Find the slope of a line given two points.
2 ▶ Find the slope from the equation of a line.
3 ▶ Use the slope to determine whether two lines are parallel, perpendicular, or neither.

We can graph a straight line if at least two different points on the line are known. A line can also be graphed by using just one point on the line if the "steepness" of the line is known.

1 ▶ Find the slope of a line given two points.

▶ One way to measure the steepness of a line is to compare the vertical change in the line (the rise) to the horizontal change (the run) while moving along the line from one fixed point to another. This measure of steepness is called the *slope* of the line.

Figure 12 shows a line with the points (x_1, y_1) and (x_2, y_2). (Read x_1 as "x-sub-one" and x_2 as "x-sub-two.") As we move along the line from the point (x_1, y_1) to the point (x_2, y_2) y changes by $y_2 - y_1$ units. This is the vertical change. Similarly, x changes by $x_2 - x_1$ units, the horizontal change. The ratio of the change in y to the change in x gives the slope of the line. We usually denote slope with the letter m. The slope of a line is defined as follows.

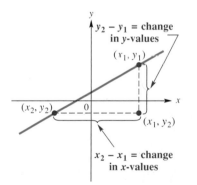

FIGURE 12

Slope Formula ___ The **slope** of the line through the points (x_1, y_1) and (x_2, y_2) is

$$m = \frac{\text{change in } y}{\text{change in } x} = \frac{y_2 - y_1}{x_2 - x_1} \quad \text{if } x_1 \neq x_2.$$

The slope of a line tells how fast y changes for each unit of change in x; that is, the slope gives the rate of change in y for each unit of change in x.

◆ C O N N E C T I O N S ◆

The idea of slope is useful in many everyday situations. For example, because $10\% = 1/10$, a highway with a 10% grade (or slope) rises one meter for every 10 meters horizontally. The highway sign shown in the first figure below is used to warn of a downgrade ahead that may be long or steep. Architects specify the pitch of a roof by indicating the slope: a 5/12 roof means that the roof rises 5 feet for every 12 feet in the horizontal direction. The slope of a stairwell also indicates the ratio of the vertical rise to the horizontal run. See the last two figures.

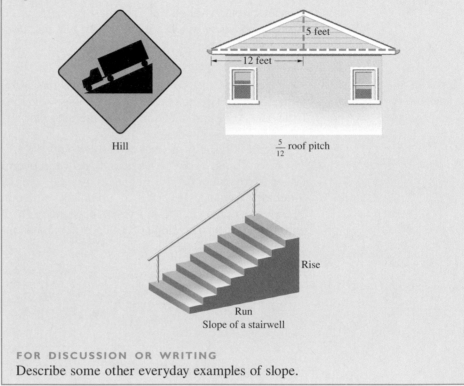

Hill

$\frac{5}{12}$ roof pitch

Slope of a stairwell

FOR DISCUSSION OR WRITING

Describe some other everyday examples of slope.

E X A M P L E   1

Finding the Slope of a Line

Find the slope of each of the following lines.

(a) The line through $(-4, 7)$ and $(1, -2)$

Use the definition of slope. Let $(-4, 7) = (x_2, y_2)$ and $(1, -2) = (x_1, y_1)$. Then

$$\text{slope} = \frac{\text{change in } y}{\text{change in } x}$$

$$m = \frac{y_2 - y_1}{x_2 - x_1}$$

$$= \frac{7 - (-2)}{-4 - 1} = \frac{9}{-5} = -\frac{9}{5}.$$

See Figure 13.

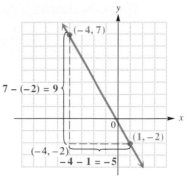

$7 - (-2) = 9$

$(-4, 7)$

$(-4, -2)$

$(1, -2)$

$-4 - 1 = -5$

FIGURE 13

(b) The line through $(12, -5)$ and $(-9, -2)$

$$m = \frac{-5 - (-2)}{12 - (-9)} = \frac{-3}{21} = -\frac{1}{7}$$

The same slope is found by subtracting in reverse order.

$$\frac{-2 - (-5)}{-9 - 12} = \frac{3}{-21} = -\frac{1}{7} \quad \blacksquare$$

CAUTION   It makes no difference which point is (x_1, y_1) or (x_2, y_2); however, it is important to be consistent. Start with the x- and y-values of one point (either one) and subtract the corresponding values of the other point.

In Example 1(a) the slope is negative and the corresponding line in Figure 13 falls from left to right. As Figure 14(a) shows, this is generally true of lines with negative slopes. Lines with positive slopes go up (rise) from left to right, as shown in Figure 14(b).

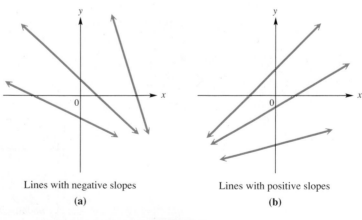

Lines with negative slopes   Lines with positive slopes

(a)   (b)

FIGURE 14

Positive and Negative Slopes

A line with positive slope rises from left to right.
A line with negative slope falls from left to right.

The next examples illustrate slopes of horizontal and vertical lines.

EXAMPLE 2

Finding the Slope of a Horizontal Line

 Find the slope of the line through $(-8, 4)$ and $(2, 4)$.

Use the definition of slope.

$$m = \frac{4 - 4}{-8 - 2} = \frac{0}{-10} = 0$$

As shown in Figure 15 by a sketch of the line through these two points, the line through the points is horizontal, with equation $y = 4$. *All horizontal lines have a slope of* 0, since the difference in y-values is always 0. ■

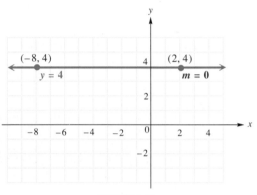

FIGURE 15

EXAMPLE 3

Finding the Slope of a Vertical Line

 Find the slope of the line through $(6, 2)$ and $(6, -9)$.

$$m = \frac{2 - (-9)}{6 - 6}$$

$$= \frac{11}{0} \qquad \text{Undefined}$$

Since division by 0 is undefined, the slope is undefined. The graph in Figure 16 shows that the line through these two points is vertical, with equation $x = 6$. All points on a vertical line have the same x-value, so *the slope of any vertical line is undefined.* ■

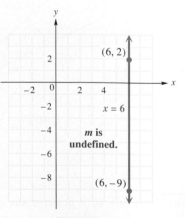

FIGURE 16

Slopes of Horizontal and Vertical Lines

Horizontal lines, with equations of the form $y = k$, **have a slope of 0. Vertical lines,** with equations of the form $x = k$, **have undefined slope.**

2 ▶ Find the slope from the equation of a line.

▶ The slope of a line also can be found directly from its equation. For example, the slope of the line

$$y = -3x + 5$$

can be found from two points on the line. Get these two points by choosing two different values of x, say -2 and 4, and finding the corresponding y-values.

If $x = -2$:	If $x = 4$:
$y = -3(-2) + 5$	$y = -3(4) + 5$
$y = 6 + 5$	$y = -12 + 5$
$y = 11.$	$y = -7.$

The ordered pairs are $(-2, 11)$ and $(4, -7)$. Now find the slope.

$$m = \frac{11 - (-7)}{-2 - 4} = \frac{18}{-6} = -3$$

The slope, -3, is the same number as the coefficient of x in the equation $y = -3x + 5$. It can be shown that this always happens, *as long as the equation is solved for y.* This fact is used to find the slope of a line from its equation.

Slope of a Line from its Equation

Step 1 Solve the equation for y.
Step 2 The slope is given by the coefficient of x.

E X A M P L E   4

Finding Slope from an Equation

Find the slope of each of the following lines.

(a) $2x - 5y = 4$

Solve the equation for y.

$$2x - 5y = 4$$
$$-5y = -2x + 4 \qquad \text{Subtract } 2x \text{ from each side.}$$
$$y = \frac{2}{5}x - \frac{4}{5} \qquad \text{Divide each side by } -5.$$

The slope is given by the coefficient of x, so the slope is

$$m = \frac{2}{5}.$$

(b) $8x + 4y = 1$

Solve the equation for y.

$$8x + 4y = 1$$
$$4y = -8x + 1 \qquad \text{Subtract } 8x \text{ from each side.}$$
$$y = -2x + \frac{1}{4} \qquad \text{Divide each side by } 4.$$

The slope of this line is given by the coefficient of x, -2. ■

3 ▶ Use the slope to determine whether two lines are parallel, perpendicular, or neither.

▶ Two lines in a plane that never intersect are **parallel.** We use slopes to tell whether two lines are parallel. For example, Figure 17 shows the graph of $x + 2y = 4$ and the graph of $x + 2y = -6$. These lines appear to be parallel. Solve for y to find that both $x + 2y = 4$ and $x + 2y = -6$ have a slope of $-1/2$. Nonvertical parallel lines always have equal slopes.

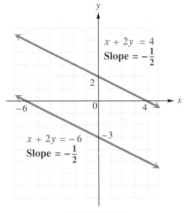

FIGURE 17

Figure 18 shows the graph of $x + 2y = 4$ and the graph of $2x - y = 6$. These lines appear to be **perpendicular** (meet at a 90° angle). Solving for y shows that the slope of $x + 2y = 4$ is $-1/2$, while the slope of $2x - y = 6$ is 2. The product of $-1/2$ and 2 is

$$\left(-\frac{1}{2}\right)(2) = -1.$$

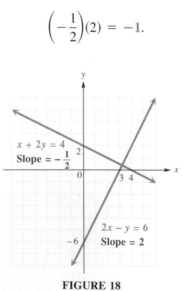

FIGURE 18

This is true in general; the product of the slopes of two perpendicular lines is always -1. This means that the slopes of perpendicular lines are negative reciprocals: if one slope is the nonzero number a, the other is $-1/a$.

Parallel and Perpendicular Lines

Two nonvertical lines with the same slope are parallel; two perpendicular lines, neither of which is vertical, have slopes that are negative reciprocals.

◆▷ **C O N N E C T I O N S** ◁◆

Because the window of a graphics calculator is a rectangle, the graphs of perpendicular lines will not appear perpendicular unless appropriate ranges are used for x and y. Graphics calculators usually have a key to select a "square" window automatically. In a square window, the x-range is about 1.5 times the y-range. The equations used in Figure 18 are graphed with the standard (nonsquare) window and then with a square window in the screens below.

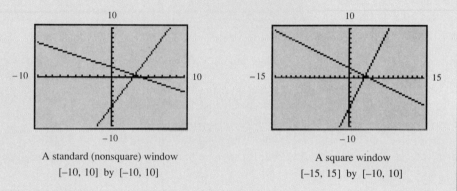

A standard (nonsquare) window
$[-10, 10]$ by $[-10, 10]$

A square window
$[-15, 15]$ by $[-10, 10]$

E X A M P L E   5

Deciding Whether Two Lines Are Parallel or Perpendicular

Decide whether the lines are *parallel, perpendicular,* or *neither.*

(a) $x + 2y = 7$
$-2x + y = 3$

Find the slope of each line by first solving each equation for y.

$$x + 2y = 7 \qquad\qquad\qquad -2x + y = 3$$
$$2y = -x + 7 \qquad\qquad\qquad y = 2x + 3$$
$$y = -\frac{1}{2}x + \frac{7}{2}$$

Slope: $-\dfrac{1}{2}$ \qquad\qquad\qquad\qquad\qquad Slope: 2

Since the slopes are not equal, the lines are not parallel. Check the product of the slopes: $(-1/2)(2) = -1$. The two lines are perpendicular because the product of their slopes is -1, indicating that the slopes are negative reciprocals.

(b) $3x - y = 4$
$6x - 2y = 9$

Find the slopes. Both lines have a slope of 3, so the lines are parallel.

(c) $4x + 3y = 6$
$2x - y = 5$

Here the slopes are $-4/3$ and 2. These two straight lines are neither parallel nor perpendicular. ∎

6.3 EXERCISES

On a pair of axes similar to the one shown, sketch the graph of a straight line having the indicated slope.

 1. negative

 2. positive

 3. undefined

 4. zero

 5. Explain in your own words what is meant by *slope* of a line.

 6. If two nonvertical lines are parallel, what do we know about their slopes? If two lines are perpendicular and neither is parallel to an axis, what do we know about their slopes?

 7. What is the slope of a line parallel to the graph of $y = 5x - 3$?

 8. What is the slope of a line perpendicular to the graph of $y = -3x + 7$?

Use the coordinates of the indicated points to find the slope of the line. See Example 1.

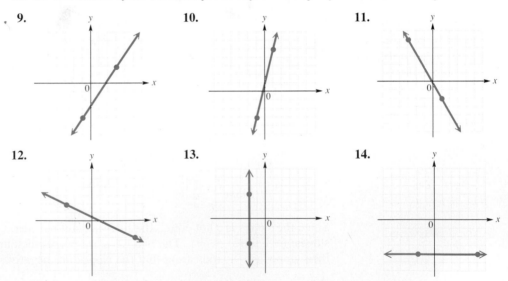

 9.

 10.

 11.

 12.

 13.

 14.

 15. Look at the graph in Exercise 9 and answer the following.
   (a) Start at the point $(-1, -4)$ and count vertically up to the horizontal line that goes through the other point plotted. What is this vertical change? (Remember: "up" means positive, "down" means negative.)
   (b) From this new position, count horizontally to the other point plotted. What is this horizontal change? (Remember: "right" means positive, "left" means negative.)
   (c) What is the quotient of the numbers found in parts (a) and (b)? What do we call this number?

16. Refer to Exercise 15. If we were to *start* at the point (3, 2) and *end* at the point (−1, −4), do you think that the answer to part (c) would be the same? Explain why or why not.

Find the slope of the line going through the pair of points. See Examples 1–3.

17. (4, −1) and (−2, −8)

18. (1, −2) and (−3, −7)

19. (−8, 0) and (0, −5)

20. (0, 3) and (−2, 0)

21. (6, −5) and (−12, −5)

22. (4, 3) and (−6, 3)

23. (−8, 6) and (−8, −1)

24. (−12, 3) and (−12, −7)

25. (3.1, 2.6) and (1.6, 2.1)

26. $\left(-\dfrac{7}{5}, \dfrac{3}{10}\right)$ and $\left(\dfrac{1}{5}, -\dfrac{1}{2}\right)$

Find the slope of the line. See Example 4.

27. $y = 2x - 3$

28. $y = 5x + 12$

29. $2y = -x + 4$

30. $4y = x + 1$

31. $y = 6 - 4x$

32. $y = 3 + 2x$

33. $-6x + 4y = 4$

34. $3x - 2y = 3$

35. $y = 4$

36. $x = 6$

The figure shows a line that has a positive slope (because it rises from left to right) and a positive *y*-value for the *y*-intercept (because it intersects the *y*-axis above the origin).

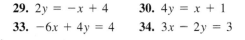

*For the line shown, decide whether (**a**) the slope is positive, negative, or zero and (**b**) the y-value of the y-intercept is positive, negative, or zero.*

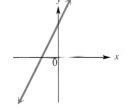

37.   **38.**   **39.**

40.   **41.**   **42.**

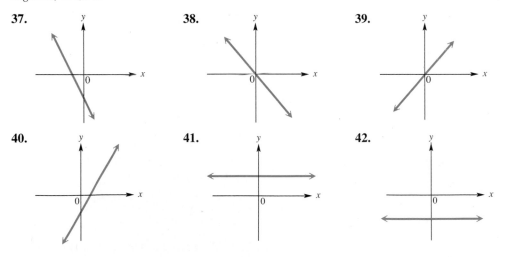

For the pair of equations, give the slopes of the lines and then determine whether the two lines are parallel, perpendicular, or neither parallel nor perpendicular. See Example 5.

43. $2x + 5y = 4$
    $4x + 10y = 1$

44. $-4x + 3y = 4$
    $-8x + 6y = 0$

45. $8x - 9y = 6$
    $8x + 6y = -5$

46. $5x - 3y = -2$
    $3x - 5y = -8$

47. $3x - 2y = 6$
    $2x + 3y = 3$

48. $3x - 5y = -1$
    $5x + 3y = 2$

49. What is the slope (or pitch) of this roof? Measurements are given in feet.

50. What is the slope (or grade) of this hill? Measurements are given in meters.

───────◆ **MATHEMATICAL CONNECTIONS** (Exercises 51–56) ◆───────

Refer to the accompanying graphs that depict the trends for student enrollment in kindergarten and grades 1–8 (Figure A), and student enrollment in grades 9–12 (Figure B) in the United States. (*Source:* U.S. National Center for Education Statistics)

Answer Exercises 51–56 in order.

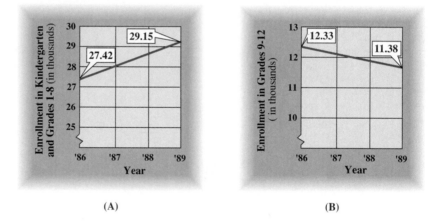

(A)   (B)

51. What is the slope of the line in Figure A?

52. The slope of the line in Figure A is _____. This means that during
 (positive/negative)
the period represented, enrollment _____ in kindergarten and grades
1–8.   (increased/decreased)

53. The slope of a line represents its *rate of change.* Based on Figure A, what was the increase in students *per year* during the period shown.

54. What is the slope of the line in Figure B?

55. The slope of the line in Figure B is _____. This means that during
 (positive/negative)
the period represented, enrollment _____ in grades 9–12.
 (increased/decreased)

56. Based on Figure B, what was the decrease in students *per year* during the period shown?

──────────────◆──────────────

57. Two views of the same line are shown in the accompanying calculator screen, along with coordinates of two points displayed at the bottoms. What is the slope of this line?

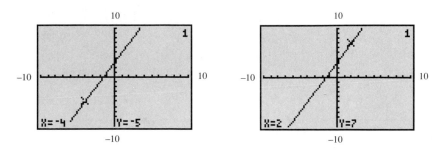

58. Repeat Exercise 57 for the line shown here.

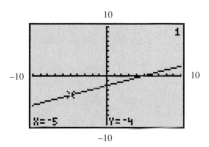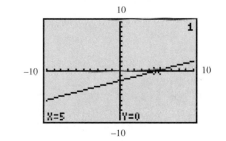

Some graphics calculators have the capability of displaying a table of points for a graph. The table shown here gives several points that lie on a line designated y_1.

59. Use any pair of points displayed to find the slope of the line.

60. What is the x-intercept of the line?

61. What is the y-intercept of the line?

62. Which one of the two lines shown is the graph of y_1?

X	Y₁
-12	-.8
-10	0
-8	.8
-6	1.6
-4	2.4
-2	3.2
0	4

X=-12

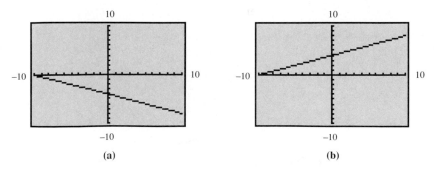

(a)   (b)

REVIEW EXERCISES

Solve for y. See Section 2.5.

63. $y - (-8) = 2(x - 4)$

64. $y - 3 = 4[x - (-6)]$

65. $y - \left(-\dfrac{3}{5}\right) = -\dfrac{1}{2}[x - (-3)]$

66. $y - \left(-\dfrac{5}{8}\right) = \dfrac{3}{8}(x - 5)$

6.4 EQUATIONS OF A LINE

FOR EXTRA HELP	OBJECTIVES
📖 **SSG** pp. 215–219 **SSM** pp. 327–334 📼 **Video** 9 💾 **Tutorial** IBM MAC	**1 ▶** Write an equation of a line given its slope and y-intercept. **2 ▶** Graph a line given its slope and a point on the line. **3 ▶** Write an equation of a line given its slope and any point on the line. **4 ▶** Write an equation of a line given two points on the line.

The last section showed how to find the slope of a line from the equation of the line. For example, the slope of the line having the equation $y = 2x + 3$ is 2, the coefficient of x. What does the number 3 represent? If $x = 0$, the equation becomes

$$y = 2(0) + 3 = 0 + 3 = 3.$$

Since $y = 3$ corresponds to $x = 0$, $(0, 3)$ is the y-intercept of the graph of $y = 2x + 3$. An equation like $y = 2x + 3$ that is solved for y is said to be in **slope-intercept form** because both the slope and the y-intercept of the line can be read directly from the equation.

Slope-Intercept Form   The slope-intercept form of the equation of a line with slope m and y-intercept $(0, b)$ is

$$y = mx + b.$$

1 ▶ Write an equation of a line given its slope and y-intercept.

▶   Given the slope and y-intercept of a line, we can use the slope-intercept form to find an equation of the line.

EXAMPLE 1
Finding an Equation of a Line

Find an equation of the line with slope 2/3 and y-intercept $(0, -1)$.
    Here $m = 2/3$ and $b = -1$, so the equation is

$$y = mx + b$$

$$y = \frac{2}{3}x - 1. \quad ■$$

◆ C O N N E C T I O N S ◆

Businesses must consider the amount of value lost, called **depreciation,** during each year of a machine's useful life. The simplest way to calculate depreciation is to assume that an item with a useful life of n years loses $1/n$ of its value each year. Historically, if the equipment had salvage value, the depreciation was calculated on the **net cost,** the difference between the purchase price and the salvage value. If P represents the purchase price and S the salvage value of an item with a useful life of n years, then the annual depreciation would be

$$D = \frac{1}{n}(P - S).$$

Because this is a linear equation, this is called **straight-line depreciation.**

However, from a practical viewpoint, it is often difficult to determine the salvage value when the equipment is new. Also, for some equipment, such as computers, there is no residual dollar value at the end of the useful life due to obsolescence. In actual practice now, it is customary to find straight-line depreciation by the simpler linear equation

$$D = \frac{1}{n}P.$$

FOR DISCUSSION OR WRITING

1. Find the depreciation using both methods for a $50,000 asset (new) with a useful life of 10 years and a salvage value of $15,000. How much is "written off" in each case over the ten-year period?
2. The depreciation equation is given in slope intercept form. What does the slope represent here? What does the y-value of the y-intercept represent? *Source:* Joel E. Halle, CPA

2 ▶ Graph a line given its slope and a point on the line.

▶ We can use the slope and y-intercept to graph a line. For example, to graph $y = \frac{2}{3}x - 1$, first locate the y-intercept, $(0, -1)$, on the y-axis. From the definition of slope and the fact that the slope of this line is 2/3,

$$m = \frac{\textbf{difference in } y\textbf{-values}}{\textbf{difference in } x\textbf{-values}} = \frac{2}{3}.$$

Another point P on the graph of the line can be found by counting from the y-intercept 2 units up and then counting 3 units to the right. We then draw the line through point P and the y-intercept, as shown in Figure 19. This method can be extended to graph a line given its slope and any point on the line.

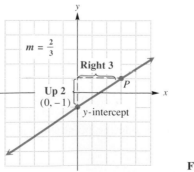

FIGURE 19

EXAMPLE 2

Graphing a Line Given a Point and the Slope

Graph the line through $(-2, 3)$ with slope -4.

First, locate the point $(-2, 3)$. Write the slope as

$$m = \frac{\text{difference in } y\text{-values}}{\text{difference in } x\text{-values}} = -4 = \frac{-4}{1}.$$

(We could have used $4/(-1)$ instead.) We locate another point on the line by counting 4 units down (because of the negative sign) and then 1 unit to the right. Finally, we draw the line through this new point P and the given point $(-2, 3)$. See Figure 20. ∎

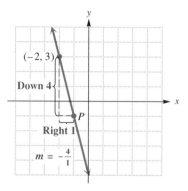

FIGURE 20

3 ▶ Write an equation of a line given its slope and any point on the line.

▶ An equation of a line also can be found from any point on the line and the slope of the line. Let m represent the slope of the line and let (x_1, y_1) represent the given point on the line. Let (x, y) represent any other point on the line. Then by the definition of slope,

$$\frac{y - y_1}{x - x_1} = m$$

or

$$y - y_1 = m(x - x_1).$$

This result is the **point-slope form** of the equation of a line.

Point-Slope Form

The point-slope form of the equation of a line with slope m going through (x_1, y_1) is

$$y - y_1 = m(x - x_1).$$

This very important result should be memorized.

EXAMPLE 3

Using the Point-Slope Form to Write an Equation

Find an equation of each of the following lines. Write the equation in the form $Ax + By = C$.

(a) Through $(-2, 4)$, with slope -3

The given point is $(-2, 4)$ so $x_1 = -2$ and $y_1 = 4$. Also, $m = -3$. Substitute these values into the point-slope form.

$$y - y_1 = m(x - x_1)$$
$$y - 4 = -3[x - (-2)]$$
$$y - 4 = -3(x + 2)$$

$y - 4 = -3x - 6$	Distributive property
$y = -3x - 2$	Add 4.
$3x + y = -2$	Add $3x$.

The last equation is in the form $Ax + By = C$.

(b) Through (4, 2), with slope 3/5

Use $x_1 = 4$, $y_1 = 2$, and $m = 3/5$ in the point-slope form.

$$y - y_1 = m(x - x_1)$$
$$y - 2 = \frac{3}{5}(x - 4)$$

Multiply both sides by 5 to clear of fractions.

$$5(y - 2) = 5 \cdot \frac{3}{5}(x - 4)$$

$$5(y - 2) = 3(x - 4)$$

$5y - 10 = 3x - 12$	Distributive property
$5y = 3x - 2$	Add 10.
$-3x + 5y = -2$	Subtract $3x$. ∎

4 ▶ Write an equation of a line given two points on the line.

▶   We can also use the point-slope form to find an equation of a line when two points on the line are known.

EXAMPLE 4

Finding the Equation of a Line Given Two Points

Find an equation of the line through the points $(-2, 5)$ and $(3, 4)$. Write the equation in the form $Ax + By = C$.

First, we find the slope of the line, using the definition of slope.

$$\text{slope} = \frac{5 - 4}{-2 - 3} = \frac{1}{-5} = -\frac{1}{5}$$

Now we use either $(-2, 5)$ or $(3, 4)$ and the point-slope form. Using $(3, 4)$ gives

$$y - y_1 = m(x - x_1)$$

$$y - 4 = -\frac{1}{5}(x - 3)$$

$5(y - 4) = -1(x - 3)$	Multiply by 5.
$5y - 20 = -x + 3$	Distributive property
$5y = -x + 23$	Add 20 on each side.
$x + 5y = 23.$	Add x on each side.

The same result would be found by using $(-2, 5)$ for (x_1, y_1). ∎

A summary of the types of linear equations is given here.

Linear Equations	$Ax + By = C$	**Standard form (A, B, and C integers, $A > 0$, $B \neq 0$)** Slope is $-A/B$. x-intercept is $(C/A, 0)$. y-intercept is $(0, C/B)$.
	$x = k$	**Vertical line** Slope is undefined. x-intercept is $(k, 0)$.
	$y = k$	**Horizontal line** Slope is 0. y-intercept is $(0, k)$.
	$y = mx + b$	**Slope-intercept form** Slope is m. y-intercept is $(0, b)$.
	$y - y_1 = m(x - x_1)$	**Point-slope form** Slope is m. Line goes through (x_1, y_1).

CAUTION The above definition of "standard form" is not the same in all texts. Also, a linear equation can be written as $Ax + By = C$ in many different (equally correct) ways. For example, $3x + 4y = 12$, $6x + 8y = 24$, and $9x + 12y = 36$ all represent the same set of ordered pairs. Let us agree that $3x + 4y = 12$ is preferable to the other forms because the greatest common factor of 3, 4, and 12 is 1.

6.4 EXERCISES

Use the geometric interpretation of slope (rise divided by run, from Section 6.3) to find the slope of the line. Then, by identifying the y-intercept from the graph, write the y = mx + b form of the equation of the line.

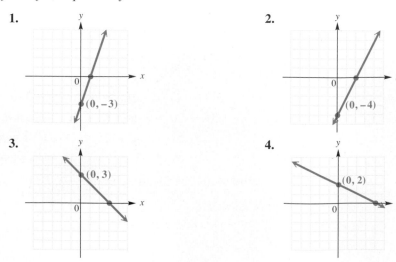

Write the equation of the line with the given slope and y-intercept. See Example 1.

5. $m = 4$, $(0, -3)$ **6.** $m = -5$, $(0, 6)$

7. $m = 0$, $(0, 3)$ **8.** $m = 3$, $(0, 0)$

9. Explain why the equation of a vertical line cannot be written in the form $y = mx + b$.

10. Match the equation with the graph that would most closely resemble its graph.

    (a) $y = x + 3$

    (b) $y = -x + 3$

    (c) $y = x - 3$

    (d) $y = -x - 3$

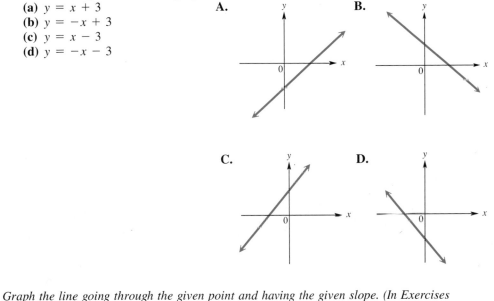

Graph the line going through the given point and having the given slope. (In Exercises 17–20, recall the types of lines having slope 0 and undefined slope.) See Example 2.

11. $(-2, 3)$, $m = \dfrac{1}{2}$ **12.** $(-4, -1)$, $m = \dfrac{3}{4}$ **13.** $(1, -5)$, $m = -\dfrac{2}{5}$

14. $(2, -1)$, $m = -\dfrac{1}{3}$ **15.** $(0, 2)$, $m = 3$ **16.** $(0, -5)$, $m = -2$

17. $(3, 2)$, $m = 0$ **18.** $(-2, 3)$, $m = 0$ **19.** $(3, -2)$, undefined slope

20. $(2, 4)$, undefined slope

21. What is the common name given to a vertical line whose x-intercept is the origin?

22. What is the common name given to a line with slope 0 whose y-intercept is the origin?

Write an equation for the line passing through the given point and having the given slope. Write the equation in the form $Ax + By = C$. See Example 3.

23. $(4, 1)$, $m = 2$ **24.** $(2, 7)$, $m = 3$ **25.** $(3, -10)$, $m = -2$

26. $(2, -5)$, $m = -4$ **27.** $(-2, 5)$, $m = \dfrac{2}{3}$ **28.** $(-4, 1)$, $m = \dfrac{3}{4}$

29. $(6, -3)$, $m = -\dfrac{4}{5}$ **30.** $(7, -2)$, $m = -\dfrac{7}{2}$

31. If a line passes through the origin and a second point whose x- and y-coordinates are equal, what is an equation of the line?

32. What point *must* the graph of $Ax + By = 0$ pass through?

Write an equation for the line passing through the given pair of points. Write the equation in the form $Ax + By = C$. See Example 4.

33. $(8, 5)$ and $(9, 6)$ **34.** $(4, 10)$ and $(6, 12)$ **35.** $(-1, -7)$ and $(-8, -2)$

36. $(-2, -1)$ and $(3, -4)$ **37.** $(0, -2)$ and $(-3, 0)$ **38.** $(-4, 0)$ and $(0, 2)$

39. $\left(\dfrac{1}{2}, \dfrac{3}{2}\right)$ and $\left(-\dfrac{1}{4}, \dfrac{5}{4}\right)$ **40.** $\left(-\dfrac{2}{3}, \dfrac{8}{3}\right)$ and $\left(\dfrac{1}{3}, \dfrac{7}{3}\right)$

The cost to produce x items is, in some cases, expressed as $y = mx + b$. The number b gives the fixed cost (that is, the cost that is the same no matter how many items are produced), and the number m is the variable cost (the cost to produce an additional item). Write the cost equation for each of the following, and answer the questions.

41. It costs \$400 to start up a business of selling ice cream cones. Each cone costs \$.25 to produce.
   (a) What will be the cost to produce 100 cones, based on the cost equation?
   (b) How many cones will be produced if total cost is \$775?

42. It costs \$2000 to purchase a copier and each copy costs \$.02 to make.
   (a) What will be the cost to produce 10,000 copies, based on the cost equation?
   (b) How many copies will be produced if total cost is \$2600?

The sales of a company for a given year can be written as an ordered pair in which the first number, x, gives the year (perhaps since the company started business) and the second number, y, gives the sales for that year. If the sales increase at the same rate each year, a linear equation for sales can be found. Sales for two years are given for two different companies.

Jeff's Rental Properties, Inc.		Star Enterprises	
Year in operation, x	*Sales in dollars,* y	*Year in operation,* x	*Sales in dollars,* y
1	4800	1	18,000
5	24,800	4	93,000

43. (a) Write two ordered pairs in the form (year, sales) for Jeff's Rental Properties, Inc.
   (b) Write the sales equation in the form $y = mx + b$.
   (c) What does the slope, m, represent in this situation?

44. (a) Write two ordered pairs in the form (year, sales) for Star Enterprises.
   (b) Write the sales equation in the form $y = mx + b$.
   (c) What does the slope, m, represent in this situation?

Two views of the same line are shown on a calculator screen. Use the displays at the bottom of the screen to find an equation of the form $y = mx + b$ for the line. Then graph the line on your own calculator to support your answer. Use the standard viewing window.

45.

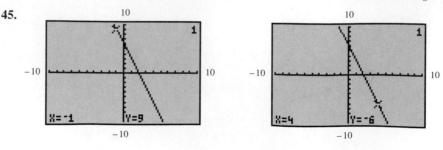

46.

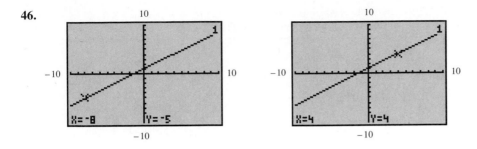

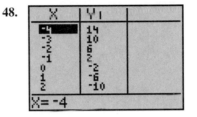

A table of points for a line y_1 is shown. Find an equation of the line, writing it in the form $y_1 = mx + b$. Then use your own calculator to support your answer.

47.

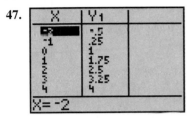

48.

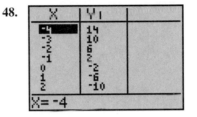

◆ **MATHEMATICAL CONNECTIONS** (Exercises 49–56) ◇

If we think of ordered pairs of the form (C, F), then the two most common methods of measuring temperature, Celsius and Fahrenheit, can be related as follows: When C = 0, F = 32, and when C = 100, F = 212. Work Exercises 49–56 in order.

49. Write two ordered pairs relating these two temperature scales.

50. Find the slope of the line joining the two points.

51. Use the point-slope form to find an equation of the line. (Your variables should be C and F rather than x and y.)

52. Write an equation for F in terms of C.

53. Use the equation from Exercise 52 to write an equation for C in terms of F.

54. Use the equation from Exercise 52 to find the Fahrenheit temperature when $C = 30$.

55. Use the equation from Exercise 53 to find the Celsius temperature when $F = 50$.

56. For what temperature is $F = C$?

REVIEW EXERCISES

Solve the inequality and graph the solutions on a number line. See Section 2.9.

57. $3x + 8 > -1$ **58.** $\frac{1}{2}x - 3 < 2$ **59.** $5 - 3x \leq -10$ **60.** $x - 4 \leq 0$

6.5 GRAPHING LINEAR INEQUALITIES IN TWO VARIABLES

FOR EXTRA HELP

SSG pp. 219–223
SSM pp. 334–339

Video
9

Tutorial
IBM   MAC

OBJECTIVES

1 ▶ Graph $\leq$ or $\geq$ linear inequalities.
2 ▶ Graph $<$ or $>$ linear inequalities.
3 ▶ Graph inequalities with a boundary through the origin.

In Section 6.2 we discussed methods for graphing linear equations, such as $2x + 3y = 6$. Now this discussion is extended to **linear inequalities in two variables,** such as

$$2x + 3y \leq 6.$$

(Recall that $\leq$ is read "is less than or equal to.")

1 ▶ Graph $\leq$ or $\geq$ linear inequalities.

▶   The inequality $2x + 3y \leq 6$ means that

$$2x + 3y < 6 \qquad \text{or} \qquad 2x + 3y = 6.$$

As we found at the beginning of this chapter, the graph of $2x + 3y = 6$ is a line. This **boundary line** divides the plane into two regions. The graph of the solutions of the inequality $2x + 3y < 6$ will include only *one* of these regions. We find the required region by solving the given inequality for y.

$$2x + 3y \leq 6$$
$$3y \leq -2x + 6 \qquad \text{Subtract } 2x.$$
$$y \leq -\frac{2}{3}x + 2 \qquad \text{Divide by 3.}$$

By this last statement, ordered pairs in which y *is less than or equal to* $(-2/3)x + 2$ will be solutions to the inequality. The ordered pairs in which y is equal to $(-2/3)x + 2$ are on the boundary line, so the pairs in which y *is less than* $(-2/3)x + 2$ will be *below* that line. To indicate the solution, shade the region below the line, as in Figure 21. The shaded region, along with the line, is the desired graph.

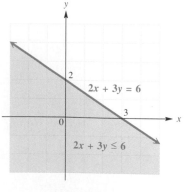

FIGURE 21

As an alternative method, a test point gives a quick way to find the correct region to shade. We choose any point *not* on the line. Because $(0, 0)$ is easy to substitute into an inequality, it is often a good choice, and we use it here. Substitute 0 for x and 0 for y in the given inequality to see whether the resulting statement is true or false. In the example above,

$$2x + 3y \leq 6 \qquad \text{Original inequality}$$
$$2(\mathbf{0}) + 3(\mathbf{0}) \leq 6 \quad ? \qquad \text{Let } x = 0 \text{ and } y = 0.$$
$$0 + 0 \leq 6 \quad ?$$
$$0 \leq 6. \qquad \text{True}$$

Since the last statement is true, shade the region that includes the test point $(0, 0)$. This agrees with the result shown in Figure 21.

2 ▶ Graph $<$ or $>$ linear inequalities.

▶   Inequalities that do not include the equals sign are graphed in a similar way.

E X A M P L E   1

Graphing a Linear Inequality

Graph the inequality $x - y > 5$.

This inequality does not include the equals sign. Therefore, the points on the line $x - y = 5$ do not belong to the graph. However the line still serves as a boundary for two regions, one of which satisfies the inequality. To graph the inequality, first graph the equation $x - y = 5$. We use a dashed line to show that the points on the line are *not* solutions of the inequality $x - y > 5$. Now we choose a test point to see which side of the line statisfies the inequality. Let us choose $(1, -2)$ this time.

$$x - y > 5 \qquad \text{Original inequality}$$
$$1 - (\mathbf{-2}) > 5 \quad ? \qquad \text{Let } x = 1 \text{ and } y = -2.$$
$$3 > 5 \qquad \text{False}$$

Since $3 > 5$ is false, the graph of the inequality is *not* the region that contains $(1, -2)$. We shade the other region, as shown in Figure 22. This shaded region is the required graph. To check that the correct region is shaded, we select a test point in the shaded region and substitute for x and y in the inequality $x - y > 5$. For example, we use $(4, -3)$ from the shaded region, as follows.

$$x - y > 5$$
$$4 - (\mathbf{-3}) > 5 \quad ?$$
$$7 > 5 \qquad \text{True}$$

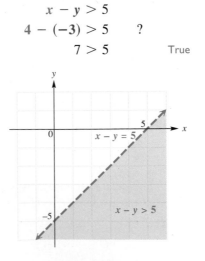

FIGURE 22

This verifies that the correct region was shaded in Figure 22.   ■

A summary of the steps used to graph a linear inequality in two variables follows.

Graphing a Linear Inequality

Step 1 **Graph the boundary.** Graph the line that is the boundary of the region. Use the methods of Section 6.2. Draw a solid line if the inequality has $\leq$ or $\geq$; draw a dashed line if the inequality has $<$ or $>$.

Step 2 **Shade the appropriate side.** Use any point not on the line as a test point. Substitute for x and y in the *inequality*. If a true statement results, shade the side containing the test point. If a false statement results, shade the other side.

EXAMPLE 2

Graphing a Linear Inequality with a Vertical Boundary Line

■ Graph the inequality $x \leq 3$.

First, we graph $x = 3$, a vertical line going through the point $(3, 0)$. We use a solid line (why?) and choose $(0, 0)$ as a test point.

$$x \leq 3 \qquad \text{Original inequality}$$
$$0 \leq 3 \qquad ? \qquad \text{Let } x = 0.$$
$$0 \leq 3 \qquad \text{True}$$

Since $0 \leq 3$ is true, we shade the region containing $(0, 0)$, as in Figure 23. ■

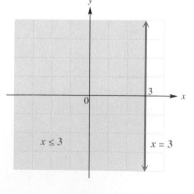

FIGURE 23

3 ▶ Graph inequalities with a boundary through the origin.

▶ The next example shows how to graph an inequality having a boundary line that goes through the origin so that $(0, 0)$ cannot be used as a test point.

EXAMPLE 3

Graphing a Linear Inequality with a Boundary Line Through the Origin

Graph the inequality $x \leq 2y$.

We begin by graphing $x = 2y$. Some ordered pairs that can be used to graph this line are $(0, 0)$, $(6, 3)$, and $(4, 2)$. Use a solid line. The point $(0, 0)$ cannot be used as a test point since $(0, 0)$ is on the line $x = 2y$. Instead, we choose a test point off the line $x = 2y$. For example, let us choose $(1, 3)$, which is not on the line.

$x \leq 2y$	Original inequality
$1 \leq 2(3)$?	Let $x = 1$ and $y = 3$.
$1 \leq 6$	True

Since $1 \leq 6$ is true, we shade the side of the graph containing the test point $(1, 3)$. (See Figure 24.) ■

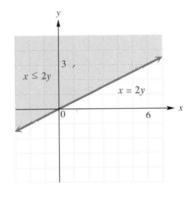

FIGURE 24

◆ **CONNECTIONS** ◆

Graphics calculators have a feature that allows us to shade regions in the plane, so they can be used to graph a linear inequality in two variables. Since all calculators are different, consult the manual for directions for your calculator. The calculator will not draw the graph as a dashed line, so it is still necessary to understand what is and is not included in the solution set.

To solve the inequalities in one variable, $-2x + 4 > 0$ and $-2x + 4 < 0$, we can use the graph of $y = -2x + 4$ (see the Connections in Section 6.2). To solve $y = -2x + 4 > 0$, we want the values of x such that $y > 0$, or *above* the x-axis. From the graph, we see that this is the case for $x < 2$. Similarly, the solution of $-2x + 4 < 0$ is $x > 2$, because $y = -2x + 4 < 0$, or *below* the x-axis, when $x > 2$.

FOR DISCUSSION OR WRITING

1. Discuss the pros and cons of using a calculator to solve a linear inequality in two variables.
2. Use a graphics calculator to solve the following inequalities in one variable from Section 2.9.
   (a) $3x + 2 - 5 > -x + 7 + 2x$ (Example 4)
   (b) $3x + 2 - 5 < -x + 7 + 2x$ (Use the result from part (a).)
   (c) $4 \leq 3x - 5 < 6$ (Example 8)

6.5 EXERCISES

Decide whether the given ordered pair is a solution of the inequality.

1. $3x - 4y < 12$
  (a) $(2, 6)$ **(b)** $(4, 0)$

2. $2x + 5y < -10$
  (a) $(4, -20)$ **(b)** $(-5, 0)$

3. $3x - 2y \geq 0$
  (a) $(4, 1)$ **(b)** $(0, 0)$

4. $6x + 3y \geq 0$
  (a) $(1, -3)$ **(b)** $(0, 0)$

In Exercises 5–14, the straight line boundary has been drawn. Complete the graph by shading the correct region. See Examples 1–3.

5. $x + y \geq 4$

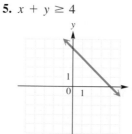

6. $x + y \leq 2$

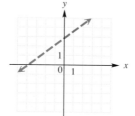

7. $x + 2y \geq 7$

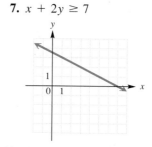

8. $2x + y \geq 5$

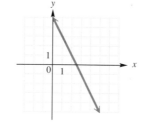

9. $-3x + 4y > 12$

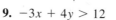

10. $4x - 5y < 20$

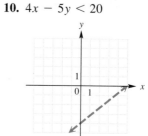

11. $x > 4$

12. $y < -1$

13. $x \leq 3y$

14. $x \geq -2y$

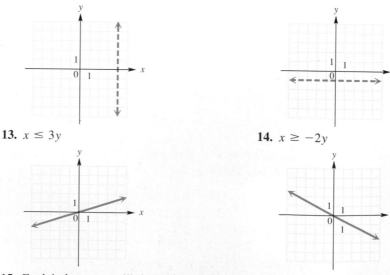

15. Explain how you will determine whether to use a dashed line or a solid line when graphing linear inequalities in two variables.

16. Explain why the point $(0, 0)$ is not an appropriate choice for a test point when graphing an inequality whose boundary goes through the origin.

Graph the linear inequality. See Examples 1–3.

17. $x + y \leq 5$ **18.** $x + y \geq 3$ **19.** $x + 2y < 4$

20. $x + 3y > 6$ **21.** $2x + 3y > -6$ **22.** $3x + 4y < 12$

23. $y \geq 2x + 1$ **24.** $y < -3x + 1$ **25.** $x \leq -2$

26. $x \geq 1$ **27.** $y < 5$ **28.** $y < -3$

29. $y \geq 4x$ **30.** $y \leq 2x$

31. Explain why the graph of $y > x$ cannot lie in quadrant IV.

32. Explain why the graph of $y < x$ cannot lie in quadrant II.

A calculator was used to generate the shaded graphs in choices A–D. Match the inequality with the appropriate choice.

33. $y \geq 3x - 7$

34. $y \leq 3x - 7$

35. $y \geq -3x + 7$

36. $y \leq -3x + 7$

A.   **B.**

C.   **D.**

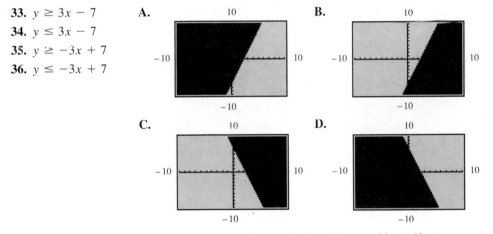

The calculator graph shown at the bottom of this page is that of $y = -2(x + 1) + 4(x - 1)$. If we solve the equation

$$-2(x + 1) + 4(x - 1) = 0,$$

we obtain the solution 3. Because the graph of the line lies *above* the x-axis for values of x greater than 3, the solution of

"y is greater than 0" means *above* the x-axis

$$-2(x + 1) + 4(x - 1) > 0$$

is the set of numbers greater than 3. On the other hand, because the graph of the line lies *below* the x-axis for values of x less than 3, the solution of

"y is less than 0" means *below* the x-axis

$$-2(x + 1) + 4(x - 1) < 0$$

is the set of numbers less than 3.

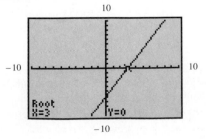

*Use the graph to solve the equation and inequalities listed in parts **(a)–(c)** below the graph.*

37. $y = -3x + 2(2x + 1)$

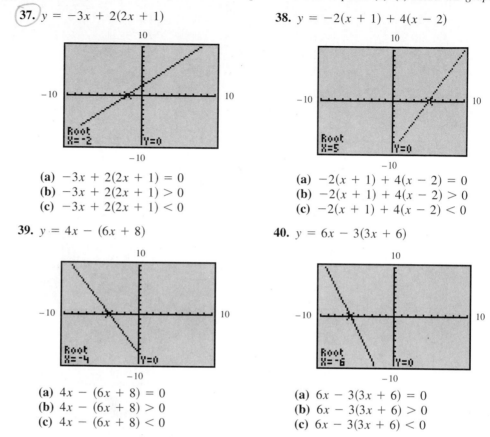

(a) $-3x + 2(2x + 1) = 0$
(b) $-3x + 2(2x + 1) > 0$
(c) $-3x + 2(2x + 1) < 0$

38. $y = -2(x + 1) + 4(x - 2)$

(a) $-2(x + 1) + 4(x - 2) = 0$
(b) $-2(x + 1) + 4(x - 2) > 0$
(c) $-2(x + 1) + 4(x - 2) < 0$

39. $y = 4x - (6x + 8)$

(a) $4x - (6x + 8) = 0$
(b) $4x - (6x + 8) > 0$
(c) $4x - (6x + 8) < 0$

40. $y = 6x - 3(3x + 6)$

(a) $6x - 3(3x + 6) = 0$
(b) $6x - 3(3x + 6) > 0$
(c) $6x - 3(3x + 6) < 0$

*For the given information **(a)** graph the inequality. Here $x \geq 0$ and $y \geq 0$, so graph only the part of the inequality in quadrant I. **(b)** Give some ordered pairs that satisfy the inequality.*

41. The Sweet Tooth Candy Company uses x pounds of chocolate for chocolate cookies and y pounds of chocolate for fudge. The company has 200 pounds of chocolate available, so that

$$x + y \leq 200.$$

42. A company will ship x units of merchandise to outlet I and y units of merchandise to outlet II. The company must ship a total of at least 500 units to these two outlets. This can be expressed by writing

$$x + y \geq 500.$$

43. A toy manufacturer makes stuffed bears and geese. It takes 20 minutes to sew a bear and 30 minutes to sew a goose. There is a total of 480 minutes of sewing time available to make x bears and y geese. These restrictions lead to the inequality

$$20x + 30y \leq 480.$$

44. Ms. Branson takes x vitamin C tablets each day at a cost of 10¢ each and y multivitamins each day at a cost of 15¢ each. She wants the total cost to be no more than 50¢ a day. This can be expressed by writing

$$10x + 15y \leq 50.$$

REVIEW EXERCISES

Find the value of $3x^2 + 8x + 5$ *for the given value of x. See Section 3.1.*

45. 0   **46.** -1 **47.** 4   **48.** $-\dfrac{5}{3}$

6.6 FUNCTIONS

FOR EXTRA HELP	OBJECTIVES
SSG pp. 223–227 **SSM** pp. 339–342	**1 ▶** Understand the definition of a relation.
Video 10	**2 ▶** Understand the definition of a function.
	3 ▶ Decide whether an equation defines a function.
Tutorial IBM MAC	**4 ▶** Find domains and ranges.
	5 ▶ Use $f(x)$ notation.

In Section 6.1, the equation $y = 25x + 250$ was used to find the cost y in cents to produce x calculators. Choosing values for x and using the equation to find the corresponding values of y led to a set of ordered pairs (x, y). In each ordered pair, y (the cost) was *related* to x (the number of calculators produced) by the equation $y = 25x + 250$.

1 ▶ Understand the definition of a relation.

▶ In an ordered pair (x, y), x and y are called the **components** of the ordered pair. Any set of ordered pairs is called a **relation.** The set of all first components in the ordered pairs of a relation is the **domain** of the relation, and the set of all second components in the ordered pairs is the **range** of the relation. Recall from Chapter 1 that sets are written with set braces, { }.

EXAMPLE 1
Using Ordered Pairs to Define a Relation

(a) The relation $\{(0, 1), (2, 5), (3, 8), (4, 2)\}$ has domain $\{0, 2, 3, 4\}$ and range $\{1, 2, 5, 8\}$. The correspondence between the elements of the domain and the elements of the range is shown in Figure 25.

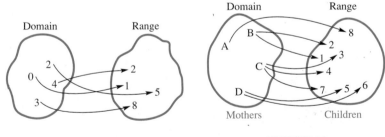

FIGURE 25   **FIGURE 26**

(b) Figure 26 shows a relation where the domain elements represent four mothers and the range elements represent eight children. The figure shows the correspondence between each mother and her children. The relation also could be written as the set of ordered pairs $\{(A, 8), (B, 1), (B, 2), (C, 3), (C, 4), (C, 7), (D, 5), (D, 6)\}$. ■

2 ▶ Understand the definition of a function.

▶ A special type of relation, called a *function,* is particularly useful in applications.

Function

A **function** is a set of ordered pairs in which each first component corresponds to exactly one second component.

By this definition, the relation in Example 1(a) is a function. However, the relation in Example 1(b) is *not* a function, because at least one first component (mother) corresponds to more than one second component (child). Notice that if the ordered pairs in Example 1(b) are reversed, with the child as the first component and the mother as the second component, the result *is* a function.

The simple relations given here were defined by listing the ordered pairs or by showing the correspondence with a figure. Most useful functions have an infinite number of ordered pairs and are usually defined with an equation that tells how to get the second component given the first component. It is customary to use an equation with x and y as the variables, where x represents the first component and y the second component in the ordered pairs.

EXAMPLE 2

Observing Examples of Functions

Some everyday examples of functions are given here.

(a) The **cost** y in dollars charged by an express mail company is a function of the **weight in pounds** x determined by the equation $y = 1.5(x - 1) + 9$.

(b) In one state, the sales tax is 6% of the price of an item. The **tax** y on a particular item is a function of the **price** x, so that

$$y = .06x. \quad ■$$

3 ▶ Decide whether an equation defines a function.

▶ Given a graph of an equation, the definition of a function can be used to decide whether the graph represents a function or not. By the definition of a function, each value of x must lead to exactly one value of y. In Figure 27 the indicated value of x leads to two values of y, so this graph is not the graph of a function. A vertical line can be drawn that cuts this graph in more than one point.

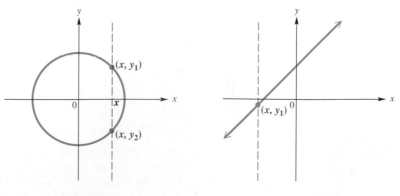

FIGURE 27   **FIGURE 28**

On the other hand, in Figure 28 any vertical line will cut the graph in no more than one point. Because of this, the graph in Figure 28 is the graph of a function. This idea gives the **vertical line test** for a function.

Vertical Line Test

If a vertical line cuts a graph in more than one point, the graph is not the graph of a function.

As Figure 28 suggests, any nonvertical line is the graph of a function. For this reason, any linear equation of the form $Ax + By = C$, where $B \neq 0$, defines a function.

E X A M P L E   3

Deciding Whether a
Relation Defines
a Function

Decide whether or not the following relations are functions.

(a) $y = 2x - 9$

This linear equation can be written as $2x - 1y = 9$. Since the graph of this equation is a line that is not vertical, the equation defines a function.

(b) Use the vertical line test. Any vertical line will cross the graph in Figure 29 just once, so this is the graph of a function.

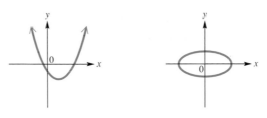

FIGURE 29   FIGURE 30

(c) The vertical line test shows that the graph in Figure 30 is not the graph of a function; a vertical line can cross the graph twice.

(d) $x = 4$

The graph of $x = 4$ is a vertical line, so the equation does not define a function.

(e) $x = y^2$

We need to decide whether any x-value corresponds to more than one y-value. Suppose $x = 36$. Then

$$x = y^2 \qquad \text{becomes} \qquad y^2 = 36.$$

The equation $y^2 = 36$ has *two* solutions: $y = 6$ and $y = -6$. Because the *one* x-value, 36, leads to *two* y-values, the equation $x = y^2$ does not define a function.

(f) $2x + y < 4$

Since this is an inequality, any choice of a value for x will lead to an *infinite* number of y-values. For example, if $x = 3$,

$$2x + y < 4$$
$$2(3) + y < 4 \qquad \text{Let } x = 3.$$
$$6 + y < 4$$
$$y < -2.$$

If $x = 3$, y can be *any* value less than -2. Since $x = 3$ leads to more than one y-value, the inequality $2x + y < 4$ does not define a function. ■

Some generalizations are suggested by Example 3. Inequalities never define functions, since each x-value leads to an infinite number of y-values. An equation in which y is squared cannot define a function because most x-values will lead to two y-values. This is true for any *even* power of y, such as y^4, y^6, and so on. Similarly, an equation involving $|y|$ does not define a function, because some x-values lead to more than one y-value.

4 ▶ Find domains and ranges.

▶ By the definitions of domain and range given for relations, the set of all numbers that can be used as replacements for x in a function is the domain of the function, and the set of all possible values of y is the range of the function.

EXAMPLE 4

Finding the Domain and Range of a Function

Find the domain and range for the following functions.

(a) $y = 6x - 9$

Any number at all may be used for x, so the domain is the set of all real numbers. Also, any number may be used for y, so the range is also the set of all real numbers.

(b) $y = x^2$

Any number can be squared, so the domain is the set of all real numbers. However, since the square of a real number cannot be negative and since $y = x^2$, the values of y cannot be negative, making the range the set of all nonnegative numbers, written as $y \geq 0$. ∎

◆ **CONNECTIONS** ◆

Here is an example of a function that is a real-life application, but is not easily expressed as an equation:

A chain-saw rental firm charges $7 per day or fraction of a day to rent a saw, plus a fixed fee of $4 for resharpening the blade. Let y represent the cost of renting a saw for x days. A portion of the graph is shown here.

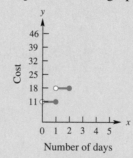

FOR DISCUSSION OR WRITING

1. Explain how the graph can be continued.
2. Explain how the vertical line test applies here.
3. What do the numbers in the domain represent here? What do the numbers in the range represent?

---

5 ▶ Use $f(x)$ notation.

▶ It is common to use the letters f, g, and h to name functions. For example, the function $y = 3x + 5$ is often written

$$f(x) = 3x + 5,$$

where $f(x)$ is read "f of x." The notation $f(x)$ is another way of writing y in a function. For the function $f(x) = 3x + 5$, if $x = 7$ then

$$f(7) = 3 \cdot 7 + 5 \qquad \text{Let } x = 7.$$
$$= 21 + 5 = 26.$$

Read this result, $f(7) = 26$, as "f of 7 equals 26." The notation $f(7)$ means the value of y when x is 7. The statement $f(7) = 26$ says that the value of y is 26 when x is 7.

To find $f(-3)$, replace x with -3.

$$f(x) = 3x + 5$$
$$f(-3) = 3(-3) + 5 \qquad \text{Let } x = -3.$$
$$f(-3) = -9 + 5 = -4$$

CAUTION The symbol $f(x)$ does *not* mean f times x. It represents the y-value that corresponds to x.

$f(x)$ **Notation**   In the notation $f(x)$, f is the name of the function,
x is the domain value,
$f(x)$ is the range value y for the domain value x.

EXAMPLE 5
Using Function Notation

For the function with $f(x) = x^2 - 3$, find the following.

(a) $f(2)$
Replace x with 2.

$$f(x) = x^2 - 3$$
$$f(2) = 2^2 - 3 \qquad \text{Let } x = 2.$$
$$f(2) = 4 - 3 = 1$$

(b) $f(0) = 0^2 - 3 = 0 - 3 = -3$

(c) $f(-3) = (-3)^2 - 3 = 9 - 3 = 6$

(d) The ordered pair with $x = -3$
From part (c), when $x = -3$, $y = f(-3) = 6$, so the ordered pair is $(-3, 6)$.

6.6 EXERCISES

Decide whether the relation is or is not a function. Give the domain and the range in Exercises 1–6. See Examples 1 and 2.

1. $\{(-4, 3), (-2, 1), (0, 5), (-2, -8)\}$

2. $\{(3, 7), (1, 4), (0, -2), (-1, -1), (-2, 5)\}$

3. $\{(-2, 3), (-1, 2), (0, 0), (1, 2), (2, -7)\}$

4. $\{(1, 5), (5, 7), (5, 9), (7, 12)\}$

5.

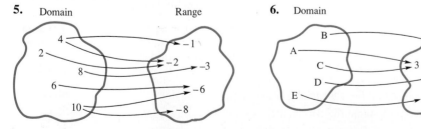

6.

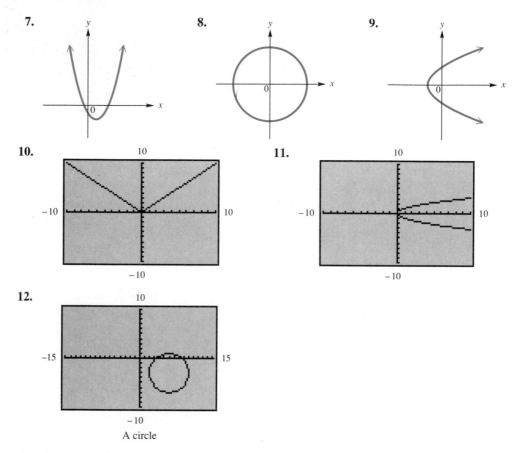

7.

8.

9.

10.

11.

12.

A circle

Decide whether the equation or inequality defines y as a function of x. (Remember that to be a function, every value of x must give one and only one value of y.) See Example 3.

13. $y = 5x + 3$ **14.** $y = -7x + 12$ **15.** $y = x^2 + 2x + 1$

16. $y = -3x^2 + x + 5$ **17.** $y = x^3$ **18.** $y = \sqrt[3]{x}$

19. $x = y^2$ **20.** $x = y^4$ **21.** $y < 3x$

22. The equation $3x - 5y = 6$ defines y as a function of x, but it is not specifically solved for y. Solve for y in terms of x.

Find the domain and the range for the function. See Example 4.

23. $f(x) = x^2$ **24.** $f(x) = x^4$ **25.** $f(x) = x + 2$

26. $f(x) = -x + 3$ **27.** $f(x) = \sqrt{x}$ **28.** $f(x) = |x|$

◆ **MATHEMATICAL CONNECTIONS** (Exercises 29–32) ◆

A function such as $f(x) = 3x - 4$ can be graphed by replacing $f(x)$ with y and then using the methods described earlier in this section. Let us assume that some function is written in the form $f(x) = mx + b$, for particular values of m and b. Answer Exercises 29–32 in order.

29. If $f(2) = 4$, name the coordinates of one point on the line.

30. If $f(-1) = -4$, name the coordinates of another point on the line.

31. Use the results of Exercises 29 and 30 to find the slope of the line.

32. Use the slope-intercept form of the equation of a line to write the function in the form $f(x) = mx + b$.

For the function f, find (a) $f(2)$, (b) $f(0)$, (c) $f(-3)$. See Example 5.

33. $f(x) = 4x + 3$ **34.** $f(x) = -3x + 5$ **35.** $f(x) = x^2 - x + 2$

36. $f(x) = x^3 + x$ **37.** $f(x) = |x|$ **38.** $f(x) = |x + 7|$

A calculator can be thought of as a function machine. We input a number value (from the domain), and then by pressing the appropriate key, we obtain an output value (from the range). Use your calculator, follow the directions, and then answer the questions.

39. Suppose we enter the value 4 and then take the square root (that is, activate the square root *function*).
  (a) What is the domain value here?
  (b) What is the range value obtained?

40. Suppose we enter the value -8 and then activate the squaring function.
  (a) What is the domain value here?
  (b) What range value is obtained?

Soft drink consumption in the United States has experienced steady increases over the past decade. The accompanying graph shows the per capita consumption for the years 1985 through 1990. We can say that consumption is a function of the year, and use the horizontal axis to represent the years and the vertical axis to represent the consumption in gallons.

41. When the domain value is 1987, what is the range value?

42. When the range value is 41.1, what is the domain value?

43. For the years 1985 through 1990, the function $f(x) = 1.54x + 35.5$ approximates the consumption in gallons, where 0 represents 1985, 1 represents 1986, and so on. Use this function to find $f(6)$, and interpret your result.

44. Discuss the relationship between the slope of the graph of $f(x) = 1.54x + 35.5$, and the trend in soft drink consumption during the period from 1985 to 1990.

The table shown was generated by a graphics calculator. The expression y_1 represents $f(x)$.

45. What is $f(3)$?

46. If $f(x) = 2$, what is the value of x?

47. The points represented in the table all lie in a straight line. What is the slope of the line?

48. What is the y-intercept of the line?

49. Write the function in the form $y = mx + b$, for the appropriate values of m and b.

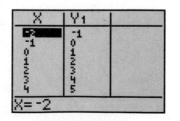

50. Give an everyday example of a function using personal relationships (for example, every person has one and only one mother).

51. Many scientists believe that rising carbon dioxide (CO_2) levels will lead to higher global temperatures and significant climate change. One study showed that the average CO_2 concentration (in appropriate units) at Mauna Loa, Hawaii, increased according to the linear function $f(x) = 1.7x + 200$, where x represents the number of years since 1900. For example, $x = 10$ represents 1910, $x = 50$ represents 1950, and so on.
(a) Find $f(60)$.   (b) Find $f(90)$.
(c) Estimate the CO_2 concentration, that is, $f(x)$, in the year 2000.

52. Compare the definitions of a relation and a function. How are they alike? How are they different?

53. In your own words, explain the meaning of domain of a function.

REVIEW EXERCISES

Add the polynomials. See Section 3.1.

54. $x - 2y$	**55.** $5x - 7y$	**56.** $9a - 5b$
$\underline{3x + 2y}$	$\underline{12x + 7y}$	$\underline{-9a + 7b}$

CHAPTER 6 SUMMARY

KEY TERMS

6.1 linear equation
ordered pair
table of values
x-axis
y-axis
rectangular
  coordinate system
quadrants
origin
coordinates
plot

6.2 graph, graphing
y-intercept
x-intercept

6.3 Slope
parallel lines
perpendicular lines

6.5 linear inequality in
  two variables
boundary line

6.6 components
relation
domain
range
function

NEW SYMBOLS

(a, b)	an ordered pair
(x_1, y_1)	x-sub-one, y-sub-one
m	slope
$f(x)$	function of x

QUICK REVIEW

CONCEPTS	EXAMPLES
6.1 LINEAR EQUATIONS IN TWO VARIABLES	
An ordered pair is a solution of an equation if it satisfies the equation.	Is $(2, -5)$ or $(0, -6)$ a solution of $4x - 3y = 18$? $4(2) - 3(-5) = 23 \neq 18$ $(2, -5)$ is not a solution. $4(0) - 3(-6) = 18$ $(0, -6)$ is a solution.

CONCEPTS	EXAMPLES
If a value of either variable in an equation is given, the other variable can be found by substitution.	Complete the ordered pair $(0, \quad)$ for $3x = y + 4$. $$3(0) = y + 4$$ $$0 = y + 4$$ $$-4 = y$$ The ordered pair is $(0, -4)$.
Plot the ordered pair $(-2, 4)$ by starting at the origin, going 2 units to the left, then going 4 units up.	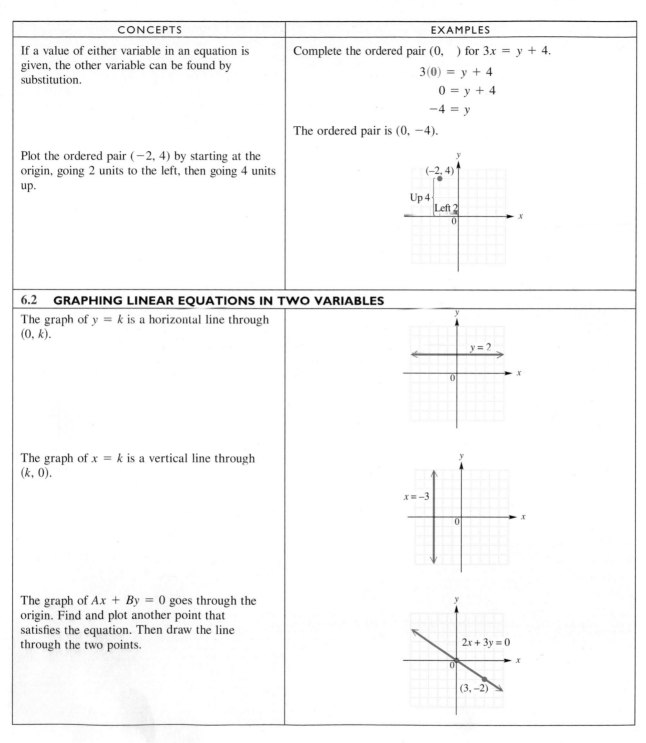

6.2   GRAPHING LINEAR EQUATIONS IN TWO VARIABLES

The graph of $y = k$ is a horizontal line through $(0, k)$.	
The graph of $x = k$ is a vertical line through $(k, 0)$.	
The graph of $Ax + By = 0$ goes through the origin. Find and plot another point that satisfies the equation. Then draw the line through the two points.	

CONCEPTS	EXAMPLES
To graph a linear equation: 1. Find at least two ordered pairs that satisfy the equation. 2. Plot the corresponding points. 3. Draw a straight line through the points.	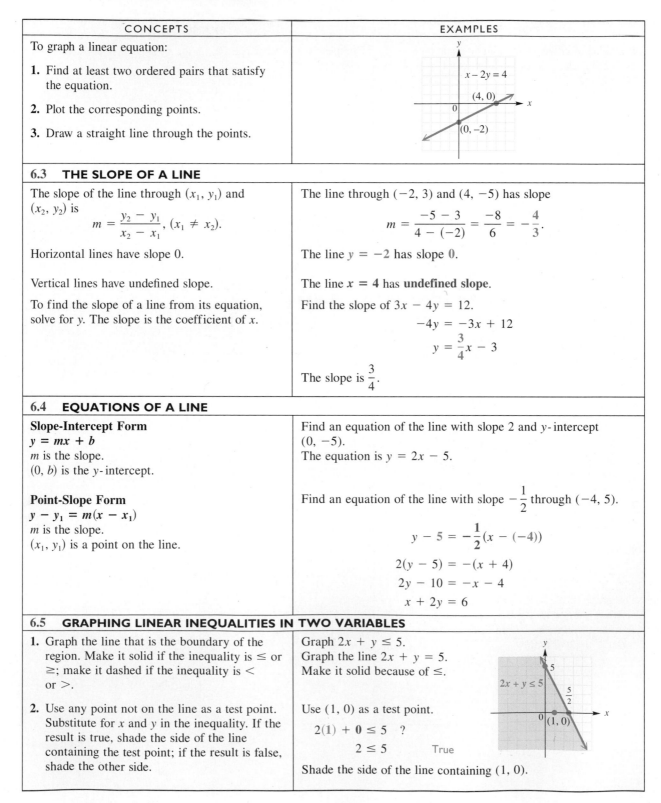

6.3 THE SLOPE OF A LINE

The slope of the line through (x_1, y_1) and (x_2, y_2) is $$m = \frac{y_2 - y_1}{x_2 - x_1}, \ (x_1 \ne x_2).$$	The line through $(-2, 3)$ and $(4, -5)$ has slope $$m = \frac{-5 - 3}{4 - (-2)} = \frac{-8}{6} = -\frac{4}{3}.$$
Horizontal lines have slope 0.	The line $y = -2$ has slope **0**.
Vertical lines have undefined slope.	The line $x = 4$ has **undefined slope**.
To find the slope of a line from its equation, solve for y. The slope is the coefficient of x.	Find the slope of $3x - 4y = 12$. $$-4y = -3x + 12$$ $$y = \frac{3}{4}x - 3$$ The slope is $\frac{3}{4}$.

6.4 EQUATIONS OF A LINE

Slope-Intercept Form $y = mx + b$ m is the slope. $(0, b)$ is the y-intercept.	Find an equation of the line with slope 2 and y-intercept $(0, -5)$. The equation is $y = 2x - 5$.
Point-Slope Form $y - y_1 = m(x - x_1)$ m is the slope. (x_1, y_1) is a point on the line.	Find an equation of the line with slope $-\frac{1}{2}$ through $(-4, 5)$. $$y - 5 = -\frac{1}{2}(x - (-4))$$ $$2(y - 5) = -(x + 4)$$ $$2y - 10 = -x - 4$$ $$x + 2y = 6$$

6.5 GRAPHING LINEAR INEQUALITIES IN TWO VARIABLES

1. Graph the line that is the boundary of the region. Make it solid if the inequality is $\le$ or $\ge$; make it dashed if the inequality is $<$ or $>$. 2. Use any point not on the line as a test point. Substitute for x and y in the inequality. If the result is true, shade the side of the line containing the test point; if the result is false, shade the other side.	Graph $2x + y \le 5$. Graph the line $2x + y = 5$. Make it solid because of $\le$. Use $(1, 0)$ as a test point. $$2(1) + 0 \le 5 \ \ ?$$ $$2 \le 5 \qquad \text{True}$$ Shade the side of the line containing $(1, 0)$.

CONCEPTS	EXAMPLES
6.6 FUNCTIONS	
Vertical Line Test If a vertical line intersects a graph in more than one point, the graph is not the graph of a function.	The graph shown is not the graph of a function. 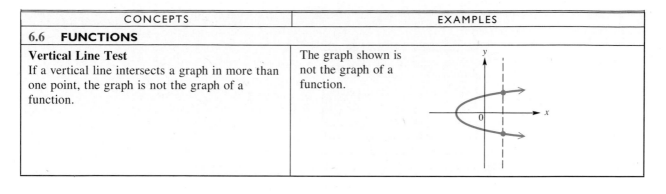

CHAPTER 6 REVIEW EXERCISES

[6.1] *Complete the given ordered pairs for the equation.*

1. $y = 3x + 2$; $(-1,\quad)(0,\quad)(\quad, 5)$
2. $4x + 3y = 6$; $(0,\quad)(\quad, 0)(-2,\quad)$
3. $x - 3y$; $(0,\quad)(8,\quad)(\quad, -3)$
4. $x - 7 = 0$; $(\quad, -3)(\quad, 0)(\quad, 5)$

Decide whether the given ordered pair is a solution of the given equation.

5. $x + y = 7$; $(2, 5)$ **6.** $2x + y = 5$; $(-1, 3)$

7. $3x - y = 4$; $\left(\dfrac{1}{3}, -3\right)$ **8.** $5x - 3y = 16$; $\left(0, -\dfrac{16}{3}\right)$

Plot the ordered pair in a rectangular coordinate system.

9. $(2, 3)$ **10.** $(-4, 2)$ **11.** $(3, 0)$ **12.** $(0, -6)$

13. If $xy > 0$, in what quadrant or quadrants must (x, y) lie?

14. On what axis does the point $(k, 0)$ lie for any real value of k?

Without plotting the point, name the quadrant in which it lies.

15. $(-3, 5)$ **16.** $(-6, -9.2)$

[6.2] *Find (a) the x-intercept and (b) the y-intercept for the line that is the graph of the equation.*

17. $y = 2x + 5$ **18.** $2x + y = -7$ **19.** $3x + 2y = 8$

Graph the linear equation.

20. $y = -2x - 5$ **21.** $x + 2y = -4$ **22.** $x + y = 0$

[6.3] *Find the slope of the line.*

23. Through $(2, 3)$ and $(-4, 6)$ **24.** Through $(0, 0)$ and $(-3, 2)$
25. Through $(0, 6)$ and $(1, 6)$ **26.** Through $(2, 5)$ and $(2, 8)$

27. $y = 3x - 4$

28. $y = \dfrac{2}{3}x + 1$

29.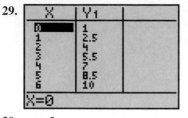

30. $y = 5$

31. $x = 0$

32. A line parallel to the graph of $y = 2x + 3$

33. A line perpendicular to the graph of $y = -3x + 3$

34. Explain why the signs of the slopes of perpendicular lines cannot be the same.

Decide whether the lines in the given pair are parallel, perpendicular, or neither.

35. $3x + 2y = 6$
$6x + 4y = 8$

36. $x - 3y = 1$
$3x + y = 4$

37. $x - 2y = 8$
$x + 2y = 8$

38. What is the slope of a line perpendicular to a line with undefined slope?

[6.4] *Write an equation for the line. Express it in the form $Ax + By = C$.*

39. $m = -1;\quad b = \dfrac{2}{3}$

40. The line in Exercise 23

41. Through $(4, -3);\quad m = 1$

42. Through $(-1, 4);\quad m = \dfrac{2}{3}$

43. Through $(1, -1);\quad m = -\dfrac{3}{4}$

44. Through $(2, 1)$ and $(-2, 2)$

45. Through $(-4, 1)$ with slope 0

46. Through $\left(\dfrac{1}{3}, -\dfrac{5}{4}\right)$ with undefined slope

[6.5] *Complete the graph of the linear inequality by shading the correct region.*

47. $x - y \geq 3$

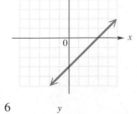

48. $3x - y \leq 5$

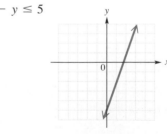

49. $x + 2y < 6$

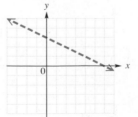

Graph the linear inequality.

50. $3x + 5y > 9$ **51.** $2x - 3y > -6$ **52.** $x - 2y \geq 0$

[6.6] *Decide whether the relation is or is not a function. In Exercises 53 and 54, give the domain and the range.*

53. $\{(-2, 4), (0, 8), (2, 5), (2, 3)\}$ **54.** $\{(8, 3), (7, 4), (6, 5), (5, 6), (4, 7)\}$

55.   **56.**

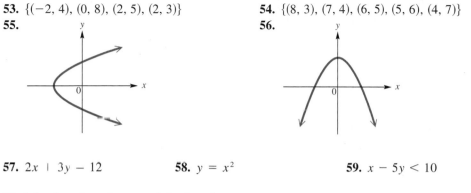

57. $2x + 3y - 12$ **58.** $y = x^2$ **59.** $x - 5y < 10$

Find the domain and range of the function.

60. $4x - 3y = 12$ **61.** $y = x^2 + 1$ **62.** $y = |x - 1|$

*Find (**a**) $f(2)$ and (**b**) $f(-1)$.*

63. $f(x) = 3x + 2$ **64.** $f(x) = 2x^2 - 1$ **65.** $f(x) = |x + 3|$

MIXED REVIEW EXERCISES

Find the intercepts and the slope of the line.

66. $11x - 3y = 4$ **67.** Through $(4, -1)$ and $(-2, -3)$

68. Through $(0, -1)$ and $(9, -5)$ **69.** $8x = 6 - 3y$

Write an equation in the form $Ax + By = C$ for the line.

70. Through $(5, 0)$ and $(5, -1)$ **71.** $m = -\dfrac{1}{4}; b = -\dfrac{5}{4}$

72. Through $(8, 6); m = -3$ **73.** Through $(3, -5)$ and $(-4, -1)$

Graph the equation or the inequality.

74. $x - 2y \leq 6$ **75.** $x + 3y = 0$ **76.** $y < -4x$

77. $y - 5 = 0$ **78.** $2x - y = 3$ **79.** $x \geq -4$

80. Find $f(2)$ if $f(x) = (3 - 2x)^2$.

81. What kind of inequality has a dashed line as its boundary? Explain why.

82. Two points determine a line. Explain why it is a good idea to plot three points on the graph of a line before drawing the line.

───────◆ **MATHEMATICAL CONNECTIONS** (Exercises 83–90) ◆───────

Use the concepts of this chapter in working Exercises 83–90 in order.

83. Plot the points $(-2, 4)$ and $(3, -11)$ and draw the line joining them.

84. Find the slope of the line in Exercise 83.

85. Find the slope-intercept form of the equation of the line in Exercise 83.

86. What is the x-intercept of this line?

87. What is the y-intercept of this line?

88. Suppose that $y = mx + b$ is the equation of the line. Graph the inequality $y > mx + b$.

89. Replace y with $f(x)$ in your answer to Exercise 85. Find $f(10)$.

90. Give the domain and the range of the function described in Exercise 89.

───────────◆───────────

CHAPTER 6 TEST

Use the graphs shown to respond to Exercises 1–3.

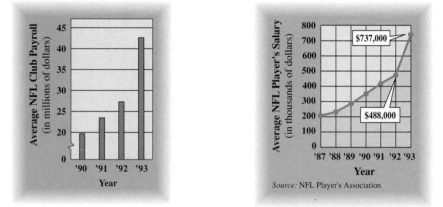

1. In what year was the average National Football League payroll $23.5 million?

2. Which one of the following is the best estimate for the average National Football League player's salary in 1989?
(a) $239,000   (b) $295,000   (c) $415,000

3. What was the percent increase in the average player's salary from 1992 to 1993 (the first year of free agency)?

Complete the ordered pairs for the given equation.

4. $y = 4x - 9$; (0,   ), (1,   ), (, -3)

5. $3x + 5y = -30$; (0,   ), (, 0), (, 3)

6. $y + 12 = 0$; (0,   ), $(-4,$ $)$, $\left(\dfrac{5}{2},$ $\right)$

7. How do you find the y-intercept and the x-intercept for a linear equation in two variables?

Graph the linear equation. Give the x- and y-intercepts.

8. $x - y = 4$ **9.** $3x + y = 6$ **10.** $y - 2x = 0$

11. $x + 3 = 0$ **12.** $y = 1$

Find the slope of the line in Exercises 13–17.

13. Through $(-4, 6)$ and $(-1, -2)$ **14.** $2x + y = 10$

15.

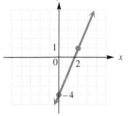

16. (These are two different views of the same line.)

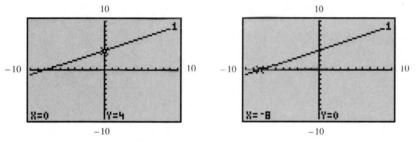

17. (a) A line parallel to the graph of $y = -4x + 6$
    (b) A line perpendicular to the graph of $y = -4x + 6$

Write an equation for the line. Express in the form $Ax + By = C$.

18. Through $(-1, 4)$ with slope 2   **19.** The line in Exercise 15

20. The line in Exercise 16

Graph the linear inequality.

21. $x + y \le 3$ **22.** $3x - y > 0$

23. If $f(x) = -4x + 8$, find $f(-3)$.

24. Decide whether the graph or the equation is that of a function. If it is, give the domain and the range.

(a)   **(b)** $y = x^2 + 2$

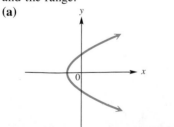

25. Why does the graph of $y = mx + b$, for real numbers m and b, satisfy the conditions for the graph of a function?

CUMULATIVE REVIEW (Chapters 1–6)

Solve the equation.

1. $-5(8 - 2z) + 4(7 - z) = 7(8 + z) - 3$

2. $A = p + prt$ for t

3. $7x^2 + 8x + 1 = 0$

4. $\dfrac{4}{x - 2} = 4$

5. $\dfrac{2}{x - 1} = \dfrac{5}{x - 1} - \dfrac{3}{4}$

Solve the inequality and graph the solutions.

6. $-2.5x < 6.5$

7. $4(x + 3) - 5x < 12$

8. $\dfrac{2}{3}y - \dfrac{1}{6}y \le -2$

Write with only positive exponents. Assume that all variables represent positive real numbers.

9. $(x^2y^{-3})(x^{-4}y^2)$

10. $\dfrac{x^{-6}y^3z^{-1}}{x^7y^{-4}z}$

11. $(2m^{-2}n^3)^{-3}$

Perform the indicated operations.

12. $2(3x^2 - 8x + 1) - 4(x^2 - 3x - 9)$

13. $(3x + 2y)(5x - y)$

14. $(x + 2y)(x^2 - 2xy + 4y^2)$

15. $\dfrac{m^3 - 3m^2 + 5m - 3}{m - 1}$

Factor the polynomial completely.

16. $y^2 + 4yk - 12k^2$

17. $9x^4 - 25y^2$

18. $125x^4 - 400x^3y + 195x^2y^2$

19. $f^2 + 20f + 100$

20. $100x^2 + 49$

Perform the indicated operation. Express the answer in lowest terms.

21. $\dfrac{3}{2x + 6} + \dfrac{2x + 3}{2x + 6}$

22. $\dfrac{8}{x + 1} - \dfrac{2}{x + 3}$

23. $\dfrac{x^2 - 25}{3x + 6} \cdot \dfrac{4x + 8}{x^2 + 10x + 25}$

24. $\dfrac{x^2 + 2x - 3}{x^2 - 5x + 4} \cdot \dfrac{x^2 - 3x - 4}{x^2 + 3x}$

25. $\dfrac{x^2 + 5x + 6}{3x} \div \dfrac{x^2 - 4}{x^2 + x - 6}$

26. $\dfrac{6x^4y^3z^2}{8xyz^4} \div \dfrac{3x^2}{16y^2}$

Simplify the complex fraction.

27. $\dfrac{\dfrac{2}{3} - \dfrac{1}{4}}{\dfrac{1}{2} + \dfrac{1}{6}}$

28. $\dfrac{\dfrac{12}{x + 6}}{\dfrac{4}{2x + 12}}$

Find the slope of the line described.

29. Through $(-4, 5)$ and $(2, -3)$

30. Horizontal, through $(4, 5)$

Find the equation of the line described. Express it in the form $Ax + By = C$.

31. Through $(4, -1)$, $m = -4$

32. Through $(0, 0)$ and $(1, 4)$

Graph the equation or the inequality.

33. $-3x + 4y = 12$ **34.** $y \le 2x - 6$ **35.** $3x + 2y < 0$

Solve the problem.

36. The audience at a concert included 36 more men than women. If there was a total of 196 men and women there, how many men and how many women attended the concert?

37. Find the measure of each angle of the triangle.

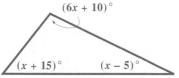

38. The length of the shorter leg of a right triangle is tripled and 4 inches is added to the result, giving the length of the hypotenuse. The longer leg is 10 inches longer than twice the shorter leg. Find the length of the shorter leg of the triangle.

39. If x varies directly as y and $x = 4$ when $y = 12$, find x when $y = 42$.

40. If a man can mow his lawn in 3 hours and his wife can do the same job in 1.5 hours, how long will it take them to do the job together?

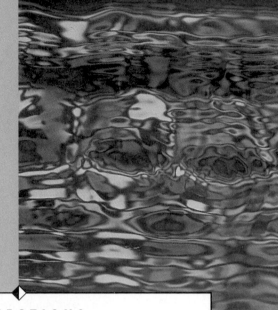

CHAPTER 7

LINEAR SYSTEMS

CONNECTIONS

In more advanced mathematics courses, systems of equations and inequalities in more than two variables are studied. Such systems have many useful applications. One such application occurred with the study of data sent back to Earth from Mars from the U.S. Viking spacecrafts. On July 20, 1976, Viking I landed on Mars and was joined by Viking II several months later. Teams of scientists obtained banks of useful information by studying photos sent back to Earth via radio beams. For example, they found the exact location of the two spacecrafts, the size of Mars, the orientation of its axis, and its rate of spin. All of these measurements were obtained using systems of equations.

As another example, in many practical problems we want to maximize or minimize some quantity, while satisfying various restrictions or constraints. *Linear programming* is used to do just that. First of all, the problem has to be clarified. Quantities involved must be designated as variables, and the objective (to be maximized or minimized) written as a function. The restrictions are usually expressed as inequalities that are rewritten as equations by introducing additional variables. The model is then treated as a large system of equations and solved by a method that is an extension of the method introduced in Section 7.3.

The procedures for solving linear programming problems were developed in 1947 by George Dantzig while he was working on a problem of allocating supplies for the Air Force in a way that minimized total cost.

FOR DISCUSSION OR WRITING

Look up the subject of linear programming to find out more about the process. Many college algebra books have short introductions to linear programming that require only a knowledge of graphing linear equations and solving systems of linear equations. Books on finite mathematics usually have at least one chapter on the topic.

When a number of equations in several variables are considered simultaneously, we have what is known as a system of equations. In this chapter we study linear systems of equations and inequalities, concentrating on systems with two variables.

7.1 SOLVING SYSTEMS OF LINEAR EQUATIONS BY GRAPHING

OBJECTIVES

1 ▶ Decide whether a given ordered pair is a solution of a system.
2 ▶ Solve linear systems by graphing.
3 ▶ Identify systems with no solutions or with an infinite number of solutions.
4 ▶ Identify inconsistent systems or systems with dependent equations without graphing.

A **system of linear equations** consists of two or more linear equations with the same variables. Examples of systems of linear equations include

$$2x + 3y = 4 \qquad x + 3y = 1 \qquad x - y = 1$$
$$3x - y = -5 \qquad -y = 4 - 2x \qquad \text{and} \qquad y = 3.$$

In the system on the right, think of $y = 3$ as an equation in two variables by writing it as $0x + y = 3$.

1 ▶ Decide whether a given ordered pair is a solution of a system.

▶ The **solution of a system** of linear equations includes all the ordered pairs that make both equations true at the same time.

E X A M P L E 1

Determining Whether an Ordered Pair Is a Solution

Is $(4, -3)$ a solution of the following systems?

(a) $x + 4y = -8$
$3x + 2y = 6$

To decide whether $(4, -3)$ is a solution of the system, we substitute 4 for x and -3 for y in each equation.

$$
\begin{array}{ll}
\quad x + 4y = -8 & \qquad\qquad 3x + 2y = 6 \\
4 + 4(-3) = -8 \quad ? & \qquad 3(4) + 2(-3) = 6 \quad ? \\
4 + (-12) = -8 \quad ? & \qquad\quad 12 + (-6) = 6 \quad ? \\
\qquad\quad -8 = -8 \quad \text{True} & \qquad\qquad\qquad 6 = 6 \quad \text{True}
\end{array}
$$

Since $(4, -3)$ satisfies both equations, it is a solution.

(b) $2x + 5y = -7$
$3x + 4y = 2$

Again, substitute 4 for x and -3 for y in both equations.

$$
\begin{array}{ll}
\quad 2x + 5y = -7 & \qquad\qquad 3x + 4y = 2 \\
2(4) + 5(-3) = -7 \quad ? & \qquad 3(4) + 4(-3) = 2 \quad ? \\
8 + (-15) = -7 \quad ? & \qquad\quad 12 + (-12) = 2 \quad ? \\
\qquad\quad -7 = -7 \quad \text{True} & \qquad\qquad\qquad 0 = 2 \quad \text{False}
\end{array}
$$

Here $(4, -3)$ is not a solution since it does not satisfy the second equation. ■

We discuss several methods of solving a system of two linear equations with two variables in this chapter.

2 ▶ Solve linear systems by graphing.

▶   One way to find the solution of a system of two linear equations is to graph both equations on the same axes. The graph of each line shows points whose coordinates satisfy the equation of that line. The coordinates of any point where the lines intersect give a solution of the system. Since two *different* straight lines can intersect at no more than one point, there can never be more than one solution for such a system.

E X A M P L E   2

Solving a System by Graphing

■ Solve the following system of equations by graphing both equations on the same axes.

$$2x + 3y = 4$$
$$3x - y = -5$$

As shown in Chapter 6, we graph these two equations by plotting several points for each line. Some ordered pairs that satisfy each equation are shown below.

$2x + 3y = 4$		$3x - y = -5$	
x	y	x	y
0	$\frac{4}{3}$	0	5
2	0	$-\frac{5}{3}$	0
-2	$\frac{8}{3}$	-2	-1

The lines in Figure 1 suggest that the graphs intersect at the point $(-1, 2)$. Check this by substituting -1 for x and 2 for y in both equations. Since $(-1, 2)$ satisfies both equations, the solution of this system is $(-1, 2)$.   ■

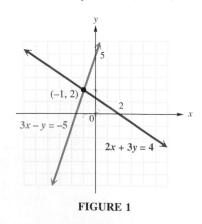

FIGURE 1

CAUTION A difficulty with the graphing method of solution is that it may not be possible to determine from the graph the exact coordinates of the point that represents the solution. For this reason, we give algebraic methods of solution later in this chapter. The graphing method does, however, show geometrically how solutions are found.

3 ▶ Identify systems with no solutions or with an infinite number of solutions.

▶ Sometimes the graphs of the two equations in a system either do not intersect at all or are the same line, as in the systems of Example 3.

E X A M P L E   3 ▪
Solving Special Systems

Solve each system by graphing.

(a) $2x + y = 2$
$2x + y = 8$

The graphs of these lines are shown in Figure 2. The two lines are parallel and have no points in common. For a system whose equations lead to graphs with no points in common, we write "no solution."

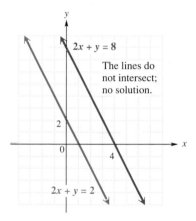

FIGURE 2

(b) $2x + 5y = 1$
$6x + 15y = 3$

The graphs of these two equations are the same line. See Figure 3. The second equation can be obtained by multiplying both sides of the first equation by 3. In this case, every point on the line is a solution of the system, and the solution is an infinite number of ordered pairs. We write "infinite number of solutions" to indicate this case. ▪

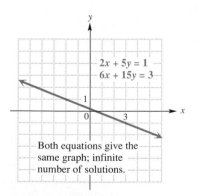

FIGURE 3

The system in Example 2 has exactly one solution. A system with a solution is called a **consistent system.** A system of equations with no solution, such as the one in Example 3(a), is called an **inconsistent system.** The equations in Example 2 are **independent equations,** equations that have different graphs. The equations of the system in Example 3(b) have the same graph. Because they are different forms of the same equation, these equations are called **dependent equations.** Examples 2 and 3 show the three cases that may occur in a system of equations with two unknowns.

Possible Types of Solutions

1. The graphs cross at exactly one point, which gives the (single) solution of the system. The **system is consistent** and the **equations are independent.**

2. The graphs are parallel lines, so there is no solution. The **system is inconsistent.**

3. The graphs are the same line. There are an infinite number of solutions. The **equations are dependent.**

---

4 ▶ Identify inconsistent systems or systems with dependent equations without graphing.

▶ Example 3 showed that the graphs of an inconsistent system are parallel lines and the graphs of a system of dependent equations are the same line. We can recognize these special kinds of systems without graphing by using slopes.

EXAMPLE 4

Identifying the Three Cases Using Slopes

Describe each system without graphing.

(a) $3x + 2y = 6$
$-2y = 3x - 5$

Write each equation in slope-intercept form by solving for y.

$$3x + 2y = 6 \qquad\qquad -2y = 3x - 5$$

$$2y = -3x + 6 \qquad\qquad y = -\frac{3}{2}x + \frac{5}{2}$$

$$y = -\frac{3}{2}x + 3$$

Both equations have a slope of $-3/2$ but they have different y-intercepts. The previous chapter showed that lines with the same slope are parallel, so these equations have graphs that are parallel lines. The system has no solution.

(b) $2x - y = 4$

$$x = \frac{y}{2} + 2$$

Again, write the equations in slope-intercept form.

$$2x - y = 4 \qquad\qquad x = \frac{y}{2} + 2$$

$$-y = -2x + 4 \qquad\qquad \frac{y}{2} + 2 = x$$

$$y = 2x - 4 \qquad\qquad \frac{y}{2} = x - 2$$

$$\qquad\qquad\qquad\qquad y = 2x - 4$$

The equations are exactly the same; their graphs are the same line. The system has an infinite number of solutions.

(c) $x - 3y = 5$
$2x + y = 8$

In slope-intercept form, the equations are as follows.

$$x - 3y = 5 \qquad\qquad 2x + y = 8$$

$$-3y = -x + 5 \qquad\qquad y = -2x + 8$$

$$y = \frac{1}{3}x - \frac{5}{3}$$

The graphs of these equations are neither parallel lines nor the same line since the slopes are different. There will be exactly one solution to this system. ■

7.1 EXERCISES

1. When a student was asked to determine whether the ordered pair $(1, -2)$ is a solution of the system

$$x + y = -1$$
$$2x + y = 4,$$

he answered "yes." His reasoning was that the ordered pair satisfies the equation $x + y = -1$: $1 + (-2) = -1$ is true. Why is the student's answer wrong?

2. Let a, b, c, d, e, and f represent six consecutive integers. (For example, if $a = 2$, then we have $b = 3$, $c = 4$, $d = 5$, $e = 6$, and $f = 7$.) You may start with *any* value for a. Then, substitute them into the system

$$ax + by = c$$
$$dx + ey = f.$$

(For example, starting with 2 we would get the equations $2x + 3y = 4$ and $5x + 6y = 7$.) Now show that $(-1, 2)$ is a solution of the system you just made up.

Decide whether the given ordered pair is a solution of the given system. See Example 1.

3. $(6, 2)$
$$3x + y = 20$$
$$2x + 3y = 18$$

4. $(3, 4)$
$$2x + y = 10$$
$$3x + 2y = 17$$

5. $(2, -3)$
$$x + y = -1$$
$$2x + 5y = 19$$

6. $(4, 3)$
$$x + 2y = 10$$
$$3x + 5y = 3$$

7. $(-1, -3)$
$$3x + 5y = -18$$
$$4x + 2y = -10$$

8. $(-9, -2)$
$$2x - 5y = -8$$
$$3x + 6y = -39$$

9. $(7, -2)$
$$4x = 26 - y$$
$$3x = 29 + 4y$$

10. $(9, 1)$
$$2x = 23 - 5y$$
$$3x = 24 + 3y$$

11. $(6, -8)$
$$-2y = x + 10$$
$$3y = 2x + 30$$

12. $(-5, 2)$
$$5y = 3x + 20$$
$$3y = -2x - 4$$

13. Which one of the ordered pairs below could possibly be a solution of the system graphed? Why is it the only valid choice?

(a) $(2, 2)$ **(b)** $(-2, 2)$
(c) $(-2, -2)$ **(d)** $(2, -2)$

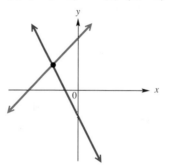

14. Which one of the ordered pairs below could possibly be a solution of the system graphed? Why is it the only valid choice?

(a) $(2, 0)$ **(b)** $(0, 2)$
(c) $(-2, 0)$ **(d)** $(0, -2)$

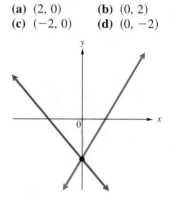

Solve the system of equations by graphing both equations on the same axes. If the system is inconsistent or the equations are dependent, say so. See Examples 2 and 3.

15. $x - y = 2$
$ x + y = 6$

16. $x - y = 3$
$ x + y = -1$

17. $x + y = 4$
$ y - x = 4$

18. $x + y = -5$
$ x - y = 5$

19. $x - 2y = 6$
$ x + 2y = 2$

20. $2x - y = 4$
$ 4x + y = 2$

21. $ 3x - 2y = -3$
$-3x - y = -6$

22. $2x - y = 4$
$ 2x + 3y = 12$

23. $2x - 3y = -6$
$ y = -3x + 2$

24. $ x + 2y = 4$
$ 2x + 4y = 12$

25. $2x - y = 6$
$ 4x - 2y = 8$

26. $2x - y = 4$
$ 4x = 2y + 8$

27. $3x = 5 - y$
$ 6x + 2y = 10$

28. $-3x + y = -3$
$ y = x - 3$

29. $3x - 4y = 24$
$$y = -\frac{3}{2}x + 3$$

30. $3x - 2y = 12$
$ y = -4x + 5$

31. Explain why a system of two linear equations cannot have exactly two solutions.

32. Explain the difficulties that may be encountered when solving a system of two linear equations using the graphing method.

33. Find a system of equations with the solution $(-2, 3)$, and show the graph.

34. Graph the system

$$2x + 3y = 6$$
$$x - 3y = 5.$$

Can you check your answer? What is the problem?

Without graphing, answer the following questions for the linear system. See Example 4.
(a) *Is the system inconsistent, are the equations dependent, or neither?*
(b) *Is the graph a pair of intersecting lines, a pair of parallel lines, or one line?*
(c) *Does the system have one solution, no solution, or an infinite number of solutions?*

35. $y - x = -5$
$x + y = 1$

36. $2x + y = 6$
$x - 3y = -4$

37. $x + 2y = 0$
$4y = -2x$

38. $4x + y = 2$
$2x - y = 4$

39. $3x - 2y = -3$
$3x + y = 6$

40. $y = 3x$
$y + 3 = 3x$

41. $5x + 4y = 7$
$10x + 8y = 4$

42. $2x + 3y = 12$
$2x - y = 4$

An application of mathematics in the study of economics deals with **supply and demand.** Typically, as the price of an item increases, the demand for the item decreases, while the supply increases. (There are exceptions to this, however.) If supply and demand can be described by straight line equations, the point at which the lines intersect determines the equilibrium supply and equilibrium demand. Suppose that an economist has studied the supply and demand for aluminum siding and has come up with the conclusion that the price per unit, p, and the demand, x, are related by the demand equation $p = 60 - \dfrac{3}{4}x$, while the supply is given by the equation $p = \dfrac{3}{4}x$. The graphs of these two equations are shown here.

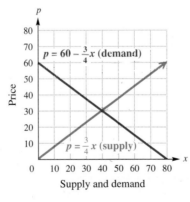

Supply and demand

Use this graph to answer the question.

43. At what value of x does supply equal demand?

44. At what value of p does supply equal demand?

45. What are the coordinates of the point of intersection of the two lines?

46. When $x > 40$, does demand exceed supply or does supply exceed demand?

◆ **MATHEMATICAL CONNECTIONS** (Exercises 47–52) ◆

In this group of exercises we show the connections between solving linear equations in one variable and solving a linear system of two equations in two variables. Work through Exercises 47–52 in order.

47. Solve the linear equation $\frac{1}{2}x + 4 = 3x - 1$ using the methods described in Chapter 2.

48. Check your solution for the equation in Exercise 47 by substituting back into the original equation. What is the value that you get for both the left and right sides after you make this substitution?

49. Graph the linear system

$$y = \frac{1}{2}x + 4$$
$$y = 3x - 1$$

and find the solution of the system.

50. How does the *x*-coordinate of the solution of the system in Exercise 49 compare to the solution of the linear equation in Exercise 47?

51. How does the *y*-coordinate of the solution of the system in Exercise 49 compare to the value you obtained in the check in Exercise 48?

52. Based on your observations in Exercises 47–51, fill in the blanks with the correct responses.

The solution of the linear equation $\frac{2}{3}x + 3 = -x + 8$ is 3. When we substitute 3

back into the equation, we get the value _____ on both sides, verifying that 3 is indeed the solution. Now if we graph the system

$$y = \frac{2}{3}x + 3$$
$$y = -x + 8,$$

the ordered pair that is the solution of the system is (_____ , _____).

◆

REVIEW EXERCISES

Solve the equation for y. See Section 2.5.

53. $3x + y = 4$ **54.** $-2x + y = 9$

55. $9x - 2y = 4$ **56.** $5x - 3y = 12$

Solve the equation. Check the solution. See Sections 2.2, 2.3, and 5.6.

57. $-2(y - 2) + 5y = 10$ **58.** $2m - 3(4 - m) = 8$

59. $p + 4(6 - 2p) = 3$ **60.** $4\left(\frac{3 - 2k}{2}\right) + 3k = -3$

61. $4x - 2\left(\frac{1 - 3x}{2}\right) = 6$ **62.** $a + 3\left(\frac{1 - a}{2}\right) = 5$

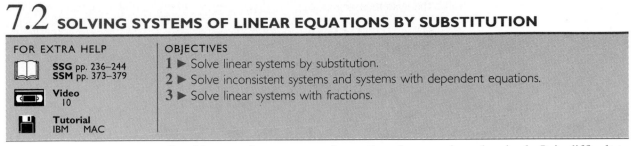

7.2 SOLVING SYSTEMS OF LINEAR EQUATIONS BY SUBSTITUTION

FOR EXTRA HELP

SSG pp. 236–244
SSM pp. 373–379

Video
10

Tutorial
IBM   MAC

OBJECTIVES

1 ▶ Solve linear systems by substitution.
2 ▶ Solve inconsistent systems and systems with dependent equations.
3 ▶ Solve linear systems with fractions.

Graphing to solve a system of equations has a serious drawback. It is difficult to estimate a solution such as $(1/3, -5/6)$ accurately from a graph (unless a graphics calculator is used).

◆ **C O N N E C T I O N S** ◆

We can use a graphics calculator to solve a linear system of equations. For instance, to solve the system of Example 1(a) in Section 7.1 with a grapher, we begin by solving each equation for y.

$$x + 4y = -8 \qquad\qquad 3x + 2y = 6$$
$$4y = -x - 8 \qquad\qquad 2y = -3x + 6$$
$$y = -\frac{1}{4}x - 2 \qquad\qquad y = -\frac{3}{2}x + 3$$

The grapher allows us to input several equations to be graphed on the same axes. To keep the two equations separate, we designate the first one y_1 and the second one y_2. We graph the two equations using a standard window. Now we use trace and zoom to find the coordinates of the point of intersection. Some graphics calculators will give the point of intersection by using a special menu, as in the graph shown below.

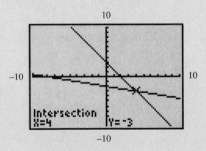

FOR DISCUSSION OR WRITING
Use a graphics calculator to solve the systems of Example 2(a) and Example 3 in Section 7.1. Compare your solutions with those shown in the text.

1 ▶ Solve linear systems by substitution.

▶   An algebraic method involving substitution is particularly useful for solving systems where one equation is solved, or can be solved easily, for one of the variables.

EXAMPLE 1

Solving a System
by Substitution

Solve the system

$$3x + 5y = 26$$
$$y = 2x.$$

The second of these two equations says that $y = 2x$. Substituting $2x$ for y in the first equation gives

$$3x + 5y = 26$$
$$3x + 5(2x) = 26 \qquad \text{Let } y = 2x.$$
$$3x + 10x = 26$$
$$13x = 26$$
$$x = 2.$$

Since $y = 2x$ and $x = 2$, $y = 2(2) = 4$. Check that the solution of the given system is $(2, 4)$. ■

A summary of the steps used in solving a system by substitution is given here.

Solving Linear Systems by Substitution

Step 1 Solve one of the equations for either variable. (If one of the variables has coefficient 1 or -1, choose it, since the substitution method is usually easier this way.)

Step 2 Substitute for that variable in the other equation. The result should be an equation with just one variable.

Step 3 Solve the equation from Step 2.

Step 4 Substitute the result from Step 3 into the equation from Step 1 to find the value of the other variable.

Step 5 Check the solution in both of the given equations.

Notice how the steps are used in the next example.

EXAMPLE 2

Solving a System
by Substitution

Use substitution to solve the system

$$2x + 3y = 8$$
$$4x + 3y = 4.$$

Step 1 The substitution method requires that an equation be solved for one of the variables. There are several ways we can begin. Let us choose the first equation of the system and solve it for x.

$$2x + 3y = 8$$
$$2x = 8 - 3y \qquad \text{Subtract } 3y \text{ from both sides.}$$
$$x = \frac{8 - 3y}{2} \qquad \text{Divide both sides by 2.}$$

Step 2 Now we substitute this value for x in the second equation of the system.

$$4x + 3y = 4$$
$$4\left(\frac{8 - 3y}{2}\right) + 3y = 4 \qquad \text{Let } x = \frac{8 - 3y}{2}.$$

Step 3   Solve this equation.

$$2(8 - 3y) + 3y = 4 \qquad \text{Divide 4 by 2.}$$
$$16 - 6y + 3y = 4 \qquad \text{Distributive property}$$
$$-3y = -12 \qquad \text{Combine terms; subtract 16.}$$
$$y = 4 \qquad \text{Divide by } -3.$$

Step 4   Find x by letting $y = 4$ in $x = \dfrac{8 - 3y}{2}$.

$$x = \frac{8 - 3 \cdot 4}{2} = \frac{8 - 12}{2} = \frac{-4}{2} = -2$$

Step 5   The solution of the given system is $(-2, 4)$. Check this solution in both equations. ■

N O T E   The addition method, explained in the next section, would require less work in Example 2. We show this example here because some systems studied in later courses *must* be solved by the substitution method.

---

2 ▶ Solve inconsistent systems and systems with dependent equations.

▶   In the previous section we solved inconsistent systems with graphs that are parallel lines and systems of dependent equations with graphs that are the same line. We can also solve these systems with the substitution method.

E X A M P L E   3 ■

Solving an Inconsistent System by Substitution

Use substitution to solve the system

$$x = 5 - 2y$$
$$2x + 4y = 6.$$

Substitute $5 - 2y$ for x in the second equation.

$$2x + 4y = 6$$
$$2(5 - 2y) + 4y = 6 \qquad \text{Let } x = 5 - 2y.$$
$$10 - 4y + 4y = 6 \qquad \text{Distributive property}$$
$$10 = 6 \qquad \text{False}$$

This false result means that the system is inconsistent and has no solution. The equations of the system have graphs that are parallel lines. See Figure 4. ■

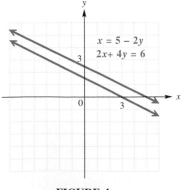

FIGURE 4

EXAMPLE 4

Solving a System with
Dependent Equations
by Substitution

Solve the following system by the substitution method.

$$3x - y = 4 \quad (1)$$
$$-9x + 3y = -12 \quad (2)$$

We can begin by solving the first equation for y to get $y = 3x - 4$. Substitute $3x - 4$ for y in equation (2) and solve the resulting equation.

$$-9x + 3(3x - 4) = -12$$
$$-9x + 9x - 12 = -12 \qquad \text{Distributive property}$$
$$0 = 0 \qquad \text{Add 12; combine terms.}$$

This true result means that every solution of one equation is also a solution of the other, so the system has an infinite number of solutions: all the ordered pairs corresponding to points that lie on the common graph. A graph of the equations of this system is shown in Figure 5. ∎

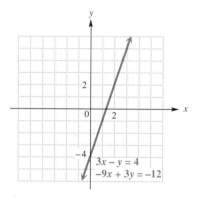

FIGURE 5

3 ▶ Solve linear systems with fractions.

▶ When a system includes equations with fractions as coefficients, we eliminate the fractions by multiplying both sides by a common denominator. Then we solve the resulting system.

EXAMPLE 5

Solving by Substitution
with Fractions
as Coefficients

Solve the system

$$4x + \frac{1}{3}y = \frac{8}{3} \quad (1)$$

$$\frac{1}{2}x + \frac{3}{4}y = -\frac{5}{2}. \quad (2)$$

We begin by eliminating fractions. Clear equation (1) of fractions by multiplying both sides by 3.

$$3\left(4x + \frac{1}{3}y\right) = 3 \cdot \frac{8}{3} \qquad \text{Multiply by 3.}$$

$$3(4x) + 3\left(\frac{1}{3}y\right) = 3 \cdot \frac{8}{3} \qquad \text{Distributive property}$$

$$12x + y = 8 \quad (3)$$

Now clear equation (2) of fractions by multiplying both sides by the common denominator 4.

$$4\left(\frac{1}{2}x + \frac{3}{4}y\right) = 4\left(-\frac{5}{2}\right) \quad \text{Multiply by 4.}$$

$$4\left(\frac{1}{2}x\right) + 4\left(\frac{3}{4}y\right) = 4\left(-\frac{5}{2}\right) \quad \text{Distributive property}$$

$$2x + 3y = -10 \qquad \qquad \text{(4)}$$

The given system of equations has been simplified to

$$12x + \ y = 8 \qquad \qquad \text{(3)}$$
$$2x + 3y = -10. \qquad \qquad \text{(4)}$$

Solve the system by the substitution method. Equation (3) can be solved for y by subtracting $12x$ on each side.

$$12x + y = 8 \qquad \qquad \text{(3)}$$
$$y = -12x + 8$$

Now substitute the result for y in equation (4).

$$2x + 3(-12x + 8) = -10 \quad \text{Let } y = -12x + 8.$$
$$2x - 36x + 24 = -10 \quad \text{Distributive property}$$
$$-34x = -34 \quad \text{Combine terms; subtract 24.}$$
$$x = 1 \quad \text{Divide by } -34.$$

Using $x = 1$ in $y = -12x + 8$ gives $y = -4$. The solution is $(1, -4)$. Check by substituting 1 for x and -4 for y in the original equations. ∎

7.2 EXERCISES

1. If you were to solve the system

$$3x + 2y = 7$$
$$5x - y = 4$$

by substitution, which variable would probably be easier to solve for in the first step? In which equation would you solve for it? Why?

2. Which one of the following systems would be easier to solve using the substitution method? Why?

$$5x - 3y = 7 \qquad 7x + 2y = 4$$
$$2x + 8y = 3 \qquad \quad y = -3x$$

Solve the system by the substitution method. Check the solution. See Examples 1–4.

3. $x + y = 12$
$y = 3x$

4. $x + 3y = -28$
$y = -5x$

5. $3x + 2y = 27$
$x = y + 4$

6. $4x + 3y = -5$
$x = y - 3$

7. $3x + 5y = 25$
$x - 2y = -10$

8. $5x + 2y = -15$
$2x - y = -6$

9. $3x + 4 = -y$
$2x + y = 0$

10. $2x - 5 = -y$
$x + 3y = 0$

11. $7x + 4y = 13$
$x + y = 1$

12. $3x - 2y = 19$
$x + y = 8$

13. $3x - y = 5$
$y = 3x - 5$

14. $4x - y = -3$
$y = 4x + 3$

15. $6x - 8y = 6$
$-3x + 2y = -2$

16. $3x + 2y = 6$
$-6x + 4y = -8$

17. $2x + 8y = 3$
$x = 8 - 4y$

18. $2x + 10y = 3$
$x = 1 - 5y$

19. $12x - 16y = 8$
$3x = 4y + 2$

20. $6x + 9y = 6$
$2x = 2 - 3y$

21. A student solves the system

$$5x - y = 15$$
$$7x + y = 21$$

and finds that $x = 3$, which is the correct value for x. The student gives the solution as "$x = 3$." Is this correct? Explain.

22. A student solves the system

$$x + y = 4$$
$$2x + 2y = 8$$

and obtains the equation $0 = 0$. The student gives the solution as $(0, 0)$. Is this correct? Explain.

23. Professor Brandsma gave the following item on a test in introductory algebra:
Solve the system

$$3x - y = 13$$
$$2x + 5y = 20$$

by the substitution method.
One student worked the problem by solving first for y in the first equation. Another student worked it by solving first for x in the second equation. Both students got the correct solution, $(5, 2)$. Which student, do you think, had less work to do? Explain.

24. When you use the substitution method, how can you tell that a system has
(a) no solutions?
(b) an infinite number of solutions?

In the system given, begin by clearing fractions and then solve the system by substitution. See Example 5.

25. $\dfrac{5}{3}x + 2y = \dfrac{1}{3} + y$

$2x - 3 + \dfrac{y}{3} = -2 + x$

26. $\dfrac{x}{6} + \dfrac{y}{6} = 1$

$-\dfrac{1}{2}x - \dfrac{1}{3}y = -5$

27. $\dfrac{x}{2} - \dfrac{y}{3} = \dfrac{5}{6}$

$\dfrac{x}{5} - \dfrac{y}{4} = \dfrac{1}{10}$

28. $\dfrac{x}{3} - \dfrac{3y}{4} = -\dfrac{1}{2}$

$\dfrac{2x}{3} + \dfrac{y}{2} = 3$

29. $\dfrac{x}{5} + 2y = \dfrac{8}{5}$

$\dfrac{3x}{5} + \dfrac{y}{2} = -\dfrac{7}{10}$

30. $\dfrac{x}{2} + \dfrac{y}{3} = \dfrac{7}{6}$

$\dfrac{x}{4} - \dfrac{3y}{2} = \dfrac{9}{4}$

The calculator screen shows the point of intersection of the lines y_1 and y_2. Substitute the values of x and y indicated at the bottom of the screen into both equations to show that the ordered pair is the correct solution.

31. $y_1 = 3x - 5$

$y_2 = -2x + 10$

32. $y_1 = -\dfrac{1}{2}x + 6$

$y_2 = x + 12$

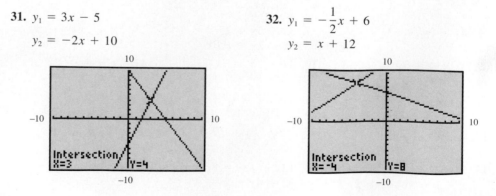

Solve the system by substitution. Then graph both lines in the same viewing window of your calculator and use the intersection feature to support your answer. (In Exercises 37 and 38 you will need to solve for y first before graphing.)

33. $y = 6 - x$
$y = 2x$

34. $y = 4x - 4$
$y = -3x - 11$

35. $y = -\dfrac{4}{3}x + \dfrac{19}{3}$
$y = \dfrac{15}{2}x - \dfrac{5}{2}$

36. $y = -\dfrac{15}{2}x + 10$
$y = \dfrac{25}{3}x - \dfrac{65}{3}$

37. $4x + 5y = 5$
$2x + 3y = 1$

38. $6x + 5y = 13$
$3x + 3y = 4$

39. If the point of intersection does not appear on your screen when solving a linear system using a graphics calculator, how can you find the point of intersection?

40. Suppose that you were asked to solve the system

$$y = 1.73x + 5.28$$
$$y = -2.94x - 3.85.$$

Why would it probably be easier to solve this system using a graphics calculator than using the substitution method?

REVIEW EXERCISES

Add. See Section 3.1.

41. $14x - 3y$
   $\underline{2x + 3y}$

42. $-6x + 8y$
   $\underline{6x + 2y}$

43. $-3x + 7y$
   $\underline{3x - 7y}$

44. What must be added to $-4x$ to get a sum of 0?

45. What must be added to $6y$ to get a sum of 0?

46. What must $4y$ be multiplied by so that when the product is added to $8y$, the sum is 0?

7.3 SOLVING SYSTEMS OF LINEAR EQUATIONS BY ADDITION

FOR EXTRA HELP

SSG pp. 244–249
SSM pp. 380–387

Video
10

Tutorial
IBM   MAC

OBJECTIVES

1 ▶ Solve linear systems by addition.
2 ▶ Multiply one or both equations of a system so that the addition method can be used.
3 ▶ Use an alternative method to find the second value in a solution.
4 ▶ Use the addition method to solve an inconsistent system or a system of dependent equations.

The graphical method and the substitution method for solving systems of linear equations were discussed in the previous sections.

1 ▶ Solve linear systems by addition.

▶   An algebraic method that depends on the addition property of equality can also be used to solve systems. As mentioned earlier, adding the same quantity to each side of an equation results in equal sums.

$$\text{If} \quad A = B, \quad \text{then} \quad A + C = B + C.$$

This addition can be taken a step further. Adding *equal* quantities, rather than the *same* quantity, to both sides of an equation also results in equal sums.

$$\text{If} \quad A = B \quad \text{and} \quad C = D, \quad \text{then} \quad A + C = B + D.$$

The use of the addition property to solve systems is called the **addition method** for solving systems of equations. For most systems, this method is more efficient than graphing.

NOTE When using the addition method, the idea is to eliminate one of the variables. (Because of this, the method is also called the *elimination method.*) To do this, one of the variables must have coefficients that are opposites in the two equations. Keep this in mind throughout the examples in this section.

EXAMPLE 1

Using the Addition Method

■ Use the addition method to solve the system

$$x + y = 5$$
$$x - y = 3.$$

Each equation in this system is a statement of equality, so, as we discussed above, the sum of the right sides equals the sum of the left sides. Adding in this way gives

$$(x + y) + (x - y) = 5 + 3.$$

Combine terms and simplify to get

$$2x = 8$$
$$x = 4. \qquad \text{Divide by 2.}$$

This result, $x = 4$, gives the x-value of the solution of the given system. To find the y-value of the solution, substitute 4 for x in either of the two equations of the system. Choosing the first equation, $x + y = 5$, gives

$$x + y = 5$$
$$4 + y = 5 \qquad \text{Let } x = 4.$$
$$y = 1.$$

The solution, $(4, 1)$, can be checked by substituting 4 for x and 1 for y in both equations of the given system.

$$x + y = 5 \qquad\qquad x - y = 3$$
$$4 + 1 = 5 \quad ? \qquad\qquad 4 - 1 = 3 \quad ?$$
$$5 = 5 \qquad \text{True} \qquad 3 = 3 \qquad \text{True}$$

Since both results are true, the solution of the given system is $(4, 1)$. ■

CAUTION A system is not completely solved until values for *both* x and y are found. Do not make the common mistake of only finding the value of one variable.

In general, we use the following steps to solve a linear system of equations by the addition method.

Solving Linear Systems by Addition

Step 1   Write both equations of the system in the form $Ax + By = C$.
Step 2   Multiply one or both equations by appropriate numbers so that the coefficients of x (or y) are opposites of each other.
Step 3   Add the two equations to get an equation with only one variable.
Step 4   Solve the equation from Step 3.
Step 5   Substitute the solution from Step 4 into either of the original equations to find the value of the remaining variable.
Step 6   Write the solution as an ordered pair and check the answer.

It does not matter which variable is eliminated first. Usually we choose the one that is more convenient to work with.

EXAMPLE 2 ■ Solve the system

Using the Addition Method

$$y = 2x - 11$$
$$4 + 5x + y = 2y + 30$$

Step 1   Rewrite both equations in the form $Ax + By = C$, getting the system

$-10 + y = -11$
$y = -1$

$-2x + y = -11$ Subtract $2x$.
$5x - y = 26$.   Subtract 4 and $2y$.

Step 2   Because the coefficients of y are 1 and -1, it is not necessary to multiply either equation by a number.

Step 3   Add the two equations. This time we use vertical addition.

$$
\begin{array}{rcl}
-2x + y &=& -11 \\
5x - y &=& 26 \\
\hline
3x &=& 15
\end{array}
$$
Add in columns.

Step 4   Solve the equation.

$$3x = 15$$
$$x = 5 \qquad \text{Divide by 3.}$$

Step 5   Find the value of y by substituting 5 for x in either of the original equations. Choosing the first gives

$$y = 2x - 11$$
$$y = 2(5) - 11 \qquad \text{Let } x = 5.$$
$$y = 10 - 11$$
$$y = -1.$$

Step 6   The solution is $(5, -1)$. Check the solution by substitution into both the original equations. Let $x = 5$ and $y = -1$.

$$y = 2x - 11 \qquad\qquad 4 + 5x + y = 2y + 30$$
$$-1 = 2(5) - 11 \ ? \qquad 4 + 5(5) + (-1) = 2(-1) + 30 \ ?$$
$$-1 = -1 \quad \text{True} \qquad\qquad 28 = 28 \quad \text{True}$$

Since $(5, -1)$ is a solution of *both* equations, it is the solution of the system. ■

In the remaining examples, we will not specifically number the steps.

2 ▶ Multiply one or both equations of a system so that the addition method can be used.

▶ In both examples above, a variable was eliminated by the addition step. Sometimes we need to multiply both sides of one or both equations in a system by some number before the addition step will eliminate a variable.

E X A M P L E   3

Multiplying Both Equations when Using the Addition Method

■ Solve the system

$$2x + 3y = -15 \qquad \textbf{(1)}$$
$$5x + 2y = 1. \qquad \textbf{(2)}$$

Adding the two equations gives $7x + 5y = -14$, which does not eliminate either variable. However, we can multiply each equation by a suitable number so that the coefficients of one of the two variables are opposites. For example, here we multiply both sides of equation (1) by 5, and both sides of equation (2) by -2.

$$
\begin{array}{ll}
10x + 15y = -75 & \text{Multiply equation (1) by 5.} \\
\underline{-10x - 4y = -2} & \text{Multiply equation (2) by } -2. \\
11y = -77 & \text{Add.} \\
y = -7 &
\end{array}
$$

Substituting -7 for y in either equation (1) or (2) gives $x = 3$. Check that the solution of the system is $(3, -7)$. ■

3 ▶ Use an alternative method to find the second value in a solution.

▶ In some cases it is easier to find the value of the second variable in a solution by using the addition method twice. The next example shows this approach.

E X A M P L E   4

Finding the Second Value Using an Alternative Method

■ Solve the system

$$4x = 9 - 3y \qquad \textbf{(1)}$$
$$5x - 2y = 8. \qquad \textbf{(2)}$$

Rearrange the terms in equation (1) so that the like terms can be aligned in columns. Add $3y$ to both sides to get the system

$$4x + 3y = 9 \qquad \textbf{(3)}$$
$$5x - 2y = 8.$$

One way to proceed is to eliminate y by multiplying both sides of equation (3) by 2 and both sides of equation (2) by 3, and then adding.

$$
\begin{array}{l}
8x + 6y = 18 \\
\underline{15x - 6y = 24} \\
23x = 42 \\
x = \dfrac{42}{23}
\end{array}
$$

Substituting 42/23 for x in one of the given equations would give y, but the arithmetic involved would be messy. Instead, solve for y by starting again with the original equations and eliminating x. Multiply both sides of equation (3) by 5 and both sides of equation (2) by -4, and then add.

$$
\begin{array}{l}
20x + 15y = 45 \\
\underline{-20x + 8y = -32} \\
23y = 13 \\
y = \dfrac{13}{23}
\end{array}
$$

Check that the solution is (42/23, 13/23). ■

When the value of the first variable is a fraction, the method used in Example 4 avoids errors that often occur when working with fractions. Of course, this method could be used in solving any system of equations.

4 ▶ Use the addition method to solve an inconsistent system or a system of dependent equations.

▶   The next example shows the result of using the addition method when the system is inconsistent or the equations of a system are dependent. To contrast the addition method with the substitution method, in part (b) we use the same system solved in Example 4 in the previous section.

E X A M P L E   5

Using the Addition Method for an Inconsistent System or Dependent Equations

Solve each system by the addition method.

(a) $2x + 4y = 5$
     $4x + 8y = -9$

Multiply both sides of $2x + 4y = 5$ by -2 and then add to $4x + 8y = -9$.

$$\begin{array}{r} -4x - 8y = -10 \\ 4x + 8y = -9 \\ \hline 0 = -19 \end{array} \quad \text{False}$$

The false statement $0 = -19$ shows that the given system has no solution.

(b) $3x - y = 4$
       $9x + 3y = -12$

Multiply both sides of the first equation by 3 and then add the two equations to get

$$\begin{array}{r} 9x - 3y = 12 \\ -9x + 3y = -12 \\ \hline 0 = 0. \end{array} \quad \text{True}$$

We saw in the previous section that this result indicates that every solution of one equation is also a solution of the other; there are an infinite number of solutions. ∎

The possible situations that may occur when using the addition method are described in the following chart.

Summary of Situations that May Occur

One of three situations may occur when the addition method is used to solve a linear system of equations.

1. The result of the addition step is a statement such as $x = 2$ or $y = -3$. The solution will be exactly one ordered pair. The graphs of the equations of the system will intersect at exactly one point.

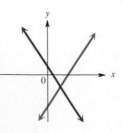

(Continued on the next page.)

2. The result of the addition step is a false statement, such as $0 = 4$. In this case, the graphs are parallel lines and there is no solution for the system.

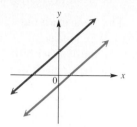

3. The result of the addition step is a true statement, such as $0 = 0$. The graphs of the equations of the system are the same line, and an infinite number of ordered pairs are solutions.

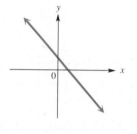

◆ C O N N E C T I O N S ◆

A determinant is a number associated with a square array of numbers written between bars. For example, the determinant of the array

$$\begin{bmatrix} a & b \\ c & d \end{bmatrix} \quad \text{is defined as} \quad \begin{vmatrix} a & b \\ c & d \end{vmatrix} = ad - cb.$$

Determinants can be used to solve linear systems. The addition method can be used to show that the solution of the system of equations

$$ax + by = m$$
$$cx + dy = n$$

is given by

$$x = \frac{D_x}{D} \quad \text{and} \quad y = \frac{D_y}{D},$$

where

$$D = \begin{vmatrix} a & b \\ c & d \end{vmatrix}, \quad D_x = \begin{vmatrix} m & b \\ n & d \end{vmatrix}, \quad D_y = \begin{vmatrix} a & m \\ c & n \end{vmatrix}.$$

For example, to solve the system $3x - 4y = 1$
$$5x + 2y = 19,$$

we find $D = \begin{vmatrix} 3 & -4 \\ 5 & 2 \end{vmatrix} = 3(2) - 5(-4) = 26$, $D_x = \begin{vmatrix} 1 & -4 \\ 19 & 2 \end{vmatrix} =$

$1(2) - 19(-4) = 78$, and $D_y = \begin{vmatrix} 3 & 1 \\ 5 & 19 \end{vmatrix} = 3(19) - 5(1) = 52$.

Thus, $x = \dfrac{D_x}{D} = \dfrac{78}{26} = 3$ and $y = \dfrac{D_y}{D} = \dfrac{52}{26} = 2$.

FOR DISCUSSION OR WRITING
1. Use determinants to solve the system of Example 4 and compare your solution with the one found there.
2. Now use determinants to solve the systems of Example 5. What happens?

7.3 EXERCISES

1. Only one of the following systems does not require that we multiply one or both equations by a constant in order to solve the system by the addition method. Which one is it?

(a) $-4x + 3y = 7$
    $3x - 4y = 4$

(b) $5x + 8y = 13$
    $12x + 24y = 36$

(c) $2x + 3y = 5$
    $x - 3y = 12$

(d) $x + 2y = 9$
    $3x - y = 6$

2. For the system
$$2x + 12y = 7$$
$$3x + 4y = 1,$$

if we were to multiply the first (top) equation by -3, by what number would we have to multiply the second (bottom) equation in order to
(a) eliminate the x terms when solving by the addition method?
(b) eliminate the y terms when solving by the addition method?

Solve the system by the addition method. Check your answer. See Examples 1, 3, and 4.

3. $x - y = -2$
   $x + y = 10$

4. $x + y = 10$
   $x - y = -6$

5. $x + y = 2$
   $2x - y = -5$

6. $3x - y = -12$
   $x + y = 4$

7. $2x + y = -5$
   $x - y = 2$

8. $2x + y = -15$
   $-x - y = 10$

9. $3x + 2y = 0$
   $-3x - y = 3$

10. $5x - y = 5$
    $-5x + 2y = 0$

11. $6x - y = -1$
    $-6x + 5y = 17$

12. $6x + y = 9$
    $-6x + 3y = 15$

13. $2x - y = 12$
    $3x + 2y = -3$

14. $x + y = 3$
    $-3x + 2y = -19$

15. $x + 3y = 19$
    $2x - y = 10$

16. $4x - 3y = -19$
    $2x + y = 13$

17. $x + 4y = 16$
    $3x + 5y = 20$

18. $2x + y = 8$
    $5x - 2y = -16$

19. $5x - 3y = -20$
    $-3x + 6y = 12$

20. $4x + 3y = -28$
    $5x - 6y = -35$

21. $3x - 2y = 22$
    $-5x + 4y = -36$

22. $-4x + 3y = -18$
    $5x - 6y = 18$

23. Explain why $(0, 0)$ *must* be a solution for the system
$$Ax + By = 0$$
$$Cx + Dy = 0$$

for any choices of A, B, C, and D.

24. Without actually solving the system, explain why

$$x + y = 1$$
$$x + y = 2$$

can have no solutions.

25. Which one of these systems would you find more difficult to solve by addition, and why?

$$
\begin{aligned}
2x + y &= 5 \\
5x + 3y &= 11
\end{aligned}
\qquad
\begin{aligned}
4x - 3y &= -7 \\
6x + 5y &= 18
\end{aligned}
$$

26. Why would it be easier to solve

$$
\begin{aligned}
8x + 3y &= -4 \\
12x + 7y &= 4
\end{aligned}
\qquad \text{than} \qquad
\begin{aligned}
4y &= 2 - 3x \\
4x &= -3y + 6
\end{aligned}
$$

by the addition method?

Solve the system by addition. See Examples 2 and 5.

27. $5x - 4y - 8x - 2 = 6x + 3y - 3$
  $4x - y = -2y - 8$

28. $2x - 8y + 3y + 2 = 5y + 16$
  $8x - 2y = 4x + 28$

29. $7x - 9 + 2y - 8 = -3y + 4x + 13$
  $4y - 8x = -8 + 9x + 32$

30. $-2x + 3y = 12 + 2y$
  $2x - 5y + 4 = -8 - 4y$

31. $2x + 5y = 7 + 4y - x$
  $5x + 3y + 8 = 22 - x + y$

32. $y + 9 = 3x - 2y + 6$
  $5 - 3x + 24 = -2x + 4y + 3$

33. $5x - 2y = 16 + 4x - 10$
  $4x + 3y = 60 + 2x + y$

34. $4 + 4x - 3y = 34 + x$
  $5y + 4x = 4y - 2 + 3x$

*In Exercises 35 and 36 (**a**) solve the system by the addition method, (**b**) solve the system by the substitution method, and (**c**) tell which method you prefer for that particular system, and why.*

35. $4x - 3y = -8$
  $x + 3y = 13$

36. $2x + 5y = 0$
  $x = -3y + 1$

Exercises 37 and 38 refer to the system

$$\frac{1}{3}x - \frac{1}{2}y = 7$$

$$\frac{1}{6}x + \frac{1}{3}y = 0.$$

37. One student solved the system by multiplying both equations by 6 to clear fractions, and another student multiplied by 12. Assuming they do all other work correctly, should they both get the same answer?

38. One student solved the system and wrote as his answer "$x = 12$," while another solved it and wrote as her answer "$y = -6$." Who, if either, was correct? Why?

Solve the system either by the addition method or the substitution method. First clear all fractions.

39. $x + \dfrac{1}{3}y = y - 2$
  $\dfrac{1}{4}x + y = x + y$

40. $\dfrac{5}{3}x + 2y = \dfrac{1}{3} + y$
  $3x - 3 + \dfrac{y}{3} = -2 + 2x$

41. $\dfrac{x}{6} + \dfrac{y}{6} = 2$
  $-\dfrac{1}{2}x - \dfrac{1}{3}y = -8$

42. $\dfrac{x}{2} - \dfrac{y}{3} = 9$

$\dfrac{x}{5} - \dfrac{y}{4} = 5$

43. $\dfrac{x}{3} - \dfrac{3y}{4} = -\dfrac{1}{2}$

$\dfrac{x}{6} + \dfrac{y}{8} = \dfrac{3}{4}$

44. $\dfrac{x}{5} + 2y = \dfrac{16}{5}$

$\dfrac{3x}{5} + \dfrac{y}{2} = -\dfrac{7}{5}$

 Refer to the system in the exercise indicated and
(a) solve the first equation for y, and call it y_1,
(b) solve the second equation for y, and call it y_2, and
(c) graph both y_1 and y_2 in the same viewing window to support the solution you obtained by the addition method. (For example, the system in Exercise 8 would be written as $y_1 = -2x - 15$, $y_2 = -x - 10$, and the graph would look like the one to the right.)

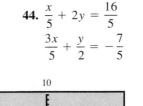
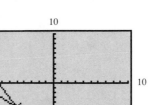

45. The system in Exercise 3

46. The system in Exercise 5

47. The system in Exercise 7

48. The system in Exercise 11

REVIEW EXERCISES

Solve the applied problem. See Sections 2.4, 2.5, 2.7, and 2.8.

49. In the 1991–1992 National Hockey League season, Mario Lemieux of Pittsburgh had a total of 131 points (goals plus assists). He had 43 more assists than goals. How many goals and how many assists did he have?

50. In the 1992 Olympic Games in Barcelona, the Unified Team earned 4 more medals than the United States. Together these two teams earned 220 medals. How many medals did each team earn?

51. The United States had the same number of gold and bronze medals, and three fewer silver medals. How many of each kind of medal did it earn? (See the answer to Exercise 50.)

52. The Unified Team had 7 more gold than silver medals, and 9 more silver than bronze medals. How many of each medal did it earn? (See the answer to Exercise 50.)

53. The perimeter of a rectangle is 46 feet. The width is 7 feet less than the length. Find the width.

54. The area of a rectangle is numerically 20 more than the width, and the length is 6 centimeters. What is the width?

55. Mark Deal, a cashier, has ten-dollar bills and twenty-dollar bills. There are 6 more tens than twenties. If there are 32 bills altogether, how many of them are twenties?

56. Michelle Johnson traveled for 2 hours at a constant speed. Because of road work, she reduced her speed by 7 miles per hour for the next 2 hours. If she traveled 206 miles, what was her speed on the first part of the trip?

7.4 APPLICATIONS OF LINEAR SYSTEMS

FOR EXTRA HELP

📖 **SSG** pp. 250–255
SSM pp. 387–395

📼 **Video**
10

💾 **Tutorial**
IBM   MAC

OBJECTIVES

1 ▶ Use linear systems to solve problems about numbers.
2 ▶ Use linear systems to solve problems about quantities and their costs.
3 ▶ Use linear systems to solve problems about mixtures.
4 ▶ Use linear systems to solve problems about rate or speed using the distance formula.

◆ C O N N E C T I O N S ◆

While national concern with automobile safety has led to improved automobiles and a reduction in drunken driving, the number of deaths caused by gunfire continues to increase.* Annual deaths per hundred thousand from motor vehicle accidents since 1965 have followed a linear pattern that can be described by the linear equation $7x + 8y = 14,066$, where x is the year. The linear equation $x + 10y = 2120$ describes the annual deaths per hundred thousand from gunfire since 1965. To see when (or if) deaths from firearms will exceed those from motor vehicles (based on these equations), we can solve a system of equations using the procedure discussed in this chapter.

If the trends described by the equations given above continue, the solution of the system indicates that their graphs will cross early in the new century.

FOR DISCUSSION OR WRITING

1. Solve the system of equations discussed above and interpret your solution.
2. When will deaths from gunfire surpass those from motor vehicles?
3. What circumstances may change this outcome?

Many practical problems are more easily translated into equations if two variables are used. With two variables, a system of two equations is needed to find the desired solution. The examples in this section illustrate the method of solving applied problems using two equations and two variables.

PROBLEM SOLVING   Recall from Chapter 2 the steps used in solving applied problems. Those steps can be modified as follows to allow for two variables and two equations.

Solving Applied Problems with Two Variables

Step 1 **Choose the variables.**
Choose a variable to represent each of the two unknown values that must be found. Write down what each variable is to represent.

Step 2 **Draw figures or diagrams.**
If figures or diagrams can help, draw them. (In some cases, they will not apply.)

---

* Based on the article "Grim Statistics" from *Scientific American,* November 1993, page 14.

Step 3 **Write two equations.**
Translate the problem into a system of two equations using both variables.

Step 4 **Solve the system.**
Solve the system of equations, using either the addition or substitution method.

Step 5 **Answer the question(s).**
Answer the question or questions asked in the problem.

Step 6 **Check.**
Check your answer by using the original words of the problem. Be sure that your answer makes sense.

1 ▶ Use linear systems to solve problems about numbers.

▶ In the first example we show how to use two variables to solve a problem about two unknown numbers.

E X A M P L E 1
Solving a Problem About Two Numbers

The sum of two numbers is 63. Their difference is 19. Find the two numbers.

Step 1 Let x = one number;
y = other number.

Step 2 A figure or diagram will not help in this problem, so go on to Step 3.

Step 3 From the information in the problem, set up a system of equations.

$$x + y = 63 \qquad \text{The sum is 63.}$$
$$x - y = 19 \qquad \text{The difference is 19.}$$

Step 4 Solve the system from Step 3. Here we use the addition method. (The substitution method works just as well.) Adding gives

$$
\begin{aligned}
x + y &= 63 \\
\underline{x - y} &= \underline{19} \\
2x &= 82.
\end{aligned}
$$

From this last equation, $x = 41$. Substitute 41 for x in either equation to find $y = 22$.

Step 5 The numbers required in the problem are 41 and 22.

Step 6 The sum of 41 and 22 is 63, and their difference is $41 - 22 = 19$. The solution satisfies the conditions of the problem. ∎

CAUTION If an applied problem asks for *two* values (as in Example 1), be sure to give both of them in your answer. Avoid the common error of giving only one of the values.

2 ▶ Use linear systems to solve problems about quantities and their costs.

▶ The next example shows how to solve a common type of applied problem involving two quantities and their costs.

PROBLEM SOLVING

Just as in Chapter 2, we can use a table or a box diagram to organize the information in order to solve an applied problem with two unknowns. We use a table in this example.

E X A M P L E   2
Solving a Problem About
Quantities and Costs

Admission prices at a football game were $6 for adults and $2 for children. The total value of the tickets sold was $2528, and 454 tickets were sold. How many adults and how many children attended the game?

Step 1 Let a = the number of adults' tickets sold;
c = the number of children's tickets sold.

Step 2 We summarize the information given in the problem in a table. The entries in the "total value" column were found by multiplying the number of tickets sold by the price per ticket.

Kind of Ticket	Number Sold	Cost of Each (in Dollars)	Total Value (in Dollars)
Adult	a	6	$6a$
Child	c	2	$2c$
Total	454	—	2528

The total number of tickets sold was 454, so

$$a + c = 454.$$

Since the total value was $2528, the final column leads to

$$6a + 2c = 2528.$$

Step 3 These two equations give the following system.

$$a + c = 454 \qquad \textbf{(1)}$$
$$6a + 2c = 2528 \qquad \textbf{(2)}$$

Step 4 We solve the system of equations with the addition method. First, multiply both sides of equation (1) by -2 to get

$$-2a - 2c = -908.$$

Then add this result to equation (2).

$$
\begin{array}{ll}
-2a - 2c = -908 & \text{Multiply (1) by } -2. \\
\underline{6a + 2c = 2528} & \\
4a = 1620 & \text{Add.} \\
a = 405 & \text{Divide by 4.}
\end{array}
$$

Substitute 405 for a in equation (1) to get

$$
\begin{array}{ll}
a + c = 454 & \qquad \textbf{(1)} \\
\textbf{405} + c = 454 & \text{Let } a = 405. \\
c = 49. & \text{Subtract 405.}
\end{array}
$$

Step 5 There were 405 adults and 49 children at the game.

Step 6 Since 405 adults paid $6 each and 49 children paid $2 each, the value of tickets sold should be $405(6) + 49(2) = 2528$, or $2528. The result agrees with the given information. ∎

3 ▶ Use linear systems to solve problems about mixtures.

▶ In Section 2.7 we learned how to solve mixture problems using one variable. Many mixture problems can also be solved using two variables. In the next example we show how to solve a mixture problem using a system of equations. The "box diagram" method first introduced in Section 2.7 is used here once again.

EXAMPLE 3

Solving a Mixture
Problem (Involving
Percent)

A pharmacist needs 100 liters of 50% alcohol solution. She has on hand 30% alcohol solution and 80% alcohol solution, which she can mix. How many liters of each will be required to make the 100 liters of 50% alcohol solution?

Step 1   Let x = the number of liters of 30% alcohol needed;
           y = the number of liters of 80% alcohol needed.

Step 2   We summarize the information using a box diagram. See Figure 6. (A chart could be used as well.)

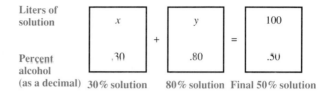

FIGURE 6

Step 3   We must write two equations. Since the total number of liters in the final mixture will be 100, the first equation is

$$x + y = 100.$$

To find the amount of pure alcohol in each mixture, multiply the number of liters by the concentration. The amount of pure alcohol in the 30% solution added to the amount of pure alcohol in the 80% solution will equal the amount of pure alcohol in the final 50% solution. This gives the second equation,

$$.30x + .80y = .50(100).$$

These two equations give the following system.

$$x + y = 100$$
$$.30x + .80y = 50 \qquad {\scriptstyle .50(100) = 50}$$

Step 4   We solve this system by the substitution method. Solving the first equation of the system for x gives $x = 100 - y$. Substitute $100 - y$ for x in the second equation to get

$$\begin{aligned}
.30(\mathbf{100} - y) + .80y &= 50 & &\text{Let } x = 100 - y. \\
30 - .30y + .80y &= 50 & &\text{Distributive property} \\
.50y &= 20 & &\text{Combine terms; subtract 30.} \\
y &= 40. & &\text{Divide by .50.}
\end{aligned}$$

Then $x = 100 - y = 100 - 40 = 60$.

Step 5   The pharmacist should use 60 liters of the 30% solution and 40 liters of the 80% solution.

Step 6   Since $60 + 40 = 100$ and $.30(60) + .80(40) = 50$, this mixture will give the 100 liters of 50% solution, as required in the original problem.

The system in this problem could have been solved by the addition method. Also, we could have cleared decimals by multiplying both sides of the second equation by 100. ■

4 ▶ Use linear systems to solve problems about rate or speed using the distance formula.

▶ Problems that use the distance formula relating distance, rate, and time, were first introduced in Section 2.8.

PROBLEM SOLVING

In some cases, these problems can be solved by using a system of two linear equations. Keep in mind that setting up a chart and drawing a sketch will help in solving such problems.

E X A M P L E   4

Solving a Problem about Distance, Rate, and Time

Two executives in cities 400 miles apart drive to a business meeting at a location on the line between their cities. They meet after 4 hours. Find the speed of each car if one car travels 20 miles per hour faster than the other.

Step 1 Let x = the speed of the faster car;
y = the speed of the slower car.

Step 2 We use the formula that relates distance, rate, and time, $d = rt$. Since each car travels for 4 hours, the time, t, for each car is 4. This information is shown in the chart. The distance is found by using the formula $d = rt$ and the expressions already entered in the chart.

	r	t	d
Faster car	x	4	$4x$
Slower car	y	4	$4y$

Find d from $d = rt$.

Draw a sketch showing what is happening in the problem. See Figure 7.

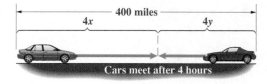

FIGURE 7

Step 3 As shown in the figure, since the total distance traveled by both cars is 400 miles, one equation is

$$4x + 4y = 400.$$

Because the faster car goes 20 miles per hour faster than the slower car, the second equation is

$$x = 20 + y.$$

Step 4 This system of equations,

$$4x + 4y = 400$$
$$x = 20 + y,$$

can be solved by substitution. Replace x with $20 + y$ in the first equation of the system and solve for y.

$$4(20 + y) + 4y = 400 \qquad \text{Let } x = 20 + y.$$
$$80 + 4y + 4y = 400 \qquad \text{Distributive property}$$

$$80 + 8y = 400 \quad \text{Combine like terms.}$$
$$8y = 320 \quad \text{Subtract 80.}$$
$$y = 40 \quad \text{Divide by 8.}$$

Step 5 Since $x = 20 + y$, and $y = 40$,

$$x = 20 + 40 = 60.$$

The speeds of the two cars are 40 miles per hour and 60 miles per hour.

Step 6 Check the answers. Since each car travels for 4 hours, the total distance traveled is

$$4(40) + 4(60) = 160 + 240 = 400$$

miles, as required. ∎

The problems in this section also could be solved using only one variable, but for many of them the solution is simpler with two variables.

CAUTION Be careful! When you use two variables to solve a problem, you must write two equations.

7.4 EXERCISES

1. Using the list of steps for solving an applied problem with two variables, write a short paragraph describing the general procedure you will use to solve the problems that follow in this exercise set.

2. Write an expression that illustrates each of the following quantities.
   (a) The monetary value of x five-dollar bills
   (b) The cost of y pounds of candy that sells for $1.30 per pound
   (c) The amount of pure acid in a solution of x liters of 40% acid
   (d) The actual speed of a plane that flies 300 miles per hour against a wind of 40 miles per hour

Write a system of equations for the problem, and then solve the system. See Example 1.

3. Find two numbers whose sum is 113 and whose difference is 71.

4. The sum of two numbers is 60, and their difference is 8. Find the numbers.

5. According to the 1990 United States census, a total of 1096 people lived in the New Mexico counties of Harding and Los Alamos. Harding County had 878 more people than Los Alamos. What was the population of each of these counties?

6. The Terminal Tower in Cleveland, Ohio, is 240 feet shorter than the Society Center, also in Cleveland. The total of the heights of the two buildings is 1656 feet. Find the heights of the buildings.

7. In a recent year London's Heathrow Airport and Tokyo's Haneda Airport serviced a total of 69,702,340 passengers. Heathrow serviced 5,348,260 more passengers than did Haneda. How many passengers did each airport service?

8. In a recent year the number of learning disabled children served in programs for the handicapped was 2,518,000 more than the number served in programs for speech impaired children. Together, a total of 6,374,000 children with these two handicaps were served. Assume none of these children had both conditions. How many learning disabled and how many speech impaired children were served?

Write a system of equations for the problem, and then solve the system. See Example 2.

9. A motel clerk counts his $1 and $10 bills at the end of a day. He finds that he has a total of 74 bills having a combined monetary value of $326. Find the number of bills of each denomination that he has.

Denomination of Bill	Number of Bills	Total Value
$1	x	
$10	y	
Totals	74	$326

10. Sabrina Avelos is a bank teller. At the end of a day, she has a total of 69 $5 and $10 bills. The total value of the money is $590. How many of each denomination does she have?

Denomination of Bill	Number of Bills	Total Value
$5	x	$5x
$10	y	
Totals		

11. At a recent soccer game, 386 tickets were sold, and a total of $523 was collected. Student tickets cost $.50 each, and nonstudent tickets cost $2.00 each. How many of each type of ticket were sold?

12. For the school band fundraiser, Deion sold license plate holders at $4.00 each and fruitcakes at $6.00 each. He sold a total of 62 items and collected $304.00. How many of each item did he sell?

13. Maria Lopez has twice as much money invested at 5% simple annual interest as she does at 4%. If her yearly income from these two investments is $350, how much does she have invested at each rate?

14. Chuck Fong invested his textbook royalty income in two accounts, one paying 3% annual simple interest and the other paying 2% interest. He earned a total of $1100 interest. If he invested three times as much in the 3% account as he did in the 2% account, how much did he invest at each rate?

15. A book collector bought some paperbacks at $.25 each and some hardbacks at $1.50 each. He bought a total of 20 books and paid $21.25. How many of each type of book did he buy?

16. A nursing home bought 38 bottles of disinfectant. Small bottles cost $5.00 each, and large ones cost $7.00 each. A total of $250.00 was spent. How many of each size bottle were bought?

Write a system of equations for the problem, and then solve the system. See Example 3.

17. A 40% dye solution is to be mixed with a 70% dye solution to get 120 liters of a 50% solution. How many liters of the 40% and 70% solutions will be needed?

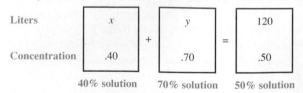

Liters	x	+	y	=	120
Concentration	.40		.70		.50
	40% solution		70% solution		50% solution

18. A 90% antifreeze solution is to be mixed with a 75% solution to make 120 liters of a 78% solution. How many liters of the 90% and 75% solutions will be used?

Liters	x	+	y	=	120
Concentration	.90		.75		.78
	90% solution		75% solution		78% solution

19. A merchant wishes to mix coffee worth $6 per pound with coffee worth $3 per pound to get 90 pounds of a mixture worth $4 per pound. How many pounds of the $6 and the $3 coffee will be needed?

Pounds	Dollars per Pound	Cost
x	6	
y	3	
90		

20. A grocer wishes to blend candy selling for $1.20 a pound with candy selling for $1.80 a pound to get a mixture that will be sold for $1.40 a pound. How many pounds of the $1.20 and the $1.80 candy should be used to get 45 pounds of the mixture?

Pounds	Dollars per Pound	Cost
x	1.20	1.20x
y	1.80	1.80y
45	1.40	63

21. How many barrels of pickles worth $40 per barrel and pickles worth $60 per barrel must be mixed to obtain 100 barrels of a mixture worth $48 per barrel?

22. The owner of a nursery wants to mix some fertilizer worth $70 per bag with some worth $90 per bag to obtain 40 bags of mixture worth $77.50 per bag. How many bags of each type should she use?

Write a system of equations for the problem, and then solve the system. See Example 4.

23. A boat takes 3 hours to go 24 miles upstream. It can go 36 miles downstream in the same time. Find the speed of the current and the speed of the boat in still water if x = the speed of the boat in still water and y = the speed of the current.

	d	r	t
Downstream	36	$x + y$	3
Upstream	24	$x - y$	3

24. It takes a boat $1\frac{1}{2}$ hours to go 12 miles downstream, and 6 hours to return. Find the speed of the boat in still water and the speed of the current. Let x = the speed of the boat in still water and y = the speed of the current.

	d	r	t
Downstream	12	$x + y$	$\frac{3}{2}$
Upstream	12	$x - y$	6

25. If a plane can travel 440 miles per hour into the wind and 500 miles per hour with the wind, find the speed of the wind and the speed of the plane in still air.

26. A small plane travels 200 miles per hour with the wind and 120 miles per hour against it. Find the speed of the wind and the speed of the plane in still air.

27. Two trains start from stations 1000 miles apart and travel toward each other. They meet after 5 hours. Find the rate of each train if one travels 20 miles per hour faster than the other.

28. Two cars start from towns 300 miles apart and travel toward each other. They meet after 3 hours. Find the rate of each car if one car travels 30 miles per hour slower than the other.

29. At the beginning of a walk for charity, Roberto and Juana are 5.5 miles apart. If they leave at the same time and walk in the same direction, Roberto overtakes Juana in 11 hours. If they walk toward each other, they meet in 1 hour. What are their speeds?

30. Mr. Abbot left Farmersville in a plane at noon to travel to Exeter. Mr. Costello left Exeter in his automobile at 2 P.M. to travel to Farmersville. It is 400 miles from Exeter to Farmersville. If the sum of their speeds was 120 miles per hour, and if they crossed paths at 4 P.M., find the speed of each.

———◆ **MATHEMATICAL CONNECTIONS** (Exercises 31–36) ◆———

A system of linear equations can be used to model the cost and the revenue of a business. Work Exercises 31–36 in order.

31. Suppose that you start a business manufacturing and selling bicycles, and it costs you $5000 to get started. You determine that each bicycle will cost $400 to manufacture. Explain why the linear equation $y_1 = 400x + 5000$ gives your *total* cost to manufacture x bicycles (y_1 in dollars).

32. You decide to sell each bike for $600. What expression in x represents the revenue you will take in if you sell x bikes? Write an equation using y_2 to express your revenue when you sell x bikes (y_2 in dollars).

33. Form a system from the two equations in Exercises 31 and 32 and then solve the system.

34. The value of x from Exercise 33 is the number of bikes it takes to *break even*. Fill in the blanks: When _____ bikes are sold, the break-even point is reached. At that point, you have spent _____ dollars and taken in _____ dollars.

35. Explain how the calculator graph shown here supports the results from Exercises 31–34.

36. Refer to the graph. When $x < 25$ what can you say about the profits? What can you say if $x > 25$?

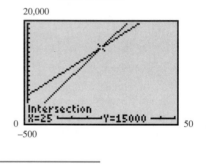

———————◆———————

REVIEW EXERCISES

Graph the linear inequality. See Section 6.5.

37. $x + y \leq 4$ **38.** $2x - y > 4$ **39.** $y \geq -3x + 2$

40. $x - 3y < 6$ **41.** $2x + 4y > 8$ **42.** $3x + 2y \leq 0$

7.5 SOLVING SYSTEMS OF LINEAR INEQUALITIES

FOR EXTRA HELP	OBJECTIVE
📖 **SSG** pp. 255–256 **SSM** pp. 395–400 📼 **Video** 11 💾 **Tutorial** IBM MAC	**1 ▶** Solve systems of linear inequalities by graphing.

Graphing the solution of a linear inequality was discussed in Section 6.5. Let us review the method. To graph the solution of $x + 3y > 12$, we first graph the line $x + 3y = 12$ by finding a few ordered pairs that satisfy the equation. Because the points on this boundary line do not satisfy the inequality, we use a dashed line.

To decide which side of the line should be shaded, we choose any test point not on the line, such as $(0, 0)$. Then we substitute 0 for x and 0 for y in the given inequality.

$$x + 3y > 12$$
$$0 + 3(0) > 12 \quad ? \quad \text{Let } x = 0, y = 0.$$
$$0 > 12 \quad \text{False}$$

Since the test point does not satisfy the inequality, we shade the region on the side of the boundary line that does not include $(0, 0)$, as in Figure 8.

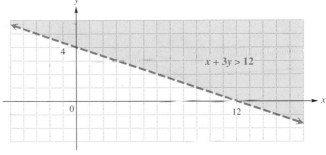

FIGURE 8

1 ▶ Solve systems of linear inequalities by graphing.

▶ The same method is used to find the solution of a system of two linear inequalities, as shown in Examples 1–3. **A system of linear inequalities** consists of two or more linear inequalities. The **solutions of a system of linear inequalities** include all ordered pairs that make all inequalities of the system true at the same time.

To solve a system of linear inequalities, use the following steps.

Graphing a System of Linear Inequalities

Step 1 Graph each inequality using the method of Section 6.5.

Step 2 Indicate the solution of the system by using dark shading on the intersection of the two graphs (the region where the two graphs overlap).

EXAMPLE 1

Solving a System of Two Linear Inequalities

Graph the solutions of the linear system

$$3x + 2y \leq 6$$
$$2x - 5y \geq 10.$$

We begin by graphing $3x + 2y \leq 6$. To do this, we graph $3x + 2y = 6$ as a solid line. Since $(0, 0)$ makes the inequality true, we shade the region containing $(0, 0)$, as shown in Figure 9.

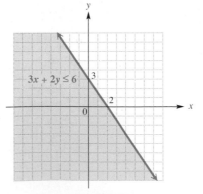

FIGURE 9

Now graph $2x - 5y \geq 10$. The solid line boundary is the graph of $2x - 5y = 10$. Since $(0, 0)$ makes the inequality false, shade the region that does not contain $(0, 0)$, as shown in Figure 10.

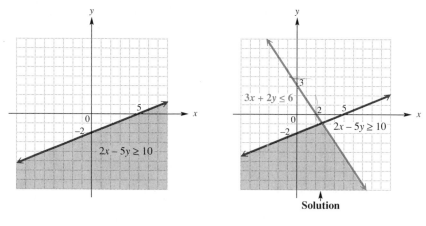

| FIGURE 10 | FIGURE 11 |

The solution of the system is given by the intersection (overlap) of the regions of the graphs in Figures 9 and 10. The solution is the shaded region in Figure 11, and includes portions of the two boundary lines. ■

NOTE In practice, we usually do all the work on one set of axes at the same time. In the following examples, only one graph is shown. Be sure that the region of the solution is clearly indicated.

E X A M P L E   2 ■
Solving a System of Two
Linear Inequalities

Graph the solutions of the system

$$x - y > 5$$
$$2x + y < 2.$$

Figure 12 shows the graphs of both $x - y > 5$ and $2x + y < 2$. Dashed lines show that the graphs of the inequalities do not include their boundary lines. The solution of the system is the shaded region in the figure. The solution does not include either boundary line. ■

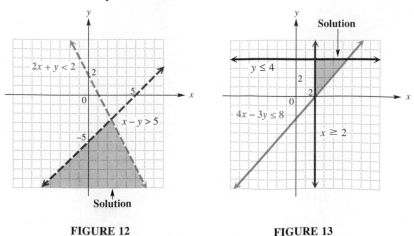

| FIGURE 12 | FIGURE 13 |

EXAMPLE 3

Solving a System of Three
Linear Inequalities

Graph the solutions of the system

$$4x - 3y \leq 8$$
$$x \geq 2$$
$$y \leq 4.$$

Recall that $x = 2$ is a vertical line through the point $(2, 0)$, and $y = 4$ is the horizontal line through $(0, 4)$. The graph of the solution is the shaded region in Figure 13. ■

◆ CONNECTIONS ◆

Linear programming, mentioned at the beginning of this chapter, is used in business, military science, and social science to find an optimum solution to certain types of problems. The quantity to be optimized must be written as a linear equation and be subject to constraints that can be written as a system of linear inequalities. The solution is found at a "corner point" of the region that contains the simultaneous solutions of the system of inequalities. For example, if the system of inequalities in Example 3 (graphed in Figure 13) represented the constraints in a linear programming problem, the corner points could be found by solving the following systems of two equations. Each equation represents a boundary line of the region.

$$4x - 3y = 8 \qquad 4x - 3y = 8 \qquad x = 2$$
$$x = 2 \qquad y = 4 \qquad y = 4$$

The solutions of these systems are, in order, $(2, 0)$, $(5, 4)$, and $(2, 4)$.

Now suppose we want to maximize the linear expression $z = 5x + 3y$, where z represents the profit (in thousands of dollars) from the sale of x units of one item and y units of another item. To decide which of the corner points has coordinates that maximize z, we substitute the values from each point into the expression for z.

corner point	$z = 5x + 3y$
$(2, 0)$	$5(2) + 3(0) = 10$
$(5, 4)$	$5(5) + 3(4) = 37 \leftarrow$ Maximum
$(2, 4)$	$5(2) + 3(4) = 22$

The results show that selling 5 units of the first item and 4 units of the second item will produce the maximum profit of $37,000.

FOR DISCUSSION OR WRITING
1. Verify the solutions of the three systems given above.
2. Which corner point produces the minimum value of z?
3. Find the maximum and minimum values of z if $z = 2x + 4y$.

7.5 EXERCISES

1. Every system of inequalities illustrated in the examples of this section has infinitely many solutions. Explain why this is so. Does this mean that *any* ordered pair is a solution?

2. Explain how you will determine in the exercises that follow in this section whether to draw a solid boundary line or a dashed boundary line.

Graph the solutions of the system of linear inequalities. See Examples 1 and 2.

3. $x + y \leq 6$
$x - y \geq 1$

4. $x + y \leq 2$
$x - y \geq 3$

5. $4x + 5y \geq 20$
$x - 2y \leq 5$

6. $x + 4y \leq 8$
$2x - y \geq 4$

7. $2x + 3y < 6$
$x - y < 5$

8. $x + 2y < 4$
$x - y < -1$

9. $y \leq 2x - 5$
$x < 3y + 2$

10. $x \geq 2y + 6$
$y > -2x + 4$

11. $4x + 3y < 6$
$x - 2y > 4$

12. $3x + y > 4$
$x + 2y < 2$

13. $x \leq 2y + 3$
$x + y < 0$

14. $x \leq 4y + 3$
$x + y > 0$

15. $-3x + y \geq 1$
$6x - 2y \geq -10$

16. $2x + 3y < 6$
$4x + 6y > 18$

17. $x - 3y \leq 6$
$x \geq -4$

18. $y \leq 2x$
$y \leq 3$

19. Which one of the following systems of linear inequalities is graphed in the figure?

(a) $x \leq 3$
$y \leq 1$

(b) $x \leq 3$
$y \geq 1$

(c) $x \geq 3$
$y \leq 1$

(d) $x \geq 3$
$y \geq 1$

20. Suppose that your friend was absent from class (because of a bad cold) on the day that systems of linear inequalities were covered, and he asked you to write a short explanation of the process. Do this in a brief, concise explanation.

Graph the solutions of the system. See Example 3.

21. $4x + 5y < 8$
$y > -2$
$x > -4$

22. $x + y \geq -3$
$x - y \leq 3$
$y \leq 3$

23. $3x - 2y \geq 6$
$x + y \leq 4$
$x \geq 0$
$y \geq -4$

24. How can we determine that the point $(0, 0)$ is a solution of the system in Exercise 21 without actually graphing the system?

In a procedure called linear programming (see the Connections in this section) it is necessary to find the "corner points" of a shaded region. For example, the system in Example 3 has corner points at $(2, 0)$, $(2, 4)$, and $(5, 4)$. See Figure 13. Find all corner points for the following systems.

25. $y \geq x - 4$
$y \leq 5$
$x \geq 4$

26. $y \leq -x + 5$
$y \geq 2$
$x \geq 1$

27. $x + y \geq 0$
$y \leq 3$
$x \leq 5$

28. $x - y \leq 0$
$x \geq -3$
$y \leq 4$

Some graphics calculators have the capability of shading above one line and below another. Of course, we must solve both inequalities for y first. For example, the graph given here is the solution of the system

$$y < x - 5$$
$$2x + y > 2.$$

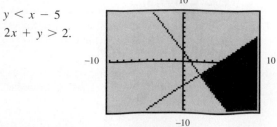

Match the system with its calculator-generated graph.

29. $y \geq x$
$y \leq 2x - 3$ **A.**

30. $y \leq x$
$y \geq 2x - 3$

31. $y \geq -x$
$y \leq 2x - 3$

32. $y \leq -x$
$y \geq 2x - 3$

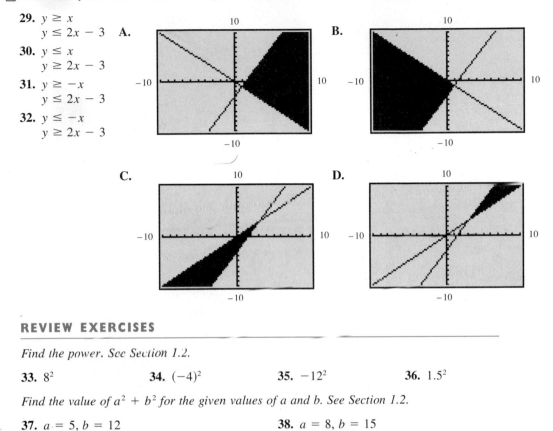

REVIEW EXERCISES

Find the power. See Section 1.2.

33. 8^2 **34.** $(-4)^2$ **35.** -12^2 **36.** 1.5^2

Find the value of $a^2 + b^2$ for the given values of a and b. See Section 1.2.

37. $a = 5, b = 12$ **38.** $a = 8, b = 15$

CHAPTER 7 SUMMARY

KEY TERMS

7.1 system of linear equations
solution of a system
consistent system

inconsistent system
independent equations
dependent equations

7.5 system of linear inequalities

solutions of a system of linear inequalities

QUICK REVIEW

CONCEPTS	EXAMPLES
7.1 SOLVING SYSTEMS OF LINEAR EQUATIONS BY GRAPHING	
An ordered pair is a solution of a system if it makes all equations of the system true at the same time.	Is $(4, -1)$ a solution of the following system? $$x + y = 3$$ $$2x - y = 9$$ Yes, because $4 + (-1) = 3$, and $2(4) - (-1) = 9$ are both true.

CONCEPTS	EXAMPLES
If the graphs of the equations of a system are both sketched on the same axes, the points of intersection, if any, are solutions of the system.	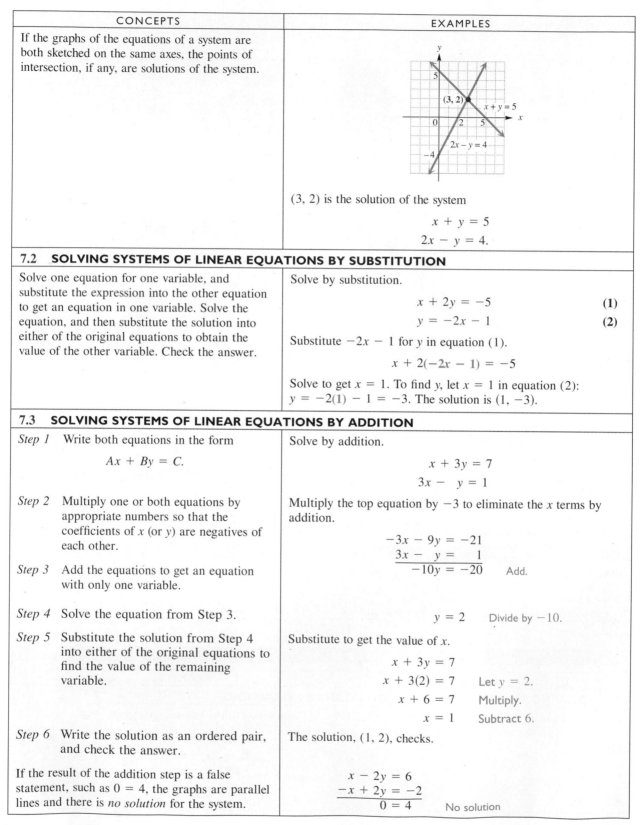 (3, 2) is the solution of the system $$x + y = 5$$ $$2x - y = 4.$$

7.2 SOLVING SYSTEMS OF LINEAR EQUATIONS BY SUBSTITUTION

Solve one equation for one variable, and substitute the expression into the other equation to get an equation in one variable. Solve the equation, and then substitute the solution into either of the original equations to obtain the value of the other variable. Check the answer.	Solve by substitution. $$x + 2y = -5 \qquad \textbf{(1)}$$ $$y = -2x - 1 \qquad \textbf{(2)}$$ Substitute $-2x - 1$ for y in equation (1). $$x + 2(-2x - 1) = -5$$ Solve to get $x = 1$. To find y, let $x = 1$ in equation (2): $y = -2(1) - 1 = -3$. The solution is $(1, -3)$.

7.3 SOLVING SYSTEMS OF LINEAR EQUATIONS BY ADDITION

Step 1 Write both equations in the form $$Ax + By = C.$$	Solve by addition. $$x + 3y = 7$$ $$3x - y = 1$$
Step 2 Multiply one or both equations by appropriate numbers so that the coefficients of x (or y) are negatives of each other.	Multiply the top equation by -3 to eliminate the x terms by addition.
Step 3 Add the equations to get an equation with only one variable.	$$\begin{array}{r} -3x - 9y = -21 \\ \underline{3x - y = 1} \\ -10y = -20 \quad \text{Add.} \end{array}$$
Step 4 Solve the equation from Step 3.	$$y = 2 \qquad \text{Divide by } -10.$$
Step 5 Substitute the solution from Step 4 into either of the original equations to find the value of the remaining variable.	Substitute to get the value of x. $$x + 3y = 7$$ $$x + 3(2) = 7 \qquad \text{Let } y = 2.$$ $$x + 6 = 7 \qquad \text{Multiply.}$$ $$x = 1 \qquad \text{Subtract 6.}$$
Step 6 Write the solution as an ordered pair, and check the answer.	The solution, $(1, 2)$, checks.
If the result of the addition step is a false statement, such as $0 = 4$, the graphs are parallel lines and there is *no solution* for the system.	$$\begin{array}{r} x - 2y = 6 \\ \underline{-x + 2y = -2} \\ 0 = 4 \qquad \text{No solution} \end{array}$$

CONCEPTS	EXAMPLES
If the result is a true statement, such as $0 = 0$, the graphs are the same line, and an *infinite number of ordered pairs are solutions*.	$x - 2y = 6$ $-x + 2y = -6$ $0 = 0$ Infinite number of solutions

7.4   APPLICATIONS OF LINEAR SYSTEMS

Step 1	Choose a variable to represent each unknown value.	The sum of two numbers is 30. Their difference is 6. Find the numbers. Let x represent one number. Let y represent the other number.
Step 2	Draw a figure or a diagram if it will help.	
Step 3	Translate the problem into a system of two equations using both variables.	$x + y = 30$ $x - y = 6$ $2x = 36$ Add. $x = 18$ Divide by 2.
Step 4	Solve the system.	
		Let $x = 18$ in the top equation: $18 + y = 30$. Solve to get $y = 12$.
Step 5	Answer the question or questions asked.	The two numbers are 18 and 12.
Step 6	Check the solution in the words of the problem.	$18 + 12 = 30$ and $18 - 12 = 6$, so the solution checks.

7.5   SOLVING SYSTEMS OF LINEAR INEQUALITIES

To solve a system of linear inequalities, graph each inequality on the same axes. (This was explained in Section 6.5.) The solutions of the system are in the overlap of the regions of the two graphs.	The shaded region shows the solutions of the system $2x + 4y \geq 5$ $x \geq 1.$

CHAPTER 7 REVIEW EXERCISES

[7.1] *Decide whether the given ordered pair is a solution of the given system.*

1. $(3, 4)$
$4x - 2y = 4$
$5x + y = 19$

2. $(1, -3)$
$5x + 3y = -4$
$2x - 3y = 11$

3. $(-5, 2)$
$x - 4y = -13$
$2x + 3y = 4$

4. $(0, 1)$
$3x + 8y = 8$
$4x - 3y = 3$

Solve the system by graphing.

5. $x + y = 4$
$2x - y = 5$

6. $x - 2y = 4$
$2x + y = -2$

7. $x - 2 = 2y$
$2x - 4y = 4$

8. $2x + 4 = 2y$
$y - x = -3$

[7.2]

9. A student solves the system $\begin{array}{l} 2x + y = 6 \\ -2x - y = 4 \end{array}$ and gets the equation $0 = 10$. The student gives the solution as $(0, 10)$. Is this correct? Explain.

10. Can a system of two linear equations in two unknowns have exactly three solutions? Explain.

Solve the system by the substitution method.

11. $3x + y = 7$
    $x = 2y$

12. $2x - 5y = -19$
    $y = x + 2$

13. $4x + 5y = 44$
    $x + 2 = 2y$

14. $5x + 15y = 3$
    $x + 3y = 2$

15. $\dfrac{1}{5}x + \dfrac{1}{8}y = 0$
    $\dfrac{1}{2}x + \dfrac{1}{4}y = \dfrac{1}{2}$

16. $\dfrac{1}{3}x + \dfrac{1}{7}y = \dfrac{52}{21}$
    $\dfrac{1}{2}x - \dfrac{1}{3}y = \dfrac{19}{6}$

17. After solving a system of linear equations by the substitution method, a student obtained the equation "$0 = 0$." He gave the solution of the system as $(0, 0)$. Was his answer correct? Why or why not?

18. Suppose that you were asked to solve the system $\begin{array}{l} 5x - 3y = 7 \\ -x + 2y = 4 \end{array}$ by substitution. Which variable in which equation would be easiest to solve for in your first step? Why?

[7.3] *Solve the system by the addition method.*

19. $2x - y = 13$
    $x + y = 8$

20. $3x - y = -13$
    $x - 2y = -1$

21. $5x + 4y = -7$
    $3x - 4y = -17$

22. $-4x + 3y = 25$
    $6x - 5y = -39$

23. $3x - 4y = 9$
    $6x - 8y = 18$

24. $2x + y = 3$
    $-4x - 2y = 6$

Solve the system by any method.

25. $2x + 3y = -5$
    $3x + 4y = -8$

26. $6x - 9y = 0$
    $2x - 3y = 0$

27. $2x + y - x = 3y + 5$
    $y + 2 = x - 5$

28. $5x - 3 + y = 4y + 8$
    $2y + 1 = x - 3$

29. $\dfrac{x}{2} + \dfrac{y}{3} = 7$
    $\dfrac{x}{4} + \dfrac{2y}{3} = 8$

30. $\dfrac{3x}{4} - \dfrac{y}{3} = \dfrac{7}{6}$
    $\dfrac{x}{2} + \dfrac{2y}{3} = \dfrac{5}{3}$

31. What are the three methods of solving systems discussed in this chapter? Choose one and discuss its drawbacks and advantages.

32. Why would a system of linear equations having the solution $\left(-\dfrac{1}{2}, \dfrac{2}{3}\right)$ be difficult to solve using the graphing method?

33. Why would it be easier to solve system A by the addition method than system B?

    A: $5x + 2y = 9$
    $3x - 2y = 12$

    B: $5x + 7y = 3$
    $3x + 3y = 14$

34. Why would it be easier to solve system B by the substitution method than system A?

    A: $-5x + 6y = 7$
    $2x + 5y = -5$

    B: $2x + 9y = 13$
    $y = 3x - 2$

[7.4] *Solve the problem by using a system of equations.*

35. The two tallest buildings in Houston, Texas, are the Texas Commerce Tower and the First Interstate Plaza. The first of these is 4 stories taller than the second. Together the buildings have 146 stories. How many stories tall are the individual buildings?

36. In the 1992 presidential election, George Bush received 100 more votes than Bill Clinton in Clark County, Idaho. Together, the two candidates received a total of 290 votes. How many votes did each receive?

37. The perimeter of a rectangle is 90 meters. Its length is $1\frac{1}{2}$ times its width. Find the length and width of the rectangle.

38. A cashier has 20 bills, all of which are $10 or $20 bills. The total value of the money is $330. How many of each type does the cashier have?

39. Candy that sells for $1.30 a pound is to be mixed with candy selling for $.90 a pound to get 100 pounds of a mix that will sell for $1 per pound. How much of each type should be used?

40. A 40% antifreeze solution is to be mixed with a 70% solution to get 90 liters of a 50% solution. How many liters of the 40% and 70% solutions will be needed?

41. Della Dawkins can buy 6 apples and 5 bananas for $5.55. She can also buy 12 apples and 2 bananas for $7.50. Find the cost of one apple and the cost of one banana.

42. A certain plane flying with the wind travels 540 miles in 2 hours. Later, flying against the same wind, the plane travels 690 miles in 3 hours. Find the speed of the plane in still air and the speed of the wind.

43. Ms. Branson invested $18,000. Part of it was invested at 3% annual simple interest and the rest was invested at 4%. Her interest income for the first year was $650. How much did she invest at each rate?

44. A textbook author receives a $2 royalty for each of his algebra books sold, and a $3 royalty for each of his calculus books sold. During one royalty period, the two books together sold 13,000 copies and he received a total of $29,000 in royalties. How many of each kind of book were sold during that period?

45. Wally bought a total of 15 compact discs. Some of them cost $12 each, and the rest cost $16 each. He paid a total of $228. How many of each did he buy?

46. Suppose that you are given a choice of methods to solve a problem, and you find that the problem can either be worked using one variable and one equation, or using two variables and two equations. Which method would you probably prefer? Why? (*Hint:* There is no right or wrong answer here. Explain your preference.)

[7.5] *Graph the solutions for the system of linear inequalities.*

47. $x + y \geq 2$
$\quad\; x - y \leq 4$

48. $\qquad y \geq 2x$
$\quad 2x + 3y \leq 6$

49. $x + y < 3$
$\quad\;\; 2x > y$

50. $3x - y \leq 3$
$\qquad x \geq -2$
$\qquad y \leq 2$

MIXED REVIEW EXERCISES

Work each of the following using the methods of this chapter.

51. $\dfrac{2x}{3} + \dfrac{y}{4} = \dfrac{14}{3}$

$\dfrac{x}{2} + \dfrac{y}{12} = \dfrac{8}{3}$

52. $x + y = 2y + 6$
$\quad y + 8 = x + 2$

53. $3x + 4y = 6$
$\quad 4x - 5y = 8$

54. $\dfrac{3x}{2} + \dfrac{y}{5} = -3$

$4x + \dfrac{y}{3} = -11$

55. $x + y < 5$
$\quad x - y \geq 2$

56. $y \leq 2x$
$\quad x + 2y > 4$

57. The perimeter of an isosceles triangle is 29 inches. One side of the triangle is 5 inches longer than each of the two equal sides. Find the lengths of the sides of the triangle.

58. Two cars leave from the same place and travel in opposite directions. One car travels 30 miles per hour faster than the other. After $2\frac{1}{2}$ hours they are 265 miles apart. What are the rates of the cars?

59. A hospital bought a total of 146 bottles of glucose solution. Small bottles cost \$2 each, and large ones cost \$3 each. The total cost was \$336. How many of each size bottle were bought?

60. Can the problem in Exercise 57 be worked using a single variable? If so, explain how it can be done, and then solve it.

61. Write each equation in the following systems in slope-intercept form. Use what you learned in Chapter 6 about slope and the y-intercept to describe the graphs of these equations.

(a) $3x + 2y = 6$ **(b)** $2x - y = 4$ **(c)** $x - 3y = 5$
     $-2y = 3x - 5$ $x = .5y + 2$ $2x + y = 8$

Explain how you can use this information to determine the solutions of certain systems.

62. Without actually graphing, determine which one of the following systems of inequalities has no solution.

(a) $x \geq 4$ **(b)** $x + y > 4$ **(c)** $x > 2$ **(d)** $x + y > 4$
     $y \leq 3$ $x + y < 3$ $y < 1$ $x - y < 3$

◆ **MATHEMATICAL CONNECTIONS** (Exercises 63–76) ◈ ─────

Consider the system

$$2x - y = 8 \tag{1}$$
$$x + 2y = 12 \tag{2}$$

and work Exercises 63–76 in order.

63. Solve the system by the substitution method.

64. Solve the system by the addition method.

65. Compare your answers in Exercises 63 and 64. What do you notice? Does it make a difference in the *method* of solution as to what answer you get?

66. Solve equation (1) for y, and call it y_1.

67. Solve equation (2) for y, and call it y_2.

68. Set the expressions for y_1 and y_2 equal to each other and solve the resulting equation. How does the solution compare to your answers in Exercises 63 and 64?

69. Check your solution from Exercise 68 back into the original equation you wrote. When you evaluate each side, you should get the same result. How does this result compare to your answers in Exercises 63 and 64?

70. What is the slope of the line in equation (1)?

71. What is the slope of the line in equation (2)?

72. Based on what you learned in Chapter 6, are the lines parallel, perpendicular, or neither?

73. The screen shows the graphs of equations (1) and (2) in the standard viewing window of a graphics calculator. How does the display at the bottom of the screen support your answers in Exercises 63 and 64? Why does it *seem* (from this view) that the lines are not perpendicular?

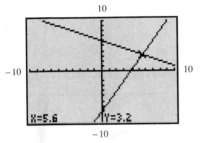

74. The screen shows the graphs of equations (1) and (2) in a square viewing window. Why is this viewing window preferable to the one shown in Exercise 73 (considering the answer to Exercise 72)?

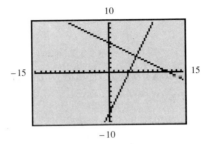

75. Graph the system of linear inequalities

$$2x - y \leq 8$$
$$x + 2y \leq 12.$$

76. Look at the "corner point" of the shaded region in your answer to Exercise 75. What are its coordinates? (*Hint:* Look at your answers in Exercises 63 and 64.)

CHAPTER 7 TEST

1. Consider the system of equations

$$2x + y = -3$$
$$x - y = -9.$$

One of the three screens below represents a graphical solution for the system. Use the x and y values displayed at the bottom to determine which one is the correct solution.

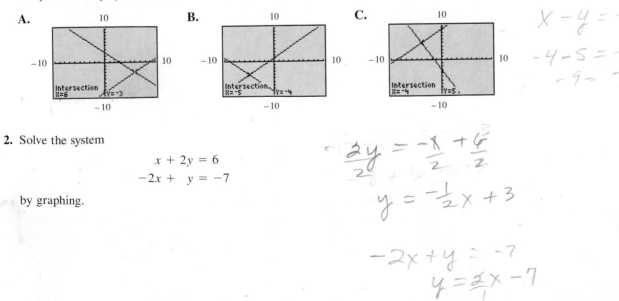

A.

B.

C.

(handwritten) $6, -3$
$-5, 4$
$-4, 5$

(handwritten)
$x - y = -9$
$-4 - 5 = -9$
$-9 = -9$

2. Solve the system

$$x + 2y = 6$$
$$-2x + y = -7$$

by graphing.

(handwritten)
$\dfrac{2y}{2} = \dfrac{-x}{2} + \dfrac{6}{2}$

$y = -\dfrac{1}{2}x + 3$

$-2x + y = -7$
$y = \dfrac{2}{1}x - 7$

Solve the system by substitution.

3. $2x + y = -4$
  $x = y + 7$

4. $4x + 3y = -35$
  $x + y = 0$

Solve the system by the addition method.

5. $2x - y = 4$
  $3x + y = 21$

6. $4x + 2y = 2$
  $5x + 4y = 7$

7. $6x + 5y = 13$
  $3x + 2y = 4$

8. $4x + 5y = 2$
  $-8x - 10y = 6$

9. $6x - 5y = 0$
  $-2x + 3y = 0$

10. $\dfrac{6}{5}x - \dfrac{1}{3}y = -20$

  $-\dfrac{2}{3}x + \dfrac{1}{6}y = 11$

11. Solve

$$8 + 3x - 4y = 14 - 3y$$
$$3x + y + 12 = 9x - y$$

by any method.

12. Suppose that the graph of a system of two linear equations consists of lines that have the same slope but different y-intercepts. How many solutions does the system have?

Write a system of two equations and solve the problem.

13. The distance between Memphis and Atlanta is 300 miles less than the distance between Minneapolis and Houston. Together the two distances total 1042 miles. How far is it between Memphis and Atlanta? How far is it between Minneapolis and Houston?

14. Admission prices for a production of *Phantom of the Opera* were $65 for orchestra seats and $50 for balcony seats. A total of 567 people attended, and $33,030 was collected. How many of each kind of seat were sold?

15. A 25% solution of alcohol is to be mixed with a 40% solution to get 50 liters of a final mixture that is 30% alcohol. How much of each of the original solutions should be used?

16. Two cars leave from Perham, Minnesota, and travel in the same direction. One car travels one and one third times as fast as the other. After 3 hours they are 45 miles apart. What are the speeds of the cars?

17. Make up a system of linear equations having $(-3, 4)$ as its only solution. (*Hint:* Start with the solution and work backward.)

Graph the solutions of the system of inequalities.

18. $2x + 7y \le 14$
  $x - y \ge 1$

19. $2x - y > 6$
  $4y + 12 \ge -3x$

20. Which one of the following is a calculator representation of the graph of the system given here?

$$y \ge x$$
$$y \le 5$$

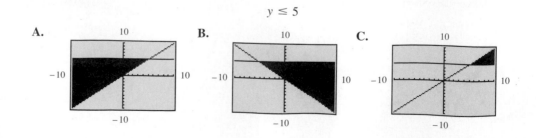

A.   **B.**   **C.**

CUMULATIVE REVIEW EXERCISES (Chapters 1–7)

1. List all integer factors of 40.

2. Find the value of the expression if $x = 1$ and $y = 5$.

$$\frac{3x^2 + 2y^2}{10y + 3}$$

Name the property that justifies the statement.

3. $5 + (-4) = (-4) + 5$

4. $r(s - k) = rs - rk$

5. $-\dfrac{2}{3} + \dfrac{2}{3} = 0$

6. Evaluate $-2 + 6[3 - (4 - 9)]$.

Solve the linear equation.

7. $2 - 3(6x + 2) - 4(x + 1) + 18$

8. $\dfrac{3}{2}\left(\dfrac{1}{3}x + 4\right) = 6\left(\dfrac{1}{4} + x\right)$

Solve the linear inequality.

9. $-\dfrac{5}{6}x < 15$

10. $-8 < 2x + 3$

11. If 15 yards of cloth are needed for 18 nurses' smocks, how much cloth would be needed for 45 smocks?

Perform the indicated operations.

12. $-3(-5x^2 + 3x - 10) - (x^2 - 4x + 7)$ **13.** $(3x - 7)(2y + 4)$

14. $\dfrac{3k^3 + 17k^2 - 27k + 7}{k + 7}$

15. Write in scientific notation: 36,500,000,000.

16. Simplify and write the answer using only positive exponents: $\left(\dfrac{x^{-4}y^3}{x^2y^4}\right)^{-1}$.

Factor completely.

17. $10m^2 + 7mp - 12p^2$

18. $64t^2 - 48t + 9$

Solve the quadratic equation.

19. $6x^2 - 7x - 3 = 0$

20. $r^2 - 121 = 0$

Perform the operation and express the answer in lowest terms.

21. $\dfrac{-3x + 6}{2x + 4} - \dfrac{-3x - 8}{2x + 4}$

22. $\dfrac{16k^2 - 9}{8k + 6} \div \dfrac{16k^2 - 24k + 9}{6}$

23. Solve the equation $\dfrac{4}{x + 1} + \dfrac{3}{x - 2} = 4$.

24. Solve the formula $P = \dfrac{kT}{V}$ for T.

Graph the linear equation or inequality.

25. $x - y = 4$

26. $3x + y = 6$

27. $x - 3y > 6$

Find the slope of the line described.

28. through $(-5, 6)$ and $(1, -2)$

29. perpendicular to the line $y = 4x - 3$

Write an equation for the line described. Express in the form $Ax + By = C$.

30. through $(2, -5)$ with slope 3

31. through the points $(0, 4)$ and $(2, 4)$

32. (a) Write an equation of the vertical line through $(9, -2)$.
　　(b) Write an equation of the horizontal line through $(4, -1)$.

33. Refer to the accompanying graph to answer the questions.

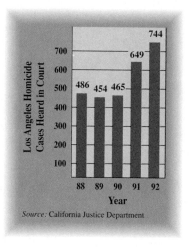

Source: California Justice Department

　　(a) What is the slope of the line segment joining the points for the years 1988 and 1989?
　　(b) Based on the answer to part (a), because the slope is

　　　　────────────── the
　　　　　(positive/negative)

　　　　number of homicide cases heard in the Los Angeles Superior Court

　　　　──────────────
　　　　　(increased/decreased)

　　　　between 1988 and 1989.

　　(c) If the points for 1988 and 1992 were joined by a single line segment, what would be its slope?

　　(d) Based on your answer to part (c), what was the average increase per year in the number of homicide cases heard from 1988 to 1992?

Solve the system by any method.

34. $2x - y = -8$
　　$x + 2y = 11$

35. $4x + 5y = -8$
　　$3x + 4y = -7$

36. $3x + 5y = 1$
　　$x = y + 3$

Use a system of equations to solve the problem.

37. The perimeter of a rectangle is 36 centimeters. The length is 8 centimeters longer than the width. Find the dimensions of the rectangle.

38. The perimeter of a triangle is 53 inches. If two sides are of equal length, and the third side measures 4 inches less than each of the equal sides, what are the lengths of the three sides?

39. The Smith family is coming to visit, and no one knows how many children they have. Janet, one of the girls, says she has as many brothers as sisters; her brother Steve says he has twice as many sisters as brothers. How many boys and how many girls are in the family?

40. In the Lopez family the number of boys is one more than half the number of girls. One of the Lopez boys, Rico, says that he has one more sister than brothers. How many boys and how many girls are in the family?

ROOTS AND RADICALS

CONNECTIONS

The Pythagorean formula was introduced in Section 4.6. The formula states that if the sides of a right triangle have lengths a, b, and c where c is the length of the longest side (the hypotenuse), then $c^2 = a^2 + b^2$. Pythagoras did not actually discover the theorem. There is evidence that the Babylonians knew the concept quite well. The first proof, however, may have come from Pythagoras. The figure on the left illustrates the theorem in a simple way, by using a sort of tile pattern. In the figure, the sides of the square along the hypotenuse measure 5 units, while those along the legs measure 3 and 4 units. If we let $a = 3$, $b = 4$, and $c = 5$, we see that the equation of the Pythagorean theorem is satisfied.

$$a^2 + b^2 = c^2$$
$$3^2 + 4^2 = 5^2$$
$$25 = 25$$

One of the topics we will consider in this chapter is how to find a, b, or c, given a^2, b^2, or c^2.

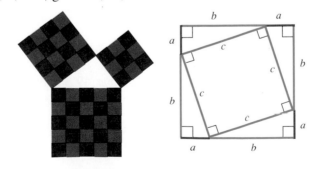

FOR DISCUSSION OR WRITING

The diagram on the right can be used to verify the Pythagorean formula. Express the area of the figure in two ways: first, as the area of the large square, and then as the sum of the areas of the smaller square and the four right triangles. Then set the areas equal and simplify the equation.

The distance a person can see to the horizon from an airplane flying at 30,000 feet is approximately $1.22\sqrt{30,000}$ feet. The number $\sqrt{30,000}$ is a *square root* or *radical expression*. In this chapter we see how to approximate a number like $\sqrt{30,000}$ and how to perform arithmetic operations with radical expressions.

Most roots and radicals represent real numbers, and the properties of real numbers studied earlier apply to them. In particular, the rules we used in Chapter 3 in our work with polynomials are used with radical expressions as well.

8.1 FINDING ROOTS

FOR EXTRA HELP	OBJECTIVES
SSG pp. 260–263 **SSM** pp. 436–440	**1** ▶ Find square roots.
	2 ▶ Decide whether a given root is rational, irrational, or not a real number.
Video 11	**3** ▶ Find decimal approximations for irrational square roots.
Tutorial IBM MAC	**4** ▶ Use the Pythagorean formula.
	5 ▶ Find higher roots.

In Section 1.2 we discussed the idea of the *square* of a number. Recall that squaring a number means multiplying the number by itself.

$$\text{If } a = 7, \text{ then } a^2 = 7 \cdot 7 = 49.$$

$$\text{If } a = 10, \text{ then } a^2 = 10 \cdot 10 = 100.$$

$$\text{If } a = -5, \text{ then } a^2 = (-5) \cdot (-5) = 25.$$

$$\text{If } a = -\frac{1}{2}, \text{ then } a^2 = \left(-\frac{1}{2}\right) \cdot \left(-\frac{1}{2}\right) = \frac{1}{4}.$$

In this chapter the opposite problem is considered.

$$\text{If } a^2 = 49, \text{ then } a = ?$$

$$\text{If } a^2 = 100, \text{ then } a = ?$$

$$\text{If } a^2 = 25, \text{ then } a = ?$$

$$\text{If } a^2 = \frac{1}{4}, \text{ then } a = ?$$

1 ▶ Find square roots.

▶ To find a in the four statements above, we must find a number that can be multiplied by itself to result in the given number. The number a is called a **square root** of the number a^2.

E X A M P L E   1

Finding all Square Roots of a Number

Find all square roots of 49.

We find a square root of 49 by thinking of a number that multiplied by itself gives 49. One square root is 7, since $7 \cdot 7 = 49$. Another square root of 49 is -7, since $(-7)(-7) = 49$. The number 49 has two square roots, 7 and -7. One is positive and one is negative. ■

Most calculators have a square root key, usually labeled $\boxed{\sqrt{x}}$, that allows us to find the square root of a number. For example, if we enter 121 and press the square root key, the display will show 11.

The positive square root of a number is written with the symbol $\sqrt{}$. For example, the positive square root of 121 is 11, written

$$\sqrt{121} = 11.$$

The symbol $-\sqrt{}$ is used for the negative square root of a number. For example, the negative square root of 121 is -11, written

$$-\sqrt{121} = -11.$$

The symbol $\sqrt{}$ is called a **radical sign** and always represents the nonnegative square root. The number inside the radical sign is called the **radicand** and the entire expression, radical sign and radicand, is called a **radical.** An algebraic expression containing a radical is called a **radical expression.**

Square Roots of a   If a is a nonnegative real number,

$$\sqrt{a} \text{ is the positive square root of } a,$$
$$-\sqrt{a} \text{ is the negative square root of } a.$$

Also, for nonnegative a,

$$\sqrt{a} \cdot \sqrt{a} = (\sqrt{a})^2 = a \text{ and } -\sqrt{a} \cdot -\sqrt{a} = (-\sqrt{a})^2 = a.$$

EXAMPLE 2

Finding Square Roots

Find each square root.

(a) $\sqrt{144}$

The radical $\sqrt{144}$ represents the positive square root of 144. Think of a positive number whose square is 144.

$$12^2 = 144, \quad \text{so} \quad \sqrt{144} = 12.$$

(b) $-\sqrt{1024}$

This symbol represents the negative square root of 1024. A calculator with a square root key can be used to find $\sqrt{1024} = 32$. Then, $-\sqrt{1024} = -32$.

(c) $\sqrt{\dfrac{4}{9}} = \dfrac{2}{3}$ **(d)** $-\sqrt{\dfrac{16}{49}} = -\dfrac{4}{7}$ ∎

As shown in the definition above, when the square root of a positive real number is squared, the result is that positive real number. (Also, $(\sqrt{0})^2 = 0.$)

EXAMPLE 3

Squaring Radical Expressions

Find the *square* of each radical expression.

(a) $\sqrt{13}$

$(\sqrt{13})^2 = 13$ Definition of square root

(b) $-\sqrt{29}$

$(-\sqrt{29})^2 = 29$ The square of a *negative* number is positive.

(c) $\sqrt{p^2 + 1}$

$(\sqrt{p^2 + 1})^2 = p^2 + 1$ ∎

2 ▶ Decide whether a given root is rational, irrational, or not a real number.

▶ All numbers that have rational number square roots are called **perfect squares.** A number that is not a perfect square has a square root that is not a rational number. For example, $\sqrt{5}$ is not a rational number because it cannot be written as the ratio of two integers. Its decimal neither terminates nor repeats. However, $\sqrt{5}$ is a real number and corresponds to a point on the number line. As mentioned earlier, a real number that is not rational is called an **irrational number.** The number $\sqrt{5}$ is irrational. Many square roots of integers are irrational.

If a is a positive number that is not a perfect square, then $\sqrt{a}$ is irrational.

Not every number has a *real number* square root. For example, there is no real number that can be squared to get -36. (The square of a real number can never be negative.) Because of this $\sqrt{-36}$ is not a real number. A calculator will show an error message in a case like this.

If a is a negative number, then $\sqrt{a}$ is not a real number.

EXAMPLE 4

Identifying Types of Square Roots

Tell whether each square root is rational, irrational, or not a real number.

(a) $\sqrt{17}$

Since 17 is a not a perfect square, $\sqrt{17}$ is irrational.

(b) $\sqrt{64}$

The number 64 is a perfect square, 8^2, so $\sqrt{64} = 8$, a rational number.

(c) $\sqrt{85}$ is irrational.

(d) $\sqrt{81}$ is rational ($\sqrt{81} = 9$).

(e) $\sqrt{-25}$

There is no real number whose square is -25. Therefore $\sqrt{-25}$ is not a real number. ■

NOTE Not all irrational numbers are square roots of integers. For example, π (approximately 3.14159) is an irrational number that is not a square root of any integer.

3 ▶ Find decimal approximations for irrational square roots.

▶ Even if a number is irrational, a decimal that approximates the number can be found by using a calculator. For example, a calculator shows that $\sqrt{10}$ is 3.16227766, although this is only a rational number approximation of $\sqrt{10}$.

EXAMPLE 5

Approximating Irrational Square Roots

Find a decimal approximation for each square root. Round answers to the nearest thousandth.

(a) $\sqrt{11}$

Using the square root key of a calculator gives $\sqrt{11} \approx 3.31662479 \approx 3.317$ rounded to the nearest thousandth, where $\approx$ means "is approximately equal to."

(b) $\sqrt{39} \approx 6.245$ **(c)** $-\sqrt{745} \approx -27.295$

(d) $\sqrt{-180}$ is not a real number, as indicated by an error message. ■

4 ▶ Use the Pythagorean formula.

▶ One application of square roots comes from using the Pythagorean formula, discussed in the Connections at the beginning of this chapter. By this formula, if c is the length of the hypotenuse of a right triangle, and a and b are the lengths of the two legs, then

$$c^2 = a^2 + b^2$$

or

$$c = \sqrt{a^2 + b^2}.$$

EXAMPLE 6
Using the Pythagorean Formula

Find the third side of each right triangle with sides a, b, and c, where c is the hypotenuse.

(a) $a = 3$, $b = 4$
We use the formula to find c^2 first.

$$c^2 = a^2 + b^2$$
$$= 3^2 + 4^2 \qquad \text{Let } a = 3 \text{ and } b = 4.$$
$$= 9 + 16 = 25 \qquad \text{Square and add.}$$

Now we find the positive square root of 25 to get c.

$$c = \sqrt{25} = 5$$

(Although -5 is also a square root of 25, the length of a side of a triangle must be a positive number.)

(b) $c = 9$, $b = 5$
Substitute the given values in the formula $c^2 = a^2 + b^2$. Then solve for a^2.

$$9^2 = a^2 + 5^2 \qquad \text{Let } c = 9 \text{ and } b = 5.$$
$$81 = a^2 + 25 \qquad \text{Square.}$$
$$56 = a^2 \qquad \text{Subtract 25.}$$

Again, we want only the positive root $a = \sqrt{56} \approx 7.483$. ■

CAUTION Be careful not to make the common mistake of thinking that $\sqrt{a^2 + b^2}$ equals $a + b$. As Example 6(a) shows,

$$\sqrt{9 + 16} = \sqrt{25} = 5 \neq \sqrt{9} + \sqrt{16} = 3 + 4,$$

so that, in general,

$$\sqrt{a^2 + b^2} \neq a + b.$$

The Pythagorean formula can be used to solve applied problems that involve right triangles.

PROBLEM SOLVING

When an applied problem involves lengths that form a right triangle, a good way to begin the solution is to sketch the triangle and label the three sides appropriately, using a variable as needed. Then use the Pythagorean formula to write an equation.

EXAMPLE 7

Solving an Application

A ladder 10 feet long leans against a wall. The foot of the ladder is 6 feet from the base of the wall. How high up the wall does the top of the ladder rest?

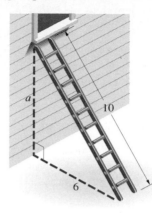

FIGURE 1

As shown in Figure 1, a right triangle is formed with the ladder as the hypotenuse. Let a represent the height of the top of the ladder. By the Pythagorean formula,

$$c^2 = a^2 + b^2$$
$$\mathbf{10}^2 = a^2 + \mathbf{6}^2 \qquad \text{Let } c = 10 \text{ and } b = 6.$$
$$100 = a^2 + 36 \qquad \text{Square.}$$
$$64 = a^2 \qquad \text{Subtract 36.}$$
$$\sqrt{64} = a$$
$$a = 8. \qquad \sqrt{64} = 8$$

Choose the positive square root of 64 since a represent a length. The top of the ladder rests 8 feet up the wall. ∎

5 ▶ Find higher roots.

▶ Finding the square root of a number is the inverse of squaring a number. In a similar way, there are inverses to finding the cube of a number, or finding the fourth or higher power of a number. These inverses are the **cube root,** written $\sqrt[3]{a}$, the **fourth root,** written $\sqrt[4]{a}$, and in general we have the following.

The *n*th root of a is written $\sqrt[n]{a}$.

In $\sqrt[n]{a}$, the number n is the **index** or **order** of the radical. It would be possible to write $\sqrt[2]{a}$ instead of $\sqrt{a}$, but the simpler symbol $\sqrt{a}$ is customary since the square root is the most commonly used root. A calculator that has a key marked $\boxed{\sqrt[x]{y}}$ or $\boxed{x^y}$ can be used to find these roots. When working with cube roots or fourth roots, it is helpful to memorize the first few *perfect cubes* ($2^3 = 8$, $3^3 = 27$, and so on), and the first few perfect fourth powers.

EXAMPLE 8

Finding Cube Roots

Find each cube root.

(a) $\sqrt[3]{8}$

Look for a number that can be cubed to give 8. Since $2^3 = 8$, then $\sqrt[3]{8} = 2$.

(b) $\sqrt[3]{-8}$

$\sqrt[3]{-8} = -2$ because $(-2)^3 = -8$. ∎

As these examples suggest, the cube root of a positive number is positive, and the cube root of a negative number is negative. *There is only one real number cube root for each real number.*

When the index of the radical is even (square root, fourth root, and so on), the radicand must be nonnegative to get a real number root. Also, for even indexes the symbols $\sqrt{}$, $\sqrt[4]{}$, $\sqrt[6]{}$, and so on are used for the *nonnegative* roots, which are called **principal roots.**

EXAMPLE 9

Finding Higher Roots

Find each root.

(a) $\sqrt[4]{16}$

$\sqrt[4]{16} = 2$ because 2 is positive and $2^4 = 16$.

(b) $-\sqrt[4]{16} = -2$

(c) $\sqrt[4]{-16}$

There is no real number that equals $\sqrt[4]{-16}$ because a fourth power of a real number must be positive.

(d) $\sqrt[3]{64} = 4$ since $4^3 = 64$.

(e) $-\sqrt[5]{32}$

First find $\sqrt[5]{32}$. The prime factorization of 32 as 2^5 shows that $\sqrt[5]{32} = 2$. If $\sqrt[5]{32} = 2$, then $-\sqrt[5]{32} = -2$. ■

8.1 EXERCISES

Answer the questions about roots.

1. How many square roots does any positive number have?
2. How many square roots does 0 have?
3. How many real number square roots does any negative number have?
4. How many real number cube roots does any positive number have?
5. How many real number cube roots does any negative number have?
6. How many cube roots does 0 have?

Find all square roots of the number. See Example 1.

7. 16
8. 9
9. 144
10. 225
11. $\dfrac{25}{196}$
12. $\dfrac{81}{400}$
13. 900
14. 1600

Find each square root that is a real number. See Examples 2 and 4(e).

15. $\sqrt{49}$
16. $\sqrt{81}$
17. $-\sqrt{121}$
18. $\sqrt{196}$
19. $-\sqrt{\dfrac{144}{121}}$
20. $-\sqrt{\dfrac{49}{36}}$
21. $\sqrt{-121}$
22. $\sqrt{-49}$

Find the square of each radical expression. See Example 3.

23. $\sqrt{100}$
24. $\sqrt{36}$
25. $-\sqrt{19}$
26. $-\sqrt{99}$
27. $\sqrt{3x^2 + 4}$
28. $\sqrt{9y^2 + 3}$

What must be true about the variable a for the statement to be true?

29. $\sqrt{a}$ represents a positive number.

30. $-\sqrt{a}$ represents a negative number.

31. $\sqrt{a}$ is not a real number.

32. $-\sqrt{a}$ is not a real number.

Write rational, irrational, or not a real number for the number. If the number is rational, give its exact value. If the number is irrational, give a decimal approximation to the nearest thousandth. Use a calculator as necessary. See Examples 4 and 5.

33. $\sqrt{25}$ **34.** $\sqrt{169}$ **35.** $\sqrt{29}$ **36.** $\sqrt{33}$

37. $-\sqrt{64}$ **38.** $-\sqrt{900}$ **39.** $-\sqrt{300}$ **40.** $-\sqrt{500}$

41. $\sqrt{-29}$ **42.** $\sqrt{-47}$

43. Explain why the answers to Exercises 17 and 21 are different.

44. Explain why $\sqrt[3]{-8}$ and $-\sqrt[3]{8}$ represent the same number.

Use a calculator with a square root key to find the root. Round to the nearest thousandth.

45. $\sqrt{571}$ **46.** $\sqrt{693}$ **47.** $\sqrt{798}$ **48.** $\sqrt{453}$

49. $\sqrt{3.94}$ **50.** $\sqrt{1.03}$ **51.** $\sqrt{.00895}$ **52.** $\sqrt{.000402}$

Use a calculator with a cube root key to find the root. Round to the nearest thousandth. (In Exercises 57 and 58, you may have to use the fact that if $a > 0$, $\sqrt[3]{-a} = -\sqrt[3]{a}$.)

53. $\sqrt[3]{12}$ **54.** $\sqrt[3]{74}$ **55.** $\sqrt[3]{130.6}$

56. $\sqrt[3]{251.8}$ **57.** $\sqrt[3]{-87}$ **58.** $\sqrt[3]{-95}$

Find the length of the unknown side of the right triangle with legs a and b and hypotenuse c. In Exercises 63 and 64, use a calculator and round to the nearest thousandth. See Example 6.

59. $a = 8, b = 15$ **60.** $a = 24, b = 10$ **61.** $a = 6, c = 10$

62. $b = 12, c = 13$ **63.** $a = 11, b = 4$ **64.** $a = 13, b = 9$

Use the Pythagorean formula to solve the problem. In Exercises 71 and 72 round the answer to the nearest thousandth. See Example 7.

65. The diagonal of a rectangle measures 25 centimeters. The width of the rectangle is 7 centimeters. Find the length of the rectangle.

66. The length of a rectangle is 40 meters, and the width is 9 meters. Find the measure of the diagonal of the rectangle.

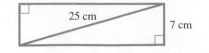

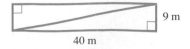

67. Margaret is flying a kite on 100 feet of string. How high is it above her hand (vertically) if the horizontal distance between Margaret and the kite is 60 feet?

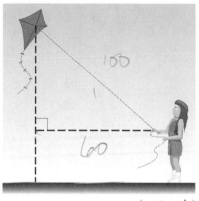

(not to scale)

68. A guy wire is attached to the mast of a short-wave transmitting antenna. It is attached 96 feet above ground level. If the wire is staked to the ground 72 feet from the base of the mast, how long is the wire?

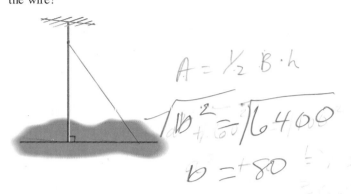

69. Two cars leave Tomball, Texas, at the same time. One travels north at 25 miles per hour and the other travels west at 60 miles per hour. How far apart are they after 3 hours?

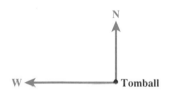

70. A boat is being pulled toward a dock with a rope attached at water level. When the boat is 24 feet from the dock, 30 feet of rope is extended. What is the height of the dock above the water?

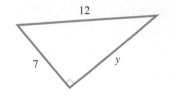

71. What is the value of x in the figure?

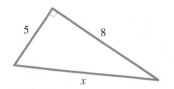

72. What is the value of y in the figure?

12

7   y

73. Use specific values for a and b different from those given in the "Caution" following Example 6 to show that $\sqrt{a^2 + b^2} \neq a + b$.

74. Why would the values $a = 0$ and $b = 1$ *not* be satisfactory in Exercise 73?

Find each root that is a real number. See Examples 8 and 9.

75. $\sqrt[3]{1000}$ **76.** $\sqrt[3]{8}$ **77.** $\sqrt[3]{125}$ **78.** $\sqrt[3]{216}$

79. $\sqrt[3]{-27}$ **80.** $\sqrt[3]{-64}$ **81.** $\sqrt[4]{625}$ **82.** $\sqrt[4]{10,000}$

83. $\sqrt[4]{-1}$ **84.** $\sqrt[4]{-625}$ **85.** $-\sqrt[5]{243}$ **86.** $-\sqrt[5]{100,000}$

◇ **MATHEMATICAL CONNECTIONS** (Exercises 87–92) ◇

One of the many proofs of the Pythagorean formula was given in the Connections at the beginning of this chapter. Here is another one, attributed to the Hindu mathematician Bhāskara. Refer to the figures and work Exercises 87–92 in order.

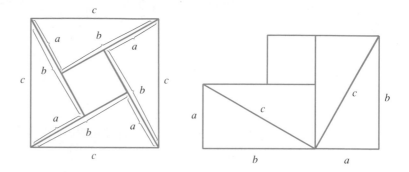

87. What is the area of the square on the left in terms of c?

88. What is the area of the small square in the middle of the figure on the left, in terms of $(b - a)$?

89. What is the sum of the areas of the two rectangles made up of triangles in the figure on the right?

90. What is the area of the small square in the figure on the right in terms of a and b?

91. The figure on the left is made up of the same square and triangles as the figure on the right. Write an equation setting the answer to Exercise 87 equal to the sum of the answers in Exercises 89 and 90.

92. Simplify the expressions you obtained in Exercise 91. What is your final result?

◇

REVIEW EXERCISES

Write the number in prime factored form. See Section 4.1.

93. 72

94. 100

95. 40

96. 242

8.2 MULTIPLICATION AND DIVISION OF RADICALS

OBJECTIVES

1 ▶ Multiply radicals.
2 ▶ Simplify radicals using the product rule.
3 ▶ Simplify radical quotients using the quotient rule.
4 ▶ Use the product and quotient rules to simplify higher roots.

In this section and the one following, we will examine how the arithmetic operations of multiplication, division, addition, and subtraction are applied to numbers involving radicals.

◆ C O N N E C T I O N S ◆

The sixteenth century German radical symbol $\sqrt{}$ we use today is probably derived from the letter *R*. The radical symbol on the left below comes from the Latin word for root, *radix*. It was first used by Leonardo da Pisa (Fibonnaci) in 1220.

The cube root symbol shown on the right above was used by the German mathematician Christoff Rudolff in 1525. The symbol used today originated in the seventeenth century in France.

FOR DISCUSSION OR WRITING
1. In the radical sign shown on the left, the *R* referred to above is clear. What other letter do you think is part of the symbol? What would the equivalent be in our modern notation?
2. How is the cube root symbol on the right related to our modern radical sign?

1 ▶ Multiply radicals.

▶ Several useful rules for finding products and quotients of radicals are developed in this section. To illustrate the rule for products, notice that

$$\sqrt{4} \cdot \sqrt{9} = 2 \cdot 3 = 6 \quad \text{and} \quad \sqrt{4 \cdot 9} = \sqrt{36} = 6,$$

showing that

$$\sqrt{4} \cdot \sqrt{9} = \sqrt{4 \cdot 9}.$$

This result is a particular case of the more general product rule for radicals.

Product Rule for Radicals

For nonnegative real numbers *x* and *y*,

$$\sqrt{x} \cdot \sqrt{y} = \sqrt{x \cdot y} \quad \text{and} \quad \sqrt{x \cdot y} = \sqrt{x} \cdot \sqrt{y}.$$

That is, the product of two radicals is the radical of the product.

CAUTION In general, $\sqrt{x + y} \neq \sqrt{x} + \sqrt{y}$. To see why this is so, let $x = 16$ and $y = 9$.

EXAMPLE 1
Using the Product Rule to Multiply Radicals

Use the product rule for radicals to find each product.

(a) $\sqrt{2} \cdot \sqrt{3} = \sqrt{2 \cdot 3} = \sqrt{6}$

(b) $\sqrt{7} \cdot \sqrt{5} = \sqrt{35}$

(c) $\sqrt{11} \cdot \sqrt{a} = \sqrt{11a}$ Assume $a \geq 0$. ■

2 ▶ Simplify radicals using the product rule.

▶ An important use of the product rule is in simplifying radicals. A radical is **simplified** when no perfect square factor remains under the radical sign. This is accomplished by using the product rule in the form $\sqrt{x \cdot y} = \sqrt{x} \cdot \sqrt{y}$. Example 2 shows how a radical may be simplified in this way.

E X A M P L E   2 ■

Using the Product Rule to Simplify Radicals

Simplify each radical.

(a) $\sqrt{20}$

Since 20 has a perfect square factor of 4,

$$\sqrt{20} = \sqrt{4 \cdot 5} \qquad \text{4 is a perfect square.}$$
$$= \sqrt{4} \cdot \sqrt{5} \qquad \text{Product rule}$$
$$= 2\sqrt{5}. \qquad \sqrt{4} = 2$$

Thus, $\sqrt{20} = 2\sqrt{5}$. Since 5 has no perfect square factor, other than 1, $2\sqrt{5}$ is called the **simplified form** of $\sqrt{20}$. Note that $2\sqrt{5}$ represents a product, where the factors are 2 and $\sqrt{5}$. Furthermore, $2\sqrt{5} \neq \sqrt{10}$.

(b) $\sqrt{72}$

We begin by looking for the largest perfect square that is a factor of 72. This number is 36, so

$$\sqrt{72} = \sqrt{36 \cdot 2} \qquad \text{36 is a perfect square.}$$
$$= \sqrt{36} \cdot \sqrt{2} \qquad \text{Product rule}$$
$$= 6\sqrt{2}. \qquad \sqrt{36} = 6$$

We could also factor 72 into its prime factors and look for pairs of like factors. Each pair of like factors produces a factor outside the radical in the simplified form.

$$\sqrt{72} = \sqrt{2 \cdot 2 \cdot 2 \cdot 3 \cdot 3} = 2 \cdot 3 \cdot \sqrt{2} = 6\sqrt{2}$$

In either case, we obtain $6\sqrt{2}$ as the simplified form of $\sqrt{72}$; however, our work is simpler if we begin with the largest perfect square factor.

(c) $\sqrt{300} = \sqrt{100 \cdot 3} \qquad \text{100 is a perfect square.}$
$$= \sqrt{100} \cdot \sqrt{3} \qquad \text{Product rule}$$
$$= 10\sqrt{3} \qquad \sqrt{100} = 10$$

(d) $\sqrt{15}$

The number 15 has no perfect square factors (except 1), so $\sqrt{15}$ cannot be simplified further. ■

Sometimes the product rule can be used to simplify a product, as Example 3 shows.

E X A M P L E   3 ■

Multiplying and Simplifying Radicals

Find each product and simplify.

(a) $\sqrt{9} \cdot \sqrt{75} = 3\sqrt{75} \qquad \sqrt{9} = 3$
$$= 3\sqrt{25 \cdot 3} \qquad \text{25 is a perfect square.}$$
$$= 3\sqrt{25} \cdot \sqrt{3} \qquad \text{Product rule}$$
$$= 3 \cdot 5\sqrt{3} \qquad \sqrt{25} = 5$$
$$= 15\sqrt{3} \qquad \text{Multiply.}$$

(b) $\sqrt{8} \cdot \sqrt{12} = \sqrt{8 \cdot 12}$ Product rule

$\qquad\qquad = \sqrt{4 \cdot 2 \cdot 4 \cdot 3}$ Factor; 4 is a perfect square.

$\qquad\qquad = \sqrt{4} \cdot \sqrt{4} \cdot \sqrt{2 \cdot 3}$ Product rule

$\qquad\qquad = 2 \cdot 2 \cdot \sqrt{6}$ $\sqrt{4} = 2$

$\qquad\qquad = 4\sqrt{6}$ Multiply. ∎

3 ▶ Simplify radical quotients using the quotient rule.

▶ The *quotient rule for radicals* is very similar to the product rule. It, too, can be used either way.

Quotient Rule for Radicals

If x and y are nonnegative real numbers and $y \neq 0$,

$$\frac{\sqrt{x}}{\sqrt{y}} = \sqrt{\frac{x}{y}} \quad \text{and} \quad \sqrt{\frac{x}{y}} = \frac{\sqrt{x}}{\sqrt{y}}.$$

The quotient of the radicals is the radical of the quotient.

E X A M P L E 4

Using the Quotient Rule to Simplify Radicals

■ Simplify each radical.

(a) $\sqrt{\dfrac{25}{9}} = \dfrac{\sqrt{25}}{\sqrt{9}} = \dfrac{5}{3}$ Quotient rule

(b) $\dfrac{\sqrt{288}}{\sqrt{2}} = \sqrt{\dfrac{288}{2}} = \sqrt{144} = 12$ Quotient rule

(c) $\sqrt{\dfrac{3}{4}} = \dfrac{\sqrt{3}}{\sqrt{4}} = \dfrac{\sqrt{3}}{2}$ Quotient rule ∎

E X A M P L E 5

Using the Quotient Rule to Divide Radicals

■ Divide $27\sqrt{15}$ by $9\sqrt{3}$.

We use the quotient rule as follows.

$$\frac{27\sqrt{15}}{9\sqrt{3}} = \frac{27}{9} \cdot \frac{\sqrt{15}}{\sqrt{3}} = 3\sqrt{\frac{15}{3}} = 3\sqrt{5} \quad ∎$$

Some problems require both the product and the quotient rules, as Example 6 shows.

E X A M P L E 6

Using Both the Product and Quotient Rules

■ Simplify $\sqrt{\dfrac{3}{5}} \cdot \sqrt{\dfrac{4}{5}}.$

$$\sqrt{\frac{3}{5}} \cdot \sqrt{\frac{4}{5}} = \frac{\sqrt{3}}{\sqrt{5}} \cdot \frac{\sqrt{4}}{\sqrt{5}} \qquad \text{Quotient rule}$$

$$= \frac{\sqrt{3} \cdot \sqrt{4}}{\sqrt{5} \cdot \sqrt{5}} \qquad \text{Multiply fractions.}$$

$$= \frac{\sqrt{3} \cdot 2}{\sqrt{25}} \qquad \text{Product rule; } \sqrt{4} = 2$$

$$= \frac{2\sqrt{3}}{5} \qquad \sqrt{25} = 5 \quad ∎$$

The properties stated earlier in this section also apply when variables appear under the radical sign, as long as all the variables represent only nonnegative numbers. For example, $\sqrt{5^2} = 5$, but $\sqrt{(-5)^2} \neq -5$.

For a real number a, $\quad \sqrt{a^2} = a \quad$ only if a is nonnegative.

EXAMPLE 7

Simplifying Radicals Involving Variables

Simplify each radical. Assume all variables represent positive real numbers.

(a) $\sqrt{25m^4} = \sqrt{25} \cdot \sqrt{m^4}$ Product rule

$\qquad\qquad = 5m^2$

(b) $\sqrt{64p^{10}} = 8p^5$ Product rule

(c) $\sqrt{r^9} = \sqrt{r^8 \cdot r}$

$\qquad\quad = \sqrt{r^8} \cdot \sqrt{r} = r^4\sqrt{r}$ Product rule

(d) $\sqrt{\dfrac{5}{x^2}} = \dfrac{\sqrt{5}}{\sqrt{x^2}} = \dfrac{\sqrt{5}}{x}$ Quotient rule ∎

4 ▶ Use the product and quotient rules to simplify higher roots.

▶ The product rule and the quotient rule for radicals also work for other roots, as shown in Example 8. To simplify cube roots, look for factors that are *perfect cubes*. A **perfect cube** is a number with a rational cube root. For example, $\sqrt[3]{64} = 4$, and since 4 is a rational number, 64 is a perfect cube. Higher roots are handled in a similar manner.

Properties of Radicals

For all real numbers where the indicated roots exist,

$$\sqrt[n]{x} \cdot \sqrt[n]{y} = \sqrt[n]{xy} \qquad \text{and} \qquad \frac{\sqrt[n]{x}}{\sqrt[n]{y}} = \sqrt[n]{\frac{x}{y}}.$$

EXAMPLE 8

Simplifying Higher Roots

Simplify each radical.

(a) $\sqrt[3]{32} = \sqrt[3]{8 \cdot 4}$ 8 is a perfect cube.

$\qquad\quad = \sqrt[3]{8} \cdot \sqrt[3]{4} = 2\sqrt[3]{4}$

(b) $\sqrt[3]{108} = \sqrt[3]{27 \cdot 4}$ 27 is a perfect cube.

$\qquad\qquad = \sqrt[3]{27} \cdot \sqrt[3]{4} = 3\sqrt[3]{4}$

(c) $\sqrt[4]{32} = \sqrt[4]{16} \cdot \sqrt[4]{2} = 2\sqrt[4]{2}$ 16 is a perfect fourth power.

(d) $\sqrt[3]{\dfrac{8}{125}} = \dfrac{\sqrt[3]{8}}{\sqrt[3]{125}} = \dfrac{2}{5}$

(e) $\sqrt[4]{\dfrac{16}{625}} = \dfrac{\sqrt[4]{16}}{\sqrt[4]{625}} = \dfrac{2}{5}$

(f) $\sqrt[3]{7} \cdot \sqrt[3]{49} = \sqrt[3]{7 \cdot 49} = \sqrt[3]{7 \cdot 7^2} = \sqrt[3]{7^3} = 7$ ∎

8.2 EXERCISES

Decide whether the statement is true or false.

1. $\sqrt{9} \cdot \sqrt{16} = \sqrt{9 \cdot 16}$ **2.** $\sqrt{9 + 16} = \sqrt{9} + \sqrt{16}$ **3.** $\sqrt{.5} = \sqrt{\dfrac{1}{2}}$

4. For nonnegative real numbers x and y, $\sqrt{xy} = \sqrt{x} \cdot \sqrt{y}$.

5. $\sqrt{(-6)^2} = -6$ **6.** $\sqrt[3]{(-6)^3} = -6$

Use the product rule for radicals to find the product. See Example 1.

7. $\sqrt{3} \cdot \sqrt{27}$ **8.** $\sqrt{2} \cdot \sqrt{8}$ **9.** $\sqrt{6} \cdot \sqrt{15}$

10. $\sqrt{10} \cdot \sqrt{15}$ **11.** $\sqrt{13} \cdot \sqrt{13}$ **12.** $\sqrt{17} \cdot \sqrt{17}$

13. $\sqrt{13} \cdot \sqrt{r}, r \geq 0$ **14.** $\sqrt{19} \cdot \sqrt{k}, k \geq 0$

15. Which one of the following radicals is simplified according to the guidelines of Objective 2?
   (a) $\sqrt{47}$ **(b)** $\sqrt{45}$ **(c)** $\sqrt{48}$ **(d)** $\sqrt{44}$

16. If p is a prime number, is $\sqrt{p}$ in simplified form? Explain your answer.

Simplify the radical according to the method described in Objective 2. See Example 2.

17. $\sqrt{45}$ **18.** $\sqrt{27}$ **19.** $\sqrt{90}$ **20.** $\sqrt{56}$

21. $\sqrt{75}$ **22.** $\sqrt{18}$ **23.** $\sqrt{125}$ **24.** $\sqrt{80}$

25. $-\sqrt{700}$ **26.** $-\sqrt{600}$ **27.** $3\sqrt{27}$ **28.** $9\sqrt{8}$

Find the product and simplify. See Example 3.

29. $\sqrt{3} \cdot \sqrt{18}$ **30.** $\sqrt{3} \cdot \sqrt{21}$ **31.** $\sqrt{12} \cdot \sqrt{48}$

32. $\sqrt{50} \cdot \sqrt{72}$ **33.** $\sqrt{12} \cdot \sqrt{30}$ **34.** $\sqrt{30} \cdot \sqrt{24}$

35. Simplify the product $\sqrt{8} \cdot \sqrt{32}$ in two different ways. First, multiply 8 by 32 and simplify the square root of this product. Second, simplify $\sqrt{8}$ and simplify $\sqrt{32}$ and then multiply. Do you get the same answer? Make a conjecture (an educated guess) about whether the correct answer can always be obtained using either method in a simplification such as this.

36. Simplify the radical $\sqrt{288}$ in two ways. First, factor 288 as $144 \cdot 2$ and then simplify. Second, factor 288 as $48 \cdot 6$ and then simplify completely, performing any additional steps as needed. Do you get the same answer? Make a conjecture concerning the quickest way to simplify such a radical.

Use the quotient rule and the product rule, as necessary, to simplify the radical expression. See Examples 4–6.

37. $\sqrt{\dfrac{16}{225}}$ **38.** $\sqrt{\dfrac{9}{100}}$ **39.** $\sqrt{\dfrac{7}{16}}$ **40.** $\sqrt{\dfrac{13}{25}}$

41. $\sqrt{\dfrac{5}{7}} \cdot \sqrt{35}$ **42.** $\sqrt{\dfrac{10}{13}} \cdot \sqrt{130}$ **43.** $\sqrt{\dfrac{5}{2}} \cdot \sqrt{\dfrac{125}{8}}$ **44.** $\sqrt{\dfrac{8}{3}} \cdot \sqrt{\dfrac{512}{27}}$

45. $\dfrac{30\sqrt{10}}{5\sqrt{2}}$ **46.** $\dfrac{50\sqrt{20}}{2\sqrt{10}}$

Simplify the radical. Assume that all variables represent nonnegative real numbers. See Example 7.

47. $\sqrt{m^2}$ **48.** $\sqrt{k^2}$ **49.** $\sqrt{y^4}$ **50.** $\sqrt{s^4}$

51. $\sqrt{36z^2}$ **52.** $\sqrt{49n^2}$ **53.** $\sqrt{400x^6}$ **54.** $\sqrt{900y^8}$

55. $\sqrt{z^5}$ **56.** $\sqrt{a^{13}}$ **57.** $\sqrt{x^6y^{12}}$ **58.** $\sqrt{a^8b^{10}}$

Simplify the radical. See Example 8.

59. $\sqrt[3]{40}$ **60.** $\sqrt[3]{48}$ **61.** $\sqrt[3]{54}$ **62.** $\sqrt[3]{135}$

63. $\sqrt[3]{128}$ **64.** $\sqrt[3]{192}$ **65.** $\sqrt[4]{80}$ **66.** $\sqrt[4]{243}$

67. $\sqrt[3]{\dfrac{8}{27}}$ **68.** $\sqrt[3]{\dfrac{64}{125}}$ **69.** $\sqrt[3]{-\dfrac{216}{125}}$ **70.** $\sqrt[3]{-\dfrac{1}{64}}$

71. $\sqrt[3]{5} \cdot \sqrt[3]{25}$ **72.** $\sqrt[4]{4} \cdot \sqrt[4]{16}$ **73.** $\sqrt[4]{4} \cdot \sqrt[4]{3}$ **74.** $\sqrt[4]{7} \cdot \sqrt[4]{4}$

75. $\sqrt[3]{4x} \cdot \sqrt[3]{8x^2}$ **76.** $\sqrt[3]{25p} \cdot \sqrt[3]{125p^3}$

77. In Example 2(a) we showed *algebraically* that $\sqrt{20}$ is equal to $2\sqrt{5}$. To give *numerical support* to this result, use a calculator to do the following:
 (a) Find a decimal approximation for $\sqrt{20}$ using your calculator. Record as many digits as the calculator shows.
 (b) Find a decimal approximation for $\sqrt{5}$ using your calculator, and then multiply the result by 2. Record as many digits as the calculator shows.
 (c) Your results in parts (a) and (b) should be the same. A mathematician would not accept this numerical exercise as *proof* that $\sqrt{20}$ is equal to $2\sqrt{5}$. Can you explain why?

78. On your calculator, multiply the approximations for $\sqrt{3}$ and $\sqrt{5}$. Now, predict what your calculator will show when you find an approximation for $\sqrt{15}$. What rule stated in this section justifies your answer?

The volume of a cube is found with the formula $V = s^3$, where s is the length of an edge of the cube. Use this information in Exercises 79 and 80.

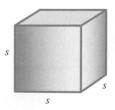

79. A container in the shape of a cube has a volume of 216 cubic centimeters. What is the depth of the container?

80. A cube-shaped box must be constructed to contain 128 cubic feet. What should the dimensions (height, width, and length) of the box be?

The volume of a sphere is found with the formula $V = (4/3)\pi r^3$, where r is the length of the radius of the sphere. Use this information in Exercises 81 and 82.

81. A ball in the shape of a sphere has a volume of 288π cubic inches. What is the radius of the ball?

82. Suppose that the volume of the ball described in Exercise 81 is multiplied by 8. How is the radius affected?

83. When we multiply two radicals with variables under the radical sign, such as $\sqrt{a} \cdot \sqrt{b} = \sqrt{ab}$, why is it important to know that both a and b represent nonnegative numbers?

84. Is it necessary to restrict k to a nonnegative number to say that $\sqrt[3]{k} \cdot \sqrt[3]{k} \cdot \sqrt[3]{k} = k$? Why?

REVIEW EXERCISES

Combine like terms. See Section 2.1.

85. $4x + 7 - 9x + 12$

86. $9x^2 + 3x^2 - 2x + 4x - 8 + 1$

87. $2xy + 3x^2y - 9xy + 8x^2y$

88. $x + 3y + 12z$

8.3 ADDITION AND SUBTRACTION OF RADICALS

FOR EXTRA HELP	OBJECTIVES
SSG pp. 266–268 **SSM** pp. 443–445	**1** ▶ Add and subtract radicals.
Video 11	**2** ▶ Simplify radical sums and differences.
Tutorial IBM MAC	**3** ▶ Simplify radical sums involving multiplication.

◆ **C O N N E C T I O N S** ◆

An interesting way to represent the lengths corresponding to $\sqrt{2}$, $\sqrt{3}$, $\sqrt{4}$, $\sqrt{5}$, and so on, is shown in the figure.

FOR DISCUSSION OR WRITING

Use the Pythagorean formula to verify the lengths in the figure.

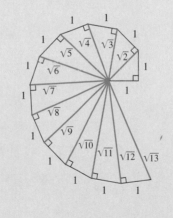

1. Which of the lengths are represented by whole numbers?

2. Is there a pattern to when such lengths occur—that is, can you predict where the next one and the one after that will occur?

1 ▶ Add and subtract radicals.

▶ We add or subtract radicals by using the distributive property. For example,

$$8\sqrt{3} + 6\sqrt{3} = (8 + 6)\sqrt{3} \quad \text{Distributive property}$$
$$= 14\sqrt{3}.$$

Also,

$$2\sqrt{11} - 7\sqrt{11} = -5\sqrt{11}.$$

Only **like radicals,** those that are multiples of the *same root* of the *same number,* can be combined this way.

E X A M P L E 1

Adding and Subtracting Like Radicals

Add or subtract, as indicated.

(a) $3\sqrt{6} + 5\sqrt{6} = (3 + 5)\sqrt{6} = 8\sqrt{6}$ Distributive property

(b) $5\sqrt{10} - 7\sqrt{10} = (5 - 7)\sqrt{10} = -2\sqrt{10}$

(c) $\sqrt[3]{5} + \sqrt[3]{5} = 1\sqrt[3]{5} + 1\sqrt[3]{5} = (1 + 1)\sqrt[3]{5} = 2\sqrt[3]{5}$

(d) $\sqrt[4]{7} + 2\sqrt[4]{7} = 1\sqrt[4]{7} + 2\sqrt[4]{7} = 3\sqrt[4]{7}$

(e) $\sqrt{3} + \sqrt{7}$ cannot be simplified further. ■

2 ▶ Simplify radical sums and differences.

▶ Sometimes we must simplify one or more radicals in a sum or difference. Doing this may result in like radicals, which we can then add or subtract.

E X A M P L E 2

Adding and Subtracting Radicals that Require Simplification

Simplify as much as possible.

(a) $3\sqrt{2} + \sqrt{8} = 3\sqrt{2} + \sqrt{4 \cdot 2}$ Factor.

$\phantom{3\sqrt{2} + \sqrt{8}} = 3\sqrt{2} + \sqrt{4} \cdot \sqrt{2}$ Product rule

$\phantom{3\sqrt{2} + \sqrt{8}} = 3\sqrt{2} + 2\sqrt{2}$ $\sqrt{4} = 2$

$\phantom{3\sqrt{2} + \sqrt{8}} = 5\sqrt{2}$ Add like radicals.

(b) $\sqrt{18} - \sqrt{27} = \sqrt{9 \cdot 2} - \sqrt{9 \cdot 3}$ Factor.

$\phantom{\sqrt{18} - \sqrt{27}} = \sqrt{9} \cdot \sqrt{2} - \sqrt{9} \cdot \sqrt{3}$ Product rule

$\phantom{\sqrt{18} - \sqrt{27}} = 3\sqrt{2} - 3\sqrt{3}$ $\sqrt{9} = 3$

Since $\sqrt{2}$ and $\sqrt{3}$ are unlike radicals, this difference cannot be simplified further.

(c) $2\sqrt{12} + 3\sqrt{75} = 2(\sqrt{4} \cdot \sqrt{3}) + 3(\sqrt{25} \cdot \sqrt{3})$ Product rule

$\phantom{2\sqrt{12} + 3\sqrt{75}} = 2(2\sqrt{3}) + 3(5\sqrt{3})$ $\sqrt{4} = 2$ and $\sqrt{25} = 5$

$\phantom{2\sqrt{12} + 3\sqrt{75}} = 4\sqrt{3} + 15\sqrt{3}$ Multiply.

$\phantom{2\sqrt{12} + 3\sqrt{75}} = 19\sqrt{3}$ Add like radicals.

(d) $3\sqrt[3]{16} + 5\sqrt[3]{2} = 3(\sqrt[3]{8} \cdot \sqrt[3]{2}) + 5\sqrt[3]{2}$ Product rule

$\phantom{3\sqrt[3]{16} + 5\sqrt[3]{2}} = 3(2\sqrt[3]{2}) + 5\sqrt[3]{2}$ $\sqrt[3]{8} = 2$

$\phantom{3\sqrt[3]{16} + 5\sqrt[3]{2}} = 6\sqrt[3]{2} + 5\sqrt[3]{2}$ Multiply.

$\phantom{3\sqrt[3]{16} + 5\sqrt[3]{2}} = 11\sqrt[3]{2}$ Add like radicals. ■

3 ▶ Simplify radical sums involving multiplication.

▶ Some radical expressions require both multiplication and addition (or subtraction). The order of operations presented earlier still applies.

EXAMPLE 3

Simplifying Radical Sums Involving Multiplication

Simplify each expression. Assume that all variables represent nonnegative real numbers.

(a) $\sqrt{5} \cdot \sqrt{15} + 4\sqrt{3} = \sqrt{5 \cdot 15} + 4\sqrt{3}$ Product rule

$= \sqrt{75} + 4\sqrt{3}$ Multiply.

$= \sqrt{25 \cdot 3} + 4\sqrt{3}$ 25 is a perfect square.

$= \sqrt{25} \cdot \sqrt{3} + 4\sqrt{3}$ Product rule

$= 5\sqrt{3} + 4\sqrt{3}$ $\sqrt{25} = 5$

$= 9\sqrt{3}$ Add like radicals.

(b) $\sqrt{2} \cdot \sqrt{6k} + \sqrt{27k} = \sqrt{12k} + \sqrt{27k}$ Product rule

$= \sqrt{4 \cdot 3k} + \sqrt{9 \cdot 3k}$ Factor.

$= \sqrt{4} \cdot \sqrt{3k} + \sqrt{9} \cdot \sqrt{3k}$ Product rule

$= 2\sqrt{3k} + 3\sqrt{3k}$ $\sqrt{4} = 2$ and $\sqrt{9} = 3$

$= 5\sqrt{3k}$ Add like radicals.

(c) $\sqrt[3]{2} \cdot \sqrt[3]{16m^3} - \sqrt[3]{108m^3} = \sqrt[3]{32m^3} - \sqrt[3]{108m^3}$ Product rule

$= \sqrt[3]{(8m^3)4} - \sqrt[3]{(27m^3)4}$ Factor.

$= 2m\sqrt[3]{4} - 3m\sqrt[3]{4}$ $\sqrt[3]{8m^3} = 2m$ and $\sqrt[3]{27m^3} = 3m$

$= -m\sqrt[3]{4}$ Subtract like radicals. ■

CAUTION Remember that a sum or difference of radicals can be simplified only if the radicals are *like radicals*. For example, $\sqrt{5} + 3\sqrt{5} = 4\sqrt{5}$, but $\sqrt{5} + 5\sqrt{3}$ cannot be simplified further. Also, $2\sqrt{3} + 5\sqrt[3]{3}$ cannot be simplified further.

8.3 EXERCISES

1. The distributive property, which says $a(b + c) = ab + ac$ and $ba + ca = (b + c)a$, provides the justification for adding and subtracting like radicals. While we usually skip the step that indicates this property, we could not make the statement $2\sqrt{3} + 4\sqrt{3} = 6\sqrt{3}$ without it. Write an equation showing how the distributive property is actually used in this statement.

2. In Example 1(e), we state that $\sqrt{3} + \sqrt{7}$ cannot be further simplified. Why is this so? Show, by using calculator approximations, that $\sqrt{3} + \sqrt{7}$ is *not* equal to $\sqrt{10}$.

Simplify and add or subtract wherever possible. See Examples 1 and 2.

3. $2\sqrt{3} + 6\sqrt{3}$

4. $6\sqrt{5} + 18\sqrt{5}$

5. $14\sqrt{7} - 19\sqrt{7}$

6. $16\sqrt{2} - 18\sqrt{2}$

7. $\sqrt{17} + 4\sqrt{17}$

8. $5\sqrt{19} + \sqrt{19}$

9. $6\sqrt{7} - \sqrt{7}$

10. $11\sqrt{14} - \sqrt{14}$

11. $\sqrt{45} + 4\sqrt{20}$

12. $\sqrt{24} + 6\sqrt{54}$

13. $5\sqrt{72} - 3\sqrt{50}$

14. $6\sqrt{18} - 5\sqrt{32}$

15. $-5\sqrt{32} + 2\sqrt{98}$

16. $-4\sqrt{75} + 3\sqrt{12}$

17. $5\sqrt{7} - 3\sqrt{28} + 6\sqrt{63}$

18. $3\sqrt{11} + 5\sqrt{44} - 8\sqrt{99}$

19. $2\sqrt{8} - 5\sqrt{32} - 2\sqrt{48}$

20. $5\sqrt{72} - 3\sqrt{48} + 4\sqrt{128}$

21. $4\sqrt{50} + 3\sqrt{12} - 5\sqrt{45}$

22. $6\sqrt{18} + 2\sqrt{48} + 6\sqrt{28}$

23. $\dfrac{1}{4}\sqrt{288} + \dfrac{1}{6}\sqrt{72}$

24. $\dfrac{2}{3}\sqrt{27} + \dfrac{3}{4}\sqrt{48}$

Perform the indicated operations. Assume that all variables represent nonnegative real numbers. See Example 3.

25. $\sqrt{3} \cdot \sqrt{7} + 4\sqrt{21}$

26. $\sqrt{13} \cdot \sqrt{2} + 7\sqrt{26}$

27. $\sqrt{6} \cdot \sqrt{2} + 9\sqrt{3}$

28. $4\sqrt{15} \cdot \sqrt{3} + 4\sqrt{5}$

29. $\sqrt{9x} + \sqrt{49x} - \sqrt{25x}$

30. $\sqrt{4a} - \sqrt{16a} + \sqrt{100a}$

31. $\sqrt{6x^2} + x\sqrt{24}$

32. $\sqrt{75x^2} + x\sqrt{108}$

33. $3\sqrt{8x^2} - 4x\sqrt{2} - x\sqrt{8}$

34. $\sqrt{2b^2} + 3b\sqrt{18} - b\sqrt{200}$

35. $-8\sqrt{32k} + 6\sqrt{8k}$

36. $4\sqrt{12x} + 2\sqrt{27x}$

37. $2\sqrt{125x^2z} + 8x\sqrt{80z}$

38. $\sqrt{48x^2y} + 5x\sqrt{27y}$

39. $4\sqrt[3]{16} - 3\sqrt[3]{54}$

40. $5\sqrt[3]{128} + 3\sqrt[3]{250}$

41. $6\sqrt[3]{8p^2} - 2\sqrt[3]{27p^2}$

42. $8k\sqrt[3]{54k} + 6\sqrt[3]{16k^4}$

43. $5\sqrt[4]{m^3} + 8\sqrt[4]{16m^3}$

44. $5\sqrt[4]{m^5} + 3\sqrt[4]{81m^5}$

45. Despite the fact that $\sqrt{25}$ and $\sqrt[3]{8}$ are radicals that have different root indexes, they can be added to obtain a single term: $\sqrt{25} + \sqrt[3]{8} = 5 + 2 = 7$. Make up a similar sum of radicals that leads to an answer of 10.

46. In the directions for Exercises 25–44, we made the assumption that all variables represent nonnegative real numbers. However, in Exercises 41 and 42, variables actually *may* represent negative numbers. Explain why this is so.

◆ **MATHEMATICAL CONNECTIONS** (Exercises 47–52) ◆

Addition and subtraction of like radicals is no different than addition and subtraction of like terms. Work Exercises 47–52 in order.

47. Combine like terms: $5x^2y + 3x^2y - 14x^2y$.

48. Combine like terms: $5(p - 2q)^2(a + b) + 3(p - 2q)^2(a + b) - 14(p - 2q)^2(a + b)$.

49. Combine like radicals: $5a^2\sqrt{xy} + 3a^2\sqrt{xy} - 14a^2\sqrt{xy}$.

50. Compare your answers in Exercises 47–49. How are they alike? How are they different?

51. In Exercises 47–49, the distributive property is essential in obtaining the correct answers. Explain how it is used.

52. Make up a problem similar to Exercise 49 whose answer is $9p^3\sqrt{2r}$.

REVIEW EXERCISES

Perform the operation. See Section 8.1.

53. $(\sqrt{6})^2$

54. $(\sqrt{25})^2$

55. $\sqrt[3]{2} \cdot \sqrt[3]{4}$

Simplify the radical. See Section 8.2.

56. $\sqrt{288}$

57. $\sqrt{7500}$

58. $\sqrt{x^2y^6}$, $x \geq 0$, $y \geq 0$

8.4 RATIONALIZING THE DENOMINATOR

FOR EXTRA HELP

📖 **SSG** pp. 268–271
SSM pp. 445–448

📼 **Video**
11

💾 **Tutorial**
IBM   MAC

OBJECTIVES

1 ▶ Rationalize denominators with square roots.
2 ▶ Write radicals in simplified form.
3 ▶ Rationalize denominators with cube roots.

1 ▶ Rationalize denominators with square roots.

▶   We found decimal approximations for radicals in the first section of this chapter. For more complicated radical expressions it is easier to find those decimals if the denominators do not contain radicals. For example, the radical in the denominator of

$$\frac{\sqrt{3}}{\sqrt{2}}$$

can be eliminated by multiplying the numerator and the denominator by $\sqrt{2}$.

$$\frac{\sqrt{3}}{\sqrt{2}} = \frac{\sqrt{3} \cdot \sqrt{2}}{\sqrt{2} \cdot \sqrt{2}} = \frac{\sqrt{6}}{2} \qquad \text{Since } \sqrt{2} \cdot \sqrt{2} = 2$$

This process of changing the denominator from a radical (irrational number) to a rational number is called **rationalizing the denominator.** The value of the number is not changed; only the form of the number is changed, because the expression has been multiplied by 1 in the form $\sqrt{2}/\sqrt{2}$.

EXAMPLE 1
Rationalizing Denominators

Rationalize each denominator.

(a) $\dfrac{9}{\sqrt{6}}$

Multiply both numerator and denominator by $\sqrt{6}$.

$$\frac{9}{\sqrt{6}} = \frac{9 \cdot \sqrt{6}}{\sqrt{6} \cdot \sqrt{6}}$$

$$= \frac{9\sqrt{6}}{6} \qquad \sqrt{6} \cdot \sqrt{6} = 6$$

$$= \frac{3\sqrt{6}}{2} \qquad \text{Lowest terms}$$

(b) $\dfrac{12}{\sqrt{8}}$

The denominator here could be rationalized by multiplying by $\sqrt{8}$. However, the result can be found more directly by first simplifying the denominator.

$$\sqrt{8} = \sqrt{4} \cdot \sqrt{2} = 2\sqrt{2}$$

Then multiply numerator and denominator by $\sqrt{2}$.

$$\frac{12}{\sqrt{8}} = \frac{12}{2\sqrt{2}}$$

$$= \frac{12 \cdot \sqrt{2}}{2\sqrt{2} \cdot \sqrt{2}} \qquad \text{Multiply by } \tfrac{\sqrt{2}}{\sqrt{2}}.$$

$$= \frac{12\sqrt{2}}{2 \cdot 2} \qquad \sqrt{2} \cdot \sqrt{2} = 2$$

$$= \frac{12\sqrt{2}}{4} \qquad \text{Multiply.}$$

$$= 3\sqrt{2} \qquad \text{Lowest terms} \quad \blacksquare$$

2 ▶ Write radicals in simplified form. ▶ A radical is considered to be in simplified form if the following three conditions are met.

___

Simplified Form of a Radical

1. The radicand contains no factor (except 1) that is a power of the root index.
2. The radicand has no fractions.
3. No denominator contains a radical.

In the following examples, radicals are simplified according to these conditions. Radicals are considered simplified only if denominators are rationalized, as shown in Examples 2–5.

___

EXAMPLE 2

Simplifying a Radical with a Fraction

■ Simplify $\sqrt{\dfrac{27}{5}}$ by rationalizing the denominator.

First we use the quotient rule for radicals.

$$\sqrt{\frac{27}{5}} = \frac{\sqrt{27}}{\sqrt{5}}$$

Now multiply both numerator and denominator by $\sqrt{5}$.

$$\frac{\sqrt{27}}{\sqrt{5}} = \frac{\sqrt{27} \cdot \sqrt{5}}{\sqrt{5} \cdot \sqrt{5}}$$

$$= \frac{\sqrt{9 \cdot 3} \cdot \sqrt{5}}{5} \qquad \sqrt{5} \cdot \sqrt{5} = 5$$

$$= \frac{\sqrt{9} \cdot \sqrt{3} \cdot \sqrt{5}}{5} \qquad \text{Product rule}$$

$$= \frac{3 \cdot \sqrt{3 \cdot 5}}{5} = \frac{3\sqrt{15}}{5} \qquad \text{Product rule} \quad \blacksquare$$

___

EXAMPLE 3

Simplifying a Product of Radicals

■ Simplify $\sqrt{\dfrac{5}{8}} \cdot \sqrt{\dfrac{1}{6}}$.

We use both the product rule and the quotient rule.

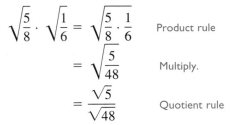

$$\sqrt{\frac{5}{8}} \cdot \sqrt{\frac{1}{6}} = \sqrt{\frac{5}{8} \cdot \frac{1}{6}} \qquad \text{Product rule}$$

$$= \sqrt{\frac{5}{48}} \qquad \text{Multiply.}$$

$$= \frac{\sqrt{5}}{\sqrt{48}} \qquad \text{Quotient rule}$$

To rationalize the denominator, first simplify, then multiply the numerator and denominator by $\sqrt{3}$ as follows.

$$\frac{\sqrt{5}}{\sqrt{48}} = \frac{\sqrt{5}}{\sqrt{16} \cdot \sqrt{3}} \qquad \text{Product rule}$$

$$= \frac{\sqrt{5}}{4\sqrt{3}} \qquad \sqrt{16} = 4$$

$$= \frac{\sqrt{5} \cdot \sqrt{3}}{4\sqrt{3} \cdot \sqrt{3}} \qquad \text{Rationalize the denominator.}$$

$$= \frac{\sqrt{15}}{4 \cdot 3} \qquad \text{Product rule}$$

$$= \frac{\sqrt{15}}{12} \qquad \text{Multiply.} \quad \blacksquare$$

EXAMPLE 4

Simplifying a Quotient of Radicals

■ Rationalize the denominator of $\dfrac{\sqrt{4x}}{\sqrt{y}}$. Assume that x and y represent positive real numbers.

Multiply numerator and denominator by $\sqrt{y}$.

$$\frac{\sqrt{4x}}{\sqrt{y}} = \frac{\sqrt{4x} \cdot \sqrt{y}}{\sqrt{y} \cdot \sqrt{y}} = \frac{\sqrt{4xy}}{y} = \frac{2\sqrt{xy}}{y} \quad \blacksquare$$

3 ▶ Rationalize denominators with cube roots.

▶ We rationalize a denominator with a cube root by changing the radicand in the denominator to a perfect cube, as shown in the next example.

EXAMPLE 5

Rationalizing a Denominator with a Cube Root

■ Rationalize each denominator.

(a) $\sqrt[3]{\dfrac{3}{2}}$

Multiply the numerator and the denominator by enough factors of 2 to make the denominator a perfect cube. This will eliminate the radical in the denominator. Here, multiply by $\sqrt[3]{2^2}$.

$$\sqrt[3]{\frac{3}{2}} = \frac{\sqrt[3]{3}}{\sqrt[3]{2}} = \frac{\sqrt[3]{3} \cdot \sqrt[3]{2^2}}{\sqrt[3]{2} \cdot \sqrt[3]{2^2}} = \frac{\sqrt[3]{3 \cdot 2^2}}{\sqrt[3]{2^3}} = \frac{\sqrt[3]{12}}{2} \qquad \text{Since } \sqrt[3]{2^3} = \sqrt[3]{8} = 2$$

(b) $\dfrac{\sqrt[3]{3}}{\sqrt[3]{4}}$

Since $4 \cdot 2 = 2^2 \cdot 2 = 2^3$, multiply numerator and denominator by $\sqrt[3]{2}$.

$$\frac{\sqrt[3]{3}}{\sqrt[3]{4}} = \frac{\sqrt[3]{3} \cdot \sqrt[3]{2}}{\sqrt[3]{2 \cdot 2} \cdot \sqrt[3]{2}} = \frac{\sqrt[3]{6}}{\sqrt[3]{2^3}} = \frac{\sqrt[3]{6}}{2} \quad \blacksquare$$

CAUTION   A common error in a problem like the one in Example 5(a) is to multiply by $\sqrt[3]{2}$ instead of $\sqrt[3]{2^2}$. Notice that this would give a denominator of $\sqrt[3]{2} \cdot \sqrt[3]{2} = \sqrt[3]{4}$. Since 4 is not a perfect cube, the denominator is still not rationalized.

8.4 EXERCISES

1. When we rationalize the denominator of an expression such as $\dfrac{4}{\sqrt{3}}$, we multiply both the numerator and the denominator by $\sqrt{3}$. By what number are we actually multiplying the given expression, and what property of real numbers justifies the fact that our result is equal to the given expression?

2. In Example 1(a), we show algebraically that $\dfrac{9}{\sqrt{6}}$ is equal to $\dfrac{3\sqrt{6}}{2}$. Give numerical support to this result by finding the decimal approximation of $\dfrac{9}{\sqrt{6}}$ on your calculator, and then finding the decimal approximation of $\dfrac{3\sqrt{6}}{2}$. Are they the same?

Rationalize the denominator. See Examples 1 and 2.

3. $\dfrac{7}{\sqrt{5}}$ **4.** $\dfrac{4}{\sqrt{3}}$ **5.** $\dfrac{8}{\sqrt{2}}$ **6.** $\dfrac{12}{\sqrt{3}}$ **7.** $\dfrac{-\sqrt{11}}{\sqrt{3}}$

8. $\dfrac{-\sqrt{13}}{\sqrt{5}}$ **9.** $\dfrac{7\sqrt{3}}{\sqrt{5}}$ **10.** $\dfrac{4\sqrt{6}}{\sqrt{5}}$ **11.** $\dfrac{24\sqrt{10}}{16\sqrt{3}}$ **12.** $\dfrac{18\sqrt{15}}{12\sqrt{2}}$

13. $\dfrac{16}{\sqrt{27}}$ **14.** $\dfrac{24}{\sqrt{18}}$ **15.** $\dfrac{-3}{\sqrt{50}}$ **16.** $\dfrac{-5}{\sqrt{75}}$ **17.** $\dfrac{63}{\sqrt{45}}$

18. $\dfrac{27}{\sqrt{32}}$ **19.** $\dfrac{\sqrt{24}}{\sqrt{8}}$ **20.** $\dfrac{\sqrt{36}}{\sqrt{18}}$ **21.** $\dfrac{-\sqrt{80}}{\sqrt{6}}$ **22.** $\dfrac{-\sqrt{24}}{\sqrt{15}}$

23. $\sqrt{\dfrac{1}{2}}$ **24.** $\sqrt{\dfrac{1}{3}}$ **25.** $\sqrt{\dfrac{13}{5}}$ **26.** $\sqrt{\dfrac{17}{11}}$

Multiply and simplify the result. See Example 3.

27. $\sqrt{\dfrac{7}{13}} \cdot \sqrt{\dfrac{13}{3}}$ **28.** $\sqrt{\dfrac{19}{20}} \cdot \sqrt{\dfrac{20}{3}}$ **29.** $\sqrt{\dfrac{21}{7}} \cdot \sqrt{\dfrac{21}{8}}$

30. $\sqrt{\dfrac{5}{8}} \cdot \sqrt{\dfrac{5}{6}}$ **31.** $\sqrt{\dfrac{1}{12}} \cdot \sqrt{\dfrac{1}{3}}$ **32.** $\sqrt{\dfrac{1}{8}} \cdot \sqrt{\dfrac{1}{2}}$

33. $\sqrt{\dfrac{2}{9}} \cdot \sqrt{\dfrac{9}{2}}$ **34.** $\sqrt{\dfrac{4}{3}} \cdot \sqrt{\dfrac{3}{4}}$

Simplify the radical. Assume that all variables represent positive real numbers. See Example 4.

35. $\sqrt{\dfrac{7}{x}}$ **36.** $\sqrt{\dfrac{19}{y}}$ **37.** $\sqrt{\dfrac{4x^3}{y}}$ **38.** $\sqrt{\dfrac{9t^3}{s}}$

39. $\sqrt{\dfrac{18x^3}{6y}}$ **40.** $\sqrt{\dfrac{24t^3}{8p}}$ **41.** $\sqrt{\dfrac{9a^2r^5}{7t}}$ **42.** $\sqrt{\dfrac{16x^3y^2}{13z}}$

43. Which one of the following would be an appropriate choice for multiplying the numerator and the denominator of $\dfrac{\sqrt[3]{2}}{\sqrt[3]{5}}$ in order to rationalize the denominator?

(a) $\sqrt[3]{5}$ (b) $\sqrt[3]{25}$ (c) $\sqrt[3]{2}$ (d) $\sqrt[3]{3}$

44. In Example 5(b), we multiply numerator and denominator of $\dfrac{\sqrt[3]{3}}{\sqrt[3]{4}}$ by $\sqrt[3]{2}$ to rationalize the denominator. Suppose we had chosen to multiply by $\sqrt[3]{16}$ instead. Would we have obtained the correct answer after all simplifications were done?

Simplify. Rationalize the denominator. Assume that variables in the denominator are nonzero. See Example 5.

45. $\sqrt[3]{\dfrac{3}{2}}$ **46.** $\sqrt[3]{\dfrac{2}{5}}$ **47.** $\dfrac{\sqrt[3]{4}}{\sqrt[3]{7}}$ **48.** $\dfrac{\sqrt[3]{5}}{\sqrt[3]{10}}$

49. $\sqrt[3]{\dfrac{3}{4y^2}}$ **50.** $\sqrt[3]{\dfrac{3}{25x^2}}$ **51.** $\dfrac{\sqrt[3]{7m}}{\sqrt[3]{36n}}$ **52.** $\dfrac{\sqrt[3]{11p}}{\sqrt[3]{49q}}$

▦ *In Exercises 53 and 54, (a) give the answer as a simplified radical and (b) use a calculator to give the answer correct to the nearest thousandth.*

53. The period p of a pendulum is the time it takes for it to swing from one extreme to the other and back again. The value of p in seconds is given by

$$p = k \cdot \sqrt{\dfrac{L}{g}},$$

where L is the length of the pendulum, g is the acceleration due to gravity, and k is a constant. Find the period when $k = 6$, $L = 9$ feet, and $g = 32$ feet per second.

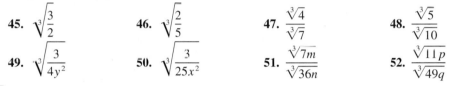

54. The velocity v of a meteorite approaching the earth is given by

$$v = \dfrac{k}{\sqrt{d}}$$

kilometers per second, where d is its distance from the center of the earth and k is a constant. What is the velocity of a meteorite that is 6000 kilometers away from the center of the earth, if $k = 450$?

REVIEW EXERCISES

Find the product. See Sections 3.3 and 3.4.

55. $(4x + 7)(8x - 3)$ **56.** $ab(3a^2b - 2ab^2 + 7)$ **57.** $(6x - 1)(6x + 1)$

58. $(r + 7)(r - 7)$ **59.** $(p + q)(a - m)$ **60.** $(3w - 8)^2$

8.5 SIMPLIFYING RADICAL EXPRESSIONS

FOR EXTRA HELP

📖 **SSG** pp. 271–274
SSM pp. 448–452

📼 **Video**
12

💾 **Tutorial**
IBM   MAC

OBJECTIVES

1 ▶ Simplify products of radical expressions.
2 ▶ Simplify quotients of radical expressions.
3 ▶ Write radical expressions with quotients in lowest terms.

It can be difficult to decide on the "simplest" form of a radical. The conditions for which a radical is in simplest form were listed in the previous section. Below is a set of guidelines to follow when you are simplifying radical expressions. Although the conditions are illustrated with square roots, they apply to higher roots as well.

Simplifying Radical Expressions

1. If a radical represents a rational number, then that rational number should be used in place of the radical.
   For example, $\sqrt{49}$ is simplified by writing 7; $\sqrt{64}$ by writing 8; $\sqrt{\dfrac{169}{9}}$ by writing $\dfrac{13}{3}$.

2. If a radical expression contains products of radicals, the product rule for radicals, $\sqrt{x} \cdot \sqrt{y} = \sqrt{xy}$, should be used to get a single radical.
   For example, $\sqrt{3} \cdot \sqrt{2}$ is simplified to $\sqrt{6}$; $\sqrt{5} \cdot \sqrt{x}$ to $\sqrt{5x}$.

3. If a radicand has a factor that is a perfect square, the radical should be expressed as the product of the positive square root of the perfect square and the remaining radical factor. A similar statement applies to higher roots.
   For example, $\sqrt{20}$ is simplified to $\sqrt{20} = \sqrt{4 \cdot 5} = \sqrt{4} \cdot \sqrt{5} = 2\sqrt{5}$; $\sqrt{75}$ to $5\sqrt{3}$; $\sqrt[3]{16} = \sqrt[3]{8 \cdot 2} = \sqrt[3]{8} \cdot \sqrt[3]{2} = 2\sqrt[3]{2}$.

4. If a radical expression contains sums or differences of radicals, the distributive property should be used to combine like radicals.
   For example, $3\sqrt{2} + 4\sqrt{2}$ is combined as $7\sqrt{2}$, but $3\sqrt{2} + 4\sqrt{3}$ cannot be further combined.

5. Any denominator containing a radical should be rationalized.
   For example, $\dfrac{5}{\sqrt{3}}$ is rationalized as $\dfrac{5}{\sqrt{3}} = \dfrac{5\sqrt{3}}{\sqrt{3} \cdot \sqrt{3}} = \dfrac{5\sqrt{3}}{3}$; $\sqrt{\dfrac{3}{2}}$ is rationalized as $\sqrt{\dfrac{3}{2}} = \dfrac{\sqrt{3}}{\sqrt{2}} = \dfrac{\sqrt{3}}{\sqrt{2}} \cdot \dfrac{\sqrt{2}}{\sqrt{2}} = \dfrac{\sqrt{6}}{2}$.

1 ▶ Simplify products of radical expressions.

▶ We begin with examples showing how to simplify radical expressions involving products.

EXAMPLE 1
Multiplying Radical Expressions

Find each product and simplify the answers.

(a) $\sqrt{5}(\sqrt{8} - \sqrt{32})$
Simplify inside the parentheses.

$$\sqrt{5}(\sqrt{8} - \sqrt{32}) = \sqrt{5}(2\sqrt{2} - 4\sqrt{2})$$
$$= \sqrt{5}(-2\sqrt{2}) \qquad \text{Subtract like radicals.}$$
$$= -2\sqrt{5 \cdot 2} \qquad \text{Product rule; commutative property}$$
$$= -2\sqrt{10} \qquad \text{Multiply.}$$

(b) $(\sqrt{3} + 2\sqrt{5})(\sqrt{3} - 4\sqrt{5})$

The products of these sums of radicals can be found in the same way that we found the product of binomials in Chapter 3. The pattern of multiplication is the same, using the FOIL method.

$$(\sqrt{3} + 2\sqrt{5})(\sqrt{3} - 4\sqrt{5})$$

$$\overset{\text{F}}{= \sqrt{3} \cdot \sqrt{3}} + \overset{\text{O}}{\sqrt{3}(-4\sqrt{5})} + \overset{\text{I}}{2\sqrt{5} \cdot \sqrt{3}} + \overset{\text{L}}{2\sqrt{5}(-4\sqrt{5})}$$
$$= 3 - 4\sqrt{15} + 2\sqrt{15} - 8 \cdot 5 \qquad \text{Product rule}$$
$$= 3 - 2\sqrt{15} - 40 \qquad \text{Add like radicals.}$$
$$= -37 - 2\sqrt{15} \qquad \text{Combine terms.}$$

(c) $(\sqrt{3} + \sqrt{21})(\sqrt{3} - \sqrt{7})$

$$(\sqrt{3} + \sqrt{21})(\sqrt{3} - \sqrt{7})$$
$$= \sqrt{3}(\sqrt{3}) + \sqrt{3}(-\sqrt{7}) + \sqrt{21}(\sqrt{3}) + \sqrt{21}(-\sqrt{7}) \qquad \text{FOIL}$$
$$= 3 - \sqrt{21} + \sqrt{63} - \sqrt{147} \qquad \text{Product rule}$$
$$= 3 - \sqrt{21} + \sqrt{9} \cdot \sqrt{7} - \sqrt{49} \cdot \sqrt{3} \qquad \text{Simplify radicals.}$$
$$= 3 - \sqrt{21} + 3\sqrt{7} - 7\sqrt{3} \qquad \begin{array}{l} \sqrt{9} = 3 \\ \text{and } \sqrt{49} = 7 \end{array}$$

Since there are no like radicals, no terms may be combined. ∎

Since radicals represent real numbers, the special products of binomials discussed in Chapter 3 can be used to find products of radicals. Example 2 uses the rule for the product that gives the difference of two squares,

$$(a + b)(a - b) = a^2 - b^2.$$

EXAMPLE 2
Using a Special Product with Radicals

Find each product.

(a) $(4 + \sqrt{3})(4 - \sqrt{3})$

Follow the pattern given above. Let $a = 4$ and $b = \sqrt{3}$.

$$(4 + \sqrt{3})(4 - \sqrt{3}) = (4)^2 - (\sqrt{3})^2$$
$$= 16 - 3 = 13 \qquad 4^2 = 16 \text{ and } (\sqrt{3})^2 = 3$$

(b) $(\sqrt{12} - \sqrt{6})(\sqrt{12} + \sqrt{6}) = (\sqrt{12})^2 - (\sqrt{6})^2$
$$= 12 - 6 \qquad (\sqrt{12})^2 = 12 \text{ and } (\sqrt{6})^2 = 6$$
$$= 6 \quad ∎$$

Both products in Example 2 resulted in rational numbers. The pairs of expressions in those products, $4 + \sqrt{3}$ and $4 - \sqrt{3}$, and $\sqrt{12} - \sqrt{6}$ and $\sqrt{12} + \sqrt{6}$, are called **conjugates** of each other.

2 ▶ Simplify quotients of radical expressions.

▶ Products of radicals similar to those in Example 2 can be used to rationalize the denominators in quotients with binomial denominators, such as

$$\frac{2}{4 - \sqrt{3}}.$$

By Example 2(a), if this denominator, $4 - \sqrt{3}$, is multiplied by $4 + \sqrt{3}$, then the product $(4 - \sqrt{3})(4 + \sqrt{3})$ is the rational number 13. Multiplying numerator and denominator by $4 + \sqrt{3}$ gives

$$\frac{2}{4 - \sqrt{3}} = \frac{2(4 + \sqrt{3})}{(4 - \sqrt{3})(4 + \sqrt{3})} = \frac{2(4 + \sqrt{3})}{13}.$$

The denominator now has been rationalized; it contains no radical signs.

Using Conjugates to Simplify a Radical Expression

To simplify a radical expression with two terms in the denominator, where at least one of those terms is a square root radical, multiply both the numerator and the denominator by the conjugate of the denominator.

EXAMPLE 3

Using Conjugates to Rationalize a Denominator

■ Simplify by rationalizing the denominator.

(a) $\dfrac{7}{3 + \sqrt{5}}$

We can eliminate the radical in the denominator by multiplying both numerator and denominator by $3 - \sqrt{5}$, the conjugate of $3 + \sqrt{5}$.

$$\frac{7}{3 + \sqrt{5}} = \frac{7(3 - \sqrt{5})}{(3 + \sqrt{5})(3 - \sqrt{5})} \qquad \text{Multiply by the conjugate.}$$

$$= \frac{7(3 - \sqrt{5})}{3^2 - (\sqrt{5})^2} \qquad (a + b)(a - b) = a^2 - b^2$$

$$= \frac{7(3 - \sqrt{5})}{9 - 5} \qquad 3^2 = 9 \text{ and } (\sqrt{5})^2 = 5$$

$$= \frac{7(3 - \sqrt{5})}{4} \qquad \text{Subtract.} \quad ■$$

(b) $\dfrac{6 + \sqrt{2}}{\sqrt{2} - 5}$

Multiply numerator and denominator by $\sqrt{2} + 5$.

$$\frac{6 + \sqrt{2}}{\sqrt{2} - 5} = \frac{(6 + \sqrt{2})(\sqrt{2} + 5)}{(\sqrt{2} - 5)(\sqrt{2} + 5)}$$

$$= \frac{6\sqrt{2} + 30 + 2 + 5\sqrt{2}}{2 - 25} \qquad \text{FOIL}$$

$$= \frac{11\sqrt{2} + 32}{-23} \qquad \text{Combine terms.}$$

$$= -\frac{11\sqrt{2} + 32}{23} \qquad \frac{a}{-b} = -\frac{a}{b} \quad ■$$

CAUTION Rationalizing the denominator in the two expressions

$$\frac{7}{\sqrt{x+5}} \quad \text{and} \quad \frac{7}{\sqrt{x}+\sqrt{5}}$$

involves different procedures. In the first, we multiply both the numerator and the denominator by $\sqrt{x+5}$. In the second, we use $\sqrt{x}-\sqrt{5}$.

◇ **CONNECTIONS** ◇

In more advanced mathematics, sometimes it is useful to change a radical expression by rationalizing the *numerator*. This is done to avoid a zero denominator. For example, the expression

$$\frac{\sqrt{x}-2}{x-4} \text{ is rewritten as } \frac{1}{\sqrt{x}+2} \text{ as follows.}$$

$$\frac{\sqrt{x}-2}{x-4} = \frac{\sqrt{x}-2}{x-4} \cdot \frac{\sqrt{x}+2}{\sqrt{x}+2} = \frac{x-4}{(x-4)(\sqrt{x}+2)} = \frac{1}{\sqrt{x}+2}$$

FOR DISCUSSION OR WRITING

1. Verify that if $x = 4$,

$$\frac{\sqrt{x}-2}{x-4} = \frac{0}{0}, \text{ which is undefined, but } \frac{1}{\sqrt{x}+2} = \frac{1}{4}.$$

2. What number makes the following expression undefined? Find an equivalent expression that is defined for that number.

$$\frac{\sqrt{x}-\sqrt{6}}{x-6}$$

---

3 ▶ Write radical expressions with quotients in lowest terms.

▶ The final example shows how to write certain quotients with radicals in lowest terms.

EXAMPLE 4
Writing a Radical Quotient in Lowest Terms

Write $\dfrac{3\sqrt{3}+15}{12}$ in lowest terms.

Factor the numerator and denominator, and then divide numerator and denominator by any common factors.

$$\frac{3\sqrt{3}+15}{12} = \frac{3(\sqrt{3}+5)}{3\cdot4} = \frac{\sqrt{3}+5}{4}$$

This technique is used in Chapter 9. ■

CAUTION A common error is to try to reduce an expression like the one in Example 4 to lowest terms before factoring. For example,

$$\frac{4+8\sqrt{5}}{4} \neq 1 + 8\sqrt{5}.$$

The correct simplification is $1 + 2\sqrt{5}$. Do you see why?

8.5 EXERCISES

Based on the work so far, many simple operations involving radicals should now be performed mentally. In Exercises 1–8, perform the operations mentally, and write the answer without doing intermediate steps.

1. $\sqrt{49} + \sqrt{36}$ **2.** $\sqrt{100} - \sqrt{81}$ **3.** $\sqrt{2} \cdot \sqrt{8}$ **4.** $\sqrt{8} \cdot \sqrt{8}$

5. $\sqrt{2}(\sqrt{32} - \sqrt{8})$ **6.** $\sqrt{3}(\sqrt{27} - \sqrt{3})$ **7.** $\sqrt[3]{8} + \sqrt[3]{27}$ **8.** $\sqrt{4} - \sqrt[3]{64} + \sqrt[4]{16}$

Simplify the expression. Use the five guidelines given in the text. See Examples 1 and 2.

9. $3\sqrt{5} + 2\sqrt{45}$ **10.** $2\sqrt{2} + 4\sqrt{18}$

11. $8\sqrt{50} - 4\sqrt{72}$ **12.** $4\sqrt{80} - 5\sqrt{45}$

13. $\sqrt{5}(\sqrt{3} - \sqrt{7})$ **14.** $\sqrt{7}(\sqrt{10} + \sqrt{3})$

15. $2\sqrt{5}(\sqrt{2} + 3\sqrt{5})$ **16.** $3\sqrt{7}(2\sqrt{7} + 4\sqrt{5})$

17. $3\sqrt{14} \cdot \sqrt{2} - \sqrt{28}$ **18.** $7\sqrt{6} \cdot \sqrt{3} - 2\sqrt{18}$

19. $(2\sqrt{6} + 3)(3\sqrt{6} + 7)$ **20.** $(4\sqrt{5} - 2)(2\sqrt{5} - 4)$

21. $(5\sqrt{7} - 2\sqrt{3})(3\sqrt{7} + 4\sqrt{3})$ **22.** $(2\sqrt{10} + 5\sqrt{2})(3\sqrt{10} - 3\sqrt{2})$

23. $(2\sqrt{7} + 3)^2$ **24.** $(4\sqrt{5} + 5)^2$

25. $(5 - \sqrt{2})(5 + \sqrt{2})$ **26.** $(3 - \sqrt{5})(3 + \sqrt{5})$

27. $(\sqrt{8} - \sqrt{7})(\sqrt{8} + \sqrt{7})$ **28.** $(\sqrt{12} - \sqrt{11})(\sqrt{12} + \sqrt{11})$

29. $(\sqrt{2} + \sqrt{3})(\sqrt{6} - \sqrt{2})$ **30.** $(\sqrt{3} + \sqrt{5})(\sqrt{15} - \sqrt{5})$

31. $(\sqrt{10} - \sqrt{5})(\sqrt{5} + \sqrt{20})$ **32.** $(\sqrt{6} - \sqrt{3})(\sqrt{3} + \sqrt{18})$

33. $(\sqrt{5} + \sqrt{30})(\sqrt{6} + \sqrt{3})$ **34.** $(\sqrt{10} - \sqrt{20})(\sqrt{2} - \sqrt{5})$

35. $(5\sqrt{7} - 2\sqrt{3})(3\sqrt{7} + 3\sqrt{3})$ **36.** $(2\sqrt{10} + 5\sqrt{2})(3\sqrt{10} - 4\sqrt{2})$

Rationalize the denominator. See Example 3.

37. $\dfrac{1}{3 + \sqrt{2}}$ **38.** $\dfrac{1}{4 - \sqrt{3}}$ **39.** $\dfrac{14}{2 - \sqrt{11}}$ **40.** $\dfrac{19}{5 - \sqrt{6}}$

41. $\dfrac{\sqrt{2}}{2 - \sqrt{2}}$ **42.** $\dfrac{\sqrt{7}}{7 - \sqrt{7}}$ **43.** $\dfrac{\sqrt{5}}{\sqrt{2} + \sqrt{3}}$ **44.** $\dfrac{\sqrt{3}}{\sqrt{2} + \sqrt{3}}$

45. $\dfrac{\sqrt{12}}{\sqrt{3} + 1}$ **46.** $\dfrac{\sqrt{18}}{\sqrt{2} - 1}$ **47.** $\dfrac{\sqrt{5} + 2}{2 - \sqrt{3}}$ **48.** $\dfrac{\sqrt{7} + 3}{4 - \sqrt{5}}$

Write the quotient in lowest terms. See Example 4.

49. $\dfrac{6\sqrt{11} - 12}{6}$ **50.** $\dfrac{12\sqrt{5} - 24}{12}$ **51.** $\dfrac{2\sqrt{3} + 10}{16}$

52. $\dfrac{4\sqrt{6} + 24}{20}$ **53.** $\dfrac{12 - \sqrt{40}}{4}$ **54.** $\dfrac{9 - \sqrt{72}}{12}$

Simplify the radical expression. Assume all variables represent nonnegative real numbers.

55. $(\sqrt{5x} + \sqrt{30})(\sqrt{6x} + \sqrt{3})$

56. $(\sqrt{10y} - \sqrt{20})(\sqrt{2y} - \sqrt{5})$

57. $(3\sqrt{t} + \sqrt{7})(2\sqrt{t} - \sqrt{14})$

58. $(2\sqrt{z} - \sqrt{3})(\sqrt{z} - \sqrt{5})$

59. $(\sqrt{3m} + \sqrt{2n})(\sqrt{5m} - \sqrt{5n})$

60. $(\sqrt{4p} - \sqrt{3k})(\sqrt{2p} + \sqrt{9k})$

61. $\sqrt[3]{4}(\sqrt[3]{2} - 3)$

62. $\sqrt[3]{5}(4\sqrt[3]{5} - \sqrt[3]{25})$

63. $2\sqrt[4]{2}(3\sqrt[4]{8} + 5\sqrt[4]{4})$

64. $6\sqrt[4]{9}(2\sqrt[4]{9} - \sqrt[4]{27})$

65. $(\sqrt[3]{2} - 1)(\sqrt[3]{4} + 3)$

66. $(\sqrt[3]{9} + 5)(\sqrt[3]{3} - 4)$

67. $(\sqrt[3]{5} - \sqrt[3]{4})(\sqrt[3]{25} + \sqrt[3]{20} + \sqrt[3]{16})$

68. $(\sqrt[3]{4} + \sqrt[3]{2})(\sqrt[3]{16} - \sqrt[3]{8} + \sqrt[3]{4})$

◆——— **MATHEMATICAL CONNECTIONS** (Exercises 69–74) ◇———

Work Exercises 69–74 in order. They are designed to help you see why a common student error is indeed an error.

69. Use the distributive property to write $6(5 + 3x)$ as a sum.

70. Your answer in Exercise 69 should be $30 + 18x$. Why can we not combine these two terms to get $48x$?

71. Repeat Exercise 22 from earlier in this exercise set.

72. Your answer in Exercise 71 should be $30 + 18\sqrt{5}$. Many students will, in error, try to combine these terms to get $48\sqrt{5}$. Why is this wrong?

73. Write the expression similar to $30 + 18x$ that simplifies to $48x$. Then write the expression similar to $30 + 18\sqrt{5}$ that simplifies to $48\sqrt{5}$.

74. Write a short paragraph explaining the similarities between combining like terms and combining like radicals.

———————————◆———————————

Solve the problem.

75. The radius of the circular top or bottom of a tin can with a surface area S and a height h is given by
$$r = \frac{-h + \sqrt{h^2 + .64S}}{2}.$$
What radius should be used to make a can with a height of 12 inches and a surface area of 400 square inches?

76. If an investment of P dollars grows to A dollars in two years, the annual rate of return on the investment is given by
$$r = \frac{\sqrt{A} - \sqrt{P}}{\sqrt{P}}.$$
First rationalize the denominator and then find the annual rate of return (in percent) if $50,000 increases to $58,320.

REVIEW EXERCISES

Solve the equation. See Section 4.5.

77. $y^2 + 4y + 3 = 0$

78. $x^2 - 6x + 9 = 0$

79. $(k - 1) = (k - 1)^2$

80. $x(x + 4) = 21$

8.6 EQUATIONS WITH RADICALS

FOR EXTRA HELP

📖 **SSG** pp. 274–278
SSM pp. 452–458

📼 **Video**
12

💾 **Tutorial**
IBM   MAC

OBJECTIVES

1 ▶ Solve equations with radicals.
2 ▶ Identify equations with no solutions.
3 ▶ Solve equations that require squaring a binomial.

◆ **C O N N E C T I O N S** ◆

The most common formula for the area of a triangle is $A = (1/2)bh$, where b is the length of the base and h is the height. What if the height is not known? What if we only know the lengths of the sides? Another formula, known as **Heron's formula,** allows us to calculate the area of a triangle if we know the lengths of the sides a, b, and c. First let s equal the semiperimeter.

$$s = \frac{1}{2}(a + b + c).$$

The area A is given by the formula

$$A = \sqrt{s(s - a)(s - b)(s - c)}.$$

For example, the familiar 3–4–5 right triangle has area

$$A = \frac{1}{2}(3)(4) = 6 \text{ square units,}$$

using the familiar formula. Using Heron's formula,

$$s = \frac{1}{2}(3 + 4 + 5) = 6.$$

Therefore,

$$A = \sqrt{6(6 - 3)(6 - 4)(6 - 5)}$$
$$= \sqrt{36} = 6$$

So $A = 6$ square units, as expected.

FOR DISCUSSION OR WRITING

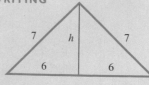

1. Use Heron's formula to find the area of a triangle with sides 7, 7, and 12.
2. The area of this triangle can be found with the formula $A = (1/2)bh$ as follows. Divide the triangle into two equal triangles as shown in the figure. Use the Pythagorean formula to find h using one of the small triangles. Note that h is the altitude of the original triangle. Now find the area using the formula $A = (1/2)bh$. Which way do you prefer?

1 ▶ Solve equations with radicals.

▶ The addition and multiplication properties of equality are not enough to solve an equation with radicals such as

$$\sqrt{x + 1} = 3.$$

Solving equations that have square roots requires a new property, the **squaring property.**

Squaring Property of Equality

If both sides of a given equation are squared, all solutions of the original equation are *among* the solutions of the squared equation.

CAUTION Be very careful with the squaring property. Using this property can give a new equation with *more* solutions than the original equation. For example, starting with the equation $y = 4$ and squaring each side gives

$$y^2 = 4^2, \quad \text{or} \quad y^2 = 16.$$

This last equation, $y^2 = 16$, has *two* solutions, 4 or -4, while the original equation, $y = 4$, has only *one* solution, 4. Because of this possibility, checking is more than just a guard against algebraic errors when solving an equation with radicals. It is an essential part of the solution process. *All potential solutions from the squared equation must be checked in the original equation.*

EXAMPLE 1

Using the Squaring Property of Equality

Solve the equation $\sqrt{p + 1} = 3$.

Use the squaring property of equality to square both sides of the equation and then solve this new equation.

$$(\sqrt{p + 1})^2 = 3^2$$
$$p + 1 = 9 \qquad (\sqrt{p + 1})^2 = p + 1$$
$$p = 8 \qquad \text{Subtract 1.}$$

Now check this answer in the original equation.

$$\sqrt{p + 1} = 3$$
$$\sqrt{8 + 1} = 3 \qquad ? \qquad \text{Let } p = 8.$$
$$\sqrt{9} = 3 \qquad ?$$
$$3 = 3 \qquad \text{True}$$

Since this statement is true, 8 is the solution of $\sqrt{p + 1} = 3$. In this case the squared equation had just one solution, which also satisfied the original equation. ■

EXAMPLE 2

Using the Squaring Property with Radicals on Each Side

Solve $3\sqrt{x} = \sqrt{x + 8}$.

Squaring both sides gives

$$(3\sqrt{x})^2 = (\sqrt{x + 8})^2$$
$$3^2(\sqrt{x})^2 = (\sqrt{x + 8})^2 \qquad (ab)^2 = a^2b^2$$
$$9x = x + 8 \qquad (\sqrt{x})^2 = x; (\sqrt{x + 8})^2 = x + 8$$
$$8x = 8 \qquad \text{Subtract } x.$$
$$x = 1. \qquad \text{Divide by 8.}$$

Check this potential solution.

$$3\sqrt{x} = \sqrt{x + 8}$$
$$3\sqrt{1} = \sqrt{1 + 8} \qquad ? \qquad \text{Let } x = 1.$$
$$3(1) = \sqrt{9} \qquad ? \qquad \sqrt{1} = 1$$
$$3 = 3 \qquad\qquad \text{True}$$

The check shows that the solution of the given equation is 1. ■

2 ▶ Identify equations with no solutions.

▶ Not all equations with radicals have a solution, as we show in Examples 3 and 4.

EXAMPLE 3

Using the Squaring Property When One Side Is Negative

Solve the equation $\sqrt{y} = -3$.
    Square both sides.

$$(\sqrt{y})^2 = (-3)^2$$
$$y = 9$$

Check this proposed answer in the original equation.

$$\sqrt{y} = -3$$
$$\sqrt{9} = -3 \qquad ? \qquad \text{Let } y = 9.$$
$$3 = -3 \qquad\qquad \text{False}$$

Since the statement $3 = -3$ is false, the number 9 is not a solution of the given equation and is said to be *extraneous*. In fact, $\sqrt{y} = -3$ has no solution. Since $\sqrt{y}$ represents the *nonnegative* square root of y, we might have seen immediately that there is no solution. ■

The steps to use when solving an equation with radicals are summarized below.

Solving an Equation with Radicals

Step 1 Arrange the terms so that one radical is alone on one side of the equation.
Step 2 Square both sides.
Step 3 Combine like terms.
Step 4 If there is still a term with a radical, repeat Steps 1–3.
Step 5 Solve the equation from Step 3 for potential solutions.
Step 6 Check all potential solutions from Step 5 in the original equation.

EXAMPLE 4

Using the Squaring Property with a Quadratic Expression

Solve $a = \sqrt{a^2 + 5a + 10}$.
    Square both sides.

$$(a)^2 = (\sqrt{a^2 + 5a + 10})^2$$
$$a^2 = a^2 + 5a + 10 \qquad (\sqrt{a^2 + 5a + 10})^2 = a^2 + 5a + 10$$
$$0 = 5a + 10 \qquad \text{Subtract } a^2.$$
$$a = -2 \qquad \text{Subtract 10; divide by 5.}$$

Check this proposed solution in the original equation.

$$a = \sqrt{a^2 + 5a + 10}$$
$$-2 = \sqrt{(-2)^2 + 5(-2) + 10} \qquad ? \qquad \text{Let } a = -2.$$

$$-2 = \sqrt{4 - 10 + 10} \qquad \text{?} \qquad \text{Multiply.}$$
$$-2 = 2 \qquad\qquad\qquad \text{False}$$

Since $a = -2$ leads to a false result, the equation has no solution.   ∎

3 ▶ Solve equations that require squaring a binomial.

▶ The next examples use the following facts from Section 3.4.

$$(a + b)^2 = a^2 + 2ab + b^2 \qquad \text{and} \qquad (a - b)^2 = a^2 - 2ab + b^2.$$

By the second pattern, for example,

$$(y - 3)^2 = y^2 - 2(y)(3) + (3)^2$$
$$= y^2 - 6y + 9.$$

EXAMPLE 5

Using the Squaring Property When One Side Has Two Terms

Solve the equation $\sqrt{2y - 3} = y - 3$.

Square each side, using the result above on the right side of the equation.

$$(\sqrt{2y - 3})^2 = (y - 3)^2$$
$$2y - 3 = y^2 - 6y + 9$$

This equation is quadratic because of the y^2 term. As shown in Section 4.5, solving this equation requires that one side be equal to 0. Subtract $2y$ and add 3, getting

$$0 = y^2 - 8y + 12.$$
$$0 = (y - 6)(y - 2) \qquad \text{Factor.}$$
$$y - 6 = 0 \qquad \text{or} \qquad y - 2 = 0 \qquad \text{Zero-factor property}$$
$$y = 6 \qquad \text{or} \qquad\quad y = 2$$

Check both of these potential solutions in the original equation.

If $y = 6$,

$$\sqrt{2y - 3} = y - 3$$
$$\sqrt{2(6) - 3} = 6 - 3 \qquad \text{?}$$
$$\sqrt{12 - 3} = 3 \qquad \text{?}$$
$$\sqrt{9} = 3 \qquad \text{?}$$
$$3 = 3. \qquad \text{True}$$

If $y = 2$,

$$\sqrt{2y - 3} = y - 3$$
$$\sqrt{2(2) - 3} = 2 - 3 \qquad \text{?}$$
$$\sqrt{4 - 3} = -1 \qquad \text{?}$$
$$\sqrt{1} = -1 \qquad \text{?}$$
$$1 = -1. \qquad \text{False}$$

Only 6 is a solution of the equation.   ∎

Sometimes we need to write an equation in a different form before squaring both sides. The next example shows why.

EXAMPLE 6

Rewriting an Equation Before Using the Squaring Property

Solve the equation $3\sqrt{x} - 1 = 2x$.

Squaring both sides gives

$$(3\sqrt{x} - 1)^2 = (2x)^2$$
$$9x - 6\sqrt{x} + 1 = 4x^2,$$

an equation that is more complicated, and still contains a radical. It would be better instead to rewrite the original equation so that the radical is alone on one side of the equals sign. To do this, we add 1 to both sides to get

$$3\sqrt{x} = 2x + 1.$$
$$(3\sqrt{x})^2 = (2x + 1)^2 \qquad\qquad \text{Square both sides.}$$

$$9x = 4x^2 + 4x + 1$$
$$0 = 4x^2 - 5x + 1 \qquad \text{Subtract } 9x.$$
$$0 = (4x - 1)(x - 1) \qquad \text{Zero-factor property}$$
$$4x - 1 = 0 \quad \text{or} \quad x - 1 = 0$$
$$x = \frac{1}{4} \quad \text{or} \quad x = 1$$

Both of these potential solutions must be checked in the original equation.

If $x = \dfrac{1}{4}$,

$$3\sqrt{x} - 1 = 2x$$
$$3\sqrt{\frac{1}{4}} - 1 = 2\left(\frac{1}{4}\right) \qquad ?$$
$$\frac{3}{2} - 1 = \frac{1}{2}. \qquad \text{True}$$

If $x = 1$,

$$3\sqrt{x} - 1 = 2x$$
$$3\sqrt{1} - 1 = 2(1) \qquad ?$$
$$3 - 1 = 2. \qquad \text{True}$$

The solutions to the original equation are 1/4 and 1. ∎

CAUTION Errors often occur when each side of an equation is squared. For instance, in Example 6 after the equation is rewritten as

$$3\sqrt{x} = 2x + 1,$$

it would be *incorrect* to write the next step as

$$9x = 4x^2 + 1.$$

Don't forget that the binomial $2x + 1$ must be squared to get $4x^2 + 4x + 1$.

Some equations with radicals require squaring twice, as in the next example.

EXAMPLE 7
Using the Squaring Property Twice

■ Solve $\sqrt{21 + x} = 3 + \sqrt{x}$.

Square both sides.

$$(\sqrt{21 + x})^2 = (3 + \sqrt{x})^2$$
$$21 + x = 9 + 6\sqrt{x} + x$$
$$12 = 6\sqrt{x} \qquad \text{Combine terms; simplify.}$$
$$(2)^2 = (\sqrt{x})^2$$
$$4 = x. \qquad \text{Square both sides again.}$$

Check the potential solution.

If $x = 4$,
$$\sqrt{21 + x} = 3 + \sqrt{x}$$
$$\sqrt{21 + 4} = 3 + \sqrt{4} \qquad ?$$
$$5 = 5. \qquad \text{True}$$

The solution is 4. ∎

8.6 EXERCISES

1. How can you tell that the equation $\sqrt{x} = -8$ has no real number solution without performing any algebraic steps?

2. Explain why the equation $x^2 = 36$ has two real number solutions, while the equation $\sqrt{x} = 6$ has only one real number solution.

Find the solution for the equation. See Examples 1–4.

3. $\sqrt{x} = 7$ 4. $\sqrt{k} = 10$ 5. $\sqrt{y + 2} = 3$

6. $\sqrt{x + 7} = 5$ 7. $\sqrt{r - 4} = 9$ 8. $\sqrt{k - 12} = 3$

9. $\sqrt{4 - t} = 7$ 10. $\sqrt{9 - s} = 5$ 11. $\sqrt{2t + 3} = 0$

12. $\sqrt{5x - 4} - 0$ 13. $\sqrt{3x - 8} = -2$ 14. $\sqrt{6y + 4} = -3$

15. $\sqrt{m - 4} = 7$ 16. $\sqrt{t + 3} = 10$ 17. $\sqrt{10x - 8} = 3\sqrt{x}$

18. $\sqrt{17t - 4} = 4\sqrt{t}$ 19. $5\sqrt{x} = \sqrt{10x + 15}$ 20. $4\sqrt{y} = \sqrt{20y - 16}$

21. $\sqrt{3x - 5} = \sqrt{2x + 1}$ 22. $\sqrt{5y + 2} = \sqrt{3y + 8}$

23. $k = \sqrt{k^2 - 5k - 15}$ 24. $s = \sqrt{s^2 - 2s - 6}$

25. $7x = \sqrt{49x^2 + 2x - 10}$ 26. $6m = \sqrt{36m^2 + 5m - 5}$

27. The first step in solving the equation $\sqrt{2x + 1} = x - 7$ is to square both sides of the equation. Errors often occur in solving equations such as this one when the right side of the equation is squared incorrectly. Why is the square of the right side *not* equal to $x^2 + 49$? What is the correct answer for the square of the right side?

28. Explain why the equation $x = 3\sqrt{x + 13}$ cannot have a negative solution.

Find the solutions for the equation. Remember that $(a + b)^2 = a^2 + 2ab + b^2$. See Examples 5 and 6.

29. $\sqrt{2x + 1} = x - 7$ 30. $\sqrt{3x + 3} = x - 5$ 31. $\sqrt{3k + 10} + 5 = 2k$

32. $\sqrt{4t + 13} + 1 = 2t$ 33. $\sqrt{5x + 1} - 1 = x$ 34. $\sqrt{x + 1} - x = 1$

35. $\sqrt{6t + 7} + 3 = t + 5$ 36. $\sqrt{10x + 24} = x + 4$ 37. $x - 4 - \sqrt{2x} = 0$

38. $x - 3 - \sqrt{4x} = 0$ 39. $\sqrt{x + 6} = 2x$ 40. $\sqrt{k + 12} - k$

Solve the equation. You will need to square both sides twice. See Example 7.

41. $\sqrt{x + 1} - \sqrt{x - 4} = 1$ 42. $\sqrt{2x + 3} + \sqrt{x + 1} = 1$

43. $\sqrt{x} = \sqrt{x - 5} + 1$ 44. $\sqrt{2x} = \sqrt{x + 7} - 1$

45. What is wrong with the following "solution"?

$$\sqrt{3x - 6} + \sqrt{x + 2} = 12$$

$3x - 6 + x + 2 = 144$	Square both sides.
$4x - 4 = 144$	Combine terms.
$4x = 148$	Add 4 on both sides.
$x = 37$	Divide by 4.

46. What is wrong with the following "solution"?

$$-\sqrt{x - 1} = -4$$

$-(x - 1) = 16$	Square both sides.
$-x + 1 = 16$	Distributive property
$-x = 15$	Subtract 1 on each side.
$x = -15$	Multiply each side by -1.

Solve the problem.

47. Police sometimes use the following procedure to estimate the speed at which a car was traveling at the time of an accident. A police officer drives the car involved in the accident under conditions similar to those during which the accident took place and then skids to a stop. If the car is driven at 30 miles per hour, then the speed at the time of the accident is given by

$$s = 30\sqrt{\frac{a}{p}}$$

where a is the length of the skid marks left at the time of the accident and p is the length of the skid marks in the police test. Find s for the following values of a and p. Round to the nearest tenth.
(a) $a = 862$ feet; $p = 156$ feet   **(b)** $a = 382$ feet; $p = 96$ feet
(c) $a = 84$ feet; $p = 26$ feet

48. A formula for calculating the distance, d, one can see from an airplane to the horizon on a clear day is

$$d = 1.22\sqrt{x}$$

where x is the altitude of the plane in feet and d is given in miles. How far can one see to the horizon in a plane flying at the following altitudes? Round to the nearest tenth.
(a) 15,000 feet   **(b)** 18,000 feet   **(c)** 24,000 feet

49. The square root of the sum of a number and 4 is 5. Find the number.

50. A certain number is the same as the square root of the product of 8 and the number. Find the number.

51. Three times the square root of 2 equals the square root of the sum of some number and 10. Find the number.

52. The negative square root of a number equals that number decreased by 2. Find the number.

REVIEW EXERCISES

Use the rules for exponents to simplify the expression. Write the answer in exponential form with only positive exponents. See Sections 3.2 and 3.5.

53. $(5^2)^3$

54. $3^{-4} \cdot 3^{-1}$

55. $\dfrac{a^{-2}a^3}{a^4}$

56. $(2x^3)^{-1}$

57. $\left(\dfrac{p}{3}\right)^{-2}$

58. $\left(\dfrac{2y^3}{y^{-1}}\right)^{-2}$

59. $\dfrac{(c^3)^2c^4}{(c^{-1})^3}$

60. $\dfrac{(m^2)^4m^{-1}}{(m^3)^{-1}}$

8.7 FRACTIONAL EXPONENTS

FOR EXTRA HELP

SSG pp. 278–280
SSM pp. 458–461

Video
12

Tutorial
IBM   MAC

OBJECTIVES

1 ▶ Define and use $a^{1/n}$.
2 ▶ Define and use $a^{m/n}$.
3 ▶ Use rules for exponents with fractional exponents.
4 ▶ Use fractional exponents to simplify radicals.

In this section we introduce exponential expressions with fractional exponents such as $5^{1/2}$, $16^{3/4}$, and $8^{-2/3}$.

1 ▶ Define and use $a^{1/n}$. ▶ How should $5^{1/2}$ be defined? We want to define $5^{1/2}$ so that all the rules for exponents developed earlier in this book still hold. Then we should define $5^{1/2}$ so that

$$5^{1/2} \cdot 5^{1/2} = 5^{1/2+1/2} = 5^1 = 5.$$

This agrees with the product rule for exponents from Section 3.2. By definition,

$$(\sqrt{5})(\sqrt{5}) = 5.$$

Since both $5^{1/2} \cdot 5^{1/2}$ and $\sqrt{5} \cdot \sqrt{5}$ equal 5,

$$5^{1/2} \text{ should equal } \sqrt{5}.$$

Similarly,

$$5^{1/3} \cdot 5^{1/3} \cdot 5^{1/3} = 5^{1/3+1/3+1/3} = 5^{3/3} = 5,$$

and

$$\sqrt[3]{5} \cdot \sqrt[3]{5} \cdot \sqrt[3]{5} = \sqrt[3]{5^3} = 5,$$

so

$$5^{1/3} \text{ should equal } \sqrt[3]{5}.$$

These examples suggest the following definition.

$\overline{a^{1/n}}$ If a is a nonnegative number and n is a positive integer,

$$a^{1/n} = \sqrt[n]{a}.$$

EXAMPLE 1

Using the Definition of $a^{1/n}$

Simplify each expression by first writing it in radical form.

(a) $16^{1/2}$

By the definition above,

$$16^{1/2} = \sqrt{16} = 4.$$

(b) $27^{1/3} = \sqrt[3]{27} = 3$ **(c)** $64^{1/3} = \sqrt[3]{64} = 4$ **(d)** $64^{1/6} = \sqrt[6]{64} = 2$ ■

2 ▶ Define and use $a^{m/n}$. ▶ Now a more general exponential expression like $16^{3/4}$ can be defined. By the power rule, $(a^m)^n = a^{mn}$, so that

$$16^{3/4} = (16^{1/4})^3 = (\sqrt[4]{16})^3 = 2^3 = 8.$$

However, $16^{3/4}$ could also be written as

$$16^{3/4} = (16^3)^{1/4} = (4096)^{1/4} = \sqrt[4]{4096} = 8.$$

The expression can be evaluated either way to get the same answer. As the example suggests, taking the root first involves smaller numbers and is often easier. This example suggests the following definition for $a^{m/n}$.

$\overline{a^{m/n}}$ If a is a nonnegative number and m and n are integers, with $n > 0$,

$$a^{m/n} = (a^{1/n})^m = (\sqrt[n]{a})^m.$$

EXAMPLE 2

Using the Definition of $a^{m/n}$

Evaluate each expression.

(a) $9^{3/2}$

Use the definition to write

$$9^{3/2} = (9^{1/2})^3 = 3^3 = 27.$$

(b) $64^{2/3} = (64^{1/3})^2 = 4^2 = 16$

(c) $-32^{4/5} = -(32^{1/5})^4 = -2^4 = -16$ ∎

Earlier, a^{-n} was defined as

$$a^{-n} = \frac{1}{a^n}$$

for nonzero numbers a and integers n. This same result applies for negative fractional exponents.

$\overline{a^{-m/n}}$ If a is a positive number and m and n are integers, with $n > 0$,

$$a^{-m/n} = \frac{1}{a^{m/n}}.$$

EXAMPLE 3

Using the Definition of $a^{-m/n}$

Write each expression with a positive exponent and then evaluate.

(a) $32^{-3/5} = \dfrac{1}{32^{3/5}} = \dfrac{1}{(32^{1/5})^3} = \dfrac{1}{2^3} = \dfrac{1}{8}$

(b) $27^{-4/3} = \dfrac{1}{27^{4/3}} = \dfrac{1}{(27^{1/3})^4} = \dfrac{1}{3^4} = \dfrac{1}{81}$ ∎

3 ▶ Use rules for exponents with fractional exponents.

▶ All the rules for exponents given earlier still hold when the exponents are fractions. The next examples show how to use these rules to simplify expressions with fractional exponents.

EXAMPLE 4

Using the Rules for Exponents with Fractional Exponents

Simplify each expression. Write each answer in exponential form with only positive exponents.

(a) $3^{2/3} \cdot 3^{5/3} = 3^{2/3+5/3} = 3^{7/3}$

(b) $\dfrac{5^{1/4}}{5^{3/4}} = 5^{1/4-3/4} = 5^{-2/4} = 5^{-1/2} = \dfrac{1}{5^{1/2}}$

(c) $(9^{1/4})^2 = 9^{2(1/4)} = 9^{2/4} = 9^{1/2} = \sqrt{9} = 3$

(d) $\dfrac{2^{1/2} \cdot 2^{-1}}{2^{-3/2}} = \dfrac{2^{1/2+(-1)}}{2^{-3/2}} = \dfrac{2^{-1/2}}{2^{-3/2}} = 2^{-1/2-(-3/2)} = 2^{2/2} = 2^1 = 2$

(e) $\left(\dfrac{9}{4}\right)^{5/2} = \dfrac{9^{5/2}}{4^{5/2}} = \dfrac{(9^{1/2})^5}{(4^{1/2})^5} = \dfrac{(\sqrt{9})^5}{(\sqrt{4})^5} = \dfrac{3^5}{2^5}$ ∎

EXAMPLE 5

Using Fractional
Exponents with Variables

Simplify each expression. Write each answer in exponential form with only positive exponents. Assume that all variables represent positive numbers.

(a) $m^{1/5} \cdot m^{3/5} = m^{1/5+3/5} = m^{4/5}$

(b) $\dfrac{p^{5/3}}{p^{4/3}} = p^{5/3-4/3} = p^{1/3}$

(c) $(x^2 y^{1/2})^4 = (x^2)^4 (y^{1/2})^4 = x^8 y^2$

(d) $\left(\dfrac{z^{1/4}}{w^{1/3}}\right)^5 = \dfrac{(z^{1/4})^5}{(w^{1/3})^5} = \dfrac{z^{5/4}}{w^{5/3}}$

(e) $\dfrac{k^{2/3} \cdot k^{-1/3}}{k^{5/3}} = k^{2/3+(-1/3)-5/3} = k^{-4/3} = \dfrac{1}{k^{4/3}}$ ∎

CAUTION Errors often occur in problems like those in Examples 4 and 5 because students try to convert the expressions to radicals. Remember that the *rules of exponents* apply here.

---

4 ▶ Use fractional exponents to simplify radicals.

▶ In fact, sometimes it is easier to simplify a radical by first writing it in exponential form.

EXAMPLE 6

Simplifying Radicals by
Using Rational Exponents

Simplify each radical by first writing it in exponential form.

(a) $\sqrt[6]{9^3} = (9^3)^{1/6} = 9^{3/6} = 9^{1/2} = \sqrt{9} = 3$

(b) $(\sqrt[4]{m})^2 = (m^{1/4})^2 = m^{2/4} = m^{1/2} = \sqrt{m}$
Here it is assumed that $m \geq 0$. ∎

◆ **CONNECTIONS** ◆

We saw earlier in this chapter that we can use the rules for adding and multiplying polynomials with radical expressions. In this section we see that radicals can be treated as exponential expressions and so the rules for exponents also apply. This should not be surprising because we know that radicals are a way to represent real numbers, and all the rules developed in this book apply to any real numbers (with exceptions in some cases). As we mentioned at the beginning of Chapter 3, polynomials are the basic algebraic expressions. Our work with radicals just extends the kind of numbers that we use in polynomials.

FOR DISCUSSION OR WRITING
Use the methods of Chapter 4 to factor the following expressions.

1. $x - 2\sqrt{x} - 3$
2. $x^2 - 10$
3. $x + 4\sqrt{x} + 4$
4. Divide the polynomial $x^2 + 3x + 5$ by $\sqrt{x}$, and simplify the result.

8.7 EXERCISES

Decide which one of the four choices is not *equal to the given expression.*

1. $49^{1/2}$ **(a)** -7 **(b)** 7 **(c)** $\sqrt{49}$ **(d)** $49^{.5}$

2. $81^{1/2}$ **(a)** 9 **(b)** $\sqrt{81}$ **(c)** $81^{.5}$ **(d)** $\dfrac{81}{2}$

3. $-64^{1/3}$ **(a)** $-\sqrt{16}$ **(b)** -4 **(c)** 4 **(d)** $-\sqrt[3]{64}$

4. $-125^{1/3}$ **(a)** $-\sqrt{25}$ **(b)** -5 **(c)** $-\sqrt[3]{125}$ **(d)** 5

Simplify the expression by first writing it in radical form. See Examples 1 and 2.

5. $25^{1/2}$ 6. $121^{1/2}$ 7. $64^{1/3}$ 8. $125^{1/3}$ 9. $16^{1/4}$

10. $81^{1/4}$ 11. $32^{1/5}$ 12. $243^{1/5}$ 13. $4^{3/2}$ 14. $9^{5/2}$

15. $27^{2/3}$ 16. $8^{5/3}$ 17. $16^{3/4}$ 18. $64^{5/3}$ 19. $32^{2/5}$

20. $144^{3/2}$ 21. $-8^{2/3}$ 22. $-27^{5/3}$ 23. $-64^{1/3}$ 24. $-125^{5/3}$

Simplify the expression. Write the answer in exponential form with only positive exponents. Assume that all variables represent positive numbers. See Examples 3–5.

25. $2^{1/2} \cdot 2^{5/2}$ 26. $5^{2/3} \cdot 5^{4/3}$ 27. $6^{1/4} \cdot 6^{-3/4}$ 28. $12^{2/5} \cdot 12^{-1/5}$

29. $\dfrac{15^{3/4}}{15^{5/4}}$ 30. $\dfrac{7^{3/5}}{7^{-1/5}}$ 31. $\dfrac{11^{-2/7}}{11^{-3/7}}$ 32. $\dfrac{4^{-2/3}}{4^{1/3}}$

33. $(8^{3/2})^2$ 34. $(5^{2/5})^{10}$ 35. $(6^{1/3})^{3/2}$ 36. $(7^{2/5})^{5/3}$

37. $\left(\dfrac{25}{4}\right)^{3/2}$ 38. $\left(\dfrac{8}{27}\right)^{2/3}$ 39. $\dfrac{2^{2/5} \cdot 2^{-3/5}}{2^{7/5}}$ 40. $\dfrac{3^{-3/4} \cdot 3^{5/4}}{3^{-1/4}}$

41. $\dfrac{6^{-2/9}}{6^{1/9} \cdot 6^{-5/9}}$ 42. $\dfrac{8^{6/7}}{8^{2/7} \cdot 8^{-1/7}}$ 43. $p^{2/3} \cdot p^{7/3}$ 44. $k^{-1/4} \cdot k^{5/4}$

45. $\dfrac{z^{2/3}}{z^{-1/3}}$ 46. $\dfrac{r^{5/4}}{r^{3/4}}$ 47. $(m^3 n^{1/4})^{2/3}$ 48. $(p^4 \cdot q^{1/2})^{4/3}$

49. $\left(\dfrac{a^{1/2}}{b^{1/3}}\right)^{4/3}$ 50. $\left(\dfrac{m^{2/3}}{n^{3/4}}\right)^{1/2}$ 51. $\dfrac{c^{2/3} \cdot c^{-1/3}}{c^{5/3}}$ 52. $\dfrac{d^{3/4} \cdot d^{5/4}}{d^{1/4}}$

Simplify the radical by first writing it in exponential form. Give the answer as an integer or a radical in simplest form. Assume that all variables represent nonnegative numbers. See Example 6.

53. $\sqrt[6]{4^3}$ 54. $\sqrt[9]{8^3}$ 55. $\sqrt[8]{16^2}$ 56. $\sqrt[9]{27^3}$

57. $\sqrt[4]{a^2}$ 58. $\sqrt[9]{b^3}$ 59. $\sqrt[6]{k^4}$ 60. $\sqrt[8]{m^4}$

◆ **MATHEMATICAL CONNECTIONS** (Exercises 61–66) ◆

The rules for multiplying and dividing radicals presented earlier in this chapter were stated for radicals having the same index. For example, we only multiplied or divided square roots, or multiplied or divided cube roots, and so on. Since we know how to write radicals with fractional exponents and from past work know how to add and subtract fractions with different denominators, we can now multiply and divide radicals having different indexes. Work Exercises 61–66 in order, to see how we can multiply $\sqrt{2}$ by $\sqrt[3]{2}$.

61. Write $\sqrt{2}$ and $\sqrt[3]{2}$ using fractional exponents.

62. Write the product $\sqrt{2} \cdot \sqrt[3]{2}$ using the expressions you found in Exercise 61.

63. What is the least common denominator of the fractional exponents in Exercise 62?

64. Repeat Exercise 62, but write the fractional exponents with the common denominator from Exercise 63.

65. Use the rule for multiplying exponential expressions with like bases to simplify the product in Exercise 64. (*Hint:* The base remains the same; do not multiply the two bases.)

66. Write the answer you obtained in Exercise 65 as a radical.

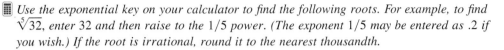

Use the exponential key on your calculator to find the following roots. For example, to find $\sqrt[5]{32}$, enter 32 and then raise to the 1/5 power. (The exponent 1/5 may be entered as .2 if you wish.) If the root is irrational, round it to the nearest thousandth.

67. $\sqrt[6]{64}$ **68.** $\sqrt[5]{243}$ **69.** $\sqrt[7]{84}$ **70.** $\sqrt[9]{16}$ **71.** $\sqrt[5]{987}$

72. $\sqrt[6]{249}$ **73.** $\sqrt[4]{19^3}$ **74.** $\sqrt[5]{27^4}$ **75.** $\sqrt[100]{2.3}$ **76.** $\sqrt[100]{9.6}$

Solve the problem.

77. A formula for calculating the distance, d, one can see from an airplane to the horizon on a clear day is

$$d = 1.22x^{1/2},$$

where x is the altitude of the plane in feet and d is given in miles. How far can one see to the horizon in a plane flying at the following altitudes? Give answers to the nearest hundredth.
(a) 20,000 feet   **(b)** 30,000 feet

78. A biologist has shown that the number of different plant species S on a Galápagos Island is related to the area of the island, A, by

$$S = 28.6A^{1/3}.$$

How many plant species would exist on such an island with the following areas?
(a) 8 square miles   **(b)** 27,000 square miles

79. Explain in your own words why $7^{1/2}$ is defined as $\sqrt{7}$.

80. Explain in your own words why $7^{1/3}$ is defined as $\sqrt[3]{7}$.

REVIEW EXERCISES

Find the square roots of the number. Simplify where possible. See Section 8.1.

81. 121   **82.** 625   **83.** 23   **84.** 160

CHAPTER 8 SUMMARY

KEY TERMS		NEW SYMBOLS	
8.1 square root	**8.3** like radicals	$\sqrt{}$	radical sign
radicand		$\approx$	is approximately equal to
radical	**8.4** rationalizing the	$\sqrt[3]{a}$	cube root of a
radical expression	denominator	$\sqrt[n]{a}$	nth root of a
perfect square		$a^{1/m}$	mth root of a
cube root	**8.5** conjugate		
fourth root			
index (order)	**8.6** squaring property		
principal root			
8.2 perfect cube			

QUICK REVIEW

CONCEPTS	EXAMPLES
8.1 FINDING ROOTS	
If a is a positive real number, $\sqrt{a}$ is the positive square root of a; $-\sqrt{a}$ is the negative square root of a; $\sqrt{0} = 0$. If a is a negative real number, $\sqrt{a}$ is not a real number. If a is a positive rational number, $\sqrt{a}$ is rational if a is a perfect square. $\sqrt{a}$ is irrational if a is not a perfect square. Each real number has exactly one real cube root.	$\sqrt{49} = 7$ $-\sqrt{81} = -9$ $\sqrt{-25}$ is not a real number. $\sqrt{\frac{4}{9}}, \sqrt{16}$ are rational. $\sqrt{\frac{2}{3}}, \sqrt{21}$ are irrational. $\sqrt[3]{27} = 3; \sqrt[3]{-8} = -2$
8.2 MULTIPLICATION AND DIVISION OF RADICALS	
Product Rule for Radicals For nonnegative real numbers x and y, $$\sqrt{x} \cdot \sqrt{y} = \sqrt{xy}$$ and $$\sqrt{xy} = \sqrt{x} \cdot \sqrt{y}.$$ **Quotient Rule for Radicals** If x and y are nonnegative real numbers and y is not 0, $$\frac{\sqrt{x}}{\sqrt{y}} = \sqrt{\frac{x}{y}} \text{ and } \sqrt{\frac{x}{y}} = \frac{\sqrt{x}}{\sqrt{y}}.$$ If all indicated roots are real, $$\sqrt[n]{x} \cdot \sqrt[n]{y} = \sqrt[n]{xy}$$ and $$\frac{\sqrt[n]{x}}{\sqrt[n]{y}} = \sqrt[n]{\frac{x}{y}}.$$	$\sqrt{5} \cdot \sqrt{7} = \sqrt{35}$ $\sqrt{8} \cdot \sqrt{2} = \sqrt{16} = 4$ $\sqrt{48} = \sqrt{16} \cdot \sqrt{3} = 4\sqrt{3}$ $\sqrt{\frac{25}{64}} = \frac{\sqrt{25}}{\sqrt{64}} = \frac{5}{8}$ $\frac{\sqrt{8}}{\sqrt{2}} = \sqrt{\frac{8}{2}} = \sqrt{4} = 2$ $\sqrt[3]{5} \cdot \sqrt[3]{3} = \sqrt[3]{15}$ $\frac{\sqrt[4]{12}}{\sqrt[4]{4}} = \sqrt[4]{\frac{12}{4}} = \sqrt[4]{3}$
8.3 ADDITION AND SUBTRACTION OF RADICALS	
Add and subtract like radicals by using the distributive property. Only like radicals can be combined in this way.	$2\sqrt{5} + 4\sqrt{5} = (2 + 4)\sqrt{5}$ $= 6\sqrt{5}$ $\sqrt{8} + \sqrt{32} = 2\sqrt{2} + 4\sqrt{2}$ $= 6\sqrt{2}$
8.4 RATIONALIZING THE DENOMINATOR	
The denominator of a radical can be rationalized by multiplying both the numerator and denominator by the same number.	$\frac{2}{\sqrt{3}} = \frac{2 \cdot \sqrt{3}}{\sqrt{3} \cdot \sqrt{3}} = \frac{2\sqrt{3}}{3}$ $\sqrt[3]{\frac{5}{6}} = \frac{\sqrt[3]{5}}{\sqrt[3]{6}} \cdot \frac{\sqrt[3]{6^2}}{\sqrt[3]{6^2}} = \frac{\sqrt[3]{180}}{6}$

CONCEPTS	EXAMPLES
8.5 SIMPLIFYING RADICAL EXPRESSIONS	
When appropriate, use the rules for adding and multiplying polynomials to simplify radical expressions.	$\sqrt{6}(\sqrt{5} - \sqrt{7}) = \sqrt{30} - \sqrt{42}$ $(\sqrt{5} - \sqrt{3})(\sqrt{5} + \sqrt{3}) = 5 - 3 = 2$
Any denominators with radicals should be rationalized.	$\dfrac{3}{\sqrt{6}} = \dfrac{3\sqrt{6}}{6} = \dfrac{\sqrt{6}}{2}$
If a radical expression contains two terms in the denominator and at least one of those terms is a radical, multiply both the numerator and the denominator by the conjugate of the denominator.	$\dfrac{6}{\sqrt{7} - \sqrt{2}} = \dfrac{6}{\sqrt{7} - \sqrt{2}} \cdot \dfrac{\sqrt{7} + \sqrt{2}}{\sqrt{7} + \sqrt{2}}$ $= \dfrac{6(\sqrt{7} + \sqrt{2})}{7 - 2}$ Multiply fractions. $= \dfrac{6(\sqrt{7} + \sqrt{2})}{5}$ Simplify.
8.6 EQUATIONS WITH RADICALS	
Solving an Equation with Radicals *Step 1* Arrange the terms so that a radical is alone on one side of the equation.	Solve $\sqrt{2x - 3} + x = 3$. $\sqrt{2x - 3} = 3 - x$ Isolate radical.
Step 2 Square each side. (By the squaring property of equality, all solutions of the original equation are *among* the solutions of the squared equation.)	$(\sqrt{2x - 3})^2 = (3 - x)^2$ Square. $2x - 3 = 9 - 6x + x^2$
Step 3 Combine like terms.	$0 = x^2 - 8x + 12$ Get one side $= 0$.
Step 4 If there is still a term with a radical, repeat Steps 1–3.	$0 = (x - 2)(x - 6)$ Factor. $x - 2 = 0$ or $x - 6 = 0$ Set each factor $= 0$.
Step 5 Solve the equation for potential solutions.	$x = 2$ or $x = 6$ Solve.
Step 6 Check all potential solutions from Step 5 in the original equation.	Verify that 2 is the only solution (6 is extraneous).
8.7 FRACTIONAL EXPONENTS	
Assume $a \geq 0$, m and n are integers, $n > 0$. $a^{1/n} = \sqrt[n]{a}$ $a^{m/n} = \sqrt[n]{a^m} = (\sqrt[n]{a})^m$	$8^{1/3} = \sqrt[3]{8} = 2$ $(81)^{3/4} = \sqrt[4]{81^3} = (\sqrt[4]{81})^3 = 3^3 = 27$

CHAPTER 8 REVIEW EXERCISES

[8.1] *Find all square roots of the number.*

1. 49   **2.** 81   **3.** 196   **4.** 121   **5.** 225   **6.** 729

Find the indicated root. If the root is not a real number, say so.

7. $\sqrt{16}$ **8.** $-\sqrt{36}$ **9.** $\sqrt[3]{1000}$ **10.** $\sqrt[4]{81}$

11. $\sqrt{-8100}$ **12.** $-\sqrt{4225}$ **13.** $\sqrt{\dfrac{49}{36}}$ **14.** $\sqrt{\dfrac{100}{81}}$

15. If $\sqrt{a}$ is not a real number, then what kind of number must a be?

16. Find the value of x:

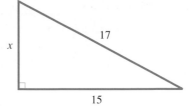

Determine whether the number is rational, irrational, *or* not a real number. *If the number is* rational, *give its exact value. If the number is* irrational, *give a decimal approximation rounded to the nearest thousandth.*

17. $\sqrt{23}$ **18.** $\sqrt{169}$ **19.** $-\sqrt{25}$ **20.** $\sqrt{-4}$

[8.2] *Use the product rule to simplify the expression.*

21. $\sqrt{2} \cdot \sqrt{7}$ **22.** $\sqrt{12} \cdot \sqrt{3}$ **23.** $\sqrt{5} \cdot \sqrt{15}$ **24.** $\sqrt{12} \cdot \sqrt{12}$

25. $-\sqrt{27}$ **26.** $\sqrt{48}$ **27.** $\sqrt{160}$ **28.** $\sqrt[3]{-125}$

29. $\sqrt[3]{1728}$ **30.** $\sqrt{12} \cdot \sqrt{27}$ **31.** $\sqrt{32} \cdot \sqrt{48}$ **32.** $\sqrt{50} \cdot \sqrt{125}$

Use the product rule, the quotient rule, or both to simplify the expression.

33. $\sqrt{\dfrac{9}{4}}$ **34.** $-\sqrt{\dfrac{121}{400}}$ **35.** $\sqrt{\dfrac{3}{49}}$

36. $\sqrt{\dfrac{7}{169}}$ **37.** $\sqrt{\dfrac{1}{6}} \cdot \sqrt{\dfrac{5}{6}}$ **38.** $\sqrt{\dfrac{2}{5}} \cdot \sqrt{\dfrac{2}{45}}$

39. $\dfrac{3\sqrt{10}}{\sqrt{5}}$ **40.** $\dfrac{24\sqrt{12}}{6\sqrt{3}}$ **41.** $\dfrac{8\sqrt{150}}{4\sqrt{75}}$

Simplify the expression. Assume that all variables represent nonnegative real numbers.

42. $\sqrt{p} \cdot \sqrt{p}$ **43.** $\sqrt{k} \cdot \sqrt{m}$ **44.** $\sqrt{r^{18}}$ **45.** $\sqrt{x^{10}y^{16}}$

46. $\sqrt{x^9}$ **47.** $\sqrt{\dfrac{36}{p^2}}, p \neq 0$ **48.** $\sqrt{a^{15}b^{21}}$ **49.** $\sqrt{121x^6y^{10}}$

50. Use a calculator to find approximations for $\sqrt{.5}$ and $\dfrac{\sqrt{2}}{2}$. Based on your results, do you think that these two expressions represent the same number? If so, verify it *algebraically.*

[8.3] *Simplify and combine terms where possible.*

51. $\sqrt{11} + \sqrt{11}$ **52.** $3\sqrt{2} + 6\sqrt{2}$ **53.** $3\sqrt{75} + 2\sqrt{27}$

54. $4\sqrt{12} + \sqrt{48}$ **55.** $4\sqrt{24} - 3\sqrt{54} + \sqrt{6}$ **56.** $2\sqrt{7} - 4\sqrt{28} + 3\sqrt{63}$

57. $\dfrac{2}{5}\sqrt{75} + \dfrac{3}{4}\sqrt{160}$ **58.** $\dfrac{1}{3}\sqrt{18} + \dfrac{1}{4}\sqrt{32}$ **59.** $\sqrt{15} \cdot \sqrt{2} + 5\sqrt{30}$

Simplify the expression. Assume that all variables represent nonnegative real numbers.

60. $\sqrt{4x} + \sqrt{36x} - \sqrt{9x}$ **61.** $\sqrt{16p} + 3\sqrt{p} - \sqrt{49p}$

62. $\sqrt{20m^2} - m\sqrt{45}$ **63.** $3k\sqrt{8k^2n} + 5k^2\sqrt{2n}$

[8.4] *Perform the indicated operations and write the answer in simplest form. Assume that all variables represent nonnegative real numbers.*

64. $\dfrac{10}{\sqrt{3}}$

65. $\dfrac{15}{\sqrt{2}}$

66. $\dfrac{8\sqrt{2}}{\sqrt{5}}$

67. $\dfrac{5}{\sqrt{5}}$

68. $\dfrac{12}{\sqrt{24}}$

69. $\dfrac{\sqrt{2}}{\sqrt{15}}$

70. $\sqrt{\dfrac{2}{5}}$

71. $\sqrt{\dfrac{5}{14}} \cdot \sqrt{28}$

72. $\sqrt{\dfrac{2}{7}} \cdot \sqrt{\dfrac{1}{3}}$

73. $\sqrt{\dfrac{r^2}{16x}},\ x \neq 0$

74. $\sqrt[3]{\dfrac{1}{3}}$

75. $\sqrt[3]{\dfrac{2}{7}}$

76. Explain how you would show, without using a calculator, that $\dfrac{\sqrt{6}}{4}$ and $\sqrt{\dfrac{48}{128}}$ represent the exact same number. Then actually perform the necessary steps.

[8.5] *Simplify the expression.*

77. $-\sqrt{3}(\sqrt{5} + \sqrt{27})$

78. $3\sqrt{2}(\sqrt{3} + 2\sqrt{2})$

79. $(2\sqrt{3} - 4)(5\sqrt{3} + 2)$

80. $(5\sqrt{7} + 2)^2$

81. $(\sqrt{5} - \sqrt{7})(\sqrt{5} + \sqrt{7})$

82. $(2\sqrt{3} + 5)(2\sqrt{3} - 5)$

83. $(\sqrt{7} + 2\sqrt{6})(\sqrt{12} - \sqrt{2})$

Rationalize the denominator.

84. $\dfrac{1}{2 + \sqrt{5}}$

85. $\dfrac{2}{\sqrt{2} - 3}$

86. $\dfrac{\sqrt{8}}{\sqrt{2} + 6}$

87. $\dfrac{\sqrt{3}}{1 + \sqrt{3}}$

88. $\dfrac{\sqrt{5} - 1}{\sqrt{2} + 3}$

89. $\dfrac{2 + \sqrt{6}}{\sqrt{3} - 1}$

Write the quotient in lowest terms.

90. $\dfrac{15 + 10\sqrt{6}}{15}$

91. $\dfrac{3 + 9\sqrt{7}}{12}$

92. $\dfrac{6 + \sqrt{192}}{2}$

[8.6] *Solve the equation.*

93. $\sqrt{m} - 5 = 0$

94. $\sqrt{p} + 4 = 0$

95. $\sqrt{k + 1} = 7$

96. $\sqrt{5m + 4} = 3\sqrt{m}$

97. $\sqrt{2p + 3} = \sqrt{5p - 3}$

98. $\sqrt{4y + 1} = y - 1$

99. $\sqrt{-2k - 4} = k + 2$

100. $\sqrt{2 - x} + 3 = x + 7$

101. $\sqrt{x} - x + 2 = 0$

102. $\sqrt{2 - x} + x = 0$

103. $\sqrt{4y - 2} = \sqrt{3y + 1}$

104. $\sqrt{2x + 3} = x + 2$

[8.7] *Simplify the expression. Assume that all variables represent positive real numbers.*

105. $81^{1/2}$

106. $-125^{1/3}$

107. $7^{2/3} \cdot 7^{7/3}$

108. $\dfrac{13^{4/5}}{13^{-3/5}}$

109. $\dfrac{x^{1/4} \cdot x^{5/4}}{x^{3/4}}$

110. $\sqrt[8]{49^4}$

MIXED REVIEW EXERCISES

Simplify the expression if possible. Assume all variables represent positive real numbers.

111. $64^{2/3}$

112. $2\sqrt{27} + 3\sqrt{75} - \sqrt{300}$

113. $\sqrt{\dfrac{121}{t^2}}$

114. $\dfrac{1}{5 + \sqrt{2}}$

115. $\sqrt{\dfrac{1}{3}} \cdot \sqrt{\dfrac{24}{5}}$ **116.** $\sqrt{50y^2}$

117. $\sqrt[3]{-125}$ **118.** $-\sqrt{5}(\sqrt{2} + \sqrt{75})$

119. $\sqrt{\dfrac{16r^3}{3s}}$ **120.** $\dfrac{12 + 6\sqrt{13}}{12}$

121. $-\sqrt{162} + \sqrt{8}$ **122.** $(\sqrt{5} - \sqrt{2})^2$

123. $(6\sqrt{7} + 2)(4\sqrt{7} - 1)$ **124.** $-\sqrt{121}$

125. $\dfrac{x^{8/3}}{x^{2/3}}$

Solve.

126. $\sqrt{x + 2} = x - 4$ **127.** $\sqrt{k} + 3 = 0$ **128.** $\sqrt{1 + 3t} - t = -3$

◆ **MATHEMATICAL CONNECTIONS** (Exercises 129–133) ◆

In Chapter 6 we plotted points in the rectangular coordinate plane. In all cases our points had coordinates that were rational numbers. However, ordered pairs may have irrational coordinates as well. Use your knowledge of the material in Chapters 6 and 8 to work Exercises 129–133 in order.
Consider the points A($2\sqrt{14}$, $5\sqrt{7}$) and B($-3\sqrt{14}$, $10\sqrt{7}$).

129. Write an expression that represents the slope of the line containing points A and B. Do not simplify yet.

130. Simplify the numerator and the denominator in the expression from Exercise 129 by combining like radicals.

131. Write the fraction from Exercise 130 as the square root of a fraction in lowest terms.

132. Rationalize the denominator of the expression found in Exercise 131.

133. Based on your answer in Exercise 132, does line *AB* rise or fall from left to right?

CHAPTER 8 TEST

On this test assume that all variables represent nonnegative real numbers.

1. Find all square roots of 196.

2. Consider $\sqrt{142}$.
   (a) Determine whether it is rational or irrational.
   ▦ **(b)** Find a decimal approximation to the nearest thousandth.

3. Simplify $\sqrt[3]{216}$.

Simplify where possible.

4. $-\sqrt{27}$ **5.** $\sqrt{\dfrac{128}{25}}$

6. $\sqrt[3]{32}$ **7.** $\dfrac{20\sqrt{18}}{5\sqrt{3}}$

8. $3\sqrt{28} + \sqrt{63}$ **9.** $3\sqrt{27x} - 4\sqrt{48x} + 2\sqrt{3x}$

10. $\sqrt[3]{32x^2y^3}$ **11.** $(6 - \sqrt{5})(6 + \sqrt{5})$

12. $(2 - \sqrt{7})(3\sqrt{2} + 1)$ **13.** $(\sqrt{5} + \sqrt{6})^2$

Solve the problem.

14. The hypotenuse of a right triangle measures 9 inches, and one leg measures 3 inches. Find the measure of the other leg.

   (a) Give its length in simplified radical form.

   (b) Round the answer to the nearest thousandth.

Rationalize the denominator.

15. $\dfrac{5\sqrt{2}}{\sqrt{7}}$

16. $\sqrt{\dfrac{2}{3x}} \; (x > 0)$

17. $\dfrac{-2}{\sqrt[3]{4}}$

18. $\dfrac{-3}{4 - \sqrt{3}}$

Solve the equation.

19. $\sqrt{x + 1} = 5 - x$

20. $3\sqrt{x} - 1 = 2x$

Simplify the expression.

21. $8^{4/3}$

22. $-125^{2/3}$

23. $5^{3/4} \cdot 5^{1/4}$

24. $\dfrac{(3^{1/4})^3}{3^{7/4}}$

25. What is wrong with the following "solution"?

$$\sqrt{2x + 1} + 5 = 0$$
$$\sqrt{2x + 1} = -5 \qquad \text{Subtract 5.}$$
$$2x + 1 = 25 \qquad \text{Square both sides}$$
$$2x = 24 \qquad \text{Subtract 1.}$$
$$x = 12 \qquad \text{Divide by 2.}$$

The solution is 12.

CUMULATIVE REVIEW (Chapters 1–8)

Simplify the expression.

1. $3(6 + 7) + 6 \cdot 4 - 3^2$

2. $\dfrac{3(6 + 7) + 3}{2(4) - 1}$

3. $|-6| - |-3|$

4. $-9 + 14 + 11 + (-3 + 5)$

5. $13 - [-4 - (-2)]$

6. $-2.523 + 8.674 - 1.928$

Solve the equation or inequality.

7. $5(k - 4) - k = k - 11$

8. $-\dfrac{3}{4}y \le 12$

9. $5z + 3 - 4 > 2z + 9 + z$

Solve the problem.

10. The perimeter of a rectangle is 56 meters. The length of the rectangle is 7 meters more than the width. Find the length and the width of the rectangle.

Simplify and write the expression without negative exponents. Assume that variables represent positive real numbers.

11. $(3x^6)(2x^2y)^2$

12. $\left(\dfrac{3^2 y^{-2}}{2^{-1} y^3}\right)^{-3}$

13. Subtract $7x^3 - 8x^2 + 4$ from $10x^3 + 3x^2 - 9$.

14. Divide: $\dfrac{8t^3 - 4t^2 - 14t + 15}{2t + 3}$.

Factor the polynomial completely.

15. $m^2 + 12m + 32$

16. $25t^4 - 36$

17. $12a^2 + 4ab - 5b^2$

18. $81z^2 + 72z + 16$

Solve the quadratic equation.

19. $x^2 - 7x = -12$

20. $(x + 4)(x - 1) = -6$

21. For what real number(s) is the expression $\dfrac{3}{x^2 + 5x - 14}$ undefined?

Multiply or divide as indicated. Express the answer in lowest terms.

22. $\dfrac{x^2 - 3x - 4}{x^2 + 3x} \cdot \dfrac{x^2 + 2x - 3}{x^2 - 5x + 4}$

23. $\dfrac{t^2 + 4t - 5}{t + 5} \div \dfrac{t - 1}{t^2 + 8t + 15}$

24. Simplify the complex fraction: $\dfrac{\dfrac{2}{3} + \dfrac{1}{2}}{\dfrac{1}{9} - \dfrac{1}{6}}$.

Add or subtract as indicated. Express the answer in lowest terms.

25. $\dfrac{y}{y^2 - 1} + \dfrac{y}{y + 1}$

26. $\dfrac{2}{x + 3} - \dfrac{4}{x - 1}$

Graph the equation or the inequality in the rectangular coordinate plane.

27. $-4x + 5y = -20$

28. $x = 2$

29. $2x - 5y > 10$

30. Find the slope of the line through the points $(9, -2)$ and $(-3, 8)$.

Solve the system of equations.

31. $\begin{aligned} 4x - y &= 19 \\ 3x + 2y &= -5 \end{aligned}$

32. $\begin{aligned} 2x - y &= 6 \\ 3y &= 6x - 18 \end{aligned}$

Solve the problem.

33. A cashier has 20 bills, all of which are ten-dollar or twenty-dollar bills. The total value of the money is $250. How many of each denomination does the cashier have?

34. The profits from a carnival are to be given to two scholarships so that one scholarship receives $\dfrac{3}{2}$ as much money as the other. If the total amount given to the two scholarships is $780, how much goes to the scholarship that receives the lesser amount?

Simplify the expression if possible. Assume all variables represent nonnegative real numbers.

35. $\sqrt{27} - 2\sqrt{12} + 6\sqrt{75}$

36. $\dfrac{2}{\sqrt{3} + \sqrt{5}}$

37. $\sqrt{200x^2y^5}$

38. $16^{5/4}$

39. $(3\sqrt{2} + 1)(4\sqrt{2} - 3)$

40. Solve the equation $\sqrt{x} + 2 = x - 10$.

CONNECTIONS

According to data from the U.S. Labor Department, the quadratic expression $.011x^2 - .097x + 4.1$ is a reasonable approximation of the number of people (in millions) in the U.S. holding more than one job. In this expression, x represents the number of years since 1970. Thus, in 1970 ($x = 0$) 4.1 million people held more than one job. To find the year in which 6 million people held more than one job, we can set the expression equal to 6, getting $.011x^2 - .097x + 4.1 = 6$, and solve for x. In this chapter we see how to solve such *quadratic equations.* Quadratic equations are good models for many applications of mathematics to everyday situations.

FOR DISCUSSION OR WRITING

Use the quadratic expression to find the number of people holding more than one job in 1980 and in 1990. Use your answers to guess in what year 6 million people held more than one job. As we see in this chapter, quadratic equations usually have two solutions. How could two solutions be interpreted in the quadratic equation we gave above for the case where 6 million people have more than one job?

A method of solving quadratic equations (by factoring) was studied in Chapter 4. In this chapter we study several other ways of solving quadratic equations, since not all quadratic equations can be solved by factoring ($x^2 - x + 1 = 0$, for instance, cannot).

9.1 SOLVING QUADRATIC EQUATIONS BY THE SQUARE ROOT PROPERTY

FOR EXTRA HELP

SSG pp. 283–286
SSM pp. 478–482

Video
12

Tutorial
IBM MAC

OBJECTIVES

1 ▶ Solve equations of the form $x^2 = $ a number.
2 ▶ Solve equations of the form $(ax + b)^2 = $ a number.

Recall that a *quadratic equation* is an equation that can be written in the form

$$ax^2 + bx + c = 0$$

for real numbers a, b, and c, with $a \neq 0$. Quadratic equations where $ax^2 + bx + c$ is factorable were solved in Chapter 4 using the zero-factor property. For example, to solve $x^2 + 4x + 3 = 0$, we begin by factoring on the left side and then set each factor equal to zero.

$$x^2 + 4x + 3 = 0$$
$$(x + 3)(x + 1) = 0 \qquad \text{Factor.}$$
$$x + 3 = 0 \qquad \text{or} \qquad x + 1 = 0 \qquad \text{Zero-factor property}$$
$$x = -3 \qquad \text{or} \qquad x = -1$$

1 ▶ Solve equations of the form $x^2 = $ a number.

▶ We can solve equations such as $x^2 = 9$ by factoring as follows.

$$x^2 = 9$$
$$x^2 - 9 = 0 \qquad \text{Subtract 9.}$$
$$(x + 3)(x - 3) = 0 \qquad \text{Factor.}$$
$$x + 3 = 0 \qquad \text{or} \qquad x - 3 = 0 \qquad \text{Zero-factor property}$$
$$x = -3 \qquad \text{or} \qquad x = 3$$

This result is generalized as the **square root property of equations.**

Square Root Property of Equations

If k is a positive number and if $a^2 = k$, then

$$a = \sqrt{k} \qquad \text{or} \qquad a = -\sqrt{k}.$$

NOTE When we solve an equation, we want to find *all* values of the variable that satisfy the equation. Therefore, we want both the positive and the negative square roots of k.

EXAMPLE 1

Solving a Quadratic Equation of the Form $x^2 = k$

Solve each equation. Write radicals in simplified form.

(a) $x^2 = 16$

By the square root property, since $x^2 = 16$, then

$$x = \sqrt{16} = 4 \quad \text{or} \quad x = -\sqrt{16} = -4.$$

An abbreviation for "$x = 4$ or $x = -4$" is written $x = \pm 4$, and is read "x equals positive or negative 4." Check each solution by substituting back into the original equation.

(b) $z^2 = 5$

The solutions are $z = \sqrt{5}$ or $z = -\sqrt{5}$, which may be written $z = \pm\sqrt{5}$.

(c) $m^2 = 8$

$$m^2 = 8$$
$$m = \sqrt{8} \quad \text{or} \quad m = -\sqrt{8} \qquad \text{Square root property}$$
$$m = 2\sqrt{2} \quad \text{or} \quad m = -2\sqrt{2} \qquad \text{Simplify } \sqrt{8}.$$
$$m = \pm 2\sqrt{2} \qquad \text{Abbreviation}$$

(d) $y^2 = -4$

Since -4 is a negative number and since the square of a real number cannot be negative, there is no real number solution for this equation. (The square root property cannot be used because of the requirement that k must be positive.)

(e) $3x^2 + 5 = 11$

First solve the equation for x^2.

$$3x^2 + 5 = 11$$
$$3x^2 = 6 \qquad \text{Subtract 5.}$$
$$x^2 = 2 \qquad \text{Divide by 3.}$$

Now use the square root property to get $x = \pm\sqrt{2}$. ■

2 ▶ Solve equations of the form $(ax + b)^2 =$ a number.

▶ In each of the equations in Example 1, the exponent 2 appeared with a single variable as its base. The square root property of equations can be extended to solve equations where the base is a binomial, as shown in the next example.

EXAMPLE 2

Solving a Quadratic Equation of the Form $(x + b)^2 = k$

Solve each equation.

(a) $(x - 3)^2 = 16$.

Apply the square root property, using $x - 3$ as the base.

$$(x - 3)^2 = 16$$
$$x - 3 = \sqrt{16} \quad \text{or} \quad x - 3 = -\sqrt{16}$$
$$x - 3 = 4 \quad \text{or} \quad x - 3 = -4 \qquad \sqrt{16} = 4$$
$$x = 7 \quad \text{or} \quad x = -1 \qquad \text{Add 3.}$$

Check both answers in the original equation.

$(x - 3)^2 = 16$		$(x - 3)^2 = 16$	
$(7 - 3)^2 = 16$?	Let $x = 7$.	$(-1 - 3)^2 = 16$?	Let $x = -1$.
$4^2 = 16$?		$(-4)^2 = 16$?	
$16 = 16$	True	$16 = 16$	True

Both 7 and -1 are solutions.

(b) $(x - 1)^2 = 6$.

By the square root property,

$$x - 1 = \sqrt{6} \qquad \text{or} \qquad x - 1 = -\sqrt{6}$$
$$x = 1 + \sqrt{6} \qquad \text{or} \qquad x = 1 - \sqrt{6}.$$

Check:

$$(1 + \sqrt{6} - 1)^2 = (\sqrt{6})^2 = 6;$$
$$(1 - \sqrt{6} - 1)^2 = (-\sqrt{6})^2 = 6.$$

The solutions are $1 + \sqrt{6}$ and $1 - \sqrt{6}$. ■

NOTE The solutions in Example 2(b) may be written in abbreviated form as

$$1 \pm \sqrt{6}.$$

If they are written this way, keep in mind that there are *two* solutions indicated, one with the $+$ sign and the other with the $-$ sign.

EXAMPLE 3

Solving a Quadratic Equation of the Form $(ax + b)^2 = k$

Solve $(3r - 2)^2 = 27$.

$$3r - 2 = \sqrt{27} \qquad \text{or} \qquad 3r - 2 = -\sqrt{27} \qquad \text{Square root property}$$
$$3r - 2 = 3\sqrt{3} \qquad \text{or} \qquad 3r - 2 = -3\sqrt{3} \qquad \sqrt{27} = \sqrt{9 \cdot 3} = 3\sqrt{3}$$
$$3r = 2 + 3\sqrt{3} \qquad \text{or} \qquad 3r = 2 - 3\sqrt{3} \qquad \text{Add 2.}$$
$$r = \frac{2 + 3\sqrt{3}}{3} \qquad \text{or} \qquad r = \frac{2 - 3\sqrt{3}}{3} \qquad \text{Divide by 3.}$$

The solutions are

$$\frac{2 + 3\sqrt{3}}{3} \quad \text{and} \quad \frac{2 - 3\sqrt{3}}{3},$$

which may be abbreviated as

$$\frac{2 \pm 3\sqrt{3}}{3}. \quad ■$$

CAUTION The solutions in Example 3 are fractions that cannot be reduced, since 3 is *not* a common factor in the numerator.

EXAMPLE 4

Recognizing a Quadratic Equation with No Real Solution

Solve $(x + 3)^2 = -9$.

The square root of -9 is not a real number. There is no real number solution for this equation. ■

9.1 EXERCISES

Decide whether the statement is true or false. If it is false, tell why.

1. If $k > 0$, then $x^2 = k$ has exactly two real solutions.

2. If $k < 0$, then $x^2 = k$ has no real solutions.

3. If $k = 0$, then $x^2 = k$ has no real solutions.

4. If k is a positive perfect square, then $x^2 = k$ has two rational solutions.

5. If k is a prime number, then $x^2 = k$ has two irrational solutions.

6. If k is a positive integer, then $x^2 = k$ must have two rational solutions.

7. When a student was asked to solve $x^2 = 81$, she wrote as her answer "9," Her teacher did not give her full credit, and the student argued, saying that because $9^2 = 81$, her answer had to be correct. Why was her answer not completely correct?

8. Explain why $x^2 = -6$ has no real solutions.

Solve the equation by using the square root property. Express all radicals in simplest form. See Example 1.

9. $x^2 = 81$ **10.** $y^2 = 121$ **11.** $k^2 = 14$ **12.** $m^2 = 22$

13. $t^2 = 48$ **14.** $x^2 = 54$ **15.** $y^2 = -100$ **16.** $m^2 = -64$

17. $z^2 = 2.25$ **18.** $w^2 = 56.25$ **19.** $3x^2 - 8 = 64$ **20.** $2t^2 + 7 = 61$

21. Explain why the square of a real number cannot be negative.

22. Does the equation $-x^2 = 16$ have real solutions? Explain.

23. Which one of these equations has exactly one real number solution?
    (a) $x^2 = 4$ **(b)** $y^2 = -4$ **(c)** $(x - 4)^2 = 1$ **(d)** $t^2 = 0$

24. Which one of the equations in Exercise 23 has no real solution?

Solve the equation by using the square root property. Express all radicals in simplest form. See Examples 2–4.

25. $(x - 3)^2 = 25$ **26.** $(y - 7)^2 = 16$ **27.** $(z + 5)^2 = -13$

28. $(m + 2)^2 = -17$ **29.** $(x - 8)^2 = 27$ **30.** $(y - 5)^2 = 40$

31. $(3k + 2)^2 = 49$ **32.** $(5t + 3)^2 = 36$ **33.** $(4x - 3)^2 = 9$

34. $(7y - 5)^2 = 25$ **35.** $(5 - 2x)^2 = 30$ **36.** $(3 - 2a)^2 = 70$

37. $(3k + 1)^2 = 18$ **38.** $(5z + 6)^2 = 75$ **39.** $\left(\dfrac{1}{2}x + 5\right)^2 = 12$

40. $\left(\dfrac{1}{3}y + 4\right)^2 = 27$ **41.** $(4k - 1)^2 - 48 = 0$ **42.** $(2s - 5)^2 - 180 = 0$

43. Johnny solved the equation in Exercise 35 and wrote his answer as $\dfrac{5 + \sqrt{30}}{2}$, $\dfrac{5 - \sqrt{30}}{2}$. Linda solved the same equation and wrote her answer as $\dfrac{-5 + \sqrt{30}}{-2}$, $\dfrac{-5 - \sqrt{30}}{-2}$. The teacher gave them both full credit. Explain why both students were correct, although their answers seem to differ.

44. In the solutions found in Example 3 of this section, why is it not valid to reduce the answers by dividing out the threes in the numerator and denominator?

▦ *Use a calculator with a square root key to solve the equation. Round your answers to the nearest hundredth.*

45. $(k + 2.14)^2 = 5.46$

46. $(r - 3.91)^2 = 9.28$

47. $(2.11p + 3.42)^2 = 9.58$

48. $(1.71m - 6.20)^2 = 5.41$

─────── ◆ **MATHEMATICAL CONNECTIONS** (Exercises 49–54) ◆ ───────

In Section 4.4 we saw how certain trinomials can be factored as the square of a binomial. Use this idea and work Exercises 49–54 in order, considering the equation

$$x^2 + 6x + 9 = 100.$$

49. Factor the left side of the equation as the square of a binomial, and write the resulting equation.

50. Write the equation from Exercise 49 as a compound statement using the word "or."

51. Solve the two individual equations from Exercise 50.

52. What are the solutions of the original equation?

53. Solve the equation $x^2 + 4x + 4 = 25$ using the method described in Exercises 49–52.

54. Solve the equation $4k^2 - 12k + 9 = 81$ using the method described in Exercises 49–52.

───────────── ◆ ─────────────

Solve the problem.

55. The number of cases commenced by the U.S. Court of Appeals per year during the period between 1984 and 1990 is approximated by the model

$$y = 68.9x^2 + 1165.3x + 31{,}676,$$

where $x = 0$ corresponds to 1984, $x = 1$ corresponds to 1985, and so on. Based on this model, approximately how many cases were commenced in 1986?
(*Source:* Administrative Office of the U.S. Courts, *Annual Report to the Director*)

56. The number of infant deaths during the past decade has been decreasing. Between 1980 and 1990, the number of infant deaths per 1000 live births each year can be approximated by the model

$$y = .0234x^2 - .5029x + 12.5,$$

where $x = 0$ corresponds to 1980, $x = 1$ corresponds to 1981, and so on. Based on this model, how many infant deaths per 1000 live births were there during 1984?
(*Source:* U.S. National Center for Health Statistics, *Vital Statistics of the United States*)

57. One expert at marksmanship can hold a silver dollar at forehead level, drop it, draw his gun, and shoot the coin as it passes waist level. The distance traveled by a falling object is given by

$$d = 16t^2,$$

where d is the distance (in feet) the object falls in t seconds. If the coin falls about 4 feet, use the formula to estimate the time that elapses between the dropping of the coin and the shot.

58. The illumination produced by a light source depends on the distance from the source. For a particular light source, this relationship can be expressed as

$$d^2 = \frac{4050}{I},$$

where d is the distance from the source (in feet) and I is the amount of illumination in foot-candles. How far from the source is the illumination equal to 50 foot-candles?

59. The area A of a circle with radius r is given by the formula

$$A = \pi r^2.$$

If a circle has area 81π square inches, what is its radius?

60. The surface area S of a sphere with radius r is given by the formula

$$S = 4\pi r^2.$$

If a sphere has surface area 36π square feet, what is its radius?

61. Becky and Brad are the owners of Cole's Baseball Cards. They have found that the price p, in dollars, of a particular Kirby Puckett baseball card depends on the demand d, in hundreds, for the card, according to the formula $p = (d - 2)^2$. What demand produces a price of $5 for the card?

62. The amount A that P dollars invested at a rate of interest r will grow to in 2 years is

$$A = P(1 + r)^2.$$

At what interest rate will $100 grow to $110.25 in two years?

REVIEW EXERCISES

Simplify all radicals, and combine like terms. Express fractions in lowest terms. See Sections 8.3 and 8.4.

63. $\dfrac{4}{5} + \sqrt{\dfrac{48}{25}}$

64. $12 + \sqrt{\dfrac{2}{3}}$

65. $\dfrac{6 + \sqrt{24}}{8}$

Factor the perfect square trinomial. See Section 4.4.

66. $y^2 - 10y + 25$

67. $x^2 - 7x + \dfrac{49}{4}$

68. $z^2 + z + \dfrac{1}{4}$

9.2 SOLVING QUADRATIC EQUATIONS BY COMPLETING THE SQUARE

FOR EXTRA HELP	OBJECTIVES
SSG pp. 286–290 **SSM** pp. 482–487	**1** ▶ Solve quadratic equations by completing the square when the coefficient of the squared term is 1.
Video 12	**2** ▶ Solve quadratic equations by completing the square when the coefficient of the squared term is not 1.
Tutorial IBM MAC	**3** ▶ Simplify an equation before solving.

1 ▶ Solve quadratic equations by completing the square when the coefficient of the squared term is 1.

▶ The properties studied so far are not enough to solve the equation

$$x^2 + 6x + 7 = 0.$$

If we could write the equation in the form $(x + b)^2 = k$, we could solve it with the square root property discussed in the previous section. To do that, we need to have a perfect square trinomial on one side. The next example shows how to rewrite the equation $x^2 + 6x + 7 = 0$ so it can be solved by that method.

EXAMPLE 1

Rewriting an Equation to Use the Square Root Property

Solve $x^2 + 6x + 7 = 0$.

Start by subtracting 7 from both sides of the equation to get $x^2 + 6x = -7$. If we are to write $x^2 + 6x = -7$ in the form $(x + b)^2 = k$, the quantity on the left-hand side of $x^2 + 6x = -7$ must be made into a perfect square trinomial. The expression $x^2 + 6x + 9$ is a perfect square, since

$$x^2 + 6x + 9 = (x + 3)^2.$$

Therefore, if we add 9 to both sides of $x^2 + 6x = -7$, the equation will have a perfect square trinomial on one side, as needed.

$$x^2 + 6x + 9 = -7 + 9 \qquad \text{Add 9 on both sides.}$$
$$(x + 3)^2 = 2 \qquad \text{Factor; combine terms.}$$

Now use the square root property to complete the solution.

$$x + 3 = \sqrt{2} \qquad \text{or} \qquad x + 3 = -\sqrt{2}$$
$$x = -3 + \sqrt{2} \qquad \text{or} \qquad x = -3 - \sqrt{2}$$

The solutions of the original equation are $-3 + \sqrt{2}$ and $-3 - \sqrt{2}$. Check this by substituting $-3 + \sqrt{2}$ and $-3 - \sqrt{2}$ for x in the original equation. ∎

The process of changing the form of the equation in Example 1 from

$$x^2 + 6x + 7 = 0 \qquad \text{to} \qquad (x + 3)^2 = 2$$

is called **completing the square.** Completing the square changes only the form of the equation. To see this, multiply out the left side of $(x + 3)^2 = 2$ and combine terms. Then subtract 2 from both sides to see that the result is $x^2 + 6x + 7 = 0$.

EXAMPLE 2

Solving a Quadratic Equation by Completing the Square

Solve the quadratic equation $m^2 - 8m = 5$.

A suitable number must be added to both sides to make one side a perfect square. Find this number as follows: recall from Chapter 3 that

$$(m + b)^2 = m^2 + 2bm + b^2.$$

In the equation $m^2 - 8m = 5$, the value of $2b$ is -8 and b^2 must be found. Set $2b$ equal to -8 to find b.

$$2b = -8$$
$$b = -4 \qquad \text{Divide by 2.}$$

Squaring -4 gives 16, the number to be added to both sides.

$$m^2 - 8m + 16 = 5 + 16 \tag{1}$$

Now the trinomial $m^2 - 8m + 16$ is a perfect square trinomial. Factor this trinomial to get

$$m^2 - 8m + 16 = (m - 4)^2.$$

Equation (1) becomes

$$(m - 4)^2 = 21.$$

Now we can use the square root property.

$$m - 4 = \sqrt{21} \qquad \text{or} \qquad m - 4 = -\sqrt{21}$$
$$m = 4 + \sqrt{21} \qquad \text{or} \qquad m = 4 - \sqrt{21}$$

Check that the solutions are

$$4 + \sqrt{21} \quad \text{and} \quad 4 - \sqrt{21}. \quad \blacksquare$$

As illustrated by Example 2, the number to add to both sides of the quadratic equation $x^2 + 2bx = k$ to complete the square is found by taking $1/2$ of $2b$, the coefficient of the x term, and squaring it. This gives $(1/2)(2b) = b$, and we add b^2 to both sides of the equation as shown in Example 2.

The steps used in solving a quadratic equation by completing the square are given below.

Solving a Quadratic Equation by Completing the Square

Step 1 If the coefficient of the squared term is 1, proceed to Step 2. If the coefficient of the squared term is not 1 but some other nonzero number a, divide both sides of the equation by a. This gives an equation that has 1 as coefficient of the squared term.

Step 2 Make sure that all terms with variables are on one side of the equals sign and that all numbers are on the other side.

Step 3 Take half the coefficient of the first degree term, and square the result. Add the square to both sides of the equation. The side containing the variables now can be factored as a perfect square.

Step 4 Apply the square root property.

2 ▶ Solve quadratic equations by completing the square when the coefficient of the squared term is not 1.

▶ The process of completing the square requires the coefficient of the squared variable to be 1 (Step 1). The next example shows how to solve a quadratic equation with this coefficient different from 1. The steps are numbered according to the summary above.

E X A M P L E 3

Solving a Quadratic Equation by Completing the Square

Solve $4y^2 + 24y - 13 = 0$.

Step 1 Divide each side by 4 to get the coefficient of y^2 to be 1.

$$4y^2 + 24y - 13 = 0$$

$$y^2 + 6y - \frac{13}{4} = 0$$

Step 2 Add $13/4$ to each side to get the variable terms on the left and the number on the right.

$$y^2 + 6y = \frac{13}{4}$$

Step 3 Now take half the coefficient of y, $(1/2)(6) = 3$, and square the result: $3^2 = 9$. Add 9 to both sides of the equation, perform the addition on the right-hand side, and factor on the left.

$$y^2 + 6y + \mathbf{9} = \frac{13}{4} + \mathbf{9} \qquad \text{Add 9.}$$

$$y^2 + 6y + 9 = \frac{49}{4} \qquad \text{Add on the right.}$$

$$(y + 3)^2 = \frac{49}{4} \qquad \text{Factor on the left.}$$

Step 4 Use the square root property and solve for y.

$$y + 3 = \frac{7}{2} \qquad \text{or} \qquad y + 3 = -\frac{7}{2} \qquad \text{Square root property}$$

$$y = -3 + \frac{7}{2} \qquad \text{or} \qquad y = -3 - \frac{7}{2} \qquad \text{Add } -3.$$

$$y = \frac{1}{2} \qquad \text{or} \qquad y = -\frac{13}{2}$$

The two solutions are $1/2$ and $-13/2$. Check by substitution into the original equation. ∎

EXAMPLE 4

Solving a Quadratic Equation by Completing the Square

Solve $4p^2 + 8p + 5 = 0$.

First divide both sides by 4 to get the coefficient 1 for the p^2 term (Step 1). The result is

$$p^2 + 2p + \frac{5}{4} = 0.$$

Add $-5/4$ to both sides (Step 2).

$$p^2 + 2p = -\frac{5}{4}$$

The coefficient of p is 2. Take half of 2, square the result, and add it to both sides. The left-hand side can then be factored as a perfect square (Step 3).

$$p^2 + 2p + 1 = -\frac{5}{4} + 1 \qquad [\tfrac{1}{2}(2)]^2 = 1$$

$$(p + 1)^2 = -\frac{1}{4} \qquad \text{Factor; combine terms.}$$

The square root of $-1/4$ is not a real number so the square root property does not apply. This equation has no real number solution. ∎

3 ▶ Simplify an equation before solving.

▶ Sometimes an equation must be simplified before applying the steps for solving by completing the square. The next example illustrates this.

EXAMPLE 5

Simplifying an Equation Before Completing the Square

Solve $(x + 3)(x - 1) = 2$.

$$(x + 3)(x - 1) = 2$$
$$x^2 + 2x - 3 = 2 \qquad \text{Use FOIL.}$$
$$x^2 + 2x = 5 \qquad \text{Add 3 to both sides.}$$
$$x^2 + 2x + 1 = 5 + 1 \qquad \text{Add 1 to get a perfect square on the left.}$$
$$(x + 1)^2 = 6 \qquad \text{Factor on the left; add on the right.}$$
$$x + 1 = \sqrt{6} \qquad \text{or} \qquad x + 1 = -\sqrt{6} \qquad \text{Square root property}$$
$$x = -1 + \sqrt{6} \qquad \text{or} \qquad x = -1 - \sqrt{6} \qquad \text{Add } -1.$$

The two solutions are $-1 + \sqrt{6}$ and $-1 - \sqrt{6}$. These can be checked by substituting into the original equation. ∎

NOTE The solutions given in Example 5 are *exact*. If we were asked to find approximations using a calculator, we would need to use the square root key, finding that $\sqrt{6} \approx 2.449$. Evaluating the two solutions we find that

$$x \approx 1.449 \quad \text{and} \quad x \approx -3.449.$$

◇ **C O N N E C T I O N S** ◇

The expression "completing the square" can be given a geometric interpretation. For example, to complete the square for $x^2 + 8x$, begin with a square with a side of length x. Add four rectangles of width 1 to the right side and to the bottom. To "complete the square" fill in the bottom right corner with 16 squares of area 1.

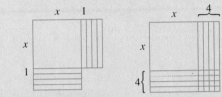

FOR DISCUSSION OR WRITING
1. What is the area of the original square?
2. What is the area of the figure after the 8 rectangles are added?
3. What is the area of the figure after the 16 small squares are added?
4. At what point did we "complete the square"?

9.2 EXERCISES

1. What is the first step that you would perform in order to solve the equation $4x^2 + 8x = 3$ by completing the square?

2. Why is it not possible to solve the equation $2x^3 - x - 1 = 0$ by completing the square?

Find the number that should be added to the expression to make it a perfect square. See Example 2.

3. $y^2 + 14y$ 4. $z^2 + 18z$ 5. $k^2 - 5k$

6. $m^2 - 9m$ 7. $r^2 + \dfrac{1}{2}r$ 8. $s^2 - \dfrac{1}{3}s$

Solve the equation by completing the square. See Examples 1 and 2.

9. $x^2 - 4x = -3$ 10. $y^2 - 2y = 8$ 11. $x^2 + 2x - 5 = 0$

12. $r^2 + 4r + 1 = 0$ 13. $z^2 + 6z + 9 = 0$ 14. $k^2 - 8k + 16 = 0$

15. Which one of the following steps is an appropriate way to begin solving the quadratic equation

$$2x^2 - 4x = 9$$

by completing the square?
(a) Add 4 to both sides of the equation.   (b) Factor the left side as $2x(x - 2)$.
(c) Factor the left side as $x(2x - 4)$.   (d) Divide both sides by 2.

16. In Example 3 of Section 4.5, we solved the quadratic equation

$$4p^2 - 26p + 40 = 0$$

by the factoring method. If we were to solve this by the method of completing the square, would we get the same solutions, $\frac{5}{2}$ and 4?

Solve the equation by completing the square. See Examples 3–5.

17. $2x^2 - 4x - 5 = 0$ **18.** $2x^2 - 6x - 3 = 0$ **19.** $4y^2 + 4y = 3$

20. $9x^2 + 3x = 2$ **21.** $2p^2 - 2p + 3 = 0$ **22.** $3q^2 - 3q + 4 = 0$

23. $3k^2 + 7k = 4$ **24.** $2k^2 + 5k = 1$ **25.** $(x + 3)(x - 1) = 5$

26. $(y - 8)(y + 2) = 24$ **27.** $-x^2 + 2x = -5$ **28.** $-r^2 + 3r = -2$

▦ *Solve the equation by completing the square. Then (a) give the exact solutions and (b) give the solutions rounded to the nearest thousandth.*

29. $3r^2 - 2 = 6r + 3$ **30.** $4p + 3 = 2p^2 + 2p$

31. $(x + 1)(x + 3) = 2$ **32.** $(x - 3)(x + 1) = 1$

Solve the problem.

33. A farmer has a rectangular cattle pen with perimeter 350 feet and area 7500 square feet. What are the dimensions of the pen? *(Hint: Use the diagram to set up the equation.)*

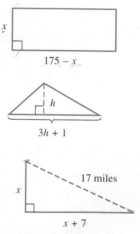

34. The base of a triangle measures 1 meter more than three times the height of the triangle. Its area is 15 square meters. Find the length of the base and the height.

35. Two cars travel at right angles to each other from an intersection until they are 17 miles apart. At that point one car has gone 7 miles farther than the other. How far did the slower car travel?

▦ **36.** Two painters are painting a house in a development of new homes. One of the painters takes 2 hours longer to paint a house working alone than the other painter. When they do the job together, they can complete it in 4.8 hours. How long would it take the faster painter alone to paint the house? (Give your answer to the nearest tenth.)

37. If an object is thrown upward from ground level with an initial velocity of 96 feet per second, its height, s, (in feet) in t seconds is given by the formula $s = -16t^2 + 96t$. How long will it take for the object to be at a height of 80 feet?

▦ **38.** How long will it take the object described in Exercise 37 to be at a height of 100 feet? Round your answers to the nearest tenth.

REVIEW EXERCISES

Write the quotient in lowest terms. Simplify the radicals. See Section 8.5.

39. $\dfrac{8 - 6\sqrt{3}}{6}$ **40.** $\dfrac{4 + \sqrt{28}}{2}$ **41.** $\dfrac{6 - \sqrt{45}}{6}$ **42.** $\dfrac{8 + \sqrt{32}}{4}$

Evaluate the expression $\sqrt{b^2 - 4ac}$ for the given values of a, b, and c. Simplify the radicals. See Sections 1.3, 8.1, and 8.2.

43. $a = 1$, $b = 2$, $c = -4$ **44.** $a = 9$, $b = 30$, $c = 25$

45. $a = 4$, $b = -28$, $c = 49$ **46.** $a = -1$, $b = -2$, $c = 4$

9.3 SOLVING QUADRATIC EQUATIONS BY THE QUADRATIC FORMULA

FOR EXTRA HELP

📖 **SSG** pp. 290–297
SSM pp. 487–497

📼 **Video**
13

💾 **Tutorial**
IBM   MAC

OBJECTIVES

1 ▶ Identify the values of a, b, and c in a quadratic equation.
2 ▶ Use the quadratic formula to solve quadratic equations.
3 ▶ Solve quadratic equations with only one solution.
4 ▶ Solve quadratic equations with fractions.
5 ▶ Use the quadratic formula to solve an applied problem.

◇ **C O N N E C T I O N S** ◇

In the Connections box in Section 5.5, we saw that $\sqrt{2}$ can be evaluated by a *continued fraction*. Now we can use the square root property to show that $\sqrt{2}$ is equivalent to the continued fraction

$$1 + \cfrac{1}{2 + \cfrac{1}{2 + \cfrac{1}{2 + \ddots}}}.$$

Let
$$x = 1 + \cfrac{1}{2 + \cfrac{1}{2 + \ddots}}.$$

Then
$$x = 1 + \cfrac{1}{1 + x}$$

$$x - 1 = \frac{1}{1 + x}$$

$$x^2 - 1 = 1.$$

$$x^2 = 2$$

$$x = \sqrt{2}.$$

(We choose the positive root since $x > 0$ here.)

FOR DISCUSSION OR WRITING

The simplest continued fraction is

$$1 + \cfrac{1}{1 + \cfrac{1}{1 + \ddots}}.$$

Since the quantity under the first numerator 1 is the fraction itself, it can be written as

$$x = 1 + \frac{1}{x}.$$

Solve this equation for the positive value of x. You should get an irrational number. Do you think continued fractions always represent irrational numbers?

Completing the square can be used to solve any quadratic equation, but the method is tedious. In this section we complete the square on the quadratic equation $ax^2 + bx + c = 0$ to get the *quadratic formula*, a formula that gives the solution for any quadratic equation. (Note that $a \neq 0$, or we would have a linear, not a quadratic, equation.)

1 ▶ Identify the values of a, b, and c in a quadratic equation.

▶ The first step in solving a quadratic equation by this new method is to identify the values of a, b, and c in the standard form of the quadratic equation.

EXAMPLE 1
Determining Values of a, b, and c in a Quadratic Equation

For each of the following quadratic equations, write the equation in standard form if necessary, and then identify the values of a, b, and c.

(a) $2x^2 + 3x - 5 = 0$

This equation is already in standard form. The values of a, b, and c are

$$a = 2, \quad b = 3, \quad \text{and} \quad c = -5.$$

(b) $-x^2 + 2 = 6x$

First rewrite the equation with 0 on the right side to match the standard form of $ax^2 + bx + c = 0$.

$$-x^2 + 2 = 6x$$
$$-x^2 - 6x + 2 = 0$$

Now identify $a = -1$, $b = -6$, and $c = 2$. (Notice that the coefficient of x^2 is understood to be -1.)

(c) $(2x - 7)(x + 4) = -23$

$$
\begin{aligned}
(2x - 7)(x + 4) &= -23 \\
2x^2 + x - 28 &= -23 \qquad \text{Use FOIL on the left.} \\
2x^2 + x - 5 &= 0 \qquad \text{Add 23 on each side.}
\end{aligned}
$$

Now, identify the values: $a = 2$, $b = 1$, $c = -5$. ■

2 ▶ Use the quadratic formula to solve quadratic equations.

▶ To develop the quadratic formula, we follow the steps for completing the square on $ax^2 + bx + c = 0$ given in the previous section. For comparison, we also show the corresponding steps for solving $2x^2 + x - 5 = 0$ (from Example 1(c)).

Step 1   Make the coefficient of the squared term equal to 1.

$$2x^2 + x - 5 = 0$$

$$x^2 + \frac{1}{2}x - \frac{5}{2} = 0 \qquad \text{Divide by 2.}$$

$$ax^2 + bx + c = 0$$

$$x^2 + \frac{b}{a}x + \frac{c}{a} = 0 \qquad \text{Divide by } a.$$

Step 2   Get the variable terms alone on the left side.

$$x^2 + \frac{1}{2}x = \frac{5}{2} \qquad \text{Add 5/2.}$$

$$x^2 + \frac{b}{a}x = -\frac{c}{a} \qquad \text{Subtract } c/a$$

Step 3   Add the square of half the coefficient of x to both sides, factor the left side, and combine terms on the right.

$$x^2 + \frac{1}{2}x + \frac{1}{16} = \frac{5}{2} + \frac{1}{16} \qquad \text{Add 1/16.}$$

$$\left(x + \frac{1}{4}\right)^2 = \frac{41}{16} \qquad \begin{array}{l}\text{Factor; add} \\ \text{on right.}\end{array}$$

$$x^2 + \frac{b}{a}x + \frac{b^2}{4a^2} = -\frac{c}{a} + \frac{b^2}{4a^2} \qquad \text{Add } \frac{b^2}{4a^2}.$$

$$\left(x + \frac{b}{2a}\right)^2 = \frac{b^2 - 4ac}{4a^2} \qquad \begin{array}{l}\text{Factor; add} \\ \text{on right.}\end{array}$$

Step 4   Use the square root property to complete the solution.

$$x + \frac{1}{4} = +\sqrt{\frac{41}{16}}$$

$$x + \frac{1}{4} = \pm\frac{\sqrt{41}}{4}$$

$$x = -\frac{1}{4} \pm \frac{\sqrt{41}}{4}$$

$$x = \frac{-1 \pm \sqrt{41}}{4}$$

$$x + \frac{b}{2a} = \pm\sqrt{\frac{b^2 - 4ac}{4a^2}}$$

$$x = -\frac{b}{2a} \pm \frac{\sqrt{b^2 - 4ac}}{2a}$$

$$x = \frac{-b \pm \sqrt{b^2 - 4ac}}{2a}$$

The final result in the column on the right is called the quadratic formula, and it is a key result that should be memorized. Notice that there are two values, one for the $+$ sign and one for the $-$ sign.

Quadratic Formula   The solutions of the quadratic equation $ax^2 + bx + c = 0$, $a \neq 0$, are

$$x = \frac{-b + \sqrt{b^2 - 4ac}}{2a} \qquad \text{and} \qquad x = \frac{-b - \sqrt{b^2 - 4ac}}{2a}$$

or, in compact form,

$$x = \frac{-b \pm \sqrt{b^2 - 4ac}}{2a}.$$

CAUTION   Notice that the fraction bar is under $-b$ as well as the radical. In using this formula, be sure to find the values of $-b \pm \sqrt{b^2 - 4ac}$ first, then divide those results by the value of $2a$.

EXAMPLE 2

Solving a Quadratic
Equation by the
Quadratic Formula

Solve $2x^2 + x - 5 = 0$ by the quadratic formula.

As found in Example 1(c), the values of a, b, and c are $a = 2$, $b = 1$, and $c = -5$. Substitute these numbers into the quadratic formula and simplify the result.

$$x = \frac{-b \pm \sqrt{b^2 - 4ac}}{2a}$$

$$x = \frac{-1 \pm \sqrt{(1)^2 - 4(2)(-5)}}{2(2)} \qquad \text{Let } a = 2, b = 1, c = -5.$$

$$x = \frac{-1 \pm \sqrt{1 + 40}}{4} \qquad \begin{array}{c}\text{Perform the operations under the}\\ \text{radical and in the denominator.}\end{array}$$

$$x = \frac{-1 \pm \sqrt{41}}{4}$$

Notice that this result agrees with the result obtained in the left column when the quadratic formula was derived. The two solutions are $\dfrac{-1 + \sqrt{41}}{4}$ and $\dfrac{-1 - \sqrt{41}}{4}$. ■

EXAMPLE 3

Rewriting an Equation
Before Using the
Quadratic Formula

Solve $x^2 = 2x + 1$.

One side of the equation must be 0 before a, b, and c can be found. Subtract $2x$ and 1 from both sides of the equation to get

$$x^2 - 2x - 1 = 0.$$

Then $a = 1$, $b = -2$, and $c = -1$. Substitute these values into the quadratic formula.

$$x = \frac{-b \pm \sqrt{b^2 - 4ac}}{2a}$$

$$= \frac{-(-2) \pm \sqrt{(-2)^2 - 4(1)(-1)}}{2(1)} \qquad \text{Let } a = 1, b = -2, c = -1.$$

$$= \frac{2 \pm \sqrt{4 + 4}}{2} = \frac{2 \pm \sqrt{8}}{2}$$

$$x = \frac{2 \pm 2\sqrt{2}}{2} \qquad \sqrt{8} = \sqrt{4} \cdot \sqrt{2} = 2\sqrt{2}$$

Write the solutions in lowest terms by factoring $2 \pm 2\sqrt{2}$ as $2(1 \pm \sqrt{2})$ to get

$$x = \frac{2(1 \pm \sqrt{2})}{2} = 1 \pm \sqrt{2}.$$

The two solutions of the given equation are $1 + \sqrt{2}$ and $1 - \sqrt{2}$. ■

3 ▶ Solve quadratic
equations with only one
solution.

▶ When the quantity under the radical, $b^2 - 4ac$, equals zero, the equation has just one rational number solution. In this case, the trinomial $ax^2 + bx + c$ is a perfect square.

EXAMPLE 4

Using the Quadratic Formula When There Is One Solution

Solve $4x^2 + 25 = 20x$.

Write the equation as $4x^2 - 20x + 25 = 0$. Here, $a = 4$, $b = -20$, and $c = 25$. By the quadratic formula,

$$x = \frac{-(-20) \pm \sqrt{(-20)^2 - 400}}{8} = \frac{20 \pm 0}{8} = \frac{5}{2}.$$

Since there is just one solution, $5/2$, we know the trinomial $4x^2 - 20x + 25$ is a perfect square. ■

4 ▶ Solve quadratic equations with fractions.

▶   The next example shows how to solve quadratic equations with fractions.

EXAMPLE 5

Solving a Quadratic Equation with Fractions

Solve the equation $\frac{1}{10}t^2 = \frac{2}{5}t - \frac{1}{2}$.

Eliminate the denominators by multiplying both sides of the equation by the common denominator, 10.

$$10\left(\frac{1}{10}t^2\right) = 10\left(\frac{2}{5}t - \frac{1}{2}\right)$$

$$t^2 = 4t - 5 \qquad \text{Distributive property}$$

$$t^2 - 4t + 5 = 0 \qquad \text{Standard form}$$

From this form identify $a = 1$, $b = -4$, and $c = 5$. Use the quadratic formula to complete the solution.

$$t = \frac{-(-4) \pm \sqrt{(-4)^2 - 4(1)(5)}}{2(1)} \qquad \text{Substitute into the formula.}$$

$$= \frac{4 \pm \sqrt{16 - 20}}{2} \qquad \text{Perform the operations.}$$

$$= \frac{4 \pm \sqrt{-4}}{2}$$

The radical $\sqrt{-4}$ is not a real number, so the equation has no real number solution. ■

5 ▶ Use the quadratic formula to solve an applied problem.

▶   Quadratic equations are needed to solve some applied problems. When we use a quadratic equation to solve an applied problem, it is important to check that the solution satisfies the physical requirements for the unknown. For example, it is impossible to have a negative length, or a fractional number of people.

EXAMPLE 6

Solving an Applied Problem Using the Quadratic Formula

If an object is thrown upward from a height of 50 feet, with an initial velocity of 32 feet per second, then its height after t seconds is given by

$$h = -16t^2 + 32t + 50, \quad \text{where } h \text{ is in feet.}$$

After how many seconds will it reach a height of 30 feet?

We must find the value of t for which $h = 30$.

$$30 = -16t^2 + 32t + 50 \qquad \text{Let } h = 30.$$
$$16t^2 - 32t - 20 = 0 \qquad \text{Rewrite in standard form.}$$
$$4t^2 - 8t - 5 = 0 \qquad \text{Divide by 4.}$$

Use the quadratic formula, with $a = 4$, $b = -8$, and $c = -5$.

$$t = \frac{-b \pm \sqrt{b^2 - 4ac}}{2a}$$

$$t = \frac{-(-8) \pm \sqrt{(-8)^2 - 4(4)(-5)}}{2(4)}$$

$$t = \frac{8 \pm \sqrt{64 + 80}}{8}$$

$$t = \frac{8 \pm \sqrt{144}}{8} = \frac{8 \pm 12}{8}$$

The two solutions of the equation are

$$\frac{8 + 12}{8} = \frac{20}{8} = \frac{5}{2} \quad \text{and} \quad \frac{8 - 12}{8} = -\frac{4}{8} = -\frac{1}{2}.$$

Since t represents time, the solution $-1/2$ must be rejected, since time cannot be negative here. The object will reach a height of 30 feet after $5/2$, or 2.5, seconds. ∎

9.3 EXERCISES

Write the equation in the standard form $ax^2 + bx + c = 0$ if it is not in this form already. Then identify the values of a, b, and c. Do not actually solve the equation. See Example 1.

1. $4x^2 + 5x - 9 = 0$

2. $8x^2 + 3x - 4 = 0$

3. $3x^2 = 4x + 2$

4. $5x^2 = 3x - 6$

5. $3x^2 = -7x$

6. $9x^2 = 8x$

7. $(x - 3)(x + 4) = 0$

8. $(x + 6)^2 = 3$

9. $9(x - 1)(x + 2) = 8$

10. $(3x - 1)(2x + 5) = x(x - 1)$

11. Why is the restriction $a \neq 0$ necessary in the definition of a quadratic equation?

12. Consider the equation $x^2 - 9 = 0$.
   (a) Solve the equation by factoring.
   (b) Solve the equation by the quadratic formula.
   (c) Compare your answers. If a quadratic equation can be solved by both the factoring and the quadratic formula methods, should you always get the same results? Explain.

13. If we were to solve the quadratic equation $-2x^2 - 4x + 3 = 0$, we might choose to use $a = -2$, $b = -4$, and $c = 3$. On the other hand, we might decide to multiply both sides by -1 to begin, obtaining the equation $2x^2 + 4x - 3 = 0$, and then use $a = 2$, $b = 4$, and $c = -3$. Show that in either case, we obtain the same solutions.

14. If we apply the quadratic formula and find that the value of $b^2 - 4ac$ is negative, what can we conclude about the solutions? *no sol*

Use the quadratic formula to solve the equation. Write all radicals in simplified form, and write all answers in lowest terms. See Examples 2–4.

15. $3x^2 + 5x + 1 = 0$

16. $6y^2 - 6y + 1 = 0$

17. $k^2 = -12k + 13$

18. $r^2 = 8r + 9$

19. $p^2 - 4p + 4 = 0$

20. $9x^2 + 6x + 1 = 0$

21. $2x^2 + 12x = -5$

22. $5m^2 + m = 1$

23. $2y^2 = 5 + 3y$

24. $2z^2 = 30 + 7z$

25. $6x^2 + 6x = 0$

26. $4n^2 - 12n = 0$

27. $7x^2 = 12x$

28. $9r^2 = 11r$

29. $x^2 - 24 = 0$

30. $z^2 - 96 = 0$

31. $25x^2 - 4 = 0$

32. $16x^2 - 9 = 0$

33. $3x^2 - 2x + 5 = 10x + 1$

34. $4x^2 - x + 4 = x + 7$

35. $-2x^2 = -3x + 2$

36. $-x^2 = -5x + 20$

37. $2x^2 + x + 5 = 0$

38. $3x^2 + 2x + 8 = 0$

39. $(x + 3)(x + 2) = 15$

40. $(2x + 1)(x + 1) = 7$

Use the quadratic formula to solve the equation. (a) Give the solutions in exact form and (b) use a calculator to give the solutions correct to the nearest thousandth.

41. $2x^2 + 2x = 5$

42. $5x^2 = 3 - x$

43. $x^2 = 1 + x$

44. $x^2 = 2 + 4x$

Use the quadratic formula to solve the equation. See Example 5.

45. $\dfrac{3}{2}k^2 - k - \dfrac{4}{3} = 0$

46. $\dfrac{2}{5}x^2 - \dfrac{3}{5}x - 1 = 0$

47. $\dfrac{1}{2}x^2 + \dfrac{1}{6}x = 1$

48. $\dfrac{2}{3}y^2 - \dfrac{4}{9}y - \dfrac{1}{3}$

49. $.5x^2 = x + .5$

50. $.25x^2 = -1.5x - 1$

51. $\dfrac{3}{8}x^2 - x + \dfrac{17}{24} = 0$

52. $\dfrac{1}{3}y^2 + \dfrac{8}{9}y + \dfrac{7}{9} = 0$

53. If an applied problem leads to a quadratic equation, what must you be aware of after you have solved the equation?

54. Suppose that a problem asks you to find the length of a rectangle, and the problem leads to a quadratic equation. Which one of the following solutions to the equation cannot be an answer to the problem if L represents the length of the rectangle?

   (a) $L = 9$ **(b)** $L = 5\dfrac{1}{4}$ **(c)** $L = \dfrac{1 + \sqrt{5}}{2}$ **(d)** $L = \dfrac{1 - \sqrt{5}}{2}$

55. Solve the formula $S = 2\pi rh + \pi r^2$ for r by first writing it in the form $ar^2 + br + c = 0$, and then using the quadratic formula.

56. Solve the formula $V = \pi r^2 h + \pi R^2 h$ for r, using the method described in Exercise 55.

Solve the problem. See Example 6.

57. For the period from 1985 to 1989, the number of female suicides by firearms in the United States can be approximated by the quadratic model

$$y = -17x^2 + 45x + 2572,$$

where $x = 0$ represents 1985, $x = 1$ represents 1986, and so on. Based on this model, in what year was the number of suicides about 2600? (*Source:* U.S. National Center for Health Statistics, *Vital Statistics of the United States*)

58. The Gross State Product in billions of (current) dollars from 1985 through 1989 can be approximated by the quadratic model

$$y = 18x^2 + 234x + 3950,$$

where $x = 0$ represents 1985, $x = 1$ represents 1986, and so on. Based on this model, in what year would the Gross State Product have been 6002 billion dollars? (*Source:* U.S. Bureau of Economic Analysis)

59. The time t in seconds under certain conditions for a ball to be 48 feet in the air is given (approximately) by

$$48 = 64t - 16t^2.$$

Solve this equation for t. Are both answers reasonable?

60. A certain projectile is located $d = 2t^2 - 5t + 2$ feet from the ground after t seconds have elapsed. How many seconds will it take the projectile to be 14 feet from the ground?

61. Karen and Jessie work at a fast-food restaurant after school. Working alone, Jessie can close up in 1 hour less time than Karen. Together they can close up in $\frac{2}{3}$ of an hour. How long does it take Jessie to close up alone?

62. In a bicycle race over a 12-mile route, Donnie finished 8 minutes ahead of Juan. Donnie pedaled 3 miles per hour faster than Juan. What was Donnie's speed?

63. A rule for estimating the number of board feet of lumber that can be cut from a log depends on the diameter of the log. To find the diameter d required to get 9 board feet of lumber, we use the equation

$$\left(\frac{d - 4}{4}\right)^2 = 9.$$

Solve this equation for d. Are both answers reasonable?

64. An old Babylonian problem asks for the length of the side of a square, given that the area of the square minus the length of a side is 870. Find the length of the side.

———◆ **MATHEMATICAL CONNECTIONS** (Exercises 65–72) ◆———

In Chapter 4 we learned how to factor trinomials. Some trinomials are not factorable using integer coefficients, however. There is a way to determine beforehand whether a trinomial of the form $ax^2 + bx + c$ can be factored. Work through Exercises 65–72 in order.

65. For the trinomial $ax^2 + bx + c$, the expression $b^2 - 4ac$ is called the discriminant. Where have you seen the discriminant before?

66. Each of the following trinomials is factorable. Find the discriminant for each one.
    (a) $18x^2 - 9x - 2$ **(b)** $5x^2 + 7x - 6$
    (c) $48x^2 + 14x + 1$ **(d)** $x^2 - 5x - 24$

67. What do you notice about the discriminants you found in Exercise 66?

68. Factor each of the trinomials in Exercise 66.

69. Each of the following trinomials is not factorable using the methods of Chapter 4. Find the discriminant for each one.
    (a) $2x^2 + x - 5$ **(b)** $2x^2 + x + 5$ **(c)** $x^2 + 6x + 6$ **(d)** $3x^2 + 2x - 9$

70. Are any of the discriminants you found in Exercise 69 perfect squares?

71. Make a conjecture (an educated guess) concerning when a trinomial of the form $ax^2 + bx + c$ is factorable.

72. Use your conjecture and a calculator to determine whether the given trinomial is factorable. (Do not actually factor.)
    (a) $42x^2 + 117x + 66$ **(b)** $99x^2 + 186x - 24$ **(c)** $58x^2 + 184x + 27$

REVIEW EXERCISES

Perform the indicated operations. See Sections 3.1, 3.3, and 3.4.

73. $(4 + 6z) + (-9 + 2z)$ **74.** $(10 - 3t) - (5 - 7t)$

75. $(4 + 3r)(6 - 5r)$ **76.** $(5 + 2x)(5 - 2x)$

SUMMARY: QUADRATIC EQUATIONS

Four methods have now been introduced for solving quadratic equations written in the form $ax^2 + bx + c = 0$. The chart below shows some advantages and some disadvantages of each method.

Method	Advantages	Disadvantages
1. Factoring	Usually the fastest method	Not all equations can be solved by factoring. Some factorable polynomials are hard to factor.
2. Square root property	Simplest method for solving equations of the form $(ax + b)^2 =$ a number	Few equations are given in this form.
3. Completing the square	Can always be used (also, the procedure is useful in other areas of mathematics)	It requires more steps than other methods.
4. Quadratic formula	Can always be used	It is more difficult than factoring because of the $\sqrt{b^2 - 4ac}$ expression.

SUMMARY EXERCISES

Solve the quadratic equation by the method of your choice.

1. $s^2 = 36$

2. $x^2 + 3x = -1$

3. $y^2 - \dfrac{100}{81} = 0$

4. $81t^2 = 49$

5. $z^2 - 4z + 3 = 0$

6. $w^2 + 3w + 2 = 0$

7. $z(z - 9) = -20$

8. $x^2 + 3x - 2 = 0$

9. $(3k - 2)^2 = 9$

10. $(2s - 1)^2 = 10$

11. $(x + 6)^2 = 121$

12. $(5k + 1)^2 = 36$

13. $(3r - 7)^2 = 24$

14. $(7p - 1)^2 = 32$

15. $(5x - 8)^2 = -6$

16. $2t^2 + 1 = t$

17. $-2x^2 = -3x - 2$

18. $-2x^2 + x = -1$

19. $8z^2 = 15 + 2z$

20. $3k^2 = 3 - 8k$

21. $0 = -x^2 + 2x + 1$

22. $3x^2 + 5x = -1$

23. $5y^2 - 22y = -8$

24. $y(y + 6) + 4 = 0$

25. $(x + 2)(x + 1) = 10$

26. $16x^2 + 40x + 25 = 0$

27. $4x^2 = -1 + 5x$

28. $2p^2 = 2p + 1$

29. $3m(3m + 4) = 7$

30. $5x - 1 + 4x^2 = 0$

31. $\dfrac{r^2}{2} + \dfrac{7r}{4} + \dfrac{11}{8} = 0$

32. $t(15t + 58) = -48$

33. $9k^2 = 16(3k + 4)$

34. $\dfrac{1}{5}x^2 + x + 1 = 0$

35. $y^2 - y + 3 = 0$

36. $4m^2 - 11m + 8 = -2$

37. $-3x^2 + 4x = -4$

38. $z^2 - \dfrac{5}{12}z = \dfrac{1}{6}$

39. $5k^2 + 19k = 2k + 12$

40. $\dfrac{1}{2}n^2 - n = \dfrac{15}{2}$

41. $k^2 - \dfrac{4}{15} = -\dfrac{4}{15}k$

42. If $D > 0$ and $\dfrac{5 + \sqrt{D}}{3}$ is a solution of $ax^2 + bx + c = 0$, what must be another solution of the equation?

43. How would you respond to this statement? "Since I know how to solve quadratic equations by the factoring method, there is no reason for me to learn any other method of solving quadratic equations."

44. How many real solutions are there for a quadratic equation that has a negative number as its radicand in the quadratic formula?

9.4 COMPLEX NUMBERS

FOR EXTRA HELP

📖 **SSG** pp. 298–304
SSM pp. 500–505

📼 **Video**
13

💾 **Tutorial**
IBM   MAC

OBJECTIVES

1 ▶ Write complex numbers like $\sqrt{-5}$ as multiples of i.
2 ▶ Add and subtract complex numbers.
3 ▶ Multiply complex numbers.
4 ▶ Write complex number quotients in standard form.
5 ▶ Solve quadratic equations with complex number solutions.

As shown earlier in this chapter, some quadratic equations have no real number solutions. For example, the solution

$$x = \frac{-3 + \sqrt{-5}}{2}$$

is not a real number because of $\sqrt{-5}$. For every quadratic equation to have a solution, we need a new set of numbers that includes the real numbers.

1 ▶ Write complex numbers like $\sqrt{-5}$ as multiples of i.

▶   This new set of numbers is defined using a new number i having the properties given below.

The Number i

$$i = \sqrt{-1} \quad \text{and} \quad i^2 = -1$$

We can now write numbers like $\sqrt{-5}$, $\sqrt{-4}$, and $\sqrt{-8}$ as multiples of i, using a generalization of the product rule for radicals, as in the next example.

EXAMPLE 1
Simplifying Square Roots of Negative Numbers

Write each number as a multiple of i.

(a) $\sqrt{-5} = \sqrt{-1 \cdot 5} = \sqrt{-1} \cdot \sqrt{5} = i\sqrt{5}$

(b) $\sqrt{-4} = \sqrt{-1} \cdot \sqrt{4} = i\sqrt{4} = i \cdot 2 = 2i$

(c) $\sqrt{-8} = i\sqrt{8} = i \cdot 2 \cdot \sqrt{2} = 2i\sqrt{2}$ ■

CAUTION It is easy to mistake $\sqrt{2}\, i$ for $\sqrt{2i}$, with the i under the radical. For this reason, it is customary to write the i factor first when it is multiplied by a radical. For example, we usually write $i\sqrt{2}$ rather than $\sqrt{2}\, i$.

Numbers that are nonzero multiples of i are *imaginary numbers*. The *complex numbers* include all real numbers and all imaginary numbers.

Complex Number

A **complex number** is a number of the form $a + bi$, where a and b are real numbers. If $b \neq 0$, $a + bi$ also is an **imaginary number.**

For example, the real number 2 is a complex number since it can be written as $2 + 0i$. Also, the imaginary number $3i = 0 + 3i$ is a complex number. Other complex numbers are

$$3 - 2i,$$

$$1 + i\sqrt{2},$$

and $$-5 + 4i.$$

In the complex number $a + bi$, a is called the **real part** and b *(not bi)* is called the **imaginary part.**

A complex number written in the form $a + bi$ (or $a + ib$) is in **standard form.** Figure 1 shows the relationships among the various types of numbers discussed in this book. (Compare this figure to Figure 4 in Chapter 1.)

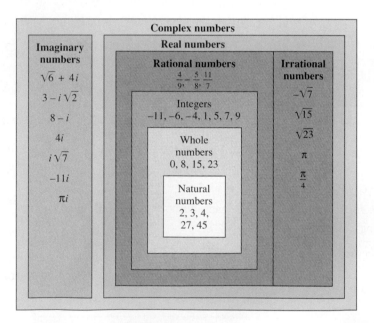

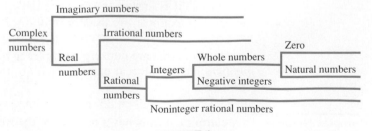

FIGURE 1

2 ▶ Add and subtract complex numbers.

▶ The operations of addition and subtraction of complex numbers are similar to these operations with binomials.

Addition and Subtraction of Complex Numbers	**1.** To add complex numbers, add their real parts and add their imaginary parts.
	2. To subtract complex numbers, change the number following the subtraction sign to its negative, and then add.

The properties of Section 1.9 (commutative, associative, etc.) also hold for operations with complex numbers.

E X A M P L E 2

Adding and Subtracting Complex Numbers

■ Add or subtract.

(a) $(2 - 6i) + (7 + 4i) = (2 + 7) + (-6 + 4)i = 9 - 2i$

(b) $3i + (-2 - i) = -2 + (3 - 1)i = -2 + 2i$

(c) $(2 + 6i) - (-4 + i)$

Change $-4 + i$ to its negative, and then add.

$$(2 + 6i) - (-4 + i) = (2 + 6i) + (4 - i) \qquad -(-4 + i) = 4 - i$$
$$= (2 + 4) + (6 - 1)i \qquad \text{Commutative, associative, and}$$
$$\text{distributive properties}$$
$$= 6 + 5i$$

(d) $(-1 + 2i) - 4 = (-1 - 4) + 2i = -5 + 2i$ ■

3 ▶ Multiply complex numbers.

▶ We multiply complex numbers as we do polynomials. Whenever i^2 appears, we replace it with -1.

E X A M P L E 3

Multiplying Complex Numbers

■ Find the following products.

(a) $3i(2 - 5i) = 6i - 15i^2 \qquad$ Distributive property
$$= 6i - 15(-1) \qquad i^2 = -1$$
$$= 6i + 15$$
$$= 15 + 6i \qquad \text{Commutative property}$$

The last step gives the result in standard form.

(b) $(4 - 3i)(2 + 5i)$

Use FOIL.

$$(4 - 3i)(2 + 5i) = 4(2) + 4(5i) + (-3i)(2) + (-3i)(5i)$$
$$= 8 + 20i - 6i - 15i^2$$
$$= 8 + 14i - 15(-1)$$
$$= 8 + 14i + 15$$
$$= 23 + 14i$$

(c) $(5 - 2i)(5 + 2i) = 25 + 10i - 10i - 4i^2$
$$= 25 - 4(-1)$$
$$= 25 + 4 = 29$$ ■

4 ▶ Write complex number quotients in standard form.

▶ The quotient of two complex numbers is expressed in standard form by changing the denominator into a real number. For example, to write

$$\frac{1 + 3i}{5 - 2i}$$

in standard form, the denominator must be a real number. As seen in Example 3(c), the product $(5 - 2i)(5 + 2i)$ is 29, a real number. This suggests multiplying the numerator and denominator of the given quotient by $5 + 2i$, as follows.

$$\frac{1 + 3i}{5 - 2i} = \frac{1 + 3i}{5 - 2i} \cdot \frac{5 + 2i}{5 + 2i}$$

$$= \frac{5 + 2i + 15i + 6i^2}{25 - 4i^2} \qquad \text{Multiply.}$$

$$= \frac{5 + 17i + 6(-1)}{25 - 4(-1)} \qquad i^2 = -1$$

$$= \frac{5 + 17i - 6}{25 + 4}$$

$$= \frac{-1 + 17i}{29} = -\frac{1}{29} + \frac{17}{29}i$$

The last step gives the result in standard form. Recall that this is the method used to rationalize certain expressions in Chapter 8. The complex numbers $5 - 2i$ and $5 + 2i$ are *conjugates*. That is, the **conjugate** of the complex number $a + bi$ is $a - bi$. Multiplying the complex number $a + bi$ by its conjugate $a - bi$ gives the real number $a^2 + b^2$.

Product of Conjugates

$$(a + bi)(a - bi) = a^2 + b^2$$

The product of a complex number and its conjugate is the sum of the squares of the real and imaginary parts.

EXAMPLE 4

Dividing Complex Numbers

Write the following quotients in standard form.

(a) $\dfrac{-4 + i}{2 - i}$

Multiply numerator and denominator by the conjugate of the denominator, $2 + i$.

$$\frac{-4 + i}{2 - i} \cdot \frac{2 + i}{2 + i} = \frac{-8 - 4i + 2i + i^2}{4 - i^2}$$

$$= \frac{-8 - 2i - 1}{4 - (-1)} \qquad i^2 = -1$$

$$= \frac{-9 - 2i}{5} = -\frac{9}{5} - \frac{2}{5}i$$

We perform the final step in order to get the complex number in standard form.

(b) $\dfrac{3 + i}{-i}$

Here, the conjugate of $0 - i$ is $0 + i$, or i.

$$\frac{3 + i}{-i} \cdot \frac{i}{i} = \frac{3i + i^2}{-i^2}$$

$$= \frac{-1 + 3i}{-(-1)} \qquad i^2 = -1, \text{ commutative property}$$

$$= -1 + 3i \quad \blacksquare$$

◆ C O N N E C T I O N S ◆

The complex number $a + bi$ is also written with the notation $\langle a, b \rangle$. (Note the similarity to an ordered pair.) This notation suggests a way to graph complex numbers on a plane in a manner similar to the way we graph ordered pairs. For graphing complex numbers, the x-axis is called the *real axis* and the y-axis is called the *imaginary axis*. For example, we graph the complex number $2 + 3i$ or $\langle 2, 3 \rangle$ just as we would the ordered pair $(2, 3)$, as shown in the figure. The figure also shows the graphs of the complex numbers $-1 - 4i$, $2i$, and -5.

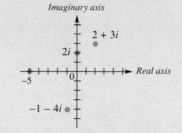

FOR DISCUSSION OR WRITING

1. Give the alternative notation for the last three complex numbers graphed above.
2. What is the real part of the complex number $-1 + 2i$? What is its imaginary part? Explain why we call the axes the real axis and the imaginary axis.

5 ▶ Solve quadratic equations with complex number solutions.

▶ Quadratic equations that have no real solutions do have complex solutions, as shown in the next examples.

EXAMPLE 5 ∎

Solving a Quadratic Equation with Complex Solutions (Square Root Method)

Solve $(x + 3)^2 = -25$ for complex solutions.

Use the square root property.

$$(x + 3)^2 = -25$$

$$x + 3 = \sqrt{-25} \qquad \text{or} \qquad x + 3 = -\sqrt{-25}$$

Since $\sqrt{-25} = 5i$,

$$x + 3 = 5i \qquad \text{or} \qquad x + 3 = -5i$$

$$x = -3 + 5i \qquad \text{or} \qquad x = -3 - 5i.$$

In standard form, the two complex solutions are $-3 + 5i$ and $-3 - 5i$. ∎

EXAMPLE 6

Solving a Quadratic Equation with Complex Solutions (Quadratic Formula)

Solve $2p^2 = 4p - 5$ for complex solutions.

Write the equation as $2p^2 - 4p + 5 = 0$. Then $a = 2$, $b = -4$, and $c = 5$. The solutions are

$$p = \frac{-(-4) \pm \sqrt{(-4)^2 - 4(2)(5)}}{2(2)}$$

$$= \frac{4 \pm \sqrt{16 - 40}}{4}$$

$$= \frac{4 \pm \sqrt{-24}}{4}.$$

Since $\sqrt{-24} = i\sqrt{24} = i \cdot \sqrt{4} \cdot \sqrt{6} = i \cdot 2 \cdot \sqrt{6} = 2i\sqrt{6}$,

$$p = \frac{4 \pm 2i\sqrt{6}}{4}$$

$$= \frac{2(2 \pm i\sqrt{6})}{4} \qquad \text{Factor out a 2.}$$

$$= \frac{2 \pm i\sqrt{6}}{2}. \qquad \text{Lowest terms}$$

In standard form, the solutions are the complex numbers

$$1 + \frac{\sqrt{6}}{2}i \quad \text{and} \quad 1 - \frac{\sqrt{6}}{2}i. \quad \blacksquare$$

9.4 EXERCISES

Decide whether the statement is true or false. If it is false, tell why.

1. $\sqrt{-5} = i\sqrt{5}$ **2.** $i^2 = -1$

3. $\sqrt{i} = -1$ **4.** $-i = 1$

5. $(-i)^2 = -1$ **6.** The conjugate of $1 + 2i$ is $-1 - 2i$.

Write the number as a multiple of i. See Example 1.

7. $\sqrt{-9}$ **8.** $\sqrt{-36}$ **9.** $\sqrt{-20}$ **10.** $\sqrt{-27}$

11. $\sqrt{-18}$ **12.** $\sqrt{-50}$ **13.** $\sqrt{-125}$ **14.** $\sqrt{-98}$

Add or subtract as indicated. See Example 2.

15. $(2 + 8i) + (3 - 5i)$ **16.** $(4 + 5i) + (7 - 2i)$

17. $(8 - 3i) - (2 + 6i)$ **18.** $(1 + i) - (3 - 2i)$

19. $(3 - 4i) + (6 - i) - (3 + 2i)$ **20.** $(5 + 8i) - (4 + 2i) + (3 - i)$

Find the product. See Example 3.

21. $(3 + 2i)(4 - i)$ **22.** $(9 - 2i)(3 + i)$ **23.** $(5 - 4i)(3 - 2i)$

24. $(10 + 6i)(8 - 4i)$ **25.** $(3 + 6i)(3 - 6i)$ **26.** $(11 - 2i)(11 + 2i)$

27. Your friend knows how to multiply binomials using the FOIL method, but does not know how to multiply complex numbers. How would you explain it?

28. Your friend knows how to rationalize the denominator of an expression like $\dfrac{1}{2 + \sqrt{3}}$ but does not know how to divide complex numbers. How would you explain it?

Write the quotient in standard form. See Example 4.

29. $\dfrac{-14 - 2i}{4 + 2i}$

30. $\dfrac{-1 + 5i}{1 + 3i}$

31. $\dfrac{40}{2 + 6i}$

32. $\dfrac{13}{3 + 2i}$

33. $\dfrac{i}{4 - 3i}$

34. $\dfrac{-i}{1 + 2i}$

—————◈ **MATHEMATICAL CONNECTIONS** (Exercises 35–40) ◈—————

When you first learned how to divide whole numbers, you probably also learned that you can check your work by multiplying your answer (the quotient) by the number doing the dividing (the divisor). For example, we know that

$$\frac{2744}{28} = 98 \text{ is true, because } 98 \times 28 = 2744.$$

This same procedure works for real numbers, other than whole numbers, as well. Does it work for complex numbers? Work Exercises 35–40 in order, so that you can determine whether it does or not.

35. Find the standard form of the quotient $\dfrac{-29 - 3i}{2 + 9i}$.

36. Multiply your answer from Exercise 35 by the divisor, $2 + 9i$. What is your answer? Is it equal to the original dividend (the numerator), $-29 - 3i$?

37. Find the standard form of the quotient $\dfrac{4 - 3i}{i}$.

38. Multiply your answer from Exercise 37 by the divisor, i. What is your answer? Is it equal to the original dividend?

39. Use the pattern established in Exercises 35–38 to determine whether the following is true: $\dfrac{14 - 5i}{4 + i} = 3 - 2i$.

40. State a rule, based on your observations, that will tell you whether your answer to a division problem involving complex numbers is correct.

—————————◆—————————

Solve the quadratic equation for complex solutions by the square root property. Write solutions in standard form. See Example 5.

41. $(a + 1)^2 = -4$

42. $(p - 5)^2 = -36$

43. $(k - 3)^2 = -5$

44. $(y + 6)^2 = -12$

45. $(3x + 2)^2 = -18$

46. $(4z - 1)^2 = -20$

Solve the quadratic equation for complex solutions by the quadratic formula. Write solutions in standard form. See Example 6.

47. $m^2 - 2m + 2 = 0$

48. $b^2 + b + 3 = 0$

49. $2r^2 + 3r + 5 = 0$

50. $3q^2 = 2q - 3$

51. $p^2 - 3p + 4 = 0$

52. $2a^2 = -a - 3$

53. $5x^2 + 3 = 2x$

54. $6y^2 + 2y + 1 = 0$

55. $2m^2 + 7 = -2m$

56. $4z^2 + 2z + 3 = 0$

57. $r^2 + 3 = r$

58. $4q^2 - 2q + 3 = 0$

Exercises 59–60 deal with quadratic equations having real number coefficients.

59. Suppose you are solving a quadratic equation by the quadratic formula. How can you tell, before completing the solution, whether the equation will have solutions that are not real numbers?

60. Refer to the solutions in Examples 5 and 6, and complete the following statement: If a quadratic equation has imaginary solutions, they are _____ of each other.

Answer true *or* false *to each of the following. If false, say why.*

61. Every real number is a complex number.

62. Every imaginary number is a complex number.

63. Every complex number is a real number.

64. Some complex numbers are imaginary.

REVIEW EXERCISES

Graph the linear equation. See Section 6.2.

65. $2x - 3y = 6$ **66.** $y = 4x - 3$ **67.** $3x + 5y = 15$

Evaluate the expression if $x = 3$. See Section 1.3.

68. $x^2 - 8$ **69.** $2x^2 - x + 1$ **70.** $(x - 1)^2$

9.5 GRAPHING QUADRATIC EQUATIONS IN TWO VARIABLES

FOR EXTRA HELP

📖 **SSG** pp. 304–310
SSM pp. 505–511

📼 **Video**
13

💾 **Tutorial**
IBM   MAC

OBJECTIVES

1 ▶ Graph quadratic equations of the form $y = ax^2 + bx + c$ $(a \neq 0)$.
2 ▶ Find the vertex of a parabola.
3 ▶ Use a graph to determine the number of real solutions of a quadratic equation.

1 ▶ Graph quadratic equations of the form $y = ax^2 + bx + c$ $(a \neq 0)$.

▶ In Chapter 6 we saw that the graph of a linear equation in two variables is a straight line that represents all the solutions of the equation. Quadratic equations in two variables, of the form $y = ax^2 + bx + c$, are graphed in this section. Perhaps the simplest quadratic equation is $y = x^2$ (or $y = 1x^2 + 0x + 0$). The graph of this equation cannot be a straight line since only linear equations of the form $Ax + By = C$ have graphs that are straight lines. However, $y = x^2$ can be graphed in much the same way that straight lines were graphed, by finding ordered pairs that satisfy the equation $y = x^2$.

EXAMPLE 1

Graphing a Quadratic Equation

■ Graph $y = x^2$.

Select several values for x; then find the corresponding y-values. For example, selecting $x = 2$ gives

$$y = 2^2 = 4,$$

and so the point $(2, 4)$ is on the graph of $y = x^2$. (Recall that in an ordered pair such as $(2, 4)$, the x-value comes first and the y-value second.) We show some ordered pairs that satisfy $y = x^2$ in a table to the side of Figure 2. If the ordered pairs from the table are plotted on a coordinate system and a smooth curve drawn through them, the graph is as shown in Figure 2. ■

x	y
3	9
2	4
1	1
0	0
−1	1
−2	4
−3	9

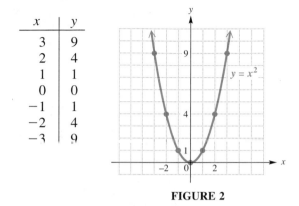

FIGURE 2

The curve in Figure 2 is called a **parabola.** The point (0, 0), the lowest point on this graph, is called the **vertex** of the parabola. The vertical line through the vertex (the y-axis here) is called the **axis** of the parabola.

The axis of a parabola is a line of symmetry for the graph. If the graph is folded on this line, the two halves will match.

Every equation of the form

$$y = ax^2 + bx + c,$$

with $a \neq 0$, has a graph that is a parabola. Because of its many useful properties, the parabola occurs frequently in real-life applications. For example, if an object is thrown into the air, the path that the object follows is a parabola (ignoring wind resistance). The cross sections of radar, spotlight, and telescope reflectors also form parabolas.

◆ **C O N N E C T I O N S** ◇

As shown in the figure, the trajectory of a shell fired from a cannon is a **parabola.** To reach the maximum range with a cannon, it is shown in calculus that the muzzle must be set at 45°. If the muzzle is elevated above 45°, the shell goes too high and falls too soon. If the muzzle is set below 45°, the shell is rapidly pulled to Earth by gravity.

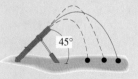

FOR DISCUSSION OR WRITING
If a shell is fired with an initial speed of 32 feet per second, and the muzzle is set at 45°, the equation of the parabolic trajectory is

$$y = x - \frac{1}{32}x^2,$$

where x is the horizontal distance in feet and y is the height in feet. At what distance is the maximum height attained and what is the maximum height? What is the maximum distance to where the shell strikes the ground?

EXAMPLE 2

Graphing a Quadratic Equation with Vertex Not at the Origin

■ Graph $y = -x^2 + 3$.

Find several ordered pairs. Begin with the intercepts. If $x = 0$,

$$y = -x^2 + 3 = -0^2 + 3 = 3,$$

giving the ordered pair $(0, 3)$. If $y = 0$,

$$y = -x^2 + 3$$
$$0 = -x^2 + 3$$
$$x^2 = 3$$
$$x = \sqrt{3} \quad \text{or} \quad -\sqrt{3},$$

giving the two ordered pairs $(-\sqrt{3}, 0)$ and $(\sqrt{3}, 0)$. Now choose additional x-values near the x-values of these three points, and complete the ordered pairs. Some ordered pairs are shown in the table of values next to Figure 3.

The vertex of this parabola is $(0, 3)$. The vertex is the *highest* point of this graph. The graph opens downward because x^2 has a negative coefficient. ■

x	y
-2	-1
$-\sqrt{3}$	0
-1	2
0	3
1	2
$\sqrt{3}$	0
2	-1

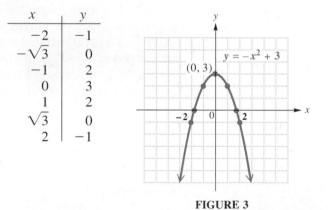

FIGURE 3

2 ▶ Find the vertex of a parabola.

▶ As the graphs above suggest, the vertex is the most important point to locate when you are graphing a quadratic equation. The next example shows how to find the vertex in a more general case.

EXAMPLE 3

Finding the Vertex to Graph a Parabola

■ Graph $y = x^2 - 2x - 3$.

We want to find the vertex of the graph. Note in Figure 3 that the x-value of the vertex is exactly halfway between the x-intercepts. If a parabola has two x-intercepts this is always the case because of the symmetry of the figure. Therefore, let us begin by finding the x-intercepts. Let $y = 0$ in the equation and solve for x.

$$0 = x^2 - 2x - 3$$
$$0 = (x + 1)(x - 3) \qquad \text{Factor.}$$
$$x + 1 = 0 \quad \text{or} \quad x - 3 = 0 \qquad \text{Set each factor equal to 0.}$$
$$x = -1 \quad \text{or} \qquad x = 3$$

There are two x-intercepts, $(-1, 0)$ and $(3, 0)$. Now find any y-intercepts. Substitute $x = 0$ in the equation.

$$y = 0^2 - 2(0) - 3 = -3$$

There is one y-intercept, $(0, -3)$.

As mentioned above, the x-value of the vertex is halfway between the x-values of the two x-intercepts. Thus, it is $1/2$ their sum.

$$x = \frac{1}{2}(-1 + 3) = 1$$

Find the corresponding y-value by substituting 1 for x in the equation.

$$y = 1^2 - 2(1) - 3 = -4$$

The vertex is $(1, -4)$. The axis is the line $x = 1$. Plot the three intercepts and the vertex. Find additional ordered pairs as needed. For example, if $x = 2$,

$$y = 2^2 - 2(2) - 3 = -3,$$

leading to the ordered pair $(2, -3)$. A table of values with the ordered pairs we have found is shown with the graph in Figure 4. ∎

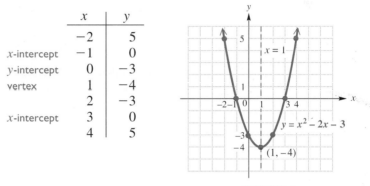

	x	y
	-2	5
x-intercept	-1	0
y-intercept	0	-3
vertex	1	-4
	2	-3
x-intercept	3	0
	4	5

FIGURE 4

We can generalize from Example 3. The x-values of the x-intercepts for the equation $y = ax^2 + bx + c$, by the quadratic formula, are

$$x = \frac{-b + \sqrt{b^2 - 4ac}}{2a} \quad \text{and} \quad x = \frac{-b - \sqrt{b^2 - 4ac}}{2a}.$$

Thus, the x-value of the vertex is

$$x = \frac{1}{2}\left(\frac{-b + \sqrt{b^2 - 4ac}}{2a} + \frac{-b - \sqrt{b^2 - 4ac}}{2a}\right)$$

$$x = \frac{1}{2}\left(\frac{-b + \sqrt{b^2 - 4ac} - b - \sqrt{b^2 - 4ac}}{2a}\right)$$

$$x = \frac{1}{2}\left(\frac{-2b}{2a}\right) = -\frac{b}{2a}.$$

For the equation in Example 3, $y = x^2 - 2x - 3$, $a = 1$, and $b = -2$. Thus, the x-value of the vertex is

$$x = -\frac{b}{2a} = -\frac{-2}{2(1)} = 1,$$

which is the same x-value for the vertex we found in Example 3. (The x-value of the vertex is $x = -\dfrac{b}{2a}$ even if the graph has no x-intercepts.) A procedure for graphing quadratic equations follows.

Graphing the Parabola
$y = ax^2 + bx + c$

Step 1 Find the y-intercept.

Step 2 Find any x-intercepts.

Step 3 Find the vertex. Let $x = -\dfrac{b}{2a}$ and find the corresponding y-value by substituting for x in the equation.

Step 4 Plot the intercepts and the vertex.

Step 5 Find and plot additional ordered pairs near the vertex and intercepts as needed, using the symmetry about the axis of the parabola.

E X A M P L E   4

Using the Steps to Graph a Parabola

Graph $y = x^2 - 4x + 1$.

Find the intercepts. Let $x = 0$ in $y = x^2 - 4x + 1$ to get the y-intercept $(0, 1)$. Let $y = 0$ to get the x-intercepts. If $y = 0$, the equation is $0 = x^2 - 4x + 1$, which cannot be solved by factoring. Using the quadratic formula to solve for x, we get

$$x = \frac{2(2 \pm \sqrt{3})}{2} = 2 \pm \sqrt{3}$$

A calculator shows that the x-intercepts are $(3.7, 0)$ and $(.3, 0)$ to the nearest tenth. The x-value of the vertex is

$$x = -\frac{b}{2a} = -\frac{-4}{2(1)} = 2.$$

The y-value of the vertex is

$$y = 2^2 - 4(2) + 1 = -3,$$

so the vertex is $(2, -3)$. The axis is the line $x = 2$.

A table of values of the points found so far, along with some others, is shown with the graph. Join these points with a smooth curve, as shown in Figure 5. ■

x	y
-1	6
0	1
$2 - \sqrt{3} \approx .3$	0
1	-2
2	-3
3	-2
$2 + \sqrt{3} \approx 3.7$	0
4	1
5	6

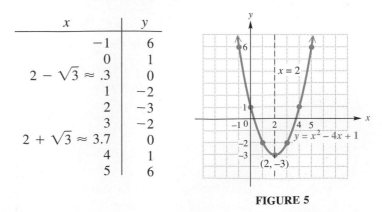

FIGURE 5

3 ▶ Use a graph to determine the number of real solutions of a quadratic equation.

▶ It can be verified by the vertical line test (Section 6.6) that the graph of an equation of the form $y = ax^2 + bx + c$ is the graph of a function. A function defined by an equation of the form $f(x) = ax^2 + bx + c$ $(a \neq 0)$ is called a **quadratic function.** The domain of a quadratic function is all real numbers; the range can be determined after the function is graphed.

Look again at Figure 5, which gives the graph of $y = x^2 - 4x + 1$. Recall that setting y equal to 0 gives the x-intercepts, where the x-values are

$$2 + \sqrt{3} \approx 3.7 \quad \text{and} \quad 2 - \sqrt{3} \approx .3.$$

The solutions of $0 = x^2 - 4x + 1$ are the x-values of the x-intercepts of the graph of the corresponding quadratic function.

**Intercepts of the
Graph of a Quadratic
Function**

The real number solutions of a quadratic equation $ax^2 + bx + c = 0$ are the x-values of the x-intercepts of the graph of the corresponding quadratic function $y = ax^2 + bx + c$.

Since the graph of a quadratic function can cross the x-axis in in two, one, or no points, this result shows why some quadratic equations have two, some have one, and some have no real solutions.

E X A M P L E 5

Determining the Number
of Real Solutions from
a Graph

(a) Figure 6 shows the graph of $y = x^2 - 3$. The equation $0 = x^2 - 3$ has two real solutions, $\sqrt{3}$ and $-\sqrt{3}$, which correspond to the x-intercepts.

(b) Figure 7 shows the graph of $y = x^2 - 4x + 4$. The equation $0 = x^2 - 4x + 4$ has one real solution, 2, which corresponds to the x-intercept.

(c) Figure 8 shows the graph of $y = x^2 + 2$. The equation $0 = x^2 + 2$ has no real solutions, since there are no x-intercepts. (The equation *does* have two imaginary solutions, $i\sqrt{2}$ and $-i\sqrt{2}$.) ∎

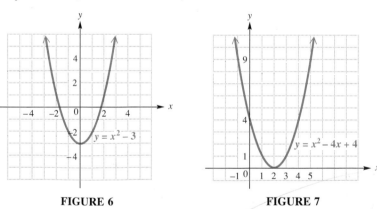

FIGURE 6   FIGURE 7

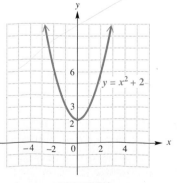

FIGURE 8

9.5 EXERCISES

1. In your own words, explain what is meant by the vertex of a parabola.

2. In your own words, explain what is mean by the line of symmetry of a parabola.

Sketch the graph of the equation and give the coordinates of the vertex. See Examples 1–4.

3. $y = 2x^2$

4. $y = 3x^2$

5. $y = x^2 + 2x + 3$

6. $y = x^2 - 4x + 3$

7. $y = x^2 - 4$

8. $y = x^2 - 6$

9. $y = 2 - x^2$

10. $y = 4 - x^2$

11. $y = (x + 3)^2$

12. $y = (x - 4)^2$

13. $y = -x^2 + 6x - 5$

14. $y = -x^2 - 4x - 3$

15. $y = -x^2 + 4x - 4$

16. $y = -x^2 - 2x - 1$

Decide from the graph how many real number solutions the corresponding equation has.
Find any real solutions from the graph. See Example 5.

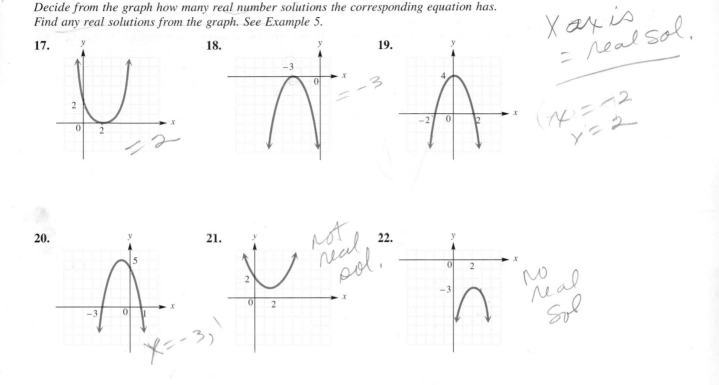

17.

18.

19.

20.

21.

22.

A quadratic equation can be solved graphically by using a graphics calculator. We enter the
equation and use the root-finding capabilities of the calculator. For example, the accompanying figures show two views of the graph of $y = x^2 - 5x - 6$. Notice that the displays
at the bottom of the screen indicate that the roots (or solutions) are $x = -1$ and $x = 6$.

*For the given quadratic equation (**a**) solve by using the quadratic formula and (**b**) graph
the left side as y_1 and use the root-finding capabilities of the calculator to support your
answer.*

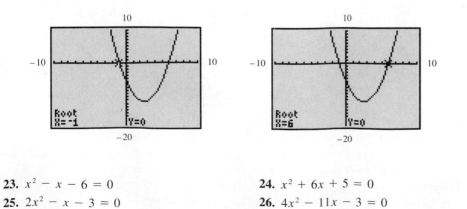

23. $x^2 - x - 6 = 0$

24. $x^2 + 6x + 5 = 0$

25. $2x^2 - x - 3 = 0$

26. $4x^2 - 11x - 3 = 0$

If a quadratic equation is not in standard form, such as $2x^2 = 9x + 5$, we can solve the equation graphically by graphing the left side as y_1, the right side as y_2, and then using the intersection-of-graphs capabilities of the calculator. The two views shown here indicate that the solutions are $-1/2$ and 5, the x-coordinates of the points of intersection. This can be determined algebraically by solving the equation using one of the methods described in this chapter, or by factoring.

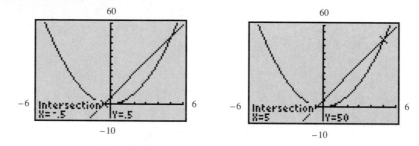

For the given quadratic equation (**a**) solve by first writing the equation in standard form and using the quadratic formula and (**b**) graph the left side as y_1 and the right side as y_2, and use the method described above to support your answer.

27. $x^2 = -2x + 8$ **28.** $x^2 = x + 6$

29. $x^2 = 6x$ **30.** $x^2 = -5x$

31. In the standard viewing window of your calculator, graph the following one at a time, leaving the previous graphs on the screen as you move along.

$$y_1 = x^2$$
$$y_2 = 2x^2$$
$$y_3 = 3x^2$$
$$y_4 = 4x^2$$

Describe the effect the successive coefficients have on the parabola.

32. Repeat Exercise 31 for the following.

$$y_1 = x^2$$
$$y_2 = \frac{1}{2}x^2$$
$$y_3 = \frac{1}{4}x^2$$
$$y_4 = \frac{1}{8}x^2$$

33. Graph the pair of parabolas $y_1 = x^2$ and $y_2 = -x^2$ on the same screen. In your own words, describe how the graph of y_2 can be obtained from the graph of y_1.

34. Graph $y_1 = -x^2$, $y_2 = -2x^2$, $y_3 = -3x^2$ and $y_4 = -4x^2$ on the same screen. Make a conjecture about what happens when the coefficient of x^2 is negative.

35. In the standard viewing window of your calculator, graph the following one at a time, leaving the previous graphs on the screen as you move along.

$$y_1 = x^2$$
$$y_2 = x^2 + 3$$
$$y_3 = x^2 - 6$$

Describe the effect that adding or subtracting a constant has on the parabola.

36. Repeat Exercise 35 for the following.

$$y_1 = x^2$$
$$y_2 = (x + 3)^2$$
$$y_3 = (x - 6)^2$$

37. This table was generated by an equation of the form $y_1 = ax^2 + bx + c$. Answer these questions by referring to the table.
(a) What is the y-intercept of the graph?
(b) What are the x-intercepts of the graph?

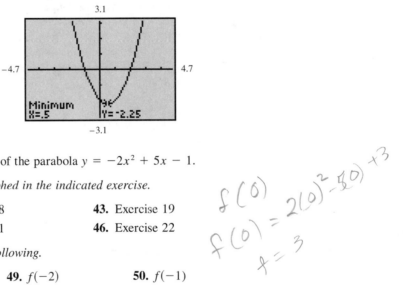

X	Y1
-4	18
-3	10
-2	4
-1	0
0	-2
1	-2
2	0

X=0

38. If we know the x-intercepts of the graph of a parabola, then we can find the x-coordinate of the vertex by finding the average of these x-coordinates. What is the x-coordinate of the vertex of the parabola that corresponds to the table in Exercise 37?

39. The graph shown here is that of the equation from Exercise 37. The calculator has the capability of finding the vertex. If the parabola opens upward, as it does in this case, the vertex is a *minimum*. The equation is $y = x^2 - x - 2$. Show *algebraically* that the point $(.5, -2.25)$ lies on the graph, by letting $x = .5$ and obtaining $y = -2.25$.

3.1

−4.7   4.7

Minimum
X=.5   Y=-2.25

−3.1

40. Use a graphics calculator to find the vertex of the parabola $y = -2x^2 + 5x - 1$.

Find the domain and range of the function graphed in the indicated exercise.

41. Exercise 17   **42.** Exercise 18   **43.** Exercise 19

44. Exercise 20   **45.** Exercise 21   **46.** Exercise 22

Given $f(x) = 2x^2 - 5x + 3$, find each of the following.

47. $f(0)$ **48.** $f(1)$ **49.** $f(-2)$ **50.** $f(-1)$

CHAPTER 9 SUMMARY

KEY TERMS		NEW SYMBOLS
9.4 complex number	**9.5** parabola	$\pm$ positive or negative
imaginary number	vertex	i $i = \sqrt{-1}$ and $i^2 = -1$
real part	axis	
imaginary part	quadratic function	
standard form (of a complex number)		
conjugate		

QUICK REVIEW

CONCEPTS	EXAMPLES

9.1 SOLVING QUADRATIC EQUATIONS BY THE SQUARE ROOT PROPERTY

Square Root Property of Equations If k is positive, and if $a^2 = k$, then $a = \sqrt{k}$ or $a = -\sqrt{k}$.	Solve $(2x + 1)^2 = 5$. $$2x + 1 = \pm\sqrt{5}$$ $$2x = -1 \pm \sqrt{5}$$ $$x = \frac{-1 \pm \sqrt{5}}{2}$$

9.2 SOLVING QUADRATIC EQUATIONS BY COMPLETING THE SQUARE

Completing the Square 1. If the coefficient of the squared term is 1, go to Step 2. If it is not 1, divide each side of the equation by this coefficient. 2. Make sure that all variable terms are on one side of the equation and all constant terms are on the other. 3. Take half the coefficient of x, square it, and add the square to each side of the equation. Factor the variable side and combine terms on the other. 4. Use the square root property to solve the equation.	Solve $2x^2 + 4x - 1 = 0$. **1.** $x^2 + 2x - \dfrac{1}{2} = 0$ **2.** $x^2 + 2x = \dfrac{1}{2}$ **3.** $x^2 + 2x + 1 = \dfrac{1}{2} + 1$ $\quad (x + 1)^2 = \dfrac{3}{2}$ **4.** $x + 1 = \pm\sqrt{\dfrac{3}{2}} = \pm\dfrac{\sqrt{6}}{2}$ $\quad x = -1 \pm \dfrac{\sqrt{6}}{2}$ $\quad x = \dfrac{-2 \pm \sqrt{6}}{2}$

9.3 SOLVING QUADRATIC EQUATIONS BY THE QUADRATIC FORMULA

Quadratic Formula The solutions of $ax^2 + bx + c = 0$, $(a \neq 0)$, are $$x = \frac{-b \pm \sqrt{b^2 - 4ac}}{2a}.$$	Solve $3x^2 - 4x - 2 = 0$. $$x = \frac{-(-4) \pm \sqrt{(-4)^2 - 4(3)(-2)}}{2(3)}$$ $$x = \frac{4 \pm \sqrt{16 + 24}}{6}$$ $$x = \frac{4 \pm \sqrt{40}}{6} = \frac{4 \pm 2\sqrt{10}}{6}$$ $$x = \frac{2(2 \pm \sqrt{10})}{6} = \frac{2 \pm \sqrt{10}}{3}$$

9.4 COMPLEX NUMBERS

The number $i = \sqrt{-1}$ and $i^2 = -1$. For the positive number b, $$\sqrt{-b} = i\sqrt{b}.$$	Simplify: $\sqrt{-19}$. $$\sqrt{-19} = \sqrt{-1 \cdot 19} = i\sqrt{19}$$

CONCEPTS	EXAMPLES
Add complex numbers by adding the real parts and adding the imaginary parts.	Add: $(3 + 6i) + (-9 + 2i)$. $(3 + 6i) + (-9 + 2i) = (3 - 9) + (6 + 2)i$ $= -6 + 8i$
To subtract complex numbers, change the number following the subtraction sign to its negative and add.	Subtract: $(5 + 4i) - (2 - 4i)$. $(5 + 4i) - (2 - 4i) = (5 + 4i) + (-2 + 4i)$ $= (5 - 2) + (4 + 4)i$ $= 3 + 8i$
Multiply complex numbers in the same way polynomials are multiplied. Replace i^2 with -1.	Multiply: $(7 + i)(3 - 4i)$. $(7 + i)(3 - 4i)$ $= 7(3) + 7(-4i) + i(3) + i(-4i)$ FOIL $= 21 - 28i + 3i - 4i^2$ $= 21 - 25i - 4(-1)$ $i^2 = -1$ $= 21 - 25i + 4$ $= 25 - 25i$
Divide complex numbers by multiplying the numerator and the denominator by the conjugate of the denominator.	Divide: $\dfrac{2}{6 + i}$. $$\frac{2}{6 + i} = \frac{2}{6 + i} \cdot \frac{6 - i}{6 - i}$$ $$= \frac{2(6 - i)}{36 - i^2}$$ $$= \frac{12 - 2i}{36 + 1}$$ $$= \frac{12 - 2i}{37}$$ $$= \frac{12}{37} - \frac{2}{37}i$$
A quadratic equation may have complex solutions. The quadratic formula will give complex solutions in such cases.	Solve for all complex solutions of $$x^2 + x + 1 = 0.$$ Here, $a = 1$, $b = 1$, and $c = 1$. $$x = \frac{-1 \pm \sqrt{1^2 - 4(1)(1)}}{2(1)}$$ $$x = \frac{-1 \pm \sqrt{1 - 4}}{2}$$ $$x = \frac{-1 \pm \sqrt{-3}}{2}$$ $$x = \frac{-1 \pm i\sqrt{3}}{2}$$ The solutions are $$-\frac{1}{2} + \frac{\sqrt{3}}{2}i \quad \text{and} \quad -\frac{1}{2} - \frac{\sqrt{3}}{2}i.$$

CONCEPTS	EXAMPLES

9.5 GRAPHING QUADRATIC EQUATIONS IN TWO VARIABLES

Graphing $y = ax^2 + bx + c$ **1.** Find the y-intercept.	Graph $y = 2x^2 - 5x - 3$. **1.** $y = 2(0)^2 - 5(0) - 3 = -3$ The y-intercept is $(0, -3)$.
2. Find any x-intercepts.	**2.** $\quad\quad\quad\quad 0 = 2x^2 - 5x - 3$ $\quad\quad\quad\quad\quad 0 = (2x + 1)(x - 3)$ $2x + 1 = 0 \quad$ or $\quad x - 3 = 0$ $\quad 2x = -1 \quad$ or $\quad\quad x = 3$ $\quad\quad x = -\dfrac{1}{2} \quad$ or $\quad\quad x = 3$ The x-intercepts are $\left(-\dfrac{1}{2}, 0\right)$ and $(3, 0)$.
3. Find the vertex: $x = -\dfrac{b}{2a}$; find y by substituting this value for x in the equation.	**3.** For the vertex: $$x = -\frac{b}{2a} = -\frac{-5}{2(2)} = \frac{5}{4}$$ $$y = 2\left(\frac{5}{4}\right)^2 - 5\left(\frac{5}{4}\right) - 3$$ $$y = 2\left(\frac{25}{16}\right) - \frac{25}{4} - 3$$ $$y = \frac{25}{8} - \frac{50}{8} - \frac{24}{8} = -\frac{49}{8} = -6\frac{1}{8}.$$ The vertex is $\left(\dfrac{5}{4}, -\dfrac{49}{8}\right)$.
4. Plot the intercepts and the vertex. **5.** Find and plot additional ordered pairs near the vertex and intercepts as needed. The number of real solutions of the equation $$ax^2 + bx + c = 0$$	**4. and 5.**
can be determined from the number of x-intercepts of the graph of $$y = ax^2 + bx + c.$$	The figure shows that the equation $$2x^2 - 5x - 3 = 0$$ has 2 real solutions.

CHAPTER 9 REVIEW EXERCISES

[9.1] *Solve the equation by using the square root property. Give only real number solutions. Express all radicals in simplest form.*

1. $y^2 = 144$

2. $x^2 = 37$

3. $m^2 = 128$

4. $(k + 2)^2 = 25$

5. $(r - 3)^2 = 10$

6. $(2p + 1)^2 = 14$

7. $(3k + 2)^2 = -3$

8. Which one of the following equations has two real solutions?
   (a) $x^2 = 0$ **(b)** $x^2 = -4$ **(c)** $(x + 5)^2 = -16$ **(d)** $(x + 6)^2 = 25$

[9.2] *Solve the equation by completing the square. Give only real number solutions.*

9. $m^2 + 6m + 5 = 0$

10. $p^2 + 4p = 7$

11. $x^2 + 5 = 2x$

12. $2y^2 - 3 = -8y$

13. $5k^2 - 3k - 2 = 0$

14. $(4a + 1)(a - 1) = -7$

Solve the problem.

15. If an object is thrown upward from a height of 50 feet, with an initial velocity of 32 feet per second, then its height after t seconds is given by $h = -16t^2 + 32t + 50$, where h is in feet. After how many seconds will it reach a height of 30 feet?

16. For the years 1987–1989, the number of computer and information science doctoral degrees conferred in the United States was given by the equation $y = 28x^2 + 26x + 374$, where $x = 0$ corresponds to 1987. How many degrees of this type were conferred in 1989? (*Source:* U.S. National Center for Education Statistics, *Digest of Education Statistics*)

17. What must be added to $x^2 + kx$ so that it will become a perfect square?

[9.3]

18. What would happen if you were to try to apply the quadratic formula to the linear equation $4x + 3 = 0$?

Solve the equation by using the quadratic formula. Give only real number solutions.

19. $x^2 - 2x - 4 = 0$

20. $3k^2 + 2k = -3$

21. $2p^2 + 8 = 4p + 11$

22. $-4x^2 + 7 = 2x$

23. $\frac{1}{4}p^2 = 2 - \frac{3}{4}p$

24. $x(5x - 1) = 1$

Solve the problem.

25. Use a calculator and the Pythagorean formula to find the lengths of the sides of the triangle to the nearest thousandth.

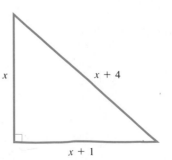

26. Use the equation in Exercise 16 to predict in what year the number of degrees would reach 926, assuming that the same trends continue.

[9.4]

27. Write an explanation of the method used to divide complex numbers.

28. Use the fact that $i^2 = -1$ to complete each of the following. Do them in order.
  (a) Since $i^3 = i^2 \cdot i$, $i^3 = $ _____ .
  (b) Since $i^4 = i^3 \cdot i$, $i^4 = $ _____ .
  (c) Since $i^{48} = (i^4)^{12}$, $i^{48} = $ _____ .

Perform the indicated operation.

29. $(3 + 5i) + (2 - 6i)$ **30.** $(-2 - 8i) - (4 - 3i)$ **31.** $(-1 + i) - (2 - i)$

32. $(4 + 3i) + (-2 + 3i)$ **33.** $(6 - 2i)(3 + i)$ **34.** $(2 + 3i)(4 - 2i)$

35. $(5 + 2i)(5 - 2i)$ **36.** $(8 - i)(8 + i)$ **37.** $\dfrac{1 + i}{1 - i}$

38. $\dfrac{19 + 13i}{3 + i}$ **39.** $\dfrac{1}{7 - i}$ **40.** $\dfrac{5 + 6i}{2 + 3i}$

41. What is the conjugate of the real number a?

42. Is it possible to multiply a complex number by its conjugate and get an imaginary product? Explain.

Find the complex solutions of the quadratic equation.

43. $(m + 2)^2 = -3$ **44.** $(x - 1)^2 = -2$ **45.** $(3p - 2)^2 = -8$ **46.** $(4p + 1)^2 = -12$

47. $3k^2 = 2k - 1$ **48.** $h^2 + 3h = -8$ **49.** $4q^2 + 2 = 3q$ **50.** $9z^2 + 2z + 1 = 0$

[9.5] *Sketch the graph of the equation and identify the vertex.*

51. $y = -3x^2$ **52.** $y = x^2 - 2x + 1$ **53.** $y = -x^2 + 5$

54. $y = -x^2 + 2x + 3$ **55.** $y = x^2 + 4x + 2$ **56.** $y = (x + 4)^2$

Decide from the graph how many real number solutions the corresponding equation has. Find any real solutions from the graph.

57.   **58.**

59.   **60.**

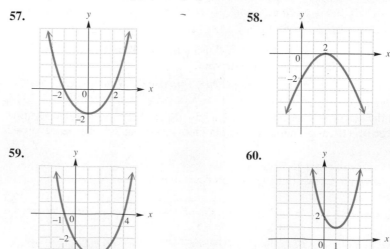

61. 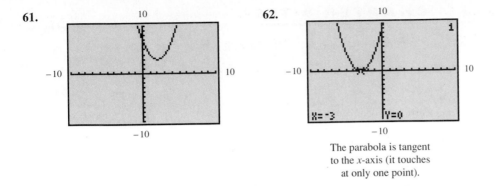 **62.**

The parabola is tangent
to the x-axis (it touches
at only one point).

MIXED REVIEW EXERCISES

Solve by any method. Give only real number solutions.

63. $(2t - 1)(t + 1) - 54$

64. $(2p + 1)^2 - 100$

65. $(k + 2)(k - 1) = 3$

66. $6t^2 + 7t - 3 = 0$

67. $2x^2 + 3x + 2 = x^2 - 2x$

68. $x^2 + 2x + 5 = 7$

69. $m^2 - 4m + 10 = 0$

70. $k^2 - 9k + 10 = 0$

71. $(3x + 5)^2 = 0$

72. $\frac{1}{2}r^2 = \frac{7}{2} - r$

73. $x^2 + 4x = 1$

74. $7x^2 - 8 = 5x^2 + 8$

◆ **MATHEMATICAL CONNECTIONS** (Exercises 75–80) ◇

*In courses such as Intermediate Algebra and College Algebra, we learn that if r and s are
solutions of the equation $x^2 + bx + c = 0$, then $x^2 + bx + c$ factors as $(x - r)(x - s)$.
For example, since 2 and 5 are solutions of $x^2 - 7x + 10 = 0$, we have*

$$x^2 - 7x + 10 = (x - 2)(x - 5).$$

*In Chapter 4 we learned various methods of factoring polynomials using integer
coefficients. Now, with the property stated above, we can factor any trinomial of the form
$x^2 + bx + c$. Work Exercises 75–80 in order.*

75. Solve the quadratic equation $x^2 - 2x - 2 = 0$ using the quadratic formula. It has two
real solutions, both of which are irrational numbers.

76. Suppose that r and s represent your solutions from Exercise 75. Write the trinomial
$x^2 - 2x - 2$ in the factored form $(x - r)(x - s)$. Use parentheses carefully.

77. Regroup the terms in the factors obtained in Exercise 76 so that in each factor, the first
two terms are grouped. These "binomials within binomials" should be the same in each
factor.

78. Multiply your factors in Exercise 77 by using the special product $(a + b)(a - b) =
a^2 - b^2$. Then simplify by using the special product $(a + b)^2 = a^2 + 2ab + b^2$.
Combine terms.

79. Compare your answer in Exercise 78 to the trinomial given in Exercise 76. They should
be the same.

80. Use the method of Exercises 75–79 to factor $x^2 - 4x - 1$. (This process is called
factoring over the real numbers.)

◆

CHAPTER 9 TEST

Items marked * require knowledge of complex numbers.

Solve by using the square root property.

1. $x^2 = 39$ **2.** $(y + 3)^2 = 64$ **3.** $(4x + 3)^2 = 24$

Solve by completing the square.

4. $x^2 - 4x = 6$ **5.** $2x^2 + 12x - 3 = 0$

6. For a quadratic equation to have two real solutions, what must be true about the quantity under the radical in the quadratic formula?

Solve by the quadratic formula.

7. $2x^2 + 5x - 3 = 0$ **8.** $3w^2 + 2 = 6w$

*9. $4x^2 + 8x + 11 = 0$ **10.** $t^2 - \dfrac{5}{3}t + \dfrac{1}{3} = 0$

Solve by the method of your choice.

11. $p^2 - 2p - 1 = 0$ **12.** $(2x + 1)^2 = 18$
13. $(x - 5)(2x - 1) = 1$ **14.** $t^2 + 25 = 10t$

Solve the problem.

15. If a ball is thrown into the air from ground level with an initial velocity of 64 feet per second, its height s (in feet) after t seconds is given by the formula $s = -16t^2 + 64t$. After how many seconds will the ball reach a height of 64 feet?

16. Which one of these equations has exactly one real number solution?
  (a) $x^2 = 8$ **(b)** $y^2 = -8$ **(c)** $(x - 8)^2 = 1$ **(d)** $t^2 = 0$

Perform the indicated operations.

*17. $(3 + i) + (-2 + 3i) - (6 - i)$ ***18.** $(6 + 5i)(-2 + i)$

*19. $(3 - 8i)(3 + 8i)$ ***20.** $\dfrac{15 - 5i}{7 + i}$

21. Find the value of x:

Sketch the graph of the equation. Identify the vertex.

22. $y = (x - 3)^2$ **23.** $y = -x^2 - 2x - 4$ **24.** $y = x^2 + 6x + 7$
25. Refer to the equation in Exercise 24, and do the following:
  (a) Determine the number of real solutions of $x^2 + 6x + 7 = 0$ by looking at the graph.
  (b) Use the quadratic formula to find the exact values of the real solutions.
  (c) Graph $y = x^2 + 6x + 7$ in the standard window of a graphics calculator, and use the root-finding capabilities of the calculator to approximate the solutions. Round to the nearest thousandth.

CUMULATIVE REVIEW (Chapters 1–9)

Note: This cumulative review exercise set may be considered as a final examination for the course.

Perform the operations.

1. $\dfrac{-4 \cdot 3^2 + 2 \cdot 3}{2 - 4 \cdot 1}$

2. $-9 - (-8)(2) + 6 - (6 + 2)$

3. $|-3| - |1 - 6|$

4. $-4r + 14 + 3r - 7$

5. $13k - 4k + k - 14k + 2k$

6. $5(4m - 2) - (m + 7)$

Solve the equation.

7. $6x - 5 = 13$

8. $3k - 9k - 8k + 6 = -64$

9. $2(m - 1) - 6(3 - m) = -4$

Solve the problem.

10. Find the measures of the marked angles.

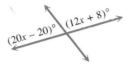

$(20x - 20)°$ $(12x + 8)°$

11. A video rental establishment displayed a rectangular cardboard standup advertisement for the movie *Forrest Gump*. The length was 20 inches more than the width, and the perimeter was 176 inches. What were the dimensions of the rectangle?

12. Solve the formula $P = 2L + 2W$ for L.   CR

Solve the inequality. Graph the solution.

13. $-8m < 16$

14. $-9p + 2(8 - p) - 6 \geq 4p - 50$ CR

Simplify the expression. Write answers with positive exponents.

quest.

15. $(3^2 \cdot x^{-4})^{-1}$ CR

16. $\left(\dfrac{b^{-3}c^4}{b^5c^3}\right)^{-2}$ CR

17. $\left(\dfrac{5}{3}\right)^{-3}$

Perform the indicated operation.

18. $(5x^5 - 9x^4 + 8x^2) - (9x^2 + 8x^4 - 3x^5)$

19. $(2x - 5)(x^3 + 3x^2 - 2x - 4)$ CR

20. $(5t + 9)^2$

21. $\dfrac{3x^3 + 10x^2 - 7x + 4}{x + 4}$ CR

Factor.

common terms

22. $16x^3 - 48x^2y$

23. $16x^4 - 1$ CR

24. $2a^2 - 5a - 3$

25. $25m^2 - 20m + 4$

factoring

factor of solving plus solving

26. Solve by factoring: $x^2 + 3x - 54 = 0$.

27. If an object is dropped, the distance d in feet it falls in t seconds is given by the formula $d = 16t^2$. How long will it take for an object to fall 100 feet?

28. Find a polynomial representing the area of the shaded region.

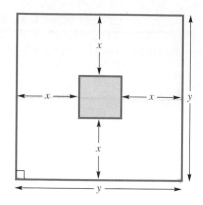

Perform the operation. Write the answer in lowest terms.

29. $\dfrac{2}{a-3} \div \dfrac{5}{2a-6}$

30. $\dfrac{1}{k} - \dfrac{2}{k-1}$

31. $\dfrac{2}{a^2-4} + \dfrac{3}{a^2-4a+4}$

32. $\dfrac{\dfrac{1}{a}+\dfrac{1}{b}}{\dfrac{1}{a}-\dfrac{1}{b}}$

33. Solve $\dfrac{1}{x+3} + \dfrac{1}{x} = \dfrac{7}{10}$.

34. The pressure exerted by a certain liquid at a given point varies directly as the depth of the point beneath the surface of the liquid. If the pressure at 10 feet is 50 pounds per square inch, what is the pressure at 20 feet?

Graph the equation or inequality.

35. $2x + 3y = 6$ **36.** $y = 3$ **37.** $2x - 5y < 10$

38. If $f(x) = -3x^2 + x - 4$, find $f(-2)$.

39. Two views of the same line are shown, and the displays at the bottom indicate the coordinates of two points on the line.
   (a) Find the slope of the line.
   (b) Explain why, when the equation of the line is written in $y = mx + b$ form, the value of b must be positive.

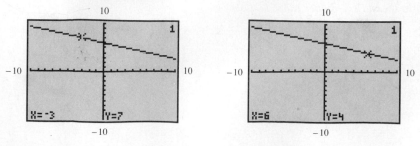

40. Write an equation of a line with slope 2 and y-intercept $(0, 3)$. Give it in the form $Ax + By = C$.

Solve the system of equations.

41. $2x + y = -4$
        $-3x + 2y = 13$

42. $3x - 5y = 8$
        $-6x + 10y = 16$

43. Based on prices in the 1995 Radio Shack catalogue, you can purchase 3 Monte Carlo phones and 2 Samantha phones for $189.95. Or you can purchase 2 Monte Carlo phones and 3 Samantha phones for $184.95. Find the price for a single phone of each model.

44. Graph the solutions of the system of inequalities.

$$2x + y \leq 4$$
$$x - y > 2$$

Simplify the expression as much as possible.

45. $\sqrt{100}$

46. $\dfrac{6\sqrt{6}}{\sqrt{5}}$

47. $\sqrt[3]{\dfrac{7}{16}}$

48. $3\sqrt{5} - 2\sqrt{20} + \sqrt{125}$

49. $\sqrt[3]{16a^3b^4} - \sqrt[3]{54a^3b^4}$

50. Evaluate $8^{2/3}$.

51. Solve the radical equation $\sqrt{x + 2} = x - 4$.

Solve the quadratic equation using the method indicated. Give only real solutions.

52. $(3x + 2)^2 = 12$ (square root property)

53. $-x^2 + 5 = 2x$ (completing the square)

54. $2x(x - 2) - 3 = 0$ (quadratic formula)

55. $(4x + 1)(x - 1) = -3$ (any method)

Solve the problem.

56. If an object is thrown upward from a height of 140 feet, with an initial velocity of 24 feet per second, its height in feet after t seconds is given by $h = -16t^2 + 24t + 140$. After how many seconds will its height be 100 feet?

*57. Write in standard form.

(a) $(-9 + 3i) + (4 + 2i) - (-5 - 3i)$ **(b)** $\dfrac{-17 - i}{-3 + i}$

*58. Find the complex solutions of $2x^2 + 2x = -9$.

59. Graph the parabola $y = x^2 - 4$ and identify the vertex.

60. If a parabola opens upward, and its equation $y = ax^2 + bx + c$ has two positive solutions, will its y-intercept be positive or negative?

Exercises designated * require knowledge of complex numbers.

APPENDIX A REVIEW OF DECIMALS AND PERCENTS

FOR EXTRA HELP

SSM pp. 542–544

Video
n/a

Tutorial
IBM   MAC

OBJECTIVES
1 ▶ Add and subtract decimals.
2 ▶ Multiply and divide decimals.
3 ▶ Convert percents to decimals and decimals to percents.
4 ▶ Find percentages by multiplication.

1 ▶ Add and
subtract decimals.

▶ A **decimal** is a number written with a decimal point, such as 4.2. The operations on decimals—addition, subtraction, multiplication, and division—are explained in the next examples.

EXAMPLE 1
Adding and Subtracting
Decimals

Add or subtract as indicated.

(a) $6.92 + 14.8 + 3.217$

Place the numbers in a column, with decimal points lined up, then add. If you like, attach zeros to make all the numbers the same length; this is a good way to avoid errors. For example,

$$
\begin{array}{r}
6.92 \\
14.8 \\
+\ 3.217 \\
\hline
24.937
\end{array}
\quad \text{or} \quad
\begin{array}{r}
6.920 \\
14.800 \\
+\ 3.217 \\
\hline
24.937.
\end{array}
$$

Decimal points lined up

(b) $47.6 - 32.509$

Write the numbers in a column, attaching zeros to 47.6.

$$
\begin{array}{r}
47.6 \\
-32.509 \\
\end{array}
\quad \text{becomes} \quad
\begin{array}{r}
47.600 \\
-32.509 \\
\hline
15.091
\end{array}
$$

(c) $3 - .253$

$$
\begin{array}{r}
3.000 \\
-\ .253 \\
\hline
2.747
\end{array}
$$

2 ▶ Multiply and
divide decimals.

▶ Multiplication and division of decimals are explained next.

E X A M P L E   2

Multiplying Decimals

Multiply.

(a) 29.3×4.52

Multiply as if the numbers were whole numbers. The number of decimal places in the answer is found by adding the numbers of decimal places in the factors.

$$
\begin{array}{r}
29.3 \\
\times 4.52 \\
\hline
5\ 86 \\
14\ 6\ 5 \\
117\ 2 \\
\hline
132.4\ 36
\end{array}
$$

1 decimal place
2 decimal places

3 decimal places in answer

(b) 7.003×55.8

$$
\begin{array}{r}
7.003 \\
\times\ 55.8 \\
\hline
5\ 602\ 4 \\
35\ 015 \\
350\ 15 \\
\hline
390.767\ 4
\end{array}
$$

3 decimal places
1 decimal place

4 decimal places in answer ■

E X A M P L E   3

Dividing Decimals

Divide: $279.45 \div 24.3$.

Move the decimal point in 24.3 one place to the right, to get 243. Move the decimal point the same number of places in 279.45. By doing this, 24.3 is converted into the whole number 243.

$$243.\overline{)2794.5}$$

Bring the decimal point straight up and divide as with whole numbers.

$$
\begin{array}{r}
11.5 \\
243\overline{)2794.5} \\
243\ \ \\
\hline
364 \\
243 \\
\hline
121\ 5 \\
121\ 5 \\
\hline
0
\end{array}
$$
■

3 ► Convert percents to decimals and decimals to percents.

► One of the main uses of decimals comes from percent problems. The word **percent** means "per one hundred." Percent is written with the sign %. One percent means "one per one hundred."

Percent

$$1\% = .01 \qquad \text{or} \qquad 1\% = \frac{1}{100}$$

EXAMPLE 4

Converting Between
Decimals and Percents

Convert.

(a) 75% to a decimal

Since 1% = .01,

$$75\% = 75 \cdot 1\% = 75 \cdot (.01) = .75.$$

The fraction form 1% = 1/100 can also be used to convert 75% to a decimal.

$$75\% = 75 \cdot 1\% = 75 \cdot \left(\frac{1}{100}\right) = .75$$

(b) 2.63 to a percent

$$2.63 = 263 \cdot (.01) = 263 \cdot 1\% = 263\% \quad \blacksquare$$

4 ▶ Find percentages by multiplication.

▶ A part of a whole is called a **percentage.** For example, since 50% represents 50/100 = 1/2 of a whole, 50% of 800 is half of 800, or 400. Percentages are found by multiplication, as in the next example.

EXAMPLE 5

Finding Percentages

Find the percentages.

(a) 15% of 600

The word *of* indicates multiplication here. For this reason, 15% of 600 is found by multiplying.

$$15\% \cdot 600 = (.15) \cdot 600 = 90$$

(b) 125% of 80

$$125\% \cdot 80 = (1.25) \cdot 80 = 100 \quad \blacksquare$$

APPENDIX A EXERCISES

Perform the indicated operations. See Examples 1–3.

1. 14.23 + 9.81 + 74.63 + 18.715

2. 89.416 + 21.32 + 478.91 + 298.213

3. 19.74 − 6.53

4. 27.96 − 8.39

5. 219 − 68.51

6. 283 − 12.42

7.
48.96
37.421
+ 9.72

8.
9.71
4.8
3.6
5.2
+8.17

9.
8.6
−3.751

10.
27.8
−13.582

11. 39.6 × 4.2

12. 18.7 × 2.3

13. 42.1 × 3.9

14. 19.63 × 4.08

15. .042 × 32

16. 571 × 2.9

17. 24.84 ÷ 6

18. 32.84 ÷ 4

19. 7.6266 ÷ 3.42

20. 14.9202 ÷ 2.43

21. 2496 ÷ .52

22. .56984 ÷ .034

Convert the percent to a decimal. See Example 4(a).

23. 53%   **24.** 38%   **25.** 129%   **26.** 174%

27. 96%   **28.** 11%   **29.** .9%   **30.** .1%

Convert the decimal to a percent. See Example 4(b).

31. .80   **32.** .75   **33.** .007   **34.** 1.4

35. .67   **36.** .003   **37.** .125   **38.** .983

Respond to the statement or question. Round your answer to the nearest hundredth if appropriate. See Example 5.

39. What is 14% of 780?

40. Find 12% of 350.

41. Find 22% of 1086.

42. What is 20% of 1500?

43. 4 is what percent of 80?

44. 1300 is what percent of 2000?

45. What percent of 5820 is 6402?

46. What percent of 75 is 90?

47. 121 is what percent of 484?

48. What percent of 3200 is 64?

49. Find 118% of 125.8.

50. Find 3% of 128.

51. What is 91.72% of 8546.95?

52. Find 12.741% of 58.902.

53. What percent of 198.72 is 14.68?

54. 586.3 is what percent of 765.4?

Solve the problem. Formulas can be found on the inside covers of this text.

55. A retailer has $23,000 invested in her business. She finds that she is earning 12% per year on this investment. How much money is she earning per year?

56. Harley Dabler recently bought a duplex for $144,000. He expects to earn 16% per year on the purchase price. How many dollars per year will he earn?

57. For a recent tour of the eastern United States, a travel agent figured that the trip totaled 2300 miles, with 35% of the trip by air. How many miles of the trip were by air?

58. Capitol Savings Bank pays 3.2% interest per year. What is the annual interest on an account of $3000?

59. An ad for steel-belted radial tires promises 15% better mileage when the tires are used. Alexandria's Escort now goes 420 miles on a tank of gas. If she switched to the new tires, how many extra miles could she drive on a tank of gas?

60. A home worth $77,000 is located in an area where home prices are increasing at a rate of 12% per year. By how much would the value of this home increase in one year?

61. A family of four with a monthly income of $2000 spends 90% of its earnings and saves the rest. Find the *annual* savings of this family.

APPENDIX B SETS

FOR EXTRA HELP

SSM pp. 544–547

Video
n/a

Tutorial
IBM   MAC

OBJECTIVES

1 ▶ List the elements of a set.
2 ▶ Learn the vocabulary and symbols used to discuss sets.
3 ▶ Decide whether a set is finite or infinite.
4 ▶ Decide whether a given set is a subset of another set.
5 ▶ Find the complement of a set.
6 ▶ Find the union and the intersection of two sets.

1 ▶ List the elements of a set.

▶ A set is a collection of things. The objects in a set are called the **elements** of the set. A set is represented by listing its elements between **set braces,** { }. The order in which the elements of a set are listed is unimportant.

EXAMPLE 1

Listing the Elements of a Set

Represent the following sets by listing the elements.

(a) The set of states in the United States that border on the Pacific Ocean = {California, Oregon, Washington, Hawaii, Alaska}.

(b) The set of all counting numbers less than 6 = {1, 2, 3, 4, 5}. ■

2 ▶ Learn the vocabulary and symbols used to discuss sets.

▶ Capital letters are used to name sets. To state that 5 is an element of

$$S = \{1, 2, 3, 4, 5\},$$

write $5 \in S$. The statement $6 \notin S$ means that 6 is not an element of S.

A set with no elements is called the **empty set,** or the **null set.** The symbols $\emptyset$ or { } are used for the empty set. If we let A be the set of all cats that fly, then A is the empty set.

$$A = \emptyset \quad \text{or} \quad A = \{ \}$$

CAUTION Do not make the common error of writing the empty set as $\{\emptyset\}$.

In any discussion of sets, there is some set that includes all the elements under consideration. This set is called the **universal set** for that situation. For example, if the discussion is about presidents of the United States, then the set of all presidents of the United States is the universal set. The universal set is denoted U.

3 ▶ Decide whether a set is finite or infinite.

▶ In Example 1, there are five elements in the set in part (a), and five in part (b). If the number of elements in a set is either 0 or a counting number, then the set is **finite.** On the other hand, the set of natural numbers, for example, is an **infinite** set, because there is no final number. We can list the elements of the set of natural numbers as

$$N = \{1, 2, 3, 4 \ldots \}$$

where the three dots indicate that the set continues indefinitely. Not all infinite sets can be listed in this way. For example, there is no way to list the elements in the set of all real numbers between 1 and 2.

EXAMPLE 2

Distinguishing Between Finite and Infinite Sets

■ List the elements of each set, if possible. Decide whether each set is finite or infinite.

(a) The set of all integers

One way to list the elements is { . . . , $-2, -1, 0, 1, 2, . . .$ }. The set is infinite.

(b) The set of all natural numbers between 0 and 5

$\{1, 2, 3, 4\}$ The set is finite.

(c) The set of all irrational numbers

This is an infinite set whose elements cannot be listed.   ■

Two sets are **equal** if they have exactly the same elements. Thus, the set of natural numbers and the set of positive integers are equal sets. Also, the sets

$$\{1, 2, 4, 7\} \qquad \text{and} \qquad \{4, 2, 7, 1\}$$

are equal. The order of the elements does not make a difference.

4 ▶ Decide whether a given set is a subset of another set.

▶   If all elements of a set A are also elements of a new set B, then we say A is a **subset** of B, written $A \subseteq B$. We use the symbol $A \not\subseteq B$ to mean that A is not a subset of B.

EXAMPLE 3

Using Subset Notation

■ Let $A = \{1, 2, 3, 4\}$, $B = \{1, 4\}$, and $C = \{1\}$. Then $B \subseteq A$, $C \subseteq A$, and $C \subseteq B$, but $A \not\subseteq B$, $A \not\subseteq C$, and $B \not\subseteq C$.   ■

The set $M = \{a, b\}$ has four subsets: $\{a, b\}$, $\{a\}$, $\{b\}$, and $\emptyset$. The empty set is defined to be a subset of any set. How many subsets does $N = \{a, b, c\}$ have? There is one subset with 3 elements: $\{a, b, c\}$. There are three subsets with 2 elements:

$$\{a, b\}, \qquad \{a, c\}, \qquad \text{and} \qquad \{b, c\}.$$

There are three subsets with 1 element:

$$\{a\}, \qquad \{b\}, \qquad \text{and} \qquad \{c\}.$$

There is one subset with 0 elements: $\emptyset$. Thus, set N has eight subsets.
The following generalization can be made.

Number of Subsets of a Set

A set with n elements has 2^n subsets.

To illustrate the relationships between sets, **Venn diagrams** are often used. A rectangle represents the universal set, U. The sets under discussion are represented by regions within the rectangle. The Venn diagram in Figure 1 shows that $B \subseteq A$.

5 ▶ Find the complement of a set.

▶   For every set A , there is a set A', the **complement** of A, that contains all the elements of U that are not in A. The shaded region in the Venn diagram in Figure 2 represents A'.

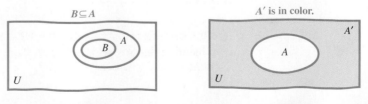

FIGURE 1   **FIGURE 2**

EXAMPLE 4

Determining the
Complement of a Set

Given $U = \{a, b, c, d, e, f, g\}$, $A = \{a, b, c\}$, $B = \{a, d, f, g\}$, and $C = \{d, e\}$, find A', B', and C'.

(a) $A' = \{d, e, f, g\}$ **(b)** $B' = \{b, c, e\}$ **(c)** $C' = \{a, b, c, f, g\}$ ∎

6 ▶ Find the union and
the intersection of
two sets.

▶   The **union** of two sets A and B, written $A \cup B$, is the set of all elements of A together with all elements of B. Thus, for the sets in Example 4,

$$A \cup B = \{a, b, c, d, f, g\}$$

and $A \cup C = \{a, b, c, d, e\}$.

In Figure 3 the shaded region is the union of sets A and B.

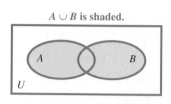

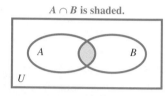

FIGURE 3   **FIGURE 4**

EXAMPLE 5

Finding the Union of
Two Sets

If $M = \{2, 5, 7\}$ and $N = \{1, 2, 3, 4, 5\}$, then

$$M \cup N = \{1, 2, 3, 4, 5, 7\}. \quad ∎$$

The **intersection** of two sets A and B, written $A \cap B$, is the set of all elements that belong to both A and B. For example if,

$$A = \{\text{Jose, Ellen, Marge, Kevin}\}$$

and $B = \{\text{Jose, Patrick, Ellen, Sue}\}$,

then $A \cap B = \{\text{Jose, Ellen}\}$.

The shaded region in Figure 4 represents the intersection of the two sets A and B.

EXAMPLE 6

Finding the Intersection
of Two Sets

Suppose that $P = \{3, 9, 27\}$, $Q = \{2, 3, 10, 18, 27, 28\}$, and $R = \{2, 10, 28\}$.

(a) $P \cap Q = \{3, 27\}$ **(b)** $Q \cap R = \{2, 10, 28\} = R$ **(c)** $P \cap R = \emptyset$ ∎

Sets like P and R in Example 6 that have no elements in common are called **disjoint sets.** The Venn diagram in Figure 5 shows a pair of disjoint sets.

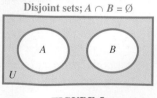

FIGURE 5

EXAMPLE 7

Using Set Operations

Let $U = \{2, 5, 7, 10, 14, 20\}$, $A = \{2, 10, 14, 20\}$, $B = \{5, 7\}$, and $C = \{2, 5, 7\}$. Find each of the following.

(a) $A \cup B = \{2, 5, 7, 10, 14, 20\} = U$ **(b)** $A \cap B = \emptyset$

(c) $B \cup C = \{2, 5, 7\} = C$ **(d)** $B \cap C = \{5, 7\} = B$

(e) $A' = \{5, 7\} = B$ ∎

APPENDIX B EXERCISES

List the elements of the set. See Examples 1 and 2.

1. The set of all natural numbers less than 8

2. The set of all integers between 4 and 10

3. The set of seasons

4. The set of months of the year

5. The set of women presidents of the United States

6. The set of all living humans who are more than 200 years old

7. The set of letters of the alphabet between K and M

8. The set of letters of the alphabet between D and H

9. The set of positive even integers

10. The set of all multiples of 5

11. Which of the sets described in Exercises 1–10 are infinite sets?

12. Which of the sets described in Exercises 1–10 are finite sets?

Tell whether the statement is true or false.

13. $5 \in \{1, 2, 5, 8\}$ **14.** $6 \in \{1, 2, 3, 4, 5\}$ **15.** $2 \in \{1, 3, 5, 7, 9\}$

16. $1 \in \{6, 2, 5, 1\}$ **17.** $7 \notin \{2, 4, 6, 8\}$ **18.** $7 \notin \{1, 3, 5, 7\}$

19. $\{2, 4, 9, 12, 13\} = \{13, 12, 9, 4, 2\}$ **20.** $\{7, 11, 4\} = \{7, 11, 4, 0\}$

Let $A = \{1, 3, 4, 5, 7, 8\}$
     $B = \{2, 4, 6, 8\}$
     $C = \{1, 3, 5, 7\}$
     $D = \{1, 2, 3\}$
     $E = \{3, 7\}$
     $U = \{1, 2, 3, 4, 5, 6, 7, 8, 9, 10\}$.

Tell whether the statement is true or false. See Examples 3, 5, 6, and 7.

21. $A \subseteq U$ **22.** $D \subseteq A$ **23.** $\emptyset \subseteq A$ **24.** $\{1, 2\} \subseteq D$ **25.** $C \subseteq A$

26. $A \subseteq C$ **27.** $D \subseteq B$ **28.** $E \subset C$ **29.** $D \not\subseteq E$ **30.** $E \not\subseteq A$

31. There are exactly 4 subsets of E.   **32.** There are exactly 8 subsets of D.

33. There are exactly 12 subsets of C.   **34.** There are exactly 16 subsets of B.

35. $\{4, 6, 8, 12\} \cap \{6, 8, 14, 17\} = \{6, 8\}$ **36.** $\{2, 5, 9\} \cap \{1, 2, 3, 4, 5\} = \{2, 5\}$

37. $\{3, 1, 0\} \cap \{0, 2, 4\} = \{0\}$ **38.** $\{4, 2, 1\} \cap \{1, 2, 3, 4\} = \{1, 2, 3\}$

39. $\{3, 9, 12\} \cap \emptyset = \{3, 9, 12\}$ **40.** $\{3, 9, 12\} \cup \emptyset = \emptyset$

41. $\{4, 9, 11, 7, 3\} \cup \{1, 2, 3, 4, 5\}$ **42.** $\{1, 2, 3\} \cup \{1, 2, 3\} = \{1, 2, 3\}$
$= \{1, 2, 3, 4, 5, 7, 9, 11\}$

43. $\{3, 5, 7, 9\} \cup \{4, 6, 8\} = \emptyset$ **44.** $\{5, 10, 15, 20\} \cup \{5, 15, 30\} = \{5, 15\}$

Let $U = \{a, b, c, d, e, f, g, h\}$
     $A = \{a, b, c, d, e, f\}$
     $B = \{a, c, e\}$
     $C = \{a, f\}$
     $D = \{d\}$.

List the elements in the set. See Examples 4–7.

45. A' **46.** B' **47.** C' **48.** D'

49. $A \cap B$ **50.** $B \cap A$ **51.** $A \cap D$ **52.** $B \cap D$

53. $B \cap C$ **54.** $A \cup B$ **55.** $B \cup D$ **56.** $B \cup C$

57. $C \cup B$ **58.** $C \cup D$ **59.** $A \cap \emptyset$ **60.** $B \cup \emptyset$

61. Name every pair of disjoint sets among A–D above.

ANSWERS TO SELECTED EXERCISES

TO THE STUDENT

If you need further help with algebra, you may want to obtain a copy of the *Student's Solution Manual* that goes with this book. It contains solutions to all of the odd-numbered exercises and all of the chapter test exercises. You also may want the *Student's Study Guide*. It has extra examples and exercises to complete, corresponding to each learning objective of the book. In addition, there is a practice test for each chapter. Your college bookstore either has these books or can order them for you.

In this section we provide the answers that we think most students will obtain when they work the exercises using the methods explained in the text. If your answer does not look exactly like the one given here, it is not necessarily wrong. In many cases there are equivalent forms of the answer that are correct. For example, if the answer section shows 3/4 and your answer is .75, you have obtained the right answer but written it in a different (yet equivalent) form. Unless the directions specify otherwise, .75 is just as valid an answer as 3/4.

In general, if your answer does not agree with the one given in the text, see whether it can be transformed into the other form. If it can, then it is the correct answer. If you still have doubts, talk with your instructor.

CHAPTER 1   THE REAL NUMBER SYSTEM

SECTION 1.1   (PAGE 8)

◆ **CONNECTIONS**   **Page 2:** Answers will vary.

EXERCISES   **1.** 3; 8   **3.** yes   **5.** product; quotient   **7.** prime   **9.** composite   **11.** composite
13. neither   **15.** $2 \cdot 3 \cdot 5$ **17.** $2 \cdot 2 \cdot 5 \cdot 5 \cdot 5$ **19.** $2 \cdot 2 \cdot 31$ **21.** 29   **23.** $\frac{1}{2}$ **25.** $\frac{5}{6}$ **27.** $\frac{1}{3}$ **29.** $\frac{6}{5}$
31. (c)   **33.** $\frac{24}{35}$ **35.** $\frac{6}{25}$ **37.** $\frac{6}{5}$ **39.** $\frac{232}{15}$ or $15\frac{7}{15}$ **41.** $\frac{10}{3}$ or $3\frac{1}{3}$ **43.** 12   **45.** $\frac{1}{16}$
47. $\frac{84}{47}$ or $1\frac{37}{47}$ **51.** $\frac{2}{3}$ **53.** $\frac{8}{9}$ **55.** $\frac{27}{8}$ or $3\frac{3}{8}$ **57.** $\frac{17}{36}$ **59.** $\frac{11}{12}$ **61.** $\frac{4}{3}$
63. $2\left(\frac{10}{100}\right) + 3\left(\frac{10}{100}\right) = \frac{50}{100}$ **65.** $81\frac{1}{2}$ dollars   **67.** $\frac{9}{16}$ inch   **69.** $618\frac{3}{4}$ feet   **71.** $5\frac{5}{24}$ inches   **73.** $4\frac{1}{2}$ cups
75. $\frac{1}{3}$ cup   **77.** (a) $\frac{9}{100}$ (b) $\frac{9}{10}$ **79.** (a) $\frac{1}{2}$ (b) $\frac{1}{4}$ (c) $\frac{1}{3}$ (d) $\frac{1}{6}$

SECTION 1.2   (PAGE 17)

◆ **CONNECTIONS**   **Page 16:** Answers will vary.

EXERCISES   **1.** false; 3^5 means $3 \cdot 3 \cdot 3 \cdot 3 \cdot 3$ **3.** true   **5.** 49   **7.** 144   **9.** 64
11. 1000   **13.** 81   **15.** 1024   **17.** $\frac{16}{81}$ **19.** .000064   **23.** 32   **25.** $\frac{49}{30}$ **27.** 12   **29.** 23.01
31. 95   **33.** 90   **35.** 14   **37.** 9   **41.** true   **43.** false   **45.** true   **47.** true   **49.** false   **51.** false
53. true   **55.** $15 = 5 + 10$ **57.** $9 > 5 - 4$ **59.** $16 \neq 19$ **61.** $2 \leq 3$ **63.** Seven is less than nineteen. True

65. Three is not equal to six. True   **67.** Eight is greater than or equal to eleven. False
69. Answers will vary. One example is $5 + 3 \geq 2 \cdot 2$.   **71.** $30 > 5$ **73.** $3 \leq 12$ **75.** is younger than
77. The inequality symbol $\geq$ implies a true statement if 12 equals 12 *or* if 12 is greater than 12.   **79.** 1989, 1992, and 1993
81. 1989, 1990, 1991, and 1992   **83.** $302.7 million   **85.** 7 miles   **86.** 5 miles   **87.** The direct distance from home to
school is less than the distance traveled via the babysitter's house.   **88.** shortest; straight line   **89.** She would have to live
directly on the line from home to school.   **90.** 9; 16; 25; 25; They are equal.

SECTION 1.3 (PAGE 24)

EXERCISES   **1.** expression   **3.** equation   **5.** equation   **7.** $2x^3 = 2 \cdot x \cdot x \cdot x$ **11.** Answers will vary. Two such
pairs are $x = 0$, $y = 6$ and $x = 1$, $y = 4$.   **13.** (a) 13 (b) 15   **15.** (a) 20 (b) 30   **17.** (a) 64 (b) 144
19. (a) $\frac{5}{3}$ (b) $\frac{7}{3}$ **21.** (a) $\frac{7}{8}$ (b) $\frac{13}{12}$ **23.** (a) 52 (b) 114   **25.** (a) 25.836 (b) 38.754   **27.** (a) 24 (b) 28
29. (a) 12 (b) 33   **31.** (a) 6 (b) $\frac{9}{5}$ **33.** (a) $\frac{4}{3}$ (b) $\frac{13}{6}$ **35.** (a) $\frac{2}{7}$ (b) $\frac{16}{27}$ **37.** (a) 12 (b) 55
39. (a) 1 (b) $\frac{28}{17}$ **41.** (a) 3.684 (b) 8.841   **43.** $12x$ **45.** $x + 7$ **47.** $x - 2$ **49.** $7 - x$ **51.** $x - 6$
53. $\frac{12}{x}$ **55.** $6(x - 4)$ **59.** yes   **61.** no   **63.** yes   **65.** yes   **67.** yes   **69.** $x + 8 = 18$; 10
71. $16 - \frac{3}{4}x = 13$; 4   **73.** $2x + 5 = 5$; 0   **75.** $3x - 2x + 8$; 8   **77.** 1.00 (billion dollars)
78. 1.27 (billion dollars)   **79.** 1.54 (billion dollars)   **80.** 1.81 (billion dollars)

SECTION 1.4 (PAGE 32)

◆ **CONNECTIONS**   **Page 27:** Answers will vary.

EXERCISES   **1.** true   **3.** true   **5.** false   **7.** true   **9.** false   **11.** (a) 3, 7 (b) 0, 3, 7 (c) $-9, 0, 3, 7$
(d) $-9, -1\frac{1}{4}, -\frac{3}{5}, 0, 3, 5.9, 7$ (e) $-\sqrt{7}, \sqrt{5}$ (f) All are real numbers.   **13.** -1760 **15.** -8 **17.** 8300
19. $-66,000,000$ (dollars)   **21.** ◆◆++++++◆+◆++→
                                              -6 -4 -2 0 2
23. ◆+◆+◆+++++◆◆→
      -6 -4 -2 0 2 4
25. $-3\frac{4}{5}$ $-1\frac{5}{8}$ $\frac{1}{4}$ $2\frac{1}{2}$
++◆+◆+◆++◆+◆+◆+→
-4 -2 0 2 4
27. It is not true. The absolute value of 0 is 0, which is not positive.   **29.** (a) A (b) A (c) B (d) B
31. (a) 2 (b) 2   **33.** (a) -6 (b) 6   **35.** (a) -3 (b) 3   **37.** (a) 0 (b) 0   **39.** $a - b$
41. -12 **43.** -8 **45.** 3   **47.** $|-3|$ or 3   **49.** $-|-6|$ or -6 **51.** $|5 - 3|$ or 2
53. true   **55.** true   **57.** true   **59.** false   **61.** true   **63.** false   **65.** gasoline in 1986   **67.** false
Answers will vary in Exercises 69–77.   **69.** true: $a = 0$ or $b = 0$ or both $a = 0$ and $b = 0$; false: Choose any values for a
and b so that neither a nor b is zero.   **71.** true: Choose a to be the opposite of b $(a = -b)$; false: $a \neq -b$
73. $\frac{1}{2}, \frac{5}{8}, 1\frac{3}{4}$ **75.** $-3\frac{1}{2}, -\frac{2}{3}, \frac{3}{7}$ **77.** $\sqrt{5}, \pi, -\sqrt{3}$

SECTION 1.5 (PAGE 39)

◆ **CONNECTIONS**   **Page 37:** **1.** $26.25   **2.** $-10°F$ Both problems require finding the sum of a negative number and a
positive number.

EXERCISES   **1.** negative   **3.** $-3, 5$ **5.** 2   **7.** -3 **9.** -10 **11.** -13 **13.** -15.9 **15.** -1
17. 13   **19.** 5   **21.** 0   **23.** -8 **25.** $\frac{1}{2}$ **27.** $-\frac{19}{24}$ **29.** $-\frac{3}{4}$ **31.** -1.6 **33.** -8.7 **35.** -25
37. -24 **39.** no   **41.** true   **43.** false   **45.** true   **47.** false   **49.** true   **51.** false   **53.** -3 **55.** 3
57. -1 **59.** -3 **61.** It must be negative.   **62.** The sum of a positive number and 5 cannot be -7.
63. It must be positive.   **64.** The sum of a negative number and -8 cannot be 2.   **65.** $-5 + 12 + 6$; 13
67. $[-19 + (-4)] + 14$; -9 **69.** $[-4 + (-10)] + 12$; -2 **71.** $[8 + (-18)] + 4$; -6 **73.** -396 million dollars
75. $47   **77.** -184 meters   **79.** 112°F   **81.** -1 yard   **83.** $286.60   **85.** 7506 million dollars

SECTION 1.6 (PAGE 46)

EXERCISES **1.** -8; -6 **3.** 7 **5.** -4 **7.** -3 **9.** -4 **11.** -10 **13.** -16 **15.** 11 **17.** 19

19. -4 **21.** 5 **23.** 0 **25.** $\frac{3}{4}$ **27.** $-\frac{11}{8}$ **29.** $\frac{15}{8}$ **31.** 13.6 **33.** -11.9 **35.** -2.8 **37.** -6.3

41. -14 **43.** -24 **45.** -16 **47.** $-\frac{17}{8}$ **49.** -48.98 **51.** Answers will vary. One example is $-8 - (-2) = -6$.

53. positive **55.** positive **57.**

```
      A               B
  |─●─┼─┼─┼─┼─┼─┼─┼─┼─●─┼─►
   -3 -1 0 1 2 3 4 5 6
```

58. 8 **59.** 8; yes

60. -8; No, because the distance is not -8, but 8. **61.** $\left| 8 \right| = \left| -8 \right| = 8$; yes **62.** subtract; absolute value
63. $4 - (-8)$; 12 **65.** $-2 - 8$; -10 **67.** $[9 + (-4)] - 7$; -2 **69.** $[8 - (-5)] - 12$; 1 **71.** -1
73. -3 **75.** -3 **77.** 50,395 feet **79.** 1345 feet **81.** $-25°F$ **83.** 14,776 feet
85. $-\$80$ **87. (a)** 74 **(b)** -22 **(c)** -42 **(d)** 14 **(e)** 7

SECTION 1.7 (PAGE 53)

◈ **CONNECTIONS** **Page 50:** \$15; $-5 \cdot 3 = -15$

EXERCISES **1.** true **3.** false **5.** true **7.** 20 **9.** -30 **11.** -28 **13.** 80 **15.** 0 **17.** $\frac{5}{6}$

19. -2.38 **21.** $\frac{3}{2}$ **23.** $-32, -16, -8, -4, -2, -1, 1, 2, 4, 8, 16, 32$

25. $-40, -20, -10, -8, -5, -4, -2, -1, 1, 2, 4, 5, 8, 10, 20, 40$ **27.** $-31, -1, 1, 31$
29. -11 **31.** -2 **33.** -60 **35.** 35 **37.** 6 **39.** -18 **41.** 67 **43.** -8 **45.** 64

49. 47 **51.** 72 **53.** $-\frac{78}{25}$ **55.** 0 **57.** -23 **59.** -9 **61.** $9 + (-9)(2)$; -9 **63.** $-4 - 2(-1)(6)$; 8

65. $7(-12) - 9$; -93 **67.** $12[9 - (-8)]$; 204 **69.** $\frac{4}{5}[-8 + (-2)]$; -8 **71.** 3 **73.** 0 **75.** -2 **77.** -2

79. -1 **81.** -3 **83.** 34.81% **85.** 21.73% **87.** positive **89.** -15
90. -15; The product of a negative number and a positive number must be a negative number.

SECTION 1.8 (PAGE 60)

EXERCISES **1.** positive **3.** less than **5.** positive **7.** $\frac{1}{11}$ **9.** $-\frac{1}{5}$ **11.** $\frac{6}{5}$ **13.** no reciprocal

15. $-\frac{7}{8}$ **17.** 2.5 **19.** (c) **21.** -3 **23.** -2 **25.** 16 **27.** 0 **29.** 25.63 **31.** $\frac{3}{2}$ **33.** 3.4 **35.** -3

37. -5 **39.** 4 **41.** 3 **43.** 7 **45.** 4 **47.** -3 **49.** $\frac{1}{2}$ **51.** 10 **53.** negative **55.** positive

57. positive **59.** $\frac{-36}{-9}$; 4 **61.** $\frac{-12}{-5 + (-1)}$; 2 **63.** $\frac{15 + (-3)}{4(-3)}$; -1 **65.** $\frac{-34(7)}{-14}$; 17 **67.** $6x = -42$; -7

69. $\frac{x}{3} = -3$; -9 **71.** $x - 6 = 4$; 10 **73.** $x + 5 = -5$; -10 **75.** 4 **77.** -4 **81.** \$1,500,000

83. -8 **85.** 4 **87.** -4 **89. (a)** 6 is divisible by 2. **(b)** 9 is not divisible by 2.
90. (a) $4 + 7 + 9 + 9 + 2 + 3 + 2 = 36$ is divisible by 3. **(b)** $2 + 4 + 4 + 3 + 8 + 7 + 1 = 29$ is not divisible by 3.
91. (a) 64 is divisible by 4. **(b)** 35 is not divisible by 4. **92. (a)** 5 is divisible by 5. **(b)** 3 is not divisible by 5.
93. (a) 2 is divisible by 2 and $1 + 5 + 2 + 4 + 8 + 2 + 2 = 24$ is divisible by 3. **(b)** While 0 is divisible by 2,
$2 + 8 + 7 + 3 + 5 + 9 + 0 = 34$ is not divisible by 3. **94. (a)** 296 is divisible by 8. **(b)** 623 is not divisible by 8.
95. (a) $4 + 1 + 1 + 4 + 1 + 0 + 7 = 18$ is divisible by 9. **(b)** $2 + 2 + 8 + 7 + 3 + 2 + 1 = 25$ is not divisible by 9.
96. (a) $4 + 2 + 5 + 3 + 5 + 2 + 0 = 21$ is divisible by 3 and 20 is divisible by 4.
(b) $4 + 2 + 4 + 9 + 4 + 7 + 4 = 34$ is not divisible by 3, and this is sufficient to show that the number is not divisible by 12.

SECTION 1.9 (PAGE 69)

EXERCISES **1.** true **3.** false **5.** true **7.** false **9.** commutative property **11.** associative property
13. commutative property **15.** associative property **17.** inverse property **19.** inverse property **21.** identity property
23. commutative property **25.** distributive property **27.** identity property **29.** distributive property **31.** identity
property **35.** $7 + r$ **37.** s **39.** $-6x + (-6) \cdot 7$; $-6x - 42$ **41.** $w + [5 + (-3)]$; $w + 2$ **43.** 11

45. 0 **47.** $-.38$ **49.** 1 **51.** Subtraction is not associative. **55.** $(5 + 1)x$; $6x$ **57.** $4t + 12$
59. $-8r - 24$ **61.** $-5y + 20$ **63.** $-16y - 20z$ **65.** $8(z + w)$ **67.** $7(2v + 5r)$ **69.** $24r + 32s - 40y$
71. $(1 + 1 + 1)q$; $3q$ **73.** $(-5 + 1)x$; $-4x$ **75.** $-4t - 3m$ **77.** $5c + 4d$ **79.** $3q - 5r + 8s$
81. for example, "putting on your socks" and "putting on your shoes" **83.** 0 **84.** $-3(5) + (-3)(-5)$ **85.** -15
86. We must interpret $(-3)(-5)$ as 15, since it is the additive inverse of -15.

CHAPTER 1 REVIEW EXERCISES (PAGE 74)

1. $\dfrac{3}{4}$ **3.** $\dfrac{9}{40}$ **5.** 625 **7.** .0000000032 **9.** 27 **11.** 39 **13.** true **15.** false **17.** $13 < 17$ **19.** 30
21. 14 **23.** $x + 6$ **25.** $6x - 9$ **27.** yes **29.** $2x - 6 = 10$; 8 **31.**

$$-\tfrac{1}{2} \quad 2.5$$
(number line from -4 to 4)

33. $-3\tfrac{1}{4} \ -1\tfrac{1}{8} \ \tfrac{5}{6} \ 2\tfrac{4}{5}$ (number line from -4 to 4) **35.** -10 **37.** $-\dfrac{3}{4}$ **39.** true **41.** true **43.** (a) 9 (b) 9

45. (a) -6 (b) 6 **47.** 12 **49.** -19 **51.** -6 **53.** -17 **55.** -21.8 **57.** -10 **59.** $(-31 + 12)$
$+ 19$; 0 **61.** $-\$8$ **63.** -2 **65.** -11 **67.** 7 **69.** 10.31 **71.** 2 **73.** $-4 - (-6)$; 2
75. $-\$29$ **79.** 36 **81.** $\dfrac{1}{2}$ **83.** -20 **85.** -24 **87.** -18 **89.** 125 **91.** $-4(5) - 9$; -29 **93.** 4
95. $-\dfrac{3}{4}$ **97.** -1 **99.** 1 **101.** $\dfrac{12}{8 + (-4)}$; 3 **103.** $8x = -24$; -3 **105.** $x - 3 = -7$; -4
107. identity property **109.** inverse property **111.** associative property **113.** distributive property
115. $(7 + 1)y$; $8y$ **117.** $3(2s + 5y)$ **121.** 16 **123.** -26 **125.** $-\dfrac{1}{24}$ **127.** 2 **129.** $-1\dfrac{1}{2}$ **131.** $-\dfrac{28}{15}$
135. $5(x + 7)$; $5x + 35$ **137.** 10,764 million dollars **139.** -2 **140.** It is less than -1. **141.** 2
142. 2 **143.** $-\dfrac{1}{2}$ **144.** yes **145.** yes **146.** yes

CHAPTER 1 TEST (PAGE 79)

[1.1] 1. $\dfrac{7}{11}$ **2.** $\dfrac{241}{120}$ **3.** $\dfrac{19}{18}$ **4.** $\dfrac{4}{25}$ **[1.2] 5.** true **6.** false **[1.4] 7.** (number line from -4 to 4)
8. $-|-8|$ (or -8) **9.** -1.277 **[1.3] 10.** $\dfrac{-6}{2 + (-8)}$; 1 **[1.5, 1.8] 11.** negative **[1.1, 1.4-1.8] 12.** 4
13. $-2\dfrac{5}{6}$ **14.** 2 **15.** 6 **16.** 108 **17.** 3 **18.** $\dfrac{30}{7}$ **[1.3] 19.** 4 **20.** -2 **[1.4–1.8] 21.** -70 **22.** 3
[1.6] 23. $177°$ F **24.** .2 million dollars; -4.4 million dollars; 4.1 million dollars **25.** (a) -5 (million)
(b) 11 (million) **(c)** -4 (million) **[1.1] 26.** approximately 11.4 billion dollars **[1.9] 27.** B **28.** D **29.** E
30. A **31.** C **32.** the distributive property **33.** (a) -18 (b) -18 (c) The distributive property assures us that the
answers must be the same, because $a(b + c) = ab + ac$ for all a, b, c.

CHAPTER 2 SOLVING EQUATIONS AND INEQUALITIES

SECTION 2.1 (PAGE 85)

EXERCISES **1.** (c) **3.** (a) **5.** $4r + 11$ **7.** $32q - 24t$ **9.** $5 + 2x - 6y$ **11.** $-7 + 3p$ **13.** 14
15. -12 **17.** 5 **19.** 1 **21.** -1 **23.** 74 **25.** Answers will vary. For example, $-3x$ and $4x$. **27.** like
29. unlike **31.** like **33.** unlike **37.** $9k - 5$ **39.** $-\dfrac{1}{3}t - \dfrac{28}{3}$ **41.** $-4.1r + 5.6$ **43.** $-2y^2 + 3y^3$
45. $-19p + 16$ **47.** $-4y + 22$ **49.** $-16y + 63$ **51.** $4k - 7$ **53.** $-23.7y - 12.6$
55. $(x + 3) + 5x$; $6x + 3$ **57.** $(13 + 6x) - (-7x)$; $13 + 13x$ **59.** $2(3x + 4) - (-4 + 6x)$; 12
61. Wording may vary. One example is "the difference between 9 times a number and the sum of the number and 2."

63. 2, 3, 4, 5   **64.** 1   **65. (a)** 1, 2, 3, 4 **(b)** 3, 4, 5, 6 **(c)** 4, 5, 6, 7   **66.** The value of $x + b$ also increases by 1 unit.
67. (a) 2, 4, 6, 8 **(b)** 2, 5, 8, 11 **(c)** 2, 6, 10, 14   **68.** m **69. (a)** 7, 9, 11, 13 **(b)** 5, 8, 11, 14 **(c)** 1, 5, 9, 13
In comparison, we see that while the values themselves are different, the number of units of increase is the same as the
corresponding parts of Exercise 67.   **70.** m **71.** -5 **73.** 15   **75.** 2   **77.** -14 **79.** 3   **81.** $-\dfrac{1}{3}$

SECTION 2.2 (PAGE 93)

◆ **CONNECTIONS**   **Page 88:** $x = 11$

EXERCISES   **1.** $x + 13$ **3.** -1 **5.** 12   **7.** 4.2   **9.** 3   **11.** -2 **13.** 4   **15.** 0   **17.** -2
19. -7 **23.** 13   **25.** 4   **27.** -4 **29.** 0   **31.** $\dfrac{7}{15}$ **33.** 7   **35.** -4 **37.** 13   **39.** 29   **41.** 18
43. 12   **47.** 6   **49.** $\dfrac{15}{2}$ **51.** -5 **53.** $-\dfrac{18}{5}$ **55.** 12   **57.** 0   **59.** -48 **61.** -12 **63.** $\dfrac{4}{7}$ **65.** 40
67. 3   **69.** 7   **71.** -5 **73.** -35 **75.** 9   **77.** $\dfrac{35}{2}$ **79.** $-\dfrac{27}{35}$ **81.** -12.2
83. Answers will vary. For example, $\dfrac{3}{2}x = -6$.   **85.** $3x = 17 + 2x$;   17   **87.** $5x + 3x = 7x + 9$;   9
89. $\dfrac{x}{-5} = 2$; -10 **91.** $24q + 32$ **93.** $-28p + 29$ **95.** $-8 + 56p$

SECTION 2.3 (PAGE 101)

EXERCISES   **3.** 2   **5.** 5   **7.** $-\dfrac{5}{3}$ **9.** -1 **11.** no solution   **13.** all real numbers   **15.** 1
17. no solution   **21.** 5   **23.** 0   **25.** $-\dfrac{7}{5}$ **27.** 120   **29.** 6   **31.** 15,000   **33.** 800
34. Yes, you will get $(100 \cdot 2) \cdot 4 = 200 \cdot 4 = 800$. This is a result of the associative property of multiplication.
35. No, because $(100a)(100b) = 10,000ab \neq 100ab$.   **36.** The distributive property involves the operation of *addition* as
well.   **37.** Yes, the associative property of multiplication is used here.   **38.** no
39. 8   **41.** $-\dfrac{13}{8}$ **43.** 0   **45.** 4   **47.** 20   **49.** all real numbers   **51.** no solution   **55.** $11 - q$
57. $x + 7$ **59.** $a + 12$; $a - 5$ **61.** $\dfrac{t}{5}$ **63.** $-6 + x$ **65.** $x - 9$ **67.** $\dfrac{-6}{x}$ **69.** $12(x - 9)$

SECTION 2.4 (PAGE 110)

◆ **CONNECTIONS**   **Page 103:** Polya's Step 1 corresponds to our Steps 1 and 2. Polya's Step 2 corresponds to our Step 3.
Polya's Step 3 corresponds to our Steps 4 and 5. Polya's Step 4 corresponds to our Step 6. Trial and error or guessing and checking
fit into Polya's Step 2, devising a plan.

EXERCISES   **1.** (c)   **5.** 3   **7.** -4 **9.** 57 Democrats, 43 Republicans   **11.** 1037   **13.** shorter piece: 15
inches;   longer piece: 24 inches   **15.** Federal Express: 9;   Airborne Express: 3;   United Parcel Service: 1   **17.** 36 million
miles   **19.** Gant: 141;   Justice: 124;   Cabrera: 3   **21.** A and B: 40 degrees; C: 100 degrees   **23.** antilock brakes: $800;
power door locks: $240   **25.** 18 prescriptions   **27.** peanuts: $22\dfrac{1}{2}$ ounces;   cashews: $4\dfrac{1}{2}$ ounces   **29.** $k - m$
31. no   **33.** $x - 1$ **35.** 80°   **37.** 26°   **39.** 55°   **41.** 68, 69   **43.** 10, 12   **45.** 101, 102   **47.** 10, 11
49. 15, 16, 17   **51.** 15, 17, 19   **53.** 1990: 1.45 billion dollars;   1991: 1.95 billion dollars;   1992: 2.20 billion dollars
55. 24   **57.** 320

SECTION 2.5 (PAGE 120)

◆ **CONNECTIONS**   **Page 115:** The maximum volume is 23,328 *cubic inches*. The volume of a box is the measure of the space
that it occupies.

EXERCISES   **3.** area   **5.** perimeter   **7.** area   **9.** area   **11.** $P = 20$ **13.** $P = 24$ **15.** $A = 70$
17. $c = 5$ **19.** $r = 40$ **21.** $I = 875$ **23.** $A = 91$ **25.** $r = 1.3$ **27.** $A = 452.16$ **29.** 4   **31.** $V = 384$
33. $V = 48$ **35.** $V = 904.32$ **37.** 1029.92 feet   **39.** perimeter: 172 inches;   area: 1785 square inches
41. $\dfrac{729}{32}$ or $22\dfrac{25}{32}$ cubic inches   **43.** 23,800.10 square feet   **45.** 107°, 73°   **47.** 75°, 75°   **49.** 139°, 139°

51. $L = \dfrac{A}{W}$ **53.** $r = \dfrac{d}{t}$ **55.** $p = \dfrac{I}{rt}$ **57.** $a = P - b - c$ **59.** $b = \dfrac{2A}{h}$ **61.** $r = \dfrac{A - p}{pt}$ **63.** $h = \dfrac{V}{\pi r^2}$

65. $m = \dfrac{y - b}{x}$ or $m = \dfrac{y}{x} - \dfrac{b}{x}$ **67. (a)** $P - 2L = 2W$ **(b)** $W = \dfrac{P - 2L}{2}$ **68. (a)** $\dfrac{P}{2} = L + W$ **(b)** $\dfrac{P}{2} - L = W$

69. (a) multiplication identity property   **(b)** An expression divided by 1 is equal to itself.   **(c)** rule for multiplication of

fractions   **(d)** rule for subtraction of fractions   **70.** $\dfrac{5T}{4} + 1$ **71.** 3   **73.** 28

SECTION 2.6 (PAGE 128)

◆ CONNECTIONS   **Page 124:** 200 mm

EXERCISES   **1.** (c)   **3.** $\dfrac{4}{3}$ **5.** $\dfrac{4}{3}$ **7.** $\dfrac{15}{2}$ **9.** $\dfrac{1}{5}$ **11.** $\dfrac{24}{5}$ **15.** true   **17.** false   **19.** true

21. 35   **23.** 27   **25.** -1 **27.** 10   **29.** $13\dfrac{1}{3}$ spears   **31.** 6.875 ounces   **33.** $706.20   **35.** $9.30   **37.** 4 feet

39. 45 heads   **41.** $338   **43.** 12,500 fish   **45.** yes: 2520;   no: 1260;   undecided: 420   **47.** 30-count size

49. 31-ounce size   **51.** 32-ounce size   **53.** 4   **55.** 1   **57. (a)**   **(b)** 54 feet   **59.** $294

61. $331   **63.** $4850   **65.** 30   **66. (a)** $5x = 12$ **(b)** $\dfrac{12}{5}$ **67. (a)** $5x = 12$ **(b)** $\dfrac{12}{5}$ **68.** They are the same. Solving

by cross products yields the same equation as multiplying by the least common denominator.   **69.** 4%   **71.** 10 years

73. 6   **75.** 4

SECTION 2.7 (PAGE 137)

◆ CONNECTIONS   **Page 133:** $\dfrac{\text{leg length}}{R} = \dfrac{100\%}{109\%}$; $R = 1.09(\text{leg length})$. For example, if the inside leg measurement is

28 inches, $R = 30.52$ inches.

EXERCISES   **1.** 35 milliliters   **3.** $350   **5.** $14.15   **7.** $533   **9.** 180%   **11.** 15   **13. (a)** Under 14:

$40,740,000;   **(b)** 18 to 24: $58,200,000;   **(c)** 45 to 64: $110,580,000   **15.** 0 liters   **17.** 160 gallons

19. $53\dfrac{1}{3}$ kilograms   **21.** $3\dfrac{1}{3}$ gallons   **23.** 4 liters   **25.** 25 milliliters   **27.** (d)   **29.** $5000 at 3%; $1000 at 5%

31. $8500 at 4%; $3500 at 5%   **33.** $40,000 at 3%; $110,000 at 4%   **35.** 25 fives   **37.** 84 fives;   42 tens

39. 50 32¢ favors   **41.** 20 pounds   **45. (a)** $.05x + .10(3400 - x) = 290$ **(b)** 1000 nickels; 2400 dimes

46. (a) $.05x + .10(3400 - x) = 290$ **(b)** $1000 at 5%; $2400 at 10%   **47.** They are the same.

48. No, you will get a different solution to the equation, but you *will* get the same answers to the problem.

51. $t = \dfrac{d}{r}$ **53.** $b = \dfrac{2A}{h}$

SECTION 2.8 (PAGE 144)

◆ CONNECTIONS   **Page 141:** 6.70 hours

EXERCISES   **1.** 3.718 hours   **3.** 1.715 hours   **5.** 7.51 meters per second   **7.** 10.08 meters per second

9. 530 miles   **13.** 10 hours   **15.** 2 hours   **17.** 5 hours   **19.** 328 feet   **21.** length: 515 feet;   width: 318 feet

23. 300 feet, 400 feet, 500 feet   **25.** 2.5 centimeters   **27.** single man: $51.58;   single woman: $40.71

29. (a) 174   **(b)** 106   **(c)** 123   **(d)** 73   **(e)** 84   **31.** 109°, 71°   **33.** 198-ounce size   **35.** A and B: 52°; C: 76°

37. 120 miles per hour   **39.** 120 pounds   **41.** $(8x - 8) + 10x + (12x + 8) = 180$;   6;   40°, 60°, 80°

42. $5x + 10 = 8x - 8$;   6;   yes   **43.** $(5x + 10) + (15x + 50) = 180$; $(15x + 50) + (8x - 8) = 180$;   6;   6;   yes

45. $<$ **47.** $<$

SECTION 2.9 (PAGE 157)

◆ **CONNECTIONS** **Page 149:** The revenue is represented by $5x - 100$. The production cost is $125 + 4x$. The profit is represented by $R - C = (5x - 100) - (125 + 4x) = x - 225$. The solution of $x - 225 > 0$ is $x > 225$. In order to make a profit, more than 225 cassettes must be produced and sold.

EXERCISES **3.** $x > -4$ **5.** $x \le 4$ **7.** (graph, at 4) **9.** (graph, at −3)

11. (graph, at 4) **13.** (graph, 8 to 10) **15.** (graph, 0 to 10) **17.** It would imply that $3 < -2$, a false

statement. **19.** $z \ge 1$ (graph, at 1) **21.** $k \ge 5$ (graph, at 5) **23.** $n < -11$ (graph, at −11)

25. It must be reversed when multiplying or dividing by a negative number. **29.** $x < 6$ (graph, at 6)

31. $y \ge -10$ (graph, at −10) **33.** $t < -3$ (graph, at −3) **35.** $x \le 0$ (graph, at 0)

37. $r > 20$ (graph, at 20) **39.** $x \ge -3$ (graph, at −3) **41.** $r \ge -5$ (graph, at −5)

43. $x < 1$ (graph, at 1) **45.** $x \le 0$ (graph, at 0) **47.** $x \ge 4$ (graph, at 4) **49.** $p < 32$ (graph, at 32)

51. $x \ge \dfrac{5}{12}$ (graph, at $\frac{5}{12}$, 0) **53.** $k > -21$ (graph, at −21) **55.** $-1 < x < 2$ **57.** $-1 < x \le 2$

59. $-1 \le x \le 6$ (graph, −1 to 6; −2 0 2 4 6) **61.** $-\dfrac{11}{6} < m < -\dfrac{2}{3}$ (graph, $-\frac{11}{6}$ to $-\frac{2}{3}$; −2 0) **63.** $1 < p < 3$ (graph; 0 1 2 3 4)

65. $-26 \le z \le 6$ (graph, −26 to 6; −30 −20 −10 0 10) **67.** $-3 \le p \le 6$ (graph, −3 to 6; −4 −2 0 2 4 6) **69.** $-\dfrac{24}{5} \le r \le 0$ (graph, $-\frac{24}{5}$ to 0; −5 −4 −2 0 2)

71. $x = 4$ (graph; 1 2 3 4 5 6 7 8) **72.** $x > 4$ The solutions are all the numbers to the right of 4. (graph; 1 2 3 4 5 6 7 8)

73. $x < 4$ The solutions are all the numbers to the left of 4. (graph; 1 2 3 4 5 6 7) **74.** The graph would be all real numbers.

75. The graph would be all real numbers. **77.** all numbers less than 3 **79.** 83 or greater **81.** It is never more than 86 degrees Fahrenheit. **83.** $x \ge 500$ **85.** 32 or greater **87.** 15 minutes **89.** 18 **91.** 63
93. $2x - 36$ **95.** $3x^3 + 11x^2 + 2$

CHAPTER 2 REVIEW EXERCISES (PAGE 164)

1. $11m$ **3.** $16p^2 + 2p$ **5.** $-2m + 29$ **7.** 6 **9.** 7 **11.** 11 **13.** 5 **15.** 5 **17.** $\dfrac{64}{5}$ **19.** all real numbers

21. all real numbers **23.** no solution **25.** Democrats: 75; Republicans: 45 **27.** Hawaii: 6425 square miles; Rhode Island: 1212 square miles **29.** 80° **31.** $A = 28$ **33.** $V = 904.32$ **35.** $h = \dfrac{2A}{b + B}$ **37.** 100°, 100°
39. diameter: approximately 19.9 feet; radius: approximately 9.95 feet; area: approximately 311 square feet

41. $\dfrac{3}{2}$ **43.** $\dfrac{3}{4}$ **45.** $\dfrac{7}{2}$ **47.** $\dfrac{25}{19}$ **49.** 36 ounces **51.** 375 kilometers **53.** 25.5-ounce size **55.** 10 fives

57. $5000 at 5%; $5000 at 6% **59.** 13 hours **61.** $\dfrac{5}{6}$ hour or 50 minutes **63.** 10 meters **67.** (graph, at 7)
69. $y \ge -3$ (graph, at −3) **71.** $x \ge 3$ (graph, at 3) **73.** $x < -5$ (graph, at −5)

75. $-2 \le m \le \dfrac{3}{2}$ (graph, −2 to $\frac{3}{2}$; −3 −2 −1 0 1 2) **77.** 88 or greater **79.** 7 **81.** $x < 2$ **83.** 70 **85.** no solution **87.** $2\dfrac{1}{2}$ cups

89. Bush: 555 votes; Dukakis: 435 votes **91.** 8 quarts **93.** 9 centimeters or less **95.** 5 meters
97. 50 meters or less **99.** $\dfrac{2}{3}(x + 63) + (.5x + 69) + (4x + 7) = 180$

100. $\frac{2}{3}(x + 63) + \left(\frac{1}{2}x + 69\right) + (4x + 7) = 180$ **101.** 6   **102.** $4(x + 63) + (3x + 414) + (24x + 42) = 1080$;
multiplication property of equality   **103.** $4x + 252 + 3x + 414 + 24x + 42 = 1080$; distributive property   **104.** 12
105. 50°, 75°, 55°   **106.** $82\frac{2}{3}$

CHAPTER 2 TEST (PAGE 169)

[2.1] 1. $21x$ **2.** $15x - 3$ **[2.2–2.3] 3.** -6 **4.** $\frac{13}{4}$ **5.** -10.8 **6.** no solution   **7.** 21   **8.** 30
9. all real numbers   **[2.4] 10.** 7   **11.** Hawaii: 4021 square miles; Maui: 728 square miles; Kauai: 551 square miles
[2.5] 12. (a) $W = \frac{P - 2L}{2}$ or $W = \frac{P}{2} - L$ **(b)** 18   **[2.6] 13.** negative   **[2.5] 14.** 100°, 80°   **15.** 75°, 75°
16. 50°   **[2.6] 17.** -29 **18.** 8 slices for \$2.19   **19.** 2300 miles   **[2.7–2.8] 20.** \$8000 at 3%; \$14,000 at 4.5%
21. 10 liters   **22.** 4 hours   **[2.9] 23.** $x \le 4$ ⟵┼──┼─┼─┼┼⟶
                                                                                        4
24. $-2 < k \le 6$ ⟵┼◦┼┼┼┼┼┼●┼⟶
                                                                                        -2 6
25. 86 or greater

CUMULATIVE REVIEW (Chapters 1–2) (PAGE 170)

1. $\frac{3}{8}$ **2.** $\frac{3}{4}$ **3.** $\frac{31}{20}$ **4.** $\frac{551}{40}$ or $13\frac{31}{40}$ **5.** 6   **6.** $\frac{6}{5}$ **7.** $2x + 12$ **8.** $3x$ **9.** $\frac{1}{2}x$ 18   **10.** $\frac{6}{x - 12}$
11. 35 yards   **12.** $4\frac{1}{6}$ cups   **13.** $\frac{5}{12}$ **14.** 12   **15.** true   **16.** true   **17.** 7   **18.** 1
19. 13   **20.** -40 **21.** -12 **22.** undefined   **23.** 6   **24.** 28   **25.** 1   **26.** 0   **27.** $\frac{73}{18}$ **28.** -64
29. -134 **30.** $-\frac{29}{6}$ **31.** distributive property   **32.** commutative property   **33.** inverse property
34. identity property   **35.** $7p - 14$ **36.** $2k - 11$ **37.** 7   **38.** -4 **39.** -1 **40.** $-\frac{3}{5}$ **41.** 2   **42.** -13
43. 26   **44.** -12 **45.** $c = P - a - b$ **46.** $s = \frac{P}{4}$ **47.** $z \le 2$ ⟵┼─┼┼─●─┼─┼⟶
                                                                                        2
48. $r \le 1$ ⟵┼─┼─┼─●─┼⟶   **49.** \$275 or more   **50.** 4 centimeters, 9 centimeters, 27 centimeters
                    1

CHAPTER 3 POLYNOMIALS AND EXPONENTS

SECTION 3.1 (PAGE 178)

◆ **CONNECTIONS** **Page 172:** Using a calculator, $\sqrt{.9} \approx .9486832981$. Using the polynomial, $\sqrt{.5} \approx .7084960938$,
$\sqrt{.7} \approx .8367460938$, $\sqrt{2} \approx 1.3984375$, $\sqrt{5} \approx -5$. Using a calculator, $\sqrt{.5} \approx .7071067812$, $\sqrt{.7} \approx .8366600265$,
$\sqrt{2} \approx 1.414213562$, $\sqrt{5} \approx 2.236067977$.   **Page 175:** For 1984, the value is about 581 billion; for 1985 it is about 580 billion.
The number of cigarettes sold in those years did not differ much, compared to the change from 1981 to 1984, for example.

EXERCISES **1.** 1; 6   **3.** 1; 1   **5.** 2; -19, -1 **7.** 2; 1, 8   **9.** $2m^5$ **11.** $-r^5$ **13.** cannot be
simplified   **15.** $-5x^5$ **17.** $5p^9 + 4p^7$ **19.** $-2y^2$ **21.** sometimes   **23.** sometimes   **25.** always   **27.** never
29. already simplified; 4; binomial   **31.** $11m^4 - 7m^3 - 3m^2$; 4; trinomial   **33.** x^4; 4; monomial
35. 7; 0; monomial   **37. (a)** 0 **(b)** 6   **39. (a)** 36 **(b)** -12 **41. (a)** 14 **(b)** -19
43. 4; \$6.00   **44.** 6; \$27   **45.** 2.5; 130   **46.** approximately 579 billion   **47.** $5m^2 + 3m$ **49.** $4x^4 - 4x^2$
51. $\frac{7}{6}x^2 - \frac{2}{15}x + \frac{5}{6}$ **53.** $6m^3 + m^2 + 12m - 14$ **55.** $15m^3 - 13m^2 + 8m + 11$ **59.** $5m^2 - 14m + 6$
61. $4x^3 + 2x^2 + 5x$ **63.** $-11y^4 + 8y^2 + y$ **65.** $a^4 - a^2 + 1$
67. $5m^2 + 8m - 10$ **69.** $8x^2 + 8x + 6$ **71. (a)** $23y + 5t$ **(b)** approximately 37.26°, 79.52°, and 63.22°   **73.** $4b - 5c$
75. $6x - xy - 7$ **77.** $-3x^2y - 15xy - 3xy^2$ **79.** $-6x^2 + 6x - 7$ **81.** $-7x - 1$ **85.** 64   **87.** 625
89. $\frac{8}{27}$ **91.** 128

SECTION 3.2 (PAGE 185)

◊ **CONNECTIONS** **Page 181:** With compound interest: $304.16; with simple interest: $300; compound interest produces more interest.

EXERCISES **1.** 3^7 **3.** $(-6)^4$ **5.** w^6 **7.** $\frac{1}{4^4}$ **9.** $(-7x)^4$ **11.** $\left(\frac{1}{2}\right)^6$ **15.** base: 3; exponent: 5; 243

17. base: -3; exponent: 5; -243 **19.** base: $-6x$; exponent: 4 **21.** base: x; exponent: 4 **25.** 5^8 **27.** 4^{12}
29. $(-7)^9$ **31.** t^{24} **33.** $-56r^7$ **35.** $42p^{10}$ **39.** $14x^4$; $45x^8$ **41.** $5a^2$; $-140a^6$ **43.** 4^6 **45.** t^{20}

47. 7^3r^3 **49.** $5^5x^5y^5$ **51.** 5^{12} **53.** -8^{15} **55.** $8q^3r^3$ **57.** $\frac{1}{2^3}$ **59.** $\frac{a^3}{b^3}$ **61.** $\frac{9^8}{5^8}$ **63.** $\frac{5^5}{2^5}$ **65.** $\frac{9^5}{8^3}$

67. $2^{12}x^{12}$ **69.** $(-6)^5p^5$ **71.** $6^5x^{10}y^{15}$ **73.** x^{21} **75.** $2^2w^4x^{26}y^7$ **77.** $-r^{18}s^{17}$ **79.** $\frac{5^3a^6b^{15}}{c^{18}}$ **83.** $12x^5$

85. $6p^7$ **87.** $125x^6$ **91.** $304.16 **93.** $1843.88 **95.** $5x + 20$ **97.** $8a + 24b$

SECTION 3.3 (PAGE 191)

EXERCISES **1.** associative; commutative; associative **3.** $40a^{14}$ **5.** $-6m^2 - 4m$ **7.** $24p - 18p^2 + 36p^4$
9. $-16z^2 - 24z^3 - 24z^4$ **11.** $14x^5y^3 + 21x^3y^2 - 28x^2y^2$ **13.** $12x^3 + 26x^2 + 10x + 1$
15. $20m^4 - m^3 - 8m^2 - 17m - 15$ **17.** $6x^6 - 3x^5 - 4x^4 + 4x^3 - 5x^2 + 8x - 3$ **19.** $5x^4 - 13x^3 + 20x^2 + 7x + 5$
21. $n^2 + n - 6$ **23.** $x^2 - 36$ **25.** $8r^2 - 10r - 3$ **27.** $9x^2 - 4$ **29.** $9q^2 + 6q + 1$ **31.** $3x^2 - 5xy - 2y^2$

33. $6t^2 + 23st + 20s^2$ **35.** $-3t^2 - 14t + 24$ **37.** $36x^2 + 24x + 4$ **41.** $6p^2 - \frac{5}{2}pq - \frac{25}{12}q^2$ **43.** $2m^6 - 5m^3 - 12$

45. $2k^5 - 6k^3h^2 + k^2h^2 - 3h^4$ **47.** $6p^8 + 15p^7 + 12p^6 + 36p^5 + 15p^4$ **49.** $-24x^8 - 28x^7 + 32x^6 + 20x^5$
51. $14x + 49$ **53.** $\pi x^2 - 9$ **55.** $30x + 60$ **56.** $30x + 60 = 600$ **57.** 18 **58.** 10 yards by 60 yards
59. $2100 **60.** 140 yards **61.** $1260 **62. (a)** $30kx + 60k$ (dollars) **(b)** $6rx + 32r$ (dollars) **63.** $9m^2$
65. $4r^2$ **67.** $16x^4$

SECTION 3.4 (PAGE 196)

◊ **CONNECTIONS** **Page 195: 1.** $b^2 - a^2$ **2.** $(b - a)^2$, $a(b - a)$, $a(b - a)$; $(b - a)^2 + 2a(b - a) =$
$(b - a)[(b - a) + 2a] = (b - a)[b - a + 2a] = (b - a)(b + a)$ **3.** Both expressions represent the same area.

EXERCISES **1. (a)** $4x^2$ **(b)** $12x$ **(c)** 9 **(d)** $4x^2 + 12x + 9$ **3.** $p^2 + 4p + 4$ **5.** $a^2 - 2ac + c^2$

7. $16x^2 - 24x + 9$ **9.** $64t^2 + 112st + 49s^2$ **11.** $25x^2 + 4xy + \frac{4}{25}y^2$ **13.** 1225 **14.** $30^2 + 2(30)(5) + 5^2$

15. 1225 **16.** They are equal. **17. (a)** $49x^2$ **(b)** 0 **(c)** $-9y^2$ **(d)** $49x^2 - 9y^2$. Because 0 is the identity element for
addition, it is not necessary to write "$+ 0$". **19.** $q^2 - 4$ **21.** $4w^2 - 25$ **23.** $100x^2 - 9y^2$ **25.** $4x^4 - 25$

27. $49x^2 - \frac{9}{49}$ **29. (a)** $4x^2 - 9$ **(b)** 55 square units **31.** $(a + b)^2$ **32.** a^2 **33.** $2ab$ **34.** b^2

35. $a^2 + 2ab + b^2$ **37.** 9999 **39.** 39,999 **41.** $399\frac{3}{4}$ **43.** $m^3 - 15m^2 + 75m - 125$ **45.** $8a^3 + 12a^2 + 6a + 1$

47. $y^4 + 16y^3 + 96y^2 + 256y + 256$ **49.** $81r^4 - 216r^3t + 216r^2t^2 - 96rt^3 + 16t^4$ **51.** $(a + b)^2 = (2 + 5)^2 = 49$;
$a^2 + b^2 = 2^2 + 5^2 = 29$; They are not equal. **55.** $\frac{1}{2}m^2 - 2n^2$ **57.** $9a^2 - 4$ **59.** $\pi x^2 + 4\pi x + 4\pi$ **61.** $\frac{1}{5}$

63. -4 **65.** 10 **67.** 3

SECTION 3.5 (PAGE 205)

◊ **CONNECTIONS** **Page 201:** Approximately .767 gram; approximately .589 gram; yes; approximately 13.1 years

EXERCISES **1.** 1 **3.** 1 **5.** -1 **7.** 0 **9.** 0 **11.** 1 **13.** 2 **15.** $\frac{1}{64}$ **17.** 16 **19.** $\frac{49}{36}$ **21.** $\frac{1}{81}$

23. $\frac{8}{15}$ **25.** negative **27.** negative **29.** positive **31.** zero **33.** 5^3 **35.** $\frac{1}{9}$ **37.** $\frac{1}{6^5}$ **39.** x^{15} **41.** 6^3

43. $2r^4$ **45.** $\frac{5^2}{4^3}$ **47.** $\frac{p^5}{q^8}$ **49.** r^9 **51.** $\frac{x^5}{6}$ **53.** 1 **54.** $\frac{5^2}{5^2}$ **55.** 5^0 **56.** $1 = 5^0$; This supports the definition for
0 as an exponent. **57.** power rules (b) and (c) **59.** A factor with a negative exponent may be moved across the fraction bar if
the sign of the exponent is changed to its opposite. Also $2^4 = 16$.

61. 7^3 or 343   **63.** $\dfrac{1}{x^2}$ **65.** $\dfrac{64x}{9}$ **67.** $\dfrac{x^2 z^4}{y^2}$ **69.** $6x$ **71.** $\dfrac{1}{m^{10}n^5}$ **73.** 5^{6r} **75.** $\dfrac{1}{x^{11a}}$ **77.** q^{5k}

79. $\dfrac{p^{3y}}{6^y}$ **81.** $2p + 1 + \dfrac{4}{p}$ **83.** $\dfrac{m^2}{3} + 3m - 2$

SECTION 3.6 (PAGE 209)

EXERCISES   **1.** $-6x^4$ **3.** $5x^7y$ **7.** $30m^3 - 10m + 5$ **9.** $5m^4 - 8m^3 + 4m^2$ **11.** $4m^4 - 2m^2 + 2m$
13. $\dfrac{m^4}{2} - 2m + \dfrac{4}{m}$ **15.** $4x^3 - 3x^2 + 2x$ **17.** $1 + 5x - 9x^2$ **19.** $\dfrac{12}{x} + 8 + 2x$ **21.** $\dfrac{4x^2}{3} + x + \dfrac{2}{3x}$
23. $9r^3 - 12r^2 - 2r + \dfrac{26}{3} - \dfrac{2}{3r}$ **25.** $-m^2 + 3m - \dfrac{4}{m}$ **27.** $4 - 3a + \dfrac{5}{a}$ **29.** $\dfrac{12}{x} - \dfrac{6}{x^2} + \dfrac{14}{x^3} - \dfrac{10}{x^4}$ **33.** 8; 13; no
35. $5x^3 + 4x^2 - 3x + 1$ **37.** $-63m^4 - 21m^3 - 35m^2 + 14m$ **39.** 1423
40. $(1 \times 10^3) + (4 \times 10^2) + (2 \times 10^1) + (3 \times 10^0)$ **41.** $x^3 + 4x^2 + 2x + 3$ **42.** They are similar in that the coefficients of powers of 10 are equal to the coefficients of the powers of x. They are different in that one is a constant while the other is a polynomial. They are equal if $x = 10$ (the base of our decimal system).   **43.** $-24k^3 + 36k^2 - 6k$
45. $-16k^3 - 10k^2 + 3k + 3$ **47.** $10x^2 - 4x + 11$

SECTION 3.7 (PAGE 214)

◆ **CONNECTIONS**   **Page 214:** **1.** -104 **2.** -104 **3.** They are both -104.   **4.** The answers should agree.

EXERCISES   **1.** $x + 2$ **3.** $2y - 5$ **5.** $p - 4 + \dfrac{44}{p + 6}$ **7.** $r - 5$ **9.** $6m - 1$ **11.** $2a - 14 + \dfrac{74}{2a + 3}$

13. $4x^2 - 7x + 3$ **15.** $2k + 5 + \dfrac{10}{7k - 8}$ **17.** $x^2 - x + 2$ **19.** $4k^3 - k + 2$ **21.** $5y^3 + 2y - 3$

23. $3k^2 + 2k - 2 + \dfrac{6}{k - 2}$ **25.** $2p^3 - 6p^2 + 7p - 4 + \dfrac{14}{3p + 1}$ **27.** (a) is correct,   (b) is incorrect.

28. (a) is incorrect,   (b) is correct.   **29.** (a) is correct,   (b) is incorrect.   **31.** $r^2 - 1 + \dfrac{4}{r^2 - 1}$ **33.** $y^2 - y + 1$

35. $a^2 + 1$ **37.** $\dfrac{3}{2}a - 10 + \dfrac{77}{2a + 6}$ **39.** $x^2 - 4x + 2 + \dfrac{9x - 4}{x^2 + 3}$ **41.** $x^3 + 3x^2 - x + 5$ **45.** $x^2 + x - 3$ units
47. $5x^2 - 11x + 14$ hours   **49.** 64,270   **51.** 1230   **53.** 3.4   **55.** .237

SECTION 3.8 (PAGE 219)

◆ **CONNECTIONS**   **Page 217: 1.** 458×10^7 **2.** 45.8×10^8 **3.** 4.58×10^9 **4.** Answers will vary.

EXERCISES   **1.** in scientific notation   **3.** not in scientific notation; 5.6×10^6 **5.** not in scientific
notation; 8×10^1 **7.** not in scientific notation; 4×10^{-3} **11.** 5.876×10^9 **13.** 8.235×10^4 **15.** 7×10^{-6}
17. 2.03×10^{-3} **19.** 750,000   **21.** 5,677,000,000,000   **23.** 6.21   **25.** .00078   **27.** .000000005134
29. 600,000,000,000   **31.** 15,000,000   **33.** 60,000   **35.** .0003   **37.** 40   **39.** .000013   **41.** 635,000,000,000
43. 1.15×10^6 **45.** 1.5×10^2; .00023; .006   **47.** .000002   **49.** 4.2×10^{42} **51.** about 15,300 seconds
53. (a) 1.75×10^6 **(b)** about $\$9.63 \times 10^{10}$ **55. (a)** 6.8×10^{17} **(b)** 1.7×10^7 **57. (a)** 8.4×10^2
(b) 2.1×10^{18} **59.** 1, 2, 3, 6, 9, 18   **61.** 1, 2, 3, 4, 6, 8, 12, 16, 24, 48

CHAPTER 3 REVIEW EXERCISES (PAGE 223)

1. $22m^2$; degree 2; monomial   **3.** already in descending powers; degree 5; none of these
5. $-8x^5 + 9x^3 - 7x$; degree 5; trinomial   **7.** $7r^4 - 4r^3 + 1$; degree 4; trinomial   **9.** $a^3 + 4a^2$
11. $-13k^4 - 15k^2 + 18k$ **13.** $7r^4 - 4r^3$ **15.** 4^{11} **17.** $-72x^7$ **19.** 19^5x^5 **21.** $5p^4t^4$ **23.** $3^3x^6y^9$
25. $6^2x^{16}y^4z^{16}$ **27.** $-63x^6$ **29.** $a^3 - 2a^2 - 7a + 2$ **31.** $5p^5 - 2p^4 - 3p^3 + 25p^2 + 15p$
33. $6k^2 - 9k - 6$ **35.** $12k^2 - 32kq - 35q^2$ **37.** $2x^2 + x - 6$ **41.** $a^2 + 8a + 16$ **43.** $4r^2 + 20rt + 25t^2$
45. $36m^2 - 25$ **47.** $25a^2 - 36b^2$ **49.** $r^3 + 6r^2 + 12r + 8$ **53.** $\dfrac{4}{3}\pi(x + 1)^3$ or $\dfrac{4}{3}\pi x^3 + 4\pi x^2 + 4\pi x + \dfrac{4}{3}\pi$ cu. in.

55. 0   **57.** $-\dfrac{1}{49}$ **59.** $\dfrac{3}{4}$ **61.** $\dfrac{1}{81}$ **63.** $\dfrac{1}{36}$ **65.** x^2 **67.** $\dfrac{r^2}{6}$ **69.** $\dfrac{1}{a^3b^5}$ **71.** $\dfrac{-5y^2}{3}$ **73.** $y^3 - 2y + 3$

75. $-2m^2n + mn + \dfrac{6n^3}{5}$ **79.** $2a^2 + 3a - 1 + \dfrac{6}{5a - 3}$ **81.** $m^2 + 4m - 2$ **83.** 2.8988×10^{10}
85. 8.24×10^{-8} **87.** 78,300,000   **89.** .00000000000995   **91.** 800   **93.** .025   **95.** $\$1.558 \times 10^6$
97. .000000001   **99.** $\$1.2 \times 10^{10}$ **101.** $\dfrac{6^3r^6p^3}{5^3}$ **103.** $\dfrac{1}{16}$ **105.** $4m + 3 + \dfrac{5}{3m - 5}$ **107.** $\dfrac{2}{3m^2}$

109. r^{13} **111.** $-y^2 - 4y + 4$ **113.** $25m^6$ **115.** $\dfrac{5}{2} - \dfrac{4}{5xy} + \dfrac{3x}{2y^2}$ **117.** $\dfrac{1}{5^{11}}$ **119.** 5^8 **121.** $\dfrac{1}{x^4y^{12}}$

122. (a) 2 **(b)** 92 **(c)** $x^3 + x^2 - 10x - 6$ **(d)** Both results are 94.

123. (a) 1 **(b)** 33 **(c)** $-x^3 - x^2 + 12x$ **(d)** Both results are -32.   **124. (a)** -6 **(b)** 12

(c) $x^4 - 2x^3 - 14x^2 + 30x + 9$ **(d)** Both results are -72.   **125. (a)** 3 **(b)** 183 **(c)** $x^2 + 4x + 1$ **(d)** Both results are 61.   **126.** Answers will vary.   **127.** We could not choose 3 because it would make the denominator equal to zero. It is the only invalid replacement for x.

CHAPTER 3 TEST (PAGE 228)

[3.1] 1. $-7x^2 + 8x$; 2; binomial   **2.** $4n^4 + 13n^3 - 10n^2$; 4; trinomial   **3.** $4t^4 + t^3 - 6t^2 - t$
4. $-2y^2 - 9y + 17$ **5.** $-21a^3b^2 + 7ab^5 - 5a^2b^2$ **6.** $-12t^2 + 5t + 8$ **[3.3] 7.** $-27x^5 + 18x^4 - 6x^3 + 3x^2$
8. $t^2 - 5t - 24$ **9.** $8x^2 + 2xy - 3y^2$ **[3.4] 10.** $25x^2 - 20xy + 4y^2$ **11.** $100v^2 - 9w^2$ **12.** $2r^3 + r^2 - 16r + 15$
13. $9x^2 + 54x + 81$ **[3.5] 14.** $\dfrac{1}{625}$ **15.** 2   **16.** $\dfrac{7}{12}$ **[3.2] 17.** $9x^3y^5$ **[3.5] 18.** 8^5 **19.** x^2y^6

20. Disagree, because $3^{-4} = \dfrac{1}{3^4} = \dfrac{1}{81}$, which is positive.   **[3.6] 21.** $4y^2 - 3y + 2 + \dfrac{5}{y}$ **22.** $-3xy^2 + 2x^3y^2 + 4y^2$

[3.7] 23. $3x^2 + 6x + 11 + \dfrac{26}{x - 2}$ **[3.8] 24.** $.00019$ **25.** 2.3 pounds

CUMULATIVE REVIEW (Chapters 1–3) (PAGE 229)

1. $\dfrac{7}{4}$ **2.** 5   **3.** $\dfrac{19}{24}$ **4.** $-\dfrac{1}{20}$ **5.** $31\dfrac{1}{4}$ cubic yards   **6.** $1836   **7.** 1, 3, 5, 9, 15, 45   **8.** positive

9. -8 **10.** 24   **11.** $\dfrac{1}{2}$ **12.** -4 **13.** associative property   **14.** inverse property   **15.** distributive property

16. 10   **17.** $\dfrac{13}{4}$ **18.** no solution   **19.** $r = \dfrac{d}{t}$ **20.** -5 **21.** -12 **22.** 20   **23.** all real numbers

24. Janet: $48; Louis: $8   **25.** 4   **26.** 81 roosters; 109 hens   **27.** $x \geq 10$ **28.** $x < -\dfrac{14}{5}$

29. $-4 \leq x < 2$ **30.** 11 feet and 22 feet   **31.** $\dfrac{5}{4}$ or $1\dfrac{1}{4}$ **32.** 2   **33.** 1   **34.** $\dfrac{2b}{a^{10}}$ **35.** 3.45×10^4
36. $11x^3 - 14x^2 - x + 14$ **37.** $18x^7 - 54x^6 + 60x^5$ **38.** $63x^2 + 57x + 12$ **39.** $25x^2 + 80x + 64$
40. $y^2 - 2y + 6$

CHAPTER 4   FACTORING AND APPLICATIONS

SECTION 4.1 (PAGE 237)

EXERCISES   1. 4   **3.** 4   **5.** 6   **7.** 1   **11.** 8   **13.** $10x^3$ **15.** $6m^3n^2$ **17.** xy^2 **19.** 6
21. yes; x^3y^2 **23.** 2   **25.** x **27.** $3m^2$ **29.** $2z^4$ **31.** $2mn^4$ **33.** $-7x^3y^2$ **35.** $12(y - 2)$ **37.** $10a(a - 2)$
39. $5y^6(13y^4 + 7)$ **41.** no common factor (except 1)   **43.** $8m^2n^2(n + 3)$ **45.** $13y^2(y^6 + 2y^2 - 3)$
47. $9qp^3(5q^3p^2 + 4p^3 + 9q)$ **49.** $a^3(a^2 + 2b^2 - 3a^2b^2 + 4ab^3)$ **51.** $(x + 2)(c - d)$
53. $(m + 2n)(m + n)$ **55.** not in factored form   **57.** in factored form   **59.** not in factored form
61. yes; $(y + 4)(18x^2 + 7)$ **63.** $(p + 4)(p + 3)$ **65.** $(a - 2)(a + 5)$ **67.** $(z + 2)(7z - a)$
69. $(3r + 2y)(6r - x)$ **71.** $(a^2 + b^2)(3a + 2b)$ **73.** $(1 - a)(1 - b)$ **75.** $(4m - p^2)(4m^2 - p)$
77. $(5 - 2p)(m + 3)$ **79.** $(6r - y)(3r + 2y)$ **81.** $(a^5 - 3)(1 + 2b)$ **83.** commutative property and associative
property   **84.** $(2xy - 8x) + (-3y) + 12$; $2x$ **85.** $(2xy - 8x) + (-3y + 12)$; yes   **86.** $2x(y - 4) - 3(y - 4)$
87. No, because it is the difference between two terms, $2x(y - 4)$ and $3(y - 4)$.   **88.** $(2x - 3)(y - 4)$; yes
89. (a) yes   **91.** $x^2 - 3x - 54$ **93.** $x^2 + 9x + 14$

SECTION 4.2 (PAGE 242)

EXERCISES   1. 1 and 12, -1 and -12, 2 and 6, -2 and -6, 3 and 4, -3 and -4;   the pair with a sum of 7 is 3 and 4.
3. 1 and -24, -1 and 24, 2 and -12, -2 and 12, 3 and -8, -3 and 8, 4 and -6, -4 and 6;   the pair with a sum of -5 is 3 and
-8.   **5.** 1 and 27, -1 and -27, 3 and 9, -3 and -9;   the pair with a sum of 28 is 1 and 27.   **7.** 1 and -48, -1 and 48, 2
and -24, -2 and 24, 3 and -16, -3 and 16, 4 and -12, -4 and 12, 6 and -8, -6 and 8;   the pair with a sum of 2 is -6

and 8.   **9.** (c)   **11.** $p + 6$ **13.** $x + 11$ **15.** $x - 8$ **17.** $y - 5$ **19.** $x + 11$ **21.** $y - 9$
23. $(y + 8)(y + 1)$ **25.** $(b + 3)(b + 5)$ **27.** $(m + 5)(m - 4)$ **29.** $(y - 5)(y - 3)$ **31.** prime
33. $(t - 4)^2$ or $(t - 4)(t - 4)$ **35.** $(r - 6)(r + 5)$ **37.** prime   **39.** $(r + 2a)(r + a)$
41. $(t + 2z)(t - 3z)$ **43.** $(x + y)(x + 3y)$ **45.** $(v - 5w)(v - 6w)$ **47.** $4(x + 5)(x - 2)$ **49.** $2t(t + 1)(t + 3)$
51. $2x^4(x - 3)(x + 7)$ **53.** $mn(m - 6n)(m - 4n)$ **57.** $a^3(a + 4b)(a - b)$ **59.** $yz(y + 3z)(y - 2z)$
61. $z^8(z - 7y)(z + 3y)$ **63.** $(a + b)(x + 4)(x - 3)$ **65.** $(2p + q)(r - 9)(r - 3)$ **67.** 1   **68.** -1
69. The sums are opposites as well.   **70.** The product is $x^2 - x - 12$. This is not correct because the middle term is incorrect.
71. The product is $x^2 + x - 12$. This is correct because we obtain the exact trinomial we were given to factor.   **72.** It is the
opposite of what it should be.   **73.** reverse the signs of the two second terms of the binomials   **74.** $(x - 5)(x + 3)$
75. $2y^2 + y - 28$ **77.** $15z^2 - 4z - 4$ **79.** $8p^2 - 10p - 3$

SECTION 4.3 (PAGE 249)

EXERCISES   **1.** (b)   **3.** (a)   **5.** (a)   **7.** $2a + 5b$ **9.** $x^2 + 3x - 4$; $x + 4, x - 1$, or $x - 1, x + 4$
11. $2z^2 - 5z - 3$; $2z + 1, z - 3$, or $z - 3, 2z + 1$

The order of the factors is irrelevant in Exercises 15–63.

15. $(2x + 1)(x + 3)$ **17.** $(3a + 7)(a + 1)$ **19.** $(4r - 3)(r + 1)$ **21.** $(3m - 1)(5m + 2)$ **23.** $(4m + 1)(2m - 3)$
25. $(4x + 3)(5x - 1)$ **27.** $(3m + 1)(7m + 2)$ **29.** $(4y - 1)(5y + 11)$ **31.** $(2b + 1)(3b + 2)$ **33.** $3(4x - 1)(2x - 3)$
35. $q(5m + 2)(8m - 3)$ **37.** $2m(m - 4)(m + 5)$ **39.** $3n^2(5n - 3)(n - 2)$ **41.** $3x^3(2x + 5)(3x - 5)$
43. $y^2(5x - 4)(3x + 1)$ **45.** $(3p + 4q)(4p - 3q)$ **47.** $(5a + 2b)(5a + 3b)$ **49.** $(3a - 5b)(2a + b)$
51. $m^4n(3m + 2n)(2m + n)$ **53.** $(5 - x)(1 - x)$ **55.** $(4 + 3x)(4 + x)$ **57.** $-5x(2x + 7)(x - 4)$
59. $-1(x + 7)(x - 3)$ **61.** $-1(3x + 4)(x - 1)$ **63.** $-1(a + 2b)(2a + b)$ **67.** $(m + 1)^3(5q - 2)(5q + 1)$
69. $(r + 3)^3(5x + 2y)(3x - 8y)$ **71.** $-4, 4$ **73.** $-11, -7, 7, 11$ **75.** $5 \cdot 7$ **76.** $(-5)(-7)$
77. The product of $3x - 4$ and $2x - 1$ is indeed $6x^2 - 11x + 4$ **78.** The product of $4 - 3x$ and $1 - 2x$ is indeed $6x^2 - 11x + 4$.
79. The factors in Exercise 78 are the opposites of the factors in Exercise 77.   **80.** $(3 - 7t)(5 - 2t)$ **81.** $-a$; $-b$
82. $-P$; $-Q$ **83.** $49p^2 - 9$ **85.** $r^4 - \dfrac{1}{4}$ **87.** $9t^2 + 24t + 16$

SECTION 4.4 (PAGE 257)

◈ **CONNECTIONS**   **Page 254:** When $x^3 + 1$ is divided by $x + 1$, the quotient is $x^2 - x + 1$. Therefore, $x + 1$ is a factor of
$x^3 + 1$. On the other hand, when $x^3 + 1$ is divided by $x - 1$, the quotient is $x^2 + x + 1$ and the remainder is 2. When the division
process leads to a remainder (other than 0), the divisor is not a factor.

EXERCISES   **1.** 1; 4; 9; 16; 25; 36; 49; 64; 81; 100; 121; 144; 169; 196; 225; 256; 289; 324;
361; 400   **3.** 1; 8; 27; 64; 125; 216; 343; 512; 729; 1000
5. (a) both of these   (b) a perfect cube   (c) a perfect square   (d) a perfect square

7. $(y + 5)(y - 5)$ **9.** $(3r + 2)(3r - 2)$ **11.** $\left(6m + \dfrac{4}{5}\right)\left(6m - \dfrac{4}{5}\right)$ **13.** $4(3x + 2)(3x - 2)$
15. $(14p + 15)(14p - 15)$ **17.** $(4r + 5a)(4r - 5a)$ **19.** prime   **21.** $(p^2 + 7)(p^2 - 7)$
23. $(x^2 + 1)(x + 1)(x - 1)$ **25.** $(p^2 + 16)(p + 4)(p - 4)$ **29.** $(w + 1)^2$ **31.** $(x - 4)^2$
33. $\left(t + \dfrac{1}{2}\right)^2$ **35.** $(x - .5)^2$ **37.** $2(x + 6)^2$ **39.** $(4x - 5)^2$ **41.** $(7x - 2y)^2$ **43.** $(8x + 3y)^2$
45. $-2(5h - 2y)^2$ **47.** $(a + 1)(a^2 - a + 1)$ **49.** $(a - 1)(a^2 + a + 1)$ **51.** $(p + q)(p^2 - pq + q^2)$
53. $(3x - 1)(9x^2 + 3x + 1)$ **55.** $(2p + 9q)(4p^2 - 18pq + 81q^2)$ **57.** $(y - 2x)(y^2 + 2yx + 4x^2)$
59. $(3a - 4b)(9a^2 + 12ab + 16b^2)$ **61.** $(5t + 2s)(25t^2 - 10ts + 4s^2)$
63. $(x^3 - 1)(x^3 + 1)$ **64.** $(x - 1)(x^2 + x + 1)(x + 1)(x^2 - x + 1)$ **65.** $(x^2 - 1)(x^4 + x^2 + 1)$
66. $(x - 1)(x + 1)(x^4 + x^2 + 1)$ **67.** The result in Exercise 64 is completely factored.   **68.** Show that $x^4 + x^2 + 1 =$
$(x^2 + x + 1)(x^2 - x + 1)$.   **69.** difference of squares   **70.** $(x - 3)(x^2 + 3x + 9)(x + 3)(x^2 - 3x + 9)$ **71.** $4mn$
73. $(m - p + 2)(m + p)$ **75.** $(x + 1)^2 - 4$ **76.** $[(x + 1) - 2][(x + 1) + 2]$ **77.** $(x - 1)(x + 3)$ **78.** $x^2 + 2x - 3$
79. $(x - 1)(x + 3)$ **80.** The results are the same. (Preference is a matter of individual choice.)   **83.** 10   **85.** 9
87. 4   **89.** $\dfrac{9}{4}$ **91.** $\dfrac{2}{3}$

SUMMARY: Exercises on Factoring (PAGE 259)

1. $(a - 6)(a + 2)$ **3.** $6(y - 2)(y + 1)$ **5.** $6(a + 2b + 3c)$ **7.** $(p - 11)(p - 6)$ **9.** $(5z - 6)(2z + 1)$
11. $(m + n + 5)(m - n)$ **13.** $8a^3(a - 3)(a + 2)$ **15.** $(z - 5a)(z + 2a)$ **17.** $(x - 5)(x - 4)$ **19.** $(3n - 2)(2n - 5)$
21. $4(4x + 5)$ **23.** $(3y - 4)(2y + 1)$ **25.** $(6z + 1)(z + 5)$ **27.** $(2k - 3)^2$ **29.** $6(3m + 2z)(3m - 2z)$
31. $(3k - 2)(k + 2)$ **33.** $7k(2k + 5)(k - 2)$ **35.** $(y^2 + 4)(y + 2)(y - 2)$ **37.** $8m(1 - 2m)$ **39.** $(z - 2)(z^2 + 2z + 4)$
41. prime   **43.** $8m^3(4m^6 + 2m^2 + 3)$ **45.** $(4r + 3m)^2$ **47.** $(5h + 7g)(3h - 2g)$ **49.** $(k - 5)(k - 6)$ **51.** $3k(k - 5)(k + 1)$

53. $(10p + 3)(100p^2 - 30p + 9)$ **55.** $(2 + m)(3 + p)$ **57.** $(4z - 1)^2$ **59.** $3(6m - 1)^2$ **61.** prime
63. $8z(4z - 1)(z + 2)$ **65.** $(4 + m)(5 + 3n)$ **67.** $2(3a - 1)(a + 2)$ **69.** $(a - b)(a^2 + ab + b^2 + 2)$
71. $(8m - 5n)^2$ **73.** $(4k - 3h)(2k + h)$ **75.** $(m + 2)(m^2 + m + 1)$ **77.** $(5y - 6z)(2y + z)$ **79.** $(8a - b)(a + 3b)$
81. $(x^3t^3 + t^3) + (2x^3t^2 + 2t^2) - (tx^3 + t) - (2x^3 + 2)$ **82.** $t^3(x^3 + 1) + 2t^2(x^3 + 1) - t(x^3 + 1) - 2(x^3 + 1)$
83. $(x^3 + 1)(t^3 + 2t^2 - t - 2)$ **84.** $(x^3 + 1)(t^2 - 1)(t + 2)$ **85.** $(x + 1)(x^2 - x + 1)(t + 1)(t - 1)(t + 2)$

SECTION 4.5 (PAGE 266)

◇ **CONNECTIONS** **Page 263: 1.** 4 seconds **2.** 64 feet

EXERCISES **1.** $4, -5$ **3.** $-\dfrac{8}{3}, -7$ **5.** $0, -4$ **7.** $0, \dfrac{4}{3}$ **13.** $-2, -1$ **15.** $1, 2$ **17.** $-8, 3$

19. $-1, 3$ **21.** $-2, -1$ **23.** -4 **25.** $-2, \dfrac{1}{3}$ **27.** $-\dfrac{4}{3}, \dfrac{1}{2}$ **29.** $-\dfrac{2}{3}$ **31.** $-3, 3$ **33.** $-\dfrac{7}{4}, \dfrac{7}{4}$

35. $-11, 11$ **37.** Another solution is -11. **39.** $0, 7$ **41.** $0, \dfrac{1}{2}$ **43.** $2, 5$ **45.** $-4, \dfrac{1}{2}$ **47.** $-12, \dfrac{11}{2}$

49. $-1, 3$ **51.** $-\dfrac{5}{2}, \dfrac{1}{3}, 5$ **53.** $-\dfrac{7}{2}, -3, 1$ **55.** $-\dfrac{7}{3}, 0, \dfrac{7}{3}$ **57.** $-2, 0, 4$ **59.** $-5, 0, 4$ **61.** $-3, 0, 5$

63. $-\dfrac{4}{3}, -1, \dfrac{1}{2}$ **65.** $-\dfrac{2}{3}, 4$ **69.** $x + 8$ **70.** $5x - 2$ **71.** $(x + 8)(5x - 2) = 0$; $-8, \dfrac{2}{5}$

72. $5x^2 + 38x - 16 = 0$; $-8, \dfrac{2}{5}$ **73.** $x(5x + 38) = 16$; $-8, \dfrac{2}{5}$ **74.** Answers will vary. One example is $5x^2 = 16 - 38x$.

75. Florida: 67 counties; California: 58 counties **77.** 7 meters

SECTION 4.6 (PAGE 271)

◇ **CONNECTIONS** **Page 271:** The diagonal of the floor should be $\sqrt{208} \approx 14.4$ ft, or about 14 feet, 5 inches. The carpenter is off by 3 inches and must correct the error to avoid major construction problems.

EXERCISES **1. (a)** $80 = (x + 8)(x - 8)$ **(b)** 12 **(c)** length: 20 units; width: 4 units
3. (a) $45 = (2x + 1)(x + 1)$ **(b)** 4 **(c)** base: 9 units; height: 5 units **5. (a)** $36\pi = \pi(x + 2)^2$ **(b)** 4 **(c)** radius: 6 units
7. length: 7 inches; width: 4 inches **9.** length: 11 inches; width: 8 inches **11.** poster: 7 feet; sign: 9 feet
13. base: 12 inches; height: 5 inches **15.** height: 13 inches; width: 10 inches **17.** 12 centimeters **19.** 8 feet
21. 6 meters **23.** 1 second **25.** 3 seconds **27.** 1 second and 3 seconds **29.** 4 seconds **31.** 9 centimeters
33. 256 feet **35.** 5.7 seconds **37.** 8.0 seconds **39.** $-3, -2$ or $4, 5$ **41.** $7, 9, 11$ **43.** $-2, 0, 2$ or $6, 8, 10$
45. 192 cubic inches **47.** 16 inches **49.** c^2 **50.** b^2 **51.** a^2 **52.** $a^2 + b^2 = c^2$; This is the Pythagorean formula.
53. $x < \dfrac{5}{2}$ **55.** $x \geq -2$ **57.** true

SECTION 4.7 (PAGE 280)

◇ **CONNECTIONS** **Page 279:** The profit is found by subtracting cost from revenue represented by $R - C$. If $R - C = x^2 - x + 12$, the solution of $R - C > 0$ is $x < -3$ or $x > 4$. Only $x > 4$ makes sense, since x must be positive in the context of the problem.

EXERCISES **1. (a)** true **(b)** true **(c)** false **(d)** true **3. (a)** false **(b)** false **(c)** true **(d)** false
5. $-3 < a < 3$ **7.** $a \leq -6$ or $a \geq 7$ **9.** $m < -3$ or $m > -2$

11. $-1 \leq z \leq 5$ **13.** $-1 < m < \dfrac{2}{5}$ **15.** $-\dfrac{1}{2} < r < \dfrac{4}{3}$

17. $1 < q < 6$ **19.** $m < -\dfrac{1}{2}$ or $m > \dfrac{1}{3}$ **21.** $-\dfrac{2}{3} < p < -\dfrac{1}{4}$

23. $m < -2$ or $m > 2$ **25.** $r < -4$ or $r > 4$

27. $-2 \le a \le \dfrac{1}{3}$ or $a \ge 4$ **29.** $r < -1$ or $2 < r < 4$

31. 3   **32.** Region A | Region B   **33.** no   **34.** yes   **35.** $x > 3$ **36.** $x > 3$ **37.** 2 seconds and 14 seconds

39. between 0 and 2 seconds or between 14 and 16 seconds   **41.** $\dfrac{25}{36}$ **43.** 2

CHAPTER 4 REVIEW EXERCISES (PAGE 285)

1. $7(t + 2)$ **3.** $35x^2(x + 2)$ **5.** $50m^2n^2(2n - mn^2 + 3)$ **7.** $(x + 3)(x + 2)$ **9.** $(q + 9)(q - 3)$
11. $(r + 8s)(r - 12s)$ **13.** $8p(p + 2)(p - 5)$ **15.** $(m + 3n)(m - 6n)$ **17.** $p^5(p - 2q)(p + q)$
19. prime   **21.** r and $6r$, $2r$ and $3r$ **23.** $(2k - 1)(k - 2)$ **25.** $(3r + 2)(2r - 3)$ **27.** $(v + 3)(8v - 7)$
29. $-3(x + 2)(2x - 5)$ **31.** (b)   **33.** $(n + 7)(n - 7)$ **35.** $(7y + 5w)(7y - 5w)$

37. prime   **39.** $(r - 6)^2$ **41.** $(m + 10)(m^2 - 10m + 100)$ **43.** $(2x - 1)(4x^2 + 2x + 1)(2x + 1)(4x^2 - 2x + 1)$
45. $-\dfrac{3}{4}, 1$ **47.** $1, 4$ **49.** $-\dfrac{4}{3}, 5$ **51.** $0, 8$ **53.** 7   **55.** $-\dfrac{2}{5}, -2, -1$ **57.** length: 10 meters;   width: 4 meters
59. length: 10 centimeters;   width: 8 centimeters   **61.** length: 7 feet;   width: 3 feet   **63.** $6, 7$ or $-5, -4$ **65.** 112 feet
67. 256 feet   **69.** 2 inches   **71.** $q < -5$ or $q > 3$ **73.** $2 \le m \le 3$ **75.** $p \le -4$ or $p \ge \dfrac{3}{2}$
77. $(z - x)(z - 10x)$ **79.** $(3m + 4p)(5m - 4p)$ **81.** $3m(2m + 3)(m - 5)$ **83.** prime   **85.** $2a^3(a + 2)(a - 6)$

87. $(10a + 3)(100a^2 - 30a + 9)$ **89.** $0, 7$ **91.** $-\dfrac{3}{4} \le x \le 1$ **93.** length: 6 meters;   width: 4 meters
95. $-6, -4$ or $6, 8$ **97.** 6 meters   **99.** 8 feet   **103.** 0   **104.** $x - 2$ **105.** $x^3 + 2x^2 + 5x + 7$
106. $(x - 2)(x^3 + 2x^2 + 5x + 7)$ **107.** Show that the product is $x^4 + x^2 - 3x - 14$.   **108.** (a) -42 (b) -33 (c) 0
Choice (c), the binomial $x - 3$, is a factor of $2x^3 - 5x - 39$.

CHAPTER 4 TEST (PAGE 289)

[4.1–4.3] **1.** (d)   [4.1–4.4] **2.** $6x(2x - 5)$ **3.** $m^2n(2mn + 3m - 5n)$ **4.** $(x + 3)(x - 8)$
5. $(x - 7)(x - 2)$ **6.** $(2x + 3)(x - 1)$ **7.** $(3x + 1)(2x - 7)$ **8.** $3(x + 1)(x - 5)$ **9.** $(5z - 1)(2z - 3)$
10. prime   **11.** prime   **12.** $(2 - a)(6 + b)$ **13.** $(3y + 8)(3y - 8)$ **14.** $(x + 8)^2$ **15.** $(2x - 7y)^2$
16. $-2(x + 1)^2$ **17.** $3t^2(2t + 9)(t - 4)$ **18.** prime   **19.** $(r - 5)(r^2 + 5r + 25)$ **20.** $8(k + 2)(k^2 - 2k + 4)$
21. It is not correct because $(p + 3)(p + 3) = p^2 + 6p + 9 \ne p^2 + 9$.   [4.5] **22.** $6, \dfrac{1}{2}$ **23.** $-\dfrac{2}{5}, \dfrac{2}{5}$ **24.** 10

25. $-3, 0, 3$ [4.6] **26.** $1\dfrac{1}{2}$ seconds and $4\dfrac{1}{2}$ seconds   **27.** 17 feet   [4.7] **28.** $-\dfrac{5}{2} < x < \dfrac{1}{3}$

29. $x \le -4$ or $x \ge 6$ [4.5] **30.** Another solution is $-\dfrac{2}{3}$.

CUMULATIVE REVIEW (Chapters 1–4) (PAGE 290)

1. 0   **2.** .05   **3.** 6   **4.** $P = \dfrac{A}{1 + rt}$ **5.** $-\dfrac{9}{5}$ **6.** antibiotics: 36; tranquilizers: 54   **7.** exports: 741 million

dollars; imports: 426 million dollars   **8.** 35 pounds   **9.** 4   **10.** $\dfrac{16}{9}$ **11.** 1   **12.** 256   **13.** $\dfrac{1}{p^2}$ **14.** $\dfrac{1}{m^6}$

15. $-4k^2 - 4k + 8$ **16.** $6m^8 - 15m^6 + 3m^4$ **17.** $3y^3 + 8y^2 + 12y - 5$ **18.** $9p^2 + 12p + 4$ **19.** $4p^2 - 9q^2$
20. $45x^2 + 3x - 18$ **21.** $4x^3 + 6x^2 - 3x + 10$ **22.** $6p^2 + 7p + 1 + \dfrac{7}{2p - 2}$ **23.** $(2a - 1)(a + 4)$
24. $(2m + 3)(5m + 2)$ **25.** $(5x + 3y)(3x - 2y)$ **26.** $(4t + 3v)(2t + v)$ **27.** $(3x + 1)^2$ **28.** $(2p - 3)^2$
29. $-2(4t + 7z)^2$ **30.** $(5r + 9t)(5r - 9t)$ **31.** $25(4x^2 + 1)$ **32.** $m(2a - 1)(3a + 2)$ **33.** $2pq(3p + 1)(p + 1)$
34. $(2x + y)(a - b)$ **35.** $\dfrac{3}{2}, -2, 6$ **36.** $-\dfrac{2}{3}, \dfrac{1}{2}$ **37.** $x \le -2$ or $x \ge \dfrac{3}{2}$ **38.** 3 centimeters
39. $-4, -2$ or $8, 10$ **40.** 5 meters, 12 meters, 13 meters

CHAPTER 5   RATIONAL EXPRESSIONS

SECTION 5.1 (PAGE 298)

◆ **CONNECTIONS** **Page 292:** One meaning of the fraction $\frac{2}{3}$ is 2 divided by 3. Other answers are possible.

Page 297: 1. $3x^2 + 11x + 8$ cannot be factored, so this quotient cannot be reduced. By long division the quotient is $3x + 5 + \frac{-2}{x + 2}$. **2.** The numerator factors as $(x - 2)(x^2 + 2x + 4)$, so by reducing, the quotient is $x - 2$. Long division gives the same quotient.

EXERCISES 1. 0 **3.** $\frac{5}{3}$ **5.** $-3, 2$ **7.** never undefined **9. (a)** 1 **(b)** $\frac{17}{12}$ **11. (a)** 0 **(b)** $-\frac{10}{3}$
13. (a) $\frac{9}{5}$ **(b)** undefined **15. (a)** $\frac{2}{7}$ **(b)** $\frac{13}{3}$ **19.** $3r^2$ **21.** $\frac{2}{5}$ **23.** $\frac{x - 1}{x + 1}$ **25.** $\frac{7}{5}$ **27.** $m - n$ **29.** $\frac{3(2m + 1)}{4}$
31. $\frac{3m}{5}$ **33.** $\frac{3r - 2s}{3}$ **35.** $\frac{z - 3}{z + 5}$ **37.** $k - 3$ **39.** $\frac{x + 1}{x - 1}$ **41.** $-\frac{x - 3}{x + 6}$ **42.** $\frac{-x + 3}{x + 6}$ **43.** $\frac{x - 3}{-x - 6}$

44. Answers will vary. For example, if the number chosen is 12, the expressions will be evaluated as $-\frac{9}{18}, \frac{-9}{18},$ and $\frac{9}{-18}$, all equal to $-\frac{1}{2}$. **45.** (d) **46.** (a) **47.** -1 **49.** $-(m + 1)$ **51.** -1 **53.** already in lowest terms **57.** $x^2 + 3$
59. $\frac{m + n}{2}$ **61.** $-\frac{b^2 + ba + a^2}{a + b}$ **63.** $\frac{z + 3}{z}$ **65.** $\frac{5}{9}$ **67.** $\frac{10}{3}$ **69.** 4

SECTION 5.2 (PAGE 306)

◆ **CONNECTIONS** **Page 305:** $\frac{2 - m}{2m(m + 1)}; \frac{-m + 2}{2m(m + 1)}$

EXERCISES 1. $\frac{3a}{2}$ **3.** $-\frac{4x^4}{3}$ **5.** $\frac{2}{c + d}$ **7.** 5 **9.** $-\frac{3}{2t^4}$ **11.** $\frac{1}{4}$ **15.** -4 causes the denominator in the first fraction to equal 0. **16.** -5 causes the denominator in the second fraction to equal 0. **17.** -7 causes the numerator in the divisor to equal 0, meaning that we would be dividing by 0. This is undefined. **18.** We *are* allowed to divide 0 by a nonzero number. **19.** $\frac{10}{9}$ **21.** $-\frac{3}{4}$ **23.** $-\frac{9}{2}$ **25.** $\frac{-9(m - 2)}{m + 4}$ **27.** $\frac{p + 4}{p + 2}$ **29.** $\frac{(k - 1)^2}{(k + 1)(2k - 1)}$ **31.** $\frac{4k - 1}{3k - 2}$
33. $\frac{m + 4p}{m + p}$ **35.** $\frac{m + 6}{m + 3}$ **37.** $\frac{y + 3}{y + 4}$ **39.** $\frac{m}{m + 5}$ **41.** $\frac{r + 6s}{r + s}$ **43.** $\frac{(q - 3)^2(q + 2)^2}{q + 1}$ **45.** $\frac{m - 8}{8}$
47. $(2r - t)(2r + t)$ **49.** $\frac{-(b - 2a)(a^2 + b^3)}{(2a + b)^2}$ or $\frac{(2a - b)(a^2 + b^3)}{(2a + b)^2}$ **51.** $y^2 - 9$ **53.** $2^4 \cdot 3$ **55.** $2^2 \cdot 3 \cdot 5$
57. 7 **59.** $18b^3$

SECTION 5.3 (PAGE 311)

EXERCISES 1. true **3.** false **5.** 60 **7.** 1800 **9.** x^5 **11.** $30p$ **13.** $180y^4$ **15.** $15a^5b^3$
17. $12p(p - 2)$ **19.** $2^3 \cdot 3 \cdot 5$ **20.** $(t + 4)^3(t - 3)(t + 8)$ **21.** The similarity is that 2 is replaced by $t + 4$, 3 is replaced by $t - 3$, and 5 is replaced by $t + 8$. **23.** $18(r - 2)$ **25.** $12p(p + 5)^2$ **27.** $8(y + 2)(y + 1)$
29. $m - 3$ or $3 - m$ **31.** $p - q$ or $q - p$ **33.** $a(a + 6)(a - 3)$ **35.** $(k + 3)(k - 5)(k + 7)(k + 8)$
37. Yes, because $(2x - 5)^2 = (5 - 2x)^2$. **39.** 7 **40.** 1 **41.** the identity property of multiplication
42. 7 **43.** 1 **44.** the identity property of multiplication **45.** $\frac{60m^2k^3}{32k^4}$ **47.** $\frac{57z}{6z - 18}$ **49.** $\frac{-4a}{18a - 36}$
51. $\frac{6(k + 1)}{k(k - 4)(k + 1)}$ **53.** $\frac{36r(r + 1)}{(r - 3)(r + 2)(r + 1)}$ **55.** $\frac{ab(a + 2b)}{2a^3b + a^2b^2 - ab^3}$ **57.** $\frac{(t - r)(4r - t)}{t^3 - r^3}$
59. $\frac{2y(z - y)(y - z)}{y^4 - z^3y}$ or $\frac{-2y(y - z)^2}{y^4 - z^3y}$ **63.** $\frac{5}{2}$ **65.** $\frac{11}{8}$ **67.** $\frac{13}{20}$ **69.** $\frac{13}{12}$

SECTION 5.4 (PAGE 318)

EXERCISES **1.** $\dfrac{5}{4}$ **2.** $\dfrac{-5}{-7+3}$ **3.** 1 **4.** $\dfrac{-5}{-4} = \dfrac{5}{4}$; The two answers are equal. **5.** Jack's answer was correct.

His answer can be obtained by multiplying Jill's correct answer by $\dfrac{-1}{-1} = 1$, the identity element for multiplication. **8. (a)** not

equivalent **(b)** equivalent **(c)** equivalent **(d)** not equivalent **(e)** not equivalent **(f)** not equivalent **9.** $\dfrac{11}{m}$ **11.** b

13. x **15.** $y - 6$ **19.** $\dfrac{3z + 5}{15}$ **21.** $\dfrac{10 - 7r}{14}$ **23.** $\dfrac{-3x - 2}{4x}$ **25.** $\dfrac{11(1 + k)}{8}$ **27.** $\dfrac{4(b + 4)}{3b}$ **29.** $\dfrac{7 - 6p}{3p^2}$

31. $\dfrac{7x - 22}{4(x + 2)(x - 2)}$ **33.** $\dfrac{43m + 1}{5m(m - 2)}$ **35.** $m - 2$ or $2 - m$ **37.** $\dfrac{-2}{x - 5}$ or $\dfrac{2}{5 - x}$ **39.** $\dfrac{-2}{y - 1}$ or $\dfrac{2}{1 - y}$

41. $\dfrac{-5}{r - y^2}$ or $\dfrac{5}{y^2 - r}$ **43.** $\dfrac{x + y}{5x - 3y}$ or $\dfrac{-x - y}{3y - 5x}$ **45.** $\dfrac{-6}{4p - 5}$ or $\dfrac{6}{5 - 4p}$ **47.** $\dfrac{-(m + n)}{2(m - n)}$

49. $\dfrac{-x^2 + 6x + 11}{(x + 3)(x - 3)(x + 1)}$ **51.** $\dfrac{-5q^2 - 13q + 7}{(3q - 2)(q + 4)(2q - 3)}$ **53.** $\dfrac{9r + 2}{r(r + 2)(r - 1)}$ **55.** $\dfrac{2x^2 + 6xy + 8y^2}{(x + y)(x + y)(x + 3y)}$ or

$\dfrac{2x^2 + 6xy + 8y^2}{(x + y)^2(x + 3y)}$ **57.** $\dfrac{15r^2 + 10ry - y^2}{(3r + 2y)(6r - y)(6r + y)}$ **59.** $\dfrac{2k^2 - 10k + 6}{(k - 3)(k - 1)^2}$ **61.** $\dfrac{7k^2 + 31k + 92}{(k - 4)(k + 4)^2}$

63. (a) $\dfrac{9k^2 + 6k + 26}{5(3k + 1)}$ **(b)** $\dfrac{1}{4}$ **65.** 6 **67.** $\dfrac{3}{25}$

SECTION 5.5 (PAGE 326)

◆ **CONNECTIONS** **Page 325:** 1.413793103; 2

EXERCISES **3.** Choice (d) is correct, because every sign has been changed in the fraction. **5.** -6 **7.** $\dfrac{1}{pq}$

9. $\dfrac{1}{xy}$ **11.** $\dfrac{2a^2b}{3}$ **13.** $\dfrac{m(m + 2)}{3(m - 4)}$ **15.** $\dfrac{2}{x}$ **17.** $\dfrac{8}{x}$ **19.** $\dfrac{a^2 - 5}{a^2 + 1}$ **21.** $\dfrac{31}{50}$ **23.** $\dfrac{y^2 + x^2}{xy(y - x)}$

25. $\dfrac{40 - 12p}{85p}$ **27.** $\dfrac{5y - 2x}{3 + 4xy}$ **29.** $\dfrac{a - 2}{2a}$ **31.** $\dfrac{z - 5}{4}$ **33.** $\dfrac{-m}{m + 2}$ **35.** It represents division.

37. $\dfrac{\frac{3}{8} + \frac{5}{6}}{2}$ **38.** $\dfrac{29}{48}$ **39.** $\dfrac{29}{48}$ **41.** $\dfrac{5}{3}$ **43.** $\dfrac{13}{2}$ **45.** $\dfrac{19r}{15}$ **47.** $\dfrac{3m(m - 3)}{(m - 1)(m - 8)}$ **49.** $12x + 2$

51. $-44p^2 + 27p$ **53.** $\dfrac{1}{2}$ **55.** -5

SECTION 5.6 (PAGE 333)

◆ **CONNECTIONS** **Page 329: 1.** 58.8; 55.2; 51.9; year 8 **2.** 44.2%. The success rates are decreasing by less and less each year, so the rate of decrease is slowing.

EXERCISES **1.** (d) **3.** 0 and 4 **5.** expression; $\dfrac{43}{40}x$ **7.** equation; $\dfrac{40}{43}$ **9.** expression; $-\dfrac{1}{10}y$

11. equation; -10 **15.** $\dfrac{1}{4}$ **17.** $-\dfrac{3}{4}$ **19.** 24 **21.** -15 **23.** 7 **25.** -15 **27.** -5 **29.** -6

31. no solution **33.** 5 **35.** 4 **37.** 1 **39.** 4 **41.** 5 **43.** $-2, 12$ **45.** no solution **47.** 3 **49.** 3

51. $-\dfrac{1}{5}, 3$ **53.** $-\dfrac{1}{2}, 5$ **55.** 3 **57.** Put all terms with k on the same side and other terms on the other side.

59. $F = \dfrac{ma}{k}$ **61.** $a = \dfrac{kF}{m}$ **63.** $R = \dfrac{E - Ir}{I}$ or $R = \dfrac{E}{I} - r$ **65.** $A = \dfrac{h(B + b)}{2}$

67. $a = \dfrac{2S - ndL}{nd}$ or $a = \dfrac{2S}{nd} - L$ **69.** $y = \dfrac{xz}{x + z}$ **71.** $z = \dfrac{3y}{5 - 9xy}$ or $z = \dfrac{-3y}{9xy - 5}$ **73.** $-\dfrac{1}{3}, 3$ **75.** $-6, \dfrac{1}{2}$

77. 6 **79.** The solution is -5. The number 3 must be rejected. **80.** The simplified form is $x + 5$. **(a)** It is the same as the actual solution. **(b)** It is the same as the rejected solution. **81.** The solution is -3. The number 1 must be rejected.

82. The simplified form is $\dfrac{x+3}{2(x+1)}$. **(a)** It is the same as the actual solution. **(b)** It is the same as the rejected solution.
85. $\dfrac{288}{t}$ miles per hour **87.** $\dfrac{289}{z}$ hours

SUMMARY: Rational Expressions (PAGE 338)

1. $\dfrac{10}{p}$ **3.** $\dfrac{1}{2x^2(x+2)}$ **5.** $\dfrac{y+2}{y-1}$ **7.** 39 **9.** $\dfrac{13}{3(p+2)}$ **11.** $\dfrac{1}{7}, 2$ **13.** $\dfrac{7}{12z}$
15. $\dfrac{3m+5}{(m+3)(m+2)(m+1)}$ **17.** no solution **19.** $\dfrac{t+2}{2(2t+1)}$

SECTION 5.7 (PAGE 344)

◆ **CONNECTIONS** **Page 341:** 54.5 miles per hour; 57.4 miles per hour; yes

EXERCISES 1. $\dfrac{1}{3}x=\dfrac{1}{6}x+2$ **3.** $\dfrac{9}{5}$ **5.** $\dfrac{1386}{97}$ **7.** $99.7 million **9.** Japan: $.14 billion; Germany: $.18 billion
11. $\dfrac{500}{x-10}=\dfrac{600}{x+10}$ **13.** $\dfrac{D}{R}=\dfrac{d}{r}$ **15.** 24.04 kilometers per hour **17.** Hussein: 12.02 miles per hour; McDermott:
8.78 miles per hour **19.** 8 miles per hour **21.** $\dfrac{37}{2}$ or $18\dfrac{1}{2}$ miles per hour **23.** $\dfrac{1}{8}x+\dfrac{1}{6}x=1$ **25.** $\dfrac{1}{10}$ job per hour
27. $\dfrac{72}{17}$ or $4\dfrac{4}{17}$ hours **29.** $\dfrac{60}{11}$ or $5\dfrac{5}{11}$ hours **31.** 3 hours **33.** $\dfrac{27}{10}$ or $2\dfrac{7}{10}$ hours **35.** $\dfrac{100}{11}$ or $9\dfrac{1}{11}$ minutes
37. 9 **39.** 125 **41.** $\dfrac{4}{9}$ **43. (a)** increases **(b)** decreases **45.** $40.32 **47.** 20 miles per hour
49. about 302 pounds **51.** 100 pounds per square inch **53.** 20 pounds per square foot **55.** 144 feet
57. (a) -7 **(b)** 23 **59. (a)** -14 **(b)** 22 **61. (a)** $-\dfrac{14}{5}$ **(b)** $-\dfrac{2}{5}$

CHAPTER 5 REVIEW EXERCISES (PAGE 352)

1. 3 **3.** $-1, 3$ **5. (a)** $-\dfrac{4}{7}$ **(b)** -16 **7. (a)** undefined **(b)** 1 **9.** $\dfrac{b}{3a}$ **11.** $\dfrac{-(2x+3)}{2}$ **13.** $\dfrac{72}{p}$ **15.** $\dfrac{2}{3m^6}$
17. $\dfrac{3}{2}$ **19.** $\dfrac{3a-1}{a+5}$ **21.** $\dfrac{p+5}{p+1}$ **23.** 96 **25.** $m(m+2)(m+5)$ **27.** $\dfrac{35}{56}$ **29.** $\dfrac{15a}{10a^4}$ **31.** $\dfrac{15y}{50-10y}$
33. $\dfrac{15}{x}$ **35.** $\dfrac{4k-45}{k(k-5)}$ **37.** $\dfrac{-2-3m}{6}$ **39.** $\dfrac{7a+6b}{(a-2b)(a+2b)}$ **41.** $\dfrac{5z-16}{z(z+6)(z-2)}$ **43. (a)** $\dfrac{a}{b}$ **(b)** $\dfrac{a}{b}$
45. $\dfrac{10}{13}$ **47.** $\dfrac{xw+1}{xw-1}$ **49.** It would cause the first (and third) denominator(s) to equal 0. **51.** -4 **53.** 3
55. $y=\dfrac{4x+5}{3}$ **57.** 12 **59.** about 6.7 hours **61.** 2 hours **63.** 4 centimeters **65.** $\dfrac{(5+2x-2y)(x+y)}{(3x+3y-2)(x-y)}$
67. $8p^2$ **69.** 3 **71.** $r=\dfrac{3kz}{5k-z}$ or $r=\dfrac{-3kz}{z-5k}$ **73.** about 17.4 million **75.** $\dfrac{36}{5}$ **77. (a)** -3 **(b)** -1
(c) $-3, -1$ **78.** $\dfrac{15}{2x}$ **79.** If $x=0$, the divisor R is equal to 0, and division by 0 is undefined. **80.** $(x+3)(x+1)$
81. $\dfrac{7}{x+1}$ **82.** $\dfrac{11x+21}{4x}$ **83.** no solution **84.** We know that -3 is not allowed, because P and R are undefined for
$x=-3$. **86.** $\dfrac{6}{5}, \dfrac{5}{2}$

CHAPTER 5 TEST (PAGE 356)

[5.1] 1. $-2, 4$ **2. (a)** $\dfrac{11}{6}$ **(b)** undefined **3.** $-3x^2y^3$ **4.** $\dfrac{3a+2}{a-1}$ **[5.2] 5.** x **6.** $\dfrac{25}{27}$ **7.** $\dfrac{3k-2}{3k+2}$
8. $\dfrac{a-1}{a+4}$ **[5.3] 9.** $150p^5$ **10.** $(2r+3)(r+2)(r-5)$ **11.** $\dfrac{240p^2}{64p^3}$ **12.** $\dfrac{21}{42m-84}$ **[5.4] 13.** 2

14. $\dfrac{-14}{5(y+2)}$ **15.** $\dfrac{-x^2+x+1}{3-x}$ or $\dfrac{x^2-x-1}{x-3}$ **16.** $\dfrac{-m^2+7m+2}{(2m+1)(m-5)(m-1)}$ **[5.5] 17.** $\dfrac{2k}{3p}$ **18.** $\dfrac{-2-x}{4+x}$

[5.6] 19. $-1, 4$ **20.** 5   **21.** no solution   **22.** $D=\dfrac{dF-k}{F}$ or $D=\dfrac{k-dF}{-F}$ **[5.7] 23.** 3 miles per hour

24. $\dfrac{20}{9}$ or $2\dfrac{2}{9}$ hours   **25.** 100 amp

CUMULATIVE REVIEW (Chapters 1–5)   (PAGE 357)

1. $\dfrac{32}{5}$ **2.** 17   **3.** $b=\dfrac{2A}{h}$ **4.** $-\dfrac{2}{7}$ **5.** $y\geq -8$ **6.** $m>4$ **7.** $\dfrac{16}{9}$ **8.** $\dfrac{1}{49}$ **9.** $\dfrac{1}{4^9}$ **10.** $\dfrac{1}{2^4 x^7}$

11. $\dfrac{1}{m^6}$ **12.** $\dfrac{q}{4p^2}$ **13.** k^2+2k+1 **14.** $72x^6 y^7$ **15.** $4a^2-4ab+b^2$ **16.** $3y^3+8y^2+12y-5$

17. $6p^2+7p+1+\dfrac{3}{p-1}$ **18.** $(4t+3v)(2t+v)$ **19.** prime   **20.** $m(2a-1)(3a+2)$ **21.** $-3, 5$ **22.** $0, 8$

23. $-\dfrac{1}{2}, \dfrac{2}{3}, 5$ **24.** -2 or -1 **25.** 6 meters   **26.** 4 inches   **27.** -2 or 4

28. (a)   **29.** (d)   **30.** $\dfrac{1}{xy^6}$ **31.** $\dfrac{4}{q}$ **32.** $\dfrac{3r+28}{7r}$ **33.** $\dfrac{7}{15(q-4)}$ **34.** $\dfrac{-k-5}{k(k+1)(k-1)}$

35. $\dfrac{7(2z+1)}{24}$ **36.** $\dfrac{195}{29}$ **37.** $\dfrac{21}{2}$ **38.** $-2, 1$ **39.** 150 miles   **40.** $\dfrac{6}{5}$ or $1\dfrac{1}{5}$ hours

CHAPTER 6   EQUATIONS AND INEQUALITIES IN TWO VARIABLES

SECTION 6.1   (PAGE 366)

◇ **CONNECTIONS**   **Page 359:** About 1977; about \$138 billion; some factors are that people are living longer and health care costs are increasing faster than anticipated.

EXERCISES   **1.** between 1989 and 1990;   262.3 thousand   **3.** 456.75 thousand;   365.4 thousand   **5.** (d)
7. Inkjet   **9.** 38.5%   **11.** \$8.38 billion   **13.** does; do not   **15.** II   **17.** 3   **19.** yes   **21.** yes   **23.** no
25. yes   **27.** yes   **29.** no   **35.** 11   **37.** 7   **39.** -5 **41.** -4 **43.** -5 **45.** -44 **49.** 4; 6; -6
51. 3; -5; -15 **53.** -9; -9; -9 **55.** -6; -6; -6 **57.–66.**   **67.** negative;

negative   **69.** positive; negative   **71.** -3; 6; -2; 4   **73.** -3; 4; -6; $-\dfrac{4}{3}$

75. -4; -4; -4; -4 **77. (a)** $(100, 5050)$ **(b)** $(2000, 6000)$ **79.** Yes, there is an approximate linear

relationship between height and weight.

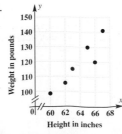

81. $(0, 1), (1, 1.25), (2, 1.50), (3, 1.75), (4, 2.00) (5, 2.25)$
83. -2 **85.** 13   **87.** $y=\dfrac{12-2x}{3}$

SECTION 6.2 (PAGE 377)

◆ **CONNECTIONS** **Page 375: 1.** -2 **2.** all real numbers **3.** no solution

EXERCISES **1.** line **3.** x **5.** horizontal **7.** 5; 5; 3

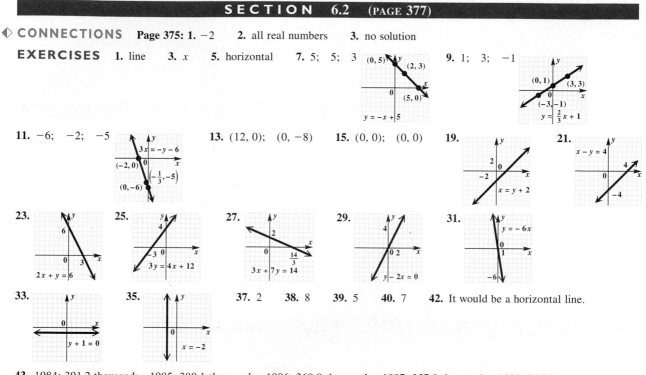

9. 1; 3; -1

11. -6; -2; -5 **13.** $(12, 0)$; $(0, -8)$ **15.** $(0, 0)$; $(0, 0)$ **19.** **21.**

23. **25.** **27.** **29.** **31.** **33.** **35.** **37.** 2 **38.** 8 **39.** 5 **40.** 7 **42.** It would be a horizontal line.

43. 1984: 391.2 thousand; 1985: 380.1 thousand; 1986: 369.0 thousand; 1987: 357.9 thousand; 1988: 346.8 thousand **45. (a)** 163.2 centimeters **(b)** 171 centimeters **(c)** 151.5 centimeters **(d)**

47. (a) \$1250 **(b)** \$1250 **(c)** \$1250 **(d)** \$6250 **49.** $\dfrac{2}{3}$ **51.** $\dfrac{1}{2}$

SECTION 6.3 (PAGE 388)

◆ **CONNECTIONS** **Page 382:** Ski slopes, "grade" on a treadmill, and slope (or "fall") of a sewer pipe are some examples.

EXERCISES **1.** Sketches will vary. The line must fall from left to right. **3.** Sketches will vary. The line must be vertical.

7. 5 **9.** $\dfrac{3}{2}$ **11.** $-\dfrac{7}{4}$ **13.** undefined **15. (a)** 6 **(b)** 4 **(c)** $\dfrac{6}{4}$ or $\dfrac{3}{2}$; slope of the line **17.** $\dfrac{7}{6}$ **19.** $-\dfrac{5}{8}$

21. 0 **23.** undefined **25.** $\dfrac{1}{3}$ **27.** 2 **29.** $-\dfrac{1}{2}$ **31.** -4 **33.** $\dfrac{3}{2}$ **35.** 0 **37. (a)** negative **(b)** negative

39. (a) positive **(b)** zero **41. (a)** zero **(b)** positive **43.** $-\dfrac{2}{5}$; $-\dfrac{2}{5}$; parallel **45.** $\dfrac{8}{9}$; $-\dfrac{4}{3}$; neither

47. $\dfrac{3}{2}$; $-\dfrac{2}{3}$; perpendicular **49.** $\dfrac{3}{10}$ **51.** .577 thousand (or 577) **52.** positive; increased **53.** 577 students

(rounded) **54.** $-.317$ thousand (or -317) **55.** negative; decreased **56.** 317 students **57.** 2 **59.** $\dfrac{2}{5}$

61. $(0, 4)$ **63.** $y = 2x - 16$ **65.** $y = -\dfrac{1}{2}x - \dfrac{21}{10}$

SECTION 6.4 (PAGE 396)

◇ **CONNECTIONS** **Page 393:** **1.** $3500, $5000; $35,000, $50,000   **2.** the loss in value each year;   the depreciation when the item is brand new ($D = 0$).

EXERCISES **1.** $y = 3x - 3$ **3.** $y = -x + 3$ **5.** $y = 4x - 3$ **7.** $y = 3$

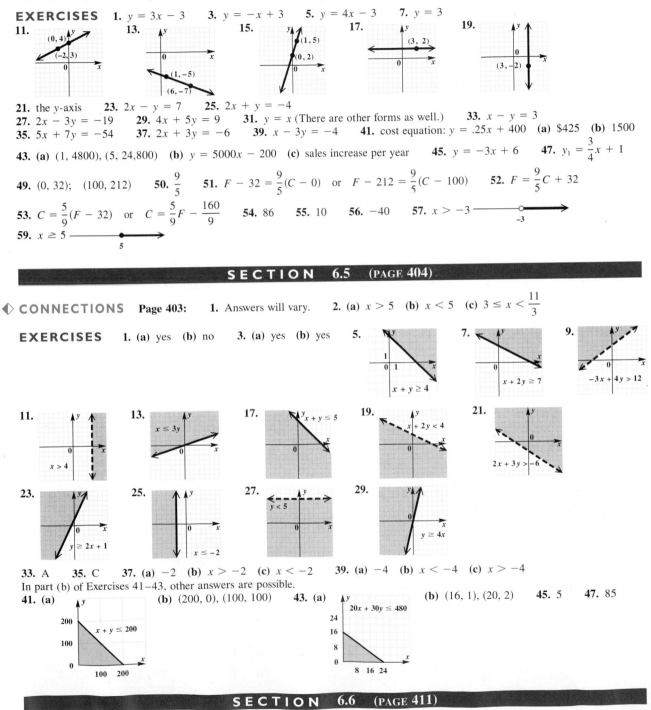

21. the y-axis **23.** $2x - y = 7$ **25.** $2x + y = -4$
27. $2x - 3y = -19$ **29.** $4x + 5y = 9$ **31.** $y = x$ (There are other forms as well.)   **33.** $x - y = 3$
35. $5x + 7y = -54$ **37.** $2x + 3y = -6$ **39.** $x - 3y = -4$ **41.** cost equation: $y = .25x + 400$ **(a)** $425   **(b)** 1500

43. (a) $(1, 4800), (5, 24,800)$ **(b)** $y = 5000x - 200$ **(c)** sales increase per year   **45.** $y = -3x + 6$ **47.** $y_1 = \dfrac{3}{4}x + 1$

49. $(0, 32)$; $(100, 212)$ **50.** $\dfrac{9}{5}$ **51.** $F - 32 = \dfrac{9}{5}(C - 0)$ or $F - 212 = \dfrac{9}{5}(C - 100)$ **52.** $F = \dfrac{9}{5}C + 32$

53. $C = \dfrac{5}{9}(F - 32)$ or $C = \dfrac{5}{9}F - \dfrac{160}{9}$ **54.** 86   **55.** 10   **56.** -40 **57.** $x > -3$

59. $x \geq 5$

SECTION 6.5 (PAGE 404)

◇ **CONNECTIONS** **Page 403:** **1.** Answers will vary.   **2. (a)** $x > 5$ **(b)** $x < 5$ **(c)** $3 \leq x < \dfrac{11}{3}$

EXERCISES **1. (a)** yes **(b)** no   **3. (a)** yes **(b)** yes

33. A   **35.** C   **37. (a)** -2 **(b)** $x > -2$ **(c)** $x < -2$ **39. (a)** -4 **(b)** $x < -4$ **(c)** $x > -4$
In part (b) of Exercises 41–43, other answers are possible.
41. (a)   **(b)** $(200, 0), (100, 100)$ **43. (a)**   **(b)** $(16, 1), (20, 2)$ **45.** 5   **47.** 85

SECTION 6.6 (PAGE 411)

◇ **CONNECTIONS** **Page 410:** **1.** The next line segment goes from $x = 2$ to $x = 3$ with $y = 4 + 3 \cdot 7 = 25$. The segments go up by 7 between consecutive x-values. Each interval has an open circle on the left and a solid one on the right.
2. No vertical line intersects more than one point on the graph, so this is the graph of a function.   **3.** days; dollars

EXERCISES **1.** not a function; domain: $\{-4, -2, 0\}$; range: $\{3, 1, 5, -8\}$ **3.** function; domain: $\{-2, -1, 0, 1, 2\}$; range: $\{3, 2, 0, -7\}$ **5.** not a function; domain: $\{4, 2, 8, 6, 10\}$; range: $\{-1, -2, -3, -6, -8\}$ **7.** function **9.** not a function **11.** not a function **13.** function **15.** function **17.** function **19.** not a function **21.** not a function **23.** domain: {all real numbers}; range: {all nonnegative real numbers} **25.** domain: {all real numbers}; range: {all real numbers} **27.** domain: {all nonnegative real numbers}; range: {all nonnegative real numbers} **29.** $(2, 4)$ **30.** $(-1, -4)$ **31.** $\dfrac{8}{3}$ **32.** $f(x) = \dfrac{8}{3}x - \dfrac{4}{3}$ **33.** **(a)** 11 **(b)** 3 **(c)** -9 **35.** **(a)** 4 **(b)** 2 **(c)** 14 **37.** **(a)** 2 **(b)** 0 **(c)** 3 **39.** **(a)** 4 **(b)** 2 **41.** 39.2 **43.** $f(6) = 44.74$, meaning that based on the past trend, in 1991 the per capita consumption would be 44.74 gallons. **45.** 4 **47.** 1 **49.** $y = x + 1$ **51.** **(a)** 302 **(b)** 353 **(c)** 370 **55.** $17x$

CHAPTER 6 REVIEW EXERCISES (PAGE 417)

1. -1; 2; 1 **3.** 0; $\dfrac{8}{3}$; -9 **5.** yes **7.** yes

9.–12. **13.** I or III **15.** II **17.** **(a)** $\left(-\dfrac{5}{2}, 0\right)$ **(b)** $(0, 5)$ **19.** **(a)** $\left(\dfrac{8}{3}, 0\right)$ **(b)** $(0, 4)$ **21.**

23. $-\dfrac{1}{2}$ **25.** 0 **27.** 3 **29.** $\dfrac{3}{2}$ **31.** undefined **33.** $\dfrac{1}{3}$ **35.** parallel **37.** neither **39.** $3x + 3y = 2$ **41.** $x - y = 7$ **43.** $3x + 4y = -1$ **45.** $y = 1$ **47.** **49.** **51.**

53. not a function; domain: $\{-2, 0, 2\}$; range: $\{4, 8, 5, 3\}$ **55.** not a function **57.** function **59.** not a function **61.** domain: {all real numbers}; range: {all real numbers greater than or equal to 1} **63.** **(a)** 8 **(b)** -1 **65.** **(a)** 5 **(b)** 2 **67.** x-intercept: $(7, 0)$; y-intercept: $\left(0, -\dfrac{7}{3}\right)$; $m = \dfrac{1}{3}$ **69.** x-intercept: $\left(\dfrac{3}{4}, 0\right)$; y-intercept: $(0, 2)$; $m = -\dfrac{8}{3}$ **71.** $x + 4y = -5$ **73.** $4x + 7y = -23$ **75.** **77.** **79.**

83. **84.** -3 **85.** $y = -3x - 2$ **86.** $\left(-\dfrac{2}{3}, 0\right)$ **87.** $(0, -2)$ **88.** **89.** -32 **90.** domain: {all real numbers}; range: {all real numbers}

CHAPTER 6 TEST (PAGE 420)

[6.1] 1. 1991 **2.** (b) **3.** 51% **4.** -9; -5; $\dfrac{3}{2}$ **5.** -6; -10; -15 **6.** -12; -12; -12 **[6.2] 7.** To find the y-intercept let $x = 0$, and to find the x-intercept let $y = 0$. **8.** $(4, 0)$, $(0, -4)$ **9.** $(2, 0)$, $(0, 6)$ **10.** $(0, 0)$, $(0, 0)$ **11.** $(-3, 0)$, no y-intercept **12.** no x-intercept, $(0, 1)$

[6.3] 13. $-\dfrac{8}{3}$ **14.** -2 **15.** $\dfrac{5}{2}$ **16.** $\dfrac{1}{2}$ **17.** **(a)** -4 **(b)** $\dfrac{1}{4}$ **[6.4] 18.** $2x - y = -6$ **19.** $5x - 2y = 8$

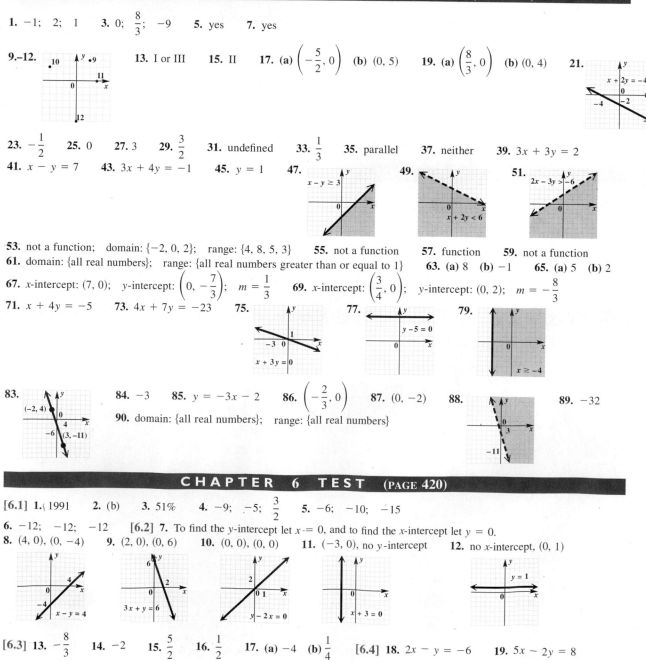

20. $x - 2y = -8$ **[6.5] 21.** 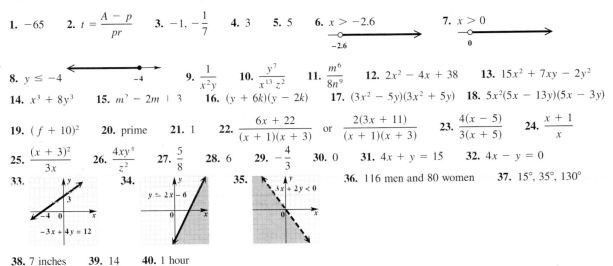   **22.**   **23.** 20   **[6.6] 24. (a)** not a function   **(b)** function;

domain: {all real numbers};   range: {all real numbers greater than or equal to 2}
25. Every vertical line intersects the graph in only one point.

CUMULATIVE REVIEW (Chapters 1–6)   (PAGE 422)

1. -65 **2.** $t = \dfrac{A - p}{pr}$ **3.** $-1, -\dfrac{1}{7}$ **4.** 3   **5.** 5   **6.** $x > -2.6$ **7.** $x > 0$

8. $y \le -4$ **9.** $\dfrac{1}{x^2 y}$ **10.** $\dfrac{y^7}{x^{13} z^2}$ **11.** $\dfrac{m^6}{8n^9}$ **12.** $2x^2 - 4x + 38$ **13.** $15x^2 + 7xy - 2y^2$
14. $x^3 + 8y^3$ **15.** $m^2 - 2m + 3$ **16.** $(y + 6k)(y - 2k)$ **17.** $(3x^2 - 5y)(3x^2 + 5y)$ **18.** $5x^2(5x - 13y)(5x - 3y)$

19. $(f + 10)^2$ **20.** prime   **21.** 1   **22.** $\dfrac{6x + 22}{(x + 1)(x + 3)}$ or $\dfrac{2(3x + 11)}{(x + 1)(x + 3)}$ **23.** $\dfrac{4(x - 5)}{3(x + 5)}$ **24.** $\dfrac{x + 1}{x}$
25. $\dfrac{(x + 3)^2}{3x}$ **26.** $\dfrac{4xy^4}{z^2}$ **27.** $\dfrac{5}{8}$ **28.** 6   **29.** $-\dfrac{4}{3}$ **30.** 0   **31.** $4x + y = 15$ **32.** $4x - y = 0$
33.   **34.**   **35.**   **36.** 116 men and 80 women   **37.** $15°, 35°, 130°$

38. 7 inches   **39.** 14   **40.** 1 hour

CHAPTER 7   LINEAR SYSTEMS

SECTION 7.1   (PAGE 429)

EXERCISES   **1.** It is not a solution of the system because it is not a solution of the second equation, $2x + y = 4$.   **3.** yes
5. no   **7.** yes   **9.** yes   **11.** no   **13.** The answer is (b), because the ordered pair must be in quadrant II.
15. $(4, 2)$ **17.** $(0, 4)$ **19.** $(4, -1)$

In Exercises 21–29, we do not show the graphs.   **21.** $(1, 3)$ **23.** $(0, 2)$ **25.** no solution (inconsistent
system)   **27.** infinite number of solutions (dependent equations)   **29.** $(4, -3)$ **33.** Answers will vary, but the lines must
intersect at $(-2, 3)$. One example is $x + y = 1, 2x + y = -1$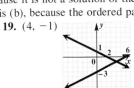

35. (a) neither   **(b)** intersecting lines   **(c)** one solution   **37. (a)** dependent   **(b)** one line   **(c)** infinite number of solutions
39. (a) neither   **(b)** intersecting lines   **(c)** one solution   **41. (a)** inconsistent   **(b)** parallel lines   **(c)** no solution   **43.** 40
45. $(40, 30)$ **47.** 2   **48.** 5   **49.** $(2, 5)$ **50.** The x-coordinate, 2, is equal to the solution of the equation.

51. The y-coordinate, 5, is equal to the value we obtained on both sides when checking.
52. 5; 3; 5   **53.** $y = -3x + 4$ **55.** $y = \dfrac{9}{2}x - 2$ **57.** 2   **59.** 3   **61.** 1

SECTION 7.2 (PAGE 437)

EXERCISES **3.** $(3, 9)$ **5.** $(7, 3)$ **7.** $(0, 5)$ **9.** $(-4, 8)$ **11.** $(3, -2)$ **13.** infinite number of solutions **15.** $\left(\frac{1}{3}, -\frac{1}{2}\right)$ **17.** no solution **19.** infinite number of solutions **25.** $(2, -3)$ **27.** $(3, 2)$ **29.** $(-2, 1)$ **31.** In both equations we get $4 = 4$. **33.** $(2, 4)$ **35.** $(1, 5)$

37. $(5, -3)$; The equations to input are $y_1 = \dfrac{5 - 4x}{5}$ and $y_2 = \dfrac{1 - 2x}{3}$.

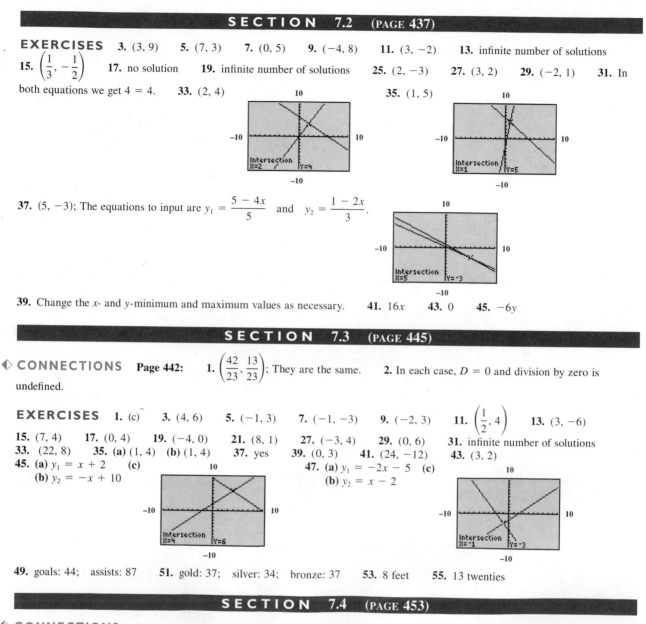

39. Change the x- and y-minimum and maximum values as necessary. **41.** $16x$ **43.** 0 **45.** $-6y$

SECTION 7.3 (PAGE 445)

◇ **CONNECTIONS** **Page 442:** **1.** $\left(\dfrac{42}{23}, \dfrac{13}{23}\right)$; They are the same. **2.** In each case, $D = 0$ and division by zero is undefined.

EXERCISES **1.** (c) **3.** $(4, 6)$ **5.** $(-1, 3)$ **7.** $(-1, -3)$ **9.** $(-2, 3)$ **11.** $\left(\dfrac{1}{2}, 4\right)$ **13.** $(3, -6)$ **15.** $(7, 4)$ **17.** $(0, 4)$ **19.** $(-4, 0)$ **21.** $(8, 1)$ **27.** $(-3, 4)$ **29.** $(0, 6)$ **31.** infinite number of solutions **33.** $(22, 8)$ **35.** (a) $(1, 4)$ (b) $(1, 4)$ **37.** yes **39.** $(0, 3)$ **41.** $(24, -12)$ **43.** $(3, 2)$ **45.** (a) $y_1 = x + 2$ (c) (b) $y_2 = -x + 10$ **47.** (a) $y_1 = -2x - 5$ (c) (b) $y_2 = x - 2$

49. goals: 44; assists: 87 **51.** gold: 37; silver: 34; bronze: 37 **53.** 8 feet **55.** 13 twenties

SECTION 7.4 (PAGE 453)

◇ **CONNECTIONS** **Page 448:** **1.** (1995, 12.5) rounded; In 1995 there will be about 12.5 deaths per hundred thousand annually from each of these sources. **2.** after 1995 **3.** Answers will vary.

EXERCISES **3.** 92 and 21 **5.** Harding: 987; Los Alamos: 109 **7.** Heathrow: 37,525,300; Haneda: 32,177,040 **9.** 46 ones; 28 tens **11.** 166 student tickets; 220 nonstudent tickets **13.** $2500 at 4%; $5000 at 5% **15.** 7 paperbacks; 13 hardbacks **17.** 80 liters of 40% solution; 40 liters of 70% solution **19.** 30 pounds at $6 per pound; 60 pounds at $3 per pound **21.** 60 barrels at $40 per barrel; 40 barrels at $60 per barrel **23.** boat: 10 miles per hour; current: 2 miles per hour **25.** plane: 470 miles per hour; wind: 30 miles per hour **27.** faster train: 110 miles per hour; slower train: 90 miles per hour **29.** Roberto: 3 miles per hour; Juana: 2.5 miles per hour **32.** $y_2 = 600x$ **33.** $y_1 = 400x + 5000$; $y_2 = 600x$; Solution: $(25, 15{,}000)$ **34.** 25; 15,000; 15,000 **36.** When $x < 25$ you are losing money. When $x > 25$ you are making money. **37.** **39.** **41.**

SECTION 7.5 (PAGE 459)

◊ **CONNECTIONS   Page 459:**   **2.** The minimum value of z is 10, produced by the point (2, 0).   **3.** The maximum value of z is 26 at (5, 4), and the minimum value is 4 at (2, 0).

EXERCISES   3.

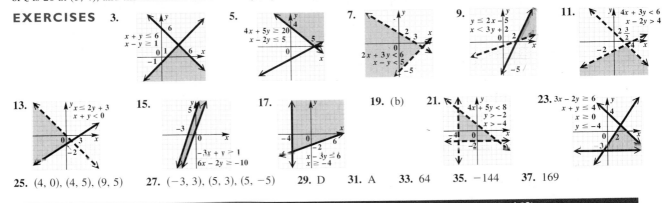

25. (4, 0), (4, 5), (9, 5)   **27.** (−3, 3), (5, 3), (5, −5)   **29.** D   **31.** A   **33.** 64   **35.** −144   **37.** 169

CHAPTER 7 REVIEW EXERCISES (PAGE 463)

1. yes   **3.** no   **5.** (3, 1)   **7.** infinite number of solutions   **11.** (2, 1)   **13.** (6, 4)   **15.** (5, −8)   **17.** His answer was incorrect since the system has infinitely many solutions (as indicated by the true statement 0 = 0).   **19.** (7, 1) **21.** (−3, 2)   **23.** infinite number of solutions   **25.** (−4, 1)   **27.** (9, 2)   **29.** (8, 9)   **33.** In system A we can eliminate the y terms by direct addition. This is not possible in system B.   **35.** Texas Commerce Tower: 75 stories; First Interstate Plaza: 71 stories   **37.** length: 27 meters; width: 18 meters   **39.** 25 pounds of $1.30 candy; 75 pounds of $.90 candy **41.** apple: $.55; banana: $.45   **43.** $7000 at 3%, $11,000 at 4%   **45.** 3 at $12; 12 at $16

47.   **49.**   **51.** (4, 8)   **53.** (2, 0)   **55.**

57. 8 inches, 8 inches, and 13 inches   **59.** 102 small bottles; 44 large bottles   **63.** $\left(\dfrac{28}{5}, \dfrac{16}{5}\right)$ **64.** $\left(\dfrac{28}{5}, \dfrac{16}{5}\right)$

65. They are the same. It makes no difference which method we use.

66. $y_1 = 2x - 8$ **67.** $y_2 = \dfrac{12 - x}{2}$ or $y_2 = 6 - \dfrac{1}{2}x$ **68.** $\dfrac{28}{5}$; The solution is the x-value found in Exercises 63 and 64.   **69.** We get $\dfrac{16}{5} = \dfrac{16}{5}$. The value we get is the y-value found in Exercises 63 and 64.   **70.** 2   **71.** $-\dfrac{1}{2}$ **72.** They are perpendicular.   **73.** The display indicates that when $x = 5.6$, $y = 3.2$. These support the values found earlier, since $\dfrac{28}{5} = 5.6$ and $\dfrac{16}{5} = 3.2$. It seems that they are not perpendicular since the viewing window is not "square."   **74.** A square window gives a truer perspective so that perpendicular lines really do appear to be perpendicular.   **75.**   **76.** $\left(\dfrac{28}{5}, \dfrac{16}{5}\right)$

$x + 2y \le 12$
$2x - y \le 8$

CHAPTER 7 TEST (PAGE 467)

[7.1] 1. C   **2.** (4, 1)   **[7.2] 3.** (1, −6)   **4.** (−35, 35)   **[7.3] 5.** (5, 6)   **6.** (−1, 3)   **7.** (−2, 5)   **8.** no solution   **9.** (0, 0)   **10.** (−15, 6)   **[7.1–7.3] 11.** infinite number of solutions   **12.** none   **[7.4] 13.** Memphis and Atlanta: 371 miles; Minneapolis and Houston: 671 miles   **14.** 312 orchestra seats; 255 balcony seats   **15.** $33\dfrac{1}{3}$ liters of 25% solution; $16\dfrac{2}{3}$ liters of 40% solution   **16.** slower car: 45 miles per hour; faster car: 60 miles per hour

[7.1] 17. Answers will vary. The ordered pair $(-3, 4)$ should satisfy the system. One example is $x + y = 1$, $x - y = -7$.
[7.5] 18.

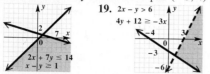

19. 2x − y > 6   4y + 12 ≥ −3x   **20.** A

CUMULATIVE REVIEW (Chapters 1–7) (PAGE 469)

1. $-1, 1, -2, 2, -4, 4, -5, 5, -8, 8, -10, 10, -20, 20, -40, 40$ **2.** 1   **3.** commutative property   **4.** distributive
property   **5.** inverse property   **6.** 46   **7.** $-\dfrac{13}{11}$ **8.** $\dfrac{9}{11}$ **9.** $x > -18$ **10.** $x > -\dfrac{11}{2}$ **11.** $37\dfrac{1}{2}$ yards
12. $14x^2 - 5x + 23$ **13.** $6xy + 12x - 14y - 28$ **14.** $3k^2 - 4k + 1$ **15.** 3.65×10^{10} **16.** x^6y
17. $(5m - 4p)(2m + 3p)$ **18.** $(8t - 3)^2$ **19.** $-\dfrac{1}{3}, \dfrac{3}{2}$ **20.** $-11, 11$ **21.** $\dfrac{7}{x + 2}$ **22.** $\dfrac{3}{4k - 3}$ **23.** $-\dfrac{1}{4}, 3$
24. $T = \dfrac{PV}{k}$ **25.**   **26.**   **27.**   **28.** $-\dfrac{4}{3}$ **29.** $-\dfrac{1}{4}$ **30.** $3x - y = 11$

31. $y = 4$ **32.** (a) $x = 9$ (b) $y = -1$ **33.** (a) -32 (b) negative; decreased   (c) 64.5   (d) 64.5 cases per year
34. $(-1, 6)$ **35.** $(3, -4)$ **36.** $(2, -1)$ **37.** length: 13 centimeters; width: 5 centimeters   **38.** 19 inches, 19 inches, 15
inches   **39.** 4 girls and 3 boys   **40.** 2 girls and 2 boys

CHAPTER 8   ROOTS AND RADICALS

SECTION 8.1 (PAGE 477)

◆ **CONNECTIONS**   **Page 471:**   The area of the large square is $(a + b)^2$ or $a^2 + 2ab + b^2$. The sum of the areas of the
smaller square and the four right triangles is $c^2 + 2ab$. Set these equal to each other and subtract $2ab$ from both sides to get
$a^2 + b^2 = c^2$.

EXERCISES   **1.** two   **3.** none   **5.** one   **7.** $-4, 4$ **9.** $-12, 12$ **11.** $-\dfrac{5}{14}, \dfrac{5}{14}$ **13.** $-30, 30$

15. 7   **17.** -11 **19.** $-\dfrac{12}{11}$ **21.** not a real number   **23.** 100   **25.** 19   **27.** $3x^2 + 4$ **29.** a must be positive.
31. a must be negative.   **33.** rational; 5   **35.** irrational; 5.385   **37.** rational; -8 **39.** irrational; -17.321
41. not a real number   **45.** 23.896   **47.** 28.249   **49.** 1.985   **51.** .095   **53.** 2.289   **55.** 5.074
57. -4.431 **59.** $c = 17$ **61.** $b = 8$ **63.** $c = 11.705$ **65.** 24 centimeters   **67.** 80 feet
69. 195 miles   **71.** 9.434   **73.** Answers will vary. For example, if $a = 2$ and $b = 7$, $\sqrt{a^2 + b^2} = \sqrt{53}$, while $a + b = 9$.
$\sqrt{53} \neq 9$.   **75.** 10   **77.** 5   **79.** -3 **81.** 5   **83.** not a real number   **85.** -3 **87.** c^2 **88.** $(b - a)^2$
89. $2ab$ **90.** $(b - a)^2 = a^2 - 2ab + b^2$ **91.** $c^2 = 2ab + (a^2 - 2ab + b^2)$ **92.** $c^2 = a^2 + b^2$ **93.** $2^3 \cdot 3^2$ **95.** $2^3 \cdot 5$

SECTION 8.2 (PAGE 485)

◆ **CONNECTIONS**   **Page 481:**   **1.** x; $\sqrt{x}$ **2.** The last part of it ($\sqrt{}$) is used as part of the radical symbol $\sqrt{}$.

EXERCISES   **1.** true   **3.** true   **5.** false   **7.** 9   **9.** $3\sqrt{10}$ **11.** 13   **13.** $\sqrt{13r}$ **15.** (a)
17. $3\sqrt{5}$ **19.** $3\sqrt{10}$ **21.** $5\sqrt{3}$ **23.** $5\sqrt{5}$ **25.** $-10\sqrt{7}$ **27.** $9\sqrt{3}$ **29.** $3\sqrt{6}$ **31.** 24   **33.** $6\sqrt{10}$
37. $\dfrac{4}{15}$ **39.** $\dfrac{\sqrt{7}}{4}$ **41.** 5   **43.** $\dfrac{25}{4}$ **45.** $6\sqrt{5}$ **47.** m **49.** y^2 **51.** $6z$ **53.** $20x^3$ **55.** $z^2\sqrt{z}$ **57.** x^3y^6
59. $2\sqrt[3]{5}$ **61.** $3\sqrt[3]{2}$ **63.** $4\sqrt[3]{2}$ **65.** $2\sqrt[4]{5}$ **67.** $\dfrac{2}{3}$ **69.** $-\dfrac{6}{5}$ **71.** 5   **73.** $\sqrt[4]{12}$ **75.** $2x\sqrt[3]{4}$

In Exercise 77, the number of displayed digits will vary among calculator models. Also, less sophisticated models may exhibit
round-off error in the final decimal place.
77. (a) 4.472135955   (b) 4.472135955   **79.** 6 centimeters   **81.** 6 inches   **85.** $-5x + 19$ **87.** $11x^2y - 7xy$

SECTION 8.3 (PAGE 489)

◆ **CONNECTIONS** **Page 487:** **1.** $\sqrt{4}, \sqrt{9}, \sqrt{16}, \sqrt{25}$, and so on **2.** There is a pattern. The differences between the numbers 4, 9, 16, 25, and so on increase by 2 each time: $9 - 4 = 5$, $16 - 9 = 7$, $25 - 16 = 9$, and so on. The next one would be $\sqrt{36}$, and the one after that $\sqrt{49}$, because $36 - 25 = 11$ and $49 - 36 = 13$.

EXERCISES **1.** $2\sqrt{3} + 4\sqrt{3} = (2 + 4)\sqrt{3} = 6\sqrt{3}$ **3.** $8\sqrt{3}$ **5.** $-5\sqrt{7}$ **7.** $5\sqrt{17}$ **9.** $5\sqrt{7}$ **11.** $11\sqrt{5}$ **13.** $15\sqrt{2}$ **15.** $-6\sqrt{2}$ **17.** $17\sqrt{7}$ **19.** $-16\sqrt{2} - 8\sqrt{3}$ **21.** $20\sqrt{2} + 6\sqrt{3} - 15\sqrt{5}$ **23.** $4\sqrt{2}$ **25.** $5\sqrt{21}$ **27.** $11\sqrt{3}$ **29.** $5\sqrt{x}$ **31.** $3x\sqrt{6}$ **33.** 0 **35.** $-20\sqrt{2k}$ **37.** $42x\sqrt{5z}$ **39.** $-\sqrt[3]{2}$ **41.** $6\sqrt[3]{p^2}$ **43.** $21\sqrt[4]{m^3}$ **45.** Answers will vary. One example is $\sqrt{36} + \sqrt[3]{64} = 6 + 4 = 10$. **47.** $-6x^2y$ **48.** $-6(p - 2q)^2(a + b)$ **49.** $-6a^2\sqrt{xy}$ **52.** Answers will vary. One example is $12p^3\sqrt{2r} + 3p^3\sqrt{2r} - 6p^3\sqrt{2r}$. **53.** 6 **55.** 2 **57.** $50\sqrt{3}$

SECTION 8.4 (PAGE 494)

EXERCISES **1.** We are actually multiplying by 1. The identity property for multiplication justifies our result. **3.** $\dfrac{7\sqrt{5}}{5}$ **5.** $4\sqrt{2}$ **7.** $\dfrac{-\sqrt{33}}{3}$ **9.** $\dfrac{7\sqrt{15}}{5}$ **11.** $\dfrac{\sqrt{30}}{2}$ **13.** $\dfrac{16\sqrt{3}}{9}$ **15.** $\dfrac{-3\sqrt{2}}{10}$ **17.** $\dfrac{21\sqrt{5}}{5}$ **19.** $\sqrt{3}$ **21.** $\dfrac{-2\sqrt{30}}{3}$ **23.** $\dfrac{\sqrt{2}}{2}$ **25.** $\dfrac{\sqrt{65}}{5}$ **27.** $\dfrac{\sqrt{21}}{3}$ **29.** $\dfrac{3\sqrt{14}}{4}$ **31.** $\dfrac{1}{6}$ **33.** 1 **35.** $\dfrac{\sqrt{7x}}{x}$ **37.** $\dfrac{2x\sqrt{xy}}{y}$ **39.** $\dfrac{x\sqrt{3xy}}{y}$ **41.** $\dfrac{3ar^2\sqrt{7rt}}{7t}$ **43.** (b) **45.** $\dfrac{\sqrt[3]{12}}{2}$ **47.** $\dfrac{\sqrt[3]{196}}{7}$ **49.** $\dfrac{\sqrt[3]{6y}}{2y}$ **51.** $\dfrac{\sqrt[3]{42mn^2}}{6n}$ **53.** (a) $\dfrac{9\sqrt{2}}{4}$ seconds (b) 3.182 seconds **55.** $32x^2 + 44x - 21$ **57.** $36x^2 - 1$ **59.** $pa - pm + qa - qm$

SECTION 8.5 (PAGE 500)

◆ **CONNECTIONS** **Page 499:** **2.** 6; For $x = 6$, $\dfrac{1}{\sqrt{x} + \sqrt{6}} = \dfrac{1}{2\sqrt{6}}$

EXERCISES **1.** 13 **3.** 4 **5.** 4 **7.** 5 **9.** $9\sqrt{5}$ **11.** $16\sqrt{2}$ **13.** $\sqrt{15} - \sqrt{35}$ **15.** $2\sqrt{10} + 30$ **17.** $4\sqrt{7}$ **19.** $57 + 23\sqrt{6}$ **21.** $81 + 14\sqrt{21}$ **23.** $37 + 12\sqrt{7}$ **25.** 23 **27.** 1 **29.** $2\sqrt{3} - 2 + 3\sqrt{2} - \sqrt{6}$ **31.** $15\sqrt{2} - 15$ **33.** $\sqrt{30} + \sqrt{15} + 6\sqrt{5} + 3\sqrt{10}$ **35.** $87 + 9\sqrt{21}$ **37.** $\dfrac{3 - \sqrt{2}}{7}$ **39.** $-4 - 2\sqrt{11}$ **41.** $1 + \sqrt{2}$ **43.** $-\sqrt{10} + \sqrt{15}$ **45.** $3 - \sqrt{3}$ **47.** $2\sqrt{5} + \sqrt{15} + 4 + 2\sqrt{3}$ **49.** $\sqrt{11} - 2$ **51.** $\dfrac{\sqrt{3} + 5}{8}$ **53.** $\dfrac{6 - \sqrt{10}}{2}$ **55.** $x\sqrt{30} + \sqrt{15x} + 6\sqrt{5x} + 3\sqrt{10}$ **57.** $6t - 3\sqrt{14t} + 2\sqrt{7t} - 7\sqrt{2}$ **59.** $m\sqrt{15} + \sqrt{10mn} - \sqrt{15mn} - n\sqrt{10}$ **61.** $2 - 3\sqrt[3]{4}$ **63.** $12 + 10\sqrt[4]{8}$ **65.** $-1 + 3\sqrt[3]{2} - \sqrt[3]{4}$ **67.** 1 **69.** $30 + 18x$ **70.** They are not like terms. **71.** $30 + 18\sqrt{5}$ **72.** They are not like radicals. **73.** Make the first term $30x$, so that $30x + 18x = 48x$; make the first term $30\sqrt{5}$, so that $30\sqrt{5} + 18\sqrt{5} = 48\sqrt{5}$. **75.** 4 inches **77.** $-3, -1$ **79.** 1, 2

SECTION 8.6 (PAGE 507)

◆ **CONNECTIONS** **Page 502:** **1.** $6\sqrt{13} \approx 21.63$ (to the nearest hundredth) **2.** $h = \sqrt{13}$; $6\sqrt{13} \approx 21.63$

EXERCISES **1.** Since $\sqrt{x}$ must be greater than or equal to zero for any replacement for x, it cannot equal -8, a negative number. **3.** 49 **5.** 7 **7.** 85 **9.** -45 **11.** $-\dfrac{3}{2}$ **13.** no solution **15.** 121 **17.** 8 **19.** 1 **21.** 6 **23.** no solution **25.** 5 **27.** The square of the right side is $x^2 - 14x + 49$. **29.** 12 **31.** 5 **33.** 0, 3 **35.** $-1, 3$ **37.** 8 **39.** 4 **41.** 8 **43.** 9 **45.** We cannot square term by term. The left side must be squared as a binomial in the first step. **47.** (a) 70.5 miles per hour (b) 59.8 miles per hour (c) 53.9 miles per hour **49.** 21 **51.** 8 **53.** 5^6 **55.** $\dfrac{1}{a^3}$ **57.** $\dfrac{3^2}{p^2}$ **59.** c^{13}

SECTION 8.7 (PAGE 512)

◇ **CONNECTIONS**   **Page 511:**   **1.** $(\sqrt{x} - 3)(\sqrt{x} + 1)$ **2.** $(x + \sqrt{10})(x - \sqrt{10})$ **3.** $(\sqrt{x} + 2)^2$
4. $x\sqrt{x} + 3\sqrt{x} + \dfrac{5\sqrt{x}}{x}$

EXERCISES   **1.** (a)   **3.** (c)   **5.** 5   **7.** 4   **9.** 2   **11.** 2   **13.** 8   **15.** 9   **17.** 8   **19.** 4   **21.** -4
23. -4 **25.** 2^3 **27.** $\dfrac{1}{6^{1/2}}$ **29.** $\dfrac{1}{15^{1/2}}$ **31.** $11^{1/7}$ **33.** 8^3 **35.** $6^{1/2}$ **37.** $\dfrac{5^3}{2^3}$ **39.** $\dfrac{1}{2^{8/5}}$ **41.** $6^{2/9}$

43. p^3 **45.** z **47.** $m^2 n^{1/6}$ **49.** $a^{2/3} b^{4/9}$ **51.** $\dfrac{1}{c^{4/3}}$ **53.** 2   **55.** 2   **57.** $\sqrt{a}$ **59.** $\sqrt[3]{k^2}$
61. $\sqrt{2} = 2^{1/2}$ and $\sqrt[3]{2} = 2^{1/3}$ **62.** $2^{1/2} \cdot 2^{1/3}$ **63.** 6   **64.** $2^{3/6} \cdot 2^{2/6}$ **65.** $2^{5/6}$ **66.** $\sqrt[6]{2^5}$ or $\sqrt[6]{32}$
67. 2   **69.** 1.883   **71.** 3.971   **73.** 9.100   **75.** 1.008   **77.** (a) 172.53 miles   (b) 211.31 miles   **81.** $-11, 11$
83. $-\sqrt{23}, \sqrt{23}$

CHAPTER 8 REVIEW EXERCISES (PAGE 515)

1. $-7, 7$ **3.** $-14, 14$ **5.** $-15, 15$ **7.** 4   **9.** 10   **11.** not a real number   **13.** $\dfrac{7}{6}$ **15.** a must be negative.
17. irrational; 4.796   **19.** rational; -5 **21.** $\sqrt{14}$ **23.** $5\sqrt{3}$ **25.** $-3\sqrt{3}$ **27.** $4\sqrt{10}$ **29.** 12   **31.** $16\sqrt{6}$
33. $\dfrac{3}{2}$ **35.** $\dfrac{\sqrt{3}}{7}$ **37.** $\dfrac{\sqrt{5}}{6}$ **39.** $3\sqrt{2}$ **41.** $2\sqrt{2}$ **43.** $\sqrt{km}$ **45.** $x^5 y^8$ **47.** $\dfrac{6}{p}$ **49.** $11 x^3 y^5$ **51.** $2\sqrt{11}$
53. $21\sqrt{3}$ **55.** 0   **57.** $2\sqrt{3} + 3\sqrt{10}$ **59.** $6\sqrt{30}$ **61.** 0   **63.** $11 k^2 \sqrt{2n}$ **65.** $\dfrac{15\sqrt{2}}{2}$ **67.** $\sqrt{5}$
69. $\dfrac{\sqrt{30}}{15}$ **71.** $\sqrt{10}$ **73.** $\dfrac{r\sqrt{x}}{4x}$ **75.** $\dfrac{\sqrt[3]{98}}{7}$ **77.** $-\sqrt{15} - 9$ **79.** $22 - 16\sqrt{3}$ **81.** -2
83. $2\sqrt{21} - \sqrt{14} + 12\sqrt{2} - 4\sqrt{3}$ **85.** $\dfrac{-2\sqrt{2} - 6}{7}$ **87.** $\dfrac{-\sqrt{3} + 3}{2}$ **89.** $\dfrac{2\sqrt{3} + 2 + 3\sqrt{2} + \sqrt{6}}{2}$ **91.** $\dfrac{1 + 3\sqrt{7}}{4}$
93. 25   **95.** 48   **97.** 2   **99.** -2 **101.** 4   **103.** 3   **105.** 9   **107.** 7^3 or 343   **109.** $x^{3/4}$ **111.** 16
113. $\dfrac{11}{t}$ **115.** $\dfrac{2\sqrt{10}}{5}$ **117.** -5 **119.** $\dfrac{4r\sqrt{3rs}}{3s}$ **121.** $-7\sqrt{2}$ **123.** $166 + 2\sqrt{7}$ **125.** x^2 **127.** no solution
129. $\dfrac{10\sqrt{7} - 5\sqrt{7}}{-3\sqrt{14} - 2\sqrt{14}}$ or $\dfrac{5\sqrt{7} - 10\sqrt{7}}{2\sqrt{14} + 3\sqrt{14}}$ **130.** $-\dfrac{\sqrt{7}}{\sqrt{14}}$ **131.** $-\sqrt{\dfrac{1}{2}}$ **132.** $-\dfrac{\sqrt{2}}{2}$ **133.** It falls from left to right.

CHAPTER 8 TEST (PAGE 518)

[8.1] 1. $-14, 14$ **2.** (a) irrational   (b) 11.916   **[8.2–8.5] 3.** 6   **4.** $-3\sqrt{3}$ **5.** $\dfrac{8\sqrt{2}}{5}$ **6.** $2\sqrt[3]{4}$ **7.** $4\sqrt{6}$
8. $9\sqrt{7}$ **9.** $-5\sqrt{3x}$ **10.** $2y\sqrt[3]{4x^2}$ **11.** 31   **12.** $6\sqrt{2} + 2 - 3\sqrt{14} - \sqrt{7}$ **13.** $11 + 2\sqrt{30}$
14. (a) $6\sqrt{2}$ inches   (b) 8.485 inches   **15.** $\dfrac{5\sqrt{14}}{7}$ **16.** $\dfrac{\sqrt{6x}}{3x}$ **17.** $-\sqrt[3]{2}$ **18.** $\dfrac{-12 - 3\sqrt{3}}{13}$ **[8.6] 19.** 3
20. $\dfrac{1}{4}, 1$ **[8.7] 21.** 16   **22.** -25 **23.** 5   **24.** $\dfrac{1}{3}$ **[8.6] 25.** 12 is not a solution. A check shows that it does not
satisfy the original equation.

CUMULATIVE REVIEW (Chapters 1–8) (PAGE 519)

1. 54   **2.** 6   **3.** 3   **4.** 18   **5.** 15   **6.** 4.223   **7.** 3   **8.** $y \geq -16$ **9.** $z > 5$ **10.** length: $17\dfrac{1}{2}$ meters;
width: $10\dfrac{1}{2}$ meters   **11.** $12x^{10} y^2$ **12.** $\dfrac{y^{15}}{5832}$ **13.** $3x^3 + 11x^2 - 13$ **14.** $4t^2 - 8t + 5$ **15.** $(m + 8)(m + 4)$
16. $(5t^2 + 6)(5t^2 - 6)$ **17.** $(6a + 5b)(2a - b)$ **18.** $(9z + 4)^2$ **19.** 3, 4   **20.** $-2, -1$ **21.** 2, -7
22. $\dfrac{x + 1}{x}$ **23.** $(t + 5)(t + 3)$ **24.** -21 **25.** $\dfrac{y^2}{(y + 1)(y - 1)}$ **26.** $\dfrac{-2x - 14}{(x + 3)(x - 1)}$

27.   **28.**   **29.**

30. $-\dfrac{5}{6}$ **31.** $(3, -7)$ **32.** infinite number of solutions **33.** 15 tens and 5 twenties **34.** $312 **35.** $29\sqrt{3}$
36. $-\sqrt{3} + \sqrt{5}$ **37.** $10xy^2\sqrt{2y}$ **38.** 32 **39.** $21 - 5\sqrt{2}$ **40.** 16

CHAPTER 9   QUADRATIC EQUATIONS

SECTION 9.1 (PAGE 525)

◆ **CONNECTIONS**   **Page 521:** 1980: 4.23;   1990: 6.56

EXERCISES **1.** true **3.** false; $x = 0$ is a solution. **5.** true **9.** $-9, 9$ **11.** $-\sqrt{14}, \sqrt{14}$
13. $-4\sqrt{3}, 4\sqrt{3}$ **15.** no real number solution **17.** $-1.5, 1.5$ **19.** $-2\sqrt{6}, 2\sqrt{6}$ **23.** (d) **25.** $-2, 8$
27. no real number solution **29.** $8 + 3\sqrt{3}, 8 - 3\sqrt{3}$ **31.** $-3, \dfrac{5}{3}$ **33.** $0, \dfrac{3}{2}$ **35.** $\dfrac{5 + \sqrt{30}}{2}, \dfrac{5 - \sqrt{30}}{2}$
37. $\dfrac{-1 + 3\sqrt{2}}{3}, \dfrac{-1 - 3\sqrt{2}}{3}$ **39.** $-10 + 4\sqrt{3}, -10 - 4\sqrt{3}$ **41.** $\dfrac{1 + 4\sqrt{3}}{4}, \dfrac{1 - 4\sqrt{3}}{4}$ **45.** $-4.48, .20$
47. $-3.09, -.15$ **49.** $(x + 3)^2 = 100$ **50.** $x + 3 = -10$ or $x + 3 = 10$ **51.** $-13;\ 7$ **52.** $-13, 7$
53. $-7, 3$ **54.** $-3, 6$ **55.** 34,282 **57.** about $\dfrac{1}{2}$ second **59.** 9 inches **61.** 424 **63.** $\dfrac{4 + 4\sqrt{3}}{5}$
65. $\dfrac{3 + \sqrt{6}}{4}$ **67.** $\left(x - \dfrac{7}{2}\right)^2$

SECTION 9.2 (PAGE 531)

◆ **CONNECTIONS** **Page 531:** **1.** x^2 **2.** $x^2 + 8x$ **3.** $x^2 + 8x + 16$ **4.** when we added the 16 squares

EXERCISES **1.** Divide both sides by 4. **3.** 49 **5.** $\dfrac{25}{4}$ **7.** $\dfrac{1}{16}$ **9.** 1, 3 **11.** $-1 + \sqrt{6}, -1 - \sqrt{6}$ **13.** -3
15. (d) **17.** $\dfrac{2 + \sqrt{14}}{2}, \dfrac{2 - \sqrt{14}}{2}$ **19.** $-\dfrac{3}{2}, \dfrac{1}{2}$ **21.** no real number solution **23.** $\dfrac{-7 + \sqrt{97}}{6}, \dfrac{-7 - \sqrt{97}}{6}$
25. $-4, 2$ **27.** $1 + \sqrt{6}, 1 - \sqrt{6}$ **29.** (a) $\dfrac{3 + 2\sqrt{6}}{3}, \dfrac{3 - 2\sqrt{6}}{3}$ (b) $2.633, -.633$ **31.** (a) $-2 + \sqrt{3}, -2 - \sqrt{3}$
(b) $-.268, -3.732$ **33.** 75 feet by 100 feet **35.** 8 miles **37.** 1 and 5 seconds **39.** $\dfrac{4 - 3\sqrt{3}}{3}$ **41.** $\dfrac{2 - \sqrt{5}}{2}$
43. $2\sqrt{5}$ **45.** 0

SECTION 9.3 (PAGE 538)

◆ **CONNECTIONS** **Page 533:** $\dfrac{1 + \sqrt{5}}{2}$

EXERCISES **1.** $a = 4, b = 5, c = -9$ **3.** $a = 3, b = -4, c = -2$ **5.** $a = 3, b = 7, c = 0$ **7.** $a = 1, b = 1,$
$c = -12$ **9.** $a = 9, b = 9, c = -26$ **11.** If a were 0, the equation would be linear, not quadratic. **13.** In either case, the
solutions are $\dfrac{-2 + \sqrt{10}}{2}, \dfrac{-2 - \sqrt{10}}{2}$ $\left(\text{or, equivalently, } \dfrac{2 - \sqrt{10}}{-2}, \dfrac{2 + \sqrt{10}}{-2}\right).$ **15.** $\dfrac{-5 + \sqrt{13}}{6}, \dfrac{-5 - \sqrt{13}}{6}$ **17.** $-13, 1$
19. 2 **21.** $\dfrac{-6 + \sqrt{26}}{2}, \dfrac{-6 - \sqrt{26}}{2}$ **23.** $-1, \dfrac{5}{2}$ **25.** $-1, 0$ **27.** $0, \dfrac{12}{7}$ **29.** $-2\sqrt{6}, 2\sqrt{6}$ **31.** $-\dfrac{2}{5}, \dfrac{2}{5}$
33. $\dfrac{6 + 2\sqrt{6}}{3}, \dfrac{6 - 2\sqrt{6}}{3}$ **35.** no real number solution **37.** no real number solution **39.** $\dfrac{-5 + \sqrt{61}}{2}, \dfrac{-5 - \sqrt{61}}{2}$
41. (a) $\dfrac{-1 + \sqrt{11}}{2}, \dfrac{-1 - \sqrt{11}}{2}$ (b) $1.158, -2.158$ **43.** (a) $\dfrac{1 + \sqrt{5}}{2}, \dfrac{1 - \sqrt{5}}{2}$ (b) $1.618, -.618$ **45.** $-\dfrac{2}{3}, \dfrac{4}{3}$
47. $\dfrac{-1 + \sqrt{73}}{6}, \dfrac{-1 - \sqrt{73}}{6}$ **49.** $1 + \sqrt{2}, 1 - \sqrt{2}$ **51.** no real number solution **53.** The solution(s) must make sense
in the original problem. (For example, a length cannot be negative.) **55.** $r = \dfrac{-\pi h \pm \sqrt{\pi^2 h^2 + \pi S}}{\pi}$ **57.** 1986
59. 1 second, 3 seconds; Both answers are reasonable. **61.** 1 hour **63.** $16, -8$; Only 16 feet is a reasonable answer.
65. It is the radicand in the quadratic formula. **66.** (a) 225 (b) 169 (c) 4 (d) 121 **67.** Each is a perfect square.
68. (a) $(3x - 2)(6x + 1)$ (b) $(x + 2)(5x - 3)$ (c) $(8x + 1)(6x + 1)$ (d) $(x + 3)(x - 8)$
69. (a) 41 (b) -39 (c) 12 (d) 112 **70.** no **71.** If the discriminant is a perfect square the trinomial is factorable.
72. (a) yes (b) yes (c) no **73.** $-5 + 8z$ **75.** $24 - 2r - 15r^2$

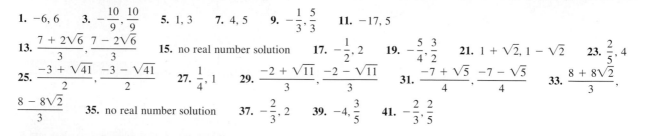

SUMMARY: Quadratic Equations (PAGE 542)

1. $-6, 6$　　**3.** $-\dfrac{10}{9}, \dfrac{10}{9}$　　**5.** $1, 3$　　**7.** $4, 5$　　**9.** $-\dfrac{1}{3}, \dfrac{5}{3}$　　**11.** $-17, 5$

13. $\dfrac{7 + 2\sqrt{6}}{3}, \dfrac{7 - 2\sqrt{6}}{3}$　　**15.** no real number solution　　**17.** $-\dfrac{1}{2}, 2$　　**19.** $-\dfrac{5}{4}, \dfrac{3}{2}$　　**21.** $1 + \sqrt{2}, 1 - \sqrt{2}$　　**23.** $\dfrac{2}{5}, 4$

25. $\dfrac{-3 + \sqrt{41}}{2}, \dfrac{-3 - \sqrt{41}}{2}$　　**27.** $\dfrac{1}{4}, 1$　　**29.** $\dfrac{-2 + \sqrt{11}}{3}, \dfrac{-2 - \sqrt{11}}{3}$　　**31.** $\dfrac{-7 + \sqrt{5}}{4}, \dfrac{-7 - \sqrt{5}}{4}$　　**33.** $\dfrac{8 + 8\sqrt{2}}{3},$

$\dfrac{8 - 8\sqrt{2}}{3}$　　**35.** no real number solution　　**37.** $-\dfrac{2}{3}, 2$　　**39.** $-4, \dfrac{3}{5}$　　**41.** $-\dfrac{2}{3}, \dfrac{2}{5}$

SECTION 9.4 (PAGE 548)

◆ **CONNECTIONS**　　**Page 547: 1.** $\langle -1, -4 \rangle, \langle 0, 2 \rangle, \langle -5, 0 \rangle$　　**2.** $-1; 2$

EXERCISES　　**1.** true　　**3.** false; $i = \sqrt{-1}$, so $\sqrt{i} = \sqrt{\sqrt{-1}}$, not -1.　　**5.** true　　**7.** $3i$　　**9.** $2i\sqrt{5}$

11. $3i\sqrt{2}$　　**13.** $5i\sqrt{5}$　　**15.** $5 + 3i$　　**17.** $6 - 9i$　　**19.** $6 - 7i$　　**21.** $14 + 5i$　　**23.** $7 - 22i$　　**25.** 45

29. $-3 + i$　　**31.** $2 - 6i$　　**33.** $-\dfrac{3}{25} + \dfrac{4}{25}i$　　**35.** $-1 + 3i$　　**36.** The product of $-1 + 3i$ and $2 + 9i$ is $-29 - 3i$, which *is* the original dividend.　　**37.** $-3 - 4i$　　**38.** The product of $-3 - 4i$ and i is $4 - 3i$, which *is* the original dividend.

39. Because $(3 - 2i)(4 + i) = 14 - 5i$ is true, the given statement is true.　　**40.** The product of the quotient and the divisor must equal the dividend.　　**41.** $-1 + 2i, -1 - 2i$　　**43.** $3 + i\sqrt{5}, 3 - i\sqrt{5}$

45. $-\dfrac{2}{3} + i\sqrt{2}, -\dfrac{2}{3} - i\sqrt{2}$　　**47.** $1 + i, 1 - i$　　**49.** $-\dfrac{3}{4} + \dfrac{\sqrt{31}}{4}i, -\dfrac{3}{4} - \dfrac{\sqrt{31}}{4}i$　　**51.** $\dfrac{3}{2} + \dfrac{\sqrt{7}}{2}i, \dfrac{3}{2} - \dfrac{\sqrt{7}}{2}i$

53. $\dfrac{1}{5} + \dfrac{\sqrt{14}}{5}i, \dfrac{1}{5} - \dfrac{\sqrt{14}}{5}i$　　**55.** $-\dfrac{1}{2} + \dfrac{\sqrt{13}}{2}i, -\dfrac{1}{2} - \dfrac{\sqrt{13}}{2}i$　　**57.** $\dfrac{1}{2} + \dfrac{\sqrt{11}}{2}i, \dfrac{1}{2} - \dfrac{\sqrt{11}}{2}i$　　**59.** If $b^2 - 4ac$ is negative, the equation will have solutions that are not real.　　**61.** true　　**63.** false;　For example, $3 + 2i$ is a complex number but it is not real.　　**65.**　　　**67.**　　　**69.** 16

SECTION 9.5 (PAGE 556)

◆ **CONNECTIONS**　　**Page 551:** The maximum height is 8 feet at 16 feet from the initial point. The maximum distance is 32 feet.

EXERCISES　　**3.** $(0, 0)$　　**5.** $(-1, 2)$　　**7.** $(0, -4)$　　**9.** $(0, 2)$　　**11.** $(-3, 0)$

13. $(3, 4)$　　**15.** $(2, 0)$　　**17.** one solution: 2　　**19.** two solutions: $-2, 2$　　**21.** no real solution

23. $-2, 3$　　　**25.** $-1, \dfrac{3}{2}$

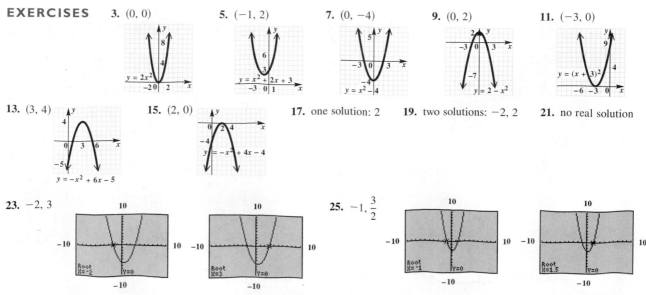

27. $-4, 2$

29. $0, 6$

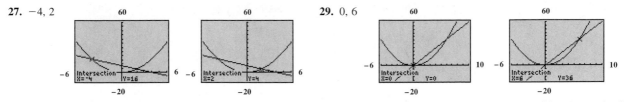

31. In each case, there is a vertical "stretch" of the parabola. It becomes narrower as the coefficient gets larger.   **33.** The graph of y_2 is obtained by reflecting the graph of y_1 across the x-axis.   **35.** By adding a positive constant k, the graph is shifted k units upward. By subtracting a positive constant k, the graph is shifted k units downward.   **37. (a)** the point $(0, -2)$ **(b)** the points $(-1, 0)$ and $(2, 0)$ **39.** $(.5)^2 - .5 - 2 = .25 - .5 - 2 = -2.25$ **41.** domain: the set of all real numbers;   range: the set of all real numbers greater than or equal to 0   **43.** domain : the set of all real numbers;   range: the set of all real numbers less than or equal to 4   **45.** domain: the set of all real numbers;   range: the set of all real numbers greater than or equal to 1
47. 3   **49.** 21

CHAPTER 9 REVIEW EXERCISES (PAGE 563)

1. $-12, 12$ **3.** $-8\sqrt{2}, 8\sqrt{2}$ **5.** $3 + \sqrt{10}, 3 - \sqrt{10}$ **7.** no real number solution   **9.** $-5, -1$

11. $-1 + \sqrt{6}, -1 - \sqrt{6}$ **13.** $-\dfrac{2}{5}, 1$ **15.** $2\dfrac{1}{2}$ seconds   **17.** $\left(\dfrac{k}{2}\right)^2$ or $\dfrac{k^2}{4}$ **19.** $1 + \sqrt{5}, 1 - \sqrt{5}$

21. $\dfrac{2 + \sqrt{10}}{2}, \dfrac{2 - \sqrt{10}}{2}$ **23.** $\dfrac{-3 + \sqrt{41}}{2}, \dfrac{-3 - \sqrt{41}}{2}$ **25.** $7.899, 8.899, 11.899$ **29.** $5 - i$ **31.** $-3 + 2i$

33. 20   **35.** 29   **37.** i **39.** $\dfrac{7}{50} + \dfrac{1}{50}i$ **41.** a (the real number itself)   **43.** $-2 + i\sqrt{3}, -2 - i\sqrt{3}$

45. $\dfrac{2}{3} + \dfrac{2\sqrt{2}}{3}i, \dfrac{2}{3} - \dfrac{2\sqrt{2}}{3}i$ **47.** $\dfrac{1}{3} + \dfrac{\sqrt{2}}{3}i, \dfrac{1}{3} - \dfrac{\sqrt{2}}{3}i$ **49.** $\dfrac{3}{8} + \dfrac{\sqrt{23}}{8}i, \dfrac{3}{8} - \dfrac{\sqrt{23}}{8}i$

51. $(0, 0)$ **53.** $(0, 5)$ **55.** $(-2, -2)$ $y = x^2 + 4x + 2$

57. two real number solutions: -2 and 2   **59.** two real number solutions: -1 and 4   **61.** no real number solution
63. $-\dfrac{11}{2}, 5$ **65.** $\dfrac{-1 + \sqrt{21}}{2}, \dfrac{-1 - \sqrt{21}}{2}$ **67.** $\dfrac{-5 + \sqrt{17}}{2}, \dfrac{-5 - \sqrt{17}}{2}$ **69.** no real number solution   **71.** $-\dfrac{5}{3}$
73. $-2 + \sqrt{5}, -2 - \sqrt{5}$ **75.** $1 + \sqrt{3}, 1 - \sqrt{3}$ **76.** $(x - (1 + \sqrt{3}))(x - (1 - \sqrt{3}))$
77. $((x - 1) - \sqrt{3})((x - 1) + \sqrt{3})$ **78.** $(x - 1)^2 - \sqrt{3}^2 = x^2 - 2x + 1 - 3 = x^2 - 2x - 2$
79. They are both $x^2 - 2x - 2$.   **80.** $(x - (2 + \sqrt{5}))(x - (2 - \sqrt{5}))$ or $((x - 2) - \sqrt{5})((x - 2) + \sqrt{5})$

CHAPTER 9 TEST (PAGE 566)

[9.1] 1. $-\sqrt{39}, \sqrt{39}$ **2.** $-11, 5$ **3.** $\dfrac{-3 + 2\sqrt{6}}{4}, \dfrac{-3 - 2\sqrt{6}}{4}$ **[9.2] 4.** $2 + \sqrt{10}, 2 - \sqrt{10}$ **5.** $\dfrac{-6 + \sqrt{42}}{2},$
$\dfrac{-6 - \sqrt{42}}{2}$ **[9.3] 6.** The quantity must be positive.   **7.** $-3, \dfrac{1}{2}$ **8.** $\dfrac{3 + \sqrt{3}}{3}, \dfrac{3 - \sqrt{3}}{3}$
[9.4] 9. $-1 + \dfrac{\sqrt{7}}{2}i, -1 - \dfrac{\sqrt{7}}{2}i$ **[9.1–9.3] 10.** $\dfrac{5 + \sqrt{13}}{6}, \dfrac{5 - \sqrt{13}}{6}$ **11.** $1 + \sqrt{2}, 1 - \sqrt{2}$
12. $\dfrac{-1 + 3\sqrt{2}}{2}, \dfrac{-1 - 3\sqrt{2}}{2}$ **13.** $\dfrac{11 + \sqrt{89}}{4}, \dfrac{11 - \sqrt{89}}{4}$ **14.** 5   **15.** 2 seconds   **16.** (d)   **[9.4] 17.** $-5 + 5i$
18. $-17 - 4i$ **19.** 73   **20.** $2 - i$ **[9.3] 21.** 6   **[9.5] 22.** $(3, 0)$ **23.** $(-1, -3)$ **24.** $(-3, -2)$

25. (a) two **(b)** $-3 + \sqrt{2}, -3 - \sqrt{2}$ **(c)** The solutions are approximately -4.414 and -1.586.

CUMULATIVE REVIEW (Chapters 1–9) (PAGE 567)

1. 15   **2.** 5   **3.** -2 **4.** $-r + 7$ **5.** $-2k$ **6.** $19m - 17$ **7.** 3   **8.** 5   **9.** 2   **10.** $100°, 80°$

11. length: 54 inches;   width: 34 inches   **12.** $L = \dfrac{P - 2W}{2}$ or $L = \dfrac{P}{2} - W$ **13.** $m > -2$

14. $p \le 4$ **15.** $\dfrac{x^4}{9}$ **16.** $\dfrac{b^{16}}{c^2}$ **17.** $\dfrac{27}{125}$ **18.** $8x^5 - 17x^4 - x^2$ **19.** $2x^4 + x^3 - 19x^2 + 2x + 20$

20. $25t^2 + 90t + 81$ **21.** $3x^2 - 2x + 1$ **22.** $16x^2(x - 3y)$ **23.** $(4x^2 + 1)(2x + 1)(2x - 1)$

24. $(2a + 1)(a - 3)$ **25.** $(5m - 2)^2$ **26.** $-9, 6$ **27.** $2\dfrac{1}{2}$ seconds   **28.** $(y - 2x)^2$ or $y^2 - 4xy + 4x^2$

29. $\dfrac{4}{5}$ **30.** $\dfrac{-k - 1}{k(k - 1)}$ **31.** $\dfrac{5a + 2}{(a - 2)^2(a + 2)}$ **32.** $\dfrac{b + a}{b - a}$ **33.** $-\dfrac{15}{7}, 2$ **34.** 100 pounds per square inch

35.   **36.**   **37.**

38. -18 **39. (a)** $-\dfrac{1}{3}$ **(b)** The line crosses the y-axis above the origin, indicating that $b > 0$, since $(0, b)$ is the y-intercept.

40. $2x - y = -3$ **41.** $(-3, 2)$ **42.** no solution   **43.** Monte Carlo: $39.99; Samantha: $34.99   **44.**

45. 10   **46.** $\dfrac{6\sqrt{30}}{5}$ **47.** $\dfrac{\sqrt[3]{28}}{4}$ **48.** $4\sqrt{5}$ **49.** $-ab\sqrt[3]{2b}$ **50.** 4   **51.** 7   **52.** $\dfrac{-2 + 2\sqrt{3}}{3}, \dfrac{-2 - 2\sqrt{3}}{3}$

53. $-1 + \sqrt{6}, -1 - \sqrt{6}$ **54.** $\dfrac{2 + \sqrt{10}}{2}, \dfrac{2 - \sqrt{10}}{2}$ **55.** no real solution   **56.** $2\dfrac{1}{2}$ seconds

57. (a) $8i$ **(b)** $5 + 2i$ **58.** $-\dfrac{1}{2} + \dfrac{\sqrt{17}}{2}i, -\dfrac{1}{2} - \dfrac{\sqrt{17}}{2}i$ **59.** $(0, -4)$ **60.** positive

APPENDIX A (PAGE 572)

1. 117.385   **3.** 13.21   **5.** 150.49   **7.** 96.101   **9.** 4.849   **11.** 166.32   **13.** 164.19   **15.** 1.344   **17.** 4.14
19. 2.23   **21.** 4800   **23.** .53   **25.** 1.29   **27.** .96   **29.** .009   **31.** 80%   **33.** .7%   **35.** 67%   **37.** 12.5%
39. 109.2   **41.** 238.92   **43.** 5%   **45.** 110%   **47.** 25%   **49.** 148.44   **51.** 7839.26   **53.** 7.39%   **55.** $2760
57. 805 miles   **59.** 63 miles   **61.** $2400

APPENDIX B (PAGE 577)

1. $\{1, 2, 3, 4, 5, 6, 7\}$ **3.** {winter, spring, summer, fall}   **5.** $\emptyset$ **7.** $\{L\}$ **9.** $\{2, 4, 6, 8, 10, \ldots\}$ **11.** The sets in
Exercises 9 and 10 are infinite sets.   **13.** true   **15.** false   **17.** true   **19.** true   **21.** true   **23.** true   **25.** true
27. false   **29.** true   **31.** true   **33.** false   **35.** true   **37.** true   **39.** false   **41.** true   **43.** false   **45.** $\{g, h\}$
47. $\{b, c, d, e, g, h\}$ **49.** $\{a, c, e\} = B$ **51.** $\{d\}$ **53.** $\{a\}$ **55.** $\{a, c, d, e\}$ **57.** $\{a, c, e, f\}$ **59.** $\emptyset$
61. B and D; C and D

INDEX

FORMULAS

FIGURE	FORMULAS	EXAMPLES
Square	Perimeter: $P = 4s$ Area: $A = s^2$	
Rectangle	Perimeter: $P = 2L + 2W$ Area: $A = LW$	
Triangle	Perimeter: $P = a + b + c$ Area: $A = \frac{1}{2}bh$	
Pythagorean Formula (for right triangles)	In a right triangle with legs a and b and hypotenuse c, $$c^2 = a^2 + b^2.$$	
Sum of the Angles of a Triangle	$A + B + C = 180°$	
Circle	Diameter: $d = 2r$ Circumference: $C = 2\pi r$ $C = \pi d$ Area: $A = \pi r^2$	
Parallelogram	Area: $A = bh$ Perimeter: $P = 2a + 2b$	
Trapezoid	Area: $A = \frac{1}{2}(B + b)h$ Perimeter: $P = a + b + c + B$	